International Federation of Automatic Control

AUTOMATIC CONTROL

WORLD CONGRESS 1990

(in six volumes)

VOLUME I

IFAC Symposia Series, 1991. Number 1

AUTOMATIC CONTROL

WORLD CONGRESS 1990

'In the service of mankind'

*Proceedings of the 11th Triennial World Congress of the
International Federation of Automatic Control
Tallinn, Estonia, USSR, 13-17 August 1990*

(in six volumes)

Edited by

Ü. JAAKSOO

*Institute of Cybernetics of the Estonian
Academy of Sciences, Tallinn, Estonia, USSR*

and

V. I. UTKIN

Institute of Problems in Control, Moscow, USSR

VOLUME I

Plenary Papers · Industrial Problems

Large Scale Systems · Linear Systems Theory

Published for the

INTERNATIONAL FEDERATION OF AUTOMATIC CONTROL

by

PERGAMON PRESS

OXFORD · NEW YORK · BEIJING · FRANKFURT
SAO PAULO · SEOUL · SYDNEY · TOKYO

UK	Pergamon Press plc, Headington Hill Hall, Oxford OX3 0BW, England
USA	Pergamon Press, Inc., Maxwell House, Fairview Park, Elmsford, New York 10523, USA
PEOPLE'S REPUBLIC OF CHINA	Maxwell Pergamon China, Beijing Exhibition Centre, Xizhimenwai Dajie, Beijing 100044, People's Republic of China
GERMANY	Pergamon Press GmbH, Hammerweg 6, D-6242 Kronberg, Germany
KOREA	Pergamon Press Korea, KPO Box 315, Seoul 110-603, Korea
AUSTRALIA	Maxwell Macmillan Pergamon Publishing Australia Pty Ltd, Lakes Business Park, 2 Lord Street, Botany, NSW 2019, Australia
JAPAN	Pergamon Press, 8th Floor, Matsuoka Central Building, 1-7-1 Nishi-Shinjuku, Shinjuku-ku, Tokyo 160, Japan

First edition 1991

Library of Congress Cataloging in Publication Data

Data applied for

British Library Cataloguing in Publication Data

A catalogue record for this title is available from the British Library

ISBN 0-08-041265-3
ISBN 0-08-040174-0 (Set)

These proceedings were reproduced by means of the photo-offset process using the manuscripts supplied by the authors of the different papers. The manuscripts have been typed using different typewriters and typefaces. The lay-out, figures and tables of some papers did not agree completely with the standard requirements: consequently the reproduction does not display complete uniformity. To ensure rapid publication this discrepancy could not be changed: nor could the English be checked completely. Therefore, the readers are asked to excuse any deficiencies of this publication which may be due to the above mentioned reasons.

The Editors

Printed in Great Britain by BPCC Wheatons Ltd, Exeter

AUTOMATIC CONTROL
11th Triennial World Congress of the International Federation of Automatic Control

Sponsored by
International Federation of Automatic Control (IFAC)

Co-sponsored by
International Association for Mathematics and Computer Simulation (IMACS)
International Federation for Information Processing (IFIP)
International Federation of Operational Research Societies (IFORS)
International Measurement Confederation (IMEKO)

Hosted by
Institute of Cybernetics of the Estonian Academy of Sciences
Tallinn Technical University

In Co-operation with
Institute of Control Sciences (Moscow)
USSR National Committee of Automatic Control

International Programme Committee
V. Utkin, USSR   (Co-Chairman)
Ü. Jaaksoo, USSR (Co-Chairman)
R. Isermann, FRG (Vice-Chairman)
R. Evans, AUS   (Vice-Chairman)

National Organizing Committee
V.A. Trapeznikov   (Chairman)
A. Work   (Vice-Chairman)
N.A. Kuznetsov   (Vice-Chairman)
I.V. Prangishvili   (Vice-Chairman)
I. Toome   (Vice-Chairman)
V.I. Venets   (Secretary)
H. Aasmäe
Ü. Jaaksoo
V.A. Lototsky
S.S. Markianov
E.D. Teryayev
V.I. Utkin
I.N. Vasilyev
M.Veiderma

SUBJECT AREAS

VOLUME I

Plenary Papers
Industrial Problems
Large Scale Systems
Linear Systems Theory

VOLUME II

Stochastic Control and State Estimation
Modelling and Identification
Biomedical Engineering
Adaptive Control

VOLUME III

Automatic Control in Aerospace
Robust Control
Nonlinear Control
Control Applications of Optimization
Distributed Parameter Systems
Theory of Discrete Event Systems

VOLUME IV

Application of Artificial Intelligence and Expert Systems of Automatic Control
Distributed Computer Control Systems
Artificial Intelligence in Real-time Control
Software Engineering for Real-time Control
Industrial Applications of Modern Control Methods
Control of Cars, Ships and Engines
Intelligent Components and Instruments for Automatic Control
Control of Electrical Drives and Power Electronics
Control System Elements

VOLUME V

Automatic Control in Manufacturing
Robot Control
Industrial Systems Engineering
Computer-aided Control System Analysis and Design
Man-machine Systems
Control of Transportation Systems
Water Resources and Environmental Systems Planning

VOLUME VI

Control of Electric Generating Plants and Power Systems
Control in Mineral, Mining and Metal Processing
Control of Chemical Processes and Processes for Natural Products like Food, Wood, Agriculture
Control System Approach to Development
Strategic Planning of Energy Systems
Modelling, Control and Decision Making in Socio-economic Systems
Automatic Control Education
Social and Cultural Aspects of Automation
Improving International Stability

OPENING SPEECH

Mr. President, Ladies and Gentelmen,

It is my honour and duty to welcome you on behalf of the International Program Committee. Our work is almost done. The scientific and technical level of the Congress will show the quality of the IPC work.

The International Program Committee decided to follow the reviewing procedure of the previous IFAC'87 Munich Congress with one very important exception - the one step reviewing procedure was approved where the acception/rejection decision was made only on the basis of the final version of the paper. It certainly caused a lot of additional work for the authors. They had to prepare the papers in the camera ready form. Still, the advantage of the one step procedure was that the quality estimation of papers was based on the final version.

The IPC received over 2700 abstracts and 1531 papers were submitted. Due to space and time limits 594 was the maximum number of papers submitted-accepted and invited by the Sub-IPC Chairmen.

Each submitted paper had to undergo a very thorough reviewing procedure, therefore the IPC believes only excellent papers to have been included into the Congress Program. The rejection rate was high indeed - only one paper of three was accepted in average. In some Subject Areas the rejection rate was even higher.

The pre-congress time showed us that in the case of hard work you will learn more about yourself and your colleagues. Now is the right time and place to thank all the authors. I would also like to thank the 37 sub-IPC Chairmen and at least three times more reviewers all over the world for the hard and responsible work. We have become aware again of how small the world actually is and how unreliable the communication with our country.

I would like to thank the IFAC Secretariat for their invaluable help in supporting the communication with the authors and the IPC members.

The four plenary papers have been invited to present the state-of-the-art and the future development of automatic control. I wish to thank our plenary speakers who are all outstanding specialists in their particular field.

An idea has cropped up concerning the plenary Sessions of the next Congress. It might be reasonable to start an IFAC Joint Project the result of which would be documented and reported as one of the plenary papers of the next Congress. This, I suppose, would promote the activities of the industrial and academic people interested in IFAC and would make the IFAC life more dynamic.

The organizers and the International Program Committee have the pleasure of welcoming you to Tallinn. We hope that the IFAC'90 World Congress will give you an opportunity to get new information about the present state and further development of automatic control in the broadest sense.

ÜLO JAAKSOO
Editor and IPC Co-Chairman

CLOSING SPEECH

Mr. President, Ladies and Gentlemen,

The 11th IFAC Congress is almost over. During 5 interesting days we have had the - at least theoretical - possibility to listen to 594 paper presentations, 4 plenaries as well as a number of case study- and discussion sessions. It is my duty as the Chairman of the IFAC Technical Board to sum up the impressions from this stimulating and rich technical program.

As you know, the technical program was prepared by 37 Sub-IPCs covering different sub-areas. More than 2700 abstracts were originally submitted, and you can imagine the amount of work that has been involved. I would like to take this opportunity to thank all the Sub-IPC members and in particular their Chairmen for an admirable work. I know that most of the work had to be done under hard time constraints. I may also add a comment about the submission of camera ready papers in the first round, since I have received several questions about it. This procedure was a necessary step to allow for a reasonable lead time in submission, due to anticipated slower mail service for this congress. It was, however, not intended to be the procedure in the future.

We may roughly divide our area into four fields. One would be the mathematically oriented tools we use for control design; another would be the non-mathematical tools for control design and implementation - this includes CAD and software and hardware at different levels · in the control system. A third field are the applications of control design and implementation in various areas. And a fourth field could be called the human aspects of control engineering; this includes education, the social and other effects of automation as well as the control field in society. With this division we have in Tallinn seen some 230 papers each in the areas of mathematical tools and of applications. About 90 papers dealt with the nonmathematical tools, and some 30 papers discussed the human aspects.

The IFAC Congress is the major international meeting in the automatic control area. It offers a unique possibility to contemplate over the status of the field and over the directions in which it is moving. It is now 30 years since the first Congress. The historic perspective of the development of a fundamental control concept was exposed in Prof. Aizerman's interesting plenary lecture. It is tempting to further dwell on the enormous developments of our field over this period, as reflected by the 11 Congress. I will not do that, but rather address a more difficult question: what is happening now? How does the 11th Congress differ from the 10th? My personal reflections based on experiences in Tallinn are the following ones:

The seemingly ever increasing interest in adaptive control appears to have stabilized. It is still, though, the largest single topic in the Congress program.

Non-linear control theory is more exciting than ever. Evidently, the new mathematical and software tools have turned out to be sharp and useful.

Discrete event dynamic systems, as evidenced in various excellent plenary lectures, is attracting more and more interest. We may now see the beginning of a "common language" for this important, but difficult area.

The exciting theoretical development in so-called H-infinity-theory has led to a tool that now is ready to be added to the control engineers' toolbox - a most valuable complement for designs where robustness to model areas are of prime importance.

These are trends in the mathematically oriented areas. The rapid development in computer hardware and software has pointed to the possibility to do sophisticated control design without traditional mathematical analysis. Many sessions at this Congress have dealt with the use of computer techniques, artificial intelligence and expert systems methodology in control systems, as well as the use of blackbox models and regulator structures on neural nets type. This is very important. Unless the control community plays an active role in this development, there is a clear risk that the major part of control systems design will be taken over by the computer science community, and that thus the broad control perspective will be lost. It is especially important that the real time aspects in the use of these tools are given adequate at tention.

In the application areas the growing interest in robot control is striking. This was without comparison the largest single application area. Clearly, these non-linear and time-varying systems represent an important challenge to us, especially regarding the next generation of less sturdy constructions. I mentioned that some 230 application papers had been presented here. A clear majority of these of course deal with the real things in terms of industrial plants, etc. - various simulated simplified models. However, it is also clear that a majority of these papers have been written by authors from universities or research institutes. Just as a sample, I checked the number of papers by authors from industry in volume 11 of the preprints (the one that deals with power plants, metal, mining and chemical applications). Out of 55 contributions, 9 are entirely industrial, most of them actually from Germany and Japan. Interesting - but not surprising - is that these papers primarily deal with what could be called model-based control: carefully using physical insight into the plant for the control design, and not so much relying upon ready-made methodologies.

Specialized research may lead to narrow perspectives. It is very important for IFAC to take all aspects and impacts of automation and control into account. At this Congress, for the first time I believe, half of the plenary lectures have been devoted to human aspects of control. In Tom Martin's enlightening survey on automation in a holistic perspective, the human was put in the centre. This is an obvious factor for anyone who deals with introducing automation in the real world, but has to be emphasized to us others over and over again. Today's most educational plenary lecture by Walter Schaufelberger really dealt with tomorrow's control engineers. This next generation will no doubt deal with interesting and difficult problems in very complex systems. It is vital how we educate them today.

The triennial IFAC Congress is, I repeat, a major international technical event in the field of control engineering. It brings together researchers, teachers, and engineers from the whole world. Here we have been 1072 people from 39 countries. The IFAC Congress is the best place not only to follow the recent developments in one's own particular interest area but also to get overviews over what happens in other fields in control engineering. The IFAC Congress is of course also a great place to meet old friends and to make new friends from all over the world. The Estonian team has done an unbelievably excellent job to provide the best possible conditions for us. We thank them all, Ülo Jaaksoo, Ants Wôrk, Boris Tamm and the whole crew behind them for their outstanding job.

LENNART LJUNG
IFAC Vice-President and Chairman
of the Technical Board

CONTENTS

Generalized State-space Systems

Algebraic Linear System Theory

Model Matching and Model Reduction

Stability of Linear Systems

State-space Methods for Control and Estimation

APPROPRIATE AUTOMATION — INTEGRATING TECHNICAL, HUMAN, ORGANIZATIONAL, ECONOMIC, AND CULTURAL FACTORS

T. Martin*, J. Kivinen, J. E. Rijnsdorp***, M. G. Rodd† and W. B. Rouse‡**

**Manufacturing Technologies Program Management Agency, Nuclear Research Centre Karlsruhe, Karlsruhe, FRG*
***Corporate Information Systems, Neste Oy, Espoo, Finland*
****Department of Chemical Technology, Twente University of Technology, Enschede, The Netherlands*
†Department of Electrical and Electronic Engineering, University College of Swansea, Swansea, UK
‡Search Technology, Inc., Norcross, Georgia, USA

Abstract. Automation technology, including digital computer and communication techniques, is being applied in an ever-increasing range of private and public spheres, also reaching third-world cultures not previously exposed to such technology. This invokes the engineer's responsibility to consider the direct and indirect effects caused by this technology.
The question is when and how to include these factors in the design and implementation process, prospectively avoiding bad effects and still fulfilling the goals imposed on the engineer by his client. To be able to carry that responsibility and make proper design decisions, the engineer must both understand 'appropriateness' within a given boundary, and have decision authority together with other parties participating in the design. Whereas sound methodologies for user-centred design are appearing, anticipating and considering the cultural effects of automation is an area going far beyond engineering. Nevertheless, engineers should get more deeply involved in comprehensive technology assessment. Encouraging experiences show how novel design approaches and consideration of comprehensive sets of requirements can lead to better overall system performance, but much research on open questions remains to be done.

Keywords. Social effects of automation; man-machine systems; human factors; organizational, economic, and cultural factors; manufacturing processes; process industries; human-centred design; socio-technical design.

INTRODUCTION

Modern technology implies the application of science in societal contexts. Automation technology, including digital computers and the communication techniques which have become such vital elements and tools of automation, is an ever-growing technology with an enormous range of subjects and applications. Its applications range from single-machine automation, through the control of industrial processes, up to computerized administrative processes and use in military intallations. New applications are continuously being developed, also in areas not foreseen earlier, for example in the private sector or in third-world cultures not previously exposed to such technology.

This enormous versatility can be achieved
only because the methods of automatic control
and computer science result in a high degree
of abstraction, as a consequence of the use
of formalized models. The engineer developing
such models tends not to be primarily concerned with the social or other consequences
of his subject. The more application-oriented
the technological implementation gets, the
more urgent the consideration of the effects
caused by this technology becomes. The question is when and how to consider the effects
in the design and implementation process.

In another IFAC Congress plenary paper,
Sheridan, Vamos, and Aida (1983) have already
raised the issue of adapting automation to
man, culture and society. In their paper, the
authors treat general issues of automation
and mainly discuss the proper extent of automation by scaling 'degrees of assistance'
that a computerized system might provide in
supervisory control. In our paper, we will
discuss criteria for appropriateness of automation, and stress issues of design and design methodology. The design considerations
will be backed up by lessons learned from
case studies in both process and manufacturing industry. We will also show examples
of cultural aspects of automated systems.
Thinking of the global effects of automation,
we as engineers can no longer be indifferent
to the effects of automation technology on
those countries which have neither the political drive and the education to utilize
this technology appropriately for their economic and cultural benefit, nor the power or
insight to prevent the negative effects of
automation technology exported to them by the
'developed' countries.

OVERALL CONSIDERATIONS OF
APPROPRIATE AUTOMATION

As in any other technical field, automation
technology must meet a given set of requirements to justify its application. These include as a minimum the criteria of adequacy
of the means (and funds) employed, and of the
reliability of the elaborated technical solution. Automation faces particular difficulties in measuring up to such requirements,
due to the interdisciplinary nature of the
problems to be solved. Many products of en-

gineering, such as electronic chips or fuel
cells, have a technical environment only. But
automation systems, methods and techniques
also have to function in a human environment.
The working population concerned must be
actively integrated in the modeling of work
processes if an automated system is to be
accepted. This requires precise determination
of the necessary interactions in the division
of labor between man and machine. Consequently, the large variety of user interfaces
(and their design considerations) has become
one of the focal points in automation technology (as will be outlined below in the
chapters on human-centred design).

Man-machine Incomparability

The rapid and broad successes of automation
technology have given rise to the hope that
it is a sort of panacea for solving complex
problems. This hope is especially entrenched
in the minds of many decision-makers, partly
as a consequence of public relations activities (and the universal trend that successes are more widely publicized than failures, although many automation projects have
failed due to insufficient automatability,
inadequate user-system-interfaces, and in-
compatibility between human needs and system
requirements).
With the advent of 'expert systems' as a new
paradigm of programming computer-based control systems, large parts of the AI establishment have in advance assumed and an-
nounced that such systems will possess almost
incredible capabilities for solving complex
tasks. Although formulation of ill-defined
problem areas by statements of facts and
rules, and heuristic searching for solutions,
will in many cases be more successful than
will conventional programming, even these
systems can only exercise supporting functions. The hope that they may render super-
fluous the responsible evaluation of results
by people is a dangerous fallacy, as expert
systems are designed to admit uncertain results without these always being recognizable
as such (Dreyfus and Dreyfus, 1986). Re-
minding ourselves of accidents that have
shaken the public globally in recent years,
like Tchernobyl (1986), the stock exchange
crash (1987), and the Iran Air airbus downing
in the Gulf (July 1988), should give reasons
enough to rethink our belief in the proper
role of automation.

Kurt Goedel showed as early as 1931 that in
any formal system there always exists a non-
decidable proposition which cannot be proved
by means of the formal system itself (Goedel's Incompleteness Theorem). An additional
problem arises from the fact that the meaning
of symbols is normally context-dependent and,
therefore, cannot be formally described in
principle. From this follows that a computer-
aided vision and pattern recognition system
cannot recognize an object for sure. Therefore, when the radar system on board the 'USS
Vincennes' recognized an 'unidentified, assumed enemy' aircraft, the order to shoot
should not have been given. The official version for the predominant cause of the accident was 'human failure'. This judgement ignores that the ship's hyper-instrumentation
led to a complexity that the persons in
charge can no longer cope with under stress
conditions (Klein, 1989). The belief in technology, having found its universal symbol in
the computer, has itself become a risk factor
(Perrow, 1984). False belief in naval automation in this case left 290 persons dead.
(In the case of Tchernobyl, however, human
failure was apparently the main reason for
the accident.)

Completely automated systems based on heuristic processes are extremely problematic, for
heuristic processes intrinsically include
wrong solutions. In situations where sur-
prises can lead to critical situations and
matters of life and death, responsible decisions cannot be left to such systems alone.
Instead, final decisions must be left to
people, because people have the ability to
master unprecedented situations. The unique
abilities of humans, as against artefacts
like control systems or computers, have been
determined by work psychologists. Recent observations reveal that skills based on empirical knowledge, such as a feeling for
machines and materials, continue to play an
important part in the work with computer-controlled machines (Böhle, Milkau, 1988). In
dealing with AI techniques for automation
purposes, we should abandon the idea of man-
computer comparability and pursue instead the
concept of complementarity. Otherwise, there
is a great danger that, e.g., expert systems
will continue to be overestimated. Then, expert systems will not primarily be adapted to
cope with the complex reality, but reality
will be reduced to that which can be for-

malized and fitted into the expert system
(Moldaschl, 1989).

Responsibility of Engineers

As outlined above, it is vital to determine
where full or even partial automation cannot
or should not be used, and where computer-
aided operations, for instance, might be appropriate instead. It needs to be emphasized
that decisions and actions in important technical, social, economic and military problem
areas must be handled not by automated systems and computers, but through the responsible actions of people.

For a person to be able to carry responsibility, two prerequisites must be fulfilled.

(1) The person needs the knowledge and skill
    to master the situation, i.e., to analyse
    it and act in a goal-oriented manner.

(2) The person must possess the proper
    authority and latitude of decision
    and action.

For the control engineer this means, firstly,
that he must be knowledgable about the social
and other nontechnical effects of automation.
The IFAC Committee on Social Effects of Automation tries hard to proliferate the available knowledge. Unfortunately, at most technical schools and universities subjects like
work science, social effects and man-machine
systems are not compulsory subjects in engineering curricula. Knowledge of that kind is,
therefore, not available to a sufficient extent. Another problem might be that the engineer, although trying to inform himself,
does not know about certain effects of his
product, e.g., the cultural effect caused by
his product in the country to which it is exported.
The second precondition is problematic, too,
because the engineer designing a product is
situated within a certain boundary beyond
which he can exert hardly any influence. So
he might be able to design a proper man-
machine interface which considers the human
effects at the work place, but unable to influence the environmental effect, or to prevent export of the product to a country where
it will cause negative cultural effects. If
we talk about responsibility we must, there-

fore, first define the boundaries within which the effects of automation can be considered.

These problems should not be an excuse, though, for engineers to avoid getting involved more actively in questions of the effects of automation. If they are still in an ivory tower where one can be in love with highly automated systems irrespective of their application, they should leave it and come to grips with human, social, economic, and cultural consequences. This involvement can range from actively including a comprehensive set of requirements in the design process, through to refusal to cooperate in a project with negative effects which are visible but not taken care of.

CRITERIA FOR APPROPRIATENESS OF AUTOMATION

The possible criteria and problem aspects to be considered for a given automation task are manifold. TABLE 1 shows a possible way of structuring the various aspects.

TABLE 1 Structure of Various Aspects of Automation

subject	actors and 'victims'	scientific field	goals and values
automation system design	automation/ control engineers	automation/ control technology	reliability; robustness; accuracy; flexibility
human inter-face design	ergonomists; system users	human factors; ergonomics	user-friendliness; efficiency
job & work organization design	personnel dept; management; workers	organization science	job satisfaction; labour pro-ductivity
coordination within the enterprise	top management; enterprise departments	organization science; information systems	flexibility; service; profit
inter-enterprise cooperation	enterprises; customers	micro-economics; computer networks	mutual benefits
society	politicians; general public	macro-ecomomics;	economic growth; employment
inter-society effects	developed and developing countries	development economics	quality of life
world	humankind; biosphere	ecology; geophysics	preservation of nature

To be able to integrate these aspects in system planning and design one must, first of all, understand them in terms of causes and effects, their dependencies and contradictions, and short- and long-term effects. Integrating these aspects is then a question of balancing different values. This will in real life be a bargaining process taking place at different levels: company level, nationally, or internationally. In the following, we will treat various aspects, but not all. (The environmental issue cannot be considered in this paper.)

THE SOCIAL DIMENSION OF AUTOMATION TECHNOLOGY

The social relevance of automation technology is pronounced, as it deals mainly with the (re-)organization of work processes ('social' being taken in a broad sense). In the design of automated systems, this focus on the work process becomes important in determining the relationship between man and machine. The machine, therefore, must always be defined in relation to the subject of the work, man. Even fully-automated systems are in no way autonomous, but are also related to real patterns of work and life, and consequently it is again people who must be determined as the subjects of these processes.

In the wake of the progressive division of labor, technical and administrative design strategies have become accepted in which humans are considered to be a potential nuisance with a high error probability, to be kept out of the immediate work process as much as possible. Such a strategy entails the danger that the use of automated systems no longer has a supporting function, but instead becomes a dominating factor. Demoting the working population from subjects of their work to machine operators kills the willingness and ability to act in a responsible way in the working environment. The subsequent effect of deskilling the workforce, together with the insufficient availability of automated systems, has been proved and documented by empirical data. The IFAC Committee on Social Effects of Automation has done its share in proliferating this knowledge and, at the same time, developing and discussing methods for the proper design of work practices in automated systems (Bibby and others, 1976; Rijnsdorp, 1978; Martin, 1984; Brödner,

1987). Jones (1984) created the notion of the
'Computer-Aided Craftsman', suggesting that
automation technology should be carefully
designed around skilled human work.
Sociologists are talking about a change of
paradigm in industrial automation (Martin,
1988). The organization and division of
labor, the question of which functions to
automate and which voluntarily not to automate, the qualification requirements and the
personnel structures are not parameters determined by developments in automation technology, but must be regarded as strategic
variables to be used deliberately by decision-makers.
With this lesson learned, how can we design
better systems?

General Guidelines for Development

One suggestion as a general guideline for
designing automated systems is that such systems be viewed from two aspects, namely that
of the social relevance and that of the tool
character (Rödiger, 1988). Since we deal with
single machines as well as with networked information systems, the notion of a 'tool'
must be seen in a broad sense. The importance
of the tool character is neglected in systems
development if people, in analogy to computers, are considered simply as information-
processing systems. The holistic, intuitive
abilities of people need the chance to establish and attain goals independently in the
work process and to acquire overviews. This,
in turn, requires latitude of action. It
further requires that fields of communicative
action be created and maintained, in order to
enable discussions and cooperation with other
people to take place.
For the individual designer involved in the
development, the question arises of how to
combine the guidelines outlined above with
the criteria and the conditions imposed by
clients. In this respect, a dialog between
developers and users is of the greatest importance in the interest of achieving these
goals.

The need to optimize both technical and human
factors suggests a parallel design approach
wherein both aspects are considered side by
side throughout the design process, to
achieve overall optimization of the system. A
scheme, shown in Fig. 1, was developed within
a project funded under the auspices of the

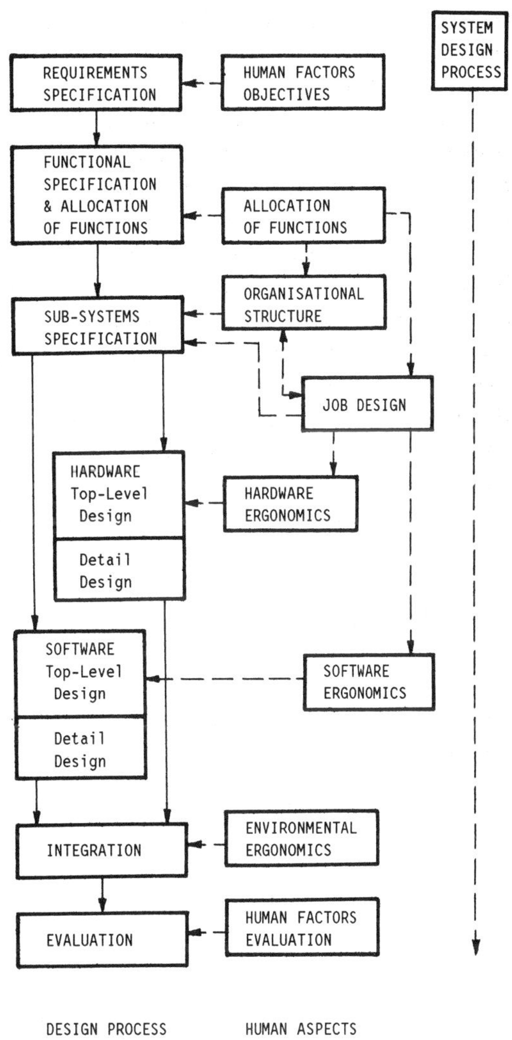

Fig. 1. Human factors mapped onto a model
of the design process

ESPRIT Programme, dealing with the design of
a flexible assembly cell (Ravden and others,
1986). It was necessary to determine which
criteria were relevant at which stages in the
design process. This suggested a hierarchical
structuring of the criteria, although in
practice design is an iterative, cyclic process. Some human aspects considered were:
Allocation of functions. This addresses the
methods by which functions (or tasks) within
the system are allocated between humans and
machines, stressing, for example, human-
machine complementarity.
Job design. By advocating factors such as
autonomy, responsibility, variety, etc., this
area concentrates on ensuring that the human
functions allocated above are designed to
maximize flexibility, responsiveness and motivation of those operating, supporting and
managing the system.

<u>Organizational structure</u>. This will possibly be the most difficult area to integrate into the design process model. Organizational factors such as self-supervision, minimum of functional specialization, etc., are included, with the overall aim of providing a supportive organizational structure.

HUMANIZATION OF TECHNOLOGY VERSUS HUMAN ENGINEERING

Within IFAC, an interesting discussion has been taking place during recent years, considering the question: "Are 'humanization of technology' and 'human engineering' for efficiency and productivity conflicting objectives, or can they support each other?" At the IFAC Congress in Budapest in 1984, Tom Sheridan suggested and then performed a delphi study with twenty experts from both the IFAC Committee on Social Effects of Automation (SOCEFF) and the Man-Machine Systems Working Group (MMS) of the Systems Engineering Committee (SECOM). The subject of the delphi study was to find out how these groups of experts differed in their views on how to create good working conditions for operators in industry. At the next Congress in Munich in 1987, Jan Forslin interpreted the results (Mårtensson, 1987). The basic issues and the differing views are given in Fig. 2, indicating how each group views each topic in the center column. Apparently, there is an underlying psychological dimension that explains the different opinions. The ensuing discussion showed a changing view of this dichotomy. As the field of man-machine systems design becomes more mature and integrated, more people seem to become competent to combine the evaluative and the analytical approach, and the design becomes objective-driven instead of technology-driven (compare chapters on human-centred design). There is a need, though, to concentrate on a common design approach. One model for this might be the layers of design shown in Fig. 3, based on the socio-technical systems design approach (Martin, Ulich, Warnecke, 1987). The suggestion is to start by designing the overall socio-technical structure, trying to anticipate the social effects of the technology used. One then works further 'from outside in', with the design of proper hardware and software at the end.

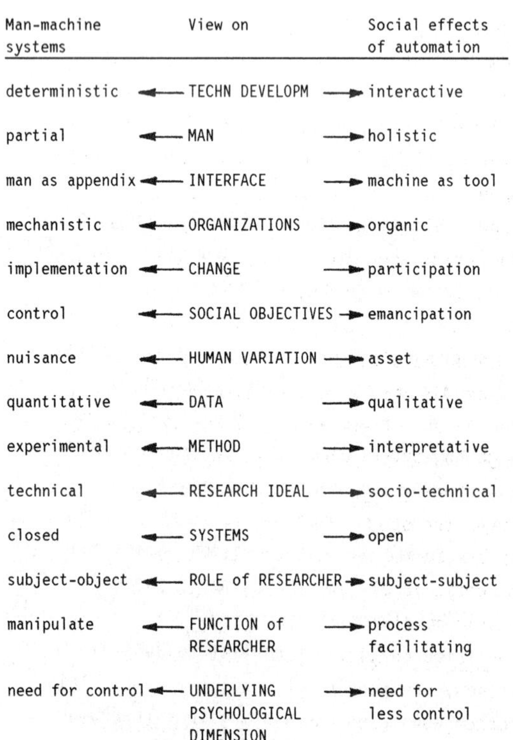

Man-machine systems	View on	Social effects of automation
deterministic ◄—	TECHN DEVELOPM	—► interactive
partial ◄—	MAN	—► holistic
man as appendix ◄—	INTERFACE	—► machine as tool
mechanistic ◄—	ORGANIZATIONS	—► organic
implementation ◄—	CHANGE	—► participation
control ◄—	SOCIAL OBJECTIVES	—► emancipation
nuisance ◄—	HUMAN VARIATION	—► asset
quantitative ◄—	DATA	—► qualitative
experimental ◄—	METHOD	—► interpretative
technical ◄—	RESEARCH IDEAL	—► socio-technical
closed ◄—	SYSTEMS	—► open
subject-object ◄—	ROLE of RESEARCHER	—► subject-subject
manipulate ◄—	FUNCTION of RESEARCHER	—► process facilitating
need for control ◄—	UNDERLYING PSYCHOLOGICAL DIMENSION	—► need for less control

Fig. 2. Conclusions from the Delphi study between experts on man-machine systems and social effects of automation (Source: J. Forslin)

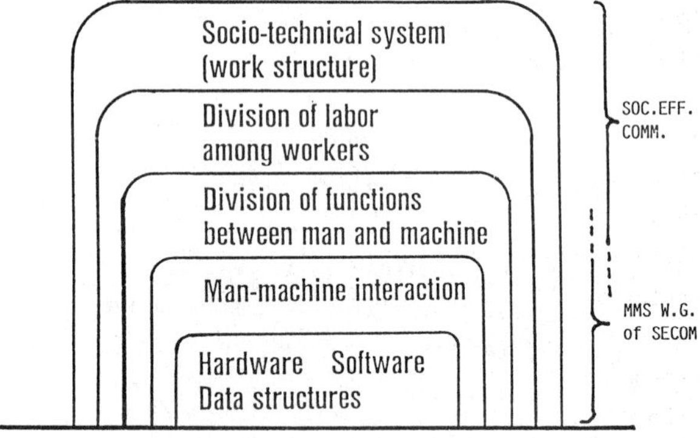

Fig. 3. Layers of design for computer-aided systems of work (trying to amalgamate the two different approaches).

HUMAN-CENTRED ISSUES OF DESIGN

The design of human-machine systems is particularly difficult (and challenging) when the context is large-scale systems. This difficulty is due to the complexity of such contexts, although this assertion is certainly not a great insight. The human-centred design issues of concern do not involve complexity in general - they involve the consequences of complexity (Rouse, 1988a).

One of these issues involves the nature of
decision making situations faced by humans in
large-scale systems. There are basically
three types of decision making situations:
- Familiar and frequent
- Familiar and infrequent
- Unfamiliar and infrequent.
We are usually pretty good at dealing with
the familiar and frequent - we know these
situations are going to arise, and they
happen often enough to justify investing in
means of dealing with them.
Familiar and infrequent situations are the
grist for the risk analyst's mill. How infrequent? What are likely consequences and
their costs? Answers to these questions determine our investment strategies for dealing
with these risks.
Unfamiliar and infrequent situations pose a
dilemma. On the one hand, perhaps we should
invest in making the unfamiliar familiar.
Then, it's back to risk analysis. On the
other hand, such a strategy ignores the in-
evitability of humans eventually having to
deal with the unfamiliar - not knowing what
has happened or what to do about it. The nonlinear, distributed nature of large-scale
systems will result in a larger number and
wider variety of unfamiliar situations. We
need to be able to design for decision making
in this type of situation.

A second issue concerns the nature of emerging large-scale systems which is resulting
in higher-order effects being increasingly
salient. Thus, for example, events in one
sector of the transportation system can affect operations in another sector. Within the
economy in general, industries affect each
other in more far-reaching ways than in the
past. Looking more broadly, economic development efforts can negatively affect the environment on a global scale. Technological
development can profoundly affect educational
requirements -and suffer due to educational
shortfalls or misdirection.
The result of these higher-order effects is
that there are a larger number and wider
variety of stakeholders in particular decision making situations. Consequently, decision makers are pulled in all directions,
decisions are made to minimize political
risks, or decisions are avoided.

A third issue concerns the roles of humans in
large-scale systems. There are two primary
reasons why humans are included as operators,
maintainers, and managers in complex systems.
One reason is people's ability to perform -
to exercise skills, judgment, and creativity.
Some pundits argue that intelligent computers
will soon eliminate the need for humans to
perform in these ways. This assessment appears to be somewhat reasonable for mani-
pulative skills, occasionally on target for
judgmental abilities, and substantially off
the mark for creativity.
Even if the soothsayers are right, perhaps
later rather than sooner, the second motivation for humans having roles in complex
systems is seldom, if ever, addressed by new
technology. Humans are included in complex
systems to be responsible for system operations and performance. The ability to accept
responsibility and, if required, to find in-
novative ways to fulfill these responsibilities is a uniquely human characteristic.
This is particularly evident when things go
wrong, sometimes unpredictably and uncon-
trollably, and some individual or group accepts responsibility and initiates efforts to
set things right again.

We raise the issue of responsibility because
we are concerned that the complexity and
technology associated with large-scale systems will lead the people involved to lose
their feelings of responsibility - to perceive that 'the system's in charge'. If this
happens, then in fact no one will be in
charge. This possibility leads to the topic
of design philosophy.

Choosing a design philosophy involves selecting objectives at their highest, value-
laden levels. From this perspective, our
design objectives involve developing the
means whereby people in complex systems can
achieve the **operational** objectives for which
they are responsible. Thus, the purpose of
the people in the system is not to 'staff'
the system - the purpose of the system is to
support the people in achieving their objectives.
In other words, our design philosophy is
simply that **people are in charge**. Based on
this philosophy, the question of whether or
not they should be in charge is open to discussion. The key question is how to support

them so that they can be in charge successfully. More specific questions concern where
in systems people should be; how they should
be trained and aided; and how we can assure
that they accept and retain feelings of being
responsible.

Adoption of the above design philosophy, as
well as the consequent design objective,
leads to three primary design goals. These
goals involve developing means for:
- Helping humans overcome their limitations
  (e.g., human error),
- Helping humans enhance their abilities
  (e.g., pattern recognition),
- Fostering user acceptance (e.g., avoiding
  the 'not invented here' attitude).

HUMAN-CENTRED DESIGN METHODOLOGY

A suitable design philosophy and design goals
are necessary for appropriate automation -
however, they are not sufficient. A design
methodology is needed that enables designers
to achieve goals in a manner consistent with
the philosophy.
The determination of human-centred functional
requirements for large-scale systems is difficult to do well. Balancing what users apparently want with what they may need, as
well as balancing technological and economic
realities, can prove to be far from straightforward. It is fairly common to discover that
the 'final' requirements are incomplete and
inconsistent. Obviously, it is better to discover this early rather than late.

We have found that a structured methodology
can ease the burden of requirements analysis,
as well as provide insights that might not
otherwise be gained (Rouse, 1987; Rouse and
Cody, 1989). This methodology is based on the
notion that the seven measurement issues
shown in Fig. 4 are central to successful
system design. If these issues are not addressed by system designers, they will eventually be addressed by system users, with perhaps unfortunate implications for sales,
liability claims, etc.
The list of questions in Fig. 4 is used in
two ways. Measurements are **executed** bottom up
- the system has to run, compute, etc. prior
to measuring actual sales. In contrast,
measurements are **planned** top down. Thus, the
eventual measurement of viability, accept-

Viability ———⟶ Are the benefits of system use sufficiently greater than its costs?

Acceptance ——⟶ Do organizations/individuals use the system?

Validation ——⟶ Does the system solve the problem?

Evaluation ——⟶ Does the system meet requirements?

Demonstration ⟶ How do observers react to the system?

Verification ——⟶ Is the system put together as planned?

Testing ———⟶ Does the system run, compute, etc?

Fig. 4. Measurement issues

ability, and validity should be planned long
before testing is considered.
From this perspective, there are two classes
of problem that can arise. The first is planning too late, where, for example, failure to
plan for assessing acceptance can preclude
measurement prior to putting a product or
system into use. The second class of problems
is execution too early where, for instance,
demonstrations are executed prior to resol-
ving test and verification issues, and potentially lead to negative initial impressions
of a product or system.

The measurement issues listed in Fig. 4 encompass a fairly diverse set of questions. If
each of these issues were pursued independently, as if they were ends in themselves,
the costs of measurement would be prohibi-
tive. Yet, each issue is important and should
not be neglected.
What is needed, therefore, is an overall approach to measurement that balances the allocation of resources among the issues, while
also integrating intermediate measurement results in a way that provides maximal benefit
to the evolution of the design product. We
have found that this can be accomplished by
viewing measurement as a process involving
four phases: 1) naturalist, 2) marketing, 3)
engineering, and 4) sales and service.

The **naturalist phase** involves understanding the domain and tasks of users, from the perspectives of individuals, the organization, and the environment. This understanding includes not only the users' activities but also the prevalent values and attitudes relative to productivity, technology, and change in general. Concerns of particular interest include identification of difficult and easy aspects of tasks, barriers to and potential avenues of improvement, and the relative leverage of various stakeholders in the organization. The results of the naturalist phase include a characterization of users' tasks and needs, as well as users' perceptions of how these needs might be met.

This characterization of users and needs serves as input to the **marketing phase**. With this understanding, one is in a position to conceptualize alternative products to support users. These product concepts are used for initial marketing to obtain users' reactions relative to validity, acceptability, and viability. In other words, one wants to determine whether or not users perceive a product concept as solving an important problem, solving it in an acceptable way, and solving it at a reasonable cost.

While one cannot resolve all of the validity, acceptability, and viability questions during the marketing phase, one can test and elaborate the plans for eventual resolution of these issues. Further, one can test and refine the hypotheses that have emerged from the naturalist phase regarding the appropriate impact of user characteristics on design choices and product concepts. Thus, the marketing phase produces both an assessment of the relative merits of the multiple product concepts that have emerged up to this point, and a preview of any particular difficulties that are likely when one later tries to ensure that users perceive the resulting product as valid, acceptable, and viable.

The **engineering phase** begins with tradeoffs between desired conceptual functionality and technological reality. Technology development will usually have been pursued prior to and in parallel with the naturalist and marketing phases. This will have at least partially ensured that the product concepts shown to potential users during the marketing phase were not technologically or economically ridiculous. However, one must now be very specific about how desired functionality is to be provided, what performance is possible, and the resources necessary to provide it. Most of the effort in this phase is associated with using various formal and informal design methodologies to transform conceptual designs to detailed design. Measurement concerns include both planning and execution of evaluation, demonstration, verification, and testing.

The **sales and service phase** begins once these four issues have been successfully resolved. The focus now shifts to validity, acceptability, and viability. At this point, one ensures that implementation conditions are consistent with the assumptions underlying the design basis of the product.

If the naturalist and marketing phases were well done, the sales and service phase should proceed easily. Nevertheless, the final marketing, sales, installation, and ongoing service of a product are the activities that should provide the validation, acceptance, and viability measurements planned earlier. Further, if the sales and service phase is well orchestrated, this will substantially lessen the investments necessary for subsequent naturalist and marketing phases for new products.

The above measurement-oriented approach provides an overall framework for human-centred design. Beyond this structure, more detailed methodological guidance is needed. During the later stages of the marketing phase, and throughout the engineering phase, methods are needed for synthesizing functionality that satisfy requirements within the scope of the design goals outlined earlier.
We have found a five-step method to be particularly useful (Rouse, 1988b). This method begins by **characterizing user-system tasks** by selecting one or more elements of a taxonomy of 13 general tasks. The next step involves **assessing relative demands on tasks**, especially to determine 'bottlenecks'. The third step concerns **identifying approaches to support** by mapping from high-demand tasks to one or more elements of a taxonomy of 17 support concepts. The fourth step involves **de-**

termining likely applications obstacles for
the candidate support concepts selected. The
last step concerns anticipating user acceptance problems by applying a 12-step procedure
to evaluate support concepts and plan their
implementation.

The methodologies outlined in this section
have been applied to a variety of complex design problems, including advanced aircraft
cockpits, power plant control rooms, several
design support systems, and a new concept for
production planning and scheduling. In all of
these applications, the philosophy, goals,
and methodologies enabled human-centred design issues to predominate, and technological
issues to be subordinated to the goals of
providing support for humans.

EXPERIENCES IN PROCESS INDUSTRIES

Fashions

As in women's and men's clothing, there have
been and still are dominant fashions in process control and automation. The sixties
showed a push for on-line quality measurement, and on-line computers for the optimization of process operation. The seventies
brought small-size control rooms with VDU-
based instrumentation for complete production
sites, handled by first-class operators (assisted by second-class operators in the plant
areas). The eighties show(ed) the impact of
adaptive control, expert systems, and statistical quality control.

Each new wave is advocated as the final solution, pushed into practice, met by difficulties, and finally subsiding into a more or
less stationary outcome. In some cases, e.g.
for on-line quality measurement, the difficulties are mainly of a technical nature.
But in other cases there are organizational
and human problems, which were (and are) not
always clearly recognized nor adequately
dealt with. So one wonders if all the potential benefits have really been reaped.

Optimization

As an example, the on-line optimization of
process operation has run into the incompati-
bility between numerical computation and
human methods of decision-making and action:

The process operator associates critical constraints with certain controller set points
('Never move temperature 703 above 435 degrees, otherwise there is excessive coking in
the flasher!'), and deals with them one at a
time. So he neither understands nor accepts
the computer output, which shows simultaneous
corrections to all set points without indicating what happens with respect to the
critical constraints. He cannot grasp, or at
least has his doubts, that the effect of
changing temperature 703 from 435 to 439 degrees is compensated for by a simultaneous
increase of flow rate 695 by 0.7 %, and a decrease of pressure 706 by 0.8 bar, etc. So,
as the operator is still held responsible for
process upsets, he will tend to play safe and
put the computer out of operation.

Evidently, optimization procedures generated
by computers have to be brought into a form
which is familiar and acceptable to the operator. One possibility is to develop near-
optimal regulatory control structures and decentralized, simplified computations, which
can also cope with variations in process conditions. Then the in-plant system needs to be
updated only infrequently, making this system
and its supervisors (the operators!) more
autonomous. In this way, appropriate technical measures can also create opportunities
for organizational decentralization.

Cooperation

A case where cooperation between operators
was vital, was the design of a major extension of a petroleum refinery (Pikaar, Lenior,
Rijnsdorp, 1985). The goal was 'whitening the
barrel': converting cheap heavy fractions
into more-valuable lighter ones (gasoline,
kerosene, light gas oil). For this purpose
the number of processing plants was to be
doubled, with many interactions due to common
energy utility systems.
These interactions would make the refinery
very vulnerable to local upsets, as these
spread rapidly through the utility systems
and so easily infect other plants. Therefore
good cooperation between control room operators was considered to be essential.

An obvious measure was to ensure eye-to-eye
contact in the control room over the VDU's on
the consoles, even between two short (seated)
operators. This led to the choice of small

VDU's (13 inch, instead of the more fashion-
able 19 inch models), and shallow keyboards
(8 inches). But organizational aspects proved
to be more problematic: the existing sub-
division of the operator team into two sub-
teams, each with its own supervisor, blocked
essential communication between control room
operators. Therefore much attention had to be
paid to integrating the total operator team
by setting up common training, and having
common supervisors. It is amazing how organizational divisions remain visible, long
after they have lost sense and purpose!

Another, unexpected problem appeared when individual plants were allocated to the different operator positions in the control
room. The object was to keep process interactions within one operator position as far
as possible, and to minimize the distance
between operators with many interactions. The
result of this optimization was a semi-circle
of 4 operator consoles (each with one operator position during normal operation, and
two operator positions during major upsets),
with a fifth console in the centre. The latter proved to be the console for the utilities, which caused an uproar in the design
team: 'What? You are going to put those
stupid utilities in the middle? Unacceptable!
Only our technologically advanced conversion
unit deserves that location!' Fortunately,
after lengthy discussions emotions were replaced by reason. Later, experience has shown
that major upsets can be dealt with only by
coordination from behind the central utilities console.

A third problem, which became evident from
observing operator actions during a major up-
set, is the variability of operator work
load. During normal operation outside office
hours, when not much is happening and there
are no maintenance technicians and staff
around, control-room operators can easily
cope with their job. But when things go
wrong, the work load can become excessive.
Unfortunately, with an increasing degree of
automation and centralization, this difference tends to become more pronounced. (Extreme differences are found in electric power
generation, where the operator's job consists
mainly of waiting for emergencies.)

Improvements can be made by training central
operators to take over some tasks from their
neighbours in the central control room, and
by providing them with the corresponding displays and controls. Evidently, organizational
and psychological problems are much harder to
recognize and to solve than technical ones,
and engineers can be quite irrational (in
this respect they prove to be as human as
other people!).

Small-Scale Plants

There is a tendency to transfer the work organization concept, as developed in the
large-scale process industry (oil refineries,
bulk chemical plants), to smaller plants. The
former works mostly with closed equipment;
hence operators in a central control room can
clearly see what happens on the VDU-screens
and other displays. Therefore, they are able
to direct the work of their field operators,
and to coordinate the overall operation effectively among themselves (see preceding
section).

In contrast, many small-scale plants have
mostly 'open' process equipment. Operators
have to see directly what is going on, and to
make local adjustments, so displaying information just in the central control room is
insufficient. Similar conditions exist in
other industries, e.g. papermaking, where
setting-up the paper machine requires monitoring both of the displays in the control
room and of the conditions in the machine
itself. In all these cases, the concept of
the central operator as the boss of the local
operators does not fit. Instead, operators
function better on equal terms, with responsibilities in the control room as well as at
the process equipment. This easily leads to
the concept of an autonomous work group with
jobs of equal status and an informal internal
organization (Rijnsdorp, Pikaar, Lenior,
1984).

Networks

Like any organization, industrial enterprises
suffer from a lack of coordination and communication. 'Multinationals' experience particular difficulties, due to their multidimensional structure: marketing offices distributed over various countries, active in

different product groups; production in other locations, each serving several marketing offices; joint ventures and co-makerships with other firms, etc. But smaller enterprises are not free from problems either. There are always conflicts of interest between sales, production, purchasing, logistics, R&D, and maintenance.

A way to improve communication and coordination is the computer network. Banks have already pioneered with this tool with very satisfactory results. People in different parts of the world with different cultural backgrounds, who have never seen each other, interleave working together with exchanging jokes!

The application of networks in industrial enterprises is probably not so straightforward. It is unlikely that conflicts can be resolved by digital data exchanges; meetings and management guidance seem essential here. But computer networks could complement direct communication by providing objective information, e.g. 'What is the cost of satisfying a very demanding customer? What is the best allocation of production runs between two factories in different countries?'

<u>Design Aspects</u>

The usual deadlines of plant design projects make it almost impossible to pay sufficient attention to the interdisciplinary aspects of automation. It is therefore desirable to set up an anticipatory phase, well ahead of the acutal project. This phase consists of:

- A situation study in a comparable running production plant.
  Here process automatability can be analyzed, and future user needs and wishes collected and evaluated.

- Developing a user participation concept, thus finding an efficient and effective way to involve the future users (or their representatives) in the design project, and encouraging acceptance by lower management and project staff.

- Allocation of tasks: a careful choice of the degree of automation for each operational function, from plant scheduling to off-normal handling. The automation 'intelligence' requires special attention, as this has an impact on job content and selection requirements.

- Structure of the work organization: setting goals for organization development, and fitting the work organization to the technical and social conditions at hand.

- Job design: within the framework of the chosen work organization, assessing the job content of production and other personnel. If results are negative, task allocation and/or work organization are modified accordingly.

- Selection, training, and career development concepts following from the previous concepts.

EXPERIENCES IN MANUFACTURING

<u>Changing Conditions and Structures of Manufacturing</u>

During the last two decades the markets for industrial consumer goods and, consequently, those for capital goods have shifted from steady expansion towards stagnation.

This lasting global trend does not yet seem to be affected by unsatisfied needs in developing countries. Terms of trade and debts hinder these countries, preventing them from turning their needs into spending power for industrial goods. Thus, the world markets are and will remain constrained to the highly industrialized areas and 'threshold' countries comprising North America, Western Europe and South-East Asia (the Comecon countries being only slightly involved).

The overall low growth rates of these limited markets mean that competition also changes its character from supplying expanding market shares, where suppliers are usually able to set the conditions, to displacing competitors, whereby customers gain the power to demand products adapted to their needs.

Under these new market conditions, price and quality of unified products are no longer the only important assets. The ability to adapt

products to customer requirements for increasing variety, and to guarantee short delivery times is becoming a more important
competitive factor.

The functional requirements that follow from
this market situation are manifold:

o Flexibility of production facilities so
   that various products, even those not yet
   known, can be manufactured.

o Simple and quick resettability of manufacturing installations in order to allow
   small lots to be manufactured.

o Reassignment of quality assurance
   functions to the productive work stations
   (quality must not be controlled, but produced).

o Rearrangement of production facilities,
   based on principles of group manufacturing, in order to shorten lead-times.

o Ability of subsystems to work autonomously
   in order to permit investments to be made
   step by step.

Trying to meet these requirements, there are
three substantial economic difficulties the
factory of today has to contend with. (The
following details are from the German machine
industry but also seem to be applicable to
other countries.) First, there is a continual
increase in capital tied up in factory equipment which compels management to make better
use of it. Second, long and varying lead-
times caused by the functional principle of
job-shop manufacturing make work-in-progress
expensive. Third, the ratio of 144 indirect
to 100 direct workers on average in the
German machine industry causes excessive personnel expenses, since well-organized firms
with comparable products have shown that a 90
to 100 ratio is sufficient. Two opposing concepts of manufacturing automation have
emerged to surmount these difficulties.

Alternative Approaches to Manufacturing
Automation

The technocentric approach. This approach
leaves the basic job-shop structure of the
production process unchanged and follows the
same fundamental objectives as in the past.

These are to reduce direct labour costs and
to gain better control over the manufacturing
process. Applied to the shopfloor, management
attempts to automate setting and operating
functions almost completely for machine tools
and handling systems. The activities focus on
automatic part and tool changes, measuring
devices and monitoring systems. They are
limited by the increasing costs of this
equipment. Although fully automatic operation
might be temporarily possible, human operators are still needed.

This concept of the 'unmanned factory' runs
into several difficulties, making its success
doubtful. First, the high cost and risk,
caused particularly by the software needed,
are too large for many small and medium-sized
firms. Despite their growing economic importance, these firms would be bypassed by
the development. Second, firms following this
strategy would suffer from relative inflexi-
bility regarding alteration of batches and
process innovation.

Third, as a sort of irony of automation,
flexible manufacturing systems are becoming
more and more complex and susceptible to
breakdowns because of automation and, consequently, increasingly dependent again on
highly qualified, skilled human labor (Bainbridge, 1982).

The human centered approach: skill-based
manufacturing. This approach regards automation (and technology in general) not as a
constraint but an option that - when properly
designed and applied - opens up the chance to
see qualified live work and automated work
not as irreconcilable contrasts, but as supplementary productive forces.
Sociologists call this new approach a change
of paradigm in industrial production.

The New Organizational Paradigm

This human centered or organizational paradigm views technology not as a thing but as a
relationship: The relationship between people
and their artefacts (van Beinum, 1988). All
technologies are tools and have meaning only
when being used and controlled by people. Introducing a new technology means introducing
a new relationship, and managing technological change means managing a relationship
which is changing. Understanding work as a

relationship between people and their tools takes on a special meaning when we consider this relationship on the organizational level.

Organizations are socio-technical systems. The concept of socio-technical systems design was developed thirty years ago by the Tavistock Institute. It arose from the consideration that any production system requires both a technical organization and a social system which creates and governs the relations among those who carry out the necessary tasks. If we design and manage the work situation so that the social system and the technical sytem are interrelated in the highest possible complementary way, we can achieve a joint optimization of the two systems, and thus of the functioning of the organization as a whole.
In the quest for the proper response to this challenge, many companies have been trying over the past few years to develop and introduce new structures of organization and labor in the manufacturing sector under the general heading of flexibility.
This can be achieved by a structure of functional blocks (products in assembly groups, productions in working groups and manufacturing cells or islands). This makes more transparent what goes on in a company and, if the control and management system is designed in a more decentralized way and the staff working there have the proper qualifications, allows even production planning and production control functions to be moved back to the operations level. This results in shorter and faster decision making pathways and fewer hierarchical steps. Delegating to the group such further activities as setting-up, programming, repair, maintenance, and quality assurance is going to further improve flexibility. The added personnel cost arising as a consequence of higher qualification will be balanced out by the increased utilization of systems (availability). In order to be able to process orders quickly enough, manufacturing plants will in this way increasingly assume the character of service companies.

This new organizational concept has recently been proved to be highly successful both in small-lot production, e.g., in the machine building industry, and in the automobile industry, which - except in Sweden - used to be organized strictly along the lines of Henry Ford.

In 1985, the Nordenham Works of F&G had 567 employees producing low voltage switch gear and accessories and electric motors. They faced serious problems typical of so many similar operations: turnover stagnated, lifespan of products became shorter (as did delivery times), lots became smaller, cost reduction on the labor side through further automation seemed infeasible. They realised that their problems could not be solved by technical means only and decided to restructure the manufacturing and information systems step-by-step. They started with one part of the mechanical workshop assigning the following tasks to the 'production island': setting of machines and tools, manufacturing the products, stock-keeping of raw materials including replenishing, routine maintenance work, quality inspection and recording, work planning within a time-span of one week (given by production control).
After one and a half years, the results were convincing, e.g., throughput times were cut by half, output per capita had gone up 25 %, and the workforce was proud of their work.
This encouraged the firm to restructure the whole mechanical workshop and the assembly departments as well.
Since the island principle had worked so well on the shop floor, they were tempted to apply the same principle to clerical work in administration where they had similar problems in long chains between customer, sales, product design, tool design, production planning etc. They, therefore, installed 'administration islands' in one room to speed up and improve their tendering services. Again, speed, quality and self-teaching team effects were overwhelming. The next steps are to restructure materials management and production control.
Automated machinery and computers were used too, of course, but - as one manager put it - 'Wherever computer application threatens to become a self-contained, self-absorbed business it should be fought like the plague!'

Summarizing General Design Aspects

The findings experienced both in the process industries and in manufacturing correlate remarkably well.

General socio-technical design aspects which
may be derived are:

- importance of overall work structure,
- team approach with job rotation (so
  operators can replace each other),
- the maximum possible degree of automation
  not necessarily being the optimum,
- importance of (re-)qualification,
- great difficulty with innovation within the
  organization (changing traditional roles
  and privileges),
- transformation of computer outputs (interactions) into forms that fit human
  criteria,
- new technologies favoring/encouraging more
  holistic work contents (tasks)
  and decentralized structures (local
  autonomy),
- autonomous work groups especially needed in
  smaller plants (SMEs), but might be the
  better approach everywhere.

ECONOMIC ASPECTS OF AUTOMATION

In industrial organizations economic aspects
of automation are considered as a part of the
business activity. The purpose of the business unit is to satisfy the needs of individual consumers, other enterprises, or the society as a whole. Sufficient profitability
and balanced cash flow are essential for the
existence of the business unit in the long
term.
Decision makers responsible for allocation of
financial resources in different business activities are first of all interested in the
efficient use of invested capital. The common
practical measure is the Return On Investment
(ROI).

Lower-level managers responsible for the production process have to decide how to use the
available financial resources. The basic reason to develop automation may be economic,
technical, human or environmental, but in
practice the responsibility of the manager is
to make technically feasible sound economic
decisions which satisfy all the requirements
he is made responsible for. Such decisions
are easy, if costs and performance of technology are deterministic and can be measured
in monetary terms. Traditional financial
measures - like ROI -are not well suited when
considering systems-properties which cannot

be evaluated in monetary terms, such as
flexibility, acceptance of the work system,
or the impact of technologies on the company's long-term competitive position, let
alone human or cultural aspects.
In recent years, methods have been developed
- e.g., by Grob (1983) - that allow for consideration of non-monetary aspects, too, in
investment decision. In general, decision
makers must learn to rely in addition on
qualitative long-term strategic decisions.

CULTURAL ASPECTS OF AUTOMATION

In developing countries, machines are expensive and people are cheap. In contrast, machines are often inexpensive relative to
people in the developed nations. This difference is crucial when we consider appropriate
technology. The key question is the meaning
of the word 'appropriate'. As we are trading
more and more with third-world and non-west-
ern nations, the meaning of this word is becoming less and less clear.
Is it inevitable that the pressure of international competition sets equal standards for
the unequal? Are there no ways for an international task division? Basic technologies
are universal, but is it possible to have
different implementations and applications of
technology which consider specific cultural
and economic needs?
In this chapter, we will try to raise some of
these issues which go far beyond the area of
engineering. Nevertheless, control engineers
who quite often accelerate the unemployment
rate with their artefacts should also play
their part in solving global problems.

Culture and Technology

An aspect which has been largely neglected by
technologists for many decades, is how amenable a particular culture is to the technology which is being introduced to (or in many
cases imposed upon) it. It has been recognized by observers in African states, for
example, that - as tough as it might appear
politically to say it - in practice it probably takes two to three generations to
change the cultural values towards, or to implement the subsequent absorption of, new
technology. The experience of many education-
alists is that while it would be desirable to
train a whole generation of engineers as ra-

pidly as possible, this has proved to be impossible in practice. In most African states
one finds, for example, that only now are in-
digenously-trained engineers emerging, and
that typically these have come from families
whose parents have at least some technological background. Often, too, the grandparents
were school teachers or at least had some
introduction to science or technology. Simply
to impose a technology on a society and believe that it will be able to cope through
extensive training has been proved not to
work. It has led to rejection of technology,
or, in the case of colonized countries, has
simply become part of the oppressor's strategy.

These comments are made on the basis of engineering, but when one considers for example
that many so-called 'underdeveloped' countries do not even have a language which is
'numeric', then one can realize the problem
sphere of introducing new technological concepts! (Here a 'numeric' language is one
which contains inherent concepts of numbers
and arithmetic functions.)

In addition to problems of language and numerics, another aspect of culture which must
be considered in terms of introducing a new
technology is the fundamental attitude of
that culture towards labour and service. A
proud culture, which has always been viol-
ently independent, will have great difficulty
in absorbing technology which demands a large
degree of adherence to sets of rigid rules.
Again, one can look at the example of many
third-world countries in which stable nations
were subjected to the imposition of colo-
nialism which in many cases depended upon
master-slave relationships. As part of these
structures, the 'masters' insisted on the
'slaves' doing certain tasks which were essentially to the well-being of the colonialists, but were denigrating to a cul-
turally proud people. For example, in many
African tribes manual work was traditionally
not done by men! However, in many parts of
Central and Southern Africa the colonialists
pursuaded men to do manual tasks - even as
far as housework, previously totally reserved
for females. Of course then come the de-colo-
nialization days and it is little wonder that
one of the first things which goes is any
form of rigidly-structured manual work. It is
no surprise that the previously proud culture

reacts and wants to attempt to return to that
state of freedom!

What, then, can one conclude about the cultures in the East? Again it is difficult to
draw any rigid guidelines, but certainly if
one looks at the 'service culture' characteristics of Japan, one sees some interesting
concepts emerging.

For many years we have been bombarded by the
fact that Japan appears to be maintaining a
low unemployment rate. Whilst some recent
studies have pointed to the fact that this
might not continue, it cannot be denied that
Japan still has one of the highest employment
rates in the world. This is, of course, despite the fact that Japan has always been
pointed to as the major consumer of automation. The concept of countries, such as
Japan, exporting unemployment, of course, is
only too well-known and there can be little
doubt that the result of Japanese high efficiency in manufacturing has been the creation of unemployment in countries who simply
cannot compete.

Many of us have also been intrigued for years
by what the Japanese actually mean when they
state that they are 'creating' a service industry. If one looks at the UK or Europe, one
sees great difficulties with the proposed involvement of more and more people in the so-
called 'service' industries. Indeed, if one
looks at the European scene, one sees that
most people have some sort of hang-up about
serving their fellow human beings! Indeed, in
the third-world circumstances, in many ex-
colonies, the whole concept of service is
looked on with much suspicion -largely a result of the master-servant relationships created by the colonists.
Although things are changing, the Japanese
tradition of service is extremely strong and
still exists. Their culture, which has had a
unique ability to absorb other cultures,
other religions and other concepts into it,
and still carry on as a unified whole, has
maintained the concept of service as a critical one. This is not just related to service
to one company, and indeed many industrial
observers have commented on the fact that the
present loyalty to one single employer will
probably change in the next few decades. (It
is already changing to some extent.) However,

the concept of service goes much further. The
ticket inspector, for example, is respected
by all whom he serves; the young lady who
opens the lift doors perennially has a smile,
and seems to find a great deal of satisfaction just doing her job properly. The person
going around a hotel in Tokyo, checking that
any marks on the floor have been removed and
that all pieces of furniture are in order
does this with much pride. As an outsider to
the culture, of course, one cannot even hope
to get close to the motivations for these
ideas, and possibly there is some form of
compulsion. However, one observes that the
concept is still very strong.

Cultural Impact on Automation

The problem is, how does all this affect
automation? It has been claimed by many that
automation has two prime effects. One is that
it undoubtedly puts people out of work - in
terms of traditional, manual labor. This appears to be 'attractive' in that it might
suit, for example, African countries to re-
vert to a situation where the jobs in the
factories are at a high level and no menial
tasks are required. However, the first problem here is that the new tasks which have
been introduced require a high degree of
training. Since, as mentioned previously, it
takes several generations to produce competent engineers, the pattern which is seen
throughout decolonialized Africa, whereby
high-technology industries simply do not
survive, seems inevitable.

The other side of automation is the claim
that the consequences are to improve the
earning power of a country supporting many of
the service functions. Again the Japanese example is worth looking at.
In discussion with Japanese industrialists,
it appears that the intention is still to decrease the human content of their operation!
Most companies state publicly that they wish
to make more use of machines, and more use of
computer integrated manufacture, and that
they aim to lower the number of people required. This is against a background of operations such as watch-manufacturing lines,
producing over 100,000 quartz crystal watches
per week, and currently manned by literally 5
or 6 people, or complete flexible manufacturing systems, with up to 10 machine tools

attached, having single operators who not
only ensure that the system keeps running but
also load up the pallets, adjust the work
pieces etc. Despite this, the companies say
that they wish to go further.

Observers are nevertheless struck by the fact
that more and more people really do seem to
be used in the service industry, not just in
serving food, but the service industry on the
railways, in farming etc. In the busiest of
Tokyo stations, where over one million people
pass through the barriers every day, the Japanese are not (at this stage?) using automatic ticket barriers such as those currently
being introduced on the London Underground.
On entering the station, one's ticket is
clipped by one inspector, and it is taken
when one leaves by another. The stations
themselves are supervised by many workers who
ensure people get into the over-full Tokyo
trains. There are always enough ticket windows open to buy tickets, and always someone
around to give you directions.

Farming is still on a small scale, despite
the tremendous demand. Japan imports vast
amounts of its food, and yet large percen-
tages of the land are turned over to agriculture. No attempt seems to be made towards
automation here, and peasant farming still
appears to be the order of the day. Throughout the countryside one sees families busy
tilling the fields or their rice paddies. Occasionally tractors are used, but in many
cases very simple, primitive farming methods
are used.

How does this all work? Somewhere within the
Japanese structure there seems to be some
careful planning to determine what can be
automated - where the value can be added to a
product - together with a clear recognition
that it is essential to provide people with
some form of employment, even though, in certain cases, particularly for internal consumption, it might appear to be more economic
to automate.

However, in the African situation with a completely different approach to the service
industry, one is left with tremendous problems. The concept of service held by a significant percentage of the population, has
been distorted by the colonialists' days, and
service is seen as being degrading. The con-

cept, then, of moving employment further into
the service sector is highly questionable.

These comments are made by technologists and
not by sociologists. The point is that the
two should get together to try to resolve the
difficulties. This could mean, on one hand,
developing technologies amenably, or appyling
them appropriately, to a particular culture,
and, on the other hand, helping a culture to
utilize the possible benefits of technology
without involuntarily giving up cultural
values. Such endeavor will remain fruitless
in the case of most under-developed countries, as long as the Western understanding
of the cultures remains minimal and in many
cases naive. One of the characteristics of
our era is still the arrogance of one culture
over another, particularly when the first
culture is one which is technologically based
and, therefore, has strength through the
technology - in the first place through its
military applications.

Working Group on Cultural Aspects of Automation

Out of the IFAC discussions between pro-
ponents of social effects research and man-
machine systems, an increased awareness has
grown that these two approaches may require a
third supplementary focus on culture. It has
been suggested that it would be a unique contribution if a learned technical institution
like IFAC took the initiative in promoting an
understanding of the larger pattern of consequences based on its own professional endeav-
ours, and considering all its complexities.
The Technical Board of IFAC has therefore
just established the Working Group on Cultural Aspects of Automation within the Social
Effects Committee. Its Scope is

- to promote an understanding of the inter-
  play between technological development,
  social conditions and effects, and
  cultural change;

- to explore new approaches for
  cross-disciplinary research;

- to develop and transfer views and methodologies that include cultural aspects to
  control engineers and interdisciplinary
  groups dealing with automation systems
  design.

CONCLUSION

If we wish engineering artefacts to be re-
conciled with aspects of life, then we must
consider comprehensive sets of requirements
in engineering design. Ignoring any impact
that an innovation might have on the environment, or the people in the environment, will
be seen by history as being negligent. Of
course, certain effects will have to be accepted, but we must, together with a wide
cross-section of the people affected, investigate all aspects before positive decisions
are made.

In view of the enormous effect of automation
technology on both the private sphere and the
world at large, a technocentric thinking that
rejects a comprehensive technology assessment
should be abandoned. It is also important to
see that a large number of examples have
shown how a holistic approach to automation
design will result in systems with improved
overall performance. Not only do the resulting systems work better, but they are
better for the world in which we all live -
now and in the future.

REFERENCES

van Beinum, H. (1988). New Technology and
Organizational Choice. QWL Focus, 6,
3-10.

Bainbridge, L. (1982). Ironies of Automation.
In G. Johannsen and J.E. Rijnsdorp
(Ed.), Proc. IFAC Symp. on Man-Machine-
Systems. Pergamon, Oxford, pp. 129-135.

Bibby, K.S., F. Margulies, J.E. Rijnsdorp,
J.M.F. Withers (1976). Man's role in
control systems. In Proc. 6th IFAC World
Congress Boston. Pergamon, Oxford,
Part 3, Paper P4.

Böhle, F., B. Milkau (1988). Computerized
Manufacturing and Empirical Knowledge.
AI and Society, 2, 235-243.

Brödner, P. (Ed.) (1987). Skill Based Automated Manufacturing. Pergamon, Oxford.

Dreyfus, H.L., S.E. Dreyfus (1986). Mind over
Machine. The Free Press, New York.

Grob, R. (1983). Erweiterte Wirtschaftlich-
keits- und Nutzenrechnung. TÜV Rhein-
land, Köln.

Jones, B. (1984). Division of Labor and Distribution of Tacit Knowledge in the Automation of Metal Machining. In T. Martin (Ed.), Design of Work in Automated Manufacturing Systems. Pergamon, Oxford. pp. 19-22.

Klein, G.A. (1989). Do decision biases explain too much? Human Factors Society Bulletin, 32, Number 5, 1-3.

Mårtensson, L. (1987). Humanization of Technology Versus Human Engineering, Record of the Discussion Session. In R. Isermann (Ed.), Proc. 10th IFAC World Congress Munich 1987. Pergamon, Oxford. pp. 111-113.

Martin, T. (Ed.) (1984). Design of Work in Automated Manufacturing Systems. Pergamon, Oxford.

Martin, T., E. Ulich, H.J. Warnecke (1987). Appropriate Automation for Flexible Manufacturing. In R. Isermann (Ed.), Proc. 10th IFAC World Congress Munich 1987. Pergamon, Oxford. pp. 265-279.

Martin, T. (1988). The Need for Human Skills in Production - the Case of CIM. Paper at the Conf. on Joint Design of Technology, Organization and People Growth, Venice, Oct. 12-14, 1988 (to appear).

Moldaschl, M. (1989). Evaluation of expert system applications in industrial and services environments. In Bernhold, T., U. Hillenkamp (Eds.), Expert Systems in Production and Services, Vol. 2. North Holland, Amsterdam (to appear).

Ortner, W.A. (1988). The Islands Principle. Paper at the Conf. on Joint Design of Technology, Organization and People Growth, Venice, Oct. 12-14, 1988 (to appear).

Perrow, C. (1984). Normal Accidents. Living with High Risk Technologies. Basic Books, New York.

Pikaar, R.N., T.M.J. Lenior, J.E. Rijnsdorp (1985). Control room design from situation analysis to final lay-out; operator contributions and the role of the ergonomist. In Mancini, G., G. Johannsen, L. Mårtensson (Eds.), Proc. 2nd IFAC Conf. on Man-Machine Systems. Pergamon, Oxford. pp. 299-303.

Rijnsdorp, J.E. (Ed.) (1978). Automation and Humanization of Work. Pergamon, Oxford.

Rijnsdorp, J.E., R.N. Pikaar, T.M.J. Lenior (1984). Process control and organization. In Gertler, J. (Ed.), Proc. 9th IFAC World Congress Budapest 1984. Pergamon, Oxford. pp. 2627-2633.

Rodd, M.G. (1988). IT in Manufacturing - the MAP to a New Society? University of Wales, Swansea, UK (Inaugural Lecture).

Rouse, W.B. (1987). On meaningful menus for measurement: Disentangling evaluative issues in system design. Information Processing and Management, 23, 593-604.

Rouse, W.B. (1988a). Designing for Human Decision Making in Large-Scale Systems. Proceedings of World Bank Workshop on Safety Control and Risk Management. World Bank, Washington.

Rouse, W.B. (1988b). Intelligent Decision Support for Advanced Manufacturing Systems. Manufacturing Review, 1, 236-243.

Rouse, W.B., W.J. Cody (1989). Information systems for design support: An approach for establishing functional requirements. Information and Decision Technologies (to appear).

Rödiger, K.H. (Ed.)(1988). Limits of a Responsible Use of Information Technology. Recommendations of Working Group 8.3.3. of the German Gesellschaft für Informatik (GI).(Working paper, to be ordered from T. Martin.)

Ravden, S.J. and others (1986). Human Factors in the Design of a Flexible Assembly Cell. In P. Brödner (Ed.), Skill Based Automated Manufacturing. Pergamon, Oxford. pp. 65-69.

Sheridan, T.B., T. Vamos, S. Aida (1983). Adapting Automation to Man, Culture and Society. Automatica, 19, 605-612.

ACKNOWLEDGEMENT

The authors wish to express their appreciation to Mrs Rodd, Swansea, for her excellent editing work and to Mrs Alter, Karlsruhe, for doing all the type-setting so well.

ANALYZING COMPLEXITY AND PERFORMANCE IN A MAN-MADE WORLD
An Introduction to Discrete Event Dynamic Systems

Yu-Chi Ho

Pierce Hall, Harvard University, Cambridge, MA 02138, USA

1. INTRODUCTION

Picture yourself with the mythical Mr. T.S.Mits (The Scientific Man In The Street) and the task of explaining to him the phenomena and workings of (i) the Gulf Stream and (ii) a computer-controlled flexible manufacturing system (FMS). Both phenomena are real and both are not completely understood. However, for task (i), you face an easy assignment since you can draw upon knowledge of calculus and differential equations to provide a succinct description of ocean currents and fluid dynamics. For task (ii), no such ready-made models are in existence[1]. One is essentially reduced to using an algorithmic description not very different from writing a computer program to simulate the FMS. In fact, modern technology has increasingly created dynamic systems which cannot be easily described by ordinary or partial differential equations. Examples of such systems are production or assembly lines, computer/communication networks, traffic systems on both the air and land side of a large airport, military C^3I (Command-Control-Communication-Intelligence) system, etc. In these instances the evolution of the system in time depends on the complex interactions of the timing of various discrete events, such as the arrival or the departure of a job, and the initiation or the completion of a task or a message. The "STATE" history of such dynamic systems is piecewise constant and changes only at the discrete instants of time at which the events occur instead of changing continuously. We shall call such man-made systems Discrete Event Dynamic Systems (DEDS) as opposed to the more familiar Continuous Variable Dynamic Systems (CVDS) in the physical world which are described by differential equations. Although systems governed by difference equations are often referred to as discrete time systems, they have more in common conceptually with CVDS than with DEDS despite the name similarity (For purpose of this paper, we shall not bother to distinguish systems modeled by differential and difference equations. They are continuous time or discrete time CVDS). To help fix ideas for DEDS, consider a Flexible Manufacturing System (FMS) with several work stations, each consisting of one or more identical machines. These stations are attended by operators and/or inspectors. Following some routing plan, parts belonging to different classes arrive at these stations via computer control. The parts queue up in the buffers according to some priority discipline until a machine and an operator are available to work on them. Stations are connected by some kind of material handling system (MHS) or automatic guided vehicles (AGV) which transport parts from station to station until they finish processing and leave the system. Typical performance criteria of interest for such an FMS are average throughput (TP), flow or wait time (WT), and work in process (WIP). Note that with some simple changes in terminology from parts to messages, work station to nodes, routes to virtual circuits, MHS to communication links, AGV to packets, fixtures to tokens, etc., the above description could be for a communication network which moves and processes information packets rather than material parts. Multi-programmed computers operating simultaneously in batch and time-shared mode, paper work flow in complex bureaucracies such as the back office of insurance or brokerage offices, and passenger, ground, and aircraft traffic at any airport are other examples fitting this description. The point here is the pervasive nature of such systems in the modern world and the relative lack of good analytically and dynamically oriented model for their description.

Conceptually, we can visualize such a DEDS as consisting of **jobs** (parts or packets) and **resources** (machines or transmitters, operators or nodal CPUs, AGVs, MHS or communication links, and buffer spaces). Jobs travel from resource to resource demanding and competing for service. The dynamics of the system are determined by the complex interactions of the timings of various discrete events associated with the jobs and the resources. In this sense, DEDS are simple. There are only two objects in DEDS, jobs and resources, which interact. When they interact, a job occupies (receives service from) a resource for a random/deterministic period of time. If we let the number of jobs waiting at a resource be a state variable, $x(t)$, and furthermore consider the approximation that there are an infinite number of jobs each infinitesimally small, then we have the differential equation

$$dx/dt = \lambda - \mu \qquad x > 0$$
$$= \lambda \qquad x = 0 \qquad (1)$$

where λ is the arrival rate of the jobs and μ is the service rate of the resource. This is a deceptively simple model of one component of a queueing system or a DEDS. The complications come because λ and μ can depend in arbitrary ways on other state variables of the DEDS. For example,

(i) admission into a queue

(ii) routing to one of two resources

(iii) order of service by a resource

may depend on the "state" of the system. When there are a multitude of such decisions interacting in a system, the resultant resource allocation problem can be very complex. Here we have an endless variety of queueing disciplines, priorities, service requirements, routing, scheduling, resource sharing, and general logical conditions that need to be met for interactions to take place. Any attempts to describe a DEDS by (1), even in cases where the approximation is appropriate, result in r.h.s. of (1) so complicated as to be useless. This description amounts to the writing of a computer program to simulate the DEDS. In fact, general purpose discrete event simulation languages provide the constructs of jobs, resources, timing of events, and logical tests while the coding in such languages produces the description of the specific systems. At this point we simply have no convenient way to capture these descriptions mathematically with the same degree of efficiency as the case of CVDS with differential equations. Thus as mentioned earlier, the workings of such DEDS are described through a system of "rules of operation" or "algorithms." This is essentially a brute force approach. However, it is important to emphasize that DEDS are nevertheless "dynamic systems" in the usually understood sense of the term, that is, it is a quintuple consisting of {input set, output set, state set, state transition map, and output map}[2]. But the specification of these five objects is far from succinct and pristine as is the case in mathematical system theory. Nevertheless the fact that we can implement <u>general purpose</u> discrete event simulation languages can be construed as testimony to the "dynamical" nature of these systems.

It is instructive to look at a typical example trajectory of a DEDS which we have illustrated in Fig.1 in terms of the operations of a simple communication system.

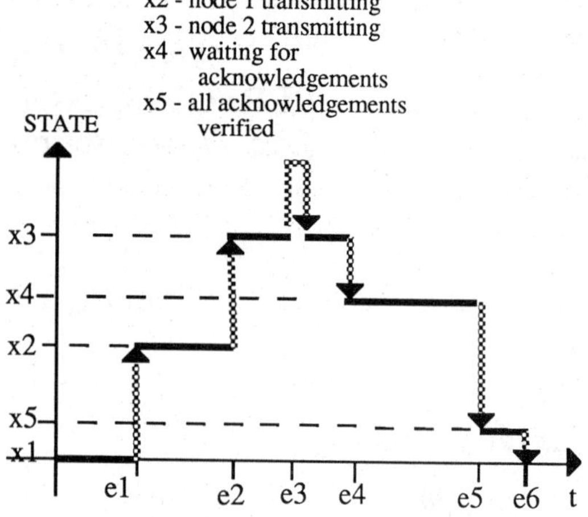

e1 - message 1 received by node 1 to transmit;
e2 - message 2 received by node 2 to transmit;
e3 - node 2 to repeat message 2;
e4 - all transmissions complete;
e5 - acknowledgements received;
e6 - shut down system.

Fig.1 An Example of a DEDS Trajectory

This is contrasted with a typical CVDS trajectory in Fig. 2 which represents a solution of some differential equations.

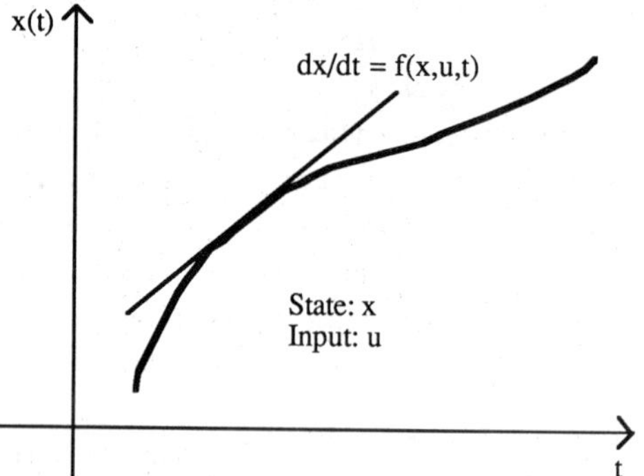

Fig.2 An Example of a CVDS Trajectory

For DEDS, the trajectory is piecewise constant and event-driven[3] . The sequence of piecewise constant segments represents the STATE SEQUENCE, and the duration of each segment, represents the HOLDING TIME in each state. While its duration is generally a continuous variable, the state takes value in a discrete set. Thus a sequence of two numbers (state and its holding time) basically characterizes the trajectory. On the other hand, a CVDS trajectory is constantly changing, with the state taking value in R^n, and is generally driven by continuous inputs. The definition of state is also different in DEDS and CVDS. In the former, it is the "physical" state (e.g., the number of messages at each node in the communication network awaiting transmission); while in the latter, it is the "mathematical" state, which is defined to be all the information required at a given time (other than external inputs) to uniquely specify the future evolution of the system. The mathematical definition of state includes the "physical" definition, but not vice versa.

Compared with CVDS, which is the underpinning of physics, DEDS is a relatively recent phenomena. It belongs in the domain of Operations Research (OR). Although a brief examination of any OR textbook tends to give the impression that OR represents a collection of useful techniques, such as decision analysis, mathematical programming, Markov chains, and queueing theory, OR can and was originally thought of as the science of operations and events of man-made systems. In fact, the father of OR, Professor P. Morse
Morse 1951], conceived OR as a parallel development to physics which is the science of nature and materials. Since it is represented primarily by man-made systems governed by "rules of operation," DEDS is in this sense properly a subject of OR. However, the developments of DEDS also received a large impetus from control system theory In particular, we submit that the concepts of dynamics, such as time constants, time and frequency responses, controllability and observability, have played and will continue to play important roles in the development of models and tools for DEDS.

1.2 MODELS OF DEDS

We cannot overemphasize the importance of a well-established modeling framework or paradigm such as the one found in Continuous Variable Dynamic Systems (CVDS) using differential equations. This framework permeates the entire scientific culture, facilitates cross disciplinary communications and influences our approach to problems. We submit that it is precisely the lack of a commonly accepted paradigm that underlies the relatively primitive analysis and synthesis effort we have with Discrete Events Dynamic Systems. While there has been no lack of attempts at constructing general models for DEDS, no consensus has been developed as to which one of the models has the potential to eventually serve as the analogy of the differential equation paradigm for CVDS. Among these modelling efforts we count

o   Markov Chains/Automaton Models -- To this group we also assign Petri nets, finite state machines.

o   Min-Max or Dioid Algebraic Models

o   Communicating Sequential Process Models

o   Queueing Network Models

o   Generalized Semi-Markov Process Models -- To this category we include all efforts relative to general discrete event simulation. These can be further classified according to Fig.3.

	Timed	Untimed
Logical	*temporal logic =>* *timed =>*	Finite State Machines; Petri Nets
Algebraic	Min-Max Algebra	Communicating Sequential Processes
Performance	Markov Chain; Queueing Networks; GSMP & Simulation models	

Stochastic--> <--Nonstochastic

Fig.3   Different Models of DEDS

The distinction between untimed and timed models of DEDS is clear. The untimed models emphasize the "state sequence" of a DEDS and de-emphasize or omit entirely the "holding time" specifications for every state. Thus in such models we primarily ask questions of a qualitative or logical nature (e.g., "yes" or "no", "true" or "false"). Timed models, on the other hand, incorporate "time" as an integral part of the model and are more suited for answering performance-related questions. Finally, algebraically based models, such as finitely recursive processes (or communicating sequential processes) aim to capture the description of the trajectories of DEDS in terms of a small set of algebraic operations on functions of states and events in very much the same way that CVDS are succinctly described by differential and algebraic operations on functions of state and inputs. However, one should be cautioned not to read too much into the classification. The boundaries of Fig.3 is by no means absolute. For example, timed Petri-nets often contain stochastic effects.

To describe adequately the features of these models would require several books. This is not the purpose of this introduction. Instead, we refer interested readers to a separate collection of papers where the working of these models are discussed in detail [Ho 1989]. In particular, in [Cassandras & Ramadge 1990] all the models are discussed in terms of a single and relatively simple example. There are several reasons for the lack of a universal modelling framework for DEDS. They can be understood by an examination of the desiderata for a DEDS model of all seasons.

The discontinuous nature of **Discrete Events**
The physical states of a DEDS are inherently discrete. Although it is possible on occasion to analyze DEDS with success by continuous approximations via diffusion models or by dealing exclusively with rates rather than explicit constituent parts, it does not seem possible to avoid the discrete nature of the problem entirely. Any kind of universal DEDS model must acknowledge this. Also, "discrete" is to be distinguished from

the commonly used adjective "digital" in CVDS where digital simply means that we do numerical analysis rather than real variable analysis.

The continuous nature of most **Performance Measures**

The performance measures for DEDS are mostly formulated in terms of continuous variables such as average throughput (the rate of occurrence of certain events) , waiting time (duration of certain events or states), utilization (% duration of certain events/states), inventory or queue length, profit, etc. In fact, "time," a fundamental performance variable, is by definition and convention a continuous variable. There is little technological or mathematical advantage in considering a discrete time or "sample-data" model of DEDS. In fact, many tools such as queueing network theory and perturbation analysis explicitly rely on the smoothing properties of the "expectation" or "average" operator to make analysis possible.

The importance of **Probabilistic Formulation**

As Murphy's law of human endeavor would have it. "Anything that can go wrong, will." Unscheduled disturbances and/or breakdowns are facts of life in DEDS operations. Almost all performance-oriented approaches to DEDS incorporate stochastic effects as integral parts of modeling. This, however, does not imply that a totally deterministic approach (e.g., Min-max algebra and Petri-nets) has no place in the scheme of things. It does mean that unless useful stochastic extensions are made, such models cannot be expected to replace probability-based models such as GSMP in performance studies.

The need for **Hierarchical Analysis.**

Many man-made operations, such as factories, operate on yearly - quarterly - monthly - weekly - daily schedules. Military commands are divided into theater - army - battalion - platoon levels. The details required for modelling and control are different for each level. Furthermore, different aspects of DEDS analysis (e.g., planning vs. operation) often take place at different time scales leading to different levels of detail. A good hierarchical DEDS model should have the property of being able to operate at any level by using the same set of constructs without the necessity of inventing new elements for each level. This is also important for aggregation and decomposition analysis.

The presence of **Dynamics**

As the late C.S. Draper of MIT was fond of saying[4] , "So damn few things in this world are static!"[3]. If we are to reach the same level of development for the dynamic control of DEDS as we have with CVDS, then it seems obvious that we must adopt models that can capture the dynamics and transient behaviors of DEDS. We submit that by ignoring the dynamical nature of DEDS and concentrating on the output analysis of the system as a black box, extant discrete event simulation literature gives up a considerable amount of valuable information inherent in the system output. Single run gradient estimation, such as likelihood ratio methods and perturbation analysis, have demonstrated this. See also the papers by Glynn and Cao in [Ho 1989]

The feasibility of the **Computational Burden.**

The number of discrete physical states of a DEDS explodes in combinatorial fashion. A brute-force enumeration of the states along the lines of a finite state machine or Markov chain without additional structure quickly leads to infeasible computational requirements even for the largest computers envisioned. Successful performance-oriented models must acknowledge this computational burden. Thus we need to distinguish between conceptual and computational models. If the computational requirement of a model grows exponentially with its size (e.g., NP-complete[5]), then the model cannot claim to have "solved" the general performance problem.

Since DEDS is primarily a man-made rather than a physical / natural system, it tends to interact more with humans than with nature. A controlled but unmanned spacecraft to Mars is interacting mostly with nature (Newton's Law of motion and gravitation, the electromagnetic spectrum, etc.). On the other hand, a Flexible Manufacturing System, even under computer control, is interacting more with human operators through display screens or actual materials handling. This has implications in practical implementation of any DEDS theory in two respects. First, AI interface and natural language-processing ability may be a much more important factor in DEDS than in CVDS.[6] In manufacturing automation, for example, DEDS are required to interface with personnel of lower skill levels than with those operating aerospace vehicles. Secondly, since DEDS are totally man-made, there are no invariant physical laws to constrain system configurations. Combinatorial explosion of complexity can be an easy occurrence. Consequently, a day when all solutions to all DEDS problems can be reduced to an algorithm is unforeseeable. Some form of rule-based solutions approach and/or fuzzy logic which tolerates imprecision, will probably be a necessary part of a general DEDS tool chest.

In summary, we see that DEDS research lies at the intersection of System Theory, Operations Research, and Artificial Intelligence - the STORAI triad in Fig. 4.

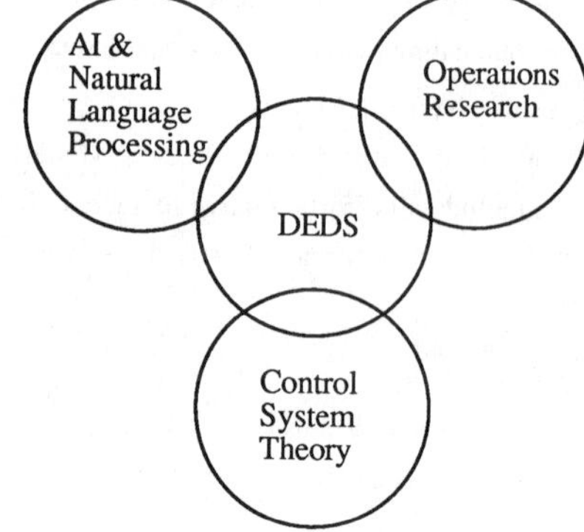

Fig.4 The System theory, Operations Research, and Artificial Intelligence Union

Finally, as **scientific** disciplines must have both experimental and theoretical components, so must DEDS . Carefully designed experiments accumulate evidence upon which a relevant theory can be built. Conversely, analytical reasoning pinpoints further experiments to be conducted for validation. To give an example, consider the area of manufacturing automation which is often touted as a prime area for the application of DEDS research. Much of the performance-oriented literature using queueing network theory and discrete event simulation deals principally with performances which are related to "resource contention and allocation." This assumes that the individual operations of a manufacturing process are sufficiently well-understood to be totally described by timing considerations. Yet problems, such as the yield percentage in semiconductor wafer manufacturing, have more to do with the physics and chemistry of materials involved in unit manufacturing operations than with resource contention. Direct experimentation and statistical quality control are integral parts of this problem. A second example is the subject of perturbation analysis itself. This tool most probably would not have been invented if it did not have the experimental evidence to support the early researches. Without this "observation - conjecture -experiment - theory - validation" cycle to inspire and to guide research, we might easily end with only a partial solution or a not so relevant model if a totally axiomatic mathematical development of DEDS is followed [Ho 1982, Hogarth 1986].

The purpose of this introduction is to advocate the rich opportunities in the field of DEDS. DEDS is used here in the narrow sense as a parallel to CVDS in the form of dx/dt=f(x,u,t). We do not address issues such as protocol, implementation, integration, interfaces, and hardware in DEDS. By this we do not imply that such issues are unimportant. But our narrower definition of DEDS is consistent with our past use of the term as well as with the convention adopted in discussions of control theory for CVDS. Conceptually, we can visualize the entire sweep of classical control-theoretic problems such as controllability and observability, estimation and identification, information and control, awaiting formulation for DEDS. Practically, many man-made systems whether they are large metropolitan airports (e.g., Heathrow international), or complex communication networks (e.g. Federal Telecommunication System 2000 for the US. government), or a modern factory (e.g. the new GM Saturn car plant) represent real opportunities for a new generation of control problems.

References.
• P. M. Morse & G. Kimball (1951) *Methods of Operations Research*, MIT Press & Wiley
• Y. C. Ho, Editor (1989) Special Issue on the Dynamics of Discrete Event Systems, *Proceedings of the IEEE*, January 1989
• Y. C. Ho, (1982) "Is it Application or os it Experimental Science? " -Editorial, *IEEE Transaction on Automatic Control*, AC-27, December 1982
• C. Cassandra & P. Ramadge, Editors (1990) Spcieal Issue on Discrete Event Dynamic Systems, *IEEE Control System Magazine*, June 1990
• R. M. Hogarth (1986) "Generalization in Decision Research: The Role of Formal Models", *IEEE Transaction on Systems, Man, & Cybernetics*, SMC-16,3, 439-449.

This paper is an updated version of an introduction written by this author for the special issue of the Proceedings of IEEE on the Dynamic of Discrete Event Systems, January 1989. The work reported here was supported by ONR contracts N00014-89-J-1023 & N00014-90-J-1093, NSF grants ECS-85-15449 and CDR-88-03012, and Army contract DAAL-03-86-K-0171.

[1] We do not regard queueing networks or automata models of an FMS as being accessible to Mr. T.S. Mits. More about these models can be found in the special issue of the Proceedings of IEEE January 1989 and the IEEE Control system Magazine June 1990.

[2] The transition and the output map for DEDS have to be interpreted to include objects such as time advance and event selection mechanism which are commonly found in discrete event simulation languages.

[3] As a possible generalization, we can replace each piecewise constant segments of the DEDS trajectory in Fig. 1 by a trajectory generated by a differential equation with a starting and ending state. This will result in a mixed CVDS and DEDS possibly modeling a large chemical plant or a power distribution system. The continuous trajectory segments can represent the system behavior under various mode of operation such as, normal, emergency, etc. while the discrete event transitions capture the high level supervisory controls. This is another example of the reason of a "dynamical" emphasis in the study of DEDS.

[4] As quoted in the video tape "Thinking about the Future" MIT Development Office 1988.

[5] NP-Complete, which stands for Nonpolynomial-Complete, is a technical term roughly indicating that the computational complexity of the problem is such that it cannot be bounded in computational time by any polynomial function of the size of the problem.

[6] As a prime example for the difficulties of natural language processing, consider the following sentence which appears on the side of a tube of well known brand of toothpaste: "for best results, squeeze tube from bottom and flatten as you go". Any child who can read can understand the meaning of this sentence. However, we are far from being able to devise an AI system with such degree of comprehension.

GENESIS OF NOTIONS "CRITERION" AND "EXTREMIZATION"

M. A. Aizerman

Institute of Control Sciences, Moscow, USSR

Abstract. The definition of the notion "criterion" is made more precise
and in short a developing of the notions "criterion" and "maximization
of criterion" is described. The classical methods of generalizing of
these notions and the three ways of their further generalization are
stated: transiting from binary relations to hyperrelations, transiting
to external description (i.e. to analysis of choice functions) and
developing of the notions of "pseudocriterion", "collection of pseudocriteria" and "pseudocriterion choice".
The basis of all the modern generalizations is the idea about influencing of context on criterial values of alternatives.

Keywords. Criterion, extremization, pseudocriterion, pseudocriterion
choice, binary relation, hyperrelation.

1. INTRODUCTORY REMARKS

Among the notions used in the control theory side by side with the notions, introduced specially for needs of this science,
interbranch notions are widely used as
well. Typical examples are "stability",
"oscillation", "equilibrium", etc. When
the notions of this kind are discussed,
the subject of theoretical analysis are
not so much   calculations and ways of
use of values or phenomena, corresponding to these notions, as discussing of the
sense put in some or other notions, its
different definitions and their development (genesis) with time.

To my mind the notion of criterion widely
used in control theory from the beginning of fifties (in connection with optimization and adaptivity approach) is a
typical example of this interdisciplinary
notion. Under some other but similar de-
nominations - "goal function", "utility
function", "scale estimate", etc. this
notion is basic for many economic, psychological, polytological and other models. Naturally, that as well as for other interdisciplinary notions, the question about significance of the term "criterion" is not evident. This notion is
"living", developing and is subjected to
discussion and analysis. Therefore in regard to the criterion there arise two
ranges of problems: 1. How to use criterion when it is specified?, and 2. What
and when must be termed as a criterion?
Almost all the literature on optimization
in control theory is looking for an answer to the question "how to find"? In the
last years more and more the question:
"what and when to search"? - becomes the
subject of a discussion. Just about this
question, i.e. strictly speaking, about
genesis of the notion "criterion", about
possible generalizations of this notion
is the text offering below to the atten-

tion of the readers.

2. WHAT IS "A CRITERION"?

Let a set A of alternatives $\{x_1, x_2, \ldots, x_n\}$ be specified (for simplicity it is
supposed further to be finite) and it be
required to "estimate" for "comparing"
and picking out from these alternatives
a subset Y_A, the properties of which are
characterized by the words "better", "worse", "middling", etc. Let the process of
picking out this subset be realized by
the scheme, presented in fig. 1: first all
the elements of A are mapped by an operator $f(\cdot)$ onto the numerical axis, and
then an operator $\mathcal{M}$, using numbers $f(x)$,
assigned to alternatives $x \in A$, picks
out from A a subset Y_A. Usually numbers $f(x)$ have a sense "worse-better",
i.e. the alternative x_i is acknowledged to
be better than the alternative x_j, if
$f(x_i) > f(x_j)$. Then the best alternatives are naturally picked out by an operator $\mathcal{M}$, realizing a process of maximization on A :

$$(1) \qquad Y_A = \arg \max_{x \in A} f(x)$$

or equivalently[1]

$$(2) \qquad Y_A = \{ y \in A : \overline{\exists x \in A}, \text{ s.t. } f(x) > f(y)\}.$$

There exist different ways of determining
the operators $\mathcal{M}$, which using the same
scale $f(x)$, specify the "middling" alternatives, etc.

Usually the term criterion is given either to the operator $f(\cdot)$, mapping A

____

[1] Here and further the bar on top signi-
fies a negation, i.e. the symbol $\overline{\exists}$ signifies "does not exist".

onto the numerical axis, or to the result of the working of this operator - the constructed scale $f(x)$, $\forall x \in A$ itself.

Though usually it is not explicitly stipulated, but it is always silently supposed that the mapping of each concrete x^* does not depend "on context": independently of the nature of the two sets A_1 and A_2, containing this x^* ($x^* \in A_1$ and $x^* \in A_2$), on mapping A_1 and A_2 alternative x^* will have identical estimates $f_{A_1}(x^*) = f_{A_2}(x^*)$. According to this proposition when not A but any its subset $X \subseteq A$, containing this x^*, is mapped by the operator $f(\cdot)$ there will be for x^* as result one and the same number $f(x^*)$. This fact, which at first sight is so obvious that it is not stipulated specially, will be further central in this text when considering nonclassical generalizations of the notion "criterion". These generalizations arose recently, and before we will transit to them let us consider the generalizations of the notion "criterion", not infringing the hypothesis "independence on context", which had been developing from the beginning to the middle of XX century and can be termed by right the classical ones.

Although the ordinary notion "criterion" in the scheme[2] of Fig. 1 permits using of different operators - "choice rules" $\mathcal{F}$ - nevertheless the extremization rule (1) is most closely connected with it historically. Both the classical and the modern nonclassical generalizations of the notion "criterion" naturally have resulted in a generalization of this extremization rule as well.

3. CLASSICAL GENERALIZATIONS OF THE NOTIONS "CRITERION" AND "EXTREMIZATION"

The first generalization of the mentioned notions is connected with a transition from a "scalar" to a "vector" understanding of the term "criterion". In the same way as in Fig. 1 a multicriterial situation is presented in Fig. 2. The set A "is processed" by m operators $\{f_1(\cdot), f_2(\cdot), ..., f_m(\cdot)\}$ attributing to each $x \in A$ m numerical estimates $f_1(x)$, $f_2(x)$,...,$f_m(x)$. Contensively this scheme is interpreted either as a presence of some set of criteria for estimation of each alternative or as a presence of a collective of m persons, each of them has one own criterion but when decision about Y_A is a collective one. But under any these interpretations it is silently supposed as before, that the principle of "independence on a context" is presented, i.e. for any two sets A_1 and A_2, containing x^*, ones and the same m numbers $f_1(x^*),...,f_m(x^*)$ are attributed to this x^*.

Used in the case of one criterion the rules $\mathcal{F}$, which process a scale into a subset of "the best" alternatives are reformulated for using in the "vector instance" $\{f_i(\cdot)\}$. Under extremization as before is understood the rule (2), if instead of $f(x)$ and $f(y)$ we shall substitute in it $f_i(x)$ and $f_i(y)$ and demand its execution for all i:

___
2) Figures and tables are given at the end of the text.

$$(3) \quad Y_A = \{y \in A: \overline{\exists} x \in A \text{ and } \overline{\exists} i \in \{1,2,...,m\}, \text{ s.t. } f_i(x) > f_i(y)\}.$$

The thus picked out subset Y_A - Pareto maximum (minimum) in A which can contain many alternatives, in particular all the set A . Because of that there were introduced many rules $\mathcal{F}$, picking out a part of the Pareto set. Typical example is a collected-extremization rule:

$$Y_A = \bigcup_i Y_A^{(i)}$$

where

$$(4) \quad Y_A^i = \{y \in A: \overline{\exists} x \in A, \text{ s.t. } f_i(x) > f_i(y)\}$$

Rules (3),(4) and their numerous analogues are natural generalizations of the extremization rule, caused by the first generalization of the notion "criterion" - by a transition to multicriterial situation.

The second, more profound generalization, is connected to the structure itself of the rules (2) and (3). These rules realize in fact pairwise comparing alternatives x and y of A , but a comparing of a special kind by value of a criterial estimate (or estimates). Naturally there arose an idea of far reaching generalization: to implement a replacing of the pairwise comparings by value of the criterial estimate in the definition of extremization rule (2) or (3) by an arbitrary binary relation of a general kind: $x \mathcal{D} y$ ("alternative x is in some relation $\mathcal{D}$ with y "), comprehending the comparings by values of criterial estimates as a particular case.

With this replacing both the rules (2) and (3) transit in the rules

$$(5) \quad Y_A = \{y \in A: \overline{\exists} x \in A, \text{ s.t. } x \mathcal{D} y\}.$$

Then instead of the criterion $f(\cdot)$ or a collection of the criteria $\{f_i(\cdot)\}$, specified on $\forall x \in A$, is introduced in consideration a more general object - an arbitrary binary relation $\mathcal{D}$ or, that is the same, a directed graph, in which vertices are alternatives, and an arc from the vertice x goes to y , if $x \mathcal{D} y$ is true. The notion $\max_{x \in A}$ is replaced then by a notion "nondominant vertices", i.e. by a set of vertices, to which arcs do not come up.

On the way of such a generalization of the notions "criterion" and "extremization" there arose, however, some difficulty: Y_A, picked out by the rules (2) or (3) is always non-empty, and Y_A picked out by the rule (5) can be empty, for example, if the relation $\mathcal{D}$ is a cycle on A . As a way out of this difficulty appeared restrictions on considered relations $\mathcal{D}$ (for example, a requirement of acyclicity of $\mathcal{D}$ is introduced), or the rule (5) is completed by a method of choice of Y_A in case of absence of vertices-dominants.

In connection with a transition from the choice of the best alternatives by criteria to choice of the best alternatives by arbitrary binary relations, there arises naturally a question: In what case are these two ways of choice equivalent?

The answer to this question is given by the adduced further Theorem 1. Let us pay our attention before to the fact that the criterial choice is always non-empty. Therefore the choice by binary relation can be equivalent to the criterial one, only if this relation is acyclic. Of course, choice not with any acyclic binary relation is equivalently reduced to criterial choice.

Theorem 1. Let an acyclic binary relation $\mathcal{P}$ be specified on A. For existing of such a collection of criteria $\{f_i(\cdot)\}$, by which for any $X \subseteq A$ with this relation $\mathcal{P}$ a rule of dominants' choice picks out a subset $Y \subseteq X$, coinciding with Pareto-maximum on X for this collection of criteria $\{f_i(\cdot)\}$, it is necessary and sufficient that $\mathcal{P}$ should be transitive. For existing one of such a criterion, it is necessary and sufficient, that $\mathcal{P}$ should be transitive and negotransitive.

Note. The relation $\mathcal{P}$ is transitive if for any triple of alternatives x, y, z of A from $x\mathcal{P}y$ and $y\mathcal{P}z$ follows $x\mathcal{P}z$. It is negotransitive, if from $x\mathcal{P}y$, $y\mathcal{P}z$ follows $x\mathcal{P}z$.

Notice now, that with transiting from "comparing by criteria" to arbitrary binary relations we imperceptibly extended the frames of the principle "independence on context": in the case of using of criteria every alternative is valued separately and in the case of the relation $\mathcal{P}$ a pare of alternatives" is valued" simultaneously and an estimate of one alternative does not have any sense and is not used anyhow. But nevertheless one circumstance, contained in "the principle of independence on context" is preserved: before, for criterial approach, for any two subsets X_1 and X_2 of the set A, containing x^{*1}, the criterial estimates x^* had one and the same value; now with replacing of criteria by a binary relation of a general type, if X_1 and X_2 contain alternatives x^* and y^*, and if $x^*\mathcal{P}y^*$ in X_1, then this relation is contained in X_2 and vice versa.

A transition from criteria to arbitrary binary relations was imagined so wide generalization of the notions "criterion" and "extremum", that to the middle of XX century the problem of development of the notions "criterion" and "maximum" seemed to be settled, to be brought to perfection. This generalization can be indeed termed the classical one.

4. INSUFFICIENCY OF THE CLASSICAL
GENERALIZATIONS. DEPENDENCE OF
ESTIMATES OF ALTERNATIVES ON
CONTEXT

The feeling of completeness of the problem was not long-term. Already in the seventies it began an accumulation of examples, demonstrating insufficiency of the classical scheme of generalization of the notions "criterion" and "extremization", and arguments, prompting in which direction the search of new generalizations should be developed. For coming up to their description we need to explain one more circumstance, which is always implied, but is not explicitly stipulated. The question is about consideration of procedures, presented by schemes in the figures 1 and 2 not as procedures, describing one act of transformation of the set A into the subset Y_A,

but as description of an algorithm, i.e. mass and effective procedure, transforming any subset $X \subseteq A$ in $Y_X \subseteq X$. This circumstance is closely linked to mentioned above principle of "independence on context".

Indeed, let us suppose only the set A to be considered, initially $Y_A \subset A$ is somehow specified and the question is put: is there a criterion $f(\cdot)$, such that maximization on $f(A)$ gives this Y_A? Positive answer on this question is evident: it is sufficient to consider as a criterion a characteristic function, equal to 1 on elements from Y_A and 0 on elements from $A \setminus Y_A$. Situation is different in case if the scheme in Fig. 1 is applied not only to A but to all $X \subset A$ as well. Let for each X initially its own $Y_X \subseteq X$ be specified, and the same question be put: is there a criterion $f(\cdot)$, such that its extremization on any X exactly determines this Y_X? It is clear, that now positive answer is possible only in exceptional situations: it can be found, that for one and the same alternative x^*, included both in X_1 and in X_2 ($X_1, X_2 \subset A$) we would to set $f(x^*) = 0$ for X_1 and $f(x^*) = 1$ for X_2, i.e. the estimates would depend on context.

Let us consider now several examples, analysis of which took an important part in a search of new generalizations of the notions "criterion" and "extremization".

1. Score sum in tournament. Let X be players or teams, A — a set of them, and a binary relation be specified by the tournament table T (for simplicity there are no draws), i.e. by a full graph on A, and the scale be constructed as a "winnings sum". Then $X \subset A$ is "subtournament", which we have when from T the lines and the columns, corresponding to $x \in A \setminus X$ are removed. It is clear that "the winnings sum in tournament" of one and the same player x^* depends on X, and in that sense the scale "winnings sum" is not a criterion (even if classical generalizations are taken into account).

2. Scale "number of votes for". Let us take notice of the scheme in Fig. 2. Let the rule $\mathcal{T}$ (for any $X \subseteq A$) attribute to each $x = x^* \in X$ an estimate equal to number from $f_1(x), \ldots, f_m(x)$, by which this x^* is situated on the first place, i.e. constructs a scale "number of the first places". It is directly seen that the estimate x^* depends on X, i.e. that choice of persons, taking "the most number of places" — is not criterial choice even in the sense of "classical generalizations". It is easy to show that the matter stands analogously, if into Y_A we will include those $x \in X$, which have turned out on the first place not less than by k criteria from $f_1(x), \ldots, f_m(x)$ for any fixed k ($1 \leq k < m$).

This example has a simple contensive interpretation in the voting systems if every i-th voter is supposed to have his criterion $f_i(\cdot)$ on A and to give his vote to that alternative (for example votee), which by this criterion is valued above the other votees from bulletin $X_1 \subset A$.

29

3. __Closedness under intersection and union__. Suppose now the rule π "to be arranged" so that: for fixed X with using of every $f_i(x)$ separately, the rule picks out the "subset of the best alternatives on X by $f_i(x)$ " and includes in Y_X the intersection of these subsets. It is easy to show that in this case as well a situation may be, when there exists no collective criterion $F(\cdot)$, by which on every X the alternatives from thus specified intersection would have by $F(x)$ the most estimate. Nevertheless in this case there exists a binary relation $\mathcal{D}$ on A such that for any X alternatives x^* from the indicated intersection on the subset X are not dominated by other alternatives ($\exists y \in X$, s.t. $y \mathcal{D} x^*$). This circumstance shows the strength and importance of classical generalizations.

Consider, nevertheless, not an operation of intersection, but an operation of union: from X into Y_A are included all alternatives, which are turned out on the first place in X even though by one criterion from $f_1(x),\ldots, f_m(x)$. It has been shown, that in this case we may have situation when binary relation does not already exist as well, in which for any X the alternatives from Y_X dominate all others from $X \setminus Y_X$, i.e. the scheme of Fig. 2 in this case does not "generate the criterion" even in the sense of classical generalizations.

4. __Picking out a part of the Pareto set.__ The last example 3 (with union of the best) is interesting by the fact that for any X in it a set Y_X is picked out, belonging the Pareto maximum on X in the space of criteria $\{f_1(x),\ldots, f_m(x)\}$. Analysis of some other rules, picking out from X not the whole Pareto subset, but its part, has shown that these rules also lead out of the frames of "classical generalizations", i.e. they realize a choice Y_X from X , which does not coincide for all X with the choice of the whole Pareto set by whatever collection of criteria.

In all these (and many other studied) cases the insufficiency of classical generalizations of the notions "criterion" and "extremization" is connected with the effect of dependence of alternatives estimate on comparing context (on X), which is only revealed, if criterial schemes of type shown in Fig. 1 and Fig. 2 are understood as "mass procedure", applying not only to the set A , but to subsets $X \subset A$ as well.

Further generalizations of the discussed notions because of that must be connected with taking into consideration of context for forming of criteria. Two ways have been outlined for solving of this problem:

1°. Immediate generalization of a criterial notation – transition to the functions, depending on x , not only on A , but on all $X \subseteq A$ as well. With using such generalizations the notion "extremization" does not change, because Y_X is determined every time by usual way for concrete X .

2°. Transition from the binary relations to the relations between the sets of alternatives. On this way the notion "domination" (as widening of the idea extremization) requires further generalizations.

It is convenient for description of two these ideas to use the conception of "input-output" description of a mapping $X \rightarrow Y_X \subseteq X$ (of a diagram of the mapping), i.e. of a choice function. Further we will first remind of ideas, connected with choice functions, then consider the way 2° introducing of hypergraph constructions instead of graph ones and only after that the way 1° (transition from criteria to pseudocriteria) is discussed.

5. EXTERNAL DESCRIPTION OF
CRITERIAL CHOICE. CHOICE
FUNCTIONS

If "at input" of the scheme, shown in Fig. 1 ("one criterion") or Fig. 2 ("m criteria") can enter not only the whole set A , but the subset $X \subseteq A$ as well, then at output the rule π picks out a part of alternatives $Y_A \subseteq A$ (or respectively, $Y_X \subseteq X$). In such a way a set of pairs $\{X, Y_X\}$ arises, i.e. the choice function $Y_X = C(X)$, specified on sets $X \subseteq A$ and having as its values the subsets $Y_X \subseteq X$. Generally speaking $Y_X = \emptyset$ is not excluded; if $\forall X \subseteq A : C(X) \neq \emptyset$, then we say in this case, that the choice function $Y_X = C(X)$ is non-empty.

Naturally, the question arose: whether all choice functions (i.e. sets of pairs $\{X, Y_X \subseteq X\}$) can be got by scheme of Fig. 1 or 2 under some choice criterion $f(x)$ or under some choice of criteria set $\{f_i(x)\}$, with π – an extremizational rule (1) (respectively (2))? If the answer is negative, then what are the choice functions, for which such a criterion-extremizational generation is possible? It turns out that to answer this question we need to introduce characteristic conditions on choice functions:

Heritance condition (H)
$$X' \subseteq X \Rightarrow C(X') \supseteq C(X) \cap X';$$
Monotonicity condition (M)
$$X' \subseteq X \Rightarrow C(X') \subseteq C(X) \cap X';$$
Constancy condition (K)
$$X' \subseteq X, \ X' \cap C(X) \neq \emptyset \Rightarrow C(X') = C(X) \cap X'$$
Concordance condition (C)
$$X' \cup X'' = X \Rightarrow C(X') \cap C(X'') \subseteq C(X)$$
Rejection condition (R)
$$C(X) \subseteq X' \subseteq X \Rightarrow C(X') = C(X).$$

It has been shown (Aizerman, 1985), that in the functional space $\mathcal{C}$ of choice functions these conditions specify domains, mutually located as it is shown in Fig. 3. It is essential that all eight intersections $H \cap C \cap O, \bar{H} \cap C \cap O, \ldots, \bar{H} \cap \bar{C} \cap \bar{O}$ are non-empty and that strict embeddings hold $M \subset C; \ M \cap H \subset K \subset H \cap C \cap O.$

Crupcial is

__Theorem 2.__ For generating of the non-empty function $C(X)$ by extremization of some criterion (or Pareto-extremization of some criteria collection) it is necessary and sufficient that it should belong to the domain K (respectively to $H \cap C \cap O$). For generating of the function $C(X)$ by a choice of dominants under some binary relation, it is necessary and sufficient that it should belong to the domain $H \cap C$.

30

Remark. The functions $C(X)$, belonging to the intersection $H \cap M$, are generated by the trivial procedure "two-valued criterion": to every alternative from A is initially ascribed estimate 0 ("not fitted") or 1 ("fitted") and from any X the alternatives with the estimate 1 are picked out, but if there are not such alternatives, then the choice is empty.

According to Theorem 2 not only criterial choice but all its classical generalizations as well generate choice functions from a comparatively narrow domain of the space $\mathcal{C}$ – from the intersection $H \cap C$. At the same time in all the examples of Section 3 (choice by score sum in tournaments, by scale "number of voices", selecting of a part of the Pareto set, etc.) choice functions sure going out of this intersection $H \cap C$ can be generated.

Now all is prepared for description of further non-classical generalizations of the notions "criterion" and "extremization". Such generalizations can be constructed as generalizations of choice by binary relations and as an immediate generalization of the notion "criterion".

6. GENERALIZATION OF NOTION
"CHOICE BY BINARY RELATION"

Let us introduce into consideration two types of hyperrelations:

1°. Relation G_1 between a subset of alternatives and individual alternatives:

$$V G_1 x_k, \quad V \subseteq X, \quad x_k \in A.$$

2°. Relations G_2 between pairs of any subsets of alternatives of type

$$V G_2 W; \quad V, W \subseteq X.$$

The images of these relations are hypergraphs, in which oriented arcs are directed from a set of vertices to individual vertices (example in Fig. 4-a) or from some sets of vertices to other sets (example in Fig. 4-b).

As a further, nonclassical generalization of the notion "maximum" we will introduce three rules of "hyperdomination":

1°. $Y = \{ y \in X \mid \exists V \subseteq X : V G_1 y \}$

2°. $Y = \{ y \in X \mid \exists x \in X \; \forall V \subseteq X : [x \in V] \Rightarrow [V G_1 y] \}$

3°. $Y = \{ X \mid \exists V \subseteq X : V G_2 Y \}.$

Of course in all three cases $Y \subseteq X$, i.e. with specifying of the hyperrelation G_1 the rule 1° (or 2°), and with specifying of hyperrelation G_2 – the rule 3° realizes a choice of a part of the alternatives from X.

Theorem 3. For any choice function $C(X)$ from domain H (or C) and only for such functions there exists a hyperrelation G_1 such that the rule of hyperdomination 1° (respectively – 2°) generates this function. For any $C(X)$ from the domain O and only for such $C(X)$ there exists a hyperrelation G_2 such that the rule of hyperdomination 3° generates this function $C(X)$.

Following examples are confirming the importance of this generalization.

Example 1. Let us consider a set of criteria $\{ f_1(x), f_2(x), \ldots, f_m(x) \}$ specified on A, and the rule: for any $X \subseteq A$ in the choice Y are included those and only those alternatives, which are the best ones in X not less that by k criteria (where $1 < k < m$). Does some specified on A criterion $F(x)$ exist, by which for any $X \subseteq A$ maximization on X realizes the same choice? If no, then does a collection of criteria exist, under which the same result is got by Pareto-optimization?

It is easy to show, that the described procedure reproduces a choice function from H, but does not always satisfy the conditions C and O. From the said above and from Theorem 2 directly follows the negative answer to these questions, and from Theorem 3 – it follows that this function realizes the described generalization: the hyperrelation G_1 and the hyperrule 1°.

Example 2. The same collection of criteria $\{ f_1(x), f_2(x), \ldots, f_m(x) \}$ is specified and the rule: on X every criterion is maximized separately and the alternatives, thus picked out by all criteria are united. For such "collected-extremal" choice the condition C may be violated, but the conditions H and O are always fulfilled. Therefore in this case as well described hyperrelations embrace choice procedures, whereas usual extremization of criteria and their classical generalizations – do not embrace them.

Described "hyper-mechanisms" are remarkable because they explicitly take into consideration "dependence on context" in forms of group influences of alternatives and that such generalization proceeds from non-classical generalization of the notions "criterion" and "extremization". The question naturally arises whether one can reach the same purpose proceeding not from the notions of "binary relation", "domination", but directly from the notions "criterion" and "extremum". Such way leads to notion "pseudocriterion" and highly general constructions, even if usual notion of the term "extremum" is retained without its generalizing.

7. PSEUDOCRITERIA (Aizerman,
Litvakov, 1989).

We will remind that the name a criterion is given to a mapping onto the numerical axis of the whole set A under which the criterial estimate $f(x^*)$ of any alternative $x^* \in A$ does not depend "on context", i.e. on what is the subset $X \subset A$ ($x^* \in X$), containing this alternative. Of course the result of criterion extremization

$$Y_X = \arg \max_{x \in X} f(x)$$

depends on X, but only because X determines the domain, over which extremization is realized, and not because criterial estimates depend on X. The situation is analogous for the collection of criteria $\{ f_i(x) \}$ and Pareto-optimization.

Let us now generalize the notion "criterion" as follows:

31

let us term as pseudocriterion on A a collection of functions[3] $\psi(x,X)$ depending (as on a parameter) on sets X, specified for all $X \subseteq A$ and taking values $0 \le \psi(x,X) \le k$ (where $k > 0$ – a fixed number) for all $x \in X$ and $\psi(x,X) \equiv 0$ for $x \notin X$. Unlike criterion $f(x)$, estimates ψ for some x^* may be different for different X, containing this x^*. In this case the usual maximization rule picks out

$$Y_X = arg \max_{x \in X} \psi(x,X)$$

depending on X both because X specifies the "maximization domain" and because the values of scale estimates themselves depend on X.

In all examples considered in Section 4 (winnings sum at a tournament, scale "number of voices for" at voting, rules picking out not the whole Pareto subset, but a part of it, etc.) is constructed not a criterion, but a pseudocriterion.

By analogy with usual criteria, one can consider collections of pseudocriteria $\{\psi_1(x,X), \psi_2(x,X), \ldots, \psi_m(x,X)\}$ and determine for each X a Pareto-maximum on this collection. We will denote via Ψ the space of all conceivable pseudocriteria on A. Of course the space of usual criteria $f(x)$ is a subset of Ψ (further it will be described by other terms), but the space $\mathcal{C}$ of all choice functions $C(X)$, which was under discussion in Section 5, is a subset of Ψ as well. Indeed, if function $Y_X = C(X)$ is specified, then one can represent it as a pseudocriterion, coinciding for each X with characteristic function Y_X:

$$\psi(x,X) = \begin{cases} 1, & \text{if } x \in Y_X = C(X) \\ 0, & \text{if } x \notin Y_X = C(X) \end{cases}$$

Thus the pseudocriterion is a very "wide" notion, generalizing both notions: "criterion" and "choice function".

We will picture, conditionally, the space Ψ and the subspace $\mathcal{C}$ as it is shown in Fig. 5. But in $\mathcal{C}$ the characteristic conditions, described in Section , specified the domains H, M, C, O and K. It is natural now to specify in the space of pseudocriteria Ψ "analogous" domains in such a way that each of them would include a domain of "usual criteria" and would be therefore their generalization, whereas in intersection with the subset $\mathcal{C}$ of choice functions it would specify the domains introduced before in $\mathcal{C}$. These domains in Ψ are introduced with depending on how $\psi(x,X)$ changes as a function of x under transition from X to $X' \subset X$.

Fig. 6 shows designations of these domains, their definition and a drawing, illustrating the character of $\psi(x,X)$ changing under transition from $\psi(x,X')$ (or $\psi(x,X')$ and $\psi(x,X'')$) to $\psi(x,X)$. Besides ψ is supposed to change as shown in Fig. 5 for all $X \subseteq A$, $X', X'' \subseteq X$.

---

3) Further $\psi(x,X)$ is conditionally used for specification both a function of two variables $x \in A$ and $X \subseteq A$ and for specification numeric values of this function with concrete variables x and X.

It holds the following

Theorem 4.1. In the space Ψ the conditions H_ψ, C_ψ and O_ψ are independent in totality (all 8 domains are non-empty: $H_\psi \cap C_\psi \cap O_\psi, \overline{H}_\psi \cap C_\psi \cap O_\psi, \ldots, \overline{H}_\psi \cap \overline{C}_\psi \cap \overline{O}_\psi$).

2. $M_\psi \subset C_\psi$, and $H_\psi \cap M_\psi = S_\psi$.

3. $S_\psi \subset K_\psi \subset H_\psi \cap C_\psi \cap O_\psi$.

4. In the intersection with the subspace $\mathcal{C}$ of choice functions the domains H_ψ, M_ψ, C_ψ, O_ψ and K_ψ transit respectively into domains H, M, C, O and K, the domain $S_\psi = H_\psi \cap M_\psi$ – into domain $H \cap M$.

Fig. 7 repeats conditionally Fig. 5 representing only the three domains: H_ψ, C_ψ and O_ψ in Ψ and their intersections with the subspace $\mathcal{C}$.

Note now, that $\psi(x,X) \in S_\psi$ is exactly what is termed "criterion": estimates $\psi(x,X)$ are identical for some $x = x^*$ with all X, containing this x^*. Note further that all above introduced domains in Ψ include S_ψ. Therefore each from such domains is peculiar generalization of the notion "criterion". At the same time every of these domains in Ψ has an exact contensive sense. In particular, the scales "score sum at the tournament", "voices sum for at voting" and "ranking places sum" belong to the domain M_ψ.

Apply now to the schemes, presented by Fig. 1, replacing in it the block "criterion $f(x)$" with a block "pseudocriterion $\psi(x,X)$" (Fig. 8). Any rule π, realized in the following block of Fig. 1, specifies $Y_X \subseteq X$. Therefore the scheme of Fig. 8 at "giving on input" all $X \subseteq A$ constructs a concrete choice function $Y_X = C(X)$ depending on the "laying" in the scheme pseudocriterion $\psi(x,X)$ and the rule π. In this sense the rule π realizes a mapping of each concrete pseudocriterion $\psi(x,X)$ into a concrete choice function, and a mapping of any domain $Q_\psi \subseteq \Psi$ into a domain in $\mathcal{C}$, i.e. projects the domain $Q_\psi \subseteq \Psi$ into the subspace $\mathcal{C}$. Such a projection depends of course on π and, in general, does not coincide with the intersection of Q_ψ and the subspace $\mathcal{C}$ (Fig. 9). There naturally arises the question: into what domains of $\mathcal{C}$ are projected the above introduced in Ψ domains M_ψ, H_ψ, O_ψ, C_ψ, K_ψ and their intersections?

Consider the three widely used (when we are dealing with usual criteria) rules π:

1°. π_{max} : $Y = arg \max_{x \in X^*} \psi(x,X^*)$

2°. π_ε : $Y = \{x \in X^*: \psi(x,X^*) \ge \max_{x \in X^*} \psi(x,X^*) - \varepsilon\}$

3°. π_d : $Y = \{x \in X^*: \psi(x,X^*) > d\}$.

where X^* – any concrete $X \subseteq A$.

With some rules for all X there certainly holds $Y_X \ne \emptyset$. Denote via $\tilde{C}(X)$ and $\tilde{\mathcal{C}}$ a choice function and a subspace of functions ($\tilde{\mathcal{C}} \subseteq \mathcal{C}$) of non-empty choice ($Y_X \ne \emptyset$, $\forall X \subseteq A$), and $\tilde{S}, \tilde{M}, \tilde{H}, \tilde{C}, \tilde{H} \cap \tilde{C}$ etc. – domains in this subspace $\tilde{\mathcal{C}}$.

1°. Maximization rule $\mathcal{J}_{max}$

Theorem 5. Each of the domains M_ψ, H_ψ and C_ψ in Ψ is projected by the rule $\mathcal{J}_{max}$ onto the whole subspace $\mathcal{C}$, the domains O_ψ, $H_\psi \cap O_\psi$ and $H_\psi \cap O_\psi \cap C_\psi$ are projected in $\mathcal{C}$ onto domains "of the same name" $\widetilde{O}$, $\widetilde{H \cap O}$ and $\widetilde{H \cap O \cap C}$ respectively, the domains K_ψ and S_ψ — onto the domain $\widetilde{K}$; the domain $H_\psi \cap C_\psi$ — into the domain $Q_1 \supset \widetilde{H \cap C}$, and $C_\psi \cap O_\psi$ — into the domain Q_2 such that $\widetilde{H \cap O} \subset Q_2 \subset \widetilde{O}$.

Remarks: 1. The fact, that any function of the non-empty choice may be realized by maximization of some pseudocriterion, is trivial — a characteristic function Y_X may be used as such a pseudocriterion. The theorem establishes more strong statement: for any function $C(X)$ can be found its own, generating this function pseudocriterion satisfying any of conditions M_ψ, H_ψ, C_ψ.

2. A projection by the rule $\mathcal{J}_{max}$ of intersection of the domain O_ψ with M_ψ and of H_ψ with C_ψ exactly coincides with the intersection of these domains with the subspace $\mathcal{C}$.

3. The question remains open: into what is projected the domain $O_\psi \cap C_\psi$ by the rule $\mathcal{J}_{max}$ in $\mathcal{C}$.

2°. Interval-extremizational rule $\mathcal{J}_\varepsilon$

The domains M_ψ, H_ψ and C_ψ are certainly projected by the rule $\mathcal{J}_\varepsilon$ onto the whole $\mathcal{C}$ — that is true even for $\varepsilon = 0$. It can be shown, that the operator $\mathcal{J}_\varepsilon$ for $\varepsilon \neq 0$ as well as for $\varepsilon = 0$, projects O_ψ and $H_\psi \cap O_\psi$ into their intersection with $\mathcal{C}$, i.e. into $\widetilde{O}$ and $\widetilde{H \cap O}$ respectively. But the domains $H_\psi \cap C_\psi \cap O_\psi$ is projected by the operator $\mathcal{J}_\varepsilon$ into a more wide domain $Q \subset \widetilde{H \cap O}$. Of course $Q \supset \widetilde{H \cap O \cap C}$. Analogously, the domains S_ψ and K_ψ are projected by the operator $\mathcal{J}_d$ not into domain $\widetilde{K}$, but into a more wide domain, containing $\widetilde{K}$, but included in $\widetilde{H \cap C \cap O}$.

For the operator $\mathcal{J}_\varepsilon$ remains unanswered the question: into what domain in $\mathcal{C}$ the domains $H_\psi \cap C_\psi$, $C_\psi \cap O_\psi$ and $M_\psi \cap O_\psi$ from Ψ are projected?

All the mentioned facts about the operators $\mathcal{J}_{max}$ and $\mathcal{J}_\varepsilon$, completed with analogous facts for the operator $\mathcal{J}_d$, are gathered in Table 1. Each column of this table corresponds to one of the domains of the space Ψ (e.g. S_ψ, K_ψ, M_ψ etc.), each line — to some rule $\mathcal{J}$, and on the intersection of the column and the line is specified the domain in $\mathcal{C}$ (or $\mathcal{C}$), onto which the rule $\mathcal{J}$, corresponding to the line, projects the domain in Ψ, corresponding to the column.

The sign $\textcircled{?}$ means, that the question about where to the corresponding domain is projected, remains open. Table 1 shows that for pseudocriteria we can ascertain facts, generalizing the facts, known for usual criteria — the first column (domain S_ψ) of Table 1 corresponds to them.

Let us transit to consideration of the case of Fig. 2, when the rule $\mathcal{J}$ is applied not to an individual pseudocriterion, but to a collection of pseudocriteria $\{\psi_i\}$. Supposing, that all $\psi_i(x, X)$ from the collection are belonging to some domain Q_ψ in Ψ, one can elucidate the question about to what a domain in $\mathcal{C}$ (or in $\mathcal{C}$) the choice functions are belonging which are obtained in consequence of applying to this pseudocriteria collection $\{\psi_i\}$ one or another rule $\mathcal{J}$, realizing in this case a "collective decision".

Three such rules were studied:

1°. $\mathcal{J}_1 : Par\ max\{\psi_i(x, X)\}$ — for each $X = X^*$ the rule constructs separately as Y_X the Pareto set for $\{\psi_i(x, X^*)\}$;

2°. $\mathcal{J}_2 : Y_X = \bigcup_i \mathcal{J}_d^i$, where $\mathcal{J}_d^i$ — the rules of over-threshold choice $\mathcal{J}_d^d$, applied to i-th pseudocriterion from the collection and

3°. $\mathcal{J}_3 : Y_X = \bigcup_i arg\ max_X \psi_i(x, X)$ — is such a collected-extremizational rule which for $X = X^*$ maximizes separately every $\psi_i(x, X^*)$ and unites the results.

In Table 2, constructed by analogy with Table 1, are gathered all the obtained till now results. Each column of Table 2 corresponds to the domain in Ψ, to which all $\psi_i(x, X)$ from the collection $\{\psi_i(x, X)\}$ belong, and the lines correspond to the rules $\mathcal{J}$ (see Fig. 2). As rules $\mathcal{J}_i$ we consider: construction of $Par\ max_{x \in X}\{\psi_i(x, X)\}$, the rule $\bigcup_i \mathcal{J}_d$, i.e. application of the rule $\mathcal{J}_d$ to each $\psi_i(x, X)$ from $\{\psi_i(x, X)\}$ separately and union of the results and the rule $\bigcup \mathcal{J}_{max}$, i.e. applying of $\mathcal{J}_{max}$ to each $\psi_i(x, X)$ separately and union of the results. In Table 2 $\textcircled{?}$ also signifies that the corresponding problem has not been solved.

We can see from Table 2, that the case is more difficult and so still little studied, when the search rule $Par\ max_X$ is used. We can see as well, how for other rules results are generalized, known before only for a collection of usual criteria (the first column of Table 2 — the domain S_ψ).

Above the extremizational rule was used conformably to pseudocriteria. A transition from usual criteria $f(x)$ to pseudocriteria $\psi(x, X)$ naturally results in further generalization of the notion "extremization".

Besides the rules 1°–3° let us consider a rule, which conditionally we write down as follows:

$$4° \quad Y = \underset{X' \subseteq X}{MAX}\ \underset{x \in X'}{max}\ \psi(x, X'),$$

where the operation $\underset{X' \subseteq X}{MAX}$ signifies picking out the subset $\underset{X' \subseteq X}{} $, on which the value $\underset{x \in X'}{max}\ \psi(x, X')$ gets its most quantity, i.e. a subset X' is looked for, on which pseudocriterion $\psi(x, X')$ has *maximum maximozum* by both arguments $x \in X'$ and $X' \subseteq X$.

By analogy, the following below rules[4] can be introduced:

$$5° \quad Y = \underset{X' \subseteq X}{MAX} \; \underset{x \in X'}{min} \; \psi(x, X)$$

$$6° \quad Y = \underset{X' \subseteq X}{MIN} \; \underset{x \in X'}{max} \; \psi(x, X)$$

and the rule of "maximization of the middling"

$$7° \quad Y = \underset{X' \subseteq X}{MAX} \; \frac{1}{|X'|} \underset{x \in X'}{\sum} \; \psi(x, X).$$

Each of these rules, applying to a concrete pseudocriterion $\psi(x, X)$, constructs a non--empty choice function, and projects any domain in the space Ψ onto the subspace of non-empty choice functions $\mathcal{C}$ .There holds[5]

Theorem 6. Any of the rules 4°,5°,6° and 7° projects accordingly the whole space Ψ in the domains $\mathcal{D}_{4°}, \mathcal{D}_{5°}, \mathcal{D}_{6°}$ and $\mathcal{D}_{7°}$, which are situated in subspace of non-empty choice functions $\mathcal{C}$ within the domain of $\mathcal{O}$.

Thus unlike usual extremization $Y = arg \underset{x \in X}{max} \psi(x, X)$, which is generally speaking "an extending rule" covering often the whole space $\mathcal{C}$ (see Table 1), generalization of extremization — the rule 4° and its analogs 5°-7° are strongly "narrowing" rules. They generate choice functions satisfying the highly natural and contensive, well interpretable condition $\mathcal{O}$.

7. CONCLUSIVE REMARKS

The notions "criterion" and "maximization" are improving and developing as well as many other inter-disciplinary notions. By the middle of XX-th century this development was linked to the generalization of the mentioned notions by introduction into consideration of binary relations of a general kind and, respectively, by replacing of the notion "maximum" with a more wide notion "domination". "Solution of von Neumann-Morgenstern" is a brilliant example of prevision of such a development.

In the last decades further generalization of these notions was continued. The leading idea became understanding of necessity to take into consideration the influence of the "context", i.e. the structure of the presented set of alternatives on the estimates of each of them. The developing of this resulted in a transition from the consideration of binary to the consideration of group relations (hyperrelations), and that naturally resulted in a further generalization of the notion "maximization" — in consideration of "hyperdomination".

The most generality and completeness has been got so far by introduction of the idea of pseudocriteria explicitly taking into account the influence of the context and comprehending as particular cases,both

usual criteria, and choice functions. It is found, that in spite of generality of this notion, it may give the possibility to specify contensive subclasses and to obtain for them nontrivial results.

Apparently, further generalizations one can expect on the way of consideration of more general structures — graphs and hypergraphs with weights on arcs under corresponding further generalization of the notion "domination", as development of the notion "maximization". In the last years publications devoted to these generalizations are appearing.

Side by side with a general search of further generalizations let us mention two more special problems.

The first problem consists in "canceling of the questions", contained in Tables 1 and 2.

The second problem requires explanation. Maximization of such widely used scales as "score sum in tournament table" and "number of votes for" at votings generate choice functions, which certainly cannot be put within some from the introduced domains (H , C or O) in $\mathcal{C}$. At the same time these scales are pseudocriteria from the domain Ψ_M . In connection with the fact that the rule of maximization projects Ψ_M onto the whole set $\mathcal{C}$, and both mentioned procedures certainly generate not the whole set,there arises a problem: to introduce (to describe) in Ψ_M subdomains, containing all pseudocriteria, generated as "score sum in tournaments" and "votes sum for at voting". One of possible approaches to this problem see in (Aizerman,Litvakov, 1989).

These two problems show to what extent the problems, which have not been still solved in the frames of introduced generalizations, are interesting.

REFERENCES

Aizerman, M.A. (1985). New problems in the general choice theory (Review of a research trend), Social Choice and Welfare, Vol. 2, pp. 235-282.

Aizerman, M.A. and B.M.Litvakov (1989). Pseudocriteria and pseudocriterial choice. Mathematical Social Sciences, Vol. 17, pp. 97-129.

Mullat, J.E. (1976). External Subsystems of Monotonic Systems, Automation and Remote Control, I. Vol. 37, No. 5, pp. 758-766; II. Vol. 37, No. 8, pp. 1286-1294; III. (1977), Vol. 38, No.1, pp. 89-97.

---

4) The operations 5° and 6° were used in the theory of monotonous systems(Mullat, 1976), where pseudocriteria from the domains Ψ_M and Ψ_H (in other terms) were considered.

5) The fact, determined by this theorem, was ascertained and the theorem was proved by V.I.Vol'skiy and B.M.Litvakov.For the domains H_ψ and M_ψ the fact was found out before by E.N.Kuznetsov(private inform.).

Table 1

Domain in Ψ / rule Π	S_φ	K_φ	M_φ	H_φ	C_φ	O_φ	$H_\varphi \cap O_\varphi$	$H_\varphi \cap C_\varphi$	$C_\varphi \cap O_\varphi$	$H_\varphi \cap C_\varphi \cap O_\varphi$	$M_\varphi \cap O_\varphi$
Π_{max}	$\widehat{K}$	$\widehat{K}$	all $\widehat{C}$	all $\widehat{C}$	all $\widehat{C}$	$\widehat{O}$	$\widehat{H\cap O}$	$Q \supset H\cap C$	$\widehat{C\cap O} \subset Q_1 \subset O$	$\widehat{H\cap C\cap O}$	(?)
Π_d	S	(?)	M	H	all C	O	$H\cap O$	$H\cap C$	$C\cap O$	$H\cap C\cap O$	$M\cap O$
Π_ε	$Q\subset \subset H\cap C\cap O$	$Q\subset H\cap \cap C\cap O$	all $\widehat{C}$	all $\widehat{C}$	all $\widehat{C}$	$\widehat{O}$	$\widehat{H\cap O}$	(?)	(?)	$Q\subset \widehat{H\cap O}$	(?)

Table 2

all φ_i from $\{\varphi_i\}$ belong to domain: / rule Π	S_φ	K_φ	M_φ	H_φ	C_φ	O_φ	$H_\varphi \cap O_\varphi$	$H_\varphi \cap C_\varphi$	$C_\varphi \cap O_\varphi$	$H_\varphi \cap C_\varphi \cap O_\varphi$	$M_\varphi \cap O_\varphi$
Par max X	$H\cap C\cap O$	$Q\subset O$, but $\not\subset H$, $\not\subset C$	(?)	(?)	(?)	(?)	(?)	(?)	(?)	(?)	(?)
$\bigcup_i \Pi_d^i$	S	(?)	M	H	all $\widehat{C}$	O	$H\cap O$	(?)	(?)	$H\cap O$	$M\cap O$
$\bigcup_i \Pi_{max}^i$	$\widehat{H\cap O}$	$\widehat{H\cap O}$	all $\widehat{C}$	all $\widehat{C}$	all $\widehat{C}$	$\widehat{O}$	$\widehat{H\cap O}$	(?)	(?)	$\widehat{H\cap O}$	(?)

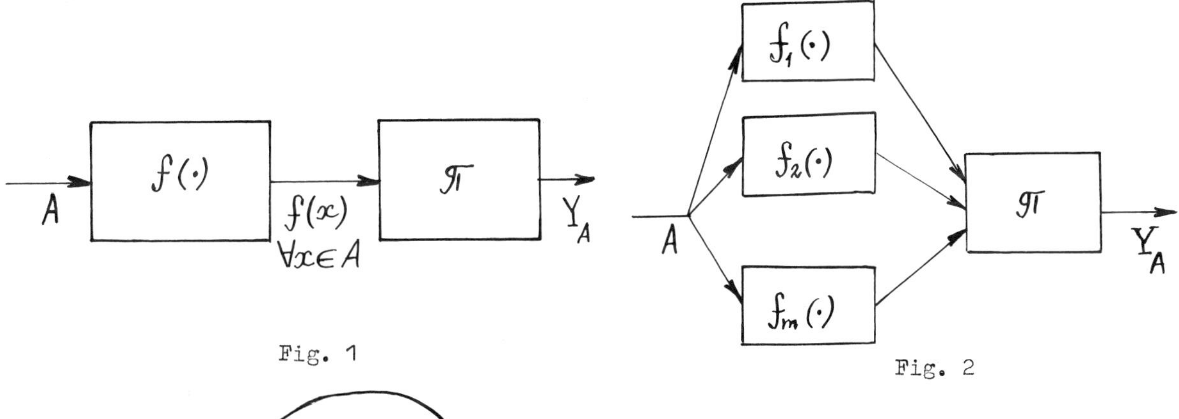

Fig. 1

Fig. 2

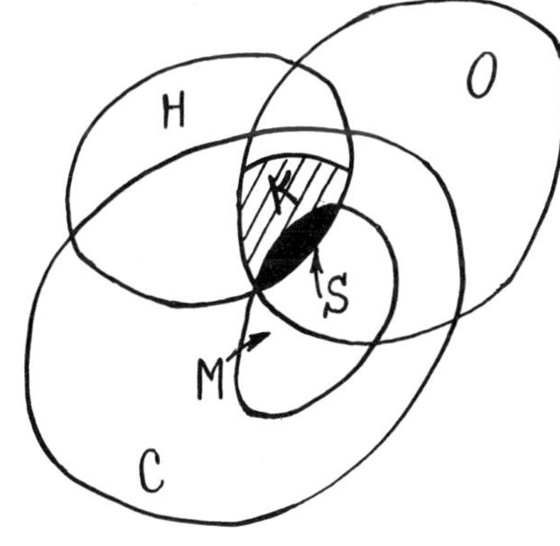

Fig. 3

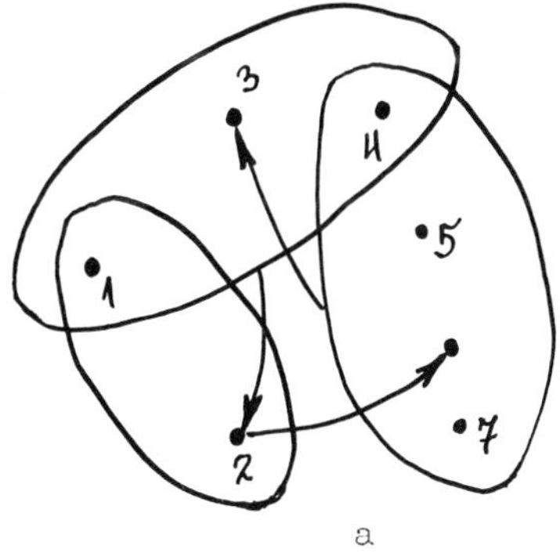

Fig. 4

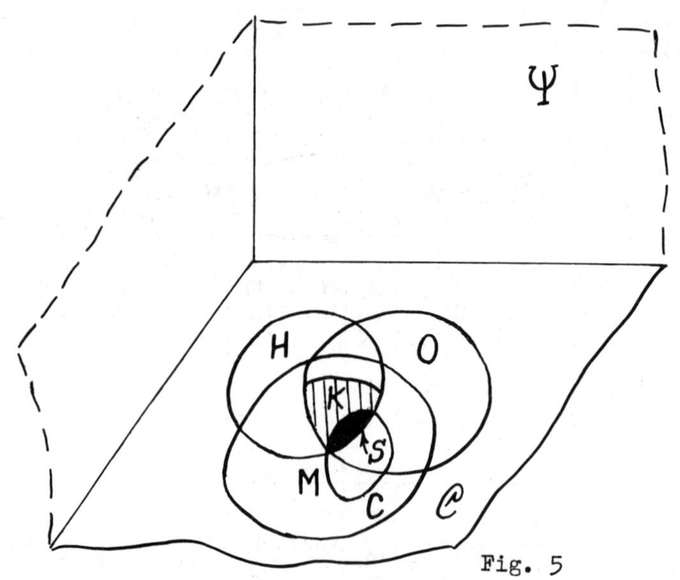

Fig. 5

name of a domain	Determining of the domain	Illustration to determining
S_ψ	$\psi(x,X') = \psi(x,X)$ for $x \in X'$	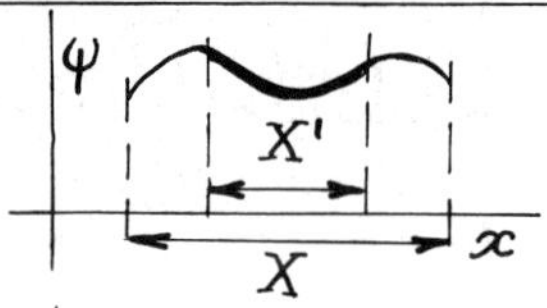
M_ψ	$\psi(x,X') \leq \psi(x,X)$	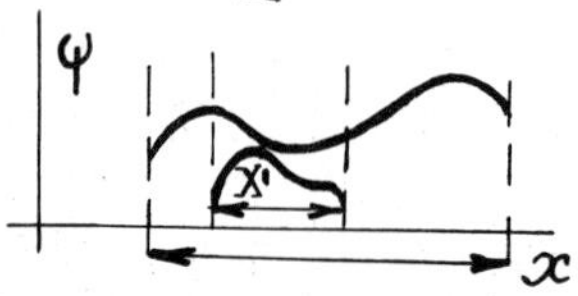
H_ψ	$\psi(x,X') \geq \psi(x,X)$	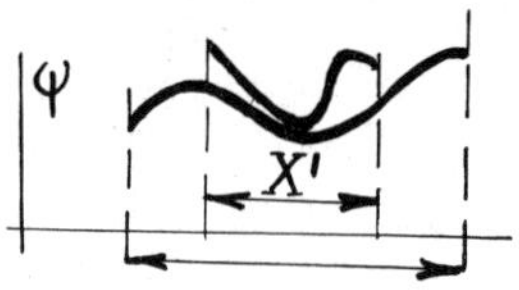
C_ψ	$min[\psi(x,X'),\psi(x,X'')] \leq \psi(x,X)$ $X', X'' \subset X$	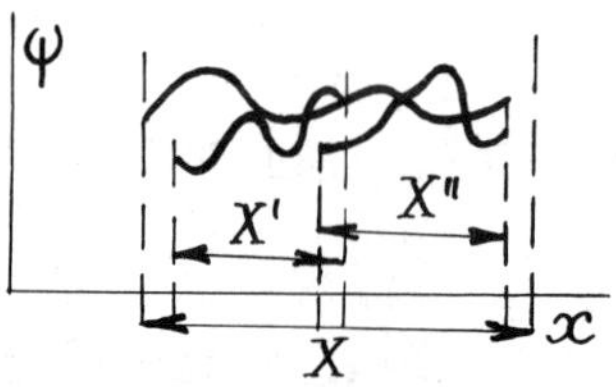
K_ψ	$\psi \in H_\psi \cap C_\psi \cap O_\psi$ and $X' \subset X,$ $\max\limits_{x \in X'} \psi(x,X) = \max\limits_{x \in X} \psi(x,X) \Rightarrow$ $\Rightarrow \psi(x,X') = \psi(x,X)$ $\forall x \in X'$	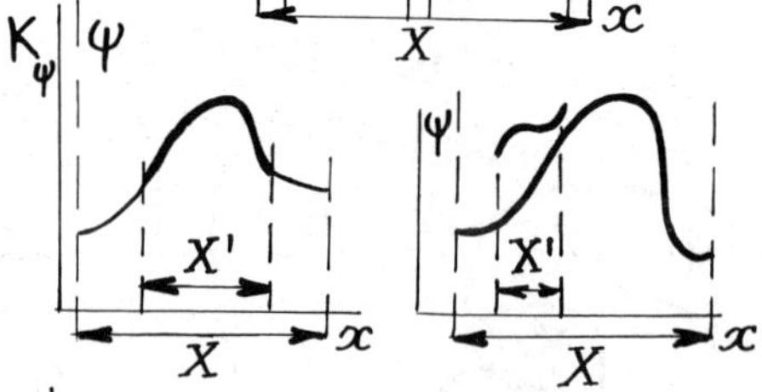
O_ψ	$\psi(x,X') = \psi(x,X)$ for $\psi(x,X) \geq \iota$ $\psi(x,X') < \iota$ for $\psi(x,X) < \iota$ where $\iota = \max\limits_{x \in X \setminus X'} \psi(x,X)$	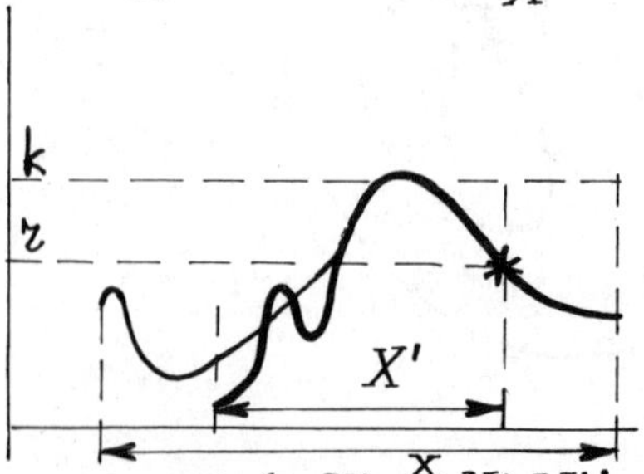 $*$ is $\max \psi(x,X)$ on $X \setminus X'$!

Fig. 6

36

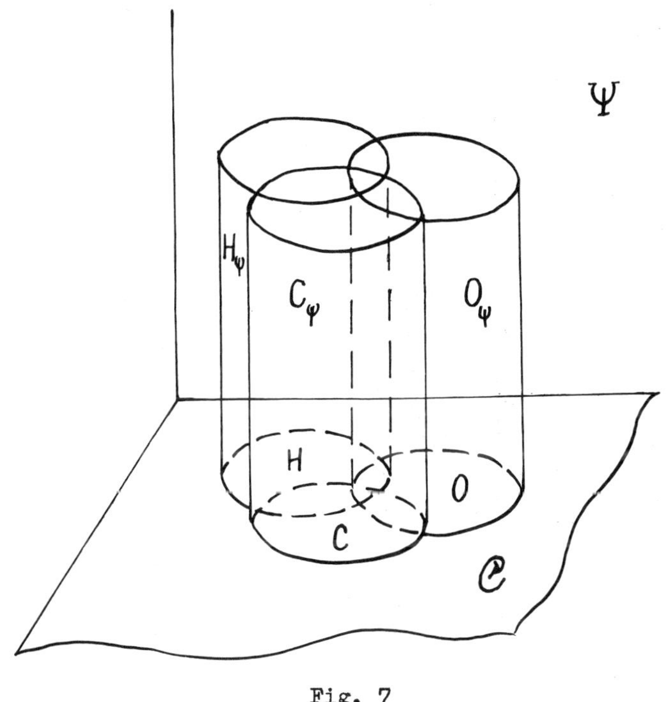

Fig. 7

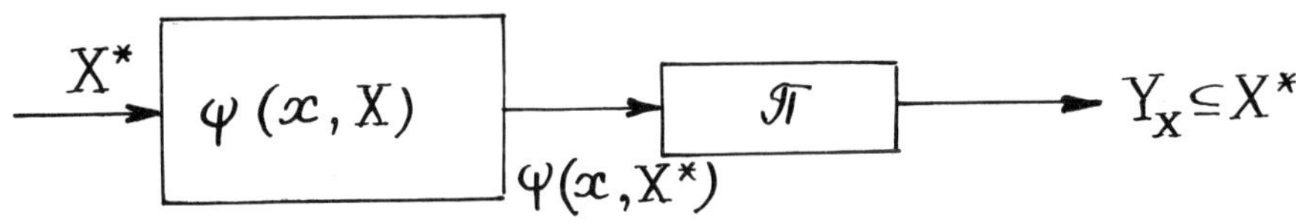

Fig. 8

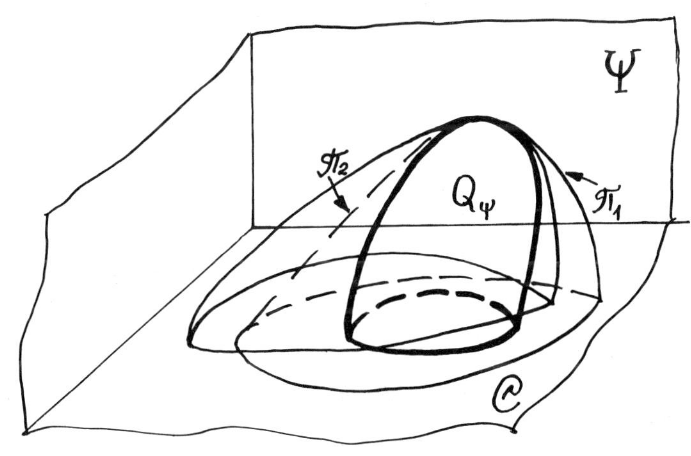

Fig. 9

EDUCATING FUTURE CONTROL ENGINEERS

W. Schaufelberger

*Project Center IDA, Swiss Federal Institute of Technology (ETH), 8092 Zurich,
Switzerland*

Abstract. The education in control theory, design and engineering undergoes considerable
changes. Requirements from Research and Industry have to be taken up and integrated into
the curricula. This situation is first analyzed briefly. Changes are then proposed that affect
the way students are taught by looking at group-teaching efforts, providing new contents such
as nontechnical subjects, discussing new ways of integrating computers into the teaching process and by showing ways for staff education.

Keywords. Computer Aided Control System Design, Computers in Control Education, Curriculum Design, Education, Group-Projects, Simulation, Staff Education, Teachware for Control,
Training programs for Control.

INTRODUCTION

Control is one of the traditional fields in almost every engineering degree education. In
some schools there are special degrees in control, in others there are not. Control has developed from beginnings in what is now called
classical control (frequency domain techniques,
Bode, Nyquist, Root Loci) through modern control (state variable feedback, observers, filters, optimal control) to sequencing, adaptive
and intelligent control. Well-known and established textbooks (Franklin et al. 1986, Unbehauen 1982) clearly indicate what the content
of a traditional introduction to control should
be as of today. Control is taught in the usual
way in courses with exercises, laboratory sessions and projects.

The attempt of the "Hütte" (Czichos 1989) to
adequately summarize the basics of engineering in one volume is worth mentioning here.
The control section by H. Unbehauen provides
an overview on control engineering for non-con-
trol engineers.

A point worth bringing up is the fact that engineers graduating today will be active for about
40 years up to the year 2030. Even if most of
them change their profession i.e. by moving into management the education must be valid for
a considerably long time. This has implica-

tions in that the educational system must look
for a careful balance between adjusting too
quickly to only temporary needs and between
not being able to adjust its basis quickly enough.

This is not a paper about the future of control.
It is assumed, that it is generally known and
accepted, that economics and ecology must grow
together in the future and that control can be of
great help for a careful use of energy and other
resources. Any education program must in some
way address these questions.

In the first part of this paper, the actual situation is assessed by looking at the present state
and at some recent developments and inquiries.
In the second part some proposals for improving
control education will be made. These affect
both the way of teaching and the contents of
the curricula.

CONTROL

The field of control has many facets. Some aspects which have influence on control education will be discussed in this section.

Theory

The theory of control is still in a rapid development. Completely new fields are coming into

focus and established fields are expanding. A recent survey of the theory committee (Ljung 1988) of IFAC shows the following areas of interest:

Linear Systems:
- Geometric theory
- Algebraic theory
- System theory
- Model reduction
- Discrete event systems

Control design and robust control:
- H∞-theory
- Robust stabilization
- Control structures
- Frequency domain methods
- Horowitz' approach
- Parametrically described uncertainties
- Variable structure regulators

Adaptive Control:
- Robust stability
- Global results
- Local results, averaging
- Chaos in adaptive control
- Switching high-gain controllers
- Stochastic adaptive control, time-varying systems
- Multivariable adaptive control, Non-minimum phase system
- Decentralized multivariable adaptive control

Distributed Parameter Systems:
- Modelling and control
- Shape design, and shape control
- Boundary control
- Distributed delays
- Actuators and sensors

System Identification:
- Mathematical tools
- Tracking properties
- Black-box identification
- Rare events and large deviations
- Change detection and model validation
- Multivariable structures

Nonlinear Control:
- "Classical nonlinear control"
- Differential geometric methods
- Inversion of systems
- Linearization and bilinearization
- Robustness
- Variable structure systems
- Chaotic motions
- Optimal control and optimization

Numerics, Software and VLSI

Nonlinear Filtering and Stochastic Control:
- Non-linear filtering
- Stochastic control
- Stochastic systems

There is no way that even specialists on control could follow the developments in all these ar-

eas and read the roughly 8000 papers produced yearly in the control field. The implication for control education can only be to produce adequate summaries and to use "courage for gaps" when putting material for courses together. This is fairly easy for the stable basis available today, but much more difficult for the many fields in rapid development.

Control System Design

A major concern in the U.S. engineering education system seems to be a perceived conflict between (control) science and (control) design (Schon 1988, Yurkovich 1989). The main argument of the designers is that control education is more often than not centered on the vast amount of theoretical results available and that young graduates do not get a sufficiently clear idea of the design process and do hardly carry out designs during their education. The design process is well understood and may consist of about the following steps (Schaufelberger, Itten 1988):

modelling
finding a suitable structure of the
model from physical equations

identification
determination of model parameters
by simulation and optimization

design
riccati, pole placement, observers,
etc.

analysis
stability, sensitivity, transient behavior
etc.

Software for most steps is available and integration of the design process into the education is therefore relatively straightforward but by its very nature time-consuming.

Control Engineering

Control engineering is seen as the practical counterpart of control theory in this paper. It deals with all the practical aspects of the design and the implementation of control systems in the many different application fields such as:

- Aerospace
- Electric Generating Plants and Power Systems
- Mineral, Mining and Metal Processing
- Chemical Processes and Processes for Natural Products like Food, Wood, Agriculture
- Cars, Ships and Engines
- Biomedical Engineering
- Electrical Drives and Power Electronics
- Strategic Planning of Energy Systems
- Socio-Economic Systems
- Manufacturing

- Robotics
- Water Resources and Environmental Systems Planning

Work in control engineering must be based on a strong theoretical background. Many aspects of how to design, implement and operate control systems are only marginally dealt with in the corresponding theory sections. An increasing demand from Industry is felt in many places for an improved education in the practical and domain specific aspects of control engineering (Drahten, Polke 1989).

CCC: Control, Computers and Communication

The rapid development of the computer has already had a strong influence on control. Some basic developments such as discrete time control would not have occurred in the same way if powerful microcomputers had not been available. Computer Aided Control Systems Design also depends crucially on computer technology. The recent development of Computer Integrated Manufacturing will influence the field of control in the near future.

A difficulty results from the fact that the education in computer science and other related areas (i.e. Electronics, CAD) is often done with considerably less mathematics and mathematical rigour than the control education. A shift to these fields in many curricula presents a severe problem because the basic education in mathematics may be affected and students find ways to earn their engineering degrees with lesser mathematics than 10 or 20 years ago.

Despite this the constant growing together of the CCC fields has a strong influence on all of them and also on the education. Computers are the design stations for control systems and at the same time important parts of the implementation. Design software must be based on reliable numerical algorithms (Ludyk 1990) and make these available through an appropriate interface (Norman 1986). Implementation software must deal with all aspects of real time (Schaufelberger et al. 1985) and also provide the man-machine interface during the system operation (Ventura, Schaufelberger 1988). Control engineers must be aware of the potential of modern computers for the design and the implementation of control systems and also of the communication systems (Rodd, Deravi 1989).

General Inquiries

Inquiries have been made recently in Industry about the quality of the control engineering education in the Federal Republic of Germany (Drahten, Polke 1989) and Switzerland (Rütter 1989). The main results are:
- The quality of the basic control education is judged as good
- efforts should be increased in
    . applications and application fields such as process control
    . nontechnical education (economics, environment)
    . team work
    . management
    . computer science and computer applications
    . problem solving techniques

SEVERAL WAYS OF CHANGING THE EDUCATION IN CONTROL

The many aspects and problems touched in the previous section can not be treated extensively in one paper. Only a partial answer to the open problems will be attempted here. Changes in a curriculum in control can be implemented in many different ways, some of the main variables being:
- the qualification and number of the teaching staff and of the students
- the number of courses and hours per course
- the course contents and notes or textbooks
- the degree and exam requirements
- the division of the time into formal lectures, exercises, laboratory assignments, projects etc.
- the use of technology such as computers
- the organization of the laboratory and project work
- the coordination and cooperation with other fields

Some fields in which the author has been active for many years will be taken as examples for possible improvement of the situation. The major topic will be the use of computers in control education which is also the theme of the recently formed WG 7.3 "Teachware for Control" of IFAC.

CONTROL SYSTEM DESIGN EDUCATION

Software as will be described in the section on computers in control education forms the basis of a design education that may be introduced in several different settings. Since the entire design process is quite time consuming, student projects are natural candidates for carrying out designs and implementations. Carefully prepared exercises may also convey the ideas of design in a short time (Schaufelberger et al. 1986). Another possibility is to devote complete courses to designs in special areas. (Schaufelberger 1983) presents such courses for process control, power systems control, machine tool control etc.

NONTECHNICAL EDUCATION

The Electrical Engineering Department of the Swiss Federal Institute of Technology offers a considerable amount of nontechnical education since 1974. For a description of the goals and the results achieved see (Davis, Schaufelberger 1986). The main courses in the compulsory program are:
- Physiology of work
- Psychology of work
- Law
- Sociology
- Technology and environment
- Economic growth and ecological equilibrium
- Management

Control engineers may profit from this program in many respects: They may gain a better understanding of their work in an overall perspective and they should get trained to foresee the implications of their work on a much larger spacial and time horizon. The program is especially useful for students who want to become entrepreneurs.

CONTINUING EDUCATION IN CONTROL

The necessity of lifelong learning and continuing education in the high technology fields such as control has been recognized for some time now. The corresponding actions are in many places rather low keyed. Except for courses and journals by the professional societies and universities which are mainly of interest to specialists there is little help for engineers that have been in professional life for some time and that do have different educational needs than young graduates. Refresher courses tuned to the needs of exactly this audience should be set up, introducing the new material in control at the level of overviews more than in depth.

GROUP-WORK OR PROJECTS

Group- or project-orientated teaching is an approach that differs considerably from the classical approach to lecturing. We use the method of the theme-centered interaction developed by R.C. Cohn (1981) and presented as a teaching and learning method by B. Eckstein (1978), Fig. 1, for our group-learning approach. In our opinion, this approach is very well tuned to the needs of technical education. Many courses and parts of courses have been taught successfully in this way during the last ten years at ETH.

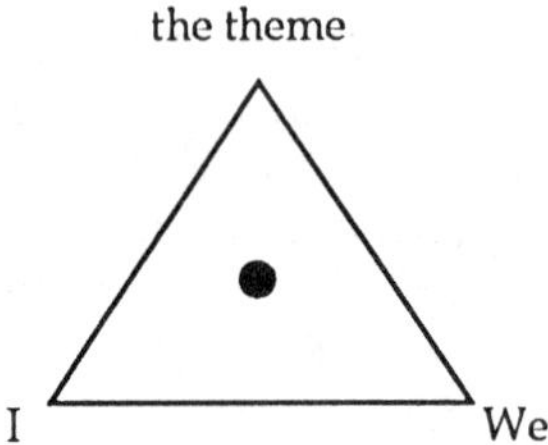

Fig. 1. Cohn triangle.
(I - the individual; We - the group; the theme - to be mastered or the problem to be solved; • working point)

Group projects and group teaching, therefore, have a long tradition in our graduate control program (Schaufelberger 1983, 1988c). Based on this, it was decided to carry out a group project in 1984/1985 in undergraduate control education. Twelve students took part in the experiment which replaced the two projects of the fourth year with a total time of about 400-500 hours.

Details of group teaching methodology used in our control education may be found in Schaufelberger 1983) and are not repeated here. Instead, a brief summary of the project is given.

Beginning of the fall term: 12 students and 4 assistants form a group. Proposed themes from the teachers side: Control of decentralized power generation systems; Control of heating systems, etc. Goal: To define problems, form small groups and organize the work within two months.

After two months: Four groups were formed:

Fuzzy control (4 students):
Goal: Apply fuzzy set theory to practical problems (fuzzy control and expert systems)

Autonomous measuring station (2 students):
Goal: Develop a measuring station for environmental measurements that will operate autonomously in mountain regions (recording temperature, wind, etc.).

Measuring system for buildings (4 students):
Goal: Develop an energy consumption measuring system with many sensors for temperature, heat-flow, etc. as required by an intermediate technology center.

Safety considerations for decentralized energy production (2 students):
Goal: Study the safety problems connected with energy production in small units, such as combined heat and electricity production.

Little influence was taken by the teachers in the formation of the groups or the selection of the themes. Each group was joined by one assistant as group-member and advisor. Weekly sessions were held throughout the year, to inform the entire group and to discuss mutual problems. Experts were invited to discuss problems, etc.

After this rather difficult initial period, the projects were carried out in a similar way as other projects with some noticeable exceptions:

- The students were aware all the time that they had chosen the subject and they therefore felt responsible and were very eager to accomplish good results.

- We all spent at least two to four hours together each week, not only to discuss technical problems, but also problems of communication, writing, reporting, etc.

- External contacts played an important role for three of the projects.

- The methodology of problem solving was applied thoroughly throughout the projects to counterbalance the sometimes "fuzzy" problem descriptions.

End of the summer term: The students hand in their reports and systems. The energy measuring system is installed as prototype in the intermediate technology center. The results obtained after two terms are comparable to the ones of other projects, if the different organizational styles are taken into consideration.

Looking back, the experience is certainly positive; unfortunately due to shortage of staff, this kind of course was not repeated since, but the style of teaching is used in other courses.

COMPUTERS IN CONTROL EDUCATION

From our experience in using personal computers in control teaching since 1985, the following four layers of software have emerged:

Layer 1: Small specialized training programs for exercises in classical and modern control.

Layer 2: Minitools for small tasks that have to be solved often such as drawing the graph of a function or a frequency response curve or solving a simple differential equation.

Layer 3: Tools such as Matlab or simulation environments for solving small to medium size problems.

Layer 4: Full grown commercial tools (professional Matlab (Little & Moler, 1985), CTRL_C, ACSL, SIMNON, Mathematica (Wolfram, 1988) etc.).

Due to licensing conditions, tools at layer 4 can usually not be made available to all students. We are convinced that software at layers 1 to 3 must be made available to all students of control engineering at minimal cost. This software must run on widely used machines available in Universities, Industry and at home. We have developed a corresponding environment consisting of about 20 programs for IBM compatibles and Macintosh that will be described in some detail in the following.

A set of examples will be shown to clarify the meaning of the four layers of software introduced above. The examples will also be used to illustrate different design techniques.

Layer 1 (specific training programs):
Figures 2 and 3 show two screens from a simple experiment for three term controller tuning. The program consists of two parts. A screen of the manual part that is organized like a textbook is shown in Fig. 2. The simulation part for doing the exercises is shown in Fig. 3. The experiment consists in studying some basic properties of classical control systems such as proportional, integral and derivative control, stability, poles and dynamic behaviour, tuning methods etc. A manual on paper about the experiment is also available for the students. All this material must be seen together with the lecture notes. Similar programs are available for: AntiWindup in three term controllers, state variable feedback and observers, adaptive control (Fig. 4), identification. Similar programs are also developed by other groups at ETH: Fig. 5 shows an animation of a two-mass-spring system used for demonstrations in a

course on mechanics and developed in the group
for Mechanics, and Fig. 6 the WorldModel developed by the Systems Analysis Group
(Fischlin et al. 1988).

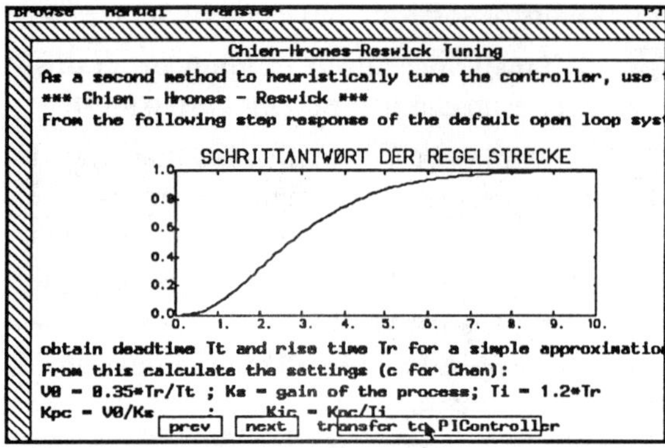

Fig. 2. Manual part of three term controller
        program. Exercises are posed in this
        part. Clicking on the transfer button
        brings up the simulation environment
        shown in Fig. 3.

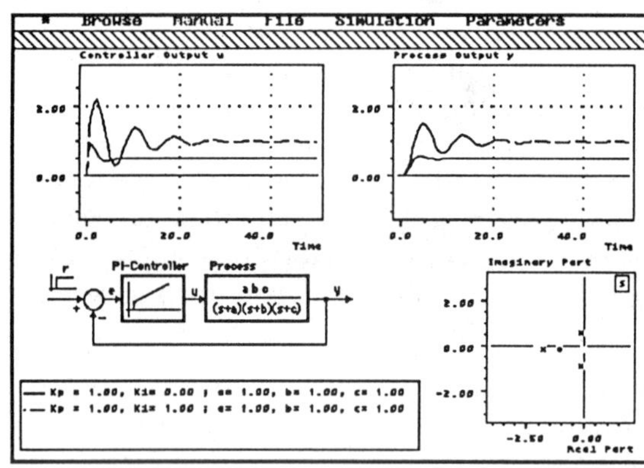

Fig. 3. Simulation part of three-term control-
        ler program. Exercises can be solved on
        this screen by changing the parameters
        of the controller and of the process. The
        relation between pole locations and
        transient behaviour can also be studied.

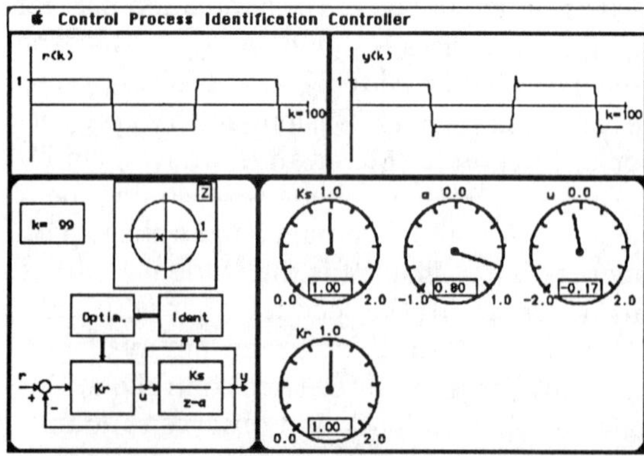

Fig. 4. Investigation of a simple adaptive con-
        trol structure with recursive least
        square identification.

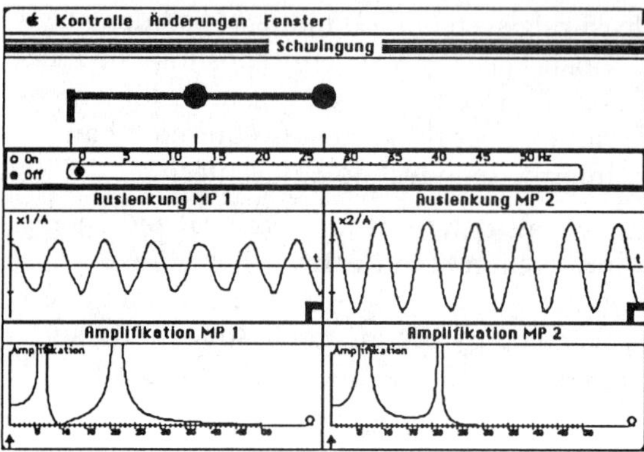

Fig. 5. Two mass spring system animation and
        simulation. The masses, spring con-
        stants and driving force may be
        changed.

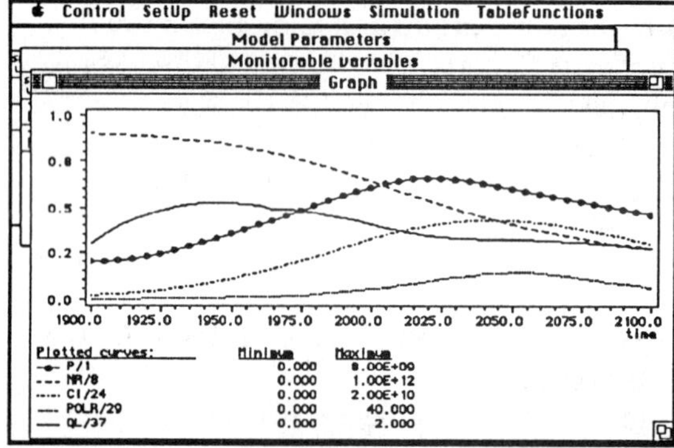

Fig. 6: Working environment for students with
        World2. The standard run of the world
        model World2 of the Club of Rome is re-
        presented. Students may change para-
        meters and compare the system beha-
        viour with their expectations. P: Popu-
        lation, NR: Natural Resources, CI: Ca-
        pital Investment, POLR: Pollution Ra-
        tio, QL: Quality of Life.

Layer 2 (minitools)

Three programs will be briefly described here.
Fig. 7 shows a screen of a program for function
graphing. Only correct sequences on the pocket
calculator keyboard are accepted. A step response is graphed in Fig. 7. This program may
find use in several courses in the area of signals, systems and control. DESolver is a program for solving differential equations of order
up to 6 in a simple and straightforward way.
Fig. 8 shows the mask for entering 4th order
equations into DESolver. The complete system
description is shown in this mask, no hidden
windows or information is needed to produce
the solution graphed in Fig. 9. A program for
frequency response analysis for systems with
order up to 6 is the last example here. Fig. 10
shows the mask for entering third-order transfer functions and Fig. 11 the four main output

windows. Small expert systems shells may also be counted here as minitools. As an example, a consultation about a controller structure is shown in Fig. 12 together with the corresponding simulation environment. Minitools have also been created for petri-net (Fig. 13) and for decision table analysis. In general, minitools should be simple and easy to understand and operate and need little or no introduction.

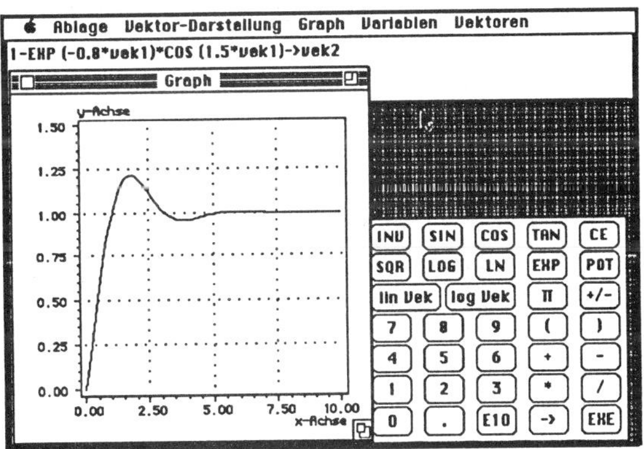

Fig. 7. A function graphing program used to calculate and draw a step response.

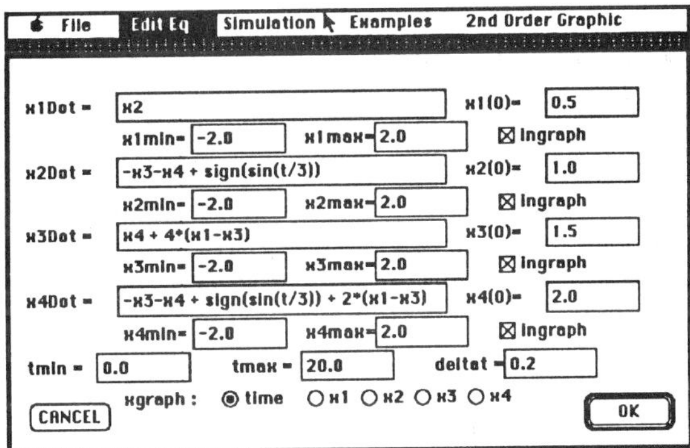

Fig. 8. Mask for entering differential equations into DESolver with equations describing a 2nd order state variable controller with observer. The complete information about the system is contained in this form.

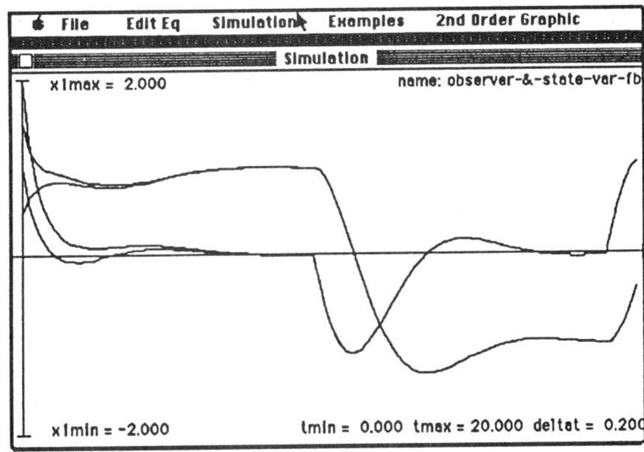

Fig. 9. Results produced by the equations of Fig. 8.

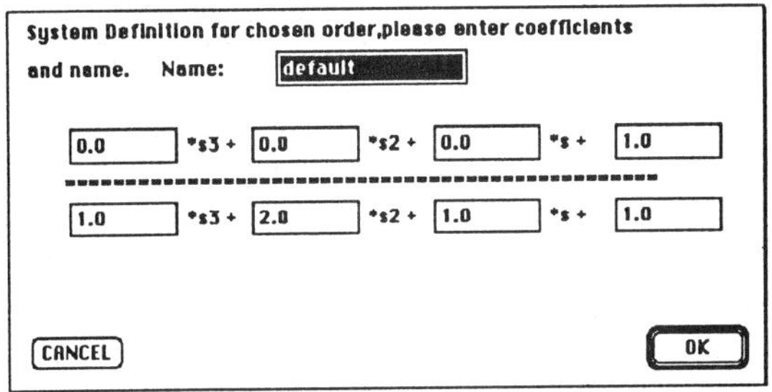

Fig. 10. A minitool for frequency response design, mask for entering the system description for a third-order transfer function.

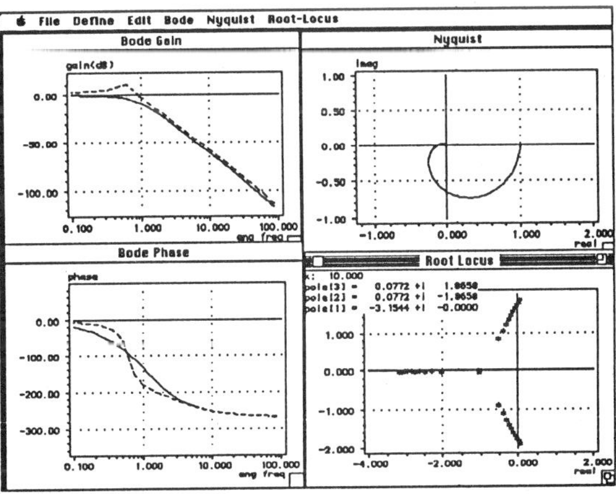

Fig. 11. A minitool for frequency response design, main output windows.

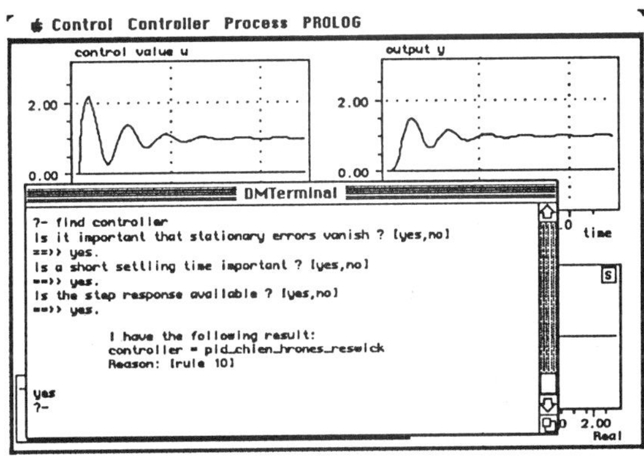

Fig. 12. Rule-based controller design. A consultation is shown in the foreground and the corresponding simulation environment in the background.

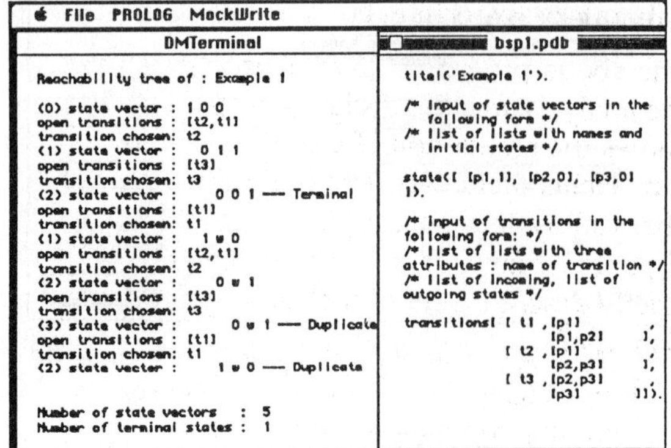

Fig. 13.   Reachability tree of a petri-net. The file describing the net is shown in the bsp1.pdb window, and the result of the analysis in the DMTerminal window.

Layer 3 (tools)
Tools are more powerful than minitools and can be used to solve reasonably sized problems. They are most often much more complex than the minitools, needing more time for an introduction. Whereas the training programs and minitools can be used efficiently even in a one hour exercise, tools usually need several hours of introduction. This is no problem if a project is carried out or if a sequence of exercises is done with the same tool. Fig. 14 shows again a state variable feedback experiment (Fischlin & Ulrich, 1988). The multi-window environment offers more possibilities than DESolver, at the same time, it is more complex to understand. The environment consists of a set of modules that provide all the services needed in continuous and discrete time simulation. The normal and well documented language, compiler and debugging facilities of Modula-2 can be used. The environment is so complex that a careful introduction on a 100 card stack for Hypercard has been prepared for the students.

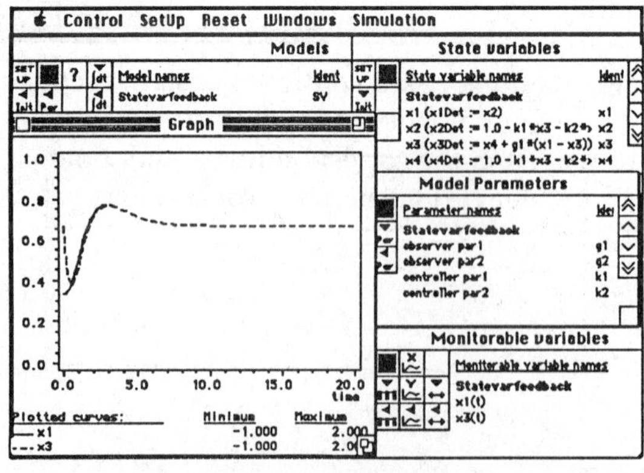

Fig. 14.   ModelWorks, a full grown simulation environment with windows for model installation and choice of integration method, initial conditions, parameters, display settings.

Fig. 15 shows our first implementation of a graphical direct manipulation interface for defining discrete-time systems by block diagrams. The corresponding execution environment allows simulations and also experiments with real processes (Maier, Schaufelberger 1990). It is based on object oriented and functional programming.

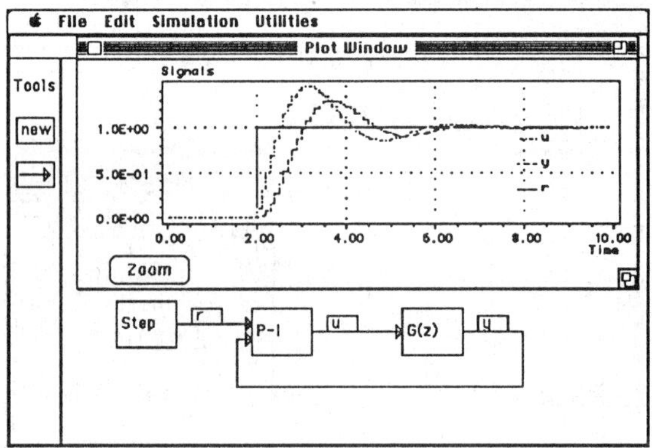

Fig. 15.   Graphic definition of control system for simulation by FPU with a direct manipulation interface.

MatETH is another tool much in use in our education. It is one of the many Matlab-descendants with graphics and toolboxes added. Fig. 16 shows the help screen of MatETH.

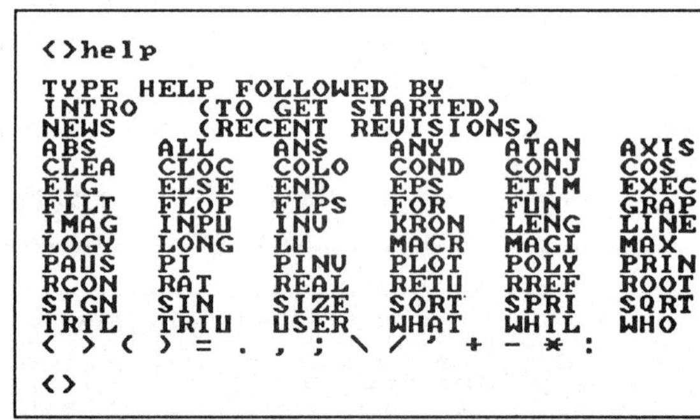

Fig. 16.   Part of the MatETH Help Screen.

Layer 4 (professional tools).
Many excellent professional programs are available for teaching i.e. PC-Matlab, Ctrl-C, MATRIXx, ACSL, Simnon, Mathematica etc. Unfortunately, the license conditions do not allow for widespread use of these programs at prices students can afford. For this reason the use of these programs is in many schools restricted to student projects, where only a relatively small number of licences must be available. The use of programs at this layer is illustrated by a bode plot from Matlab (Little, Moler 1985), Figs. 17, 18, and by a three dimensional representation of a surface from a simple identifica-

tion task in Mathematica (Fig. 19) (Wolfram, 1988).

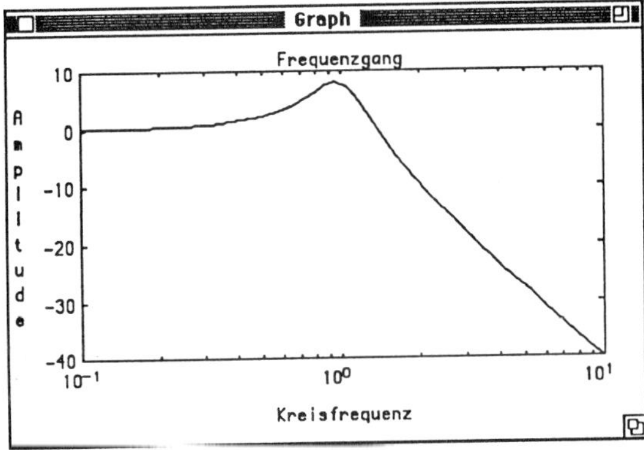

Fig. 17. Magnitude plot for a second-order system (Bode).

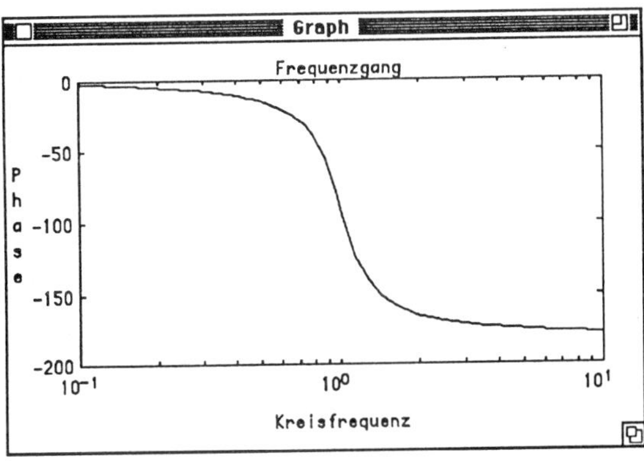

Fig. 18. Phase plot for a second-order system (Bode).

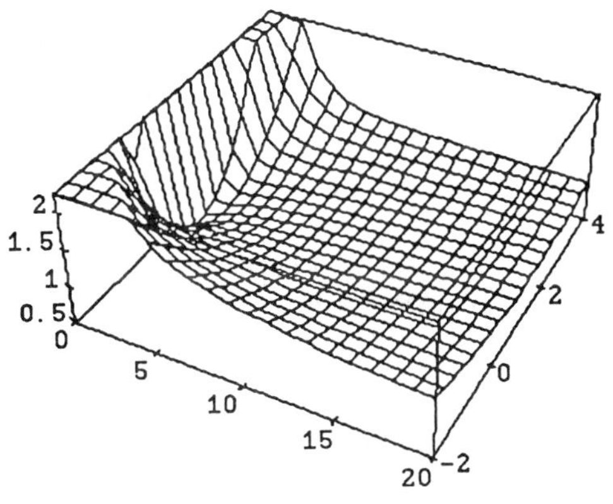

Fig. 19. Three-dimensional representation of the performance index of a simple identification problem produced with the professional program "Mathematica".

Laboratory Experiments

The above programs have mainly illustrated the analysis and the design process. Fig. 20 shows the realization of a control system for our "Helicopter" two degree of freedom mechanical system as a typical laboratory assignment. Mansour and Schaufelberger 1989 briefly describe a set of such assignments consisting of:

- Traffic Lights
- High-Bay Warehouse
- Coin-Exchanging Machine
- Heating and Ventilation Control System
- Three-Basin Level Control System
- "Helicopter" Model
- Three-Mass-Spring-System
- DC-Servo
- Inverted Pendulum
- Model Railway
- Speed and Tension Control in a Tape Transporting Machine
- Electric Power Network Model
- Manufacturing Cell

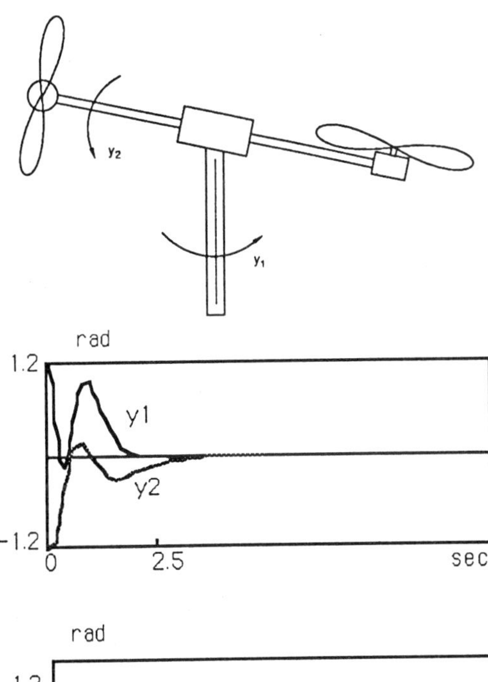

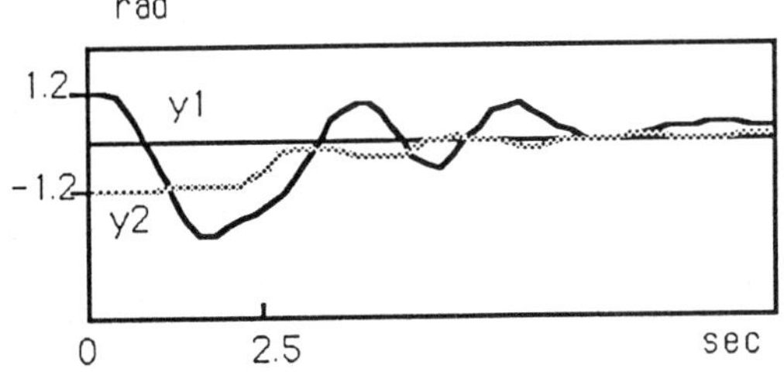

Fig. 20. Two-degree of freedom "helicopter" model design example.
Top: Physical model. Middle: Simulation. Bottom: Measurements on the real plant.

This wide range of different physical processes allows experimentation with classical and modern control techniques and sequencing control. From the programs developed in our group FPU (Maier, Schaufelberger 1990) is a discrete time simulator that can be used for online control by including access to A/D and D/A converters and to the clock of an IBM compatible PC. In this way, programming can be reduced to a minimum during laboratory sessions, where Matlab is used for the design of the control systems.

Design and Implementation of Teachware

By using modern methods of software engineering the development time for programs as described in this paper has been considerably shortened. We use a consistent approach based on the "user's model of the system" by Norman (1986) to design our programs for ease of use and an authoring system (DialogMachine: Fischlin 1988, Keller 1989) based on Modula-2 for implementation on IBM compatibles and on Macintoshes. From our experience, it takes a few days to implement a small training program and a few weeks for a minitool. We spend about one third of the total development time for the basic design, about one third for the documentation and only one third for actual programming. Beside Modula-2 we use Hypercard, cT (the former Carnegie Tutor (Sherwood 1988)), and public domain versions of Lisp (Xlisp) and Prolog. Lisp and Prolog are especially well suited for minitools with simple language input because they contain an interpreter. Our minitools for sequencing control (decision tables, petri-nets) and expert system shells have been implemented in this way (Schaufelberger 1988a).

Student projects are very well suited for the development of teachware, our function graphing environment (Fig. 7) and the graphical interface for FPU (Fig. 15 (Kolb, Rickli 1989)) have been developed by students.

Availability of the programs described in this paper

All programs developed at the Project Center IDA are available at a handling charge from the centre for educational use. They may be freely copied and distributed among interested staff members and students and may also be used in continuing education. For some programs a runtime license (GEM desktop) is needed,

others can only be used together with a special compiler. A leaflet is available from the IDA centre describing the available software, the distribution charge and the requirements for using the programs on either Macintosh or IBM compatible computers. Some of the programs are accompanied by manuals with graphics. This prevents us from sending our software through electronic mail or bulletin boards. Tests with our own board have shown that distributing software in this way is also very time consuming and that software is still best distributed by sending disks and manuals by ordinary mail.

Call for Cooperation

Implementing software as described in this paper remains a considerable effort despite the use of modern techniques of software engineering. We therefore propose to join forces with other groups and to make software available to students through a common effort. If many groups were willing to cooperate by contributing their own software to the control community, by developing software for this purpose or by translating existing software (i.e. from German to English) we would certainly be able to put together an IFAC control education software package of considerable power in a short amount of time. Interested colleagues are invited to contact the Project Center IDA at ETH Zürich.

STAFF EDUCATION

Staff promotion is usually strongly based on research achievements (publish or perish) and many academics act accordingly by devoting as much of their time to research as possible. From this it follows that especially young staff members often have little or no training in how to teach.

To improve this situation, the University of Zürich together with ETH offers a staff education program for both Universities. Courses usually are given to a small number of participants (about 12) and have a duration of about three days. The program for 1989/90 contains the following courses:

- My First Course
- New Views of Things
- Rhetorics
- Microteaching
- Media-training

- Individualization of Teaching
- Leading Discussion Groups
- Discussion Training
- Teaching at Universities (Goals and Reality)
- Oral Exams
- Methods of Knowledge Transfer
- Scientific Presentations
- Language and Oral Presentations in University Teaching
- Computers in University Education
- Computer Workshop
- Experiences of Woman Staff Members
- How to Make a Movie for Teaching
- How to Write
- Course for Assistants
- Course Visits by Experts
- Didactics à la Carte
- Curriculum Design

In addition, at ETH we have a long-standing tradition of "Weekend-Seminars". Every one or two years, all staff members are invited for a Friday-Saturday discussion on a selected topic which is introduced by internationally known experts. Among the topics of the past are:

- Classical and Modern Ways of Teaching
- Relations between Teaching Goals and Didactics
- Exams
- Participation of Students
- How to Solve Problems
- Student Motivation
- Visualization in Teaching
- Student Problems
- Group-Work (possibilities and limits)
- Technological Development why and where ? - Implications for Teaching
- Feedback
- Teaching Large Courses and Small Seminars
- Teaching in the Year 2000
- Exams
- Long-term Effects of Education in Natural Sciences

The attendance to these events is usually 30 to 70 out of a staff of 1000. These approaches are seen as one of the possibilities to improve teaching in our school.

CONCLUSIONS

The rapid changes in technology have wide ranging influences on control education. Several quite different aspects of this education have been elaborated on in this paper. Changing the content by expanding it or by adding complementary material such as nontechnical subjects is only one way of adjusting the curriculum. The form of teaching may also be affected and adapted by educating staff members and by using using new approaches to teaching such as group-work. The computer has a dominating role in the entire discussion and computer assisted education in control has correspondingly been looked at in some detail.

Improvements in these areas are seen as vital for the further development of control as an engineering discipline.

REFERENCES

Allenspach, H., and W. Schaufelberger (1985). The Use of a Home Computer for Computer Control Experiments. *IFAC/IFORS Conf. on Control Science and Technology for Development.* Beijing, PRC.

Åstrom, K. J. , B. Wittenmark (1984). *Computer Controlled Systems.*

Åstrom, K. J., B. Wittenmark (1989). *Adaptive Control.* Addison-Wesley.

Battersby, G.A., D.E.P. Jenkins , and W. Schaufelberger (1984). Horizons in Engineering Education - a systems approach to the situation in Western Europe. *IFAC World Congress,* Budapest.

Cohn, R.C. (1981). *Von der Psychoanalyse zur themenzentrierten Interaktion.* Klett-Cotta, Stuttgart.

Czichos, H. Hrsg. (1989). *Hütte: Die Grundlagen der Ingenieurwissenschaften.* Springer.

Davis, J.S., and W. Schaufelberger, (1986). Non-Technical Education in the Electrical Engineering Department of the Swiss Federal Institute of Technology. *Europ. J. of Eng. Educ., 11,* pp. 5-12.

Drahten, H., and M. Polke, (1989). Anforderungsprofil und Qualifikationsprofil als Basis zur Ermittlung des Weiterbildungsbedarfs - Ergebnis einer Umfrage. In G. Schmidt, and H. Steusloff H. (Eds.), *Mit vernetzten, intelligenten Komponenten zu leistungsfähigen Mess- und Automatisierungssystemen,* INTERKAMA, Düsseldorf, pp. 643-651.

Eckstein, B. (1978). *Einmaleins der Hochschullehre.* München, Kösel., Englewood Cliffs N.J.

Fischlin, A. (1988). *The 'DialogMachine' for the Macintosh.* Projekt-Zentrum IDA, ETH Zürich.

Fischlin, A., and W. Schaufelberger (1987). Arbeitsplatzrechner im technisch-naturwissenschaftlichen Hochschulunterricht. *Bulletin SEV/ VSE, 78,* pp. 15-21.

Fischlin, A., and M. Ulrich (1987). Interaktive Simulation schlechtdefinierter Systeme auf modernen Arbeitsplatzrechnern - die Modula-2 Simulationssoftware ModelWorks. *ASIM-Tagung,* Zürich.

Fischlin, A., and M. Ulrich (1988). ModelWorks: an Interactive Modula-2 Simulation Environment. *Int. Report No. 4,* Projekt-Zentrum IDA, ETH Zürich.

Fischlin, A., T. Blanke, D. Gyalistras, M. Baltensweiler, and M. Ulrich (1988). *Unterrichtsprogramm Weltmodell 2*, ETH Zürich, 25 p.

Franklin, G.F., J.D. Powell, and A. Emami-Naeini (1986). *Feedback Control of Dynamic Systems*. Addison-Wesley.

Keller, D. (1989). Introduction to the DialogMachine. *Int. Report No. 5*, Projekt-Zentrum IDA, ETH Zürich.

Kolb, P., and M. Rickli (1989). *Bedienoberfläche zur funktionalen Programmirumgebung*. Nachdiplomstudienarbeit, ETH Zürich.

Little, J., and C. Moler (1985). *PC-MATLAB User's Guide*. The MathWorks, Portola Valley, CA.

Ljung, L. (1988). Control Theory 1984-1986. A Progress Report from IFAC's Technical Committee on Theory. *Automatica*, 24/4, pp. 573-583.

Ludyk, G. (1990). *CAE von Dynamischen Systemen*. Springer.

Maier, G.E., and W. Schaufelberger (1990). Simulation and Implementation of Discrete-Time Control Systems on IBM-compatible PCs by FPU. *Submitted to IFAC World Congress*, Tallinn.

Mansour, M., and W. Schaufelberger (1981). Digital Computer Control Experiments in the Control Group of ETH Zürich. *IFAC World Congress*. Kyoto, Japan.

Mansour, M., and W. Schaufelberger (1989). Software and Laboratory Experiments Using Computers in Control Education. *IEEE Control Systems Magazine*, 9, No. 3, 19-24.

Mansour, M., W. Schaufelberger, F.E. Cellier, G.E. Maier, and M. Rimvall (1984). The Use of Computers in the Education of Control Engineers at ETH Zürich. *Europ. J. of Eng. Educ.*, 9, 135-151.

Nievergelt, J., A. Ventura, and H. Hinterberger (1986). *Interactive Compuer Programs for Education. Philosophy, Techniques, and Examples*. Addison-Wesley.

Norman, D.A., and S.W. Draper (Eds.) (1986). *User-Centered Systems Design*. Lawrence Erlbaum Ass., London.

Rodd M.C., and F. Deravi (1989). *Communication Systems for Industrial Automation*, Prentice Hall.

Rütter, H. (1989). *Bedürfnisse für die Aus- und Weiterbildung von ETH Ingenieuren/innen 1989*. Informationsstelle "Ingeniure für die Schweiz von Morgen".

Schaufelberger, W. (1983). Experiences with Project-Orientated Teaching Courses in a Graduate Control Programme. *Europ. J. of Eng. Educ.*, 8, 79-87.

Schaufelberger, W. (1987). Teachware for Control. *10th IFAC World Congress*, Munich.

Schaufelberger, W. (1988a). Experiences in the Teaching of Sequencing Control. *IFAC/ TRIC-MED*, Swansea.

Schaufelberger, W. (1988b). Teachware for Control. *American Control Conference*, Atlanta.

Schaufelberger, W. (1988c). Experiences with Group-Projects and Teachware in Control Engineering Education at ETHZ. *ASEE Conference*, Portland.

Schaufelberger, W. (1989a). Simple Models of Co-operation and Conflict. *IFAC/SWIIS*, Budapest.

Schaufelberger, W. (1989b). Design and Implementation of Interactive Programs for Education in Engineering and Natural Sciences. In H. Maurer (Ed.). *Computer Assisted Learning*. Springer Lecture Notes CS 360, pp. 468-479.

Schaufelberger, W. (1989c). Low cost control education software for IBM PC's. *IFAC/ LCA '89*, Milano.

Schaufelberger, W. (1989d). Einsatz von Rechnern im Unterricht der Automatisierungstechnik, *INTERKAMA*, Düsseldorf, pp. 662-670.

Schaufelberger, W., and A. Itten (1988). Low-Cost CACSD on the Apple Macintosh. *IMACS World Congress*, Paris.

Schaufelberger, W., H. Good, and A. Itten (1986). Education for Microprocessor Application in Control. *IFAC Symp. on Microprocessor Application in Process Control*. Istanbul, Turkey, 22.-25.7.86.

Schaufelberger, W., M. Mansour, A. Fischlin, and M. Rimvall (1986). Simulation and Computer Aided Control System Design in Engineering Education. *IFAC/IMACS Int. Symp. on Simulation of Control Systems*. Vienna, Austria, 22.-29.9.86.

Schaufelberger, W., P. Sprecher, and P. Wegmann (1985). *Echtzeitprogrammierung bei Automatisierungssystemen*. Teubner Studienbücher Elektrotechnik.

Schon, D.A. (1988). Marrying Applied Science and Artistry in Engineering Education: Recovering from the last Reform. *ASEE Annual Conference*, Portland (Mini-Plenary).

Sherwood B.A., Sherwood J.N. (1988). *The cT Language Stipes*.

Unbehauen H. (1982, 83, 85) *Regelungstechnik I .. III*. Vieweg.

Ventura A., and W. Schaufelberger (1988). Benützungsmodelle und Oberflächen: Vermittler zwischen Mensch und Maschine. *Bulletin SEV/VSE*, 79, pp. 921-925.

Ventura, A., W. Schaufelberger, and A. Fischlin (1988). The Use of Apple Macintosh Computers in Teaching at ETH. *Wheels Europe*, No. 2, 19-26.

Wolfram, St. (1988). *Mathematica*. Addison-Wesley.

Yurkovich, S. (1989). Control Education. *IEEE Control Systems Magazine*, p. 3 (Editorial).

TECHNOLOGY TRANSFER: NOT JUST
A MATTER OF DOING
Contribution to Technology Transfer Panel

H. Bolk

InterVisie, Schipholweg 88, NL-2316 XD Leiden, The Netherlands

Keywords: Technology Transfer; Implementation of CIM;
Organisation; Management.

Recent developments in management science point at interesting relationships beteen the individual, the organisation and technology.

Traditionally these three concepts have been studied separately from each other. Technology was the domain of the engineer, the organisation was the domain of the economist or expert on administration, the individual was the domain of the psychologist.
It was only until recently that it was found that the three concepts in practice needed to be linked. First rather primitive efforts were undertaken to establish a link, for instance in theories like the Human Resource point of view, in which some sort of interaction between the human and the organisation is incorporated. Or for instance in early ergonomics, where the man-machine interface is dealt with.

Looking at developments in management science, one could say that many moments in its history are characterized by attempts at bringing the three concepts together. One may note the attempts at understanding the organisation from the technologists angle of incidence, in which the lay-out of the factory and the materials-flow and connected information flow are seen as the sole conditions for understanding <u>and</u> designing organisations. One may also note the attempts at understanding the organisation solely from the mutual interactions between work-groups, in which conflict and harmony are seen as the key issues for understanding what goes on. Finally, attempts have been made in understanding individuals as being exclusively moulded by the organisational norms and values and the organisational structure.

Summarizing these 'attempts' one could say that perhaps each of the points-of-view may be relevant in one way or another, but at the same time one must confess that too many attempts pretend(ed) to tell the one and only exclusive truth about the relationships between the three concepts.

Management scientist have become careful, therefore. Not one aspect of organisational life, expecially when dealing with advanced technological implementations, can be understood in mono-causal, uni-dimensional, relationships.
As I stated in an earlier paper: the technological issue has become a very complex one, so has the human and organisational issue ("Management Science's Role in CIM", ESPRIT-Conference, Athens, 1989).

Let us take a closer look at technology and organisation and the role of the human being and see if we can learn something from recent management science with respect to technology transfer.

Although many times it has been tried to establish some sort of explanation for the link between the human being and technology it is still the case that technology and 'technique' are seen as something that can be known objectively. It is often perceived as just 'being there'.
Of course, when walking around in a factory or modern office, the machinery cannot be overlooked. Even after the last shift has left the machines are still there, uninfluenced as they may seem by the human being that operates them.
<u>But</u>, these machines were once designed, they were built according to specifications, certain products were to be manufactured by them, certain other products were not. Selections have been made as to what would be the degrees of freedom for operating them. Decisions were taken about data-specifications, operating equipment,

Therefore, conscious <u>as well as unconscious</u> decisions are embedded in the machinery. Technology is, therefore, also a matter of choice. Technology is, in other words, very much related to the human being of the designing engineer(s) and others, including their own points-of-view <u>and prejudices</u> about organisations and e.g. operators and how they should function.

Technology is not 'just there', it has been defined, choosen, installed according to individual preferences of several people involved.

Much too often technology is being attempted to be implemented and transferred on the basis of the conviction that "I know the technology, I know what benefits it brings, I know what is best for the organisation of for others, so do as I say!"

Needless to say, in the light of what I said before, that many automation or knowledge transfer projects encounter severe obstacles. The reason for this is that one 'forgets' that in every situation the technology <u>needs redefining, reassessment</u> taking the new organisational context in which it is to be embedded. And in these processes of redefinition and reassessment one must involve every person that has to work with the new technology (of course using accountable modern management instruments to prevent situations of everyone needing to talk to everyone!).

So, technology transfer is not 'just a matter of doing', it is not just a case of telling others what to do and how to do it. Rather, in my view, it is a matter of continuous and continued definition and reassessment.
The scope of this contribution does not allow me to go into detail with respect to the way this defining and reassessment strategy can be operationalised. Let us hope that the conference gives us the opportunity.

AN INCLUSIVE MODEL FOR EFFECTIVE TECHNOLOGY TRANSFER

F. Emspak

Center for Applied Technology, Boston, MA 02108, USA

Abstract: Technology transfer depends on involving those who actually use the technology in its elaboration and implementation. Thus structures that involve working people in the design and implementation of technologies have to be developed. Technology transfer will not take place successfully if coupled with job degradation.

Key Words: Technology transfer; skills; worker involvement

Technology transfer is often treated as if it is some arcane artform. Various models are conceived of, technopoles are organized and battalions of experts are mobilized in order to get technical ideas transferred from large to small firms or from the university to industry.

However too little thought is given to those who are the users of the new technologies. The result is extreme difficulty and expense in transferring new technologies as well as wasted time. This will become clear if we look at the mechanisms in place in most developed capitalist economies for technology transfer.

Generally a new idea or technology is identified by the owners of a firm or its engineers. They may get the idea from a trade publication, conference or university outreach. Most of these transfer methods are not available to working people.

The owners or technical staff make a decision as to wether or not to employ the new technology. They buy the technology or technique. In some cases the technology development is carried on by the firm - almost always physically separated from the point of use. At this point workers, the users of the new technologies, may be brought into the process to discuss implementation. Often the actual users do not enter into meaningful participation until the day the techniques are to be deployed on the shop floor or in the office.

We should contrast this model to the historical model. Contact between craftsmen was the most common form of technology transfer during most of human history. Universities played an important role in the advance of science but were not the arbitrators of technology until the end of the 1920ies.

The users were the principle inventors and through their intercommunication transferred new techniques. Migration was one of the key methods of transferring skills from one place to another- for example the migration of skilled German metal workers to the US from about 1870 on. Guilds with strict apprenticeship rules were one of the more important European methods for restricting technology transfer to another group.

At this moment in history technology transfer is dominated by a new exclusionary model. The process of technology transfer makes the user the object. He/she is a recipient not an actuator of knowledge. The owner of the technology is the subject. The owner, manager or engineer conceives and someone else executes. Direct numerical control of the machining process is probably the most extreme example of this type of separation of execution and conception.

New technologies have been associated with negative as well as positive factors. In many advanced countries technology transfer is often accompanied by unease concerning employment, wages and skill. This has probably always been true, but now the negative factors tend to weigh most heavily on the purported users of the technologies and the positive aspects are most evident to those who buy and control the technologies. This disjuncture contributes to the problems one has with transferring technologies.

A new approach is needed. From a logical point of view it is clear that the user must be the subject not the object of technology transfer organization. The user, not necessarily only the owner, must play an enhanced role in the process. Otherwise technology transfer continues to be something imposed from outside the work place or the culture of the worker.

Two ideas to overcome this problem suggest themselves. One is organizational. Just as engineers and managers are encouraged to go out of their workshop and factory to learn new ideas and concepts, so must we develop the forms to enable workers to exchange ideas and concepts. Exposure of working people to new ideas should not be something workers can do on their spare time but part

of the normal work day. Concomitant with the exposure to ideas must be the training and education needed to assimilate the ideas.

The second concept is more philosophical in nature. It has to do with design and thus control of technologies. Technology design and development goes on outside the work place and without worker participation. There are few or no structures that enable knowledgeable workers to participate in or even review technology development projects. Thus an action item to improve technology transfer would be to experiment with user friendly structures of organization so that the user can participate in development decisions in some meaningful way. Naturally if the idea of participation is accepted it connotes that the users will also have influence over the design criteria of a particular technology. Thus as time goes on, one would expect the technologies to be designed in such a way as to take into account the needs and aspirations of the users as well as the owners.

What would be some of the expected consequences of a user centered technology transfer model? We would expect technology transfer to be more rapid, because the users would be part of the design and implementation process. We would expect the cost to be less, because more use would be made of the tacit

knowledge of the work force and less of expensive computer models. We would expect the new technologies to embody fewer social costs because they would be designed with skill and health and safety concerns in mind. In other words we would expect more rapid transfer of appropriate technologies.

TECHNOLOGY TRANSFER FROM UNIVERSITY TO INDUSTRY IN THE NETHERLANDS

G. Honderd* and J. J. A. M. van Eijk**

**Delft University of Technology, Department of Electrical Engineering, Control Laboratory, P.O. Box 5031, 2600 GA Delft, The Netherlands*
***Delft University of Technology, Transfer Point, P.O. Box 5048, 2628 CN Delft, The Netherlands*

Abstract
The transfer of technology, especially between technological universities and industry is a growing area of interest to both parties. The development in the last decades indicates a much stronger relation than could have been foreseen before. From experiences with so-called transfer offices in universities where industry could put their questions and problems and where possible knowledge transfer was stimulated and organized, now a new style "Industrial Liaison Office" is created. Some experiences and a look into the future are presented in this paper, centered around the Delft University of Technology in the Netherlands.

Keywords. Transfer of technology, organisational and financial aspects, coordinated university-industry research, interfaculty cooperative agreements.

1. Introduction

The transfer of technology to industry is attracting a great deal of attention. This is expressed, for example, in the encouragement the national government -- in The Netherlands as well as elsewhere -- is giving to this phenomenon: an active policy on such transfers of technology has been the subject of discussion since 1979.

The principal reason to formulate an active policy lies in the desire to more often and more efficiently put high-quality research and development to work on behalf of the technological renewal of industry, especially small and medium sized enterprises.

A central point here is the proposition that sufficient technological knowledge is available in The Netherlands, but that this knowledge is insufficiently transmitted between the links in the chain of knowledge: from fundamental to applied research, and then to industry.

The important areas of consideration in the governmental policy are:

-- education and training [proceeding from the idea that people are the most important medium for the transfer and dissemination of knowledge];

-- publicly financed research [striving for greater openness and market orientation among research institutions];

-- broadening and increasing R&D-activities in the market sector [this clearly increases the receptiveness of this sector to technological information;

-- national and international technology programs and the acquisition of technology from other countries;

-- additionally, a number of specific stimuli for the transfer to and use of technology by medium-sized and small firms are included.

Among the specific stimuli for such transfer and use of technology by medium-sized and small firms was the institution [in 1980] of the 'Liaison Offices' or 'Transfer Bureaus', which are centers for scientific advice and information, located in universities and research institutions. The task of the Liaison Offices is primarily to improve the access of the target group to the knowledge available within such institutions.
Additional stimuli include:

-- promotion of temporary placement in smaller firms [including trainees] of those who have completed some degree of higher education;

-- promotion of temporary positions for entrepreneurs in public technological and educational institutions;

-- promotion of transfer of technological and organizational knowledge from multinationals and major corporations to smaller firms via subcontractor/supplier relationships [co-development and co-makership]; promotion of 'contact days' for businesses;

-- promotion of cooperation within sector or trade associations [national technology programs];

-- the creation of a national infrastructure for advice and information: the institution of regional Innovation Centers [see also Section 4].

Due to the diversity of their development, university Liaison Offices in 1989 have quite variable characters and required tasks. These sometimes differ considerably from the original design, but all still nevertheless fall within the intended theme.

The subject of this report is the transfer of knowledge from universities and research institutions to the business community. These topics will be briefly covered: historical development [section 2]; the original design for the objectives and procedures of the university Liaison Offices, plus the results [3]; the changing perception of the transfer of technology from the university, after the partial loss of governmental support [4]; implementation of such changes at Delft University of Technology [TU Delft] [5]; the financial aspects of transfer of technology [6]; and, finally, future trends with respect to the transfer of technology from universities to the business community.

2. Historical development

Universities, and in particular the technical universities, have worked for many years with the business community. At Delft University of Technology, the training program for engineers has had a close relationship with research and industry ever since the University was founded in 1852. The mutual dependence of the University and the business community is such that it is impossible to ignore each other for long without adverse effects. How can engineers be prepared for their tasks in industry and society if TU Delft does not get involved in the problems of industry and society? How can research be linked up with recent [industrial] developments if the University is hermetically sealed off from the business community?

For many years we have been familiar with the professor who works part-time at the university: he or she gives lectures, but also has a position in the business community or in government. Such a person effects a two-way transfer of knowledge: the knowledge available within the university can stream through to the firm, and at the same time a business component is brought into education an research. Such contacts are very valuable to TU Delft.

Further, many full-time professors at the university are originally coming from industry and the business community. With the experience acquired in these earlier positions, they bring the social components into scientific education and research.

At the same time, cooperative ties regularly arise between industry and the departments of the university, due to a growing recognition of the expertise available in these departments.

In these ways relations have been and are established 'naturally' between on the one hand, the TU Delft, and, on the other hand, commercial firms -- that is to say, principally, major concerns and multinationals, since only they have operated at a high level of technology, using engineers, for many years.

In contrast, small and medium-sized enterprises have had, and have, few or no academics in general [and engineers in particular] as employees. Therefore contacts with universities have not been established, or have come about only with great difficulty, even though these firms too could certainly profit from university contacts.

Consciousness of this problem was expressed by the government in the 1979 'Memorandum on Innovation',

and led to the institution of the university Liaison Offices, subsidized by the Ministry of Economic Affairs and the Ministry of Education and Science.

The main objective of these Liaison Offices is 'the promotion of the transfer of scientific knowledge from centers of knowledge to small and medium-sized enterprises, in order to increase the capacity of these enterprises for innovation.' Thus the Liaison Offices actively contribute to making publicly financed centers of knowledge more accessible to the society. [In 1985, such a contribution to increased accessibility was added to the Law on Scientific Education as a third major task of the universities, in addition to education and research.]

The first Liaison Offices, including the ones within the technical universities, began their work in 1981. By 1985, a national network of twelve Liaison Offices had become a reality.

Over time [arising from the first partial, and, after 1988, complete loss of government support], various subsidiary objectives have been formulated, including:

-- the strengthening of [regional] economic structures;

-- attention to the demand side of the process of technology transfer i.e., the promotion of what the various centers of scientific knowledge have to offer [for example, at trade fairs];

-- making the universities and research institutes themselves familiar with business problems, to promote more practically oriented education and research;

-- guiding the development of spin-off from scientific research;

-- contributing to a better attunement of education to the labor market.

We will return to this in Section 4. New philosophies on transfer of technology.

3. Industrial Liaison Office;
original operational plans and results

Transfer of technology from universities to the business community is of great importance. The role of the Liaison Offices in this process can be characterized as good but still modest. It is modest because in most cases such transfers are accomplished bilaterally -- that is, they are arranged directly between the firm seeking the technological knowledge and the university department which has knowledge to offer, without a third party as intermediary. It is also modest in extent: on average, a Liaison Office employs five people, of whom three are liaison officers.

This information and the data below are the results of a study of the functioning of the university Liaison Offices [source: Buck Consultants International, Nijmegen, The Netherlands, 1988]. These data primarily relate to the work of the Liaison Offices in the period 1985-1987, when they were operating under the original objectives: generating questions relevant to the intended target group and getting them answered, either within the university where the Liaison Office is located or elsewhere.

The position of the Liaison Office in the university organization differs from one university to another. The large universities [including TU Delft] use a decentralized model, in which a network of faculty members [one within each faculty] works with the Liaison Office; in smaller universities, Liaison Office functions are more centralized. Almost all Liaison Offices have an advisory board, whose task is to give both advice and direction. Members of these boards are drawn from the university and outside [for example, from the business community]. In practice, most Liaison Offices have reasonable access to the University Council, in spite of the variety of ways in which they may fit into the university structure. Such access is important because the Liaison Offices can suggest to the university which research areas must be seen as important [i.e., they have a signal function].

Illustration 1:
Total 1986 number of questions from industry per Liaison Office, showing the number of medium-sized and small firms included in the total

Liaison Offices
[number of questioners]

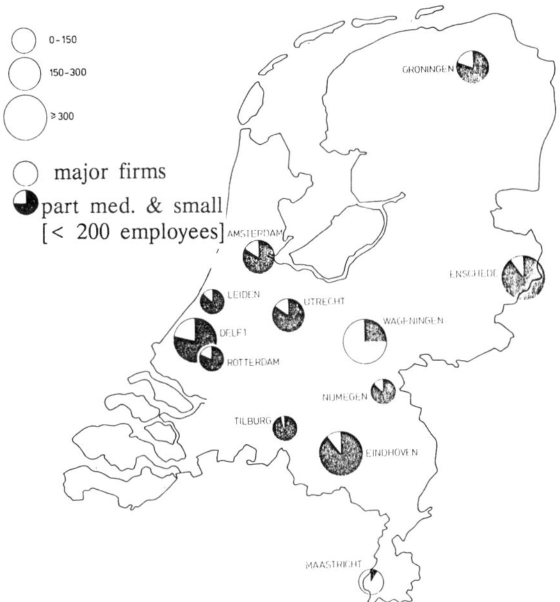

The twelve university Liaison Offices in The Netherlands handle more than 3,300 questions each year. The majority come from small and medium-sized businesses [those with a maximum of 200 employees]. TU Delft accounts for about 500 of these questions [see Illustration 1].

Approximately one third of those who ask questions appear to do so on behalf of industrial firms; 23 % of questions concern requests from university departments for help with specific projects [e.g. drawing up contracts].

In general, those who ask for information are located in the same region as the university and its Liaison Office. However, TU Delft is active nationally. For the Liaison Offices as a whole, about 14 % of questions come from other countries.

On average more than 50 % of questions are handled by an academic or staff department within the university; about 20 % are resolved within the Liaison Office, while nearly one-fourth are referred elsewhere. In about 15 % of cases, replies are worked up by students, while about 70 % are dealt with by academic staff. The remainder of questions are answered by personnel brought in on a project basis.

In 1986 the average Liaison Office worked with a budget of [somewhat] more then a half million guilders (1 Dutch guilder equals approximately 1 US dollar). In 1985, more than 60 % of this came from the government. Financial reimbursement by those who made use of Liaison Office services was not requested.

In more than half of the cases, information which already existed within the university was transferred [consultancies]. Knowledge newly generated by a university department was involved in 30 % of the questions [contract research]. The remaining questions involved making facilities or equipment available, doing literature searches, and so forth. Approximately 50 % of the projects involving a transfer of knowledge led to some form of innovation in the firms involved.

A few examples of cases:

***** Development of a fly ash monitor:**
A manufacturer of measuring instruments and control devices had problems with the further development of an idea for a device which would continually monitor the quality of the fly ash in a coal burning power plant.

***** Filtering of smoke stack gases from heating systems and storage of CO2:**
This question came from a company growing products in greenhouses. The exhaust gases were so heavily contaminated that it was not possible to measure their CO2 content accurately. The addition of CO2 increases the company's yields.

***** Recovery of oil particles from granular material:**
A business active in industrial cleaning approached TU Delft asking for help in the development and operation of an installation which would make it possible to remove and recover the oil particles released from granular material during the extraction of oil and gas.

***** Development of a very high capacity sprinkler system:**
A producer of sprinkler system equipment, e.g. for extinguishing fires, had designed a very high capacity system [up to 6,000 liters per minute]; tests using the prototype were not acceptable, because the water flow was not constant.

***** 'Healthy versus sick buildings':**
The user of an office building reported scattered complaints of illness, and wondered to what extent these might be connected with the climate within the building.

4. New philosophies on transfer of technology

As indicated by the above discussion, in 1987 the Liaison Offices were fulfilling their obligations; and, based on the original operational plan, they still do so. Naturally there are many businesses that still make no use of the Liaison Offices. However, the number of projects [completed] appears not inconsiderable, both quantitatively and qualitatively.

Further, the Liaison Offices are also a positive factor for the universities. That is not so much a matter of bringing in the so-called 'third stream' of financing for research [income derived from research contracts with industry and government]; the worth of a Liaison Office lies more in establishing a positive image, in giving substance to the concept of increasing the access of the society to centers of scientific knowledge, and in the stimulus it gives to research and education.

The foregoing sections suggest a favorable picture of the work of the Liaison Offices. Nevertheless, for several reasons the place and task of these organizations has been a matter of debate since 1987. Among the reasons:

-- universities have been confronted with financial cutbacks, which has decreased the capacity of the departments to help [often on an ad hoc-basis] with technology transfer projects;

-- in practice, little importance is attached to the third major task of the universities [increasing the access of the society to centers of scientific knowledge];

-- governmental technology policy regarding small and medium-sized enterprises has been reinforced by the establishment of local Innovation Centers. Since 1989 these have served as central desks and as a place to ask questions for small and medium-sized firms. They are located not in the universities but instead with Chambers of Commerce;

-- at the same time, government financing has ended, and the universities have been obliged to assume the full financing of the Liaison Offices.

These developments have led to a reorientation of the university Liaison Offices. In general, the attempt to let the direction of the work be determined by the inquiries received will decrease. The Liaison Offices will then, together with the university departments, undertake more contract research generating activities. The result may be a more commercial arrangement that leaves less room for requests from smaller enterprises that cannot cover the costs of the departments involved.

5 Transfer of technology at the Delft University of Technology; the new style Industrial Liaison Office

At Delft University of Technology, the new environment has meant modifications within the organization [see Illustration 2: The new style Liaison Office].

a. on the one hand, the ability to respond to questions is being maintained, even while making cutbacks. This will require more selective guidance in referring questions from businesses. [These may come directly to the Liaison Office or, for example, via an Innovation Center]. Transfer of technology from a specific university department may take place in the form of, for example, a short-term research contract plus, if relevant, a contract for education; a brief consultancy; a study performed by a graduate student as a degree requirement; a literature search; or the use of equipment.

b. on the other hand, a new 'technology coordination' function has been created: the organization of inter-faculty [and sometimes inter-university] cooperative agreements, in order to deal with multi-disciplinary and collective inquiries from businesses. Transfer of technology then takes place via medium to long-term contract research; in these cases it is primarily fundamental research which is in order.

The task of the technology coordinator will consist of handling the projects that require a 'tailor-made' project organization within TU Delft. This can involve:

-- where concentrations of knowledge are clearly available, reinforcing existing structures while assuring that the capacity available is always somewhat exceeded by the demand for services. Some examples:

*** **Delft Center for Phase Equilibria and Thermodynamic Properties** [Faculty of Chemical Engineering and Materials Science]

-- where the knowledge required is available but is diffuse and widely dispersed over several faculties, establishing the necessary linkages. For example:

*** **Delft Advanced Control**
[departments within the Faculties of Electrical Engineering, Mechanical Engineering, Technical Mathematics and Informatics, Applied Physics, Chemical Engineering and Materials Science];

*** **Delft Center for Adhesives Technology**
[departments within the Faculties of Chemical Engineering and Materials Science, Aerospace Engineering, Civil Engineering, Industrial Design Engineering];

-- taking the initiative to establish cooperative agreements to deal with questions which will require work over periods between six months and three years: that is, too long for an M.Sc. level student and too short for a doctoral thesis;

-- drawing the attention of the university to blank spots, i.e. areas in which knowledge is not available within the [technical] universities in The Netherlands, despite the fact that questions about this area come in regularly from the business community.

The objective of the requisite cooperative agreements is to open up financially interesting markets for the university.

The organizational form of the intended cooperative agreements should be dictated by the market situation: it is not pre-determined.
In general, internal work groups can be formed by cooperating departments, represented externally by a coordinating Project Bureau that operates on a commercial basis and provides a clearly recognizable point for communication. This bureau would use the title 'Institute' and would have the legal form of a foundation.

This 'new style' Liaison Office at TU Delft is a part of the Internal and External Communication Service of the University administration.

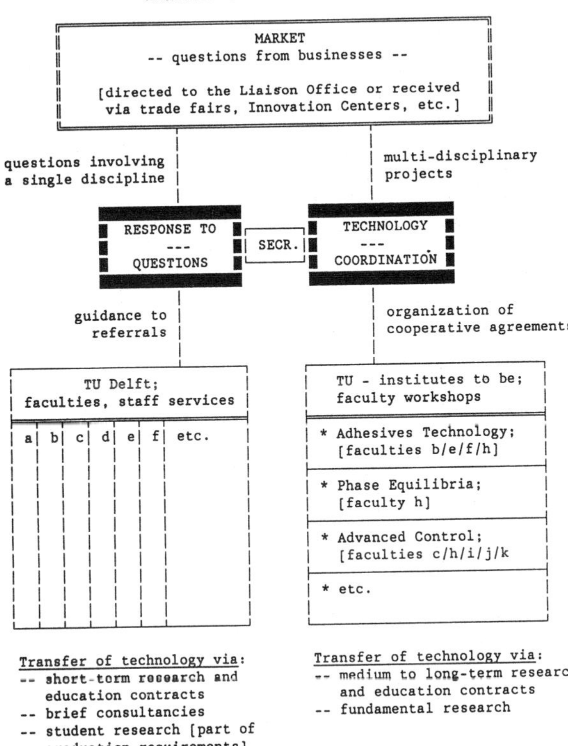

6. Financial aspects of transfer of technology

Three 'streams of financing' are available to universities in The Netherlands as sources of funding for their activities. These are:

-- the first stream [government funding: Ministry of Education and Science];
-- the second stream [grants from ZWO, the Foundation for Pure Scientific Research; and STW, the Foundation for Technical Sciences];
-- the third stream [income from contract research for industry and governmental institutions].

The first stream is by far the largest; the amount received is related to the number of students. For TU Delft, this amount is some 90 % of the total budget. The use of income from the second stream is strictly confined to the work outlined in the grant proposal. The third stream consists of income received from working cooperatively with industry or with governmental agencies. Such contract research is steadily increasing. For an overview of the extent of contract research at Delft University of Technology, see the following table.

At the moment, universities feel they must increase their income from sources other than the first and second streams, among other reasons because of financial cutbacks by the national government. Only in the third stream are increases possible. One of the obstacles, however, is that such contract research needs to be compatible with the on-going research funded by the first stream. Scientific interests should take first priority.

These and other conditions are important in seeking a suitable organizational form for cooperative agreements among university departments and faculties. These are

Table 1: Income from contract research at TU Delft [excluding medium and short-term projects]; the total budget of TU Delft in 1988 was approximately 485 million guilders.

year [-]	income [mill. fl.]	portion industry [%]	portion government [%]
1983	16.9	27	73
1984	17.9	23	77
1985	21.8	25	75
1986	23.7	32	68
1987	29.6	25	75
1988	31.4	36	64

now being established, with consideration for cooperation with and transfer of technology to industry. As noted above, it appears that coordinating and commercially oriented Project Bureau would be in order here. Such a bureau would have a clear externally directed task [providing a clearly recognizable point for communication, negotiating in the name of the university, concluding contracts] plus a clear internally directed task [staff communications to bring potentially interested parties together, general monitoring and support of projects, and monitoring of how money is spent within university. See Illustration 3.

The surplus from the activities carried out for this market sector must be deployed as a contribution to the preservation of the infrastructure of the university and the support of fundamental scientific research. Further, bringing together applied and fundamental research will be mutually beneficial.

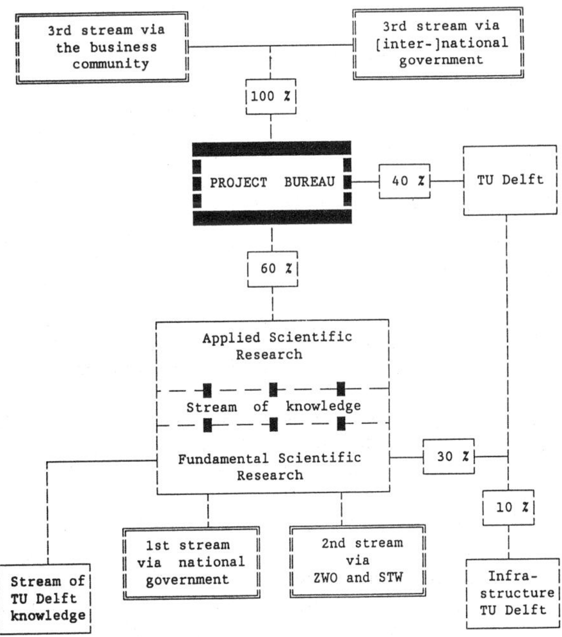

7. Future trends

The activities being developed by the new style Industrial Liaison Office at Delft University of Technology are quite in line with developments and trends in The Netherlands and elsewhere. The following trends are clearly discernable:

-- members of the business community [following the Japan example] are often getting together to collectively seek cooperative agreements with research organizations for short-term contracts;

-- the speed of innovation and introduction of new technologies is increasing;

-- a shift from 'natural resource economy' to a 'human resource economy' is taking place;

-- other universities in The Netherlands are also seeking both internal and external cooperative agreements;

-- reapportionment of research tasks among universities and institutes in The Netherlands [as in West Germany in 1988] is becoming a necessity;

-- the European integration in 1992 is already casting its shadow; research organizations in other countries [such as the German Forschungsinstituten] look upon our country and others as an interesting new market area.

The place and task of the Liaison Office in the future follow from these points. At the moment ways are being sought to bring together all of the experience and know-how available within the university regarding the second and third streams of financing. This may mean that in the [near] future the Liaison Office will be merged with, for example, staff who seek other external funding, experts on the European Community, financial experts, lawyers, and so forth.

References

Buck Consultants International, Nijmegen, the Netherlands, 1984

Annual Report Delft University of Technology, 1987

APPENDIX

Case study of a long-term project in ship control systems

In cooperation with the Royal Netherlands Navy, research institutes and private companies, TU Delft has developed several ship control systems.

An adaptive autopilot for seagoing ships has been developed which is able to carry out better manoeuvres and reduces the fuel consumption. The research has mainly been done by staff members of the university between 1972 and 1983. Full-scale trials have been carried out on board navy ships. The autopilot is commercially available.

Techniques similar to those applied in the adaptive autopilot have been used to design an autopilot which not only controls the yaw motion (cf. heading angle), but also reduces the roll motion (cf. heeling angle). It has appeared to be possible to control the heading and to reduce the roll by only using the rudder. The Rudder Roll Stabilization (RRS) project has been initiated by the university. The Dutch company which presently manufactures the RRS autopilot has detached a research worker at the university between 1983 and 1987 and has provided necessary funds. The RRS autopilot will be used on board the new M-frigates of the Royal Netherlands Navy.

The Netherlands Technology Foundation (STW) has supported a Ph. D. project for the development of an adaptive track predictor for ships. The track predictor uses a simple model of the ship while adaptation provides for sufficient accuracy of the prediction and for the insensitivity to varying circumstances. Experiments at the manoeuvring simulator of the TNO Institute of Perception indicated the usefulness of the track predictor as a manoeuvring aid.

A small company which manufactures electronic and hydraulic systems for inland ships has asked TU Delft for support with the development of an autopilot for inland ships. Initially TU Delft provided algorithms for a simple digital autopilot. Later on funds have been obtained from the Netherlands Technology Foundation to do more exhaustive research in a 2-year project. An adaptive autopilot which is able to control the rate-of-turn of the inland ship as well as its heading will be on the market in 1990.

Recently TU Delft has been asked to assist in the development of the control system for a surface-effect ship (SES).

TRANSFER OF TECHNOLOGY FROM RESEARCH TO PRODUCTION

T. Huhtelin

Valmet Automation Inc., Head Office, Tampere, Finland

Abstract. Technology transfer as an industrial problem is a quite
fascinating but a difficult subject to discuss. There are different
opinions how to handle this problem. It must be known how the company
will work while trying to obtain new technology. In this short statement I have taken 'the two-step approach' principle in R&D of an
industrial enterprise. In the first step all relevant information
including technologies will be obtained. After this information has
been treated properly and a decision has been made to develop a
product, no more information and technology is needed. It is true
that most companies do not work in this way yet.

Keywords. Automation; economics; technology forecasting; strategy;
product development.

R&D MANAGEMENT OPINION

INTRODUCTION

Management of a strategic business unit or a
company must take care of the future competitiveness. The main tool for this work is strategic
management. There are alternative ways available to
implement strategic management. One of the main
components is strategic R&D through strategic
development projects. This is a path from great
uncertainty to success or failure. At the same time
this is a path from research to development and
to production.

There are alternative strategic options available
for each company. The selection of a product to be
developed must be based on market needs, defined
business strategy, and available technologies. At
the same time strengths and weaknesses of the
company and available distribution channels must
be considered. The decision-making process is
specific for each company and also specific to the
business. However, some general rules can be described.

The decision to develop a product must be based on
a project plan. In this plan the development
process and the product are to be described. Timing
is a very important issue. The product must be
developed in the shortest possible time. Any
failures or mistakes are not allowed during this
development project. Only successful development
projects are allowed today.

The planning before the development project is a
key to success. During this planning process three
main topics must be studied:

> What are the market needs?
> What are the selected technologies?
> What is the product concept?

At the time the above mentioned information is
available the financial figures can be studied in
more detail:

> What are the investments in
> development, production,
> distribution channels, and
> marketing?
> What are accumulated income
> figures from sales?
> What are accumulated profit
> figures?

In many companies it is not enough to make these
plans. These plans must be compared to other
competing proposals of other businesses. The plan
must win the funding.

MARKET NEEDS

Key issue in defining market needs is the understanding of the market window. Each product has
a certain time span to be introduced. If the
product introduction happens too early or too
late, the market will not accept the product.

A second key issue is price and quality of the
product compared to the value of the product for
the customer. These parameters must be studied
very carefully, and they create the framework for
development.

In this area also the following questions must be
studied:

> Who is the customer and what are
> his opinions?
> What are the 'fashion needs'?
> What are the competitors doing
> right now in R&D?

PRODUCT CONCEPT

The product concept must be based on one leading
idea. In the beginning there will be plenty of
uncertainty. As stated before, it is not allowed

to make any failures during the development project.
This must be a guideline during the work done in
this phase. The information on market needs and
available technologies must be combined to the
product concept.

The product concept must take care of production
facilities and distribution channels. All these
must match together. If this part of the work is
done properly, the development is just straightforward work according to available plans.

TECHNOLOGY

Every product is composed of a number of distinct
and identifiable technologies. Here the meaning of
technology is the practical application of science
to address a particular product or manufacturing
need, or to an area of specialized expertise.

Every business unit must deploy and manage an
array or a portfolio of technologies. In many
cases there will be up to 200 technologies to be
managed. Some technologies have the greatest impact
on competitive performance at a certain particular
time. These technologies are called key technologies.

The development of a technology follows an
'S-shaped' curve, rising from first experimental
attempts to full achievement of performance
potential. Technology maturity can be classified
in four phases: embryonic, growth, mature, and
aging.

Of each selected technology the following factors
should be defined:

> How far are we from theoretical maximum?
> How much money and time is needed to
> reach it?
> What are the available alternative
> technologies?

Each company may have different needs concerning
the utilization of technologies. Companies can
act as:

> Innovators
> Early adaptors
> Late adaptors
> Laggards

TECHNOLOGY TRANSFER

As described before, the company will have the need
to obtain new technology as a part of a development project planning phase. The time to obtain
new technology is clearly dictated by other factors
than the availability and interests of the technology sources. This fact can also create some
problems, because there is also a so-called
technology window. This technology window is open
only for a relatively short period of time, and
after that it is more difficult to obtain information of new technologies. The company trying to
obtain new technologies will ask the following
questions:

> Who owns this technology?
> Who knows it in detail?
> Who is still interested in
> working with us?
> Who can make theoretical
> calculations?
> Who can make comparisons
> between related technologies?
> How reliable and accurate is
> the available information?

CONCLUSION

It has been described how an industrial enterprise
can operate in R&D. What are the key elements to
drive R&D decision-making and what is the role of
technology in this process. For an industrial
enterprise technology is a must. But it is a must
only during the short period of time when it is
being obtained. This is a key issue for any
technology source to be considered. Technology
sources must obtain more cabability to assist and
advise customers in the selection of the right
technology.

PANEL DISCUSSION ON THE TOPIC OF TRANSFER OF TECHNOLOGY FROM RESEARCH TO PRODUCTION

Y. H. Pao

Department of Electrical Engineering and Computer Science, Case Western Reserve University and AI Ware, Inc., Cleveland, OH 44106, USA

Keywords: Industrial Production Systems, Technological Forecasting, Management Systems.

I participate in this panel discussion with a sense of anticipation, believing that the sharing of perspectives and experiences will result in a richer knowledge base for the guidance of our actions in the future.

The subject matter is challenging and tantalizing and also very complex, perhaps as complex as that of trying to understand human nature itself.

In approaching this discussion, I draw on a set of varied experiences. These experiences include years of soliciting funds for the support of research at University, years of administering and managing research program for a government agency engaged in the support of university research, years as a researcher and professor, as a university research center director, as consultant for industry and also as the president of a start-up enterprise.

In my discussion my emphasis, as always, is on people, rather than events. Accordingly, I describe the activity of interest to us in terms of the individuals who in an ideally structured society would characteristically gravitate to the corresponding types of activities involved in the chain events in the overall "transfer of technology".

I characterize these groups to be

- the basic researchers, the "truth-seekers",
- the inventors
- the entrepreneurs
- the engineers
- the managers
- the "ministers"

All of these are essential to the successful implementation of a meaningful new product. All of these also pay a part in the transfer of technology from research to production, but to different degrees and in different manners.

We can and will discuss these matters at greater length, but it suffices, in this brief statement, to point out that it is crucial that would be "central planners" understand the nature of the chain of events. It should also be understood, that in any one environment it is not necessary that all the various activities be supported within that one environment. There is

evidence that the inputs and outputs from the various individual activities in that crucial chain of events can be imported and exported from individual environments to the benefit of all. This is true not only vis-a-vis academia and industry, but also from state to state.

Understanding of these opportunities may lead to a better utilization of human talents and of material resources. In our presentation we will cite various different scenarios to provide illustrations of how different mechanism operate to faciliate and maintain transfer of technology.

EXAMPLE OF TECHNOLOGY TRANSFER: FROM UNIVERSITY TO INDUSTRY

M. Piirto and H. Koivo

Tampere University of Technology, Control Engineering Laboratory, PO Box 527, SF-33101 Tampere, Finland

Abstract. In Finland cooperation between the industrial and academic sectors is becoming more common. Those from the academic sector help industry by providing new technology. The industrial sector for its part offers academia real and often quite challenging problems. In the academic sector one main problem in this cooperation is: How to serve industry practically? A case study of a portable data acquisition and control station is presented here to indicate how academic research can be quite rapidly transferred to industry.

Keywords: control equipment, real-time systems, technology transfer.

1 INTRODUCTION

Industry is increasingly interested in using academic know-how. There are three main ways the academic sector can transfer its know-how to industry:

1. By entering into cooperative research projects.

2. By motivating students to pursue higher degrees before going to industry.

3. By providing continuing education courses for those already in industry.

Of all the universities in Finland, the Tampere University of Technology (TUT) has been the most successful in obtaining research funds from industry. The Control laboratory of TUT is one of the most active in providing not only these research"services" [1], but also in providing the training of items 2 and 3 above. The Control laboratory has succeeded in providing these services in part because of the acquisition and development of good tools. One such tool, which can be used in a direct fashion for technology transfer and has been used in all three services mentioned above, is described here.

2 DATA ACQUISITION AND CONTROL (DAAC) SYSTEM

Previously, measurements were done on-site using instrument recorders, which were then taken to another location, maybe faraway from the test site. The rapid development of computer hardware and software has now made it possible to do on-site analysis and measurement monitoring during experiment. In this way the testing time can be significantly reduced, experiments repeated, and operation continued without long delays. In many cases it is also advantageous to experiment with modern control algorithms. By using the DAAC system

it is possible for example to compare normal PID-controllers to modern ones. The DAAC system provides the environment which can easily be connected to the I/O cards of the process station. A brief description of the developed system follows [2].

Hardware: The DAAC system (Fig.1) hardware consists of a portable HP 9000/350 microcomputer having the computational power of a VAX 11/780, an HP 3852A datalogger and an Opto 22 unit for outputing control signals. The data logger sends measurements to the computer via an HP-IB bus. Control signals are sent to the Opto 22 unit via an RS-232 interface. Measurements and other data from the controller routines are stored on hard disk.

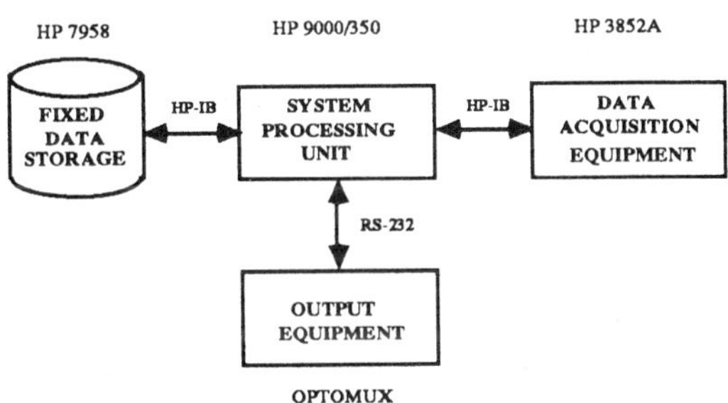

Fig. 1. Data acquisition and control station.

Software: The software is written in C. It can be divided into two parts: a) real-time programs, and b) off-line programs. In both the real-time and off-line programs the user is presented with a series of menus. The user may use simply the keyboard or a combination of mouse+keyboard in selecting items from the menu and entering data.

Real-time programs are responsible for measurement and control tasks. The software has been made so that it is easy to add new control algorithms. At this moment the control algorithms implemented in the software are single-input single-output PID-controllers & STCs and a multivariable PID & a multivariable self-tuning PID-controller.

Measurements and control data saved can be drawn, analyzed, and processed using the off-line programs. For example the analysis part includes provision for calculating of auto-correlations, cross-correlation and power spectra.

3   USE OF THE DAAC SYSTEM

The DAAC system has been designed primarily as a tool for performing control tests and for the collection of accurate process data in industry, but it has been used also for education and research in conjunction with the pilot processes of the laboratory. In industry it has been used for example on paper machines and in power plants.

In this example it is shown how to compare different control algorithms. The DC-motor's speed is measured with a tachometer and controlled by changing the current to the motor. In Fig. 2 results are shown for an experiment where a change was made from a PID controller to a STC at about t=57s. (Note that the setpoint only goes to that controller which is actually controlling the process, hence the setpoint for the STC remains constant during the first set of up-and-down setpoint changes.) The change from PID controller to STC occurs without disturbance.

The standard controller implemented in industrial automation systems is a digital PI(D)-controller. By feeding appropriate parameters into the DAAC system's own PID-controller(s), the DAAC system can mimic the performance a plant's own automation system. With DAAC system one can rapidly switch from the PID-controller(s) to, for example, self-tuning controller(s). Thus it is possible to find out quickly if there is any improvement between original, "industrial" controllers and modern, "academic" controllers.

4   CONCLUSION

It has been shown that the DAAC system is a practical tool to transfer know-how from university to industry. There are many different research and teaching teams in the control laboratory which are frequent users of DAAC system. In addition a very direct technology transfer has taken place when a number of industrial companies have purchased the system software: a power company, a pulp and paper company, and a company manufacturing chemicals.

Here is the summary of the advantages of the DAAC system:

- portability
- data analysis on-site
- ability to compare different control algorithms
- ability to test modern control algorithms in industry

ACKNOWLEDGEMENT

We wish to thank MSc Frank Cameron for proofreading the paper.

REFERENCES

[1]   KOIVO, H. (1987). Panel Session User/ Supplier/Academic Relationship Experiences At Tampere University of Technology, IFAC 10th World Congress on Automatic Control, Munich, Federal Republic of Germany, Vol.11, pp. 67.

[2]   PIIRTO, M., H. KOIVO, and P. VIRTANEN (1989). Portable Data Acquisition and Control Station for Self-Tuning Control Applications, 1989 IEEE International Conference on Control and Applications, Jerusalem, Israel.

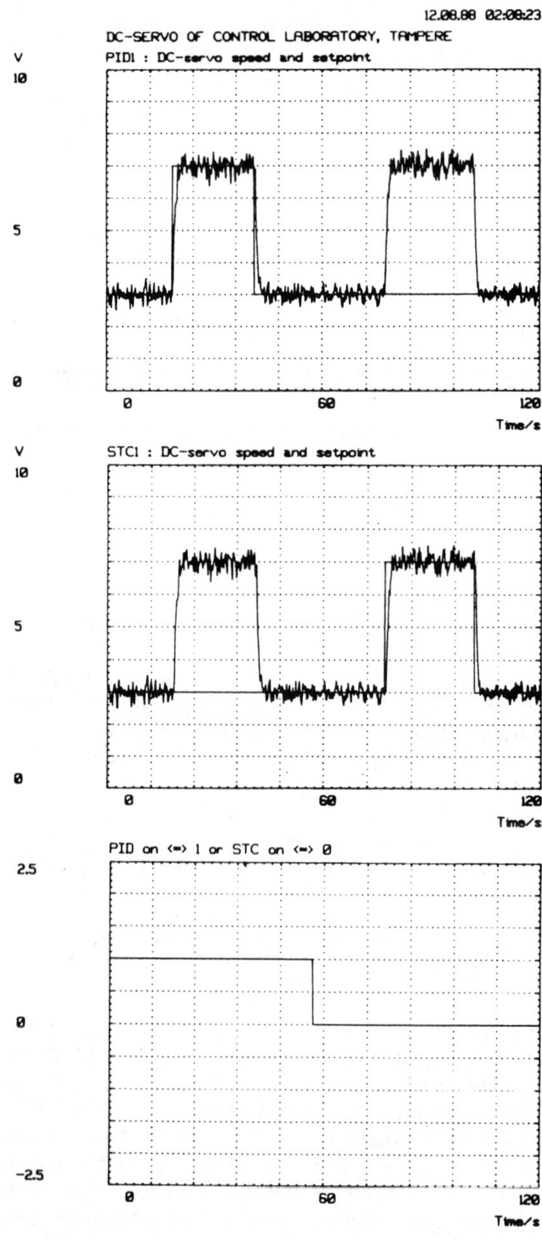

Fig. 2. DC-servo motor test with PI and STC in succession.

ADAPTIVE SWITCHING CONTROLLER
A Case Study in Technology Transfer from University to Industry

H. Rake and P. Schmitz

*Institut für Regelungstechnik, RWTH Aachen, Steinbachstr. 54,
D-5100 Aachen, FRG*

Abstract. The development of a controller from the basic idea up to the technology transfer to an industrial manufacturer of control equipment is described. For marketing the comparison of the performance of the controller with that of existing products is essential. The experiences with patenting efforts show that the timing of publications and the application for a patent can be of crucial importance.

Keywords. Adaptive control; on-off control; technology transfer.

INTRODUCTION AND TIMETABLE

Technology transfer is sometimes regarded as transfer of readily applicable knowledge from one partner to the other in the same manner as goods are transferred by sale "off the shelf". This picture is misleading in most cases. In this contribution the development of an idea and the efforts to transfer it to possible users will be described. The following table shows the time it took the idea to mature.

1978
- Conception of basic idea
- First implementation on process computer
- Tests with model of polymer extruder

1980
- Presentation of controller on process computer at INTERKAMA Fair
- Improvement of concept
- Implementation with 8 Bit-microcomputer

1983
- Filing of national patent application
- Presentation of dedicated 8 channel microcomputer-based controller at INTERKAMA Fair
- After intensive discussions with several instrument manufacturers agreement on cooperation with an international company
- Filing of international patent applications
- Building of prototype controller, setting-up of extensive documentation of hard- and software, training of personnel for industrial partner
- Tests of prototype against proprietory product on the premises of industrial partner

1986
- Presentation of prototype of an industrial controller at INTERKAMA Fair
- Discussions with national Patent Office with regard to patentability of claims, priorities of publications etc.
- Further improvement of controller concept

THE BASIC IDEA

The basic idea of an adaptive switching controller (ASC) is to employ a model of the process to fore-cast the future process output as resulting from possible process input sequences. For a limited prediction horizon and a discrete-time controller the number of input sequences is bounded and thus an optimal process output can be selected from a bounded number of candidates. The first element of the process input sequence corresponding to the optimal process output is executed in the succeeding time interval. The procedure of evaluating all possible process input sequences and selecting an optimal one is repeated at each and every sampling instant. Adaptation of the process model via on-line process identification gives the controller adaptive capabilities. Fig. 1 shows a bloc diagram of the basic ASC.

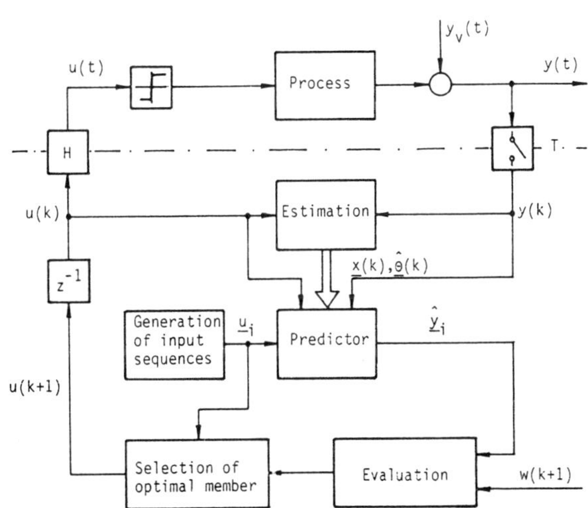

Fig. 1: Adaptive Switching Controller

MARKETING EFFORTS

From the very beginning of the development of the ASC efforts to compare its performance with that of conventional switching controllers (Hoffmann, 1984) and contacts to manufactures and users of these controllers were sustained. An excellent opportunity to present the idea and its possibilities to the control community was seen in the participation of

the institute in the INTERKAMA Fairs, the world's
largest exhibition of control-related hardware and
services. As given in the timetable the ASC has
been presented three times on this fair. The second
one resulting in an agreement on cooperation with
an international manufacturer of control equipment.

PATENTING EFFORTS

Most scientists after conceiving an idea first try
to convince themselves of its viability and then
endeavour to do the same with colleagues and members of the scientific community e.g. by publications. Only much later in the process of development they realize that the idea might be of commercial value and should be protected by claiming patent rights.

The case described here is no exeption from the
rule given. When stating patent claims for filing
with the national patent office two publications
(Breddermann, 1980; Hoffmann, 1981), exhibiting
major features of the idea had to be taken into
account. All features given in the publications
mentioned could not be protected by the laid-open
patent (1984). This resulted not only in lengthy
discussions with patent officials but also in a
significant reduction of the number of claims stated originally. It is hoped that finally a patent
will be granted before the IFAC Congress.

In parallel to the efforts sketched above the industrial partner stated patent claims in many countries using the priority of the national claims
mentioned and requesting and receiving help on
short notice in discussions with patent offices all
over the world.

COOPERATION WITH EQUIPMENT MANUFACTURER

This act of technology transfer into a foreign
country-industrial environment was performed by

- building and supplying a prototype controller,
- writing an extensive documentation of concept.
  software and hardware in English,
- training personnel of the industrial partner
  in the institute,
- sending personnel of the institute to the premises of the industrial partner for further
  training of personnel and tests of the controller supplied,
- long-term cooperation on introduction and enhancement of the concept.

CONCLUSION

Technology transfer is a very tough business. In
most cases demands on resourcefulness, enthusiasm
and even funds of both sides are greater than originally expected. Reliability on both sides is an
absolute must for a successful transfer. The benefits are much less certain than the costs. However
the alternative of industry and university working
side by side without much interaction is worse than
many transfer efforts with partial success only.

REFERENCES

Breddermann, R. (1980). Realization and application
of a self-tuning on-off controller. _Lecture
Notes in Control and Information Sciences_ 24,
Springer Verlag, pp. 74-83

Hoffmann, U. and Breddermann, R. (1981). Entwick-
lung und Erprobung eines Konzepts zur adaptiven
Zweipunktregelung. _Regelungstechnik_ 29, 6 , pp.
212-213

Hoffmann, U., Müller, U., Schürmann, B. and Rake,
H. (1984). An on-off self-tuner development,
real-time application and comparison to conventional on-off controllers. _Proc. 9th IFAC World
Congress_, Budapest, Hungary, pp. 2777-2782

Laid-open German patent no. DE 3418501 A1 (1984).
Vorrichtung zur prädiktiven zeitdiskreten Ein-
Aus-Regelung zeitkontinuierlicher Prozesse mit
binär wirkenden Stellelementen.

TRANSFER OF TECHNOLOGY FROM ACADEMIA TO INDUSTRY

N. Rozsenich

Bundesministerium für Wissenschaft und Forschung, Sektion Forschung und Technologie, Freyung 1, A-1014 Wien, Austria

1. The role of the Federal Ministry for Science and Research (BMWF):

Activities of the BMWF in the area of technology transfer started in the early eighties. Based upon several studies potential areas for innovation in the Austrian economy were investigated and existing barriers between the scientific community and industry identified. As a result numerous measures have been taken in order to improve the flow of know-how originating from research and its subsequent transformation into products.

On the policy level the promotion of R&D in a wide range of economically promising areas was intensified and a number of specific programmes - prominent among those a programme for the promotion of microelectronics and information processing - introduced. In 1989 a government document on technology policy ("Technologiepolitisches Konzept der Bundesregierung") was issued which underlines the importance of strengthening and modernising applied R&D with a view to improving the international competitiveness of the Austrian economy, of intensifying the cooperation between science and industry and of the strategic role of the public sector in respect of planning and implementation of large scale projects with long term effects.

The promotion of research in Austria is conducted by two funding organisations, the fund for the promotion of scientific (FWF) and the research promotion fund for small and medium sized enterprises (FFF). While the FWF is mainly concerned with basic research carried out by researchers, respectively research teams, usually from the universities, the FWF aims at supporting research undertaken by enterprises, increasingly in cooperation with the university sector. The BMWF is responsible for the allocation of the annual government grant to the funds and supervises the activities of the funding organisations.

In view of the structural features of the Austrian economy in addition to R&D incentives specific efforts for furthering the transfer of technology have to be made to ensure quick and systematic transformation of R&D results into products for which demand in world markets exists. Therefore various schemes have been put into operation and several initiatives started, which will be described in more detail.

2. Transfer of research personnel:

Under this heading two schemes run by the BMWF should be mentioned:

"Scientists for Industry" and "Company start-up for Scientists". Up to now more than 100 scientists have made use of the possibility to spend a period of up to two years in industry. Under the terms of the scheme a return to the previous post is guaranteed and the participating company receives a subsidy. The success of this scheme gave rise to the complementary company start-up scheme.

3. Technology transfer activities of the universities:

The basic organisational unit of an Austrian university is the "Institut". (Altogether there are more than 800 technology transfer activities is the domain of the so called "Außeninstitute", which in addition to acting as liaison centres to the business community are also active in the area of continuing education.

4. Technology (incubation) centres and research parks:

From 1986 onwards a steadily growing number of technology (incubation) centres started their operation. Due to the federal structure of the country these developments take place in close cooperation with provincial and local authorities, chambers of commerce and - of course - federal institutions. The

BMWF takes an active interest in the activities of the centres and supports in manifold ways the progress of this dynamic section of the regional economy.

While size, organisational and ownership structure of these centres vary widely, the cooperation and exchange of information between them is well organised under the umbrella of a countrywide organisation ("Vereinigung der Technologiezentren Österreichs").

The secretariate of this association is currently set up in a government sponsored agency ("Innovationsagentur"), which itself is active in the field of technology transfer.

5. Examples of successful technology
   transfer:

The effect of the combined effort by government and the private sector, by science and industry, can best be demonstrated by a selection of successful cases, which will be presented during the panel discussion at the 1990 IFAC World Congress.

RESEARCH, INDUSTRY AND TECHNOLOGY TRANSFER AT THE NIST AMRF

D. A. Swyt

Precision Engineering Division, US National Institute of Standards and Technology, Gaithersburg, MD 20899, USA

Abstract. This paper, which deals with the Automated Manufacturing Research Facility (AMRF) at the U.S. National Institute of Standards and Technology, outlines briefly the nature of AMRF research, interactions with industry, and principal mechanisms of technology transfer to large-, medium- and small-size firms.

Keywords. Automation, computer interfaces, flexible manufacturing, machine tools, robotics, standards, technology transfer.

INTRODUCTION

Within its Automated Manufacturing Research Facility (AMRF), the U.S. National Institute of Standards and Technology (NIST), formerly the National Bureau of Standards (NBS), is addressing issues of measurement-based quality control and standards-based system integration within the flexibly-automated "factory of the future".

In the conduct of its forefront research on this emerging technology, NIST researchers already work directly with counterparts from large R&D-oriented industries and universities.

To facilitate transfer of research results to small- and medium-size firms without such an R&D orientation, a new mechanism of technology transfer through state-based technology centers is being initiated.

THE RESEARCH

The NIST AMRF consists of robots, computers and machine tools from a variety of domestic manufacturers being integrated under a generic control system architecture; the purpose of AMRF research is to support development by industry of voluntary standards for the integration of such multi-vendor systems of flexible automation [Simpson 1982; Nanzetta 1984].

The graduate-level work within the AMRF consists of: (1) research on real-time sensory-feedback control; (2) development, based on that research, of a standard control-system architecture; (3) implementation, under that architecture, of a laboratory facility in the form of a flexibly-automated-manufacturing system; and (4) demonstration, by means of that facility, of the feasibility of such an architecture as the basis of commercial multi-vendor systems of flexible automation.

The results of this work within the AMRF are forms of knowledge transferred to industry by a wide variety of mechanisms.

INTERACTIONS WITH INDUSTRY

Considered to be the most important means for the NIST AMRF program to interact with U.S. industry is the Industrial Research Associate Program [ORTA 1989]. Under this program, technical people from domestic industrial firms work with NBS counterparts in what is, in effect, co-generation of public-domain knowledge. Over the last eight years, 50 firms (and a similar number of universities) have had Research Associates in the AMRF.

In addition to joint public-domain research under the auspices of the RA program, major NIST interactions with industry have included: demonstrations of system capabilities to 3500 attendees at public test runs of AMRF systems, technical briefings to 20,000 visitors to the facility and off-site showings of AMRF videotapes to 100,000 viewers. Further, technical communications include nearly one hundred publications as journal articles and conference proceedings and an estimated 2500 technical talks.

TECHNOLOGY TRANSFER

Technology transfer from the AMRF to industry occurs in hard and soft forms. The broadly-successful "soft" form conveys information on the nature and implications of flexible automation to technical, managerial, educational and public-policy audiences; the narrowly-successful "hard" form conveys specific technical results to industry for potential use in development of commercial products and processes.

In terms of the recipients of its technology transfer in the "hard" form, NIST has been more effective with certain types of industrial firms than its has been with others.

One type of manufacturing enterprise for which there has been very effective technology transfer is the large firm with commitment to its own internal development of process technologies based on advanced robotics, machine tools, software or sensors. A second type for which technology transfer has been effective is the production facility of a federal agency with a mission need for advanced automation. A third type is the producer-goods company, independent of size, which has product lines associated with robotics, machine tools, software or sensors.

However, a fourth type of manufacturing enterprise-- one for which there has been no

effective direct transfer of results from the
AMRF -- is that of the local small manufacturer.
That NIST cannot itself deal directly with this
type of firm has been indicated by the report of
the reactions of small shop owners to the AMRF
[MSB 1985].

Given the notable successes of the NIST AMRF in
dealing with large R&D-oriented industrial
firms, a problem remains regarding such direct
transfer of research results to small- and
medium-size firms which operate at the local
level.

Because of the importance of such transfer to
the nation's international competitiveness as
well as to local economic development, Federal
legislation has been enacted to address the
problem of facilitating transfer of results of
NIST's advanced research to small- and medium-
size manufacturing firms; in addition to
changing the name of the National Bureau of
Standards to the National Institute of Standards
and Technology, the legislation provides for
Federal match-funding of state-based technology
centers [CUS 1988].

Activities of each center, three of which have
been established with 50% Federal matching
funds, are expected to include:

1) informing and educating the industrial
    firms in its region about advanced
    manufacturing techniques;

2) demonstrating the applicability of
    advanced technology to these firms;

3) actively assisting firms in evaluating
    their requirements;

4) assisting with the implementation of
    desired applications;

5) supporting work-force training and
    retraining; and

6) communicating technology transfer
    experiences to a wide national audience.

Given the impetus to create such technology
centers, understanding of the means to transfer
of results of research from Federal laboratories
to local industries still poses major conceptual
as well as practical difficulties.

To address this problem, a model system of
technology transfer has been proposed, one in
which a Federal laboratory -- the NIST AMRF --
is linked to small manufacturers through state-
and community-based technology centers [Swyt,
1988].

In the model, the four types of institutions in
the system are matched along each of four
dimensions:

        -- governmental level,
        -- innovation stage,
        -- educational counterpart, and
        -- capital-revenue base.

In summary, the institutions in the model system
are:

(1) Federal laboratories, with are public-
     funded, with PhD-level workers carrying
     out a research function:

(2) a first type of technology center, --
     state-based, non-profit, with master's-
     degree-level workers, carrying out a
     development function;

(3) a second type of technology centers,
     community-based, not-for-profit, with
     bachelor's degree-level workers, carrying
     out an engineering-application function;
     and

(4) small manufacturers, private, for-profit,
     with associate degree-level workers,
     carrying out a production function.

This model for a four-tier system linking
Federal laboratories to small manufacturers --
which involves two types of technology centers
as intermediaries rather than a one -- is
helping shape the system of new centers now
being initiated.

CONCLUSION

Within the NIST Automated Manufacturing Research
Facility, public-domain research on the
measurements and standards issues of flexibly
automated manufacturing is being conducted in
collaboration with larger and/or R&D-orientated
industrial firms. To complement that type of
government-industry interaction, new technology
centers are also being established within the
states to provide local support to small and
medium size firms.

REFERENCES

Congress of the United States, Technology
Competitiveness Act (a subsection of the Omnibus
Trade and Competitiveness Act), Washington DC,
August 1988.

Manufacturing Studies Board, National Research
Council, "Reactions of Small Machine Shop Owners
to the Automated Manufacturing Research Facility
at the National Bureau of Standards", National
Academy Press, Washington DC, 1985.

Office of Research and Technology Applications,
National Institute of Standards and Technology,
Gaithersburg Maryland 20899.

Simpson, J. A. et al. "The NBS Automated
Manufacturing Facility", Journal of
Manufacturing Systems, Society of Manufacturing
Engineering, 1(1):17-32, 1982; and Nanzetta, P.
, "NBS Research Facility Addresses Problems of
Set-Up in Small Batch Manufacturing", Industrial
Engineering, Vol. 16, No. 6, June 1984.

Swyt, D.A., "Transferring NBS Technology to
Small Manufacturers Through State and Local
Centers," Journal of Technology Transfer, Vol.
13, No. 1, (1988).

COMPLEXITY IN MAN–MACHINE SYSTEMS

G. Johannsen

*Laboratory for Man–Machine Systems, University of Kassel (GhK),
D-3500 Kassel, FRG*

Abstract. Technical, behavioural, and task perspectives of complexity are explained. The subjective aspect of task complexity is mentioned. Possibilities for coping with complexity are outlined. They deal with training, transparency, and decision support.

Keywords. Man-machine systems; complexity; decision support system; training.

DIMENSIONS OF COMPLEXITY

Complexity is a term frequently used for characterising many of today's technical systems as well as economic, social and other systems. Generally, complexity can be defined as the whole composed of a number of parts, features or other entities. Being complex is the opposite of being simple but does not necessarily contain the meaning of being complicated.

The complexity aspects become even more important when dealing with man-machine systems which comprise man (one or several persons) and machine (in the broadest sense, i.e., all kinds of technical systems) with their interactions. Thus, technical, behavioural, and task perspectives of complexity can be distinguished in man-machine systems.

The technical perspectives include structural and functional complexity. Any technical system can be broken down into possibly several levels of subsystems and components. They may consist of hardware as well as software. The whole of all the subsystems and components with their interconnections determine the structural complexity of a system. Possible measures for structural complexity may be the number of subsystems and components or the number and pattern of interconnections. Software complexity is sometimes dealt with separately because not only the program structure but also such issues as program length and software quality may be of concern.

The functional complexity expresses the multitude and combination of functioning within a system. Particularly, dynamic aspects of the system and their inter-relationships are described. The number of functions under different operational conditions in normal and failure situations as well as functions of reliability and maintainability contribute to the degree of functional complexity. It can be measured by the number of these functions and their relationships as well as by reaction times and accuracy measures.

The behavioural perspectives of complexity are concerned with human abilities and limitations in man-machine systems. Henneman and Rouse (1986) differentiate between perceptual complexity and problem-solving complexity. This distinction relates to the visual and cognitive aspects of human behaviour, respectively. Perceptual complexity deals with the human ability of recognising and interpreting displayed information. This leads to the possibility of manipulating some characteristics of information presentation in order to improve the speed and efficiency of man-machine interactions.

Rouse and Rouse (1979) found, however, that problem-solving complexity is more relevant than perceptual complexity in fault diagnosis tasks. Probably even in pure monitoring tasks, this result may be found because the underlying human understanding of what is going on in the system is essential. The problem-solving complexity relates to human reasonsing abilities and problem-solving skills. The number of relevant relationships between possible causes and symptoms in failure situations can be used as a measure of problem-solving complexity. Individual differences and attained skill levels play an important roll in perceptual and problem-solving complexity. If human cognitive limits (e.g.,of human short-term memory) are exceeded, some system failures may not be diagnosable (as stated by Wohl, after Henneman and Rouse, 1986).

Task complexity is the most predominant dimension of complexity in man-machine systems. It relates the human abilities and limitations to the structural and functional properties of the technical system. Both technical and behavioural perspectives determine the tasks to be performed (Leplat, 1988). One can distinguish between prescribed tasks or work assignments on the one hand and perceived tasks on the other hand. The perceived tasks are characterised by mental representations of the prescribed ones and are, thus, influenced by subjectives factors. The degree of the individual understanding of structural and functional characteristics of the technical system is of this subjective nature, based on the acquired skill level of the human operator or the maintenance personnel. Their knowledge about the technical system is not only formal knowledge about structures and functions of the technical system but also task-dependent knowledge about how to utilise the technical system. Low levels of skill and knowledge can lead to wrong mental representations of the tasks whereby the real systems and task complexity are not appropriately grasped.

COPING WITH COMPLEXITY

The question is now how to deal with the different dimensions of complexity in man-machine systems. One of the main goals of man in interacting with technical systems is to make their complexity manageable. As Woods (1988) stated, one standard tactic of humans is to bound the world to be considered. He continues: "Thus, one might address only a single time slice of a dynamic process or only a subset of the interconnections between parts. However, it is not at all clear that simple aggregation of the results of bounded investigations will capture all the relevant aspects of the whole." This way of reducing complexity seems to be a risky one. Other approaches for coping with complexity need to be pursued. A major goal in these efforts will be to achieve transparency.

One way towards transparency is training (Leplat, 1988). Task complexity effectively reduces when a human operator or maintenance person built up better skills through training, that means that he or she developed a clearer mental representation of the prescribed tasks with a better understanding of the structural and functional complexity behind these tasks. Then, structures, functions, and tasks appear more transparent and possibly allow higher human performance with lower mental workload.

Another approach tries to reduce task complexity by technical means. During the design process, more functions will be assigned to the technical system by an appropriate function and task allocation which may even be adaptive to different operational conditions and failure situations. Assigning more functions to the technical system creates higher levels of structural and functional complexity. To understand this equally well can increase task complexity correspondingly for the human. If the human limitations are compensated by the function allocation, however, then the effective task complexity may be reduced. This can be achieved either by completely autonomous technical systems functions which are safe enough because of built-in redundancy, or by decision support systems which aid the human in problem solving tasks. Although decision support systems become part of the technical system and, thus, increase their structural and functional complexity, they can lead to better transparency. It is obvious that this improved transparency as well as the related reduced task complexity can only be achieved when the task demands and human information needs are clearly analysed in a human-centred design approach.

Examples of decision support systems within dynamic technical systems are described by Johannsen (1990) at this IFAC World Congress. Results from the ESPRIT-GRADIENT project on "Graphics and Knowledge Based Dialogue for Dynamic Systems" as well as more relevant references are given in that paper. The suggested decision support systems are of two types, namely based on user models or application models. User-model based decision support systems can evaluate human errors or even recognise human plans. Such systems are currently investigated in several research and development projects. They will contribute to more error-tolerant man-machine systems with less stringent task demands. An expert system for diagnosising technical faults is an example for a decision support system of the application-model type. Such systems are particularly developed for making the structural and functional complexity more transparent.

Concluding this short discussion on complexity in man-machine systems, it is pointed out that the effective task complexity may be reduced by means of new automation and information technologies, thereby at the same time increasing the structural and functional complexity. Human-centred and task-oriented approaches are needed in order to deal appropriately with this design complexity. However, it must be admitted that not all problems mentioned in this article, particularly about human problem solving and decision support systems, have been solved sufficiently in order to be immediately applicable in industrial settings. Further progress can be achieved with joint research and development projects between industrial companies and universities, such as within the ESPRIT-programme of the European Communities.

REFERENCES

Henneman, R.L., and W.B. Rouse (1986). On measuring the complexity of monitoring and controlling large-scale systems. IEEE Trans. Systems, Man, Cybernetics, SMC-16, 193-207.

Johannsen, G. (1990). Towards a new quality of automation in complex man-machine systems. Preprints, 11th IFAC World Congress Automatic Control. Tallinn.

Leplat, J. (1988). Task complexity in work situations. In L.P. Goodstein, H.B. Andersen, and S.E. Olsen Francis, London, pp.105-115.

Rouse, W. B., and S. H. Rouse (1979). Measures of complexity of fault diagnosis tasks. IEEE Trans. Systems, Man, Cybernetics, SMC-9, 720-727.

Woods, D.D. (1988). Coping with complexity: the psychology of human behaviour in complex systems. In L.P. Goodstein, H.B. Andersen, and S.E. Olsen (Eds.), Tasks, Errors and Mental Models. Taylor & Francis, London, pp.128-148.

THREE RELIABILITY FACTORS IN AUTOMATIC CONTROL SYSTEMS

H. J. Leśkiewicz

Warsaw Technical University, Poland

Abstract. Hardware reliablity, software reliability and human operator realiability
were discused as three main reliability factors in automatic control systems. The relation between those three reliability factors and failure modes were presented. The
problem of optimal participation of hardware, software and human operator in automatic
control systems was approached from the point of view of reliability.

Keywords. Probability; reliability; safety; process control; redundancy; human operator;
man-machine systems.

INTRODUCTION

Reliability of an automatic control system is expressed by the probability that this system will
work satisfactory in a given period of time.
Among technical sciences reliability is distinguished by the need of using a probabilistic approach to technology and not a deterministic one which
the majority of engineers are accustomed to.

In 1713 Bernoulli avrote his "Ars Conjectandi" which
may be regarded as the beginning of a probabilistic
approach to nature. It had soon become evident that
this approach was necessary in some problems of nuclear physics, agriculture, biology, medicine, military strategies, and public opinion testing. Reliability is based on a probabilistic approach as well.

Efficiency in massproduction causes permanent increase in the value of products in a given automatically controlled production system or production line
in a time unit. This fact is responsible for the
recently observed tendency to increase the reliability of automatic control systems consisting the
economical factor. The other reason can not be expressed in economical terms as it is connected with
the increasing number of disasters destroying human beings and environment.

Any technical problem is complex enough to promote
a simplified model suitable for research.The below
presented three reliability factors in automatic
control systems consist such a model. The factors
are as followes: hardware realiability, software
reliability and human operator´s reliability. The
problem is the reliability of system.

HARDWARE RELIABILITY

Hardware of an automatically controlled production
process or production line is composed from two main parts. The first is the hardware which is directly involved in production, and the second one
                is the hardware which controls it automa-
tically. In some cases the division between these
two parts is not very well defined. Automatic control system´s designers and producers concentrate
more on hardware reliability of their systems, but
both hardware parts have an inpact on the reliability of production process or production line.

It was hardware that reliability considerations in
technology started with. Primarily the classical
methods of reliability evaluation were in use; they
were based either on known or on assumed reliability of separate blocs forming the hardware system.
Then some data - based methods were introduced additionally. Probability density function and probability distribution function started to be used as well
failure density and hazard rate function. The hazard rate function is very useful in presenting differences in the three lifeperiods from the reliability point of view of nearly any hardware and namely, the period immediately following installation,
the period of quasi-stabilized performance and the
period of decay.

Hardware failures may be divided in to four failure
modes: poor quality fabrication, design error,
overload of the component and wear-out. All attempts
to minimize those failures and thus to increase the
hardware reliability were not satisfactory. The new
way of increasing reliability of a hardware system
not changing the reliability of its components started to be developed in 1956 initiated by the paper
by E.F.Moore and C.E.Shannon: "Reliable Circuits
Using Less Reliable Relays". The paper presented a
possibility of creating a more reliable system from
less reliable components. The methods making use of
redundancy, majority voting and artificial intelligence in order to in crease the hardware reliability followed.

SOFTWARE RELIABILITY

Programmability of a majority of modern automatic
control components and instruments is now growing.
The software used is sharing responsability with
hardware and human operator for the reliability of
the whole production process or production line,
including safety. Software reliability has become
a very important technical problem. The price of
software almost equals hardware. Out of the four
above mentioned failure modes only wear-out does
not appear in software since software does not
wear-out.

Failure mode called poor quality fabrication in case of software may mean, for instance, a wrong version of subroutine used in the program or a typo-
graphical error. Failure mode called design error
may mean, for example, that we failed to clear all
registers returning from a subroutine to the main
program or we overlooked that the used series expansion of a calculated function does not converge

in that range. Another failure mode, called overload of a component, is rare in soffware for automatic control equipment, but it might appear in a text-editing system when a typist is too quick.

HUMAN OPERATOR RELIABILITY

Basing on life experience it is known how unpredictable human behaviour can be. This should be kept in mind when using human operator as a part of an automatic control system. The probability of proper behaviour or proper functioning of human operator is called a human operator´s reliability. Regarding human operator as a component of an automatic control system usual methods may be applied to evaluate the reliability of such a component.

Considered as a component, human operator is a very complex biological system and thus it is particularly difficult to be investigated. Taking it into consideration, in order to increase a human operator´s reliability it is not only necessary to be sure that a man or a woman in question is rested, not drunk, and not under any special stress, but a proper man-machine relalionship has to be established. This aspect is nowdays developed and investigated. It deals with such problems as where and how to sit, what to look at what kind of movements to perform.

RELIABILITY AS OPTIMALIZATION PARAMETER

In order to optimize a technical project the optimalization parameter must be defined. Deterministic nature of the majority of those parameters does not exclude a probabilistic one, as e.g., reliability. The growing importance of reliability is leading to reliability-oriented design of automatic control systems and to regard reliability as optimalization parameter of such systems.

The project stage owns its importance to a possibility of influencing task distribution among hardware, software and human operator; it may increase reliability of an entire system. The next step is a maximum increase of the three already discussed reliability factors.

To effect progress in reliability it would be very desirable to achieve a much wider international standarization of reliability of automatic control components and systems.

CONCLUSIONS

The purely economical pressure and the fear of disasters with financially unmeasurable implications are promoting all efforts to increase the reliability of automatic control systems. Having in mind the reliability increase, possible changes in a project of an automatic control system have to be considered. From the three factors, and namely, hardware reliability, software reliability and human operator´s reliability, it seems that the last one is the most difficult to be improved.

REFERENCES

Adams, E.C. (1980). Minimizing cost impact of software defects. RC 8228, IBM T.J.Watson Research Labs., Yorktown Heights, N.Y., April 11.

Asada, H., and Slotine, J.J.E. (1986). Robot Analysis and Control. Wiley, New York.

Bruce, R.C. (1986). Software Debugging of Microcomputers. Reston Publishing Company, Reston, Va.

Crossman, E.R.F.W.,J.E. Cookeeand R.J. Beishon. (1974). Visual attention and the sampling of displayed information in process control. In E. Edwards and F.P. Lees (Eds). The Human Operator in Process Control. Taylor and Francis, London.

Leskiewicz, H.J. (1985). Cascade voting redundancy in automatic control. IFAC Conference on Control Science and Technology for Development, Beijing, China. Vol. I, pp.445 - 454.

Leskiewicz, H.J. (1988). Reliability aspects of structural design of automatic control systems. Seminar ASR´88 Automatizace Inzenyrskich Praci, VSB Ostrawa, pp.55-61.

Littlewood, B. (1980). Theories of software reliability: how good are they and how can they be improved? IEEE Trans. Software Eng. Vol. SE6, no 5, Sept., pp. 489-500.

Moore, E.F., and Shannon, C.E. (1956). Reliable circuits using less reliable relays. J. Franklin Inst., Vol. 262, Sept., pp. 191-208 amd Vol. 262, Oct., pp.281-297.

Rosenborough, J.B. and T.B. Sheridan. (1986). Aiding human operators with states estimates. Man-Machine Systems Lab. Rep.,MIT.,Cambridge, Mass.

Schick, G.J., and R.W. Wolverton. (1978). Analysis of competing software reliability models. IEEE Trans. Software Eng., Vol. SE-4, no.2, pp.104-120, March.

Sharpe, J.J.E. (1988). Technical and human operational requirements for skill transfer in teleoperations. In C.A.Mason (Ed.), Teleoperation and Control Symposium. The Ergonomics Society, 12,15 July.

Sheridan, T.B. (1989). Telerobotics. IFAC Automatica, July, Vol. 25, no 4, pp.487-507.

Shooman, M.L. (1987). Software Engineering. McGraw-Hill, pp. 683.

Thayer, T.A. (1978). Software Reliability: a Study of Large Project Reality. North Holland Publishing Company, New York.

Tomovic, R. (1969). On man-machine control. Automatica 5, pp. 401-404.

SUMMARY: INDUSTRIALIST AND AUTOMATION

Chairman: J. Scrimgeour (Canada)
Co-chairman: J. Paiuk (Argentina)

Summary

As one in the series of industrial problem sessions, which are important to maintaining a healthy linkage between IFAC and industry, five short invited papers and one contributed paper were obtained for this session — all six were contained in the preprints and four were presented. Additional experts added to the panel for discussion purposes were W. Miller (U.S.A.) and P. Uronen (S.F.).

The paper by Yoshitani lists many problem areas in modern society and identifies some key areas in which automatic control technology is contributing to advances and solutions. Emphasis in the paper is placed on the more efficient use of natural resources, with data given on reductions which have been achieved in the energy consumption of the steel industry relative to industry output. For example, by 1987, the index of real specific energy consumption for most major steel making countries lay in the range of 114–141 compared to a base value of 100 for Japan. While it is difficult to identify the proportion of reductions that are derived specifically from automatic control systems, these are significant, such as the large reduction in thermal losses achieved with direct rolling implemented with computer control and scheduling systems. In addition the existence of down-stream energy savings external to the steel industry was noted. For example, much of the reduction in gasoline consumption world-wide has resulted from weight reductions in automobiles, achieved in part through the use of lighter gauge steels, whose production in turn has been enhanced or enabled through improved automatic control systems.

The paper by Swyt describes research, industry and technology transfer at the National Institute of Standards and Technology in the U.S.A. and the work of the Automated Manufacturing Research Facility (AMRF) within that institution. A strong program of interaction with industry is maintained. Research Work includes the development of CIM/Control system architectures, quality-in-automation and contributions to the forthcoming product data exchange standards PDES/STEP. As part of a nation wide network of research, development, technology transfer and application activities at the federal, state, local government and industrial firm level, three regional manufacturing technology centres have been established in a program intended later to reach ten to twelve centres.

While production automation is still in its infancy, the paper contributed by H.J. Warnecke from the Fraunhofer Institute for Manufacturing and Automation in Germany stresses that the use of modern sensor systems can contribute much to the widening of its scope and application. Attention is also drawn to a trend in flexible production automation towards the use of multi-purpose machining centres and flexible machining cells, rather than flexible manufacturing systems, although use of the latter is also known to be increasing. Software is identified as the critical key element in FMS operation, which currently tends to make their use expensive, particularly for small users. Obstacles or problems to the greater use of sensor-guided robot systems requiring attention and development are identified.

The paper by D.G. Fisher (Canada), presented by S. Shah, notes the progress made in computer based industrial automation and stresses that in the coming decade greater emphasis will be placed on the integration of process control and management techniques. In this context an entire process or production facility will be regarded as an entity for control, rather than emphasis on the control of individual process loops. This will require a re-organization of people as well as technical processes, and will lead to a greater involvement by management in both systems design and operation. While large computer systems for process control are now taken as the norm, with less emphasis therefore required on economic justification, there is an on-going need for individual practitioners and corporations to keep abreast of new technology. There is also a need for users to show leadership and to become more pro-active in their relationship with vendors with regard to defining user and system requirements.

Additional contributions include the paper by J.O. Gray (U.K.) which provides an overview of initiatives taken by the government in that country to promote advanced robotics research in both the academic and industrial community, including the Advanced Robotics Research Centre (ARRC) located at the University of Salford. Total funding over five years is in the order of £45 million.

An additional paper by Kotter (FRG) describes a new approach to production automation involving greater emphasis on the design of human work tasks in which a new interdisciplinary approach in factory planning is required. Worker participation in systems and work design is seen as a forward step in making CIM factories both productive and socially acceptable.

PANEL SESSION: "THE INDUSTRIALIST AND AUTOMATION"

D. G. Fisher

*Department of Chemical Engineering, University of Alberta, Edmonton, Alberta,
Canada*

Abstract. There is no doubt that computers have had, and will continue to have, a
profound effect on industrial automation. In the past decade the results have been
primarily in the area of new means of implementation for existing process control and
management techniques. In the next decade computer-based automation will be regarded
as an established tool and the focus will shift towards: a) new control techniques that
consider an entire process unit or production facility rather than a single loop; b)
integration of what were formerly regarded as separate technologies, e.g. process
control, global on-line databases, maintenance and production planning; c) a
reorganization of the plant management and operations personnel-structure so that top
management become more involved in "on-line, real-time decisions" and the operating
personnel assume increased responsibility with the help of computer-assisted systems.

INTRODUCTION

The discussion in this paper is primarily
from the perspective of automation in the
(continuous) process industries, such as
petrochemicals, food processing and mineral
upgrading, rather than discrete parts manufacturing
or production line systems. In the process
industries large computer-based automation systems
are now the norm rather than the exception. The
suitability and applicability of digital systems
for industrial applications is now a proven and
acknowledged fact and is accelerating the
historical trend of transferring functions and
responsibilities from humans to automation systems.

Consider the historical trends in automation.
Within the lifetime of many current practitioners
we have seen process operation switch from manual
operation to one of increasing automation. The
process operator implementing feedback control
using field-mounted indicators and manual final-
control-elements has been largely replaced by
automatic control instrumentation. Miniature,
board-mounted control instrumentation combined with
schematic control panels gave operators the ability
to oversee an entire production unit and the use of
computer-based control systems and electronic video
displays accelerated this trend towards centralized
control. Operators became "supervisors" rather
than "implementors" of control technology and
became increasingly removed from direct contact
with the process equipment. From another point of
view operators now have less involvement with basic
control tasks, e.g. how to manipulate one variable
to bring another to the desired value, and more
responsibility for "supervisory" functions, e.g.
what should the setpoints be and what exceptions
have to be dealt with. More generally, operator
functions moved up the "process control hierarchy"
as automation took over the lower levels.

Justification for Automation. Even if we describe
the progress of the past decade (oversimply and
narrowly) as simply a change from discrete analog
or pneumatic components to larger, more integrated
systems based on digital electronics, we still must

acknowledge strong economic justifications

• the greater consistency of automatic
control over manual control has improved quantity,
quality and safety of production

• digital technology has made it easier
and cheaper to implement larger, more complex
control systems with on line plant databases so
that management and production planning have become
more effective

• digital technology, including
measurement sensors, field buses, man-machine
interfaces, etc., have reduced the cost of
implementation and maintenance. (For example,
"self-testing" by built in microprocessors;
"automatic" recalibration and/or re-scaling;
centralized access from a maintenance or engineer's
console; noise immunity and greater bandwidth of
digital field buses).

To summarize, one could say that digital technology
is functionally and physically better than earlier
alternatives and permits easier, higher-level
access to a more reliable and more complete plant
database.

TRENDS IN TECHNOLOGY

Some of the current trends in industrial
automation can be conveniently summarized under the
titles, control technology, digital and
communication technology, and management
technology.

Control Technology. In the past, the approach of
the control engineer to a specific application was
restricted by: a) limitations on available hardware
functionality; b) the limited educational
background of engineers and plant personnel in
control technology and; c) a conservative
industrial practice that was largely limited to
"conventional" multiloop control. Today, digital
technology provides an almost unlimited capability
for the implementation of more advanced control
techniques; universities and other post-secondary
institutions are offering specialized programs in

computer process control; and industry, pushed by
the necessity of improved process control to
maintain competitiveness in global markets and re-
assured by the success of existing digital process
control systems, has demonstrated a willingness to
use more advanced techniques.

Modern control technology has been extended
to handle multivariable systems, hard constraints
and control of process trajectories rather than
simply point values. Not only has the available
control technology improved but the number of
functions that are being implemented on-line has
also increased. Two specific examples are self-
tuning and adaptive controllers.

In summary, control technology is becoming
more sophisticated, broader in scope, and is
increasingly being moved from the "human" to the
"computer" domain via computer assisted design
systems and more integrated on-line computer
systems. This trend will continue and we will see
more debate about "computer-assisted" versus
"computer-automated" control systems.

<u>Digital and Communication Technology</u>. Digital
hardware technology and the equally important
software technology have revolutionized the design,
implementation and operation of process control
systems. The impact of current and future
developments on industrial automation will be just
as significant.
 • Computers with multiple processing
units (whether distributed or centralized) will
offer the advantages of: incremental, on-line
expansion; "fail-soft" operation; parallel system
monitoring and functional validation; and more
"generic" (rather than proprietary) computing
capability. Relative to process and field
instrumentation costs, computing power and memory
will be "free" and the challenge will be how to
harness it for process functions.
 • Hardware, software and communication
standards are of increasing importance. Two of the
most important are a standardized "digital field
bus" (e.g., to replace the analog 4-20 ma type
standard) to permit multivendor interconnection of
equipment in a real-time process environment and
communication standards to permit interconnection
of process control computers, management level
computer systems, local area networks and
international networks. The Open-Systems
Interconnect (OSI) standards by the International
Standards Organization (ISO) are a good start but
rapid progress and practical implementations are
essential.
 • In the software area the most important
requirements are in the area of software
portability and improved development environments.
The requirements of industrial automation systems
are often unique to a single application. However,
it is impractical to develop and maintain a
complete software system for a single application
and/or manufacturer-specific implementation
platform. Operating systems like UNIX and
languages like C prove the portability is possible.
Object-oriented programming and Computer Assisted
Software Engineering (CASE) systems will facilitate
software development and maintenance. Standardized
graphical users interfaces (GUI) and packaged
subsystems, such as databases and network support,
will also speed up system development.
 • digital hardware technology will
probably evolve faster than it can be harnessed for
process control. The primary challenge for
industry will be to ensure that evolving standards
and product developments meet the needs for on-
line, real-time process management and control
rather than simply the less demanding requirements
of the broader business community. The second

challenge will be to make effective use of the new
digital and communications technology for improved
process automation.

<u>Management Technology</u> Process automation is not
just a job for technical specialists and people on
the plant floor. It requires increasing support
and involvement by the highest levels of plant
management. The time for decision making will
shorten. The scope and significance of decisions
by plant management will increase. For the same
reasons that we hear increasingly about the "global
(electronic) village" we will experience a much
more integrated approach to process automation and
management. We already have examples of computer
networks that integrate raw material suppliers +
manufacturers + customers. Suppliers assume
responsibility for "just-in-time-delivery" using
on-line access to the manufacturer's latest
production schedules. Customers access the
"statistical quality control and production
records" of the manufacturer to ensure that their
requirements are met. Manufacturers track the
sales performance and forecasts of their customers
via international computer communications. Thus
"automation' is expanded-from the equipment level
to the corporate level. Any discussion of the
"Industrialist" and "Automation" must be in this
broader context.

Perhaps the most obvious management level
change could be describes as "integration" of what
most people today consider to be separate functions
or departments e.g. production operations,
maintenance, inventory control, design, accounting
etc. These different functions or departments
utilize various computer-based tools, e.g.
databases, process control systems, spreadsheets,
management information systems, etc. Often these
tools are from different vendors and run on
different computers so the potential for
integration is not obvious. However, we already
have large computer systems that run multiple
applications (often on "separate" virtual computers
running under their own operating system and
locally defined (virtual) environment) and are
"networked" to systems that can provide everything
from real-time process data to the results of
management planning simulations. With proper
"integration" the potential exists for "industrial
automation" which begins with global monitoring of
actual sales, adds supply data based on satellite
monitoring of current resources (e.g. agricultural)
and then works down through the various levels of
the management and control hierarchy, to adjust
current product specifications and production
rates.

<u>Conclusions</u>. The technology for industrial
automation, which includes hardware, software and
theoretical concepts, will continue to expand and
develop at a rate which will challenge individual
practitioners and industrial corporations. Both
must keep abreast of current technology; influence
the development of standards and automation
products; expand the conventional and continuing
educational programs; and be both flexible and
aggressive enough to evaluate and implement the new
technology as it becomes available. The need is
for leaders. Followers will falter.

HUMAN ORIENTED AUTOMATION (HOA)

Y. Yoshitani

Nagaoka University of Technology, Japan

Abstract

There are many problems in modern society such as environmential pollution and increase of aged people. Automation has many possibilities to solve those problems. Paper proposed Human oriented automation (HOA) rather than fully automated manufacturing.

Keywords

Automation, social effect of Automation

1. Introduction

It has been passed half century since word "Automation" had been used in the society. Meaning and concepts of automation has been expanded along with the progress of technology. Previously, the automation had been used mainly in the mass production plants. Today, the automation has been applied in every corner of the modern society, which are simbolized as the new words as office automation(OA), home automation(HA) and etc.

In spite of the progress of technology, there are many problems in the modern societies. Main problems are the followings.

1. Environmential pollution.
2. Long life time (problem of aged people).
3. Higher education (less young labor force).
4. Desire to have individual life styles.
5. Globarization.

Drastic changes has been going on around the industries in the past decade. Recent trends are summerised as follows.

1. Strong competition in both inside and outside.
2. Shorter products life.
3. Faster delivery.
4. Small volume production with more products variety.
5. Reguest of better working environment.

Huge energy and resources has been consumed by the advanced countries of the world. First step towards pollution contrl is the reduction of energy and resource consumptions. Automation could promote both energy and resource savings.

Because of higher education, people prefer to work in more intellectual jobs. There are big needs of automation for dirty, dengerous, dark jobs, in the modern society. Automation is becomming a key technology for the solution of those problems.

However, the people still have wrong image of automation which Chales Chapline showed in his film "Modern Times". Today, the automation created many new industries and emplyments. Working environment of the factries has been improved far better than the age of Modern Times.

Today, the interests of industry has been oriented towards computer integrated manufacturing. However, it is a huge systems and takes more man-power, more time to builed and require big investments. Only excellent companies could build CIM systems today. Majority of manufacturing factories are either medium and small, which have not enough engineers and funds for the automation. There are many manufacturing industries which are still difficult to automate, such as foundry, sweing and etc. Also there are many manufacturing processes which are still depend on human skills.

For the sound progress of automation, we have to promote better understanding of automation among the society. We have to show the actual exsamples which the

automation will really help promoting both natural and human enviropments. Not orienting complete un-maned production but rather building a system which will harmonize men and machines.

2. Benefit of automaton

Automation had been applied for various improvements such as productivity, quality, and man-power savings. However, one of the big benefits of automation would be a problem finding through the studies of production system. Before introducing automation to the production,it always reguire studies of production system. Those studies include material flow, material handling, individual processings and inspection of products. Through those studies you will find many problems and faced to solve those before the automation. Sometimes you will improve your production by introducing the ideas of improved processing or handling.
Key of the managements is the full use of resources of the firm including human resource. Characteristics of men and machine are quit opposite as shown in the Table 1. Also there are diverse difference in the individual skills and apptitude. Engineers and designer tend to concentrate on machine side, and do not pay much attentions on who will going to operate and maintain them.

Table 1. Comparison of man and
machine

	Man (Soft)	Machine (Hard)
	Intermi- ttent	Continous
Motion	Back and forth	Rotation
	Delicate	Monotonous
	Complicate	Simple

There are the steps for the automating production system. For automate production system, it is preferable to reform stream line processing or assembling in order to reduce handling cost and make it easier to adapt automatic system. There are many problems in introducing stream line operation, such as the sychronization of individual processings. First step is to automate individual unit operations, then the second step is line automation. The last step is the full integration of entire jobs in the factory. There are many studies and improvements require to proceed those steps. Those studies are very important for the training of both operaters and engineers for the automation.
Any automatic system has to be operable, maintenable and able to up-date the system within the avariable employees. If the system level is so high for both operaters and maintenance crews, you could not gain any profits of automation. If the system could not update for the changes of market or products change, system will be obsolite within few years. Automation is not just introducing computer, control devices and robots to the factory but rather the reform of production system which will allow full use of men and machines. Therefore, I would like to redefine the automation as the creation of new production system for given factory environments.
Another big influence brough by the automation is the integration of engineering skills. Previously the design of the products had not much concern about how to produce them. Production side had a responsibility to produce them according to the design specifications. If the products composed of less parts and materials which are easy to fabricate, it will be easy to produce.
By the introduction of FMS, importance of better communication between design and production, sales and production has been realized. Review of products design revealed the possibilites of the reduction of processing costs.
One of the good exsample of this is the success of the Toyota Motor's New production system(NPS). Toyota started NPS system developments in relatively early(1950). You will find many operaters on their assembly plants. The NSP system looks far behind of the CIM. However, they successed the world competitive production system(1)(2). It has been taken so much time for the development of automation of individual processings for synchronize operations, and developments of quick change devices for the change of products(such as quick press dice change).
Technolgy used to create by-products as automobile and IC created various by-products. Automation has also created such benefitial by products mentioned here. Also automation will creat new industrial possibilities.

3.Environmential control

Environmential problem becomes big issue of the world. Pollution control is becoming a vital issue of the survival of industries, especially process industries. Environmential control is both important for better human working environments but also for relieability and maintenance of delicate automatic devices.
First steps towards better environment control is to reduce energy and resource consumptions. Remarkable success has been found in the development of pollution control in the Japanese iron and steel industry(3).
Of cource, it is the results of various developments of pollution abatement equipments and introduction of new energy savings and recovery systems. Nearly half of the energy savings had came from the improvements of products yield (due to less resource) by the close control of production. Development of power electronics also promoted efficient use of power (See figures).
On-stream analyzer for water and atmospher has been remarkably advanced since 1965, which is one of the by-product of pollution control.
Development of those on-line instrumentations also stimulated the on-line quality instrumentations.

Investiments for the pollution control will not directly help either productivity or quality. However, it has indirect benefits through the better working environments and social acceptance of the industry.

4.Direction of Automation

Two type of labour problems are exist in modern society. One is the shortage of young labour force. Automation will offer more intellectual jobs such as research and development, design and soft ware productions. Also the development of automation of dirty, dangerous and dark jobs are requested in the modern society.
Another problem would be the longer life time. We need to use more aged people. We need to develope better interface for those aged people.
Man and machine are quit opposit in nature. We have to use full capability of human being. Therefore, the automation should promote better human working environments.

5.Conclusion

Automation has a possibility to solve many problems of modern society, if the industrialists wish to build better human and natural environment.Also the direction of Robotics is not trying to develop human like robot but rather working towards developments of robots for dirty, dangerous job.

Reference

1.M.Imai.   "Kaizen"
  Randam House, N.Y(1986)
2.M.Katagiri. Just-in-time
  Edit,J.Motimer,(1986)IFS.VK.
3.The Japan Iron and Steel Federation
  "The steel industry Japan 1989"

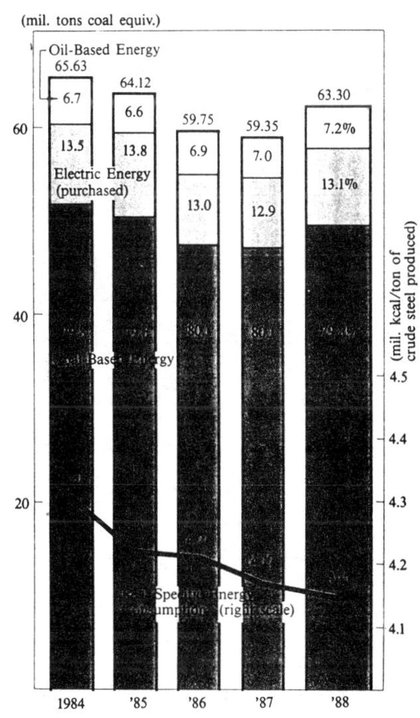

Energy Consumption by the Steel Industry, 1984—1988

Note: These figures represent energy consumption by the steel industry, excluding the coke-making and ferroalloy-producing sectors.
**Total energy consumption by the steel industry divided by crude steel production.*

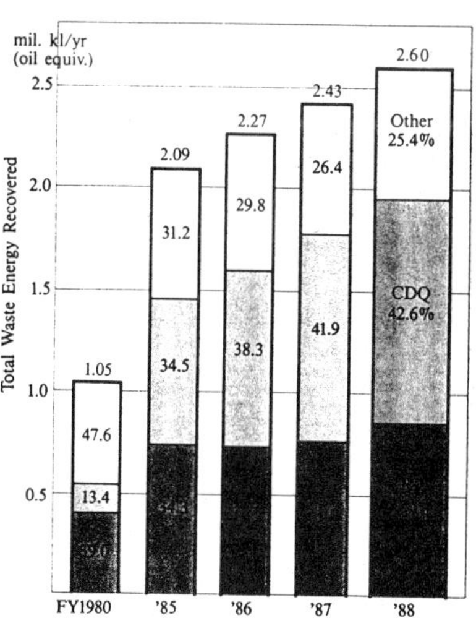

Recovery of Waste Energy Generated by Integrated Steelworks, Fiscal 1980—1988

Electricity Consumption by the Steel Industry and Generation by Its Captive Power Plants, 1980—1988

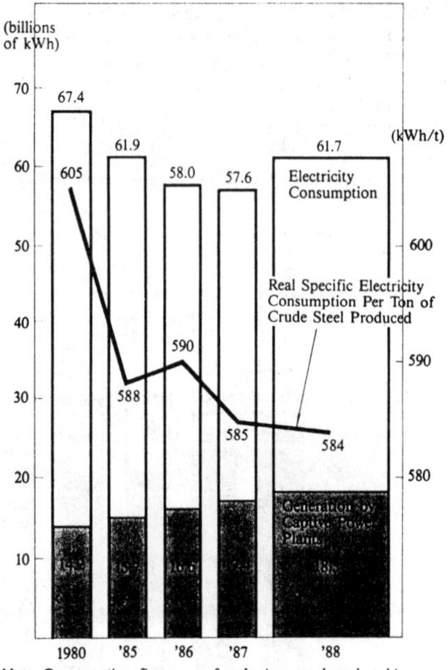

Note: Consumption figures are for the iron- and steel-making sector only.

Real Specific Energy Consumption in Major Steelmaking Countries, 1987

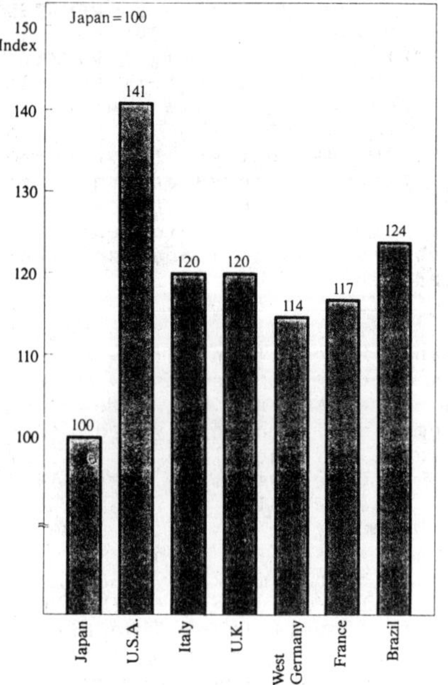

Source: estimated from IISI data.

CLOSING SUMMARY: INDUSTRIAL PROBLEM SESSIONS

P. Uronen

*Helsinki University of Technology, Department of Chemical Engineering, SF-02150
Espoo, Finland*

The Industrial Problem Session part in the 11th
IFAC World Congress consisted of three subsessions
dealing with selected topics and thereafter a
concluding session was held. The topics discussed
in various subsessions were IPS-1&2: Technology
Transfer (Chairman prof. Haase), IPS-3: Managing
the Complexity and safety in highly automated
systems (Chairman prof. Gertler) and IPS-4:
Industrialist and automation (Chairman Dr. J.
Scrimgeour). The number of people attended in
these sessions varied from 15-40 and the panelists
represented both industry and academia;
unfortunately most panelists in IPS-4 were
absent and therefore this session partially
failed.

From the topics selected it can be said that
"Technology transfer" and "Complexity and Safety"
were very actual and interesting to the audience;
the time devoted for the sessions was not long
enough. On the contrary the theme "Industrialist
and automation" was too general and undefined
and it did not raise very much discussion.

In concluding session the central findings of these
panels were summarized as follows:

Technology transfer:

1. Problems are different in different countries
   and in different industries. The speed of innovation and introduction of new technologies is
   increasing.
2. A structure and orgnization for technology
   transfer is needed.
3. Technology/science centers are used and closer
   confacts between business community and
   research organizations are needed.
4. Liason offices and "gateway people" are
   important.
5. More research and education in technology
   transfer is needed.
6. Technology transfer is a multidisciplinary
   complex process; therefore also technology
   assesment is important.
7. Theoretical research is of vital importance as
   feeding surface for technology transfer.
8. The large multinational programs can be bery
   useful, one problem here is how the small and
   medium sized enterprises can benefit from
   these.

Complexity and safety

1. The complexity of systems is increasing
   rapidly.
2. There is lack of definitions and terminology
   in this area.
3. The role of human engineering in this
   context must be stressed. More research and
   education in these areas are needed.
4. How to define and measure the observability
   and controllability of safety?
5. The possibilities to use sensor-based
   condition monitoring systems should be
   evaluated.
6. Finally the economical aspects of safety will
   set the limits.

Industrialist and automation

1. Computer technology has given new potential
   to industry.
2. Leadership in applying new technologies is
   haunted. Also risks are higher.
3. Production chains are becoming longer and
   more complex; therefore decision makers
   are not eager to take risks in applying
   new automation technology.
4. The basic question which remains to be
   answered: The economics of automation.

THE CONCEPT OF RELEGATION FOR DECENTRALIZED CONTROL

Ü. Özgüner and D. Schoenwald

*Department of Electrical Engineering, The Ohio State University, Columbus,
OH 43210, USA*

Abstract. Relegation is the assignment of control tasks and information channels in view of control effectiveness, on-line computational complexity, controller capabilities, and physical and structural constraints. In this paper one particular form of relegation, namely *decentralized set–point relegation* will be introduced.

The decentralized control utilized here consists of proportional state feedback and integral output error feedback for each input/output channel. The choice of output set-points is constrained by a given function which has to be satisfied while minimizing the overall cost criterion of the linear quadratic regulator. We consider the cases of uncoupled and coupled subsystems, linear and nonlinear constraint equations.

Keywords. Decentralized control, large scale systems, hierarchical decision making.

I. Introduction

In the control of large systems the notion of decentralization plays an important role. The decentralization constraint enters into large scale systems because it may be impractical or even impossible to communicate signals from one controller to another. Moreover, decentralization may be required by the control designer to achieve reliability and a degree of redundancy, and it may impose a structure to the control implementation by relegating control authority to separate channels.

This paper introduces and analyzes the concept of relegation for the design of decentralized controllers for large systems.

Relegation is the assignment of control tasks and information channels in view of control effectiveness, on-line computational complexity, controller capabilities, and physical and structural constraints [1,2]. In this paper one particular form of relegation, namely *decentralized set–point relegation* will be introduced.

The decentralized control utilized here consists of proportional state feedback and integral output error feedback for each input/output channel. The feedback gain matrix is obtained via the solution of two coupled Lyapunov equations. The optimal cost will then contain a term which is quadratic in the output set-points, and it is the minimization of this term subject to a linear or nonlinear constraint function which leads to the solution of the optimal choice of output set-points. In Section II we briefly consider the Decentralized Quadratic Regulator problem and introduce the idea of

Decentralized Set-Point Relegation. In Section III we point out how set–points are incorporated into the quadratic regulator setting, we then indicate how these can be assigned to uncoupled subsystems while further minimizing the cost. In Section V we outline the interconnected system configuration and advocate a series expansion method (further detailed in the Appendix) for calculation of the control. Finally we introduce an iterative linearization approach for use when the constraint equations are nonlinear.

II. The Decentralized Quadratic Regulator

Consider the basic *decentralized quadratic regulator problem*:

$$\dot{x} = Ax + \sum_{i=1}^{N} B_i u_i \qquad\qquad ; x(0) = x_0 \qquad (1)$$

$$y_i = C_i x \qquad\qquad ; i = 1, \ldots, N. \qquad (2)$$

where N is the number of input/output channels, m_i is the dimension of u_i, and p_i is the dimension of y_i. We wish to minimize the cost criterion with cost

$$J = \int_0^{\infty} (x^T Q x + + \sum_{i=1}^{\nu} u_i^T R_i u_i) \, dt \qquad (3)$$

and the following feedback structure constraint:

$$u_i = K_i y_i \qquad ; i = 1, \ldots, N. \qquad (4)$$

87

It can be shown [3,4,5,6] that the necessary conditions for minimizing J given by (3) with the controller structure (4) imply the solution of the following system of nonlinear algebraic equations:

$$\begin{cases} A_c^T P + P A_c + \bar{Q} = 0 \\ A_c L + L A_c^T + X_0 = 0 \end{cases}$$

and

$$\nabla_{K_i} J = B_i^T P L C_i^T + R_i K_i C_i L C_i^T = 0$$

where

$$A_c = A + \sum_{i=1}^{N} B_i K_i C_i$$
$$\bar{Q} = Q + \sum_{i=1}^{N} C_i^T K_i^T R_i K_i C_i$$
$$X_0 = x_0 x_0^T.$$

Now consider the situation when each control channel has a set of variables y_{ri} to be regulated and these have been assigned output set-points $y_i^d \ i = 1,..,N$. Note that we differentiate here between the local measurements and the local variables to be regulated. If the set-points are predetermined and fixed, fairly straight forward extensions to the decentralized regulator problem are possible to find the associated controllers. However, in many problems of interest, in large space structures, robotics and manufacturing, these set points jointly satisfy a constraint equation. Their one-to-one assignment to channels is not obvious. We shall call this *the decentralized set-point relegation problem* (See Figure 1). Depending on the type of constraints the set-points need to satisfy, there are a number of distinct special cases of interest in decentralized set-point relegation:

1. Linear constraints

2. Nonlinear constraints

3. Time–varying constraints

In this paper we shall consider the first two cases above.

III. Incorporation of Set–Points

We first outline the incorporation of set–points into the standard quadratic regulator problem. Consider the system:

$$\begin{aligned} \dot{x} &= Ax + Bu \\ y_r &= Cx \quad, \end{aligned} \tag{5}$$

under the cost criterion

$$J = \int_o^\infty (z^T Q z + \dot{u}^T R \dot{u}) dt \quad, \tag{6}$$

where $z(t)$ is defined as

$$z(t) \stackrel{\triangle}{=} \begin{bmatrix} \dot{x} \\ \Delta y \end{bmatrix} \tag{7}$$

$$\Delta y(t) \stackrel{\triangle}{=} y_r(t) - y^d \quad, \tag{8}$$

where y^d, a constant set point, has been specified.

The state equations for the system with $\dot{u}$ as input and $z()$ as state vector can now be written. If $Q = block\ diag.\{Q_1, Q_2\}$ then the solution of the above problem takes the form

$$u = K^1 x + K^2 \int_o^t (y_r(\tau) - y^d) d\tau \tag{9}$$

where the $\{K^1, K^2\}$ pairs are calculated from the associated Riccati equation.

Let the Riccati matrix be partitioned as,

$$P = \begin{bmatrix} P_1 & P_3 \\ P_3^T & P_2 \end{bmatrix}$$

and let,

$$S = BR^{-1}B^T$$

Then the Riccati equation decouples into three equations:

$$-P_3^T S P_3 + Q_2 = 0$$

$$P_3^T A + P_2 C - P_3^T S P_1 = 0$$

$$A^T P_1 + C^T P_3^T + P_1 A + P_3 C - P_1 S P_1 + Q_1 = 0$$

As can be seen, P_3 can be calculated from the first equation. Then,

$$P_2 = P_3^T (A - S P_1) C^T (C C^T)^{-1} \tag{10}$$

and P_1 can be calculated from the Riccati equation,

$$A^T P_1 + P_1 A - P_1 S P_1 + (Q + C^T P_3^T + P_3 C) = 0 \tag{11}$$

The optimal cost is given by

$$J^* = z^T(0) P z(0) \tag{12}$$

Since $z(0)$ has y^d in it, the "optimal cost" J^* is a function of the value of the set–point.

IV. Relegation for Uncoupled Systems

Consider the set of *dynamically uncoupled* linear subsystems,

$$\begin{aligned} \dot{x}_i &= A_i x_i + B_i u_i \\ y_{ri} &= C_i x_i \quad i = 1, 2, \ldots, N \quad, \end{aligned} \tag{13}$$

under the cost criterion

$$J = \sum_{i=1}^{N} \int_o^\infty (z_i^T Q_i z_i + \dot{u}_i^T R_i \dot{u}_i) dt \quad, \tag{14}$$

$$z_i(t) \stackrel{\triangle}{=} \begin{bmatrix} \dot{x}_i \\ \Delta y_i \end{bmatrix} \tag{15}$$

$$\Delta y_i(t) \stackrel{\triangle}{=} y_{ri}(t) - y_i^d \quad, \tag{16}$$

where y_i^d, a constant set point, has been specified.

The solution of the above problem takes the form

$$u_i = K_i^1 x_i + K_i^2 \int_o^t (y_{ri}(\tau) - y_i^d) d\tau \quad , \quad i = 1, 2, \ldots, N \quad (17)$$

where neither the relative feedback nor the solution for $\{K_i^1, K_i^2\}$ are coupled. In fact, the $\{K_i^1, K_i^2\}$ pairs are calculated from uncoupled Riccati equations which provide the matrix pairs $\{P_1^i, P_2^i\}$ such that the optimal cost is given by

$$J^* = \sum_{i=1}^{N} z_i^T(0) P^i z_i(0) \qquad (18)$$

We now assume that one step higher on the hierarchy there exists a *relegator* (See Figure 1) which must specify the $\{y_i^d\}$ set points, constrained with a set of linear, static equations of the general form

$$\sum_{i=1}^{N} F_i y_i^d = G \qquad (19)$$

The question now is whether the relegator can pick a specific set $\{y_i^d\}$ that satisfies the constraints (19) while further minimizing J^*. Thus, we consider the problem

$$\min_{y_i^d} J^* = \min_{y_i^d} \sum_{i=1}^{N} (y_{ri}(0) - y_i^d)^T P_2^i (y_{ri}(0) - y_i^d) \qquad (20)$$

such that (19) holds. This is equivalent to minimizing

$$\hat{J} = \sum_{i=1}^{N} (y_i^d)^T P_2^i y_i^d - 2 y_{ri}^T(0) P_2^i y_i^d + \Lambda^T (F_i y_i^d) - \Lambda^T G \quad , \quad (21)$$

where Λ is a vector of Lagrange multipliers. In such a minimization exercise, we obtain

$$\frac{\partial \hat{J}}{\partial y_i^d} = 2 P_2^i y_i^d - 2 P_2^i y_{ri}(0) + F_i^T \Lambda = 0 \qquad (22)$$

for $i = 1, \ldots, N$. Thus, in view of this and (19),

$$y_i^d = y_{ri}(0) - 0.5 (P_2^i)^{-1} F_i^T \Lambda \quad , \qquad (23)$$

and

$$\Lambda = 2 [\sum_{i=1}^{N} F_i (P_2^i)^{-1} F_i^T]^{-1} [\sum_{i=1}^{N} F_i y_{ri}(0) - G] \qquad (24)$$

to give

$$y_i^d = y_{ri}(0) - (P_2^i)^{-1} F_i [\sum_{i=1}^{N} F_i (P_2^i)^{-1} F_i^T]^{-1} [\sum_{i=1}^{N} F_i y_{ri}(0) - G] \quad (25)$$

for $i = 1, \ldots, N$.

<u>Example</u>

A simple example involves two motors with mirrors mounted on them, as shown in Figure 2. The desired slew angle of the ray of light is specified as θ^d. It can be easily shown that the individual angles of rotation for each motor $y_{ri} = \theta_i$ are related to the ray slew angle through the equation,

$$\theta_1^d - \theta_2^d = \theta^d \qquad (26)$$

Thus the feedback controls for the individual motors can be derived in conjunction with the specified individual desired angles.

V.   Relegation for Coupled Systems

We shall consider the special case of interconnected systems with local state measurements available for feedback. We shall furthermore assume that a linear transformation of local states is to be regulated to a given set point.

Consider the set of dynamically coupled linear subsystems,

$$\dot{x}_i = \sum_{j=1}^{N} A_{ij} x_j + B_i u_i \qquad (27)$$
$$y_{ri} = C_i x_i, \quad i = 1, 2, \ldots, N$$

where N is the number of input/output channels, n_i is the dimension of x_i, m_i is the dimension of u_i, and p_i is the dimension of y_{ri}. We wish to minimize the cost criterion

$$J = \frac{1}{2} \sum_{i=1}^{N} \int_0^{\infty} (z_i^T Q_i z_i + \dot{u}_i^T R_i \dot{u}_i) \, dt \qquad (28)$$

where $z_i(t)$ is the $n_i + p_i$ dimension vector defined as

$$z_i(t) = \begin{bmatrix} \dot{x}_i \\ \triangle y_i \end{bmatrix} \qquad (29)$$

$$\triangle y_i = y_{ri} - y_i^d \qquad (30)$$

$$z_i(0) = \begin{bmatrix} \dot{x}_i(0) \\ \triangle y_{ri}(0) \end{bmatrix} = z_{i0} \qquad (31)$$

with Q_i positive semi-definite and R_i positive definite matrices of appropriate dimensions.

The linear system can be redefined in terms of $z(t)$ as

$$\dot{z} = \begin{bmatrix} A_{11} & 0 & A_{12} & 0 & \ldots & A_{1N} & 0 \\ C_1 & 0 & 0 & 0 & \ldots & 0 & 0 \\ A_{21} & 0 & A_{22} & 0 & \ldots & A_{2N} & 0 \\ 0 & 0 & C_2 & 0 & \ldots & 0 & 0 \\ & & & \vdots & & & \\ A_{N1} & 0 & A_{N2} & 0 & \ldots & A_{NN} & 0 \\ 0 & 0 & 0 & 0 & \ldots & C_N & 0 \end{bmatrix} z + \begin{bmatrix} B_1 & \ldots & 0 \\ 0 & \ldots & 0 \\ 0 & \ldots & 0 \\ 0 & \ldots & 0 \\ & \vdots & \\ 0 & \ldots & B_N \\ 0 & \ldots & 0 \end{bmatrix} \dot{u}$$

or more compactly as,

$$\dot{z} = \hat{A} z + \hat{B} \dot{u} \qquad (32)$$

with decentralized feedback

$$\dot{u}_i = K_i z_i = K_i^1 \dot{x}_i + K_i^2 \triangle y_i, \quad K_i = [K_i^1 \; K_i^2] \qquad (33)$$
$$\dot{u} = K z, \quad K = \text{Block-diag}[K_1, K_2, \ldots, K_N]$$

such that each channel input has the form

$$u_i = K_i^1 x_i + K_i^2 \int_0^t (y_{ri}(\tau) - y_i^d) \, d\tau, \quad i = 1, 2, \ldots, N. \quad (34)$$

The optimal solution for the decentralized feedback gain matrix K to the minimization of the quadratic cost criterion J is known to involve the iterative solution of two coupled Lyapunov equations as mentioned before. Due to the special structure of the system matrices here we shall advocate the utilisation of a series expansion in terms of a coupling parameter [7,8]. The Appendix gives the details of the expansion.

It is well known that the *cost* associated with any stabilizing feedback is, in fact,

$$z^T(0)Pz(0)$$

where P is the solution of the related Lyapunov equation

$$A_c^T P + P A_c + \bar{Q} = 0 \tag{35}$$

Thus the series expansion will also give an approximation to the weighting factors $\{P_2^i\}$ to be used in the set–point distribution.

VI.   Nonlinear Constraints

At this point we assume there exists a *relegator* whose function is to specify the y^d set-points subject to the following set of **nonlinear** constraints

$$H_i(y^d) = 0 , \quad i = 1, 2, \ldots, N \tag{36}$$

where $H_i(y^d)$ is a scalar function of the output set-points vector y^d. Because general nonlinear functions are difficult to solve, we shall expand $H_i(y^d)$ about a point $\bar{y}$ that represents the current estimate of y^d. The power series is carried to quadratic terms in y^d as follows

$$H_i(y^d) = H_i(\bar{y}) + \frac{\partial H_i(\bar{y})}{\partial y^d}(y^d - \bar{y}) + \frac{1}{2}(y^d - \bar{y})^T \frac{\partial^2 H_i(\bar{y})}{\partial(y^d)^2}(y^d - \bar{y}) \tag{37}$$

for $i = 1, 2, \ldots, N$, where the gradients are defined as follows

$$\frac{\partial H_i(\bar{y})}{\partial y^d} = \begin{bmatrix} \frac{\partial H_i(\bar{y})}{\partial y_1^d} & \frac{\partial H_i(\bar{y})}{\partial y_2^d} & \cdots & \frac{\partial H_i(\bar{y})}{\partial y_N^d} \end{bmatrix} , \tag{38}$$

for $i = 1, 2, \ldots, N$ and the Hessians are defined as

$$\frac{\partial^2 H_i(\bar{y})}{\partial(y^d)^2} = \begin{bmatrix} \frac{\partial^2 H_i(\bar{y})}{\partial(y_1^d)^2} & \cdots & \frac{\partial^2 H_i(\bar{y})}{\partial y_N^d y_1^d} \\ \vdots & & \vdots \\ \frac{\partial^2 H_i(\bar{y})}{\partial y_1^d y_N^d} & \cdots & \frac{\partial^2 H_i(\bar{y})}{\partial(y_N^d)^2} \end{bmatrix} . \tag{39}$$

Thus, we wish to minimize the following cost criterion with respect to y^d

$$\hat{J} = y^{d^T} P_2 y^d - 2 y_r(0)^T P_2 y^d + \sum_{i=1}^{N} \Lambda_i [H_i(\bar{y})$$
$$\frac{\partial H_i(\bar{y})}{\partial y^d}(y^d - \bar{y}) + \frac{1}{2}(y^d - \bar{y})^T \frac{\partial^2 H_i(\bar{y})}{\partial(y^d)^2}(y^d - \bar{y})] . \tag{40}$$

We proceed by setting the gradient of $\hat{J}$ equal to zero and solving for y^d thus obtaining

$$y^d = [2P_2 + \sum_{i=1}^{N} \frac{\partial^2 H_i(\bar{y})}{\partial(y^d)^2} \Lambda_i]^{-1}$$

$$\cdot [2P_2 y_r(0) - \sum_{i=1}^{N}(\frac{\partial H_i(\bar{y})^T}{\partial y^d} - \frac{\partial^2 H_i(\bar{y})}{\partial(y^d)^2} \bar{y}) \Lambda_i] \tag{41}$$

where Λ_i, $i = 1, 2, \ldots, N$, are solved for by substituting (41) into (37). This motivates the following algorithm to solve for y^d:

1.) Choose $\bar{y} = 0$ initially.
2.) Evaluate $H_i(\bar{y})$, $\frac{\partial H_i(\bar{y})}{\partial y^d}$, and $\frac{\partial^2 H_i(\bar{y})}{\partial(y^d)^2}$ for $i = 1, 2, \ldots, N$.
3.) Solve for Λ_i, $i = 1, 2, \ldots, N$ by substituting (41) into (37).
4.) Using above, solve for y^d by solving (41) for its p components.
5.) If $\| y^d - \bar{y} \| \leq \delta$ then stop; otherwise set $\bar{y} = y^d$ and go to step 2.

It should be noted that with some sacrifice in accuracy one can linearize (37) instead of expanding to second order and obtain the following simpler equations for Λ_i and y^d

$$\frac{\partial H_i(\bar{y})}{\partial y^d} P_2^{-1} \sum_{i=1}^{N} \frac{\partial H_i(\bar{y})^T}{\partial y^d} \Lambda_i = 2[H_i(\bar{y}) + \frac{\partial H_i(\bar{y})}{\partial y^d}(y_r(0) - \bar{y})] , \tag{42}$$

for $i = 1, 2, \ldots, N$. Finally,

$$y^d = y_r(0) - 0.5 P_2^{-1} \sum_{i=1}^{N} \frac{\partial H_i(\bar{y})^T}{\partial y^d} \Lambda_i . \tag{43}$$

The algorithm above can still be followed to find y^d except Hessian information is no longer needed.

VII.   Conclusion

A number of comments can be made regarding the problem outlined:

1. Note that y_i^d depends on $y_{ri}(0)$ which may be assumed to be known. Also, y_i^d can be solved for assuming $y_{ri}(0) = 0$.

2. Equation (19) may be generalized to (slowly) time varying constraints.

3. The quadratic regulator problem can be generalized to accomodate frequency weighting. Note that the interesting and intuitively obvious conclusion that a band limited actuator will be given a "closer" set point to achieve.

In conclusion, it can be observed that *decentralized setpoint relegation* provides the hierarchical framework that couples a version of resource allocation with the dynamics of the controlled system. This is accomplished in a globally optimal manner, that is, both the high–level resource allocation (or target allocation) and the lower level dynamic control problem have a single goal. We believe this to be avery general framework that can be utilized in many large scale system problems, an example may be found in [2].

VIII.   Acknowledgement

Research sponsored by the Air Force Office of Scientific Research (AFSC) under Contract F49620–89–C–0046. The

IX.   References

[1] U. Özgüner, "Decentralized/relegated control," in *Proc. ICES-88*, Atlanta, Georgia, April 1987.

[2] E. Barbieri, Ümit Özgüner, and S. Yurkovich, "Vibration compensation in optical tracking systems," *Journal of Guidance and Control*, vol. 12, no. 3, , May 1989.

[3] W. S. Levine, T. L. Johnson, and M. Athans, "Optimal limited state variable feedback controllers for linear systems," *IEEE Transactions on Automatic Control*, vol. AC–16, no. 6, pp. 785–792, December 1971.

[4] J. C. Geromel and J. Bernussou, "Optimal decentralized control of dynamic systems," *Automatica*, vol. 18, no. 5, pp. 545–557, 1982.

[5] F. Khorrami, S. Tien, and Ümit Özgüner, "DOLORES: A software package for analysis and design of optimal decentralized control," in *Procdings of the 40th National Aerospace and Electronics Conference*, Dayton, OH, May 1988.

[6] F. Khorrami and U. Özgüner, "Frequency–shaped cost functionals for decentralized systems," in *Proc. 25th CDC*, Athens, Greece, Dec. 1988.

[7] P.V. Kokotovic, W.R. Perkins, J.B. Cruz and G. D'Ans, "ε–coupling method for near optimum design of large scale linear systems," *Proc. IEE*, vol. 116, , May 1969.

[8] U. Özgüner and W. R. Perkins, "A series solution to the Nash strategy for large scale interconnected systems," *Automatica*, vol. 13, pp. 313–315, 1977.

X.   Appendix

In this Appendix we use the series expansion (ϵ-coupling) method to generate the local feedback solution for interconnected (decentralized) systems where the states are available at the subsystem level.

We shall embed a coupling parameter ϵ between the subsystems, so that the equations satisfying the necessary condition of optimality are as given below:

<u>General</u>

$$\bar{A}_c = \begin{bmatrix} A_{11} + B_1 K_1 & \epsilon A_{12} \\ \epsilon A_{21} & A_{22} + B_2 K_2 \end{bmatrix}$$

$$\bar{Q}_c = \begin{bmatrix} Q_1 + K_1^T R_1 K_1 & 0 \\ 0 & Q_2 + K_2^2 R_2 K_2 \end{bmatrix}$$

Partition P and L as

$$P = \begin{bmatrix} P_1 & P_3 \\ P_3^T & P_2 \end{bmatrix}$$

$$L = \begin{bmatrix} L_1 & L_3 \\ L_3^T & L_2 \end{bmatrix}$$

The gradients are then,

$$\begin{aligned}
\nabla_{K1} J &= 2[B_1^T(P_1 L_1 + P_3 L_3^T) + R_1 K_1 L_1] = 0 & (44a) \\
\nabla_{K2} J &= 2[B_2^T(P_3 L_3 + P_2 L_2^T) + R_2 K_2 L_2] = 0 & (44b)
\end{aligned}$$

<u>P equations</u>

$$\begin{aligned}
(A_{11} + B_1 K_1)^T P_1 + \epsilon A_{21}^T P_3^T + \epsilon P_3 A_{21} + P_1(A_{11} + B_1 K_1) \\
+ (Q_1 + K_1^T R_1 K_1) = 0 & \quad (45a) \\
(A_{22} + B_2 K_2)^T P_2 + \epsilon A_{12}^T P_3^T + \epsilon P_3 A_{12} + P_2(A_{22} + B_2 K_2) \\
+ (Q_2 + K_2^T R_2 K_2) = 0 & \quad (45b)
\end{aligned}$$

$$(A_{11} + B_1 K_1)^T P_3 + \epsilon A_{21}^T P_2 + \epsilon P_1 A_{12} + P_3(A_{22} + B_2 K_2) = 0 \quad (45c)$$

<u>L equations</u>

$$\begin{aligned}
(A_{11} + B_1 K_1) L_1 + \epsilon A_{12} L_3^T + \epsilon L_3 A_{12}^T + L_1(A_{11} + B_1 K_1)^T \\
+ X_{01} = 0 & \quad (46a) \\
(A_{22} + B_2 K_2) L_3 + \epsilon A_{21} L_3^T + \epsilon L_3 A_{21}^T + L_2(A_{22} + B_2 K_2)^T \\
+ X_{02} = 0 & \quad (46b)
\end{aligned}$$

$$(A_{11} + B_1 K_1) L_3 + \epsilon A_{12} L_2 + \epsilon L_1 A_{21}^T + L_3(A_{22} + B_2 K_2)^T = 0 \quad (46c)$$

Consider the power series expansion in terms of ϵ for P and L:

$$P = P^0 + \epsilon P^1 + \frac{1}{2}\epsilon^2 P^2 + \cdots$$

$$L = L^0 + \epsilon L^1 + \frac{1}{2}\epsilon^2 L^2 + \cdots$$

resulting in the power series for K.

$$K = K^0 + \epsilon K^1 + \frac{1}{2}\epsilon^2 K^2 + \cdots$$

We shall henceforth use the superscript to indicate both the order of the derivative or the term in the series expansion.

<u>o'th order terms:</u>

Setting $\epsilon = 0$ Eq. (45c) becomes

$$(A_{11} + B_1 K_1^0)^T P_3^0 + P_3^0(A_{22} + B_2 K_2^0) = 0 \quad (47)$$

We temporarily assume the following assertion to be true:

Assertion I : K_1^0 and K_2^0 asymptotically stabilize the subsystems $\{A_{11}, B_1\}$ and $\{A_{22}, B_2\}$

Under Assertion I for the unique solution for Eq. (47) (and similarly from Eq. (46c)) we obtain

$$P_3^0 \equiv 0 , \quad L_3^0 \equiv 0$$

With $\epsilon = 0$ Eq. (46a) is a Lyapunov equation and under Assertion I admits a positive definite solution for L_1^0. Thus Eq. (44a) simplifies to

$$B_1^T P_1^0 + R_1 K_1^0 = 0$$

Substituting into Eq. (45a) at $\epsilon = 0$ we obtain the standard subsystem Ricatti equation:

$$A_{11}^T P_1^0 + P_1^0 A_{11} - P_1^0 B_1 R_1^{-1} B_1^T P_1^0 + Q_1 = 0$$

Similar arguments result in the Ricatti equation for the second subsystem. Therefore, if $\{A_{11}, B_1\}$ and $\{A_{22}, B_2\}$ are stabilizable pairs, Assertion I is indeed true.

1'st order term

Under Assertion I and the solution of the 0'th order equation we can find:

$$K_1^1, K_2^1, P_1^1, P_2^1, L_1^1, L_2^1 \text{ all zero.}$$

The terms P_3^1 and L_3^1 can be calculated from the linear equations:

$$(A_{11} + B_1 K_1^0)^T P_3^1 + P_3^1 (A_{22} + B_2 K_2^0) + A_{21}^T P_2^0 + P_1^0 A_{12} = 0$$

$$(A_{11} + B_1 K_1^0) L_3^1 + L_3^1 (A_{22} + B_2 K_2^0)^T + A_{12}^T L_2^0 + L_1^0 A_{21}^T = 0$$

2nd order terms

P_1^2 is solved from:

$$(A_{11} + B_1 K_1^0)^T P_1^2 + P_1^2 (A_{22} + B_1 K_1^0) + A_{21}^T P_3^{1T} + P_3^1 A_{21} = 0$$

Then K_1^2 is

$$K_1^2 = -R_1^{-1} B_1^T P_1^2 - 2R_1^{-1} P_3^1 L_3^{1T} (L_1^0)^{-1}$$

The following terms can then be sequentially generated. Note that the odd order terms will give no local feedback, yet one intermediate set of linear matrix equations will still have to be solved. After the o'th term the local feedbacks will be calculated from Lyapunov equations.

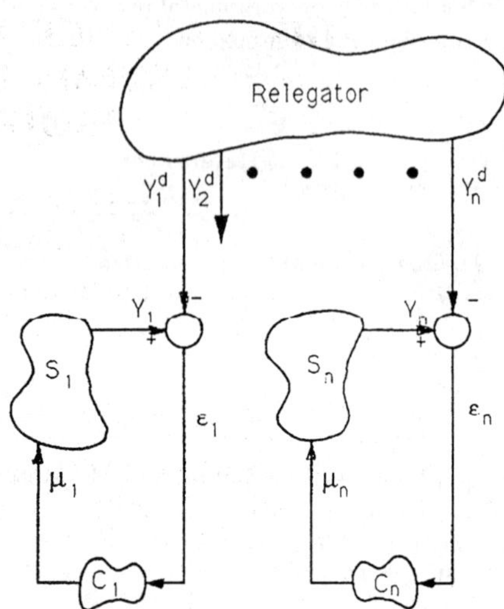

Figure 1 : Set Point Relegation

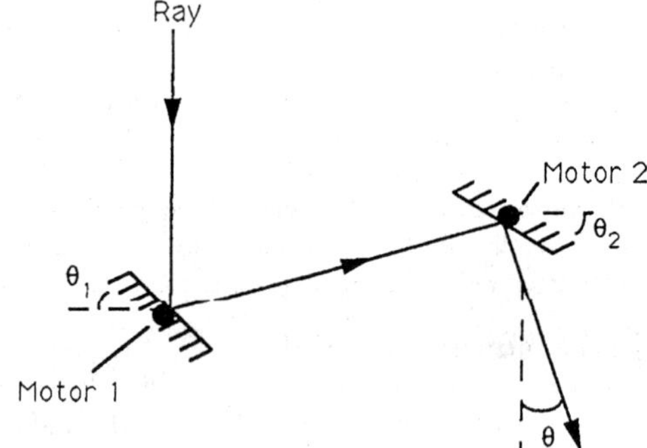

Figure 2 Relegation Example for Ray Path Determination

DISTURBANCE REJECTION IN LARGE SCALE SYSTEMS USING DECENTRALIZED PROJECTIVE CONTROLS*

J. V. Medanić, **W. R. Perkins**, **R. A. Ramaker**,
F. A. Latuda and **Z. Uskoković***

**University of Illinois, Urbana, IL 61801, USA*
****University of Titograd, Yugoslavia*

Abstract. We formulate a two stage design problem to achieve disturbance rejection using low-order decentralized controllers. First, using the H_∞-optimal state feedback solution as the reference, we determine desired closed-loop poles to be retained in the final design. Then, restricting admissible controllers to decentralized projective controllers we create fixed modes at desired closed-loop poles and we parameterize all decentralized controllers of a given order retaining these poles in the closed-loop system. Then, using a Frobenius-Hankel norm minimization approach we determine the free parameters in the parametrization to place the remaining zeros and poles for disturbance rejection. The procedure is illustrated on a 7th order example where two first order controllers were designed.

Key Words. Decentralized control, disturbance rejection, control system design

INTRODUCTION

The disturbance rejection problem arises frequently in system design and serves as a paradigm for other control problems (Francis and Doyle, 1987). The generic formulation of the disturbance rejection problem is shown in Fig. 1 where

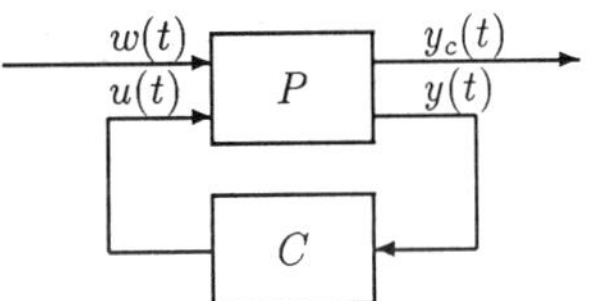

Fig. 1.

$w(t)$ is the disturbance input, $u(t)$ is the control input, $y(t)$ is the measured output and $y_c(t)$ is the regulated output. The disturbance rejection problem is to choose the feedback controller, C, such that the system response $y_c(t)$ to the disturbance, $w(t)$, is sufficiently small as measured by a selected performance criterion.

The choice of the criterion distinguishes the various approaches used in solving the disturbance rejection problem. If the closed-loop transfer function between $w(s)$, and $y_c(s)$ is denoted by $G(s)$, then it is well known that, if $G(s)$ is stable,

$$\|y_c\|_2 \leq \|G(s)\|_\infty \|w(s)\|_2 \qquad (1)$$

where $\| \cdot \|_2$ is the L_2 norm and

$$\|G(s)\|_\infty = \max_\omega \sigma_{max}(\omega), \qquad (2)$$

σ_{max} being the maximal singular value of $G(-j\omega)^T G(j\omega)$. So it is clear that the smaller $\|G\|_\infty$, the greater the suppression of the disturbance in the output. Consequently, the disturbance minimization problem is often reduced to minimizing, over the class of admissible controllers, the H_∞ norm

of the transfer function between the disturbance input and the regulated output.

Unfortunately, minimization of the H_∞ norm, or the determination of a controller that guarantees an H_∞-norm bound results in a controller of relatively high order, often higher than the order of the system itself. Recent results on the algebraic Riccati equation based approach (Petersen, 1987; Doyle and others, 1988; Bernstein and Haddad, 1989) have shown that there exist state feedback optimal and suboptimal controllers in addition to optimal and suboptimal output feedback controllers of same order as the system. All controllers providing a given bound on the H_∞-norm have now been parameterized (Glover and Doyle, 1988). These results do not, however, shed new insight into the construction of low order dynamic controllers with the same properties since it is not known which members of the parameterized family, if any, will result in a low order controller. Consequently, the determination of suboptimal H_∞ controllers of low order remains computationally involved. Also, until recently, there have been no theoretical results available on **decentralized** H_∞ optimal or suboptimal controllers. Recent work (Veillette, Medanić and Perkins, 1989) has shown that decentralized controllers, each of same order as the plant, can be constructed to guarantee a given H_∞-norm bound if there exists a positive semi-definite solution to an algebraic Riccati-like equation. No similar constructive procedure however, exists for low order decentralized controllers.

To overcome computational difficulties in the design of low order controllers for the disturbance rejection problem other approaches have been proposed, and in particular the use of the Frobenius-Hankel (FH) norm (Young, 1987; Medanić and Perkins, 1987; Ramaker, Medanić and Perkins, 1989). As will be seen in the next section, the minimization of this norm also guarantees bounds on the H_∞ norm, and leads, in a wide class of problems (Young, 1987; Paz and Medanić, 1989; Paz and Medanić, 1989) to a minimization problem that is computationally much simpler to solve than the H_∞ optimization problem. The problem reduces to the minimization of a trace function subject to two Lyapunov equations as constraints. Moreover, in many instances the free

*This work was supported in part by the Flight Dynamics Laboratory, Air Force Wright Aeronautical Laboratory (AFCS), USAF, by the Joint Services Electronics Program under Contract N00014-84-C-0149, by University of Illinois and by the Fulbright Foundation.

design parameters appear linearly in the coefficient matrices of the Lyapunov equations. This enables the use of algorithms developed for solution of various constrained regulator design problems (Ramaker, Medanić and Perkins, 1989), as well as the use of a new algorithm based on the iteration of algebraic Riccati equations (Paz and Medanić, 1989), to resolve initialization and convergence problems and efficiently compute FH-norm optimal controllers.

Motivated by the capability to computationally handle FH norm minimization problems for systems linear in the free design parameters we formulate a **decentralized** disturbance rejection problem and solve it using decentralized controllers based on the projective control approach. By default to one channel, the results apply to the problem of centralized design of a low order controller. For simplicity we will consider the case of two controllers where the same number of outputs, r, is available to each controller, and each controllers is of the same order, p. Extensions to additional controllers and controllers with different numbers of available measurements and different orders are straightforward.

The essence of the projective controls approach is that it parameterize the entire family of decentralized controllers that retain a subset of the dominant eigenvalues (and associated invariant subspaces) of a reference state feedback design. In the case of dynamic controllers each controller is individually parametrized via a $p \times r$ matrix of free parameters (Uskoković and Medanić, (1983). The projective controls parametrization generates a new decentralized structure in which the selected eigenvalues become decentralized fixed modes placed at desired locations. The idea, then, is to restrict admissible controllers to the family of projective controllers and so place in the first phase of design the selected modes at desired locations, and then to choose in the second phase of design the free parameters to achieve optimal disturbance rejection using the FH norm. If only strictly proper controllers are used p modes can be retained and the closed-loop system is linear in the free parameters. If proper controllers are used $r + p$ modes can be retained but the obtained system representation is not linear in the free parameters. The problem is resolved by identifying a similarity transformation that reduces the system to a linear-in-the-free-parameters (LIFP) representation, thus rendering the problem computationally tractable by the FH-norm approach.

THE FH NORM AND DISTURBANCE ATTENUATION

The Frobenius-Hankel (FH) norm approach to disturbance attenuation is based on minimizing the sum of squares of the Hankel singular values of the system using the available design freedom. By minimizing the Frobenius norm defined on the Hankel singular values, the disturbance rejection problem leads to a trace minimization problem subject to Lyapunov equation constraints. Given the stable system with transfer function $G(s)$ and a corresponding minimal and balanced realization (Moore, 1981) $H(sI - A_c)^{-1}G$, the Hankel singular values associated with the system are the eigenvalues of the Hankel matrix (Glover, 1984), and

$$\Sigma = \begin{bmatrix} \sigma_1 & & & \\ & \sigma_2 & & \\ & & \ddots & \\ & & & \sigma_n \end{bmatrix}, \quad \sigma_i \geq \sigma_{i+1} \tag{3}$$

is the balanced observability/controllability Grammian satisfying (Moore, 1981)

$$\begin{aligned} A^T\Sigma + \Sigma A + H^T H &= 0 \\ A\Sigma + \Sigma A^T + GG^T &= 0. \end{aligned} \tag{4}$$

The product of the controllability Grammian and the observability Grammian is invariant under similarity transformation, that is,

$$\begin{aligned} \|G(s)\|^2_{FH} &= \text{Trace } \Sigma^2 = \text{Trace } T^{-1}\Sigma^2 T \\ &= \text{Trace } T^{-1}\Sigma T^{-T}T^T\Sigma T \\ &= \text{Trace } PQ \end{aligned} \tag{5}$$

where P, Q are the controllability and observability Grammians, respectively, for an arbitrary state space representation. Therefore, the problem of minimizing the FH norm reduces to minimizing the trace of the product of the controllability Grammian with respect to the disturbance input, and the observability Grammian with respect to the regulated output, with the feedback closed between the measured output and the control input. Thus, the optimal solution for the disturbance rejection problem can be found by minimizing with respect to the free controller parameters

$$J(K) = \text{Trace } PQ \tag{6}$$

subject to

$$\begin{aligned} A_c^T Q + QA_c + H^T H &= 0 \\ A_c P + PA_c^T + GG^T &= 0 \end{aligned} \tag{7}$$

where in general A_c, H and G depend on the free design parameters.

This resulting problem is a finite dimensional constrained minimization problem. By defining the Lagrange multipliers, $L \in R^{n \times n}$, $M \in R^{n \times n}$, $L^T = L$, $M^T = M$, the problem can be reduced to an unconstrained minimization problem. This is accomplished by defining the augmented criterion

$$\begin{aligned} J(K, P, Q, L, M) &= \text{Trace } [PQ + L(A_c^T Q + QA_c + H^T H) \\ &\quad + M(A_c P + PA_c^T + GG^T)]. \end{aligned} \tag{8}$$

The necessary conditions for an optimal solution can then be written as

$$\partial J/\partial P = A_c^T M + MA_c + Q = 0 \tag{9a}$$

$$\partial J/\partial Q = A_c L + LA_c^T + P = 0 \tag{9b}$$

$$\partial J/\partial L = A_c^T Q + QA_c + H^T H = 0 \tag{9c}$$

$$\partial J/\partial M = A_c P + PA_c^T + GG^T = 0 \tag{9d}$$

$$\frac{\partial J}{\partial K} = 0 \tag{9e}$$

with the form of (9e) depending on how A_c, G, H depend on the free design parameters.

A number design problems can be reduced to the above generic disturbance rejection problem in which the admissible controllers are static controllers, or dynamic controllers of fixed order (Young, 1987). For the static controller problem and other problems equivalent to it, $u = Ky$ and the

$$\begin{aligned} \dot{x} &= Ax + Bu + Gw \\ y &= Cx \\ y_c &= Hx. \end{aligned} \tag{10}$$

So, G and H are independent of K while $A_c = A + BKC$. Then

$$\frac{\partial J}{\partial K} = 2B^T(QL + MP)C^T \tag{11}$$

For dynamic controllers of the form

$$\begin{aligned} \dot{z} &= K_1 z + K_2 y \\ u &= K_3 z + K_4 y \end{aligned} \tag{12}$$

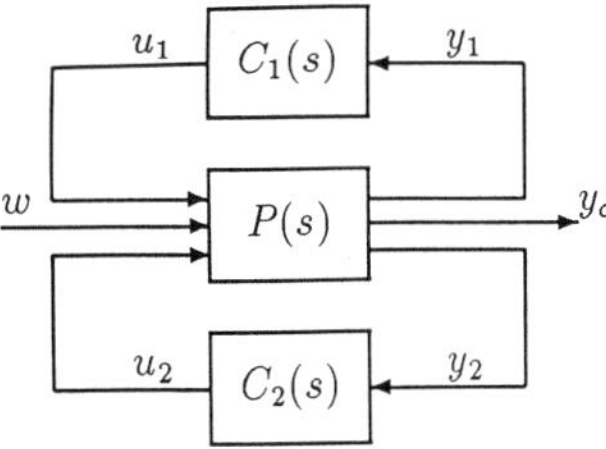

Fig. 2.

the same representation is obtained when

$$K = \left[\begin{array}{cc} K_1 & K_2 \\ K_3 & K_4 \end{array} \right]$$

is introduced and A_c, G, H are considered to represent the closed-loop system consisting of the system and controller.

The necessary conditions (9), with $\partial J/\partial K$ given by (11) can serve as a paradigm in discussing the computational aspect of the FH norm minimization problem. Due to the decoupled nature of equations (9) many computational approaches are possible. The feasible direction algorithm has been used with success and will be discussed in Section 5. A new algorithm based on iterating algebraic Riccati equations, which converge to relevant Lyapunov equations, has been developed for the discrete FH-norm minimization problem, and can be used for certain continuous problems (Paz and Medanić, 1989; Paz and Medanić, 1989).

The justification for using the FH norm, beyond the computational advantages, lies in its relation to the H_∞ norm. Recalling the definition of the Hankel norm

$$\|G(s)\|_H = \sigma_1 \tag{13}$$

the Trace norm

$$\|G(s)\|_T = \sum_{i=1}^{n} \sigma_i \tag{14}$$

and the FH norm (5), it has been established that (Glover, 1984)

$$\|G(s)\|_H^2 \leq \|G(s)\|_\infty^2 \leq 2\|G(s)\|_T^2 \tag{15}$$

while it follows from properties of matrix norms and the Cauchy-Schwarz inequality that

$$\|G(s)\|_H^2 \leq G(s)\|_{FH}^2 \leq \|G(s)\|_T \|G(s)\|_H \tag{16}$$

From this follows, in particular, that if $\|G(s)\|_{FH} = \epsilon$ then

$$\frac{1}{\sqrt{n}}\epsilon \leq \|G(s)\|_\infty \leq 2\sqrt{n}\epsilon \tag{17}$$

where n is the dimension of the closed-loop system. Hence, as the value of the FH norm is reduced, the bounds on the H_∞-norm contract and the interval of uncertainty of the H_∞ norm is reduced.

DISTURBANCE ATTENUATION VIA DECENTRALIZED PROJECTIVE CONTROLS

Consider the decentralized system of Fig. 2. The state space description of this system can be written as

$$\begin{array}{rcl} \dot{x} & = & Ax + B_1 u_1 + B_2 u_2 + Gw \\ y_1 & = & C_1 x \\ y_2 & = & C_2' x \\ y_c & = & Hx \end{array} \tag{18}$$

where $x \in R^n$, $u_1, u_2 \in R^m$, $y_1, y_2 \in R^r$, $y_c \in R^s$, and $w \in R^q$. The goal is to determine $C_1(s)$ and $C_2(s)$ to achieve certain performance and disturbance rejection goals.

Reference Controller

It has been shown (Uskoković and Medanić, 1987) that if $u = K^0 x$, leading to

$$u_i = K_i^0 x, \quad K_2^0 = \left[\begin{array}{c} K_1^0 \\ K_2^0 \end{array} \right] \tag{19}$$

is a stable feedback control producing the closed-loop system

$$\begin{array}{rcl} \dot{x} & = & Fx + Gw \\ y_r & = & Hx. \end{array} \tag{20}$$

with $F = A - BK^0$ then there exist projective controllers such that the resulting closed-loop system retains a subset of the eigenvalues of the reference system together with the associated eigenvectors.

If transient performance is of primary concern, then the reference state feedback solution can be determined by solving an LQ optimization problem, and projective controls will then retain in the closed-loop system the dominant poles of the reference solution that define acceptable transient response. The free parameters are then determined by solving an auxiliary problem to shape the residual dynamics.

If disturbance rejection and transient response are of concern then the reference solution can be determined to minimize the H_∞ norm (Ramaker, Medanić, and Perkins, 1989) resulting in $K^0 = K^\infty$, where

$$K^\infty = -B^T M \tag{21}$$

where $M > 0$ is the solution of the ARE

$$A^T M + MA - MBB^T M + \frac{1}{\gamma^2}MGG^T M + H^T H = 0 \tag{22}$$

with $B = \left[\begin{array}{cc} B_1 & B_2 \end{array} \right]$ and γ the minimal value for which $M > 0$ solving (22) exists. For details see, for example (Doyle and others, 1988). The use of (21) guarantees

$$\|G_c(s)\|_\infty = \|H(sI - F)^{-1}G\|_\infty \leq \gamma \tag{23}$$

In this case, projective controls will fix the dominant poles of the system at locations determined by the reference solution. The free parameters P_1, P_2 (see below) sare then to be determined to shape the residual dynamics to achieve disturbance rejection. It will be shown that when strictly proper controllers are used in the closed loop system representation is linear in the free parameters P_1, P_2. To efficiently apply the FH norm minimization approach to the disturbance rejection problem when proper controllers are used, a transformation developed in (Uskoković and Medanić, 1987) that reduces the closed-loop to a linear-in-the-free-parameters (LIFP) form will be applied.

Static Controllers

For completeness, consider first the use of static output feedback. When projective controls are used, r eigenvalues can be retained by using the controller

$$u_i = K_i y_i \tag{24}$$

where

$$K_i = K_i^0 X_r (C_i X_r)^{-1} \tag{25}$$

where X_r spans the eigenspace associated with the r retained eigenvalues, Λ_r.

The transfer function of the closed loop system is

$$G(s) = H(sI - A_c)^{-1}G \qquad (26)$$

where

$$A_c = A + B_1 K_1 C_1 + B_2 K_2 C_2 \qquad (27)$$

with K_1, K_2 fixed. Hence, with static controls, there is no freedom available for the next design phase within the class of projective controls. This solution, however, serves as a foundation for expanding the admissible controllers by allowing dynamic controllers of given order. The advantages are twofold: more eigenvalues can be retained and free parameters are introduced to shape the residual dynamics to achieve disturbance rejection. We treat separately proper controllers and strictly proper controllers.

Strictly Proper Dynamic Controllers

Assume strictly proper controllers of the form

$$\begin{aligned} \dot{z}_i &= H_i z_i + D_i y_i \\ u_i &= N_{di} z_i \end{aligned} \qquad (28)$$

where $z_i \in R^p$ are used in all channels. Then the family of all projective controllers of order p that retain by simultaneous action the subspectrum Λ_p is parameterized by

$$\begin{aligned} H_i &= \Lambda_p - P_i C_i X_p \\ D_i &= P_i \\ N_{di} &= K_i X_p \end{aligned} \qquad (29)$$

where $P_i \in R^{p \times r}$ are free parameter matrices.

The transfer function of the closed loop system is then

$$\bar{G}(s) = \bar{H}(sI - \bar{A}_c)^{-1}\bar{G} \qquad (30)$$

with

$$\bar{A}_c = \begin{bmatrix} A & B_1 K_1 X_p & B_2 K_2 X_p \\ P_1 C_1 & \Lambda_p - P_1 C_1 X_p & 0 \\ P_2 C_2 & 0 & \Lambda_p - P_2 C_2 X_p \end{bmatrix}$$

$$\qquad (31)$$

$$\bar{G} = \begin{bmatrix} G \\ 0 \\ 0, \end{bmatrix} \qquad \bar{H} = \begin{bmatrix} H & 0 & 0 \end{bmatrix}$$

Note that this system is linear in the free parameters and can be represented by the triplet

$$\begin{aligned} \bar{A}_c &= \bar{A}_o + \bar{B}_1 P_1 \bar{C}_1 + \bar{B}_2 P_2 \bar{C}_2 \\ \bar{G} &= \bar{G}_o \\ \bar{H} &= \bar{H}_o \end{aligned} \qquad (32)$$

Viewing the system characterized by (32) as a new decentralized structure it is noted that eigenvalues in Λ_p are decentralized fixed modes of the system. Hence projective controls offer the possibility of placing fixed modes in decentralized systems at desired locations. The freedom in the free parameters P_1 and P_2 can be used to shape the residual dynamics and in the disturbance suppression problem to shape the system's transfer function.

It is instructive to look at the single disturbance, single regulated output case (SISO transfer function) for an interpretation of the two phase design. The dominant poles as selected from the reference state feedback controlled system are placed in the first phase. By restricting admissible controllers to projective controllers, these poles become fixed modes and in the second phase of design, adjustment of

P_1, P_2 places system zeros and residual poles to achieve disturbance attenuation. The LIFP representation (32) enables the efficient application of the FH norm approach.

Proper Dynamic Controllers

Assume finally that proper controllers of the form

$$\begin{aligned} \dot{z}_i &= H_i z_i + D_i y_i \\ u_i &= N_{di} z_i + K_{di} y_i \end{aligned} \qquad (33)$$

where $z_i \in R^p$ are used in all channels. It has been shown (Uskoković and Medanić, 1983) that in this case the family of projective controllers of order p that retain $r + p$ eigenvalues, Λ_r and Λ_p of the reference solution are parameterized by

$$\begin{aligned} H_i &= \Lambda_p + P_i F_{12}^i B_o^i \\ D_i &= P_i F_r^i - \Lambda_p P_i - P_i F_{12}^i B_o^i P_i \\ N_{di} &= K_{2i}^o \\ K_{di} &= K_i - N_{di} P_i \end{aligned} \qquad (34)$$

where again $P_1, P_2 \in R^{p \times r}$ are free parameters while F_{12}^i, B_o^i, F_r^i (as well as K_{2i}^o, K_i) are all defined by the reference solution and invariant eigenvalues. See (Uskoković and Medanić, 1983) for a complete definition.

The transfer function of the closed loop system is now

$$\tilde{G}(s) = \tilde{H}(sI - \tilde{A}_c)^{-1}\tilde{G} \qquad (35)$$

where

$$\tilde{A}_c = \begin{bmatrix} A_d & -B_1 N_{d1} & -B_2 N_{d2} \\ (P_1 F_r^1 - H_1 P_1)C_1 & \Lambda_p + P_1 F_{12}^1 B_o^1 & 0 \\ (P_2 F_r^2 - H_2 P_2)C_2 & 0 & \Lambda_p + P_2 F_{12}^2 B_o^2 \end{bmatrix}$$

$$\tilde{G} = \begin{bmatrix} G \\ 0 \\ 0 \end{bmatrix}, \qquad \tilde{H} = \begin{bmatrix} H & 0 & 0 \end{bmatrix} \qquad (36)$$

$$A_d = A_c + B_1 N_{d1} P_1 + B_2 N_{d2} P_2$$

Transforming as in (Uskoković and Medanić, 1987) the closed loop system by the similarity transformation

$$T = \begin{bmatrix} I_n & 0 & 0 \\ P_1 C_1 & I_p & 0 \\ P_2 C_2 & 0 & I_p \end{bmatrix},$$

$$\qquad (37)$$

$$T^{-1} = \begin{bmatrix} I_n & 0 & 0 \\ -P_1 C_1 & I_p & 0 \\ -P_2 C_2 & 0 & I_p \end{bmatrix}$$

yields a representation of the closed loop system

$$\begin{aligned} \tilde{A}_{ce} &= T^{-1}\tilde{A}T = \tilde{A}_e + \tilde{B}_{1e} P_1 \tilde{C}_{1e} + \tilde{B}_{2e} P_2 \tilde{C}_{2e} \\ \tilde{G}_e &= T^{-1}\tilde{G} = G_e - \tilde{B}_{1e} P_1 C_1 G - \tilde{B}_{2e} P_2 C_2 G \\ \tilde{H}_e &= \tilde{H}T = H_e \end{aligned} \qquad (38)$$

Thus, the expressions derived for $\tilde{A}_{ce}$, $\tilde{G}_e$, and $\tilde{H}_e$ all exhibit a **linear** dependence on the free parameter matrices P_1 and P_2. ($\tilde{H}_e$ is in fact independent of the free parameter matrices.) This linear dependence can now be utilized to determine suitable P_1 and P_2 (and thus the dynamic controllers) to achieve disturbance rejection by minimizing the FH norm. The advantage here is that $r + p$ fixed modes have been created and the FH norm approach can now be applied to achieve disturbance attenuation.

FH NORM MINIMIZATION

The FH norm minimization now reduces to the minimization of $J_e = \text{Trace } P_e Q_e$ subject to the constraints:

$$\tilde{A}_{ce}^T Q_e + Q_e \tilde{A}_{ce} + \tilde{H}^T \tilde{H} = 0 \qquad (39)$$

$$\tilde{A}_c P_e + P_e \tilde{A}_c^T + \tilde{G}\tilde{G}^T = 0 \qquad (40)$$

where P_e, A_e are the controllability and observability Gramians, respectively, of (32), or (38). By defining the Lagrange multipliers, $L_e = L_e^T$, $M_e = M_e^T$, the problem can again be reduced to an unconstrained minimization with the augmented criterion written as described earlier. Since $\tilde{A}_c$, $\tilde{G}$, and $\tilde{H}$ are all linear functions of the free parameters, only minor changes are introduced in the necessary conditions (9). In particular, condition (9e) here has reduced to the two conditions

$$\partial J/\partial P_1 = 2\tilde{B}_{1e}^T(M_e P_e + Q_e L_e)\tilde{C}_{1e}^T - 2\tilde{B}_{1e}^T M_e \tilde{G}_e G^T C_1^T \quad (41a)$$

$$\partial J/\partial P_2 = 2\tilde{B}_{2e}^T(M_e P_e + Q_e L_e)\tilde{C}_{2e}^T - 2\tilde{B}_{2e}^T M_e G_e G^T C_2^T. \quad (41b)$$

The feasible direction algorithm can be applied to solve the problem. Initially, the free parameter matrices, P_1 and P_2, are set to zero, (although arbitrary values can be used). If the resulting closed-loop system matrix, $\tilde{A}_{ce}$ given by (35) has unstable eigenvalues, J_e is not defined. In this case, an embedding parameter, ρ, is chosen such that

$$\rho > \max_i(Re\ \lambda_i(A)) \qquad (42)$$

and $\tilde{A}_{ce}$ modified to $\tilde{A}_{ce} - \rho I$. If the resulting $\tilde{A}_{ce}$ is stable, the embedding parameter is then zero. This leads to a modified extension J_{ea} with an expanded region of definition encompassing the initial P_1, P_2.

This resulting algorithm applicable to an arbitrary number of controllers can be summarized as follows:

1. Initialize P_i^1, $i = 1, ..., k$. Evaluate the resulting closed-loop system matrix $\tilde{A}_{ce}$, based on the current iterative value of P_i, $i = 1, ..., k$. If $\tilde{A}_{ce}$ is stable, proceed to step 5.

2. Choose an embedding parameter, ρ, such that $\rho > \max_i(Re\Lambda_i(A))$.

3. Solve the associated minimization problem recursively until ρ can be set equal to zero. (If $\tilde{A}_{ce}$ cannot be stabilized, increase the order of the dynamic controllers to be used and start the algorithm over.)

4. Solve for P_e, Q_e, L_e and M_e from the necessary conditions for a minimum Trace $P_e Q_e$.

5. Use the partial derivative equations with respect to the free parameter matrices to calculate gradient directions.

6. Set $P_{i+1}^j = \Delta P_i + P_i^j$ where $\Delta P_i = -\bar{h}\partial J/\partial P_i$, $i = 1, ..., k$, with

$$\bar{h} = \arg\min_{h>0} J_{ea}(P_1 - h\frac{\partial J}{\partial P_1}, P_2 - h\frac{\partial J}{\partial P_2}) \qquad (43)$$

7. Repeat until P_i, $i = 1, ..., k$ converges to their optimal values.

8. Use calculated P_i, $i = 1, ..., k$, to determine controller parameters based on the parameterization given by (25) to complete the controller design.

Using the embedding parameter method the maximal eigenvalue of $\tilde{A}_{ce}$ can be successively moved towards the imaginary axis. The parameter ρ can then be decreased and ultimately when the system is stabilized ρ can be set to zero resulting in the minimization of the original criterion J_e. However, it may not be possible to move all unstable eigenvalues into the left-half plane without simultaneously forcing previously stable eigenvalues into the right-half plane. If this situation occurs, the order of the dynamic controller must be increased to provide additional design freedom needed to stablize the system. Expanding controllers if necessary is simple in view of the way controllers are parameterized (see Uskokovic and Medanić, (1987)) and expanded free parameter matrices can utilize the latest iterates of P_1 and P_2 to simply continue the combined stabilization and optimization process.

EXAMPLE

In order to illustrate the approach, consider a seventh-order system with two decentralized dynamic controllers to be designed so as to minimize the effects of a disturbance input on the regulated outputs. A system of the form (18) will be used, characterized by the following matrices:

$$A = \begin{bmatrix} -2 & 1 & 0 & 0 & -1 & 1 & 0 \\ -2 & -3 & 1 & 0 & 0 & 1 & 1 \\ -2 & -3 & -2 & 0 & -1 & -1 & -1 \\ 0 & 0 & 1 & -3 & -1 & 0 & 0 \\ -1 & 0 & 1 & 0 & -2 & 1 & 0 \\ 0 & 2 & -1 & -1 & -2 & 1 & 1 \\ -1 & 0 & -3 & 0 & -2 & -2 & -4 \end{bmatrix} \quad B_1 = \begin{bmatrix} 0 \\ 0 \\ 1 \\ 0 \\ 0 \\ 0 \\ 0 \end{bmatrix} \quad B_2 = \begin{bmatrix} 0 \\ 0 \\ 0 \\ 0 \\ 0 \\ 0 \\ 1 \end{bmatrix}$$

$$C_1 = \begin{bmatrix} 1 & 0 & 0 & 0 & 0 & 0 & 0 \\ 0 & 1 & 0 & 0 & 0 & 0 & 0 \end{bmatrix}$$

$$C_2 = \begin{bmatrix} -1 & 0 & 0 & 1 & 0 & 0 & 0 \\ 0 & 1 & 0 & 0 & 1 & 0 & 0 \end{bmatrix}$$

$$G = \begin{bmatrix} 0 \\ 0 \\ 0 \\ 1 \\ 1 \\ 1 \\ 0 \end{bmatrix} \quad H = \begin{bmatrix} 1 & 0 & 0 & 0 & 0 & 0 & 0 \\ 0 & 1 & 0 & 0 & 0 & 0 & 0 \end{bmatrix}$$

$$Q = diag\{100, 10, 100, 0, 0, 0, 0\} \qquad R_{11} = R_{22} = 1.$$

The open-loop system is unstable, with the spectrum,

$$\Lambda(A) = \{-2.80 \pm j2.41, -4.48, -3.71, .83, -.68, -1.26\}.$$

The reference system (here an LQ solution, i.e., $\gamma \to \infty$) is characterized by the optimal spectrum,

$$\Lambda(F) = \{-10.09, -1.62 \pm j1.63, -1.15, -4.5, -3.10 \pm j.87\}.$$

Since in this problem $r_1 = r_2 = 2$, two modes of the reference solution can be retained with static projective controls. In addition, by using two first-order dynamic controllers, one additional mode can be retained. The dominant modes are chosen for retention; thus,

$$\Lambda_r = \begin{bmatrix} -1.62 + j1.63 & 0 \\ 0 & -1.62 - j1.63 \end{bmatrix} \qquad \Lambda_p = [-1.15].$$

The initial choices for P_1 and P_2 are

$$P_1 = [0\ \ 0], \qquad P_2 = [0\ \ 0]$$

producing the closed-loop spectrum,

$$\Lambda(\tilde{A}_{ce}) = \{-3.65 \pm j2.16, -3.89, .36, -1.62 \pm j1.64, -.92, -1.15, -1.15\}.$$

Note that this choice of P_1 and P_2 fails to stabilize the resulting closed loop system. Consequently, the embedding parameter method must be used initially until a stable A_{ce} is achieved, or it is determined that the order of the dynamic controllers must be increased. For the example at hand, first order controllers did produce a stable system; thus, the order need not be increased.

The feasible direction algorithm is then employed to yield the optimum parameters for P_1 and P_2 for disturbance minimization. These optimal values are found to be

$$P_1 = [-2.41 \quad 0.71]$$

$$P_2 = [-3.65 \quad 1.58]$$

with an optimal value of the cost criterion of

$$J = \text{Trace } P_e Q_e = 1.430E^{-2}.$$

(Notice that for the initial choice, $J = \infty$, since the system was unstable.) The closed loop spectrum, $\tilde{A}_{ce}$, produced by the free parameter matrices is now

$$\Lambda(\tilde{A}_{ce}) = \{-5.77 \pm 3.36j, -.58 \pm 1.72j,$$
$$-4.26, -3.42, -1.62 \pm 1.63j, -1.15\}$$

Once P_1 and P_2 are determined, the controller parameters can then be determined from (25) to be

$$\begin{aligned}
H_1 &= -1.93, & H_2 &= -7.92 \\
D_1 &= [-5.48 \quad -4.58], & D_2 &= [-25.17 \quad 17.05] \\
K_{d1} &= [-3.43 \quad -1.95], & K_{d2} &= [-34.17 \quad 20.48] \\
N_{d1} &= -0.95 & N_{d2} &= -8.87,
\end{aligned}$$

thus completing the design.

CONCLUSIONS

A methodology is presented for the disturbance rejection problem using decentralized projective controls. It is shown that a decentralized system with dynamic controllers can be parameterized by a set of free parameter matrices. Furthermore, the system can be transformed such that the system parameters exhibit a linear dependence on the design free parameters. The minimization of the Frobenius-Hankel norm is used since it yields necessary conditions with a significant degree of decoupling which are therefore computationally feasible to solve and provides bounds on the H_∞ norm. A seventh-order problem illustrates the procedure and results. The optimal values for the free parameter matrices are determined first, and the controller parameters are then easily calculated from the given parameterizations.

It is of interest to note that the disturbance minimization problem is solved in two steps. In the first step, the dominant modes of the system are selected and retained based on the system's transient response and disturbance input affects using projective controls. In the second step, the disturbance minimization problem is resolved by positioning the residual poles and zeros. That is, the two-stage approach to the problem can be viewed as first fixing the dominate nodes, determined by the system's transient response, and then manipulating the system's zeros and additional poles to minimize the effects of the disturbance inputs.

REFERENCES

Bernstein, D., W. Haddad (1989). LQG control with an H_∞ performance bound: A Riccati equation approach, *IEEE Trans. Auto. Control*, AC-34.

Doyle, J., K. Glover, P. Khargonekar, B. Francis (1988). State-space solutions to standard H_2 and H_∞ control problems, *Proc. Amer. Control Conf.*, Atlanta, GA, pp. 1691-1696.

Francis, B. A., J. C. Doyle (1987). Linear Control Theory with an H_∞ Optimality Criterion, *SIAM J. Contr. Optim.*, Vol. 25, pp. 815-844.

Glover, K. (1984). All Optimal Hankel-Norm Approximations of Linear Multivariable Systems and Their L^∞ Error Bounds, *Int. J. Contr.*, Vol. 39, pp. 1115-1193.

Glover, K., J. Doyle (1988). State-space formulae for all stabilizing controllers that satisfy an H_∞ norm bound and relations to risk sensitivity, *Systems and Control Letters*, Vol. 11, pp. 967-172.

Medanić, J. V., W. R. Perkins (1987). Frobenius-Hankel norm framework for disturbance rejection and low order decentralized control design, *Proc. 5th AFORS Workshop on Control of Space Structures*, Monterey, CA.

Moore, B. C. (1981). Principal Components Analysis in Linear Systems: Controllability, Observability and Model Reduction, *IEEE Trans. Automatic Control*, Vol. AC-26, No. 1, pp. 17-31.

Paz, R. A., J. V. Medanić (1989). Design of disturbance attenuating controllers: a Riccati-based FH norm optimization algorithm, submitted for publication.

Paz, R. A., J. V. Medanić (1989). A Riccati-Based FH Norm Optimization Algorithm for Robust Low-Order Controller Design, submitted for publication.

Petersen, I. R. (1987). Disturbance attenuation and H_∞ optimization: a design method based on the algebraic Riccati equation, *IEEE Trans. Auto. Control*, AC-32, pp. 427-429.

Ramaker, R. A., J. Medanić, W. R. Perkins (1989). Strictly Proper Projective Controllers for Disturbance Attenuation, to appear *Int. J. Control*.

Uskoković, Z., J. Medanić (1983). Design of Centralized Static and low order dynamic regulators for large scale systems, *Proc. 22nd IEEE Conf. on Decision and Control*, San Antonio, TX, December.

Uskoković, Z., J. V. Medanić (1987). New Parametrization of Decentralized Dynamic Regulators Designed by Projective Controls, *IEEE Proceedings, 26th Conference on Decision and Control*, Los Angeles, CA, pp. 2289-2294, December.

Veillette, R. J., J. V. Medanić and W. R. Perkins (1989). Robust Stabilization and Disturbance Rejection for Uncertain Systems by Decentralized Control, to appear in *Proc. Workshop on Control of Uncertain Systems*, Bromen, June, Series "Progress in Systems and Control Theory", Birkhauser Publ. Co.

Young, K. David (1987). Control System Research for Directed Energy Weapons FY87 Annual Technical Report, Lawrence Livermore National Laboratory, December.

DECENTRALIZED ESTIMATION AND CONTROL WITH OVERLAPPING INPUT, STATE, AND OUTPUT DECOMPOSITION

A. Iftar

*Department of Electrical Engineering, University of Toronto, Toronto, Ontario
M5S 1A4, Canada*

Abstract: The extension principle, which was introduced recently, is generalized to the case of input, state, and output expansion. Estimator and controller design problems are considered within the framework of the extension principle. Decentralized estimator and controller design with overlapping decompositions is also discussed within the same framework. It is shown that, if the extension principle is used, then any estimator or controller designed in the expanded spaces is contractible to the original spaces for implementation. Furthermore, it is shown that, if an estimator designed for the expanded system achieves good–estimation, then the contracted estimator also achives good–estimation for the original system. Similarly, it is also shown that, if a controller designed for the expanded system achieves stability and good–performance, then the contracted controller achieves stability and good–performance for the original system.

Keywords: Large–scale systems; control theory; estimation theory; decentralized control; decentralized estimation; overlapping decompositions; linear systems.

I. INTRODUCTION

Many of today's technological and social problems involve so complex systems that it is very costly if not impossible to handle such large dimensional systems as a whole. For the purposes of estimation and control, it is usually necessary to decompose the given system into a number of interconnected subsystems. Once a decomposition is available, each subsystem is to be considered independently and the solutions to the subproblems are to be combined to obtain a solution for the original problem.

In order to obtain a useful decomposition, it is essential to identify the parts of a system that are *weakly* interconnected. However, many large scale systems may consist of subsystems which are strongly connected through certain dynamics (*"the overlapping part"*), but weakly connected otherwise. Electric power systems [1], socio-economic systems [2], large flexible structures [3], and freeway traffic regulation [4] are examples of such behaviour. For such systems, disjoint decompositions may easily fail to produce useful results. However, it has been demonstrated that the recently introduced *overlapping decompositions* [5] may produce useful solutions in such cases (*e.g.*, see [3] or [6]).

The estimator or controller design approach within the framework of the *inclusion principle* [7] starts with expanding certain spaces (*e.g.*, state, input, and/or output) of a dynamic system with overlapping subsystems into larger spaces such that the subsystems appear as disjoint. Then, decentralized estimators or decentralized controllers are designed in the expanded spaces and finally contracted to the smaller spaces for implementation on the original system.

The earlier results on the inclusion principle, however, were restricted to the expansions and contractions of the state space only. Expansions and contractions of the input and output spaces were first considered in [8] and [9]. Later, in [10], controller design with state and input inclusion was considered. In [10], a special case of inclusion,

called *extension* was introduced; it was shown that, if the extension approach is used, then any control law designed in the expanded spaces is contractible to the original spaces for implementation. Contractibility is required in order to preserve the desired relations between the expanded and the original systems following the application of appropriate controllers or estimators (*e.g.*, see [10]).

In the present paper, the extension principle is generalized to the case where the output space is also expanded besides the state and the input spaces. Estimator and controller design problems are considered within the framework of the extension principle. Decentralized estimator and controller design with overlapping decompositions is also discussed within the same framework. It is shown that, if the extension principle is used, then any estimator or controller designed in the expanded spaces is contractible to the original spaces for implementation. Note that, this property may not hold if some other inclusion approach is employed (*e.g.*, see [8]). Furthermore, it is shown that, if an estimator designed for the expanded system achieves *good–estimation*, then the contracted estimator also achieves good–estimation for the original system. Similarly, it is also shown that, if a controller designed for the expanded system achieves stability and *good–performance*, then the contracted controller also achieves stability and good–performance for the original system.

Throughout the paper, $\mathbf{R}^k$ denotes the k–dimensional real vector space, I_k denotes the identity operator on $\mathbf{R}^k$, I denotes the identity operator on the appropriate vector space, and $(\cdot)^T$ denotes the transpose of $(\cdot)$.

II. ESTIMATOR AND CONTROLLER DESIGN WITH EXTENSION

In this section a special case of inclusion, called extension, is introduced; estimator and controller design within the framework of this principle is also discussed. Consider the following linear time–invariant (LTI) systems:

$$\Sigma: \quad \begin{aligned} \dot{x} &= Ax + Bu \\ y &= Cx \end{aligned} \qquad (1)$$

and

$$\tilde{\Sigma}: \quad \begin{aligned} \dot{\tilde{x}} &= \tilde{A}\tilde{x} + \tilde{B}\tilde{u} \\ \tilde{y} &= \tilde{C}\tilde{x} \end{aligned} \qquad (2)$$

Here $x \in \mathbf{R}^n$, $u \in \mathbf{R}^m$, and $y \in \mathbf{R}^l$ are, respectively, the state, input, and output vectors of the system Σ. Sim*i*larly, $\tilde{x} \in \mathbf{R}^{\tilde{n}}$, $\tilde{u} \in \mathbf{R}^{\tilde{m}}$, and $\tilde{y} \in \mathbf{R}^{\tilde{l}}$, are the state, input, and output vectors of the system $\tilde{\Sigma}$. It is assumed that $\tilde{n} \geq n$, $\tilde{m} \geq m$, and $\tilde{l} \geq l$. In the sequel the system Σ is refered to as the *original system* and $\tilde{\Sigma}$ is refered to as the *expanded system*. The state, input, and output spaces $\mathbf{R}^n$, $\mathbf{R}^m$, and $\mathbf{R}^l$ of Σ are called, respectively, the *original* state, input, and output spaces; similarly, the spaces $\mathbf{R}^{\tilde{n}}$, $\mathbf{R}^{\tilde{m}}$, and $\mathbf{R}^{\tilde{l}}$ of $\tilde{\Sigma}$ are called the *expanded* state, input, and output spaces.

Consider the transformations:

$$T : \mathbf{R}^n \rightarrow \mathbf{R}^{\tilde{n}} \quad , \qquad rank(T) = n \ , \qquad (3a)$$

$$R : \mathbf{R}^{\tilde{m}} \rightarrow \mathbf{R}^m \quad , \qquad rank(R) = m \ , \qquad (3b)$$

$$S : \mathbf{R}^l \rightarrow \mathbf{R}^{\tilde{l}} \quad , \qquad rank(S) = l \ , \qquad (3c)$$

and

$$T^{\#} : \mathbf{R}^{\tilde{n}} \rightarrow \mathbf{R}^n \quad , \qquad T^{\#}T = I_n \ . \qquad (3d)$$

Definition 1: *The system $\tilde{\Sigma}$ is an extension of the system Σ, and Σ is a disextension of $\tilde{\Sigma}$, if there exist transformations as in (3a)–(3c) such that for any initial state $x_o \in \mathbf{R}^n$ and any input $\tilde{u}(t) \in \mathbf{R}^{\tilde{m}}$, $0 \leq t < \infty$, the choice*

$$\tilde{x}_o = Tx_o \qquad (4a)$$

$$u(t) = R\tilde{u}(t) \quad \forall t \geq 0 \qquad (4b)$$

implies that

$$\tilde{x}(t; \tilde{x}_o, \tilde{u}) = Tx(t; x_o, u) \ , \quad \forall t \geq 0 \qquad (5a)$$

and

$$\tilde{y}(t; \tilde{x}) = Sy(t; x) \ , \quad \forall t \geq 0. \qquad (5b)$$

Remark 1: *The extension defined above is a generalization of the extension first defined in [10] to the case where the output space is also expanded besides the state and the input spaces. It is a special case of inclusion defined in [8]. In fact, it is a generalization of unrestriction first discussed in [7] to the case of state, input, and output inclusion. However, it is different than the unrestriction defined in [8]. In [8] unrestriction is defined for an arbitrary input $u(t)$ in the original input space, and the input in the expanded space is obtained by a transformation: $\tilde{u}(t) = \tilde{R}u(t)$. Here, on the other hand, the extension is defined for an arbitrary input $\tilde{u}(t)$ in the expanded input space, and the input in the original space is obtained by the transformation given in (4b). Therefore, for the unrestriction the allowable set of inputs for $\tilde{\Sigma}$ at any time is only an m–dimensional subset of $\mathbf{R}^{\tilde{m}}$, but for the extension it is $\mathbf{R}^{\tilde{m}}$.*

The conditions of extension are provided by the following theorem:

Theorem 1: *The system $\tilde{\Sigma}$ is an extension of the system Σ if and only if there exist transformations as in (3a)–(3c) such that*

$$TA = \tilde{A}T \ , \qquad (6a)$$

$$TBR = \tilde{B} \ , \qquad (6b)$$

and

$$SC = \tilde{C}T \ . \qquad (6c)$$

Proof: The responses of the two systems (1) and (2) are given by

$$x(t) = e^{At}x_o + \int_0^t e^{A(t-\tau)}Bu(\tau)d\tau \qquad (7)$$

and

$$\tilde{x}(t) = e^{\tilde{A}t}\tilde{x}_o + \int_0^t e^{\tilde{A}(t-\tau)}\tilde{B}\tilde{u}(\tau)d\tau \qquad (8)$$

respectively. By substituting (4a) and (4b) for $\tilde{x}_o$ and $u(\tau)$, and using the series expansion for the matrix exponentials, it can be shown that (5a) holds for any x_o and any $\tilde{u}(t)$ if and only if

$$\tilde{A}^i T = TA^i \qquad (9a)$$

and

$$\tilde{A}^i \tilde{B} = TA^i BR \qquad (9b)$$

for all $i \in \{0, 1, 2, ...\}$. For $i = 0$ (9a) holds trivially. For $i = 1$ (9a) holds if and only if (6a) holds; for $i = 0$ (9b) holds if and only if (6b) holds. Forthermore, if (6a) holds, then (9a) holds for $i \geq 2$ and (9b) holds for $i \geq 1$.

Similarly, (5b) holds for any x_o and any $\tilde{u}(t)$ if and only if

$$\tilde{C}\tilde{A}^i T = SCA^i \qquad (10a)$$

and

$$\tilde{C}\tilde{A}^i \tilde{B} = SCA^i BR \qquad (10b)$$

for all $i \in \{0, 1, 2, ...\}$. Using (6a) and (6b), however, (10a) and (10b) both reduce to (6c). Hence the result follows. $\square$

Under the transformations (3a)–(3d), the matrices of Σ and $\tilde{\Sigma}$ can be related as:

$$\tilde{A} = TAT^{\#} + M \ , \qquad (11a)$$

$$\tilde{B} = TBR + N \ , \qquad (11b)$$

and

$$\tilde{C} = SCT^{\#} + L \ , \qquad (11c)$$

where M, N, and L are constant complementary matrices. In order $\tilde{\Sigma}$ to be an extension of Σ, these complementary matrices must satisfy certain conditions provided by the following theorem:

Theorem 2: *The system $\tilde{\Sigma}$ is an extension of the system Σ if and only if*

$$MT = 0 \ , \qquad (12a)$$

$$N = 0 \ , \qquad (12b)$$

and

$$LT = 0 \ . \qquad (12c)$$

Proof: By substituting (11a)–(11c) into (6a)–(6c) and using $T^{\#}T = I$, we obtain: $TA + MT = TA$, $TBR + N = TBR$, and $SC + LT = SC$; these relations imply (12a)–(12c). $\qquad\square$

Next, we discuss designing LTI estimators and controllers by using extension. A LTI estimator or controller for the system Σ can be described by:

$$\Gamma : \quad \begin{aligned} \dot{z} &= Fz + Gy + EBu \\ v &= Hz + Ky \,, \end{aligned} \qquad (13)$$

where $z \in \mathbf{R}^p$ is the state and $v \in \mathbf{R}^k$ is the output of the estimator/controller Γ. In the case of an estimator, the output v is an *estimate* of

$$w = Dx \qquad (14)$$

which is a k–dimensional ($k \le n$) linear combination of the state x of the system Σ. In the case of a controller, the output v of the controller Γ is applied to the input of the system Σ in order to control it; *i.e.*:

$$u = v \,. \qquad (15)$$

In the latter case, the dimension of v is assumed to be the same as the dimension of u ($k = m$).

Similarly, a LTI estimator or controller for the system $\tilde{\Sigma}$ can be described by:

$$\tilde{\Gamma} : \quad \begin{aligned} \dot{\tilde{z}} &= \tilde{F}\tilde{z} + \tilde{G}\tilde{y} + \tilde{E}\tilde{B}\tilde{u} \\ \tilde{v} &= \tilde{H}\tilde{z} + \tilde{K}\tilde{y} \,, \end{aligned} \qquad (16)$$

where $\tilde{z} \in \mathbf{R}^{\tilde{p}}$ is the state and $\tilde{v} \in \mathbf{R}^{\tilde{k}}$ is the output of the estimator/controller $\tilde{\Gamma}$. In the case of an estimator, the output $\tilde{v}$ is an *estimate* of

$$\tilde{w} = \tilde{D}\tilde{x} \qquad (17)$$

which is a $\tilde{k}$–dimensional ($\tilde{k} \le \tilde{n}$) linear combination of the state $\tilde{x}$ of the system $\tilde{\Sigma}$. In the case of a controller, the output $\tilde{v}$ of the controller $\tilde{\Gamma}$ is applied to the input of the system $\tilde{\Sigma}$ for control purposes:

$$\tilde{u} = \tilde{v} \,, \qquad (18)$$

in which case $\tilde{k} = \tilde{m}$.

We also assume that the dimension of Γ does not exceed the dimension of $\tilde{\Gamma}$ (*i.e.*, $\tilde{p} \ge p$); this assumption may be justified since Σ is a part of $\tilde{\Sigma}$, and thus should not require a larger estimator or controller. In the case of an estimator, we further assume that the dimension of $\tilde{w}$ is not less than the dimension of w (*i.e.*, $\tilde{k} \ge k$).

In addition to the transformations (3a)–(3c), we define the following transformation:

$$P : \mathbf{R}^{\tilde{p}} \to \mathbf{R}^p \,, \qquad rank(P) = p \,. \qquad (19)$$

We now introduce contractibility:

Definition 2: *The estimator/controller (16) is contractible to the estimator/controller (13), if there exist transformations as in (3a)–(3c) and (19) such that for any initial state $x_o \in \mathbf{R}^n$ of the system Σ, for any input $\tilde{u}(t) \in \mathbf{R}^{\tilde{m}}$, $0 \le t < \infty$ of the system $\tilde{\Sigma}$, and for any initial state $\tilde{z}_o \in \mathbf{R}^{\tilde{p}}$ of the estimator/controller $\tilde{\Gamma}$, the choice*

$$\tilde{x}_o = Tx_o \,, \qquad (20a)$$

$$u(t) = R\tilde{u}(t) \,, \qquad \forall t \ge 0 \,, \qquad (20b)$$

and

$$z_o = P\tilde{z}_o \qquad (20c)$$

implies that

$$z(t; z_o, y, u) = P\tilde{z}(t; \tilde{z}_o, \tilde{y}, \tilde{u}) \,, \qquad \forall t \ge 0 \qquad (21a)$$

and

$$v(t; z, y) = Q\tilde{v}(t; \tilde{z}, \tilde{y}) \,, \qquad \forall t \ge 0 \,, \qquad (21b)$$

where in the case of an estimator $Q : \mathbf{R}^k \to \mathbf{R}^k$ satisfies

$$D = Q\tilde{D}T \,, \qquad (22)$$

and in the case of a controller

$$Q = R \,. \qquad (23)$$

It is important to satisfy contractibility in order to preserve the desired relations (such as stability, *good-performance*, and *good-estimation*) between the expanded and the original systems following the application of the appropriate estimators or controllers. Given two vector functions $a(t)$ and $b(t)$, which are of the same dimensions, we say that $a(t)$ is *close* to $b(t)$ with respect to a given linear transformation W and a given scalar function g, and write $a \sim_g^W b$, if $(a(t) - b(t))^T W^T W (a(t) - b(t)) \le g(t)$, $\forall t \in \mathcal{T}$, where $\mathcal{T}$ is the common domain of $a(\cdot)$, $b(\cdot)$, and $g(\cdot)$. Given a system (such as Σ), we say that an estimator (such as Γ) achieves *good-estimation*, with respect to a given linear transformation W and a given scalar function g, if the the output of the estimator is close to the variable which is to be estimated with respect to W and g (if $v \sim_g^W w$). Given a system (such as Σ), we say that a controller (such as Γ) achieves *good-performance*, with respect to a given *performance function* $f(t)$, a given linear transformation W, and a given scalar function g, if the state of the controlled system is close to f with respect to W and g (if $x \sim_g^W f$). Now we can prove the following result:

Theorem 3: *Suppose that $\tilde{\Sigma}$ is an extension of Σ and that $\tilde{\Gamma}$ is contractible to Γ. Moreover, suppose that the estimator/controller Γ is applied to the system Σ and the estimator/controller $\tilde{\Gamma}$ is applied to the system $\tilde{\Sigma}$. Then*

a) in the case of an estimator, $\tilde{v} \sim_g^{WQ} \tilde{w}$ implies that $v \sim_g^W w$, where Q satisfies (22); i.e., if $\tilde{\Gamma}$ achieves good-estimation with respect to WQ and g, then Γ achieves good-estimation with respect to W and g.

b) in the case of a controller, stability (asymtotic stability) of the expanded closed-loop system (obtained by applying $\tilde{\Gamma}$ to $\tilde{\Sigma}$) implies the stability (asymptotic stability) of the original closed-loop system (obtained by applying Γ to Σ); furthermore, if $\tilde{x} \sim_g^{WT^{\#}} \tilde{f}$, where $\tilde{f}(t)$ is a given performance function and $T^{\#}$ is as in (3d), then $x \sim_g^W f$, where $f(t) \triangleq T^{\#}\tilde{f}(t)$; i.e., *if $\tilde{\Gamma}$ achieves stability and good–performance with respect to $\tilde{f}$, $WT^{\#}$, and g, then Γ achieves stability and good–performance with respect to $T^{\#}\tilde{f}$, W, and g.*

Proof: a) Note that, $\tilde{v} \sim_g^{WQ} \tilde{w}$ implies $Q\tilde{v} \sim_g^W Q\tilde{w}$. However, by (21b): $Q\tilde{v} = v$ and by (17), (5a), (22), and (14): $Q\tilde{w} = Q\tilde{D}\tilde{x} = Q\tilde{D}Tx = Dx = w$; thus $v \sim_g^W w$.

b) First note that if $\tilde{\Gamma}$ is contractible to Γ, then by (21b) $u(t) = v(t) = R\tilde{v}(t) = R\tilde{u}(t)$; thus (20b) is satisfied. If the expanded closed–loop system is stable (asmptotically stable), then $[\tilde{x}(t) \; ; \; \tilde{z}(t)]$ is bounded (and $\lim_{t\to\infty}[\tilde{x}(t) \; ; \; \tilde{z}(t)] = 0$); then, however, $[x(t) \; ; \; z(t)] = [T^{\#}\tilde{x}(t) \; ; \; P\tilde{z}(t)]$ (by (5a) and (21a)) is also bounded (and $\lim_{t\to\infty}[x(t) \; ; \; z(t)] = 0$); thus the original closed–loop system is also stable (asmptotically stable). If $\tilde{x} \sim_g^{WT^{\#}} \tilde{f}$, then $T^{\#}\tilde{x} \sim_g^W T^{\#}\tilde{f}$; furthermore, since $T^{\#}T = I$, by (5a): $T^{\#}\tilde{x} = x$; thus $x \sim_g^W T^{\#}\tilde{f}$. $\qquad\square$

Remark 2: *Recall that the estimator or the controller is to be designed in the expanded spaces and then contracted for implementation. Thus, in the case of an estimator, if good–estimation with respect to W and g is desired for the original system, then the estimator $\tilde{\Gamma}$ must be designed to achive good–estimation with respect to WQ and g. Similarly, in the case of a controller, if closed–loop (asymptotic) stability and good–performance with respect to f, W, and g are desired for the original system, then the controller $\tilde{\Gamma}$ must be designed to achive closed–loop (asymptotic) stability and good–performance with respect to Tf, $WT^{\#}$, and g.*

The conditions of contractibility are given by the following theorem:

Theorem 4: *The estimator/controller (16) is contractible to the estimator/controller (13), if and only if there exist transformations as in (3a)–(3c) and (19) such that*

$$FP = P\tilde{F} \, , \qquad\qquad (24a)$$
$$GCA^j = P\tilde{G}\tilde{C}\tilde{A}^j T \, , \qquad \forall j \in \{0,1,2,...\} \, , \qquad (24b)$$
$$GCA^j BR = P\tilde{G}\tilde{C}\tilde{A}^j \tilde{B} \, , \quad \forall j \in \{0,1,2,...\} \, , \qquad (24c)$$
$$EBR = P\tilde{E}\tilde{B} \, , \qquad\qquad (24d)$$
$$HP = Q\tilde{H} \, , \qquad\qquad (24e)$$
$$KCA^j = Q\tilde{K}\tilde{C}\tilde{A}^j T \, , \qquad \forall j \in \{0,1,2,...\} \, , \qquad (24f)$$

and

$$KCA^j BR = Q\tilde{K}\tilde{C}\tilde{A}^j \tilde{B} \, , \quad \forall j \in \{0,1,2,...\} \, , \qquad (24g)$$

where in the case of an estimator Q satisfies (22), and in the case of a controller $Q = R$.

Proof: The responses of the two estimators/controllers (13) and (16) are given by:

$$z(t) = e^{Ft}z_o + \int_0^t e^{F(t-\tau)}(Gy(\tau) + EBu(\tau))d\tau \qquad (25)$$

and

$$\tilde{z}(t) = e^{\tilde{F}t}\tilde{z}_o + \int_0^t e^{\tilde{F}(t-\tau)}(\tilde{G}\tilde{y}(\tau) + \tilde{E}\tilde{B}\tilde{u}(\tau))d\tau \qquad (26)$$

respectively. By substituting $y(\tau) = Cx(\tau)$ and $\tilde{y}(\tau) = \tilde{C}\tilde{x}(\tau)$, by using (7) and (8) for $x(\tau)$ and $\tilde{x}(\tau)$, by substituting (20a)–(20c) for $\tilde{x}_o$, $u(\tau)$, and z_o, and using the series expansions for the matrix exponentials, it can be

shown that (21a) holds for any x_o, for any $\tilde{u}(t)$, and for any $\tilde{z}_o$ if and only if

$$F^i P = P\tilde{F}^i \, , \qquad\qquad (27a)$$
$$F^i GCA^j = P\tilde{F}^i \tilde{G}\tilde{C}\tilde{A}^j T \, , \qquad\qquad (27b)$$
$$F^i GCA^j BR = P\tilde{F}^i \tilde{G}\tilde{C}\tilde{A}^j \tilde{B} \, , \qquad\qquad (27c)$$

and

$$F^i EBR = P\tilde{F}^i \tilde{E}\tilde{B} \, , \qquad\qquad (27d)$$

for all $j \in \{0,1,2,....\}$ and for all $i \in \{0,1,2,....\}$. For $i = 0$ (27a) holds trivially; for $i = 1$ (27a) holds if and only if (24a) holds, and given that (24a) holds, (27a) holds for $i \geq 2$. By using (24a): (27b), (27c), and (27d) reduce to (24b), (24c), and (24d) respectively.

Similarly, (21b) holds for any x_o, for any $\tilde{u}(t)$, and for any $\tilde{z}_o$ if and only if

$$HF^i P = Q\tilde{H}\tilde{F}^i \, , \qquad\qquad (28a)$$
$$HF^i GCA^j = Q\tilde{H}\tilde{F}^i \tilde{G}\tilde{C}\tilde{A}^j T \, , \qquad\qquad (28b)$$
$$HF^i GCA^j BR = Q\tilde{H}\tilde{F}^i \tilde{G}\tilde{C}\tilde{A}^j \tilde{B} \, , \qquad (28c)$$
$$HF^i EBR = Q\tilde{H}\tilde{F}^i \tilde{E}\tilde{B} \, , \qquad\qquad (28d)$$
$$KCA^j = Q\tilde{K}\tilde{C}\tilde{A}^j T \, , \qquad\qquad (28e)$$

and

$$KCA^j BR = Q\tilde{K}\tilde{C}\tilde{A}^j \tilde{B} \, , \qquad\qquad (28f)$$

for all $j \in \{0,1,2,....\}$ and for all $i \in \{0,1,2,....\}$. By using (24a): (28a) reduces to (24e); by using (24a) and (24e): (28b) reduces to (24b), (28c) reduces to (24c), and (28d) reduces to (24d). Hence the result follows. $\qquad\square$

The above conditions reduce to simpler ones if the expanded system is an extension of the original system:

Corollary 1: *Given that the system $\tilde{\Sigma}$ is an extension of the system Σ, the estimator/controller (16) for the system $\tilde{\Sigma}$ is contractible to the estimator/controller (13) for the system Σ, if and only if there exist transformations as in (3a)–(3c) and (19) such that*

$$FP = P\tilde{F} \, , \qquad\qquad (29a)$$
$$GC = P\tilde{G}SC \, , \qquad\qquad (29b)$$
$$EBR = P\tilde{E}TBR \, , \qquad\qquad (29c)$$
$$HP = Q\tilde{H} \, , \qquad\qquad (29d)$$

and

$$KC = Q\tilde{K}SC \, , \qquad\qquad (29e)$$

where in the case of an estimator Q satisfies (22), and in the case of a controller $Q = R$.

Proof: By using (6a)–(6c): (24b) and (24c) both reduce to (29b), (24d) reduces to (29c), and (24f) and (24g) both reduce to (29e). $\qquad\square$

Since the estimator/controller is to be designed in the expanded spaces and then contracted for implementation, it is important that any estimator/controller designed in the expanded spaces be contractible. In fact, if $\tilde{\Sigma}$ is an extension of Σ then such a property holds:

Corollary 2: *If the system $\tilde{\Sigma}$ is an extension of the system Σ, then any estimator or controller of the form (16) for the system $\tilde{\Sigma}$ is contractible to an estimator/controller of the form (13) for the system Σ with:*

$$F = \tilde{F} \, , \qquad\qquad (30a)$$

$$G = \tilde{G}S \,, \tag{30b}$$

$$E = \tilde{E}T \,, \tag{30c}$$

$$H = Q\tilde{H} \,, \tag{30d}$$

and

$$K = Q\tilde{K}S \,, \tag{30e}$$

where in the case of an estimator Q satisfies (22), and in the case of a controller $Q = R$.

Proof: With $P = I$, (30a)–(30e) imply (29a)–(29e). $\square$

III. OVERLAPPING DECOMPOSITIONS

In this section decentralized estimator and controller design with overlapping decompositions is discussed within the framework of extension. Consider the system Σ given in (1). Suppose that the state, the input, and the output are partitioned as:

$$x = (x_1^T \,,\ x_2^T \,,\ x_3^T)^T \,, \qquad x_i \in \mathbf{R}^{n_i} \,, \tag{31a}$$

$$u = (u_1^T \,,\ u_2^T \,,\ u_3^T)^T \,, \qquad u_i \in \mathbf{R}^{m_i} \,, \tag{31b}$$

and

$$y = (y_1^T \,,\ y_2^T \,,\ y_3^T)^T \,, \qquad y_i \in \mathbf{R}^{l_i} \,. \tag{31c}$$

Here it is assumed that x_2, u_2, and y_2 correspond to the overlapping parts of the state, input, and output spaces respectively. Suppose that the system matrices in (1) are also partitioned compatibly:

$$A = \begin{bmatrix} A_{11} & A_{12} & A_{13} \\ A_{21} & A_{22} & A_{23} \\ A_{31} & A_{32} & A_{33} \end{bmatrix} \,, \quad B = \begin{bmatrix} B_{11} & B_{12} & B_{13} \\ B_{21} & B_{22} & B_{23} \\ B_{31} & B_{32} & B_{33} \end{bmatrix} \,,$$

and

$$C = \begin{bmatrix} C_{11} & C_{12} & C_{13} \\ C_{21} & C_{22} & C_{23} \\ C_{31} & C_{32} & C_{33} \end{bmatrix} \,.$$

Then, an extension of Σ is obtained by choosing:

$$S = \begin{bmatrix} I_{l_1} & 0 & 0 \\ 0 & I_{l_2} & 0 \\ 0 & I_{l_2} & 0 \\ 0 & 0 & I_{l_3} \end{bmatrix} \,, \quad T = \begin{bmatrix} I_{n_1} & 0 & 0 \\ 0 & I_{n_2} & 0 \\ 0 & I_{n_2} & 0 \\ 0 & 0 & I_{n_3} \end{bmatrix} \,,$$

$$R = \begin{bmatrix} I_{m_1} & 0 & 0 & 0 \\ 0 & I_{m_2} & I_{m_2} & 0 \\ 0 & 0 & 0 & I_{m_3} \end{bmatrix} \,,$$

$$T^{\#} = \begin{bmatrix} I_{n_1} & 0 & 0 & 0 \\ 0 & \frac{1}{2}I_{n_2} & \frac{1}{2}I_{n_2} & 0 \\ 0 & 0 & 0 & I_{n_3} \end{bmatrix} \,,$$

$$M = \begin{bmatrix} 0 & \frac{1}{2}A_{12} & -\frac{1}{2}A_{12} & 0 \\ 0 & \frac{1}{2}A_{22} & -\frac{1}{2}A_{22} & 0 \\ 0 & -\frac{1}{2}A_{22} & \frac{1}{2}A_{22} & 0 \\ 0 & -\frac{1}{2}A_{32} & \frac{1}{2}A_{32} & 0 \end{bmatrix} \,, \quad N = 0 \,,$$

and

$$L = \begin{bmatrix} 0 & \frac{1}{2}C_{12} & -\frac{1}{2}C_{12} & 0 \\ 0 & \frac{1}{2}C_{22} & -\frac{1}{2}C_{22} & 0 \\ 0 & -\frac{1}{2}C_{22} & \frac{1}{2}C_{22} & 0 \\ 0 & -\frac{1}{2}C_{32} & \frac{1}{2}C_{32} & 0 \end{bmatrix} \,.$$

The resulting expanded system matrices are:

$$\tilde{A} = \begin{bmatrix} A_{11} & A_{12} & \vdots & 0 & A_{13} \\ A_{21} & A_{22} & \vdots & 0 & A_{23} \\ \cdots & \cdots & \cdots & \cdots & \cdots \\ A_{21} & 0 & \vdots & A_{22} & A_{23} \\ A_{31} & 0 & \vdots & A_{32} & A_{33} \end{bmatrix} \triangleq \begin{bmatrix} \tilde{A}_1 & \tilde{A}_{12} \\ \tilde{A}_{21} & \tilde{A}_2 \end{bmatrix} \,,$$

$$\tilde{B} = \begin{bmatrix} B_{11} & B_{12} & \vdots & B_{12} & B_{13} \\ B_{21} & B_{22} & \vdots & B_{22} & B_{23} \\ \cdots & \cdots & \cdots & \cdots & \cdots \\ B_{21} & B_{22} & \vdots & B_{22} & B_{23} \\ B_{31} & B_{32} & \vdots & B_{32} & B_{33} \end{bmatrix} \triangleq \begin{bmatrix} \tilde{B}_1 & \tilde{B}_{12} \\ \tilde{B}_{21} & \tilde{B}_2 \end{bmatrix} \,,$$

and

$$\tilde{C} = \begin{bmatrix} C_{11} & C_{12} & \vdots & 0 & C_{13} \\ C_{21} & C_{22} & \vdots & 0 & C_{23} \\ \cdots & \cdots & \cdots & \cdots & \cdots \\ C_{21} & 0 & \vdots & C_{22} & C_{23} \\ C_{31} & 0 & \vdots & C_{32} & C_{33} \end{bmatrix} \triangleq \begin{bmatrix} \tilde{C}_1 & \tilde{C}_{12} \\ \tilde{C}_{21} & \tilde{C}_2 \end{bmatrix} \,.$$

If an estimator is to be designed, we also partition the variable to be estimated as follows:

$$w = (w_1^T \,,\ w_2^T \,,\ w_3^T)^T \,, \qquad w_i \in \mathbf{R}^{k_i} \,. \tag{32}$$

The matrix D in (14) is also partitioned compatibly:

$$D = \begin{bmatrix} D_{11} & D_{12} & D_{13} \\ D_{21} & D_{22} & D_{23} \\ D_{31} & D_{32} & D_{33} \end{bmatrix} \,.$$

Then, the corresponding variable to be estimated for the expanded system is given by (17), where

$$\tilde{D} = \begin{bmatrix} D_{11} & D_{12} & \vdots & 0 & D_{13} \\ D_{21} & D_{22} & \vdots & 0 & D_{23} \\ \cdots & \cdots & \cdots & \cdots & \cdots \\ D_{21} & 0 & \vdots & D_{22} & D_{23} \\ D_{31} & 0 & \vdots & D_{32} & D_{33} \end{bmatrix} \triangleq \begin{bmatrix} \tilde{D}_1 & \tilde{D}_{12} \\ \tilde{D}_{21} & \tilde{D}_2 \end{bmatrix} \,.$$

Note that (22) is satisfied with

$$Q = \begin{bmatrix} I_{k_1} & 0 & 0 & 0 \\ 0 & \frac{1}{2}I_{k_2} & \frac{1}{2}I_{k_2} & 0 \\ 0 & 0 & 0 & I_{k_3} \end{bmatrix} \,.$$

Next, consider the *decoupled subsystems*:

$$\tilde{\Sigma}_i^D : \quad \begin{aligned} \dot{\tilde{x}}_i &= \tilde{A}_i\tilde{x}_i + \tilde{B}_i\tilde{u}_i \\ \tilde{y}_i &= \tilde{C}_i\tilde{x}_i \end{aligned} \,, \quad i = 1,2. \tag{33}$$

Suppose that local estimators or controllers described by:

$$\tilde{\Gamma}_i^D : \quad \begin{aligned} \dot{\tilde{z}}_i &= \tilde{F}_i \tilde{z}_i + \tilde{G}_i \tilde{y}_i + \tilde{E}_i \tilde{B}_i \tilde{u}_i \\ \tilde{v}_i &= \tilde{H}_i \tilde{z}_i + \tilde{K}_i \tilde{y}_i \end{aligned} \quad , \quad i = 1, 2, \quad (34)$$

are designed for the decoupled subsystems $\tilde{\Sigma}_i^D$, for $i = 1, 2$, respectively. These local estimators/controllers are designed such that the overall decentralized estimator/controller described by (16), with

$$\tilde{F} = \text{blockdiag}\{\tilde{F}_1 , \tilde{F}_2\}, \quad (35a)$$
$$\tilde{G} = \text{blockdiag}\{\tilde{G}_1 , \tilde{G}_2\}, \quad (35b)$$
$$\tilde{H} = \text{blockdiag}\{\tilde{H}_1 , \tilde{H}_2\}, \quad (35c)$$
$$\tilde{E} = \text{blockdiag}\{\tilde{E}_1 , \tilde{E}_2\}, \quad (35d)$$

and

$$\tilde{K} = \text{blockdiag}\{\tilde{K}_1 , \tilde{K}_2\}, \quad (35e)$$

satisfies the design requirements (such as stability, good–performance, or good–estimation) when applied to the expanded system $\tilde{\Sigma}$. Let us partition the matrices of the local estimators/controllers as follows:

$$\tilde{G}_i = \begin{bmatrix} G_1^i & G_2^i \end{bmatrix}, \quad \tilde{E}_i = \begin{bmatrix} E_1^i & E_2^i \end{bmatrix},$$
$$\tilde{H}_i = \begin{bmatrix} H_1^i \\ H_2^i \end{bmatrix}, \quad \text{and} \quad \tilde{K}_i = \begin{bmatrix} K_{11}^i & K_{12}^i \\ K_{21}^i & K_{22}^i \end{bmatrix},$$

for $i = 1, 2$, where $G_1^1 \in \mathbf{R}^{p_1 \times l_1}$, $G_2^1 \in \mathbf{R}^{p_2 \times l_3}$, $E_1^1 \in \mathbf{R}^{p_1 \times n_1}$, $E_2^1 \in \mathbf{R}^{p_2 \times n_3}$, $H_1^1 \in \mathbf{R}^{k_1 \times p_1}$, $H_2^1 \in \mathbf{R}^{k_3 \times p_2}$, $K_{11}^1 \in \mathbf{R}^{k_1 \times l_1}$, $K_{22}^2 \in \mathbf{R}^{k_3 \times l_3}$, p_i is the dimension of $\tilde{z}_i$ ($i = 1, 2$), and in the case of a controller $k_i = m_i$ ($i = 1, 2, 3$). The contraction of the above described overall decentralized estimator/controller to the original spaces is given by (13), where

$$F = \tilde{F}, \quad (36a)$$
$$G = \begin{bmatrix} G_1^1 & G_2^1 & 0 \\ 0 & G_1^2 & G_2^2 \end{bmatrix}, \quad (36b)$$
$$E = \begin{bmatrix} E_1^1 & E_2^1 & 0 \\ 0 & E_1^2 & E_2^2 \end{bmatrix}, \quad (36c)$$
$$H = \begin{bmatrix} H_1^1 & 0 \\ H_2^1 & H_1^2 \\ 0 & H_2^2 \end{bmatrix}, \quad (36d)$$

and

$$K = \begin{bmatrix} K_{11}^1 & K_{12}^1 & 0 \\ K_{21}^1 & K_{22}^1 + K_{11}^2 & K_{12}^2 \\ 0 & K_{21}^2 & K_{22}^2 \end{bmatrix}. \quad (36e)$$

Assuming that the designed decentralized estimator/controller $\tilde{\Gamma}$ achieves stability, good–performance, and/or good–estimation when applied to the expanded system $\tilde{\Sigma}$, by Theorem 3, the above described contracted estimator/controller Γ achieves stability, good–performance, and/or good–estimation when applied to the original system Σ.

IV. CONCLUSIONS

The extension principle, first introduced in [10], has been generalized to the case where the output space is also expanded besides the state and the input spaces. This principle is conceptually close to the unrestriction defined in [8]. However, in [8] the unrestriction was defined for an arbitrary input in the original input space, and the input in the expanded space was obtained by a transformation. This fact brings in the restriction that a particular estimator or controller designed for the expanded system by the approach of [8] may not be contractible to the original spaces. It has been shown, however, that with the approach undertaken here any estimator or controller designed in the expanded spaces is always contractible to the original spaces.

Furthermore, it has been shown that, if an estimator designed for the expanded system achieves good–estimation, then the contracted estimator also achieves good–estimation for the original system. Similarly, if a controller designed for the expanded system achieves stability and good–performance, then the contracted controller also achieves stability and good–performance for the original system.

Finally, decentralized estimator and controller design with overlapping decompositions has been discussed within the framework of extension. Although systems with only two subsystems have been presented for notational simplicity, the extension of the results to systems with more than two subsystems is straightforward.

REFERENCES

[1] D. D. Šiljak, *Large–Scale Dynamic Systems: Stability and Structure*. New York: North–Holland, 1978.

[2] M. Aoki, "On decentralized stabilization and dynamic assignment problems," *J. International Economics*, vol. 6, pp. 143–171, 1976.

[3] Ü. Özgüner, F. Khorrami, and A İftar, "Two controller design approches for decentralized systems," in *Proceedings of the AIAA Guidance, Navigation, and Control Conference*, Minneapolis, MN, 1988.

[4] L. Isaksen and H. J. Payne, "Suboptimal control of linear systems by augmentation with application to freeway traffic regulation," *IEEE Transactions on Automatic Control*, vol. AC–18, pp. 210–219, 1973.

[5] M. Ikeda and D. D. Šiljak, "Overlapping decompositions, expansions, and contractions of dynamic systems," *Large scale systems*, vol. 1, pp. 29–38, 1980.

[6] A. İftar and Ü. Özgüner, "Local LQG/LTR controller design for decentralized systems," *IEEE Transactions on Automatic Control*, vol. AC–32, pp. 926–930, 1987.

[7] M. Ikeda, D. D. Šiljak, and D. E. White, "An inclusion principle for dynamic systems," *IEEE Transactions on Automatic Control*, vol. AC–29, pp. 244–249, 1984.

[8] M. Ikeda and D. D. Šiljak, "Overlapping decentralized control with input, state, and output inclusion," *Control Theory and Advanced Technology*, vol. 2, pp. 155–172, 1986.

[9] Y. Ohta, D. D. Šiljak, and T. Matsumoto, "Decentralized control using quasi-block diagonal dominance of transfer function matrices," *IEEE Transactions on Automatic Control*, vol. AC–31, pp. 420–430, 1986.

[10] A. İftar and Ü. Özgüner, "Contractible controller design and optimal control with state and input inclusion," *Automatica*, 1989. (to appear).

NETWORK MODEL OF TRAFFIC UPDATES AND ITS APPLICATION TO ROUTING OPTIMIZATION

J. Filipiak* and P. Chemouil**

**The Teletraffic Research Centre, The University of Adelaide, G.P.O. Box 498,
Adelaide, S.A. 5001, Australia*
***Centre National d'Etudes des Télécommunications, 92131 Issy Les Moulineaux,
France*

Abstract: The real time control of traffic routing in large communication networks is investigated. An algorithm is derived which updates routing tables according to changing load conditions. Traffic measurements are described by means of linear difference equations. The model incorporates real control and information structures. The theorem is given which asserts that the network performance objectives can be achieved by means of the state dependent shortest route algorithm. Implementation issues related to traffic estimation and prediction are addressed. An actual version of the algorithm is evaluated by means of simulations.

Keywords: Large scale systems, Communication networks, Traffic control, Routing, Traffic estimation and prediction

Introduction

An analysis and synthesis of traffic management and control in communication networks poses a complex problem. Control rules must be find for a large set of geographically dispersed local controllers acting under stringent time constraints in a random environment. Different implementation issues must be taken into account, which are implied by hardware parameters, quality of network services and PTT regulations.

Currently, routing in the network is done by means of routing tables. For each destination code the routing table specifies a sequence of overflow directions. Usually, a direct route is first tried. If all trunks on the direct route are busy alternative routes are attempted in a prespecified order. The sequence of overflow is updated from a keyboard on a time horizon of weeks or months after an analysis of long term traffic measurements. In contrast to that the on-line routing control consists of updating routing tables with a short cycle length so as to profit from instantaneous traffic fluctuations.

A control system is appended to a node which measures a number of trunks busy on all outgoing links and uses that information to determine a sequence of overflow routes, see, e.g., [15] and [10]. Practical update times range from several seconds up to several minutes depending on a make of the telephone exchange.

Our previous research on modelling and control of telephone traffic concentrated on the fields trials and simulation experiments. Hardware and software elements which we used to test an adaptive routing in a real network are described in [10]. We also constructed an analytical model of on-line traffic measurements and used it for time series analysis of sequential traffic records, cf. [7]. Performance of heuristic algorithms was analysed in [4] and [5].

In this paper we intent to determine a routing rule in a systematic way. Several optimization problems are formulated. An attempt is made to take into account constraints on control imposed by hardware. The network model is based on realistic patterns of information about the network state.

In the next section link and network models are presented. formulate problems of sequential and random routing. The control problem is formulated and solved in Section 3. Implementation issues and simulation results are considered in Section 4. In Section 5 some comments on properties of the proposed algorithms are given.

Routing problem

In this section a model of on-line traffic measurements is first introduced. The routing scheme is then presented, which defines traffic distribution at a network node. Link and node descriptions are combined to give a network model. The sequential routing problem and its equivalent, the state dependent routing problem, are formulated.

In circuit switching machines with stored program control traffic records are collected by online sampling of all trunks going out from an exchange. Since from the point of view of switching processor this is a low priority task an updating period Δt can be long. The update cycle length depends on a type of exchange. A number $x(\kappa)$ of trunks busy at time t_κ and a number $u(\kappa)$ of calls accepted in the time period $(t_{\kappa-1}, t_\kappa)$ are usually recorded. κ denotes the measurement epoch. Measurement time instants t_κ as well as the update cycle length Δt can vary from node to node.

The following model of traffic measurements on a sligthly loaded trunk group was derived in [7]:

$$X(\kappa + 1) = aX(\kappa) + bX(\kappa + 1) + X(\kappa + 1), \quad \kappa = 1 \ldots K, \quad (1a)$$

It was shown that under an assumption of Poisson arrivals and exponential holding times coefficients a and b have the following form:

$$a = \exp(-T) \qquad (1b)$$

$$b = \frac{1}{T}[1 - \exp(-T)] \qquad (1c)$$

where T is an update cycle length normalized by a mean holding time τ: $T = \Delta t/\tau$. Random variables $X(\kappa)$, $U(\kappa)$, and $V(\kappa)$ are Poisson.

Assume that the considered process is stationary and denote by ϱ an intensity of offered traffic: $\varrho = \bar{u}/T$. In the teletraffic engineering ϱ is measured in Erlangs. We have: $\bar{x} = E[X(\kappa)] = \varrho$, $\sigma_x^2 = var[X(\kappa)] = \varrho$, and $\sigma_u^2 = var[U(\kappa)] = T\varrho$. Variable $V(\kappa)$ can be treated as a noise with a zero mean and the variance

$$\sigma_v^2 = (1 - a^2 - Tb^2)\varrho \tag{2}$$

An autocorrelation function of the process $\{X(\kappa), \ \kappa = 1..K\}$ decays exponentially:

$$\alpha_k = \exp(-kT) \tag{3}$$

It can be shown, cf. [6], that Eq.(1) is valid for loss systems with nonexponential service times.

The link model defined by dependencies (1) to (3) is attractive for control applications. We shall use it in the sequel to construct the network model.

Routing control scheme

Routing at network nodes is done according to routing tables. Denote by N a set of network nodes: $N = \{i, j, k \ldots\}$, and by L a set of network links: $L = \{(i,j), (j,k) \ldots\}$. We shall admit routes composed of two links at most.

Denote by R_{jk}^l a path from node j to node l through node k: $R_{jk}^l = \{(j,k), (k,l)\}$. An algorithm of updating routing tables consists of the following steps.

Algorithm 1.

1. Update an information $\Xi(\kappa)$ about a state of all links:

$$\Xi(\kappa) = \{x_{jk}(\kappa), u_{jk}(\kappa) : (j,k) \in L\} \tag{4}$$

2. Estimate traffic congestion $z_{jk}(\kappa + 1)$ on link (j,k) in the approaching time interval:

$$z_{jk}(\kappa + 1) = \Lambda_{jk}\{\Xi(\iota), \ \iota = 1 \ldots \kappa\}, \ \forall (j,k) \in L \tag{5}$$

3. Determine qualities $\Omega_{jk}^l[\mathbf{z}(\kappa)]$ of routes R_{jk}^l, $k \in N$, $k \neq j$, from j to l, for all origin-destination pairs $j, l \in N$.

4. For each origin j and destination l determine a sequence $\Psi_j^l(\kappa)$ of overflow routes:

$$\Psi_j^l(\kappa) = \{R_{jk_m}^l(\kappa), \ m = 1 \ldots M_j^l, \ k_m \in N, \ k_m \neq j\} \tag{6}$$

by arranging them according to their quality.

We shall denote by $\Psi(\kappa)$ a set of overflow sequences:

$$\Psi(\kappa) = \{\Psi_j^l(\kappa) : \ j, l \in N\} \tag{7}$$

Algorithm 1 suits well a centralized execution. One could easily reformulate it so that each node could independently calculate overflow sequence of routes starting from it. In both cases an information about a state of some links may be obsolete due to communication delays and an asynchronous mode of traffic measurements.

We stress that the structure of Algorithm 1 is imposed by the hardware. One could change it only at the expense of significant investments into a redesign of actually exploited systems. The fixed structure of control reduces the design of traffic routing to a choice of measures of trunk group congestion z_{jk} and route length Ω_{jk}^l. Though the Algorithm 1 is presented here as a heuristic algorithm we shall demonstrate in Section 3 that it finds a theoretical support.

Sequential routing

Assume that the traffic which enters node $j \in N$ in time interval $(t_\kappa, t_{\kappa+1})$ from the outside of the network and is destined to node $l \in N$ is Poisson with intensity $\varrho_j^l(\kappa+1)$. According to a definition of traffic intensity:

$$\bar{u}_j^l(\kappa + 1) = T\varrho_j^l(\kappa + 1), \quad j, l \in N \tag{8}$$

Under an assumption of Poisson distribution $\bar{u}_j^l(\kappa)$ fully describes $U_j^l(\kappa)$. Variables $\mathbf{r}(\kappa+1) = \{\varrho_j^l(\kappa+1), \ j, l \in N\}$, $\kappa = 1 \ldots K-1$, define the network load pattern.

The links going out from node j are attempted by calls to l in the prespecified order $\Psi_j^l(\kappa + 1)$. This means that all traffic entering node j is first forwarded to the link (j, k_1):

$$U_{jk_1}^l(\kappa + 1) = U_j^l(\kappa + 1), \quad \forall j, l \in N \tag{9a}$$

The overflow $W_{jk_1}^l(\kappa + 1)$ from the link (j, k_1) is directed to the second choice link (j, k_2): $U_{jk_2}^l(\kappa+1) = W_{jk_1}^l(\kappa+1)$, and so forth. For the $m - th$ choice link we have:

$$U_{jk_m}^l(\kappa + 1) = W_{jk_{m-1}}^l(\kappa + 1), \quad \forall j, l \in N, \ m = 2 \ldots M_j^l \tag{9b}$$

Assume next that the following dependencies are given:

$$W_{jk_m}^l(\kappa + 1) = \Gamma_{jk_m}^l[X_{jk_m}(\kappa), U_{jk_m}^l(\kappa + 1)] \tag{10}$$

for all $j, l \in N$, $m = 1 \ldots M_j^l - 1$. Denote by $I(j)$ a set of links incoming to node j: $I(j) = \{(i,j) : i \in N\}$, and by $O(j)$ a set of links going out from j: $O(j) = \{(j,k) : \ k \in N\}$. Link (j,k) of the route R_{jk}^l will be called an *upstream* link, and link (k,l) a *downstream* link.

The total traffic offered to the link (j,k) is composed of a direct, upstream, and downstream traffic:

$$U_{jk}(\kappa+1) = \sum_{(i,j)\in I(j)} U_{ij}^k(\kappa+1) + U_{jk}^k(\kappa+1) + \sum_{(k,l)\in O(k)} U_{jk}^l(\kappa+1) \tag{11}$$

for $(j,k) \in L$. Eq.(1a) written for the link (j,k) has the form:

$$X_{jk}(\kappa+1) = aX_{jk}(\kappa) + bU_{jk}(\kappa+1) + V_{jk}(\kappa+1), \quad (j,k) \in L, \tag{12}$$

for $\kappa = 1 \ldots K$. Eqs.(8) to (12) define *Model 1* of the network with state-dependent routing. Note that the model parameters a and b in (12) does not depend on a link number. This is true if the mean holding time τ and observation period Δt are the same on all links. Though this assumption is not necessary for further derivations we take it because in many cases it is both valid and simplifies notation.

A quality of the network operation is described by the functional:

$$J = E\{\sum_{\kappa=1}^{K-1} \Phi[\mathbf{X}(\kappa + 1)]\} \tag{13}$$

where $\mathbf{X}(\kappa) = \{X_{jk}(\kappa) : (j,k) \in L\}$.

The problem of routing control synthesis can be formulated as follows.

Problem 1. *Sequential routing optimization.* Given the network load pattern $\{\mathbf{r}(\kappa+1), \ \kappa = 1 \ldots K\}$ and the Model 1 of network operation find a set of overflow sequences $\{\Psi(\kappa+1), \ \kappa = 1 \ldots K-1\}$ which is optimal in a sense of the performance index J defined by Eq.(13).

This problem can be solved only with a difficulty. It is suggested in [9] that the set of the optimal overflow sequences is found by means of the NP-complete algorithm. Therefore, we reformulate the Problem 1 so as to make it amenable to solution.

State dependent routing

Define the random routing variable θ_{jk}^l as a proportion of the input traffic U_j^l, which is forwarded to the link (j,k):

$$U_{jk}^l(\kappa+1) = \theta_{jk}^l(\kappa)U_j^l(\kappa+1), \quad (j,k) \in L, \; l \in N \qquad (14)$$

We have

$$\theta_{jk_m}^l(\kappa)U_j^l(\kappa+1) = W_{jk_{m-1}}^l(\kappa+1) - W_{jk_m}^l(\kappa+1) \qquad (15)$$

for $m = 1 \ldots M_j^l$, $j,l \in N$. where $W_{jk_0}^l(\kappa+1) = U_j^l(\kappa+1)$.

It follows from the introduced definitions that

$$\theta_{jk}^l(\kappa) \geq 0, \quad j,k,l \in N \qquad (16a)$$

and

$$\sum_{(j,k)\in O(j)} \theta_{jk}^l(\kappa) \leq 1, \quad j,l \in N \qquad (16b)$$

Using (14) in (11) and then substituting (11) in (12) gives the model

$$X_{jk}(\kappa+1) = aX_{jk}(\kappa) + b\{ \sum_{(i,j)\in I(j)} \theta_{ij}^k(\kappa)U_i^k(\kappa+1)$$
$$+\theta_{jk}^k(\kappa)U_j^k(\kappa+1) + \sum_{(k,l)\in O(k)} \theta_{jk}^l(\kappa)U_j^l(\kappa+1)\} + V_{jk}(\kappa+1)$$

for $(j,k) \in L$. In a matrix notation

$$\mathbf{X}(\kappa+1) = \mathbf{A}\,\mathbf{X}(\kappa) + \mathbf{B}\,\mathbf{U}(\kappa+1)\,\theta(\kappa) + \mathbf{V}(\kappa+1) \qquad (17)$$

where $\mathbf{X}(\kappa+1)$, $\mathbf{X}(\kappa)$, and $\mathbf{V}(\kappa+1)$ are random vectors, $\mathbf{A}$ and $\mathbf{B}$ are matrices of parameters, $\mathbf{U}(\kappa+1)$ is a matrix of random elements, and $\theta(\kappa)$ is a control vector. The network performance index has the form

$$J = E\{ \sum_{\kappa=1}^{K-1} \Phi[\mathbf{X}(\kappa+1), \theta(\kappa)]\} \qquad (18)$$

Calculating averages in the state equations yields

$$x_{jk}(\kappa+1) = ax_{jk}(\kappa) + b\{ \sum_{(i,j)\in I(j)} \theta_{ij}^k(\kappa)u_i^k(\kappa+1)$$
$$+[1 + \theta_{jk}^k(\kappa)]u_j^k(\kappa+1) + \sum_{(k,l)\in O(k)} \theta_{jk}^l(\kappa)u_j^l(\kappa+1)\}$$

for $(j,k) \in L$, where $x_{pq} = E[X_{pq}]$ for $\forall(p,q) \in L$ and $u_r^s = E[U_r^s]$ for $\forall r,s \in N$. Moreover, we have redefined routing variables θ_{jk}^k on direct links (j,k): $\theta_{jk}^k(\kappa) := 1 + \theta_{jk}^k(\kappa)$, to stress that all traffic is first forwarded to the direct routes. The control constraints now have the form:

$$1 + \theta_{jl}^l(\kappa) \geq 0, \qquad (19a)$$

$$\theta_{jk}^l(\kappa) \geq 0, \quad (j,k) \in O(j), k \neq l, \qquad (19b)$$

$$\sum_{(j,m)\in O(j)} \theta_{jm}^l(\kappa) = 0 \qquad (19c)$$

for all origin-destination pairs $j,l \in N$. The above equations define the *Model 2* of the network.

Assume that the function Φ is separable and does not explicitly depend on $\theta(\kappa)$:

$$Q = \sum_{\kappa=1}^{K-1} \sum_{(p,q)\in L} \phi_{pq}[x_{pq}(\kappa+1)] \qquad (20)$$

We are interested in a solution of the following problem.

PROBLEM 2. *Synthesis of a state-dependent routing control.* Given the Model 2 of the network, the initial state $\mathbf{x}(1)$, and the load pattern $\{u_j^l(\kappa+1) : j,l \in N, \kappa = 2 \ldots K$, find the routing control variables $\theta_{jk}^l(\kappa)$, $j,k,l \in N$, such that the index Q defined by (22) takes the minimum value.

The above formulation does not impose any constraints on the final network state for the communication networks are operated over an infinite time horizon.

The following theorem defines the *shortest route property* of the optimum solution to the Problem 2.

THEOREM 1. Routes R_{jk}^l which are used by the traffic from origin j to destination l have the same length

$$\Omega_{jk}^l(\kappa) = \sum_{(u,v)\in R_{jk}^l} \omega_{uv}(\kappa), \quad (j,k) \in O(j) \qquad (21)$$

and are not longer than the unused routes. The function $\omega_{uv}(\kappa)$ is defined by

$$\omega_{uv}(\kappa) = \frac{d\phi_{uv}[x_{uv}(\kappa+1)]}{dx_{uv}(\kappa+1)}, \quad (u,v) \in L \qquad (22)$$

and may be interpreted as the state dependent link length.

This solution can be implemented by means of the Algorithm 1. Several comments may now be in order. First of all we note that the Theorem 1 provides a theoretical substantiation for the shortest route algorithms. That is, it asserts that this type algorithms can accomplish some global network objectives if the route length function is properly chosen. The second comment refers to the decentralized execution of the Algorithm 1 in the fully connected network. Note that each network node can independently execute the algorithm steps for the traffic originating from that node. The problem is that to do so each node must have a full information about the state of all network links. Providing all nodes with the same information pattern $\Xi(\kappa)$ is not practical. Therefore the traffic control is excercised in a network control center.

To fully specify the algorithm steps we need to decide a form of the objective function. Moreover, the traffic estimation procedure must be made precise.

Consider first the form of the function $\phi_{uv}[x_{uv}(\kappa+1)]$. Assume that the network links have finite capacities m_{uv}, $\forall(u,v) \in L$. The routing control consits of avoiding the link saturation, which is equivalent to a maximization of the function

$$\phi_{uv}[x_{uv}(\kappa+1)] = [m_{ux} - x_{uv}(\kappa+1)]^2 \qquad (23)$$

This corresponds to the following form of the route length function for the alternate routes

$$\Omega_{jk}^l(\kappa) = -[\omega_{jk}(\kappa) + \omega_{kl}(\kappa)] \qquad (24a)$$

For the direct routes we obtain:

$$\Omega_{jl}^l(\kappa) = -\omega_{jl}(\kappa) \qquad (24b)$$

The link length function is given by

$$\omega_{uv}(\kappa) = m_{uv} - x_{uv}(\kappa+1), \quad \forall(u,v) \in L \qquad (25)$$

The next problem consists of finding an estimate $z_{jk}(\kappa+1)$ of the trunk group congestion $x_{jk}(\kappa)$ given the set of network measurements. The natural approach is to calculate $x_{jk}(\kappa)$ directly from the state equations of the Model 2. However, in order to on-line evaluate $x_{jk}(\kappa+1)$ at time κ, we need to know the forecasts of the traffic $u_p^q(\kappa)$, which will be offered to the network nodes in the approaching time interval. The trouble is that in some networks the total number of calls offered to a link is recorded, rather than the number of calls which are sent from a given node to a specific destination. In a general case, different estimates $z_{jk}(\kappa+1)$ of $x_{jk}(\kappa+1)$ are needed, which take into account the available information on the network state.

Implementation issues and simulation results

Different techniques of traffic estimation and prediction for real time traffic management and control were investigated previously, cf. [5] and [3]. We briefly discuss the methods which can be adopted in the case under scrutiny. Performance of the proposed solution is then investigated by means of simulation.

Traffic estimation and prediction

The state equation for the link (j,k) has the form

$$X_{jk}(\kappa+1) = aX_{jk}(\kappa)+bU_{jk}(\kappa+1)+V_{jk}(\kappa+1), \quad \forall(j,k) \in L, \quad (26)$$

where $U_{jk}(\kappa)$ is given by Eq.(11). We assume that the number $u_{jk}(\kappa)$ of calls connected on each link is measured while the values of traffic elements $u_p^q(\kappa)$ are not known. Under that assumption, in order to predict $x_{jk}(\kappa+1)$, the equation (26) is more usefull than the state equations of the Model 2.

Taking averages in Eq.(26) gives

$$\tilde{x}_{jk}(\kappa+1) = ax_{jk}(\kappa) + b\tilde{u}_{jk}(\kappa+1), \quad \forall(j,k) \in L \quad (27)$$

We shall distinguish between three cases.

- The updating period Δt is small as compared to the mean holding time τ. Then $T \approx 0$ and according to Eqs.(1) $a \approx 1$ and $b \approx 0$. It follows from Eq. (27) that the reasonable estimate $z_{jk}(\kappa+1)$ of $x_{jk}(\kappa+1)$ can be obtained by taking

$$z_{jk}(\kappa+1) = x_{jk}(\kappa), \quad \forall(j,k) \in L, \quad (28)$$

  where $x_{jk}(\kappa)$ is the last measurement of the link state. The measurements of the number of accepted calls are not needed in this case.

- The updating period Δt has a value somewhere between $.5\tau$ and 1.5τ. Both coefficients in Eq.(27) are then different from zero which means that we must take

$$z_{jk}(\kappa+1) = \tilde{x}_{jk}(\kappa+1), \quad \forall(j,k) \in L \quad (29)$$

  where $\tilde{x}_{jk}(\kappa+1)$ is calculted from (27). The forecasts $\tilde{u}_{jk}(\kappa+1)$, which are needed in (27), are made under an assumption that the arrivals can be described by the IMA process, cf. [5]:

$$\tilde{u}_{jk}(\kappa) = \sigma u_{jk}(\kappa) + (1-\sigma)\tilde{u}_{jk}(\kappa), \quad \forall(j,k) \in L \quad (30)$$

- The updating period is much bigger than the mean holding time. In this case $a \approx 0$ and $b \approx 1/T$. From (27) we have $\tilde{x}_{jk}(\kappa+1) \approx \tilde{u}_{jk}(\kappa+1)/T$, which means that the estimate $z_{jk}(\kappa+1)$ of the trunk group congestion in the approaching time interval is equal to the forecast of the carried traffic:

$$z_{jk}(\kappa+1) = \tilde{u}_{jk}(\kappa+1)/T, \quad \forall(j,k) \in L \quad (31)$$

  $\tilde{u}_{jk}(\kappa+1)$ can be estimated by using the prediction formula (30). The measurements of the trunk group state are not needed.

Combining definitions of route length with traffic prediction techniques gives a multiplicity of possible algorithms. The choice of the traffic prediction method to a great extend is determined by the available data on the network state and the length of the updating period.

Simulation results

Performance of the proposed algorithms was evaluated by means of simulations. We simulated the network which was used during the adapting routing field trials as described in [8] and [5]. New assumptions include a nonhierarchical network structure, bidirectional use of the network links, and the crankback signalling.

The performance of the two routing algorithms using different route length functions is compared in Table 1. In the Case A the function $\Omega_{jk}^l(\kappa)$ had the form (24). In the Case B the route length was defined as follows:

$$\Omega_{jk}^l(\kappa) = -\min\{\omega_{jk}(\kappa), \omega_{kl}(\kappa)\}, \quad \forall i,j,k \in N$$

This heuristic measure of trunk group congestion was previously used in [10] and [2]. In both cases the link length $\omega_{jk}(\kappa)$ has been calculated according to (25). Since in the investigated network the number of trunks busy was only measured we used the estimate (28) of $x_{jk}^l(\kappa+1)$ for all values of the update cycle length.

To obtain results presented in Table 1 we assumed that the direct link is always tried as the first choice route, and that only one alternate route, the shortest one, can be attempted. The network grade of service was measured as a proportion of lost calls. It seen that for the long update times $\Delta t \geq 120s$ the Algorithm A outperformed the Algorithm B. The mean holding time τ was equal to 240 seconds.

We note an interesting fact that the multiple overflows did not improve the efficiency of the investigated network. This is illustrated in Table 2. To obtain these results the *state protection mechanism* was additionally implemented. The state protection consists of prohibiting the use of direct links by an overflow traffic if at the instant of measurement the number of free circuits is smaller than the threshold value. In our case the threshold was equal to 2. During simulations we assumed that the calls discarded on the downstream link are sent back to the orginating office so that the next alternate route can be tried.

TABLE 1. Network grade of service: % of lost calls

Δt [sec]	Algorithm A	Algorithm B
7.5	1.8199	1.8709
15	1.9645	2.0835
30	2.0750	2.0750
60	2.0920	2.0580
120	2.5342	2.5853
180	2.6193	2.8829
240	2.4747	3.2164
360	2.7979	3.4257

TABLE 2. Grade of service of the network with multiple overflows

Δt [sec]	Number of overflows					
	1	2	3	4	5	6
15	1.9474	1.9474	2.0750	2.3301	2.6193	2.6533
30	1.8794	1.9740	2.0750	2.3301	2.6193	2.6533
60	1.9304	1.9474	2.0750	2.3301	2.6193	2.6533
120	2.2366	2.3012	2.5767	2.6193	2.6958	2.7043
180	2.5257	1.9474	2.0750	2.3301	2.6193	2.6533
240	2.6363	2.7553	2.7213	2.4407	3.1040	3.4272

After analysis of this particular implementation we illustrate some further problems involved in a modelling and synthesis of state-dependent routing algorithms.

Feedback control law

The shortest route property defined in the Theorem 1 can be used to find the feedback control law for the Problem 3. Assume that standing at time κ we know the actual network state $\mathbf{x}(\kappa)$ and the load $\mathbf{u}(\kappa+1)$ in the approaching time epoch. The vectors $\mathbf{x}(\kappa)$ and $\mathbf{u}(\kappa+1)$ are defined as follows: $\mathbf{x}(\kappa) = \{x_{jk}(\kappa) : (j,k) \in L\}$ and $\mathbf{u}(\kappa+1) = \{u_j^l(\kappa+1) : j,l \in N\}$. The problem consists of finding the vector $\theta(\kappa) = \{\theta_{jk}^l(\kappa) : j,k,l \in N\}$.

If the routes $R_{jk_1}^l$ and $R_{jk_2}^l$ are to be used by the traffic from j to l in the time epoch $\kappa+1$ then according to the Theorem 1

$$\Omega_{jk_1}^l(\kappa) = \Omega_{jk_2}^l(\kappa) \tag{32}$$

Assume that the objective function is given by (27) which means that $\omega_{uv}(\kappa) = m_{uv} - x_{uv}(\kappa+1)$. Using this in Eq.(32) and substituting from the state equations for $x_{uv}(\kappa+1)$ gives

$$
\begin{aligned}
m_{jk_1} &- \{ax_{jk_1}(\kappa) + b\{ \sum_{(i,j)\in I(j)} \theta_{ij}^{k_1}(\kappa)u_i^{k_1}(\kappa+1) \\
&+ [1 + \theta_{jk_1}^{k_1}(\kappa)]u_j^{k_1}(\kappa+1) + \sum_{(k_1,l)\in O(k_1)} \theta_{jk_1}^l(\kappa)u_j^l(\kappa+1)\}\} \\
&+ m_{k_1 l} - \{ax_{k_1 l}(\kappa) + b\{ \sum_{(m,k_1)\in I(k_1)} \theta_{mk_1}^l(\kappa)u_m^l(\kappa+1) \\
&+ [1 + \theta_{k_1 l}^l(\kappa)]u_{k_1}^l(\kappa+1) + \sum_{(l,n)\in O(l)} \theta_{k_1 l}^n(\kappa)u_{k_1}^n(\kappa+1)\}\} \\
&= m_{jk_2} - \{ax_{jk_2}(\kappa) + b\{ \sum_{(i,j)\in I(j)} \theta_{ij}^{k_2}(\kappa)u_i^{k_2}(\kappa+1) \\
&+ [1 + \theta_{jk_2}^{k_2}(\kappa)]u_j^{k_2}(\kappa+1) + \sum_{(k_2,l)\in O(k_2)} \theta_{jk_2}^l(\kappa)u_j^l(\kappa+1)\}\} \\
&+ m_{k_2 l} - \{ax_{k_2 l}(\kappa) + b\{ \sum_{(m,k_2)\in I(k_2)} \theta_{mk_2}^l(\kappa)u_m^l(\kappa+1) \\
&+ [1 + \theta_{k_2 l}^l(\kappa)]u_{k_2}^l(\kappa+1) + \sum_{(l,n)\in O(l)} \theta_{k_2 l}^n(\kappa)u_{k_2}^n(\kappa+1)\}\}
\end{aligned}
$$

Such equations can be written for all used routes. Taking into account the constraints (19c) we obtain a set of linear equations which uniquely determine the control variables $\theta_{jk}^l(\kappa)$:

$$\mathbf{F}\mathbf{x}(\kappa) + \mathbf{G}\theta(\kappa) = \mathbf{H} \tag{33}$$

We have assumed in (33) that the offered traffic is stationary and that the forecasts of $u_j^l(\kappa+1)$ do not depend on time. From (33) we obtain the following feedback control law

$$\theta(\kappa) = -\mathbf{G}^{-1}\mathbf{F}\mathbf{x}(\kappa) + \mathbf{G}^{-1}\mathbf{H} \tag{34}$$

Since due to the high system dimension it may be difficult to calculate the inverse matrix $\mathbf{G}^{-1}$ Eqs.(33) can be solved by using the Gauss-Seidel method or other well known techniques. The trouble is that in most of cases we do not know *a priori* which routes are to be used by the traffic. This means that we do not know for which routes to write the above equations.

Another approach consists of using the following relaxation algorithm.

ALGORITHM 2.

1. Take the measurements. Update the network state $\mathbf{x}(\kappa)$ and the load $\mathbf{u}(\kappa+1)$. Set $\theta_{jk}^l(\kappa) = 0 \ \forall j,k,l \in N$.

2. Given $\mathbf{x}(\kappa)$, $\mathbf{u}(\kappa+1)$, and the vector $\theta(\kappa)$ of control variables calculate $\mathbf{x}(\kappa+1)$ and the lengths $\Omega_{jk}^l(\kappa)$ of all possible routes.

3. If there is the unused route $R_{jk_e}^l$ which is shorter than the longest used route shift a traffic increment $\delta\theta_{jk}^l(\kappa)u_j^l(\kappa+1)$ from that route to the route $R_{jk_e}^l$. Repeat that for all origin-destination pairs $j \to l$.

4. If for a given origin destination pair $j \to l$ all unused routes are longer than the longest used route than shift the traffic increment $\delta\theta_{jk}^l(\kappa)u_j^l(\kappa+1)$ from the longest used route to the shortest used route. Check it for all origin destination pairs $j \to l$.

5. Memorize new values $\theta_{jk}^l(\kappa)$, $\forall j,k,l \in N$.

6. If there is at least one source destination pair $j \to l$ for which the used routes have different lengths go to the Step 2. Otherwise stop.

The convergence of this type algorithms was investigated in [7] and [1].

Let us examine in more detail how the Algorithm 2 is operated in a real environment. At the begining of each epoch κ the measurements are taken, the load $\mathbf{u}(\kappa+1)$ is evaluated, and the Algorithm 2 is run to determine $\theta(\kappa)$:

$$A2 : \ \mathbf{x}(\kappa), \mathbf{u}(\kappa+1) \to \theta(\kappa) \tag{35}$$

The control variables must be calculated at the very beginning of the period Δt because they are supposed to be effected in that time interval. Keeping in mind that the length of Δt is about several seconds up to one minute, and that several hundreds routing variables must be calculated, we see that the time constraints are really stringent. Moreover, as we have mentioned previously, the knowledge about actual values of $\theta(\kappa)$ is not sufficient to uniquely determine the routes of individual calls. This is due to the fact that $\theta_{jk}^l(\kappa)$ defines the proportion of calls which must be sent along the route R_{jk}^l. An additional algorithm is needed to decide which actual calls must be directed to R_{jk}^l to obtain the traffic intensity $\theta_{jk}^l(\kappa)u_j^l(\kappa+1)$. Such *flow realization algorithms* are time consuming and can hardly be implemented by means of routing tables cf. [4].

Assume, however, that in spite of the foregoing difficulties, the control law (34) has been determined. Substituting from Eq.(34) for $\theta(\kappa)$ in Eq.(17) gives

$$\mathbf{X}(\kappa+1) = \mathbf{A}\mathbf{X}(\kappa) + \mathbf{B}\mathbf{U}(\kappa+1)[-\mathbf{G}^{-1}\mathbf{F}\mathbf{X}(\kappa) + \mathbf{G}^{-1}\mathbf{H}] + \mathbf{V}(\kappa+1) \tag{36}$$

This equation describes a time evolution of the network with state-dependent routing.

Concluding remarks

We have derived the algorithm which updates routing tables in circuit switching networks according to the changing traffic conditions. The algorithm is based on realistic assumptions about the structure of control and available system status information.

We first described the network by means of stochastic difference equations and formulated the problem of sequential routing optimization. Then we have discussed simplifications and modifications needed to obtain the practical solution. We have indicated how our problem and its solution relate to the real time traffic estimation and prediction. We took into account information patterns available in contemporery nonhierarchical networks and suggested the centralized implementation of the control algorithm.

In this paper we have formulated more problems than we have solved. In particular, it would be useful to solve the Problem 2 of random routing synthesis. Finding a set of control variables $\theta(\kappa)$ from (33) is a challenging task as well. From the point of view of network design a method is needed to evelute traffic statistics from (36).

Appendix: Proof of the Theorem 1

Define the Lagrange function:

$$
\begin{aligned}
F = \sum_{\kappa=1}^{K-1} \{ &\sum_{(j,k)\in L} \phi_{jk}[x_{jk}(\kappa+1)] - \sum_{j,l\in N} \lambda_j^l(\kappa) \sum_{(j,m)\in O(j)} \theta_{jm}^l(\kappa) \\
&- \sum_{(j,k)\in L} \omega_{jk}(\kappa)\{x_{jk}(\kappa+1) - ax_{jk}(\kappa)
\end{aligned}
$$

$$-b\{ \sum_{(i,j)\in I(j)} \theta_{ij}^k(\kappa)u_i^k(\kappa+1)$$

$$+[1+\theta_{jk}^k(\kappa)]u_j^k(\kappa+1) + \sum_{(k,l)\in O(k)} \theta_{jk}^l(\kappa)u_j^l(\kappa+1)\}\}$$

$$- \sum_{j,l\in N} \beta_{jl}^l(\kappa)[1+\theta_{jl}^l(\kappa)] - \sum_{j,l\in N} \sum_{(j,k)\in O(j),\,k\neq l} \beta_{jk}^l(\kappa)\theta_{jk}^l(\kappa)\}$$

The Lagrange multipliers λ_j^l correspond to the equality constraints (19c). The multipliers β_{jl}^l and β_{jk}^l relate to constraints (19a) and (19b), respectively, while the conjugate variables ω_{jk} refer to the equality constraints imposed by the state equations.

Optimality conditions for the nonlinear program under scrutiny are as follows:

$$\frac{\partial F}{\partial \theta_{jl}^l(\kappa)} = 0, \quad \forall j,l \in N \qquad (A.1a)$$

$$\frac{\partial F}{\partial \theta_{jk}^l(\kappa)} = 0, \quad \forall j,k,l \in N, \ k \neq l, \qquad (A.1b)$$

$$\frac{\partial F}{\partial x_{jk}(\kappa+1)} = 0, \quad \forall (j,k) \in L, \qquad (A.1c)$$

$$\frac{\partial F}{\partial \lambda_j^l(\kappa)} = 0, \quad \forall j,l \in N, \qquad (A.1d)$$

$$\frac{\partial F}{\partial \omega_{jk}(\kappa)} = 0, \quad \forall (j,k) \in L, \qquad (A.1e)$$

$$\beta_{jl}^l(\kappa) \geq 0, \quad \beta_{jl}^l(\kappa)[1+\theta_{jl}^l(\kappa)] = 0, \quad j,l \in N \qquad (A.1f)$$

$$\beta_{jk}^l(\kappa) \geq 0, \quad \beta_{jk}^l(\kappa)\theta_{jk}^l(\kappa) = 0, \quad \forall j,k,l \in N, \ k \neq l \qquad (A.1g)$$

for $\kappa = 1 \ldots K-1$. Calculating derivatives in (A.1e) and (A.1d) we obtain the state equations and the control constraints (19a). From (A.1a) we get

$$-\lambda_j^l(\kappa)+b\omega_{jl}(\kappa)u_j^l(\kappa+1)-\beta_{jl}^l(\kappa) = 0, \quad \forall j,l \in N, \ j \neq l \quad (A.2a)$$

Eqs.(A.1b) yields

$$-\lambda_j^l(\kappa)+b\omega_{jk}(\kappa)u_j^l(\kappa+1)+b\omega_{kl}(\kappa)u_j^l(\kappa+1)-\beta_{jk}^l(\kappa) = 0 \quad (A.2b)$$

for all $j,k,l \in N, \ k \neq l$. From (A.1c) we obtain

$$\omega_{jk}(\kappa+1) = -\frac{1}{a}\frac{d\phi_{jk}}{dx_{jk}(\kappa+1)} + \frac{1}{a}\omega_{jk}(\kappa), \quad (j,k) \in L \qquad (A.3)$$

Assume now that the alternate routes $R_{jk_e}^l = \{(j,k_e),(k_e,l)\}$, $e = 1 \ldots E_j^l$, are used by the traffic from j to l in the time epoch $\kappa+1$:

$$\theta_{jk_e}^l(\kappa) > 0, \quad e = 1 \ldots E_j^l \qquad (A.4)$$

According to (A.1g) this means that

$$\beta_{jk_e}^l(\kappa) = 0, \quad e = 1 \ldots E_j^l \qquad (A.5)$$

Using this in (A.2b) yields

$$\omega_{jk_e}(\kappa) + \omega_{k_el}(\kappa) = \frac{\lambda_j^l(\kappa)}{bu_j^l(\kappa+1)} \qquad (A.6a)$$

for used routes, and

$$\omega_{jk}(\kappa) + \omega_{kl}(\kappa) \geq \frac{\lambda_j^l(\kappa)}{bu_j^l(\kappa+1)} \qquad (A.6b)$$

for the unused routes for which $\theta_{jk}^l(\kappa) = 0$. Using similar arguments one can check that if the alternate routes are used then for the direct route $R_j^l = \{(j,l)\}$

$$\omega_{jl}(\kappa) = \frac{\lambda_j^l(\kappa)}{bu_j^l(\kappa+1)} \qquad (A.6c)$$

Note that according to the definition (21) the left hand sides of Eqs.(A.6) are the route lengths $\Omega_{jk}^l(\kappa)$. The right hand sides in

all equations are the same and do not depend on the tandem node k. This proves the shortest route property.

To derive the definition (22) of the link length note that except of (A.3) there are no additional constraints on the conjugate variables $\omega_{uv}(\kappa)$, $\forall (u,v) \in L$. Thus, we can take the simple solution

$$\omega_{uv}(\kappa+1) = \omega_{uv}(\kappa), \quad (u,v) \in L \qquad (A.7)$$

Using this in (A.3) yields

$$\omega_{uv}(\kappa) = \frac{1}{1-a}\frac{d\phi_{uv}[x_uv(\kappa+1)]}{dx_{uv}(\kappa+1)} \qquad (A.8)$$

Since the parameter a is the same for all links then Eq.(A.8) gives the same classification of routes as the definition (22), which completes the proof.

References

[1] D.P. Bertsekas, E.M. Gafni, and R.G. Gallager. Second derivative algorithms for minimum delay distributed routing in networks. *IEEE Trans. Commun.*, 32(8):911–191, 1984.

[2] W. H. Cameron, P. Galloy, and W. J. Graham. Report on the Toronto advanced routing concept trial. In *Proc. NETWORKS '80 Conf.*, Paris, 1980.

[3] P. Chemouil, J. Filipiak, and P. Gauthier. Analysis and control of traffic routing in circuit-switched networks. *Computer Networks and ISDN Systems*, 11:203 – 217, 1986.

[4] F.R.K. Chung, R.L. Graham, and F.K. Hwang. Efficient realization techniques for network flow patterns. *Bell Syst. Tech. J.*, 60(8):1771–1786, 1981.

[5] J. Filipiak and P. Chemouil. Modelling and prediction of traffic fluctuations in telephone networks. *IEEE Trans. Commun.*, COM-35(9):931–941, 1987.

[6] J. Filipiak and P. Chemouil. Time series analysis of traffic updates in loss systems. *Submitted for publication.*

[7] R.G. Gallager. A minimum delay routing algorithm using distributed computation. *IEEE Trans. Commun.*, 25:73–85, 1977.

[8] P. Gauthier and P. Chemouil. A system for testing an adaptive routing in France. In *Proc. IEEE GLOBECOM'87 Conference*, Tokyo, 1987.

[9] A. Girard and Y. Cote. Sequential routing optimization for circuit switched networks. *IEEE Trans. Commun.*, COM-32:1234–1243, 1984.

[10] E. Szybicki and A.E. Bean. Advanced traffic routing in local telephone networks: Performance of proposed call routing algorithms. In *Proceedings of the 9th International Teletraffic Congress*, Torremolinos, 1979.

A DECENTRALIZED CONTROL STRATEGY
WITH DISCRETE-TIME UPDATING FOR
DYNAMIC ROUTING

A. Iftar and E. J. Davison

*Department of Electrical Engineering, University of Toronto, Toronto, Ontario
M5S 1A4, Canada*

Abstract: A decentralized controller for dynamic routing in multi-destination data-communication networks is presented. It is assumed that each node of the network corresponds to a different control agent. The controls at different nodes are updated at different discrete instants. A dynamic model, which can incorporate arbitrary, different processing delays at different nodes, is developed to describe the network dynamics. The structure of the controller is motivated by an optimal control problem. It is shown that the controller can clear the queues in finite time in the absence of external flows.

Keywords: Data communication networks; computer communication; large-scale systems, decentralized control; discrete time systems; optimal control.

I. INTRODUCTION

An important problem in the operation of data communication networks is the routing of messages. Typically, a data communication network consists of many nodes which are connected through a number of links. The routing problem is to direct messages from one node to another, through such links, until they reach their desired destination.

Since in a typical situation the amount of messages entering a network at various nodes may vary from time to time, a dynamic routing strategy, which can adopt to such variations is required. Furthermore, it is often the case that the number of nodes in a network is large; in this case the vast number of different possible paths from one node to another, makes it virtually impossible to implement a centralized controller. Centralized controllers are also vulnerable to failures in the network and introduce a large communication overhead on the network. Thus, decentralized controllers, which can be implemented locally at individual nodes, and which require a minimum amount of information from the other nodes, are desirable to implement in practice.

Early routing algorithms, such as those implemented in the ARPANET [1] and TYMNET [2], were based on finding the shortest path from the initial node to the destination node [3]. In these algorithms, the length of a path is usually taken to be proportional to the message flow rate on that path. Most of these algorithms could be implemented in a decentralized way in the sense that the computations can be done locally; however, for dynamic routing, they require excessive information transfer inside the network (all the nodes must be informed about the changing link lengths). Other algorithms have been proposed to improve the network performance by minimizing a cost function related to the link congestion (message flow rate on a link relative to the capacity of that link) [4]-[6]. However, for dynamic routing, these algorithms also require excessive information exchange; furthermore, it has been argued that the link congestion does not actually reflect the network performance [3]. It should also be noted that the shortest path algorithms and the algorithms based on link congestion minimization are suitable only when the total input flow to a network is small compared to its total capacity; these algorithms usually require separate flow control algorithms to deal with situations where congestion occur.

The problem of determining a routing controller which minimizes a measure of the total queue length of the network was studied in [7]; it is clear that such a measure reflects the overall performance better than a measure based on link congestion. Furthermore, algorithms based on minimizing such measures can also work well under congestion. In [7], a conceptual algorithm was given to compute the optimal control for minimizing a measure of the queue length of a network. However, it was stated that the implementation of such an algorithm may not be possible to achieve due to the computational complexity required and a number of other problems. A decentralized routing algorithm, based on minimizing a measure of the queue length and the total travel time required (traffic networks, rather than communication networks were considered; hence the travel time is non-zero), was considered in [8]. The algorithm, however, is valid only for single destination networks. Multi-destination networks were considered within the same context in [9]. However, in [9] it was assumed that the total flow rate entering a node (including the flow coming from the upstream nodes) is constant, and nodes with only two outgoing links were considered. A local optimization approach for decentralized routing and "signalization" for a particular class of traffic networks was considered in [10].

An optimal control problem, which minimizes a measure of queue lengths was considered in [11]. Motivated by the structure of the optimal controller, a decentralized controller for solving the routing problem was proposed. In the present paper, a new decentralized controller, motivated by the same optimal control problem, is proposed. The basic difference between the controller of [11] and the present proposed controller is that here it is assumed that the controls are updated only at discrete instants at each node (each node corresponds to a different control agent). Thus, the present controller is more realistic to implement. Furthermore, the proposed controller also incorporates a "most preferred path" algorithm; however, unlike the earlier shortest path algorithms of [1]-[3], the present algorithm does not require information exchange to occur between different nodes. With these changes, we can prove that the proposed controller clears the queues of a network in a finite time, if there are no external input flows. It is to be noted that the controller of [11] can only guarantee asymptotic clearance of the network in this situation

A dynamic model is developed in Section II to describe the network dynamics; the developed model can incorporate arbitrary, different processing delays at different nodes. The control strategy is developed in Section III. Some properties of this strategy are discussed in Section IV where it is shown that the controller can clear the queues in finite time in the absence of external input flows. Simulation results for a five node network with processing delays and with two different external input flow rates are presented in Section V. The simulation results confirm that the proposed controller can perform reasonably well in practical situations.

II. MODELLING NETWORK DYNAMICS

Consider a data communication network consisting of N nodes. The nodes are connected through directed links on which messages can be transmitted. Each node receives messages from both up-stream nodes inside the network and from outside the network. Each message has a destination which is a node in the network. Messages are absorbed as soon as they arrive to their destination. Messages arriving to a node other than their final destination are put into a queue (or "buffer") and eventually are sent out to a down-stream node. However, it is assumed that a certain fixed amount of processing delay τ_i occurs at each node before an arriving message can be put into a queue and sent out. Furthermore, it is assumed that the rate of messages being sent out (controls) are updated only at discrete time instants $t_1^i < t_2^i < t_3^i < \cdots$ at node i ($i=1,...,N$). Note that each node may have a different processing delay and different control update instants (i.e., the network is not necessarily synchronous). It is assumed that messages travel along a link instantaneously (the travel time is usually negligible compared to processing delays at the nodes). Under these assumptions, the queue dynamics at each node can be modelled as shown in Figure 1, where:

r_i^k is the volume of messages with destination node k, presently being processed at node i ($i \neq k$),

w_i^k is the volume of processed messages with destination node k, waiting at node i (queue length) ($i \neq k$),

$q_i^k \triangleq w_i^k + r_i^k$ is the total volume of messages at node i with a destination node k ($i \neq k$),

f_i^k is the input flow rate of messages with destination node k entering the network at node i ($i \neq k$),

u_{ij}^k is the flow rate of messages with destination node k, sent out from node i to the downstream node j along the link i to j ($i \neq j, i \neq k$), and

$\mathbf{U}(i)$ and $\mathbf{D}(i)$ are respectively the sets of adjacent upstream and downstream nodes of node i, i.e.,

$\mathbf{U}(i) \triangleq \{ j \mid \text{there exists a link from } j \text{ to } i \}$

$\mathbf{D}(i) \triangleq \{ j \mid \text{there exists a link from } i \text{ to } j \}$.

By examining Figure 1, we can obtain the differential equation governing the total queue dynamics at node i for messages with destination node k:

$$\dot{q}_i^k = \dot{w}_i^k + \dot{r}_i^k$$
$$= f_i^k + \sum_{j \in \mathbf{U}(i)} u_{ji}^k - \sum_{j \in \mathbf{D}(i)} u_{ij}^k \qquad (1)$$

Note that, due to a cancellation, the delay nature of the system does not appear in the total queue length dynamics. However, if w_i^k and r_i^k were considered individually, then a set of delay-differential equations would be used in place of (1). The above equations can be compactly written as:

$$\dot{q} = f + Bu \qquad (2)$$

where $q \triangleq (q_1^2, q_1^3, \ldots, q_1^N, q_2^1, q_2^3, \ldots, q_N^{N-1})^T$ is the vector of total queue lengths, $f \triangleq (f_1^2, f_1^3, \ldots, f_1^N, f_2^1, f_2^3, \ldots, f_N^{N-1})^T$ is the vector of input message flow rate entering the network, u is the vector of message flow rates $u_{ij}^k, \forall i, \forall k \neq i, \forall j \in \mathbf{D}(i)$, and B is a matrix consisting of 1's corresponding to u_{ji}^k's with $j \in \mathbf{U}(i)$, -1's corresponding to u_{ij}^k's with $j \in \mathbf{D}(i)$ on the row corresponding to $\dot{q}_i^k$, and 0's elsewhere. For the purpose of routing control, q can be viewed as the state-vector of the system, and u can be viewed as the control input vector. The

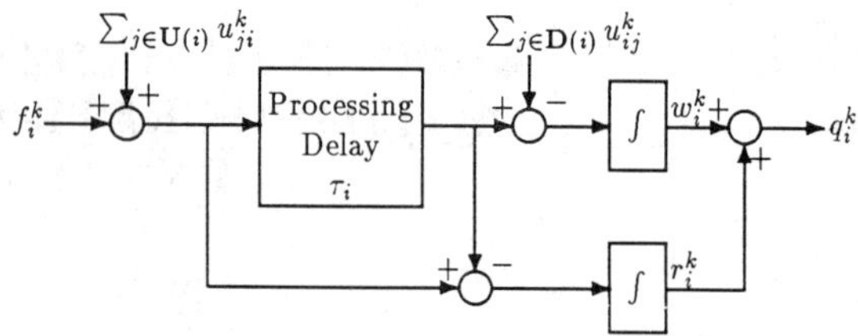

Figure 1: Queue dynamics at node i for messages with destination node k

aim of routing control is to determine a suitable controller for u (i.e., to control the flow rate along the links) to regulate the queue lengths q in the presence of f.

Certain constraints must be considered:

(a) The message flow rate into the network must be non-negative:

$$f_i^k \geq 0 \qquad \forall i, \forall k \neq i, \qquad (3)$$

(b) The messages being processed at a node can not be negative:

$$r_i^k \geq 0 \qquad \forall i, \forall k \neq i, \qquad (4)$$

(c) The messages waiting at a node can not be negative:

$$w_i^k \geq 0 \qquad \forall i, \forall k \neq i, \qquad (5)$$

(d) The message flow rate along any link can not be negative:

$$u_{ij}^k \geq 0 \qquad \forall i, \forall k \neq i, \forall j \in \mathbf{D}(i), \qquad (6)$$

(e) The total message flow along a link can not exceed the capacity $c_{ij} > 0$ of that link:

$$\sum_{\substack{k=1 \\ k \neq i}}^{N} u_{ij}^k \leq c_{ij} \qquad \forall i, \forall j \in \mathbf{D}(i). \qquad (7)$$

Note that constraint (a) is satisfied naturally (no one can insert a negative amount of message volume into the network). Constraint (b) is also automatically satisfied as long as constraints (a) and (d) are satisfied, since:

$$r_i^k(t) = \int_{t_0}^{t} (\overline{f}_i^k(v) - \overline{f}_i^k(v - \tau_i)) dv \quad \text{where} \quad \overline{f}_i^k \triangleq f_i^k + \sum_{j \in \mathbf{U}(i)} u_{ji}^k \quad \text{and}$$

$\overline{f}_i^k(t) = 0$ for $t < t_0$, where t_0 denotes the start-up time. Hence a control strategy must be chosen such that constraints (c), (d) and (e) are satisfied at all times. Furthermore, since the controls can be updated only at discrete instants, the control law must also satisfy:

$$u_{ij}^k(t) = u_{ij}^k(t_n^i), \quad \forall t \in [t_n^i, t_{n+1}^i), \quad n = 1, 2, \ldots,$$

where $[a, b)$ denotes the semi-closed interval of the real line from (including) a to (excluding) b. We will denote the time between the two updating instants t_n^i and t_{n+1}^i by δ_n^i, i.e., $\delta_n^i \triangleq t_{n+1}^i - t_n^i$.

To ease the difficulty of controller design, we will replace constraint (c) by:

$$(c') \qquad \sum_{j \in \mathbf{D}(i)} u_{ij}^k(t_n^i) \leq \frac{w_i^k(t_n^i)}{\delta_n^i} \qquad (8)$$

which is expressed in terms of the control inputs rather than states. Note that if (8) is satisfied, then (5) is also satisfied.

Finally we note that in a practical situation, there may also be constraints imposed on the volume of messages than can be processed and buffered (i.e., upper constraints on r_i^k and w_i^k). However, in this paper, we assume that such limits are sufficiently high so that such constraints are not violated.

In the rest of the paper we will restrict our discussion to connected networks. A network is said to be a "connected network" if every node of the network can be reached from every other node, through a number of directed links with positive capacity. The results, however, can be directly extended to non-connected networks, which have the property that $f_i^k=0$ for all node pairs (i,k) with the property that node k can not be reached from node i. The results presented in this paper are also valid when some or all nodes in a network are delay-free; in this case we simply set $\tau_i=0$, $r_i^k=0$, and $q_i^i=w_i^k$.

III. ROUTING CONTROL STRATEGY

The structure of the proposed controller is motivated by the solution of the following optimization problem:

Find controls for the system described by (2) such that constraints (4)-(7) are satisfied when (3) is satisfied, and such that the cost function

$$J = \frac{1}{2} q(t_f)^T H q(t_f) + \frac{1}{2} \int_{t_0}^{t_f} q(t)^T Q q(t) dt , \qquad (9)$$

where H and Q are positive semi-definite matrices and t_0 and t_f are initial and final times, is minimized.

It is shown in [11], that to minimize (9), the optimal controls u_{ij}^{k*} must satisfy the following necessary conditions:

$$-p_i^j u_{ij}^{j*} \leq -p_i^j u_{ij}^j , \quad \forall i, \forall j \in \mathbf{D}(i) \qquad (10a)$$

and

$$-(p_i^k-p_j^k)u_{ij}^{k*} \leq -(p_i^k-p_j^k)u_{ij}^k , \qquad (10b)$$

$$\forall i, \forall j \in \mathbf{D}(i), \ \forall k \neq j, i ,$$

for all admissible controls u_{ij}^k satisfying (6), (7) and (8), where p_i^k is the "co-state" corresponding to q_i^k. In [11], it has been demonstrated that the optimal co-state (and hence the solution to the optimal control problem) in general depends on future in-coming external flows and controls. However, under certain assumptions (see [11]), the co-state vector $p \triangleq (p_1^2, p_1^3, \ldots, p_1^N, p_2^1, p_2^3, \ldots, p_N^{N-1})^T$ can be obtained as a feedback from the state vector q. Using this observation for motivation, we will propose a decentralized controller to solve the routing problem, which satisfies the conditions (10), where p is obtained as:

$$p = Kq \qquad (11)$$

where K is a diagonal matrix with positive elements on its diagonal.

Note that conditions (10), together with the applicable constraints, imply that to implement an optimal controller one should apply $u_{ij}^k=0$, if $p_i^k-p_j^k<0$, and let u_{ij}^k be as large as possible if $p_i^k-p_j^k>0$. However, these conditions do not tell us how to distribute the available capacities among the different u_{ij}^k's with positive $p_i^k-p_j^k$. To develop a strategy to achieve a sensible distribution, define at each node i an "ordering" R_i of the set $\{1,2,\ldots,i-1,i+1,\ldots,N\}$, i.e., R_i is an ordered set consisting of all the elements of $\{1,2,\ldots,i-1,i+1,\ldots,N\}$. Also define a non-cyclic path, called the "most preferred path" from each node i $(i=1,2,\ldots,N)$ to each other node k $(k=1,2,\ldots,i-1,i+1,\ldots,N)$ of the network, such that if the nodes $i=i_0,i_1,\ldots,i_{m-1},i_m=k$ form the most preferred path from node i to node k, then $i_n,i_{n+1},\ldots,i_{m-1},i_m=k$ form the most preferred path from node i_n $(n=0,1,\ldots,m-1)$ to node k.

The above sequences are formed such that there exists a link with positive capacity from each node to the successive node in this sequence (i.e., $i_{n+1} \in \mathbf{D}(i_n)$, $n=0,1,\ldots,m-1$), and that none of the nodes are repeated (i.e., $i_l \neq i_r$ if $l \neq r$, $l,r=0,1,\ldots,m$). Here, m denotes the number of nodes on the most preferred path from i to k and is a function of i and k; we omit to explicitly show this functional dependence for the

sake of clarity. Next define an ordering S_i^k of the set $\mathbf{D}(i)$ for each destination $k=1,2,\ldots,i-1,i+1,\ldots,N$, at each node $i=1,2,\ldots,N$, such that if $i=i_0,i_1,\ldots,i_{m-1},i_m=k$ is the most preferred path from i to k, then i_1 is the first element of S_i^k; other elements of $\mathbf{D}(i)$ may be ordered arbitrarily in S_i^k.

Now we are ready to present the following controller for routing:

Decentralized routing controller: Defining $p_j^j=0$ and $\sum_{i=1}^{0}(\cdot)=0$, for notational convenience, the proposed decentralized routing controller is described as follows:

$$u_{ij_s}^{k_r}(t_n^i) = \begin{cases} 0 & , \text{ if } p_i^{k_r}(t_n^i)-p_{j_s}^{k_r}(t_n^i) \leq 0 \\ \min\left[\dfrac{w_i^{k_r}(t_n^i)}{\delta_n^i} - \sum_{l=1}^{s-1} u_{ij_l}^{k_r}, c_{ij_s} - \sum_{l=1}^{r-1} u_{ij_s}^{k_l}(t_n^i) \right] & , \text{ otherwise} \end{cases} \qquad (12)$$

for $s=1,2,\ldots,\text{size}(S_i^{k_r})$, $r=1,2,\ldots,N-1$, $n=1,2,3,\ldots$, and $i=1,2,\ldots,N$, where $\text{size}(\cdot)$ denotes the number of elements of the set $(\cdot)$. Here j_s is the sth element of $S_i^{k_r}$ and k_r denotes the rth element of R_i. Note that j_s is in fact a function of i and k_r, and k_r is a function of i. We omit to explicitly show these functional dependences for the sake of clarity.

Note that, the above controller can be implemented in a decentralized way at each node. The only information required at node i from the rest of the network, during the operation, is the value of p_j^k for $j \in \mathbf{D}(i)$ and $k=1,2,\ldots,N$, $k \neq i$, $k \neq j$. Thus, each node is required to pass its "co-state" values only to adjacent upstream nodes. The presence of a central authority is not required except for determining the most preferred paths, which would be done "off-line" before starting the actual operation.

The proposed control strategy has the tendency to send out messages with destination k, first to the downstream nodes which are ranked more preferable according to S_i^k. It also has the tendency to clear the queue q_i^k before the queue q_i^l if k precedes l in the ordering defined by R_i. The first property makes the most preferred path from node i to node k, more preferable for messages travelling from node i to node k. The messages are directed through other paths only if the most preferred path is congested. The second property may be desirable in certain applications where messages with a certain destination are considered as having a higher priority. However, in some other applications this property may be undesirable. In such a case, one could alter R_i from time to time in order to minimize this type of effect.

IV. FINITE TIME QUEUE CLEARANCE

In this section we will show that the proposed decentralized controller enjoys an important property. Namely, given a connected network, the proposed control strategy will clear all the queues of the system in a finite time, if the input flow to the network is zero.

Consider the sets R_i and R_j which are orderings of $\{1,\ldots,i-1,i+1,\ldots,N\}$ and $\{1,\ldots,j-1,j+1,\ldots,N\}$ respectively. We say that R_i and R_j are "comparable" if the ordered set R_i excluding the element j is the same as the ordered set R_j excluding the element i. We also say that more than two such sets are "comparable" if all possible pairs of these sets are comparable. In the case when all the sets $R_1,R_2,\ldots,R_N$ are comparable, we define a "global ordering" R, which is an ordering of the set $\{1,2,\ldots,N\}$ such that the ordered set R excluding the element i is the same as the ordered set R_i for $i=1,2,\ldots,N$.

For technical simplicity, we will first assume that all the sets $R_1,R_2,\ldots,R_N$ are comparable and fixed. We will also assume that controls are updated periodically, such that the time interval between any successive updating instants t_n^i and t_{n+1}^i at node i is the same, and is denoted by $\delta^i > 0$. Note that different nodes may still have a different updating frequency.

Let the most preferred path from node i to node k be $i = i_0, i_1, ..., i_{m-1}, i_m = k$. Define the "total delay" from node i to node k by:

$$\tau_i^k \triangleq \sum_{r=0}^{m-1} (\tau_{i_r} + \delta^{i_r}) \; ; \tag{13}$$

then the "maximum delay" in the network is defined as:

$$\tau_{\max} \triangleq \max_{i,k} (\tau_i^k) \; . \tag{14}$$

Define the set $\overline{D}(i) \subset D(i)$ as the set of adjacent downstream nodes of i, such that any one of these nodes is the first element of S_i^k for at least one k, i.e.:

$$\overline{D}(i) \triangleq \{ j \mid j \text{ is the first element of } S_i^k$$

$$\text{for at least one } k \in \{1, 2, ..., i-1, i+1, ..., N\} \} \; .$$

Also let us define $\sum_{i,k} (\cdot)$ as a short-hand notation for $\sum_{i=1}^{N} \sum_{\substack{k=1 \\ k \ne i}}^{N} (\cdot)$.

We will now prove the following result:

Theorem 1: Given a connected network as described in section II, assume that the decentralized controller (12) is applied, where the "co-state" p is obtained as described in (11) and the controls at node i are updated periodically with period $\delta^i > 0$. Assume that $f(t) = 0$, $\forall t \geq T_0$ for some T_0, and that the sets $R_1, R_2, ..., R_N$ are comparable and remain fixed for $t > T_0$. Then

$$q(t) = 0, \quad \forall t \geq T_0 + T_c \tag{15}$$

where

$$T_c \triangleq \frac{\tau_{\max} \sum_{i,k} q(T_0)}{\min_{\substack{i \in \{1,2,...,N\} \\ j \in \overline{D}(i)}} (\delta^i c_{ij})} + \sum_{i,k} \tau_i^k \; . \tag{16}$$

Proof: It has been shown in [11] that when $f = 0$

$$\sum_{i,k} \dot{q}_i^k = - \sum_{i=1}^{N} \sum_{k \in D(i)} u_{ik}^k \; . \tag{17}$$

Since $u_{ij}^k \geq 0$, $\forall i, j, k$, this implies that $\sum_{i,k} \dot{q}_i^k \leq 0$. Thus if $\sum_{i,k} q_i^k(T) = 0$ for some $T \geq T_0$, this implies that $q(t) = 0$, $\forall t \geq T$. Note that $\sum_{i,k} q_i^k$ is the sum of the components of q which are non-negative. Suppose $\sum_{i,k} q_i^k(T) > 0$ for some $T \geq T_0$, and let k be chosen such that there exists an i with $q_i^k(T) > 0$ and such that $q_l^r(T) = 0$ for all $l = 1, 2, ..., N$ and for all r which precedes k in the global ordering R. Let $i = i_0, i_1, ..., i_{m-1}, i_m = k$ be the most preferred path from i to k. Then there exists an updating instant

$$t_{n_0}^{i_0} \in [T, T + \tau_{i_0} + \delta^{i_0}) \tag{18}$$

such that $w_i^k(t_{n_0}^{i_0}) > 0$. Suppose that

$$w_i^k(t_{n_0}^{i_0}) \geq \delta^{i_0} c_{i_0 i_1} \; . \tag{19}$$

Then, assuming $q_{i_1}^k(t_{n_0}^{i_0}) = 0$ (if this is not true take $i = i_1$), this implies that $p_{i_1}^k(t_{n_0}^{i_0}) = 0$ and $u_{i_0 i_1}^k(t_{n_0}^{i_0}) = c_{i_0 i_1}$; thus there exists an updating instant $t_{n_1}^{i_1}$ (for node i_1) satisfying

$$t_{n_1}^{i_1} \in [t_{n_0}^{i_0}, t_{n_0}^{i_0} + \tau_{i_1} + \delta^{i_1}) \subset [T, T + \tau_{i_0} + \delta^{i_0} + \tau_{i_1} + \delta^{i_1})$$

such that $w_{i_1}^k(t_{n_1}^{i_1}) \geq \delta^{i_0} c_{i_0 i_1}$. Then assuming $q_{i_2}^k(t_{n_1}^{i_1}) = 0$ (if this is not true take $i = i_2$), this implies that

$$u_{i_1 i_2}^k(t_{n_1}^{i_1}) \geq \min \left[\frac{\delta^{i_0} c_{i_0 i_1}}{\delta^{i_1}}, c_{i_1 i_2} \right] .$$

By continuing this procedure, we obtain that there exists an updating instant

$$t_{n_{m-1}}^{i_{m-1}} \in [T, T + \tau_i^k)$$

such that

$$u_{i_{m-1} i_m}^k(t_{n_{m-1}}^{i_{m-1}}) \geq \min \left[\frac{\delta^{i_0} c_{i_0 i_1}}{\delta^{i_{m-1}}}, ..., \frac{\delta^{i_{m-2}} c_{i_{m-2} i_{m-1}}}{\delta^{i_{m-1}}}, c_{i_{m-1} i_m} \right]$$

$$\geq \frac{\min_{\substack{i \in \{1,2,...,N\} \\ j \in \overline{D}(i)}} (\delta^i c_{ij})}{\delta^{i_{m-1}}} \; .$$

Thus, by (17), $\sum_{i,k} \dot{q}_i^k \leq - \dfrac{\min_{\substack{i \in \{1,2,...,N\} \\ j \in \overline{D}(i)}} (\delta^i c_{ij})}{\delta^{i_{m-1}}}$ for a duration of $\delta^{i_{m-1}}$; hence

$$\sum_{i,k} q_i^k(T + \tau_{\max}) - \sum_{i,k} q_i^k(T) \leq - \min_{\substack{i \in \{1,2,...,N\} \\ j \in \overline{D}(i)}} (\delta^i c_{ij}) \tag{20}$$

Furthermore, (20) is valid as long as there exists i, k such that (19) is true.

Now suppose there exists no updating time (18) for which (19) is true. Then, however, the total amount of messages $q_i^k(T)$ is transferred to node i_1 within $\tau_i + \delta^i$ time units. This implies that either $u_{i_l i_{l+1}}^k(t_{n_l}^{i_l}) = c_{i_l i_{l+1}}$ for some i_l on the most preferred path so that (20) holds, or that the total amount of messages $q_i^k(T)$ is transferred to node k within τ_i^k units of time. Combining this observation with (20), one concludes that (15) is true.

$\square$

Remark 1: The assumption that the updating intervals δ_n^i ($n = 1, 2, ...$) are fixed, is not used in the above proof, except in the derivation of the upper bound in the time it takes for a particular message to reach its destination. Therefore, the same proof applies even if these intervals are not fixed, but are bounded, i.e., if there exists $\delta_{\min}^i$ and $\delta_{\max}^i$ ($i = 1, 2, ..., N$) such that $0 < \delta_{\min}^i \leq \delta_n^i \leq \delta_{\max}^i$ ($n = 1, 2, ...$). In this case we simply use $\delta_{\max}^i$ instead of δ^i in the derivation of the aforementioned upperbound. The lower bound $\delta_{\min}^i$ is needed since the control strategy (and the proof of Theorem 1) involves division by δ_n^i.

Remark 2: The assumption that the sets R_i ($i = 1, 2, ..., N$) are comparable and fixed may also be removed. In particular, suppose that these sets are not necessarily comparable nor fixed, and suppose that in the proof of Theorem 1, we choose an arbitrary pair (i, k) such that $q_i^k(T) > 0$. Now, it is possible that at some node i_l ($l = 1, 2, ..., m-1$) on the most preferred path from i to k, there may exist a queue $q_i^r(t_{n_{l-1}}^{i_{l-1}}) > 0$, where r precedes k in the ordering defined by R_{i_l} at time $t_{n_{l-1}}^{i_{l-1}}$, thus, the flow of messages along the most preferred path from node i to node k may be blocked for some time. However, during that time, there must exist at least one other pair (i', k'), such that the messages flow from i' to k' along the most preferred path from i' to k' as described in the proof of Theorem 1. Furthermore, even if the messages flowing from node i to node k are blocked at node i_l, there must exist a time t' at which $q_{i_l}^r(t') = 0$, for all r preceeding k in the ordering defined by R_{i_l} at time t'; hence, the flow of messages from node i to node k eventually continues.

Combining Remarks 1 and 2, we obtain the following result:

Corollary 1: Given a connected network as described in section II, assume that the controller (12) is applied, where the "co-state" p is obtained as described in (11) and the controls at node i are updated with time intervals δ_n^i, $n=1,2,...$, satisfying $0<\delta_{min}^i \leq \delta_n^i \leq \delta_{max}^i$ for fixed δ_{min}^i and δ_{max}^i. Assume that $f(t)=0$, $\forall t \geq T_0$ for some T_0. Then this implies that $q(t)=0$, $\forall t \geq T_0+T_c$ where T_c is given by (16) with δ^i replaced by δ_{max}^i in (13) and (16).

Remark 3: From the proof of Theorem 1, it can also be deduced that the proposed control strategy can also keep the queue lengths bounded under sufficiently small message input rates. In particular, if

$$\sum_{i,k} f_i^k(t) \leq \frac{\min_{\substack{i \in \{1,2,...,N\} \\ j \in \overline{D}(i)}} (\delta^i c_{ij})}{\tau_{max}}, \quad \forall t \geq T_0 \qquad (21)$$

then this implies that

$$\sum_{i,k} q_i^k(t) \leq \tau_{max} \max_{T_0 \leq \tau \leq t} \sum_{i,k} f_i^k(\tau) \leq \max_{\substack{i \in \{1,2,...,N\} \\ j \in \overline{D}(i)}} (\delta^i c_{ij}), \quad (22)$$

$$\forall t \geq T_0+T_c$$

where T_c is given by (16).

Remark 4: Note that (16) is simply an upper bound on the time required to clear all the queues. The proposed controller may clear the queues in a faster time than predicted by (16). In fact, tighter bounds can be developed if the individual initial queue lengths $q_i^k(T_0)$ and the actual network topology is taken into account. Similarly, the controller may keep the queue lengths bounded for input rates which are much larger than predicted by (21).

V. EXAMPLES

In this section we consider the network shown in Figure 2. The capacity of each link is indicated next to the link in the figure. Note that the network is connected. Each node has a processing delay of $\tau_i=1$ sec ($i=1,2,...,5$). The decentralized controller (12) is used to control the routing for the system. The "co-state" p is obtained as described in (11) with $K=I$, where I is the identity matrix.

The controls are assumed to be updated each 0.1 second. The ordering of the down-stream nodes are assumed to be as follows:

$$S_1^2 = \{2,5,3\}, \quad S_1^3 = \{3,2,5\}, \quad S_1^4 = \{3,2,5\}, \quad S_1^5 = \{5,3,2\}$$

$$S_2^1 = \{4,3\}, \quad S_2^3 = \{3,4\}, \quad S_2^4 = \{4,3\}, \quad S_2^5 = \{4,3\}$$

$$S_3^1 = S_3^2 = S_3^4 = S_3^5 = \{4\}$$

$$S_4^1 = \{1,5\}, \quad S_4^2 = \{5,1\}, \quad S_4^3 = \{1,5\}, \quad S_4^5 = \{5,1\}$$

$$S_5^1 = \{1,2\}, \quad S_5^2 = \{2,1\}, \quad S_5^3 = \{2,1\}, \quad S_5^4 = \{2,1\}$$

It is also assumed that there are no messages being processed or waiting at time zero (i.e., $q(0)=0$).

Case 1: First we assume that the ordering of the destination nodes are fixed and comparable where the global ordering is $R=\{1,2,3,4,5\}$. We consider the following input flow rate:

$$f_i^k(t) = \begin{cases} 1, & 0 \leq t < 5 \\ 0, & t \geq 5 \end{cases}, \quad \forall i, \forall k \neq i \qquad (23)$$

On applying the proposed decentralized controller (12), the resulting queue lengths at node 1 (i.e., q_1^k, $k=1,2,3,4$) are plotted in Figure 3. Due to space limitations, the queue lengths at the other nodes are not shown; however, they behave in a similar manner. The total queue length $\sum_{i,k} q_i^k$ is plotted in Figure 4. It is observed that the queue lengths

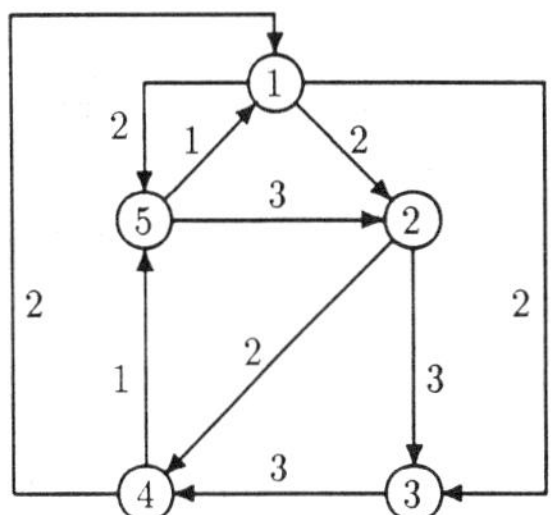

Figure 2: Example network.

increase with constant rate for $t \in [0,1]$. During this time, the incoming messages are processed at each node and all the link flows are zero. During the period $t \in (1,5)$, the queues continue to increase on average, but the rate of increase is slower, which indicates that a portion of the incoming messages actually reach their destination and leave the network. After the shut-off time $t=5$ sec., the queue lengths start decreasing on average, and all the queues are cleared by $t=22$ sec. We have to note that during the early phase of the time interval $[1,22]$, many links operate at their full capacity; this fact indicates that the link capacity constraints are a limiting factor in dictating how fast the queues can be cleared. The link from node 4 to 5, for example, operates at its full capacity for $t \in [1,16.2]$.

Case 2: Next, we consider a constant input flow rate:

$$f_i^k(t) = 0.3, \quad t \geq 0; \quad \forall i, \forall k \neq i. \qquad (24)$$

The queue lengths at node 1 are plotted in Figure 5. The total queue length is plotted in Figure 6. It is observed that by the time $t=15$ sec. occurs, a steady-state is reached where the total queue length fluctuates about 15.3. The controller in this case keeps the queue lengths bounded.

Under this external flow-rate some of the links (such as the link from node 4 to node 5) operate at their maximum capacities most of the time for $t>1$ sec. Some other links (such as the link from node 1 to node 2) operate at levels less that their maximum capacities for most of the time, and they exhibit a periodic steady-state type of behaviour for $t>1$ sec.

Case 3: Finally, we assume that the ordering of the destination nodes are "shifted-around" once at each updating instant at each node (e.g., $R_1=\{2,3,4,5\}$ becomes $R_1=\{3,4,5,2\}$ at the next updating instant). The input flow rate is assumed to be given by (23). In this case the queue lengths at node 1, and the total queue length are plotted in Figures 7 and 8 respectively. On comparing Figures 3 and 7, it is seen that using the latter strategy results in the queue lengths for different destinations being closer to each other at a particular node. On comparing Figures 4 and 8, however, it is seen that the overall performances of the two strategies are very similar.

VI. CONCLUSIONS

The routing control problem in multi-destination data communication networks has been considered in this paper. The dynamic model developed to describe the network dynamics is a generalization of previous models, e.g. [7], in that processing delays can be incorporated. Based on this model, a decentralized controller for routing control is proposed. The structure of the proposed controller has been motivated by an optimal control problem (the 'optimal' controller obtained in the optimal control problem cannot be directly implemented since it requires the unrealistic assumption that all future behaviour of the systems inputs be known in advance).

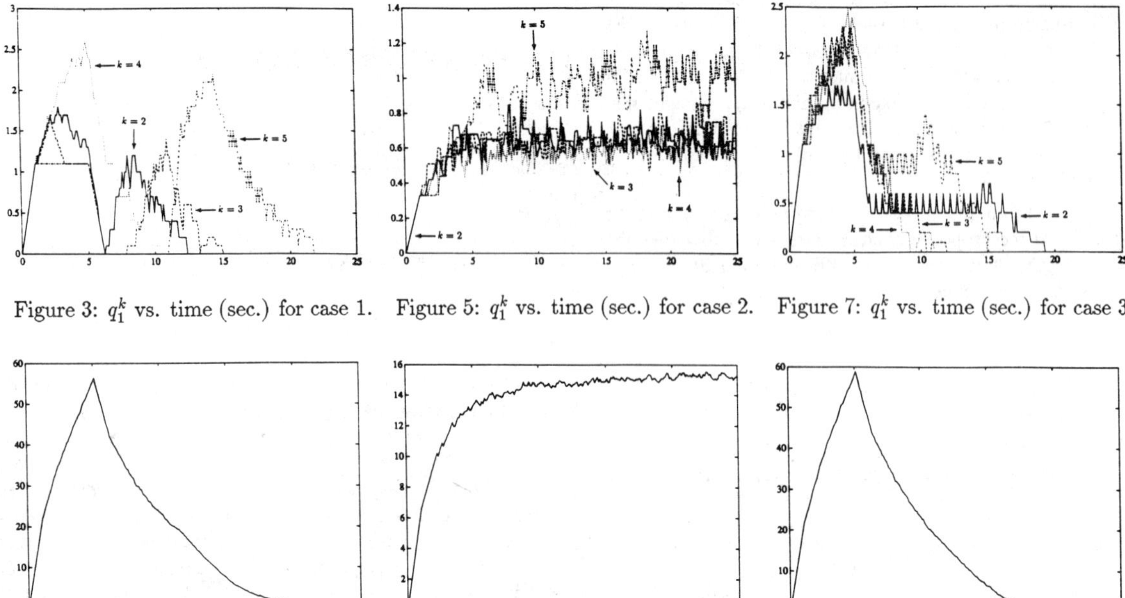

Figure 3: q_1^k vs. time (sec.) for case 1. Figure 5: q_1^k vs. time (sec.) for case 2. Figure 7: q_1^k vs. time (sec.) for case 3.

Figure 4: $\sum_{i,k} q_i^k$ vs. time for case 1. Figure 6: $\sum_{i,k} q_i^k$ vs. time for case 2. Figure 8: $\sum_{i,k} q_i^k$ vs. time for case 3.

The proposed controller has a number of desirable properties. It is decentralized in the sense that all the necessary computations to determine the flow rates on a link can be done at the node where the link originates. Furthermore, the only necessary information required for these computations to be done are that the total queue lengths at the two end nodes of that link be known. The only communication overhead on the network is that each node has to pass the co-state values, which depend on the queue lengths at that node, to its adjacent upstream nodes. We have shown that, in the absence of external input flows, the controller clears all the queues in finite-time. This property also suggests that the queue lengths can be kept bounded when the external input flow rate is sufficiently small. In fact, via simulation, we have verified that the queue lengths remain at reasonable values when the input flow rates are "moderate" compared to the link capacities.

The control strategy can also cope easily with modifications, expansions, and failures in the network. For example, suppose that the capacity of a link is modified, a new link is added, or an existing link has failed; then, it is only required to modify the link capacity information at the nodes where such a link originates. (Set it to its new value, or in the case of a failure, set it to zero.) If the link from node i to node j fails, then node j can be removed from the set $\mathbf{D}(i)$ and the ordered sets S_i^k, ($k=1,2,...,i-1,i+1,...,N$). If a new link is added from node i to node j, then the set $\mathbf{D}(i)$ and the ordered sets S_i^k ($k=1,2,...,i-1,i+1,...,N$) must be modified to include node j. If a node fails, then the computations, at the upstream nodes, can be done by assuming that the link to the failed node has failed. If a new node is added, then the ordered sets R_i ($i=1,2,...,N$) are modified to include this node. Assuming that the number of link failures and additions are bounded in a given finite time interval and that the network remains connected, the basic properties of the controller continues to hold: the queues are cleared in finite time in the absence of external flows, and the queue lengths are bounded if the external flow rates are sufficiently small.

REFERENCES

[1] J.M. McQuillan, I. Richer and E.C. Rosen, "The new routing algorithm for the ARPANET", *IEEE Trans. Communications,* COM-28, pp. 711-719, 1980.

[2] L. Tymes, "Routing and flow control in TYMNET", *IEEE Trans. Communications,* COM-29, pp. 392-398, 1981.

[3] D. Bertsekas and R. Gellager, *Data Networks,* Prentice-Hall Inc., Englewood Cliffs, N.J., 1987.

[4] A. Ephremides, "The routing problem in computer networks", in *Communications and Networks,* I.F. Blake and H.V. Poor (eds.), Springer-Verlag, New York, 1986.

[5] T.P. Vum, "The design and analysis of a semidynamic deterministic routing rule", *IEEE Trans. Communications,* COM-29, pp. 498-504, 1981.

[6] J.S. Meditch and J.C. Mandojana, "A decentralized algorithm for optimal routing in data-communication networks", *Large Scale Systems,* vol. 1, pp. 149-158, 1980.

[7] F.H. Moss and A. Segall, "An optimal control approach to dynamic routing in networks", *IEEE Trans. Automatic Control,* AC-27, pp. 327-339, 1982.

[8] P.E. Sarachik and U. Ozguner, "On decentralized dynamic routing for congested traffic networks", *IEEE Trans. Automatic Control,* AC-27, pp. 1233-1238, 1982.

[9] P.E. Sarachik, "An effective local dynamic strategy to clear congested multi-destination networks", *IEEE Trans. Automatic Control,* AC-27, pp. 510-513, 1982.

[10] F. Khorrami and U. Ozguner, "Simultaneous routing and signalization in traffic networks", in *Proceedings of the 22nd Allerton Conference on Communication, Control and Computing,* pp. 211-220, Monticello, Illinois, 1984.

[11] A. Iftar and E.J. Davison, "A decentralized control strategy for dynamic routing", in *Proceedings of the 28th Conference on Decision and Control,* Tampa, Florida, 1989 (to appear).

STABILITY ANALYSIS OF DESYNCHRONIZED SYSTEMS

E. A. Asarin*, M. A. Krasnoselskii*, V. S. Kozyakin** and
N. A. Kuznetsov**

*Institute of Control Sciences, Profsojuznaja 65, Moscow, 117806, USSR
**NPO ASU "Moskva", Bakhrushina 18, Moscow, 113054, USSR

Abstract. Multi-component systems consisting of components working at discrete time instants are considered. It is supposed that instants of different components functioning do not always coincide. Such systems are referred to as desynchronized. Multi-processor systems and computer systems with asynchronous computations may serve as typical examples. The influence of instants of components acting over the system stability is studied. Three situations, substantially differing by both methods and results of investigation, are considered. The former deals with the case when the system's components work periodically, the periods in general not coinciding. The second situation embraces the case when the components switching instants are random. The latter situation concerns the case when we know nothing about laws describing instants of components work. The results of the investigation demonstrate that the type of desynchronization can essentially affect the dynamics of the system. Sometimes desynchronized systems possess some properties not possessed by synchronized systems.

Keywords. Mathematical system theory; stability; discrete time systems; stochastic systems; linear systems.

INTRODUCTION

Control systems including sampled data elements are traditionally studied in the control theory (Tsypkin, 1961; Kalman, Falb and Arbib 1969). Analysis of such systems is quite difficult in the case that some elements or subsystems work nonsynchronously with other ones. Recently the interest for such desynchronized (or asynchronous) systems had increased considerably, mainly due to the progress in multiprocessor computing systems (Beletskii, 1986; Bertsekas and Tsitsiklis, 1988). Classical mathematical methods turned out to be to a great extent unfit for analysis of desynchronized systems. New approaches and methods were to be developed. From the beginning of the 80-th questions of analysis of desynchronized systems' dynamics engross the attention of a group of mathematicians at Institute of Control Sciences and NPO ASU "Moskva" (M.A.Krasnoselskii, N.A.Kuznetsov, A.F.Kleptsyn, V.S.Kozyakin, A.M.Krasnoselskii and E.A.Asarin). In this report some results established by the group are presented. Most of these results hold for both linear and nonlinear systems. Nevertheless we will discuss only results concerning linear desynchronized systems. As we got to know recently analogous investigations had been conducted by some other authors: Baudet (1978), Bertsekas and Tsitsiklis (1988). They obtained some similar results.

DESYNCHRONIZED SYSTEMS

Consider a system W consisting of components (elements, subsystems, parts etc.) $W_1,\ldots,W_N$. Let the state of any component W_i be described by a vector $x_i \in \mathbb{R}^{n_i}$, $n_i \geq 1$, and be changed at some discrete instants according to the rule

$$x_{i\ new} = a_{i1}x_1 + a_{i2}x_2 + \ldots + a_{iN}x_N + f_i, \qquad (1)$$

where a_{ij} are matrices of corresponding dimensions, f_i is a so called vector of external influences. Let us denote instants of change of the state of W_i by $\ldots < T_{i0} < T_{i1} < \ldots < T_{in} < \ldots$. Then changes of the variable state x_i of the component W_i can be expressed by the equation

$$x_i(T_{i\,n+1}) = a_{i1}x_1(T_{in}) + \ldots + a_{iN}x_N(T_{in}) + f_i(T_{in}), \qquad (2)$$

where the function $x_i(t)$ is assumed to be constant on any interval $T_{in} < t \leq T_{i\,n+1}$. From physical reasons it is natural to assume that $T_{in} \to \infty$ when $n \to \infty$. The moments T_{in} will be referred to as the switching times of the component W_i.

If all the components are switched simultaneously, i.e.

$$T_{1n} = T_{2n} = \ldots = T_{Nn} \qquad (-\infty < n < \infty),$$

then the system W is referred to as *synchronized*. Otherwise it is called *desynchronized*. For example the system is desynchronized if the switching times of its components are expressed by the following formula:

$$T_{in} = \tau_i n + \varphi_i \qquad (i=1,2,\ldots,N; \ -\infty < n < \infty).$$

Such systems will be called *phase and frequency desynchronized* (Kleptsyn and colleagues, 1983-1984, 1984b). In some cases it is reasonable to assume that the components of the system W are switched at random instants. Then probabilistic methods can be used to investigate the dynamics of the system. Eventually situations are possible when no information concerning rules of switching of the components of the system W is available. This situation (if the researcher is interested in stability problem for W) leads to the idea of absolute stability (cf. Aizerman and Gantmakher, 1963) of the system W under some class of desynchronizations.

Sometimes a discrete model of desynchronized system is more convenient then (2) (see also Baudet, 1978). Generally it is possible that one, two or more components are switched simultaneously. Let $\omega \subseteq \{1,2,\ldots,N\}$ be a set of their indices. Denote by A_ω the block matrix defined by the following conditions. If $i \in \omega$ then the i-th string of A_ω coincides with the corresponding string of the matrix $A=(a_{ij})$. Otherwise it coincides with the corresponding string of the identity matrix I. For example in the case when $\omega=\{i\}$ the matrix A_ω is as follows

$$A_\omega = A_i = \begin{bmatrix} I & \ldots 0 & \ldots 0 \\ & \cdot\cdot\cdot & \\ a_{i1} & \ldots a_{ii} & \ldots a_{iN} \\ & \cdot\cdot\cdot & \\ 0 & \ldots 0 & \ldots 0 \end{bmatrix}.$$

Denote by X the state space of the system W, i.e. the set of vectors $x=(x_1,x_2,\ldots,x_3)$ where $x_i \in \mathbb{R}^{n_i}$. Denote by X_ω the linear subspace of the space X consisting of all the vectors $x=(x_1,x_2,\ldots,x_3)$ with $x_i=0$ for any $i \notin \omega$. In the general case the change of the state of the system W is described by the vector equation

$$x_{new}=A_\omega x+F_\omega \quad \text{where} \quad F_\omega=(f_1,f_2,\ldots,f_N) \in X_\omega.$$

Let $\ldots <T_0 <T_1 <\ldots <T_n <\ldots$ be all the switching times of all the components of the system W. Denote the state vector of W at the time T_n by $x(n)$; denote the set of the indices of all the components switching at the instant T_n by $\omega(n)$. Then the following equation describes the dynamics of W

$$x(n+1)=A_{\omega(n)}x(n)+F(n), \quad F(n) \in X_{\omega(n)}. \quad (4)$$

If W is a synchronized system then $\omega=\{1,2,\ldots,N\}$. The dynamics of the synchronized system W is described by the equation

$$x(n+1)=Ax(n)+F(n). \quad (5)$$

Suppose now that there is no external influence on the system W (i.e. $F(n)\equiv 0$). Then the question of stability of the equilibrium $x=0$ naturally arises. It is the main question of this report.

The question of stability of a synchronized system W can be solved in terms of eigenvalues of its matrix A. The classic theorems providing such a solution can be found e.g. in (Tsypkin, 1963).

PHASE AND FREQUENCY DESYNCHRONIZATION

Let all the values of τ_i in the formula (3) be equal. Then

$$T_{in}=\tau n+\varphi_i \quad (i=1,2,\ldots,N; \ -\infty<n<\infty) \quad (6)$$

and the system is called *phase desynchronized*. Let

us note that in this case the equation

$$x(n+1)=A_{\omega(n)}x(n) \quad (7)$$

is N-periodic in n. Hence, as it is well known (e.g. Tsypkin, 1963) the question about stability of W can be solved in terms of eigenvalues of the matrix

$$C=A_{\omega(N)}A_{w(N-1)}\cdots A_{\omega(1)}. \quad (8)$$

Provided the values of phase shifts φ_i in (6) satisfy the condition

$$0\leq\varphi_1<\varphi_2<\ldots<\varphi_N<\tau,$$

there is no need to compute the matrix C because its eigenvalues coincide with the roots of the equation

$$\det \begin{bmatrix} a_{11}-\lambda & a_{12} & \ldots & a_{1N} \\ \lambda a_{21} & a_{22}-\lambda & \ldots & a_{2N} \\ & \cdot\cdot\cdot & & \\ \lambda a_{N1} & \lambda a_{N2} & & a_{NN}-\lambda \end{bmatrix} = 0.$$

Note that from the computational viewpoint the iterative procedure (7) is similar to the Gauss-Seidel's procedure of solving the equation $x=Ax$.

In general the matrix C differs from A. It is important to stress that each of the matrices A and C can be stable or unstable (in the sense of stability of the corresponding system) independently from the stability of another. Note also that values of phase shifts φ_i do not affect the matrix C. This matrix depends only on their mutual positioning.

Thus when analysing desynchronized systems one must take into account the influence of arbitrarily small desynchronization onto the stability of the synchronized system. A stable system can become unstable and vice versa.

If the system (7) is phase and frequency desynchronized, i.e. not all the values of τ_i in (3) coincide, then the analysis of its stability becomes a problem of considerable difficulty. The authors do not know any effective necessary and sufficient condition of stability of such systems.

Brilliant results concerning stability of two-component systems W were obtained by A.F.Kleptsyn (1985). The step k of Kleptsyn's algorithmical procedure consists in computation of a number $\lambda(k)$. This number is an estimate of the mean rate of decrease of the norms of vectors $x(n)$ in the equality (7). It turned out that any two-component desynchronized system W with non-commensurable switching periods τ_1 and τ_2 of its components is asymptotically stable if and only if at some step of the Kleptsyn's algorithm the inequality $\lambda(k)<1$ holds.

The initial proof of the algorithm was rather complicated. A simpler proof heavily based on ideas of the symbolic dynamics (see Birkhoff, 1927) is due to one of the authors of the report. This new proof made possible to obtain the following assertion.

Theorem 1. *If a two-component phase and frequency desynchronized system W with a non-degenerate matrix A is asymptotically stable with the set of parameters $\tau_1,\tau_2,\varphi_1,\varphi_2$ where τ_1 and τ_2 are non-commensurable then W is asymptotically stable with any set of parameters $\tau_1,\tau_2,\varphi_1,\varphi_2$ with τ_i*

118

close to τ_i and arbitrary (!) values of phase shifts φ_i.

Unfortunately the fruitful work of A.F.Kleptsyn was stopped by his tragic death.

By now it is not clear whether an analogy of the important theorem 1 holds for multi-component systems or not. One of possible ways to analyze a multi-component phase and frequency desynchronized system is as follows. Let us divide the real axis by intervals of equal lengths, for example $[nH,(n+1)H)$, where $H>0$, $-\infty<n<\infty$. Denote by D_n the transition matrix from the state of the system W at the instant nH to its state at the instant $(n+1)H$. In the situation considered there are only a finite set $\{F_1,F_2,\ldots,F_L\}$ of different matrices D_n. Denote by f_{lk} the number of matrices F_l in the sequence of matrices $D_0,D_1,\ldots,D_k$. Then for any $l=1,2,\ldots,L$ there exists a finite limit

$$p_l = \lim_{k\to\infty} (f_{lk}/k)$$

This limit can be treated as the mean frequency of occurrences of the matrix F_l in the sequence $\{D_k\}_{k=0}^{\infty}$. Provided the periods $\tau_1,\tau_2,\ldots,\tau_N$ are non-commensurable, the matrices F_l and the frequencies p_l can be evaluated effectively (Kleptsyn and colleagues, 1984b) using geometric considerations based on ergodic theory.

Theorem 2. *Let* $\|F_1\|^{p_1}\|F_2\|^{p_2}\ldots\|F_L\|^{p_L}<1$ *and the periods* $\tau_1,\tau_2,\ldots,\tau_N$ *of a phase and frequency desynchronized system W are noncommensurable. Then the system W is asymptotically stable.*

STOCHASTICALLY DESYNCHRONIZED SYSTEMS

In some situations a model with random switching times of components is more adequate. Consider a system W with possibly multidimensional state vectors of components. Assume that for any $i=1,2,\ldots N$ the sequence $\{T_{in}\}$ of switching instants of the i-th component forms a simplest flow of events with intensity λ_i, these flows being independent for different i. Such a system is referred to as *stochastically desynchronized.*

Denote by $D(t)$ the random transition matrix from the state of W at time 0 to its state at time t. It can be found from the equality

$$D(t)=A_{i_k} A_{i_{k-1}} \ldots A_{i_1},$$

where $i_1,i_2,\ldots,i_k$ stand for numbers of components switched during the interval $(0,t]$. According to an ergodic theorem due to Furstenberg and Kesten (1960) there exists a non-random number μ satisfying with probability 1 the following equality

$$\mu= \lim_{t\to\infty} t^{-1}\ln\|D(t)\|.$$

The number μ is called *Lyapunov index* of the system W. Let us say the system W is *L-stable* iff $\mu<0$.

Three methods of estimation of Lyapunov indices were developed. The first one consists in straightforward computer modelling of the desynchronized system (more precisely, of its discrete time analogy).

The second method has a much less computational complexity than the first one. It is based on the following algebraic construction. Consider a matrix

$$Q=\sum_{i=1}^{N} \lambda_i A_i^{\otimes 2} \Big/ \sum_{i=1}^{N} \lambda_i,$$

where $A_i^{\otimes 2}$ stands for the Kroneker square (Lancaster, 1969) of the matrix A_i. Then the following estimate of the Lyapunov index holds.

Theorem 3. *The inequality*

$$\mu\le \ln \rho(Q) \sum_{i=1}^{N} \lambda_i /2,$$

$\rho(Q)$ *standing for the spectral radius of Q, holds for any stochastically desynchronized system.*

Thus the inequality $\rho(Q)<1$ is a sufficient condition of L-stability of the system W. If this condition holds one can estimate the rate of convergence of $\|D(t)\|$ to zero in probability.

Theorem 4. *Let* $\rho(Q)<\sigma<1$. *Then there exists a number q such that for any* $t,\varepsilon>0$ *the following inequality holds*

$$\mathbb{P}\{\|D(t)\|>\varepsilon\} < q\varepsilon^{-2}\exp\left[-(1-\sigma)t\sum_{i=1}^{N}\lambda_i\right].$$

The computational experiment has demonstrated that the estimate given by theorem 3 is not very accurate. However it is much more accurate than the trivial estimate

$$\mu\le \sum_{i=1}^{N} \lambda_i \ln\|A_i\|,$$

the latter estimate not taking into account the system's geometry. The same experiment has revealed effects of synchronized (or phase desynchronized) systems gaining or loosing stability under stochastical desynchronization.

The third method of checking the L-stability and estimating the Lyapunov index can be used if the system satisfies some special conditions and geometric characteristics of the system's dynamics can be investigated with more details. The next part of the report contains such an investigation.

REGULAR STOCHASTICALLY DESYNCHRONIZED SYSTEMS

By now the authors have succeeded in understanding the behaviour of stochastically desynchronized systems satisfying the following conditions.

A. The system W is two-dimensional and two-component.

B. All the elements of the matrix

$$A=\begin{bmatrix} a_{11} & a_{12} \\ a_{21} & a_{22} \end{bmatrix}$$

are positive.

C. The matrix A is "far" from the identity in the following sense. The inequality

$$a_{21}a_{12}>f(a_{11},a_{22})$$

holds, where

$$f(x,y)=\begin{cases} 1 & \text{if } x\ge1, \ y\ge1 \\ 1-x(1-y) & \text{if } x\ge1, \ y<1 \\ 1-y(1-x) & \text{if } x<1, \ y\ge1 \\ \max\{xy,(1-x)(1-y)\} & \text{if } x<1, \ y<1 \end{cases}$$

119

Systems W satisfying conditions A-C are referred to as *regular*. It is convenient to pass to the polar coordinates (ρ,φ) for the analysis of dynamics of regular systems. It turns out that for any regular system W there exists an invariant measure $\nu(\varphi)$ on the interval $[0,\pi/2]$. The distribution law of the polar angle of the state vector of the system tends to $\nu(\varphi)$ when $t\to\infty$, for any initial distribution law (at t=0). The support $\mathfrak{M}$ of the measure ν is nowhere dense, its Hausdorf dimension (see e.g. the survey by Farmer, Ott and Yorke, 1983) is fractal. A quickly converging algorithm (resembling the well-known construction of the Cantor's ladder) of computation of the measure ν is proposed.

When knowing the measure ν it is easy to evaluate the Lyapunov index of the system W this index being equal to the integral of some explicit function of φ by the measure ν. It is possible to answer other natural questions about the behaviour of regular stochastically desynchronized systems. It is worth mentioning that it turned out possible to investigate the dynamics of such systems, including the above-stated properties of the measure ν and its "strange" support $\mathfrak{M}$, at the mathematical level of strictness (cf. Farmer, Ott and Yorke, 1983).

ABSOLUTE STABILITY OF DESYNCHRONIZED SYSTEMS

Assume that while analyzing the stability of desynchronized systems one knows nor the amount of components switched at any time instant neither the law of alternation of switching times of different components. Then one naturally comes to the notion of absolute stability (cf, Aizerman and Gantmakher, 1963) of desynchronized systems under some class of desynchronizations. Let us note that unlike classical situations the problem of absolute stability of desynchronized systems is non-trivial even for linear systems.

The formal definition shall be given in terms of the equation (7). Denote by $\mathfrak{P}(A)$ the set of all the matrices A_ω with $\omega\subseteq\{1,2,\ldots,N\}$. Let $\mathfrak{J}$ be a non-empty subset of the set $\mathfrak{P}(A)$. The equation (7) (or the system W) is referred to as *absolutely stable under the class $\mathfrak{J}$ of desynchronizations* if and only if all its solutions, corresponding to various sequences of matrices $A_{\omega(n)}\in\mathfrak{J}$ and to various initial values, are bounded. Let us say that the sequence of matrices $A_{\omega(n)}\in\mathfrak{J}$ switches any component of the system W infinitely many times if any number $i=1,2,\ldots,N$ belongs to infinitely many sets $\omega(n)$ ($n\geq0$). The equation (7) (or the system W) is called *absolutely asymptotically stable under the class $\mathfrak{J}$ of desynchronizations* if all its solutions, corresponding to various initial values and various sequences of matrices $A_{\omega(n)}\in\mathfrak{J}$ switching any component of the system infinitely many times, tend to zero when $n\to\infty$. It is clear that the problem of absolute asymptotical stability has some sense only if there exists at least one sequence of matrices $A_{\omega(n)}\in\mathfrak{J}$ switching any component of the system W infinitely many times. Therefore assume that the set $\mathfrak{J}\subseteq\mathfrak{P}(A)$ contains some matrices $A_{\omega_1},A_{\omega_2},\ldots,A_{\omega_k}$ satisfying the condition

$$\omega_1\cup\omega_2\cup\ldots\cup\omega_k=\{1,2,\ldots n\}.$$

Such a set $\mathfrak{J}$ is called *generating*. For example the set $\mathfrak{P}_k(A)$ (with any $k=1,2,\ldots,N$) of all the matrices A_ω such that the set ω contains no more

than k elements.

Various authors (Kleptsyn and colleagues, 1984a; Bertsekas and Tsitsiklis, 1988; Beletskii, 1988) proposed sufficient conditions of absolute stability of systems W with scalar component states and symmetric or non-negative matrix A. Let us formulate some more general results about necessary and sufficient conditions of absolute stability of such systems.

Theorem 5. *Let the matrix* A *of a system* W *be scalar (i.e. its blocks are scalars) and symmetrical. Then the system* W *is absolutely asymptotically stable under the class* $\mathfrak{P}_k(A)$ *of desynchronizations iff all the eigenvalues of* A *are less than 1 and all the eigenvalues of any its diagonal submatrix of the order* k *are greater than* -1.

In particular this theorem implies that no desynchronization of an asymptotically stable synchronized system with a symmetric matrix leads to the loss of stability.

Theorem 6. *Let the elements of a matrix* $A=(a_{ij})$ *be non-negative. Then the system* W *is absolutely asymptotically stable under a generating class* $\mathfrak{J}$ *of desynchronizations if and only if the spectral radius* $\rho(A)$ *of the matrix* A *is less than 1.*

The theorem 6 implies an important for applications sufficient condition of absolute stability of desynchronized systems with arbitrary block matrices $A=(a_{ij})$. In this case any matrix a_{ij} can be treated as a linear map from some space X_j into X_i. Let some norms $\|\cdot\|_i$ be fixed in the spaces X_i ($i=1,2,\ldots,N$). Put

$$\|a_{ij}\| = \sup_{x\in X_j,\, x\neq0} \frac{\|a_{ij}x\|_i}{\|x\|_j}.$$

Theorem 7. *If* $\rho(S)<1$, *the scalar matrix* S *being defined by the equality* $S=(\|a_{ij}\|)$, *then the system* W *with the matrix* A *is absolutely asymptotically stable under any generating class of desynchronizations.*

It is important to note that for any solution $x(n)$ of a an absolutely stable desynchronized system W the following estimate holds

$$\|x(n)\|\leq C\|x(0)\| \qquad (n=1,2,\ldots),$$

where the constant C does not depend on the sequence of matrices $A_{\omega(n)}$ (in the equation (7)) corresponding to the solution $x(n)$. If a desynchronized system W is absolutely asymptotically stable, then for any solution of the equation (7) the following inequality holds

$$\|x(n)\| \leq Cq^{-\mathfrak{æ}(\{\omega(\cdot)\},n)}\|x(0)\| \qquad (q<1)$$

where the constants C and q do not depend on the sequence of matrices $A_{\omega(n)}$ corresponding to the solution $x(n)$. The expression $\mathfrak{æ}(\{\omega(\cdot)\},n)$ denotes the maximum number of disjoint subintervals of the interval $[0,n]$, on each of which the sequence of matrices $A_{\omega(n)}$ switches all the components of W.

CORRECTNESS OF ABSOLUTE STABILITY

It is important to know whether the property of absolute asymptotical stability of a linear desynchronized system W is conserved under little perturbation of the system's matrix. The answer to the analogous question about synchronized systems is positive and the proof is trivial (see, e.g. Tsypkin, 1963). The answer for desynchronized systems turned out to be positive too. However the proof of this fact is complicated and requires special mathematical constructions.

Theorem 8. *Let a desynchronized system W with a matrix A be absolutely asymptotically stable under the class of all the desynchronizations. Then any desynchronized system with a matrix A close to the matrix A is absolutely asymptotically stable.*

The proof of this theorem is essentially based on the following deep fact. The property of absolute asymptotical stability of a system is equivalent to the property of its absolute stability under constantly acting perturbations. We say that a desynchronized system W is absolutely stable under constantly acting perturbations and a class of desynchronizations $\mathfrak{F}$ if any solution of the non-homogeneous equation (4) with matrices $A_{\omega(n)} \in \mathfrak{F}$ and $F(n)$ coordinated with the matrices: $F(n) \in X_{\omega(n)}$, $\|F(n)\| \leq 1$, is bounded when $n \geq 0$.

Let α be a set of integers $i_1 < i_2 < \ldots < i_k$ from the interval $[1,N]$. Consider a subsystem W^{α} of the system W, consisting of its components W_i with numbers $i \in \alpha$. In this case items $a_{ij} x_j$ with indices $j \notin \alpha$ in the equation (1) describing the changes of state of the component W_i can be considered as "external influences". Identify the state space of the system W^{α} with X_{α} and denote by A^{α} the result of replacement by zero of all the elements a_{ij} of the block matrix A with $i \notin \alpha$ or $j \notin \alpha$. Then the equation of dynamics of the subsystem W^{α} takes the form analogous to (4)

$$x(n+1) = (A^{\alpha})_{\omega(n)} x(n) + F(n),$$

where $\omega(n) \subseteq \alpha$, $x(n) \in X_{\alpha}$, $F(n) \in X_{\omega(n)} \subseteq X_{\alpha}$.

The following theorem expresses an important property of absolutely asymptotically stable desynchronized systems without analogies for synchronized systems.

Theorem 9. *Let a desynchronized system W be absolutely asymptotically stable under the class of all the desynchronizations. Then any its subsystem W^{α} is absolutely asymptotically stable under the class of all the desynchronizations.*

The method of proof of theorems 8 and 9 was found to be helpful to state a new condition of absolute stability of desynchronized systems. Let the states of all the components of a system W be scalars. If the matrix A of the system W is symmetric and its eigenvalues belong to the interval $[-\rho,\rho]$ where $\rho<1$, then by theorem 5 the system W is absolutely asymptotically stable under the class of all the desynchronizations. Now let A=B+C, the matrix B being symmetric with eigenvalues in the interval $[-\rho,\rho]$ where $\rho<1$, the matrix C being antisymmetric. Denote the spectral radius of the matrix C by r.

Theorem 10. *Let*

$$r < \rho \sqrt{\frac{\rho-1}{\rho+1}} \left(\frac{1}{\sqrt{1-(1-\rho^2)^N}} - 1 \right). \qquad (9)$$

Then the system W with the matrix A=B+C is absolutely asymptotically stable under the class of all the desynchronizations.

If $\rho=0$, i.e. the matrix A is antisymmetric, then the condition (9) takes the form $r < N^{-1/2}$. This condition is rough enough even for N=2.

CONCLUSION

Many important for applications phenomena, situations and processes can be described in terms of desynchronized systems. Many properties of synchronized systems hold for desynchronized systems too. However the authors' experience has demonstrated that proofs of the most "obvious" properties of desynchronized systems offer major difficulties and require special mathematical technics and non-traditional approaches. It is necessary to stress that desynchronized systems can possess some important properties not possessed by synchronized ones. This implies the possibility to solve applied problems by means of intentional desynchronization of the work of a system's components. The mathematical theory of desynchronized system by no means can be considered as formed - it is at the very beginning of its development. We hope that our report will attract attention to this interesting and promising, at our opinion, area of control science.

REFERENCES

Aizerman, M.A., and F.R.Gantmakher (1963) *Absolute Stability of Remote-control Systems.* AN SSSR, Moscow (in Russian).

Baudet, G.M. (1978) Asynchronous iterative algorithms for multiprocessors. *J. Assoc. Comput Mach.*, **22**, 226-244.

Beletskii, V.N. (1988). *Multiprocessor and Parallel Structures with Asynchronous Computations Organization.* Naukova Dumka, Kiev (in Russian).

Bertsekas, D.P., and J.N. Tsitsiklis (1988). *Parallel and Distributed Computation. Numerical Methods.* Prentice Hall, Englewood Cliffs.

Birkhoff, G.D. (1927) *Dynamical Systems.* AMS, Providence, Rhode Island.

Farmer, J.D., E.Ott, and J.A.Yorke (1983). The dimension of chaotic attractors. *Phys.D.*, **7D**, 153-180.

Furstenberg, H., and H.Kesten (1960). Products of random matrices. *Ann. Math. Stat.*, **31**, 457-469.

Kalman, R.E., P.L.Falb, and M.A.Arbib (1969). *Topics in Mathematical System Theory.* McGraw-Hill Book Company, New York.

Kleptsyn, A.F., V.S.Kozyakin, M.A.Krasnoselskii, and N.A.Kuznetsov (1983-1984). On the effect of small synchronization errors on stability of complex systems. I-III. *Avtomatika i Telemekhanika*, 1983: **7**, 44-50; 1984: **3**, 42-47; **8**, 63-67(in Russian, Engl. transl. in *Automation and Remote Control*).

Kleptsyn, A.F., V.S.Kozyakin, M.A.Krasnoselskii, and N.A.Kuznetsov (1984a) Stability of desynchronized systems. *Dokl. AN SSSR*, **274**, 1053-1056 (in Russian, Engl. transl. in *Sov. Math. Dokl.*).

Kleptsyn, A.F., M.A.Krasnoselskii, N.A.Kuznetsov, and V.S.Kozjakin (1984b). Desynchronization of linear systems. *Math. and Comput. in Simulation*, **26**, 423-431.

Lankaster, P. (1969). *Theory of Matrices.* Academic Press, New York.

Tsypkin, Ja.Z. (1963). *Theory of Linear Sampled-data Systems.* Fizmatgiz, Moscow (in Russian).

ON GENERATING VARIABLE STRUCTURE ARCHITECTURES FOR DISTRIBUTED INTELLIGENCE SYSTEMS[1]

J. J. Demaël and A. H. Levis

Laboratory for Information and Decision Systems, Massachusetts Institute of Technology, Cambridge, MA 02139, USA

ABSTRACT. A quantitative approach for modeling and generating variable structure distributed intelligence systems is presented. In these systems, the interactions between the objects can change depending on the task being processed. Colored Petri Nets are used as the appropriate mathematical framework for representing design requirements and for modeling the variable structures. The set of variable structures that satisfy both the design requirements and some generic constraints is characterized, and an algorithm for solving the design problem is described. An application to a non trivial example for an Air Traffic Control System is outlined.

Keywords. Distributed Intelligence Systems, Colored Petri Nets.

INTRODUCTION

Recent developments in the theory of distributed intelligence systems have addressed the problem of developing architectures whose performance would meet specific requirements (Levis, 1988). Two major issues have been identified, the system design problem and the system control problem. The former refers to the task of designing a system given some constraints or requirements, as imposed by the users and by the state of technology. The latter refers to the need to control the system once a design has been chosen, so as to reach a desired level of performance. Issues related to the design and control of fixed structure architectures are becoming well understood. In fixed structure systems, the interactions between intelligent nodes are fixed and well defined. To meet ever increasing requirements from users for reliability and reconfigurability, systems that adapt their structure of interactions between components to the task being processed are being considered. As some patterns of interactions may be more suitable for processing a given input than others, a properly designed variable structure system with an efficient control can be expected to achieve a higher overall performance, provided that it adapts its structure to the most appropriate interactions for each type of input.

Theoretical and practical evidence point to the fact that the control problem can be more easily solved if the design of the plant has included some desirable properties. A good plant design, for example, may eliminate some difficult control problems. However, there are very few references in the literature on methodologies for the generation in some orderly manner of plant designs that are not just variants of a single structure, given the constraints relevant to the problem being investigated. There is indeed need for a framework that addresses quantitatively the generation of system architectures. In generating distributed system architectures, designers face the problem that the computation of *all* the systems that satisfy the constraints of the design is intractable, even for systems with a small number of resources. An indirect approach to the generation of fixed structure organizations has been developed (Remy, 1988) in which Petri Net theory was used to express their problem in mathematical terms. Then, an algorithm was given for generating all feasible architectures and characterizing them in terms of lattices of fixed structures that satisfy the designer's requirements. The computational requirements are very modest.

This paper presents a major extension of the earlier work by addressing the problem of designing *variable structure* distributed intelligence systems (Demaël, 1989) for a well defined set of systems, those that process deterministically and repetitively a set of simultaneous observations. An appropriate mathematical framework is defined based on the theory of Colored Petri Nets (Jensen, 1987), an extension of Petri Net theory (Reisig, 1985). As for Ordinary Petri Nets, a Colored Petri Net is a graph with two sets of nodes, places and transitions. A transition, denoted by a bar node, describes a process, while a place, denoted by a circle node, models a (communication) buffer between two processes. The precedence relations between places and transitions are described by the links or arcs of the net. In CPN theory, the tokens represent messages and have an identity, i.e., attributes which contain some information about the message. The variable interactions between processes are described by annotating the links of the net; some tokens are allowed to be carried by an link, while some others cannot pass. The annotation of the links is based on the language of Linear Algebra.

In the next sections, the key elements of the methodology are presented. First, a framework is described for representing variable structure as well as fixed structure systems. The third section describes properties of variable structures in the language of Colored Petri Nets. The fourth section describes the constraints that must be verified by a variable structure to make physical sense for a particular design problem. In the fifth section, an algorithm that characterizes the set of solutions is outlined, while in the sixth section a non-trivial example of the methodology is presented.

MATHEMATICAL MODEL

A Distributed Intelligence System (DIS) is seen as an information processing system that must perform several functions to accomplish its mission (Minsky, 1986; Levis, 1988). The functions are divided into individual tasks, the *roles*. Each role is a series of repetitive procedures that are prescribed by the requirements of the mission, so that each object's activity contributes a little to each of the several functions. Each role is performed by a human decision maker or an intelligent computerized node. The inputs to the system are the observations made by the *sensors*. These items of information are transmitted to the proper destinations within the system, they are analyzed, and the selected response is implemented by the effectors. The model is restricted to observations that are *temporally consistent*, they refer to the same temporal origin, i.e., to an event with a specific time of occurrence (Grevet, 1988). It is further assumed that the processing of one set of simultaneous observations is *deterministic*,

---

[1] This work was carried out at the MIT Laboratory for Information and Decision Systems with support provided by the U.S. Office of Naval Research under contract no. N00014-84-K-0519.

it is achieved while involving a unique set of interactions.

A DIS has a *variable structure* if the interactions between roles can vary. Conversely, a system for which the interactions cannot vary is said to have a *fixed structure*. This paper is restricted to systems whose variability is triggered exclusively by the information contained in the sensors' observations. The goal of the methodology is to create a Colored Petri Net (CPN) model of the flow of data from the sources to the roles, the exchange of information between roles, and the communication of messages to the effectors. A CPN can then be used to assess the effectiveness of a structure, using the System Effectiveness Analysis methodology as described in Monguillet (1988).

Sensors: A DIS processes data from N sources of information, i.e., N sensors labeled Sensor 1 to Sensor N. Sensor n can output one signal or symbol from its associated set of possible signals, its output alphabet $Xn = \{xn_1, ..., xn_{|Xn|}\}$, which contains $|Xn|$ elements. In the Petri Net formalism, each independent Sensor is modeled by a place, as represented in Fig. 1. A transition models the communication of the sensor's observations. The temporal consistency of the observations is embedded in the fact that all sensors are the output of a single process. This process has a single input place p0, which is called the external place.

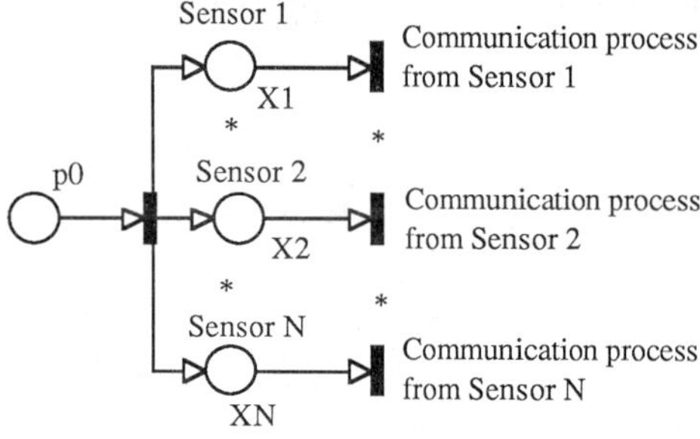

Fig. 1 Sensors

From the system point of view, temporal consistency also implies that the input to the system is an N-dimensional vector: $x = (x1, x2, ..., xN)$. This vector has as components the N independent observations and belongs to the alphabet X, cross-product of the source alphabets: $X = X1 \times X2 \times ... \times XN$.

CPN Representation of an Interaction

As indicated before, an interaction is characterized by its pattern of activation over the set of inputs. Therefore, every interaction is described by a *diagonal* $|X| \times |X|$ matrix L.

- $L_{ii} = 1$ if the i-th input in the lexicographic ordering of X activates the interaction.
- $L_{ii} = 0$ if the i-th input in the lexicographic ordering of X does not activate it.

In the Colored Petri Net model, an interaction is represented by a link between two transitions t1 and t2, as depicted in Fig. 2. The link indicates that the output of process t1 is an input to process t2. The place that belongs to the link is an interactional place, which models a communication buffer. The links are annotated by the matrix L.

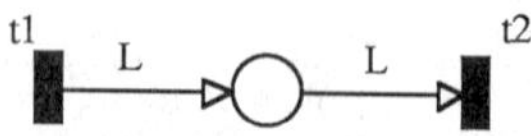

Fig. 2 Link

The tokens in the CPN have an identity and represent messages. A token's identity belongs to X and $x = <x1,...,xN>$ describes a message that has been generated by the set of simultaneous observations $<x1,...,xN>$. The matrix L attached to an interaction (a link) describes the set of tokens that can go through that link. A transition is enabled if and only if all its input places contain messages (tokens) as described by the annotation of its input-links. When a transition fires, tokens of the type indicated by the annotation in the links are taken from the input places and tokens are generated in the output places.

There are three basic types of interactions in a variable structure.
- The *permanent* links. These are the links for which L is the $|X| \times |X|$ identity matrix. Every input requires this interaction to be processed. By convention, these links are depicted without annotation on a CPN model of the system.
- The *inadmissible* links. These are the links for which L is the $|X| \times |X|$ null matrix. No input requires this interaction to be processed. These links are never depicted on a CPN model of the system
- The *variable* links. These are the links for which L has 0s and 1s on the diagonal. Some inputs require this interaction in order to be processed, while some do not.

Permanent links, as well as inadmissible links, are not key elements in generating variable structures. If a link is permanent or inadmissible, the existence of the interaction is not subordinated to the information content of the input while for variable links the decision to interact or not is based on the information content of the input x. The alphabet Xi of Sensor i is said to be *effective*, if the decision to interact is based, in part or in whole, on the output of Sensor i. More formally:

Proposition 1 Given a variable interaction described by a diagonal matrix L, the alphabet Xi is an effective alphabet of the interaction if and only if there exist two signals in Xi, xi_1 and xi_2, such that there exists an input $x = (x1, x2, ..., xN)$ in X that activates the interaction, with $xi = xi_1$ and there exists an input $x' = (x'1, x'2, ..., x'N)$ in X that does not activate the interaction, with $x'i = xi_2$.

Interactions: Each role is modeled by a subnet with four transitions and three internal places, as shown in Fig. 3 (Levis, 1988). The four stage decision making process consists of four algorithms SA, IF, CI, and RS. In Figure 3, x represents an input signal from an external source of information or from the rest of the organization, i.e., from another role. The *Situation Assessment (SA)* algorithm processes the incoming signal to formulate an assessment of the situation. The assessed situation z may be transmitted to other roles. Concurrently, the role may incorporate one or several signals z" from other parts of the system. The signals z and z" are fused together in the *Information Fusion* stage *(IF)* to produce the final situation assessment z'. The next algorithm, the *Command Interpretation* algorithm *(CI)* receives and interprets possible commands (v') from other roles, which restrict the set of responses that can be generated. The CI stage outputs a command v, which is used in the *Response Selection* algorithm *(RS)* together with the assessed situation z to produce the response of the role, the output y. This output can be sent to the effectors and/or to other roles.

The input stage of a role may be SA, IF or CI; these are the stages that can accept external inputs. The final output stage must be RS, the stage in which the role selects its response.

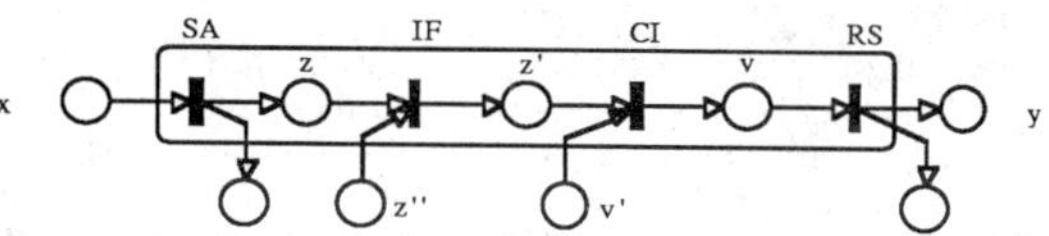

Fig. 3 Four stage model of a role

Every role might not have access to all the sensors' observations. It might base its situation assessment on a restricted number of observations. Fig. 4 depicts schematically the interactions between sensors and roles. S_{ij} models a link between Sensor i and Role j.

Only certain types of interactions between roles make sense within the model (Remy, 1988). They are depicted on Fig. 5. For the sake of clarity, only the links from the i-th role to the j-th role have been represented. The symmetrical links from i to j are valid interactions as well.

The place s_i models the case in which the i-th role communicates the response it has selected to the external environment through the effectors. If Role i sends its response to the effectors, then there exists a link between the RS stage of Role i and the output transition. This output transition has a unique output place, which is called the sink. The place F_{ij} denotes the interaction that occurs when the situation assessment which is produced as an output of the SA stage is sent to the j-th role to be fused with the assessment of the j-th role, and/or assessments from other roles. G_{ij} depicts the case where the response selected by the i–th role is the input of the j-th role. H_{ij} shows the sharing of a result, Role i informs Role j of its final decision. The j-th role may or may not take this information into account. Finally, C_{ij} has been introduced to model hierarchies between roles. It describes the possibility of role i sending a command to role j.

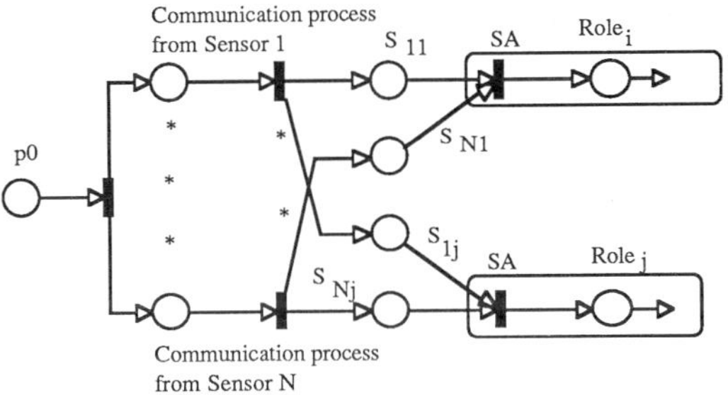

Fig. 4 Interactions roles-sensors

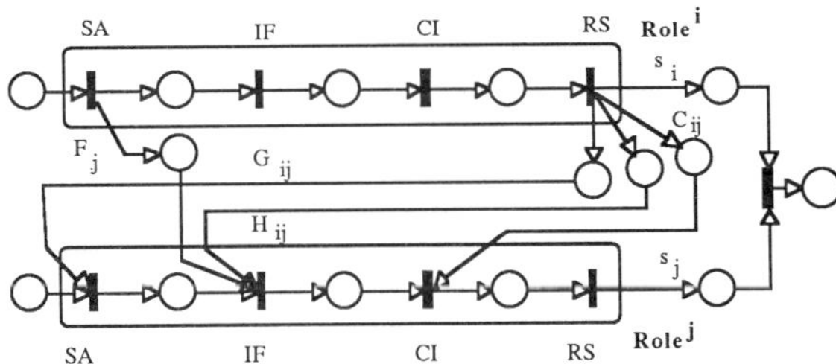

Fig. 5 Allowable interactions between two roles

Matrix Representation of Interactions: The model of interactions in a variable structure DIS can be represented in matrix form. Suppose that a variable structure contains R roles, N sensors, and that its input alphabet is X = X1 × X2 ×...× XN. Then, a DIS is completely determined by the six-tuple:
$$\Pi = (S,\ s,\ F,\ G,\ H,\ C)\ .$$

- S is a N × R block array, which depicts the flow of information from the external environment to the DIS.
- s is a 1 × N block array that depicts the flow of information from the DIS to the effectors.
- F, G, H, C are four R × R block arrays which model the interactions between roles within the DIS.
  - F_{ij} models the link from the SA stage of Role i to the IF stage of Role j.
  - G_{ij} models the link from the RS stage of Role i to the SA stage of Role j.
  - H_{ij} models the link from the RS stage of Role i to the IF stage of Role j.
  - C_{ij} models the link from the RS stage of Role i to the CI stage of Role j.
- Every block in S, s, F, G, H, C is a |X| × |X| diagonal matrix L.
  - $L_{ii} = 1$ if the i-th input in the lexicographic ordering activates the link.
  - $L_{ii} = 0$ if the i-th input in the lexicographic ordering does not activate the link.

For fixed dimensions X, N, and R, the set of all six-tuples Π of dimensions (X, N, R) is called V, the set of Well Defined Variable Structures (WDVS). Let us now turn briefly to fixed structures, which have been studied more extensively in Remy (1988). A fixed structure is determined by two parameters, the number of roles, R, and the number of sensors, N. A fixed structure of dimensions (N, R) can be represented in this model by the six-tuple:
$$\Sigma = (S',\ s',\ F',\ G',\ H',\ C')$$

- S' is an N × R array
- s' is a 1 × N array
- F', G', H', C ' are four R × R arrays.
- Their entries are in {0,1}
  - 1 if the interaction is present
  - 0 if the interaction is not present.

Here again, if N and R are fixed, the set of all six-tuples S of dimensions (N, R) is called W, the set of Well Defined Fixed Structures (WDFS).

PROPERTIES

Some properties of the sets of variable structures, V, and of the set of fixed structures, W, are stated in this section. First, the mathematical framework of the previous section is related to the language of the Petri Net theory, which is used as the formalism within which the design problem is articulated.

Proposition 2 Each element of V can be equivalently described in a matrix form Π or by a Colored Petri Net.

There exists a one to one relationship between the representation of a structure in matrix form and a CPN model of the structure. One can thus work with the language that is most appropriate for one's needs. The proof is as follows. One transition is created for each process in the system, and a link is drawn between any two transitions that interact. This information is provided by the matrix form Π. Note, however, that a DIS contains also internal links, which describe the continuous flow of information within one role, and are not embedded in the matrix form Π. The key proposition is that the internal links are completely determined by the activation of interactional links: a link between two internal processes t1 and t2 within a role exists if and only if t1 has at least one input link - internal or interactional.

Proposition 3 Each element of W can be equivalently described in a matrix form Σ or by an Ordinary Petri Net.

The proof for fixed structures is similar to the proof for variable structures. Next, the theory of variable structures is related to the theory of fixed structures through Proposition 4.

Proposition 4 Any variable structure corresponds to a mapping $\qquad \Pi: X \to W \qquad x \to \Pi(x)$ which associates with each input in X one and only one fixed structure.

In other terms, the Colored Petri Net model of a variable structure can be represented as a mapping from X into the set of Ordinary Petri Nets W. In the Petri Net literature, the decomposition of a CPN into a mapping is referred to as "unfolding" (Genrich and Lautenbach, 1981), and the translation of a mapping into a CPN is called "folding", because it yields a more compact representation. Finally, the sets V and W can be investigated using some partial orderings, which allow one to sort elements in a set, as described by Propositions 5 and 6.

Proposition 5 The set V of variable structures is ordered by the binary relation ACT, where
$$\Pi\ ACT\ \Pi'$$
is equivalent to every input that activates an interaction in Π activates the same interaction in Π', i.e., each interaction in Π has fewer or equal number of activations than in Π'.

The elements of V can be ordered from the ones with the least activation to the ones with the largest activation. Similarly, it is easy to prove Proposition 6.

Proposition 6 The set W of fixed structures is ordered by the binary relation SUB, where
$$\Sigma\ SUB\ \Sigma'$$
means that every interaction in Σ is present in Σ', i.e. the Ordinary Petri Net that represents Σ is a subnet of the Ordinary Petri Net that represents Σ'.

The elements of W can be ordered from the ones that are least connected to the ones that are maximally connected (Remy, 1988).

CONSTRAINTS

Generic Constraints: Any WDVS might not model a DIS that makes physical sense. Some generic constraints must be defined on V, to restrict the class of variable structures to those that are admissible. The generic constraints on V can be divided into two classes. The first class relates the properties of variable structures to the properties of fixed structures, as described in Remy (1988). The second class is specific to variable structures. The set of variable structures that satisfy the generic constraints is called AV, the set of Admissible Variable Structures (AVS).

Let Π be a variable structure. For any x in X, the fixed structure $\Pi(x)$ must satisfy
(R1) (a) The Ordinary Petri Net that corresponds to $\Pi(x)$ should be connected, i.e., there should be at least one (undirected) path between any two nodes in the Net,
(b) A directed path should exist from the external place to every node and from every node to the sink.
(R2) The Ordinary Petri Net that corresponds to $\Pi(x)$ should have no loops, i.e., the structure must be acyclic.
(R3) In the Ordinary Petri Net that corresponds to $\Pi(x)$, there can be at most one link from the RS stage of a role i to another role j, i.e., for each i and j, only one element of the triplet $\{G(x)_{ij}, H(x)_{ij}, C(x)_{ij}\}$ can be non-zero.
(R4) Information fusion can take place only at the IF and CI stages. Consequently, the SA stage of a role can either receive observations from sensors, or receive one and only one response sent by some other role.
(R5) There cannot be one link from the SA stage of role i to the IF stage of role j and a link from the RS stage of role i to the SA stage of role j.

Constraint R1(a) eliminates data flow structures that do not represent a single structure. Constraint R1(b) insures that the flow of information is continuous within the organization from the sensors to the effectors. Constraint R2 allows acyclical fixed dataflow structures only. This restriction is imposed to avoid deadlocks and infinite circulation of messages within the organization. Constraint R3 indicates that it does not make sense to send the same output to the same role at several stages. It is assumed that once the output has been received by a role, this output is stored in its internal memory and can be accessed at later stages. Constraints R4 and R5 ensure that the IF stage is indeed a stage at which items of information coming from different sources are fused and that deadlocks are avoided.

(R6) If the first stage of a role is SA, then each input link in S and G is permanent
(R7) If the first stage of a role i is IF, then each input link F_{ji}, H_{ji} for j in [1..R] is permanent.
(R8) If the first stage of a role i is CI, then each input link C_{ji} for j in [1..R] is permanent.

(R9) Let L be a variable link between two stages t1 and t2, and let us suppose that Xi is an effective alphabet of the variable interaction.
• If t1 is a SA stage and t2 is a IF stage, then there must be in every $\Pi(x)$ a directed path from the place Sensor i to t1, and a directed path from the place Sensor i to the SA stage of the role that contains t2.
• If t1 is a RS stage and t2 is a IF stage, then there must exist in every $\Pi(x)$ a directed path from the place Sensor i to t1, and a directed path from Sensor i to the SA stage of the role that contains t2.
• If t1 is a RS stage and t2 is CI stage, then there must exist in every $\Pi(x)$ a directed path from the place Sensor i to t1, and a directed path from the place Sensor i to the IF stage of the role that contains t2.

Constraints R6, R7, and R8 proceed from a common rationale. They state that a role at its input stage could not have any knowledge about the input to the system and, therefore, cannot exhibit a variable interaction. Constraints R7 and R8 incorporate the fact that the input stage of a role can be the SA, IF, CI stages. Constraint R9 states that a variable interaction between two stages t1 and t2 must be based on sources of information that are accessed jointly by the roles that interact. The stage t1 must determine, based on some information it has accessed, whether or not it has to send a message to t2. Similarly, the role that contains t2 must infer from some of the information it has already received, whether or not it must wait for a message from t1 before initiating process t2. The condition that the information can be accessed is formulated by checking that there is a flow of information (a directed path) from a source that is effective to the stages at which the information contained in the sensor observa-

tion is needed. In other words, an effective source of information must be *accessible*. Constraints R6 to R9 lead to the introduction of a new concept, the accessible pattern of a variable structure, which keeps track of the sensors that are accessible for each potentially variable interaction. Let Π be any AVS. Its *accessible pattern* is a set of arrays:

$$E(\Pi) = [\, F_e(\Pi), H_e(\Pi), C_e(\Pi), s_e(\Pi)\,].$$

• $F_e(\Pi)$, $H_e(\Pi)$, $C_e(\Pi)$ are three $R \times R$ arrays that correspond to the interactions between roles that can be variable.
• $s_e(\Pi)$ is an $R \times 1$ array that corresponds to the links from the roles to the effectors, which can be variable.
• Each array contains accessible alphabets for this interaction.

User - Defined Constraints: A designer can also introduce constraints that reflect his knowledge about the structure under study. He may rule in or rule out some links, force a certain pattern of variability, or express hierarchical relationships between the roles. For example, due to a particular expertise, one might like to indicate that a certain set of observations can only be processed by some roles, while the processing of other sets of observations can be carried out by any role (Stabile, 1984). The designer can translate his knowledge by filling 0s and 1s at the appropriate places in the arrays S, s, F, G, H, C. The other elements will remain unspecified, and constitute the degrees of freedom of the design.

A designer can impose two types of conditions, fixed constraints or colored constraints. Fixed constraints are the constraints that are valid for any input x in X. They can be of two types, ruling in or ruling out. A link is ruled out by putting the $|X| \times |X|$ null matrix O in the appropriate entry of Π. A link is set to be a permanent link by putting the $|X| \times |X|$ identity matrix I in the appropriate entry of Π. Colored constraints are used to rule out a link for some set of observations and rule it in for some other set of inputs. This can be done by filling the appropriate matrix L in Π.

COMPUTATION OF THE SOLUTIONS

The design problem is to determine all the Admissible Variable Structures that satisfy the user-defined requirements. Demaël (1989) proves that this task can be carried out within acceptable computational limits. First, it is proven that the task of determining the set of solutions can be carried out quantitatively using the formalism of the CPN theory. Then, it is shown how to characterize the set of solutions without having to do a computationally expensive, and practically infeasible, exhaustive enumeration. Indeed, the set can be determined from its boundaries, i.e., its

minimal and maximal elements. A solution Π_0 is *minimal* if it is not possible to have a WDVS Π, with Π ACT Π_0, without violating one of the constraints R1 to R9. A solution Π^0 is *maximal* in V if it is not possible to have a WDVS Π, Π^0 ACT Π, without violating one of the constraints R1 to R9. The next propositions lead step by step to a characterization of the set of solutions that can be translated into a design algorithm.

Proposition 7 Consider the fixed structure, called the Universal Net, which contains all the interactions that have not been ruled out at the design specifications stage. Then, for every x, $\Pi(x)$ must be a subnet of the Universal Net.

The rationale is that any $\Pi(x)$ cannot have a link that has been ruled out for any input to the system. Then, the colored constraints must be analyzed to determine the correlation of the links whose variability has been set explicitly by the designer. Indeed, for each variable link that has been specified, the set of inputs X is divided into two subsets, as is the set W. The inputs that activate the variable link L (the inputs in AC) must be assigned to fixed structures that contain the fixed link L (fixed structures in W^1), and the inputs that do not activate the variable link L (the inputs in DC) must be assigned to fixed structures that do not contain the fixed link L (fixed structures in W^0). The fact that the activation of links is correlated is illustrated in Fig.6, two links may not induce the same partitions of X and W.

This analysis yields:
• A partition of X into k *elementary sets of inputs* EX_i.
• A characterization of k disjoint subsets in W, W^i, i = 1..k.

- The condition that each input in EX_i must be assigned to a unique structure in W^i, Σ^i.

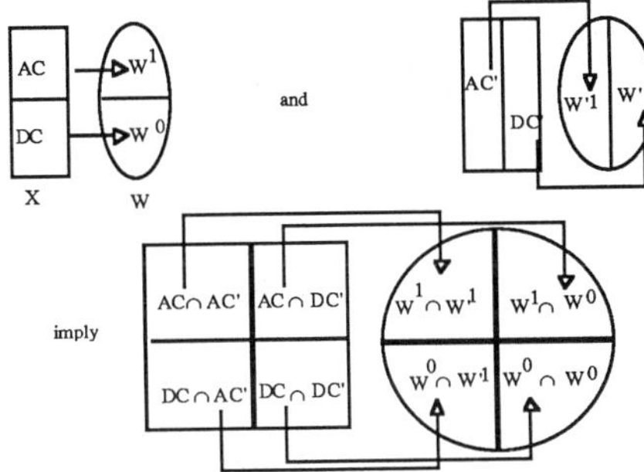

Fig. 6 Correlation of colored constraints

The elements of every subset W^i, $i = 1..k$, can be characterized using *simple information flow paths*. A simple information flow path is a directed path without loops from the external place p0 of the Universal Net to the sink.

Proposition 8 Σ is a fixed structure that belongs to W^i, $i = 1..k$, if and only if Σ is a union of simple information flow paths of the Universal Net and lies between a maximal element and a minimal element in W^i:

$$\exists\ \Sigma_1\ \text{and}\ \Sigma_2\ \text{such that}\ \Sigma_1\ \text{SUB}\ \Sigma\ \text{SUB}\ \Sigma_2$$

The proof comes from Remy (1988). Each W^i is uniquely determined by its minimal and maximal elements. Furthermore, the unit leading in each W^i from a structure Σ to a structure that is immediately above Σ, according to the partial ordering SUB, is a simple information flow path. Furthermore, the Lattice Algorithm of Remy can compute the boundaries in each W^i.

Let a *candidate solution* be an AVS that satisfies all constraints of the design but R6 to R9. Proposition 9 characterizes candidate solutions.

Proposition 9 An AVS verifies all constraints of the design but R6 to R9 if and only if Π lies between a maximal candidate solution and a minimal candidate solution :

$$\exists\ \Pi_1\ \text{and}\ \Pi_2\ \text{such that}\ \Pi_1\ \text{ACT}\ \Pi\ \text{ACT}\ \Pi_2$$
Π_1 is such that for every i, $i = 1..k$,

$EX_i \dashrightarrow \Sigma^i$, with Σ^i minimal element in W^i
Π_2 is such that for every i, $i = 1..k$,

$EX_i \dashrightarrow \Sigma^i$, with Σ^i maximal element in W^i

The proof is a combination of Proposition 4 and Proposition 8. The set of the AVSs that satisfy all constraints of the design but R6 to R9 is thus completely determined by its boundaries. Unfortunately, all AVSs that satisfy Proposition 9 do not fulfill constraints R6 to R9. The first reason is that candidate minimal and candidate maximal solutions may not fulfill one of the constraints R6 to R9. The second reason is that variable structures between the candidate maximal and minimal solutions have variable links in which some effective alphabets are not accessible. However, the set of solutions is completely determined by its minimal and maximal solutions, as described by Proposition 10 and illustrated on Fig. 7.

Proposition 10 The set of solutions can be partitioned into families with each family containing all the structures that have the same permanent input links.
- One family corresponds to a layer of partially ordered sets. In each layer, the AVSs have the same accessible patterns.
- Within one family, Π is a solution if and only if
- Π fulfills Proposition 9.
- Π is bounded by at least one minimal and one maximal solutions that have the same accessible pattern.

$$\Pi_1\ \text{ACT}\ \Pi\ \text{ACT}\ \Pi_2.$$
These accessible patterns are sorted in increasing number of accessible alphabets.
- The boundaries of each layer are determined by the maximal and minimal elements of the family.

A minimal solution to the problem is called a VMINO, whereas a maximal solution to the problem is called a VMAXO. The constraints that are specific to variable structures require that the set of solutions be divided into subsets of solutions with the same input links (Constraints R6, R7, and R8). The VMINOs and VMAXOs determine completely the set of solutions within one family. Between a VMINO and a VMAXO there are several layers of solutions. Each layer has boundaries, and any AVS: 1) between the boundaries, 2) whose variability is based on the corresponding accessible pattern, is a solution to the design problem. Finally, the boundaries are determined by adding selected links of the maximal solution to the minimal solution.

Figure 7 depicts one possible configuration of a set of solutions. There are two families of solutions. Each subset is characterized by some VMINOs and VMAXOs. In the first family, there are two VMINOs and two VMAXOs. From VMINO2 to VMAXO 2, there exist three layers of solutions with the same effective pattern, which correspond to the effective patterns E1, E2, E3.

The computation that allows one to reduce the set of candidate solutions to the set of solutions can be done as follows. First, one determines the number of input links that are allowed by the degrees of freedom of the design. One family of solutions will be determined for each combination. Then, for each family, the computation considers its minimal candidate solutions and its maximal candidate solutions, and searches for solutions that verify constraint R9, which is checked (i) by determining for each variable link its effective alphabets; (ii) if Xi is an effective alphabets, by checking if there exist simple information flow paths, from Sensor i to the stages that interact in each fixed structure Σ^k. If one variable link does not fulfill R9, the structure is rejected.

VMINOs are computed by checking first all the candidate minimal solutions. If some minimal candidate solutions verify R9, then the computation stops. Otherwise, the search continues inductively on the variable structures that are immediately above, until a solution is found. Symmetrically, the VMAXOs are computed by scanning first all the candidate maximal solutions. If some such structures fulfill R9, the search stops. Otherwise, it continues inductively by checking the set of candidate solutions that are immediately below the ones just scanned.

Finally, the intermediate boundaries between one VMINO and one VMAXO are determined as follows. One computes the effective pattern of the VMINO and the VMAXO. If they are equal the search is over. If they are not, the interactions at which the VMAXO has more accessible alphabets are determined. Then, for each pair (interaction I, accessible alphabet Xi) one determines the combinations of simple information flow paths that make Xi accessible at I. These information flow paths, when added to the VMINO, generate the set of intermediate boundaries.

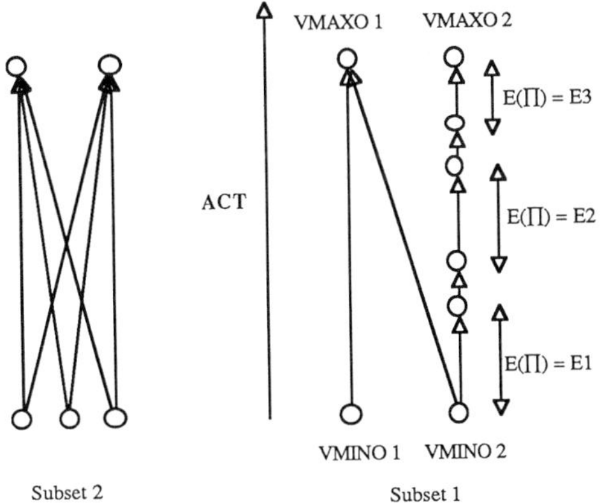

Fig. 7 Structure of a set of solutions

To summarize the computational procedure, we have

127

Step 1:   Given the set of constraints to the design, determine the Universal Net, the elementary sets of inputs EX_i, $i = 1..k$, and the constraints on W^i, $i = 1..k$.

Step 2:   Apply the Lattice Algorithm of Remy to compute the minimal and maximal elements of each W^i, $i = 1..k$.

Step 3:   Use Proposition 4 to obtain candidate minimal solutions and candidate maximal solutions.

Step 4:   Sort the families of solutions. Search for VMINOs and VMAXOs.

Step 5:   Determine the intermediate boundaries of each family.

APPLICATION

This section outlines an application of the methodology to an hypothetical design of a Airport Surface Traffic Control system (ASTC) for Logan Airport in Boston, Massachusetts. ASTC is broadly defined as the portion of the Air Traffic Control system that is responsible for traffic on the runways and taxiways of an airport. The system consists of two control positions, Local Control and Ground Control, which are stationed in the tower cab. One Local Controller handles the traffic on the runways and in the airspace in the immediate vicinity of the airport, while two Ground Controllers handle the traffic on the taxiways. In the example, as depicted in Fig.8, the airport has three terminals and two runways, Runways A and B. The utilization of the runways depends on the direction of the wind, the guiding principle being that landings and takeoffs are done "against the wind." Both runways can be used if the speed of the wind is below a certain limit, however large planes land and take off from Runway A exclusively. Runway B is used solely for general aviation.

It is assumed that the division of Ground Control between two Ground Controllers is done on a geographic basis. One GC, hereafter called GC 1, is responsible for the southern sector of the airport, while the other GC, GC 2, monitors the northern sector. Five sources of information can trigger variable patterns of interaction in the system.

- Sensor 1 : Wind. Wind is a parameter that influences the direction of landings and takeoffs. This source can take five values: $X1 = \{0, E, N, S, W\}$. N (S, E, W respectively) indicates that the wind comes from the North (South, East, West respectively). 0 models low wind conditions in which both runways can be used.

- Sensor 2 : Runway Status. This source of information models the movements on the runways. Note that one and only one runway is used if the output of Sensor 1 is 0, E, N, S or W. Both runways can be used under low wind conditions but the safety standards require that there cannot be at the same time one movement on the Runways A and B.

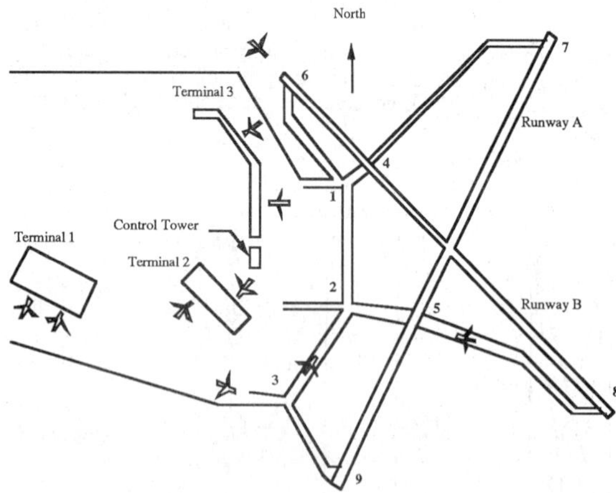

Fig. 8 Plan of Logan airport

- Sensor 3 : Terminals 1 and 2. This source models a plane leaving Terminal 1 or Terminal 2. Its output alphabet is $X3 = \{DP1, NDP1\}$ where DP1 models a departure from Terminal 1 or Terminal 2, and NDP1 indicates that no plane is departing. Note that a departing plane must be directed to a runway. If several runways are used at the same time,

Ground Controller 1 must ask the Local Controller where to direct the plane.

- Sensor 4 : Terminal 3. Similarly, this source models a plane leaving Terminal 3. Its output alphabet is $X4 = \{DP3, NDP3\}$, where DP3 models the case in which one plane is leaving, and NDP3 the case in which no plane leaves. GC2 must ask the Local Controller where to direct the plane in case of low wind conditions.

- Sensor 5 : Conflicts. This source of information models the existence of conflicts at the boundaries of the northern and the southern sectors.

Constraints are expressed to restrict the set of solutions to the design problem from AV to those AVSs that are relevant for ASTC. These constraints are translated into the language of Colored Petri nets on Fig.9. To ease the readability of the net, a link that has been ruled out has not been represented, a link that is permanent is drawn with a bold line without annotations, a link whose variability is imposed has been drawn with a bold line, and a degree of freedom of the design is depicted by a plain line. The matrix L1 indicates an interaction that is activated if and only if $x1 = N$ or W or 0. Matrix L2 annotates an interaction activated for $x1 = S$ or E or 0, while matrix L3 represents an interaction that is activated if and only if $x1 = 0$ and $x4 = DP3$. Finally, matrix L4 annotates an interaction that is activated if and only if $x1 = 0$ and $x3 = DP1$.

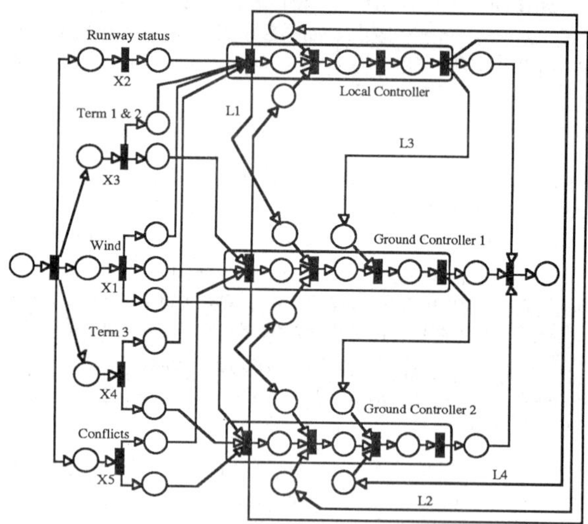

Fig. 9 Colored Petri Net model of the constraints

There are six elementary sets of inputs, which correspond to different patterns of activations for Link 1, the link from SA of LC to IF of GC1; Link 2, the link from SA of LC to IF of GC2; Link 3, the link from RS of LC to CI of GC 1; and Link 4, the link from RS of LC to CI of GC2.

- EX_1 contains all the inputs with $x1 = N$ or $x1 = W$. For those inputs, Link 1 must be activated and Links 2, 3, 4 must not be activated.
- EX_2 contains all the inputs with $x1 = E$ or $x1 = S$. For those inputs, Link 2 must be activated, and Links 2, 3, 4 must not be activated.
- EX_3 contains all the inputs with $x1 = 0$, $x3 = DP1$, and $x4=DP3$. For those inputs, Links 1, 2, 3, 4, must be activated.
- EX_4 contains all the inputs with $x1 = 0$, $x3 = DP1$, and $x4=NDP3$. For those inputs, Links 1, 2, 4, must be activated, while Link 3 is not activated.
- EX_5 contains all the inputs with $x1 = 0$, $x3 = NDP1$, and $x4 = DP3$. For those inputs, Links 1, 2, 3, must be activated, while Link 4 is not activated.
- EX_6 contains all the inputs with $x1 = 0$, $x3 = NDP1$, and $x4= NDP3$. For those inputs, Links 1, 2, must be activated, while Links 3, 4 are not activated.

There are four families for the ASTC system. Only one family is presented here, because of length requirements. This family has a unique VMINO represented in Fig. 10, and a unique VMAXO depicted in Fig. 11. Furthermore there is one and only one intermediate boundary shown in Fig. 12. Five links constitute the degrees of freedom from the minimal solution to the maximal solution: Link A, the link from the SA stage of GC1 to the IF stage of GC 2; Link B, the link from the SA stage of GC2 to the IF stage of GC1; Link C, the link from the RS stage of GC1, to the IF stage of GC2; Link D, the link from the RS stage of GC 1 to the IF stage of LC; and Link E, the link from the RS stage of GC 2 to the IF stage of LC. The highest layer is composed of those structures that are above the intermediate boundary, while the lowest layer is composed of those that do not.

- If Π is above the intermediate boundary, then
  - X1 is the only alphabet accessible for Links A and B.
  - X1, X3, X4 are accessible at Link C.
  - X1, X3 are accessible at Link D.
  - X1, X4 are accessible at Link E.
- If Π is above the minimal solution without Intermediate Boundary ACT Π.
  - X1 is the only alphabet accessible for Links A, B, and C.
  - X1, X3 are accessible at Link D.
  - X1, X4 are accessible at Link E.

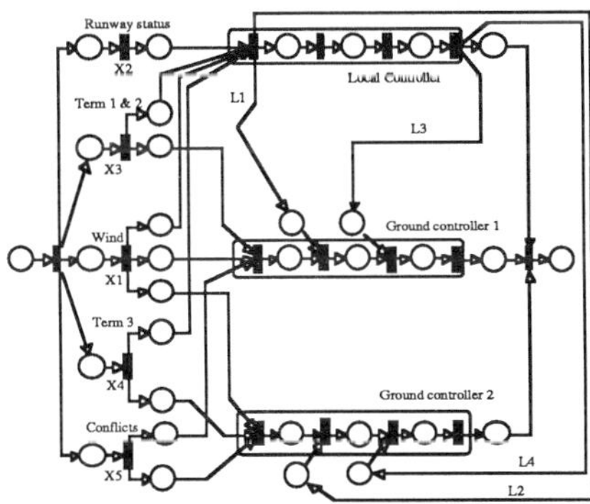

Fig. 10 VMINO

This family of solutions to the ASTC design problem can be interpreted as follows. If Local Control knows when planes leave terminals, then there are many ways to design an organization that coordinates its tasks correctly, as indicated by the degrees of freedom on the activation of Links A to E. Each GC can have a variable interaction with LC that is based both on the Wind direction and on the departures at the appropriate terminal. Similarly, the GCs can have many interactional patterns. Furthermore, the existence of two layers shows that the issuance of an advisory regarding a command by GC1 to GC2 can be elaborated either on one sensor, or three sensors. The framework quantifies precisely, through the layers, the sources of information on which variable interactions can be based. Finally, it must be noted that the methodology leaves open to the designer the choice of a particular structure in this family. It describes explicitly the parameters (VMINOs, VMAXOs, intermediate boundaries) that characterize the set of solutions.

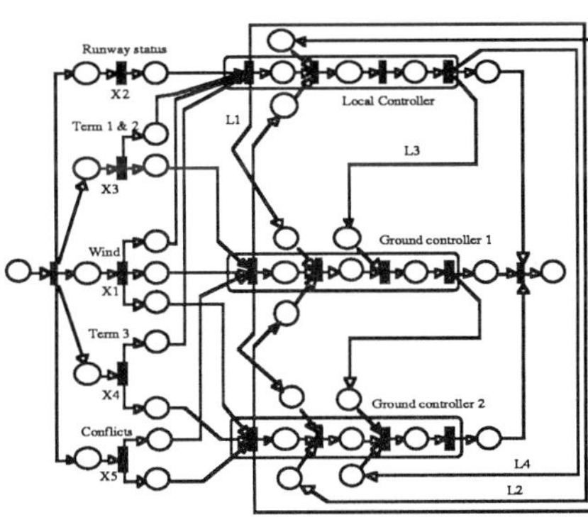

Fig. 11 VMAXO

CONCLUSION

In this paper a quantitative methodology for modeling variable structure distributed intelligence systems and for characterizing the set of solutions to a particular organizational design problem is presented. The class of structures considered process deter-ministically a set of simultaneous observations. This methodology models variable structures with Colored Petri Nets, which are used as the basic mathematical framework to generate the set of structures that satisfy design requirements. The designer of a system can describe his degree of knowledge about the requirements in a matrix form that is translated into the language of Colored Petri Nets. Then, an algorithm that reduces the problem to a tractable level has been outlined. This methodology provides a basic step towards the design and analysis of more realistic systems, because the optimal decision problem can now be formulated in a structured hierarchical way so that the optimum decision strategies for each configuration can be obtained.

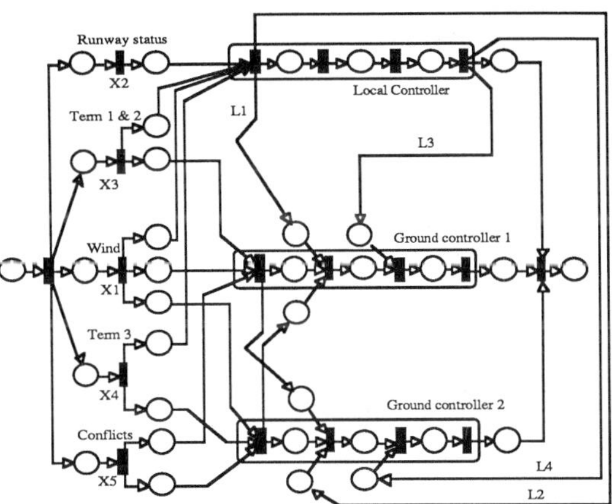

Fig. 12 Intermediate Structure

REFERENCES

Demaël, J.J. (1989). "On the Generation of Variable Structure Distributed Architectures". LIDS-TH-1869. Laboratory for Information and Decisions Systems, MIT, Cambridge.

Genrich, H.J. and Lautenbach, K (1981). "System modeling with High-Level Petri Nets," *Theoretical Computer Science*, No. 13, pp. 109-136.

Grevet, J.-L, and Levis, A. H. (1988). "Coordination in Organizations with Decision Support Systems". *Proc. 1988 Symposium on Command and Control Research.* Science Applications International Corporation, McLean, Virginia.

Jensen, K. (1987). " Colored Petri Nets " in *Advanced Course on Petri Nets 1986*, Lecture Notes in Computer Science, Springer-Verlag, Berlin, Germany.

Levis, A. H. (1988) ."Human Organizations as Distributed Intelligence Systems." *Proc. IFAC Symposium on Distributed Intelligence Systems.* Pergamon Press, Oxford, England.

Minsky, M. (1986). *The Society of Mind.* Simon and Schuster, New York.

Monguillet, J.-M., and Levis, A. H. (1988). " Modeling and Evaluation of Variable Structure Command and Control Organizations." *Proc. 1988 Symposium on Command and Control Research.* Science Applications International Corporation, McLean, Virginia.

Reisig, W. (1985). *Petri Nets, An Introduction.* Springer-Verlag, Berlin, Germany.

Remy, P. and Levis, A. H. (1988). " On the Generation Of Organizational Architectures Using Petri Nets. " in *Advances in Petri Nets 1988*, Lecture Notes in Computer Science, G.Rozenberg Ed. Springer-Verlag, Berlin, Germany.

Stabile, D.A. and Levis, A.H. (1984) "The Design of Information Structures: Basic Allocations Strategies for Organizations." *Large Scale Systems*, Vol. 6, No. 2, North-Holland.

TWO-LEVEL OPTIMAL RESOURCE
ALLOCATION IN A LARGE SCALE SYSTEM

R. Gessing and Z. Duda

*Institute of Automatic Control, Silesian Technical University, Pstrowskiego 16,
44-101 Gliwice, Poland*

Abstract. In the paper the resource allocation problem in a large-scale
system , at the presence of the resource shortages is considered. Two-
level control structure with the coordinator and the local controllers
having different information is considered. The so called elastic constraint (Gessing, 1985) is used for coordination . It is shown that
the control laws of the local controllers have an analitical form, while
that of the coordinator results from a numerical calculations . On the
basis of the simulation researches the influence of the demands variance,
the lenght of the moving horizon and the reserve capacity in the storage
reservoir   on the quality of control is analized .

Keywords. Large scale systems; hierarchical systems ; stochastic
control ; optimal control ; discrete-time systems .

INTRODUCTION

In the literature there exist many papers
concerning the large-scale control systems. The deterministic (Findeisen, 1974;
Findeisen and others, 1980) as well as
the systems with uncertainty are considered in which the different decision makers have the same or different information (Chong and Athans, 1971 ; Ho, 1980,
Ho and Chu, 1972). The different information structures (Chu, 1974 ; Ho and Chu,
1974) are analized and their influence on
the form of the solution is researched.

In the present paper the problem formulation is most closely related to Gessing
(1985, 1987) where the discrete-time model for the resource allocation problem
in the two-level control structure at the
presence of uncertainty is considered. It
is assumed that the coordinator has at
disposal the information which is essential for the whole system, while the local
controllers - more detailed information
which is essential for the particular subsystems .

The present paper differs from Gessing
(1985, 1987) in this, that the resource
allocation problem for the local controllers is static, while the system dynamics results from the storage reservoir
capacity and is taken into account by the
coordinator . The elastic constraint introduced in (Gessing, 1985) is also utilized.

The contribution of the paper is in showing by the way of simulation researches
that it is profitable to assign a part of
the system storage reservoir capacity as
a reserve capacity designed for accumulation of the resources reserves . These
reserves make it possible to realize the
considered decentralized control. This
means that the detailed information of the
local controllers significantly improves
the quality of the control.

MODEL OF A SYSTEM

We consider a static linear large scale
system composed of M independent subsystems (receivers). The receivers take water from a storage reservoir with a given
capacity . The reservoir is supplied with
an inflow and it is possible to let out
some water from the reservoir out of the
system. We assume that in the considered
system it is a deficit of water in some
periods of time and then the demands of the
receivers can't be fully satisfied .

We consider a discrete model of the system
in which the discrete time n is related to
the basic period of discretization (e.g.
month, week). The variables d_n and z_n^i
denote the inflow to the storage reservoir
and the demand of the i-th subsystem, respectively in the n-th period of time. We
assume that d_n and z_n^i , i=1,2, ..., M,
are the stochastic processes defined by
a probability distribution .

We assume that decisions about the resource
allocation are made in a two-level hierarchical control structure consisting of multiple decision-makers with different infor-
mation.The decisions on the control u_n^i
(the resource allocation) for the i-th
subsystem are made by the i-th lower level
local controller on the basis of its information, about the demands from the appropriate subsystem and a value of the coordinating variable p_n^i from the coordinator .
The coordinator takes decisions on the
coordinating variables p_n^i using information about the estimation of the inflow,the
amount of water in the reservoir and the

mean values of demands . These coordinating variables are transmitted to the local controllers .

The performance index defining the losses in the whole system which will be minimized has the form :

$$I = E \sum_{n=0}^{N} \sum_{i=1}^{M} (u_n^i - z_n^i)^2 \qquad (1)$$

where E denotes the mean operation and N denotes the stopping time. The goal of the control is the determination the coordinating variables p_n^i - by the coordinator and the resource allocations u_n^i - by the local controllers so as to minimize the losses (1).

In order to solve the problem we introduce some simplifying assumptions. We assume that the pipelines are well designed and it is no necessity (in the analitical calculations) to take into account the inequality constraints concerning u_n^i . The constraints $u_n^i \geqslant 0$, $p_n^i \geqslant 0$ will be taken into account only in the numerical calculations. The fulfilment of these constraints mostly results from the appropriately prepared given data. The unique inequality constraint taken into account concerns the amount of the water in the storage reservoir and has the form

$$h_{min} \leqslant h_n \leqslant h_{mx} \quad n=1,2, \ldots, N+1 \qquad (2)$$

where h_n is the amount of the water in the beginning of the i-th period of discretization, h_{min}, h_{mx} - the minimal and maximal amount of the water in the recervoir. The occurence of the constraint (2) causes the necessity of performing the numerical calculations .

We assume that z_n^i , z_n^j are white noises mutually (for $i{\neq}j$) independent with some given mean values $\bar{z}_n^i$, i=1,2, ..., M , n=0,1, ... N. We also assume that the relation between u_n^i and p_n^i is determined by the so called elastic constraint (Gessing, 1985) which will be formulated in the next section .

Our goal is to perform the calculations by the analitical way so far as it will be possible. It means that the numerical calculations should be reduced to a minimum. It will be possible owing to the analitical way of solving the static problem of the resource allocation which will be shown in the next section .

TWO-LEVEL STRUCTURE FOR
THE OPTIMAL RESOURCE ALLOCATION

Let us introduce a variable e_n described by the equation

$$e_n = \sum_{i=1}^{M} p_n^i \qquad (3)$$

Now, we assume that the variable e_n is given and the decisions p_n^i, u_n^i are related by the equation

$$p_n^i = E (u_n^i \mid p_n^i) \qquad (4)$$

where E denotes the conditional mean, given p_n^i . The relation plays the role of the so called elastic constraint of the control variable (Gessing, 1985) .

A part of the performance index (1) related to discrete time n has the form

$$I_n = E \sum_{i=1}^{M} (u_n^i - z_n^i)^2 \qquad (5)$$

For the considered system, the optimal control laws $u_n^i = a_n^{io}(z_n^i, p_n^i)$ and $p_n^i = b_n^{io}(e_n)$, i = 1,2, ... , M , are to be found, for which the performance index (5) with the constraints (3) and (4) is minized.

From the assumptions related to z_n^i ,it results that the function a_n^{io} may be found by solving the problem

$$I_n^i = \underset{a_n^i}{Min} \; E \; (u_n^i - z_n^i)^2 \qquad (6)$$

with the elastic constraint (4).

After using the method of the Lagrange multipliers we can find the law a_n^{io} in the form .

$$u_n^i = p_n^i + (z_n^i - \bar{z}_n^i) \qquad (7)$$

where $\bar{z}_n^i = E (z_n^i)$.

Substituting (7) into (6) and (5) we have

$$I_n^i = E (p_n^i - z_n^i)^2 \qquad (8)$$

$$I_n = E \sum_{i=1}^{M} (p_n^i - z_n^i)^2 \qquad (9)$$

The function $b_n^{io}(e_n)$ can be determined by solving the minimization problem (9) with the constraint (3) .

After using the method of the Lagrange multipliers we can find the law b_n^{io} in the form

$$p_n^i = \bar{z}_n^i + (e_n - \sum_{i=1}^{M} \bar{z}_n^i)/M \qquad (10)$$

Substituting (10) into (9) we obtain

$$I_n = E \sum_{j=1}^{M} (e_n - \sum_{i=1}^{M} z_n^i)^2/M^2 \qquad (11)$$

132

or $\quad I_n = E \left[(e_n - \sum_{i=1}^{M} \bar{z}_n^i)^2 \right] / M \qquad (12)$

The performance index (12) related to discrete time n will be used for calculation of the variable e_n .

CALCULATION OF THE VARIABLE e_n

The decisions on the coordinating variables p_n^i , $i=1,2, \ldots M$, depend on the variable e_n (see (10)). The optimal value of the variable e_n results from minimization of the performance index

$$I = E \sum_{n=0}^{N} (e_n/M - \sum_{i=1}^{M} z_n^i)^2 + k' q_n^2 \qquad (13)$$

with the constraint (2) .

Last term in (13) represents the controllable outflow on the outside of the system .

The constraint (2) related to the time $(n+1)$ has the form

$$h_{min} \leqslant h_n + d_n - \sum_{i=1}^{M} u_n^i - q_n \leqslant h_{mx} \qquad (14)$$

Substituting (7) into (14) and taking into account (10) we obtain

$$h_{min} \leqslant h_n + d_n - e_n - \sum_{i=1}^{M} (z_n^i - \bar{z}_n^i) - q_n \leqslant h_{mx} \qquad (15)$$

From (7) and (10) it results that the total water allocation ($\sum_{i=1}^{M} u_n^i$) determined by the local controllers may be greated than the value e_n determined by the coordinator. In order to realize the control u_n^i in the two-level structure we must accumulate in a magazine of the system some additional resource reserves for satisfying some randomly increased resource demands . The resource reserves must be accumulate in some "reserve capacity" . As the reserve capacity may be used some special buffer reservoirs or a part of the storage reservoir. In the paper the last case will be considered. In this case, the coordinator will have at his disposal a smaller controlled capacity . It seems that the creation of the reserve capacity in the storage reservoir is resonable but it is very difficult to motivate this suggestion by an analitical way . We will discuss this problem in the next chapter .

Determination of the optimal solution of the minimization problem (13) with the constraint (2) is impossible, even numerically. Then we propose to solve a suboptimal problem basing on the open-loop feedback (OLF) control.

In accordance with this idea , the coordinator minimizes the performance index

$$I_k = E \sum_{n=k}^{k+N'} (e_n - \sum_{i=1}^{M} \bar{z}_n^i)/M + k' q_n^2 \qquad (16)$$

with respect to $e_k, e_{k+1}, \ldots e_{k+N'}, q_k, q_{k+1} \ldots q_{k+N'}$, and with the constraints

$$h_{min} + \Delta h \leqslant \hat{h}_{k+j|k} \leqslant h_{mx} \quad j = 1,2, \ldots N'+1 \quad (17)$$

$$e_n \geqslant 0 \quad q_n \geqslant 0 \quad n = k, k+1, \ldots k+N' \qquad (18)$$

where $\hat{h}_{k+j|k}$ - the estimation of the variable h_{k+j} , given information available at the time k , N'- horizon of the control. The estimation $\hat{h}_{k+j|k}$ may be determined from the equation

$$\hat{h}_{k+j|k} = h_k + \sum_{i=k}^{k+j-1} \hat{d}_{i|k} - \sum_{i=k}^{k+j-1} (e_i + q_i) \leqslant h_{mx}$$
$$(19)$$

where $\hat{d}_{i|k}$ - the estimation of the inflow d_i, given information at the time k .

At the time k the coordinator knows h_k and $\hat{d}_{i|k}$ and determines the values of the variables $e_k \ldots e_{k+N'}$, $q_k, q_{k+N'}$ which minimize (16) with the constraints (17), (18). Then the variables p_k^i, $i=1,2, \ldots M$ are determined and transmitted to the local controllers. The local controller transmits the decision on u_k^i to the i-th subsystem. The coordinator repeats this alghoritm at each time $k=0,1,2, \ldots$.

Let us notice that in the case of inactive constraints (17) the optimal value of p_n^{io} equals $\bar{z}_n^i$ and $u_n^{io} = z_n^i$. In the case of active constraints (17) the water allocation is less than the demand and results from (7) .

NUMERICAL INVESTIGATIONS

Let us consider the system composed of five subsystems (M=5). The consideration of a large system creates no additional difficulties .

We assume that the demands of the receivers are stationary gaussian white noises defined by $N(\bar{z}_n^i, \delta^2)$, where $\bar{z}_n^i = 15$, $i=1,2, \ldots 5$.

The storage reservoir is feeded by an periodical inflow d_n with the period 12. It may be interpreted as the period of 12 months . We assume that the predictions of the inflow are determined by the means $\bar{d}_n = E(d_n)$ and that the variations d_n with respect to $\bar{d}_n$ can be neglected. The values of the inflow are as follows : $d_i = 100$, $i=1,2, \ldots 6$, $d_i = 40$, $i=7,8, \ldots 12$.

The values of z_n^i, d_n and h_n represent some conventional units . It is easy to notice that the mean year inflow (840) is smaller than the mean year demands (900) . We assume that N=480 (months i.e. 40 years).

The coordinator determines the values of the variable e_n in accordance with the idea presented in the previous chapter. The Wolf's method is used to numerical calculations .

Let us notice that the difference between p_n^i and u_n^i for the i-th local controller has the form

$$u_n^i - p_n^i = z_n^i - \bar{z}_n^i \tag{20}$$

and for the whole system

$$w_n = \sum_{i=1}^{M} (u_n^i - p_n^i) = \sum_{i=1}^{M} (z_n^i - \bar{z}_n^i) \tag{21}$$

The variable w_n is gaussian random variable defined by $N(\,0,\ \sum_{i=1}^{M} \delta_i^2\,)$.
The full realization of the optimal control is possible when the probability $P\,(w_n \leqslant \Delta h) = 1$. It can be fulfilled for the variable of limited distribution. In our case only a suboptimal control can be realized. As the reserve capacity Δh is taken as the part of the storage reservoir, the quantity of Δh influences the quality of control . This influence was investigated numeracilly . The results are presented on the Fig. 1 . for the data $N'=1, h_{min}=0$ and z_n^i with the parameters $N(15,45)$.

From the numerical investigations it results that the "optimal" capacity of the storage reservoir is $h_{mx}=260$. In this case, the storage reservoir can accumulate the surplus of water resulting from the increased inflow or the decreased demands . For this capacity, the minimal losses are for $\Delta h=30$. It gives about 11% of the whole capacity . The losses increase for Δh below $\Delta h=30$. For $\Delta h=0$ (the lack of reserve capacity) the losses are greater about 30% in comparison to the minimal ones .

On the Fig. 1 . are also presented the results of calculations for the storage reservoirs with a smaller capacity (180, 160, 130) . We can see, that the losses are higher than in the case discussed above.

The value of the reserve capacity depends on variance of the demands. On the Fig. 2. an influence Δh on the control equality is presented for $h_{mx}=260$ and for different variance of the demands $N(15,20)$, $N(15,45)$, $N(15,80)$.

On the Fig. 3. an influence of the horizon N' on the control quality is presented, for $h_{mx}=260$ and z_n^i defined by $N(15,45)$

It is easy to notice that the influence of the horizon is essential. It shows that the coordinator plays useful role in the above structure. We must emphasize that the influence of the horizon on the control quality depends on estimation of the inflow. Let us notice that the optimal value of the reserve capacity Δh does not change for $N'=0,1,2,$

Sometimes it is possible that the value of the variable u_n^i resulting from (7) may be negative . In this case we assume that $u_n^i=0$. For considered system the number of suboptimal decisions was not great (about 2% in comparison to all of the decisions).

The values z_n^i were determined by a random-number generator $N(15, \delta^2)$. When proportion $\delta/15$ increases then the probability of the "great" demand increases, too. In the case of unlimited distributions, sometimes a great demand can be generated giving the decision (7) which can't be realized. Practically, the minimal and maximal demands can be estimated and the model with a limited distribution is then more proper .

FINAL CONCLUSIONS

The essential point of this paper is the assumption that particular decision makers have different information . The coordinator has the information about mean values of demands for the whole system, while the local controllers have the information about real demands .

In the case of the considered problem the optimal control laws of the local controllers and partially of the coordinator, take an analitical form .

The elastic constraint admits some freedom in taking decisions for the local controllers, owing to this they can better use its more detailed information. Owing to the elastic constraint used for the coordination and the assumptions concerning the information structure, it is possible to decompose the calculations and to realize the decentralized optimal control .

For realization of the control in the two-level structure, some water reserve is needed in order to satisfy some randomly increased demands . From numerical investigations it results that it is profitable to use a part of the storage reservoir as the reserve capacity . The reserve capacity is not controlled by the coordinator and is only used by the local controllers. Though, the reserve capacity dimineshes the capacity of the storage reservoir controlled by the coordinator, nevertheless it improves the quality of the control . It means that the control decentralization with using more detailed information of the local controllers gives the better results than the centralized control with less detailed information of the coordinator and the bigger storage reservoir capacity . The more detailed information of the local controllers plays here a crucial role and speaks for the decentralization .

The solution of the control resource allocation problem presented in this paper may

be used for more complex   systems with
many reservoirs, inflows and transfers
between different parts of the system.

ACKNOWLEDGMENT

The paper was supported by the Program
CPBP-03.09 coordinated by the Institute of
Geophysics of the Polish Academy of Scien-
ces .

REFERENCES

Chong, C.Y. and M.A.Athans (1971). On the
    Stochastic Control of Linear Systems
    with Different Information Sets.
    I.E.E.E. Trans. autom.Control, 16,15,
    pp.423-430.
Chu, K.C. (1974). Team Decision Theory
    and information structures in optimal
    control problems. Automatica, Vol.
    10, 4 .
Findeisen, W. (1974). Multilevel control
    systems, Warsaw .
Findeisen W., F.N. Bailey, M.Brdyś, K.
    Malinowski, P.Tatjewski, and A.Woźniak
    (1980). Control and Coordination in
    hierarchical systems .
Gessing,R. (1985). Two-level hierarchical
    control for stochastic optimal resource
    allocation. Int. J.Control 41, 1,
    pp.161-175.
Gessing,R. (1987). Optimal control laws
    for two-level hierarchical resource
    allocation. Large Scale Systems 12.
Ho, Y.C. (1980). Team Decision Theory
    and Information Structures. Proc.Int.
    elect. Engrs., 68, 6, pp.644-654 .
Ho, Y.C. and K.C. Chu (1972). Team deci-
    sion theory and information structu-
    res in optimal control problems.Part I.
    I.E.E.E. Trans. autom.Control,23,2,
    pp.108-128

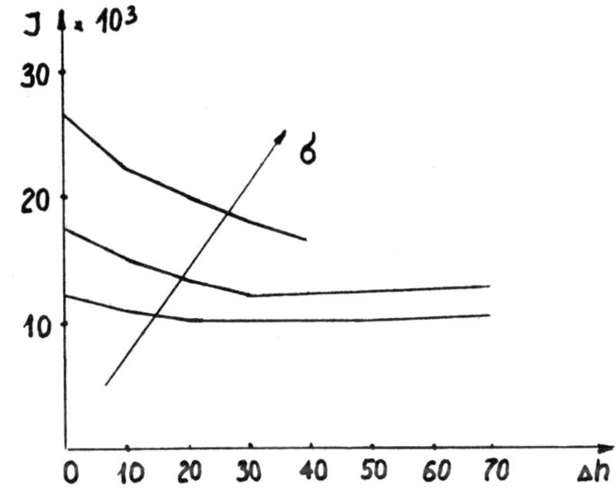

Fig. 2.   The influence of the variance
          on the control quality .

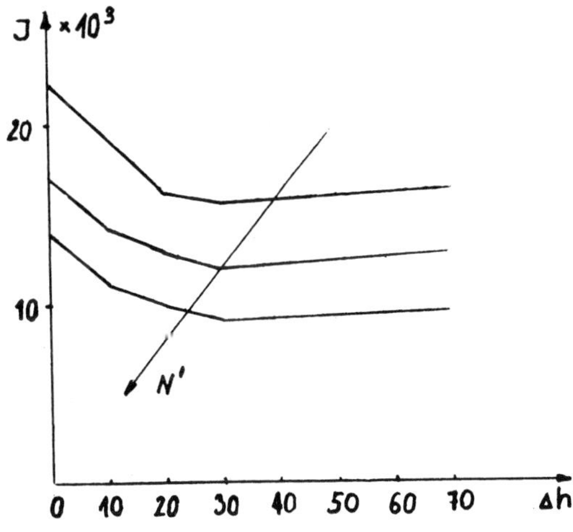

Fig. 3.   The influence of the horizon
          on the control quality .

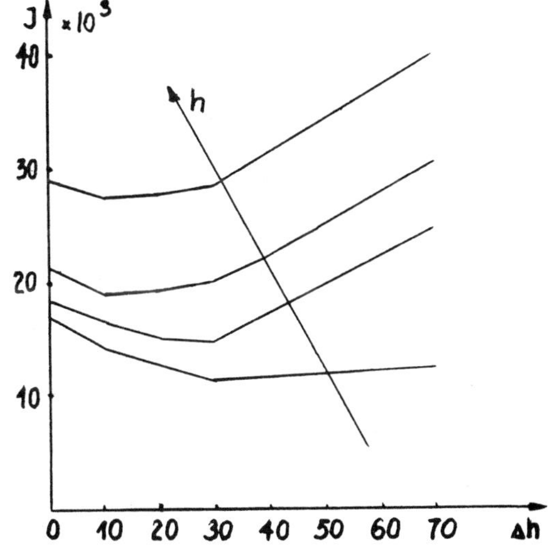

Fig. 1.   The influence of the reserve
          capacity on the control quality .

HIERARCHICAL CONTROLLED COMPONENT SYNTHESIS OF LARGE SPACE STRUCTURES

K. D. Young

*Lawrence Livermore National Laboratory, University of California, Livermore,
CA 94550, USA*

Abstract. This paper describes a new framework called *Controlled Component Synthesis* for the design
of multi-level decentralized controllers for large flexible structures. In contrast to the existing decentralized
control approach in which the design begins with a given dynamic model of the flexible structure, *controlled
components* are built and assembled into a controlled flexible structure that meets performance specifications. A simple truss structural control problem is employed to illustrate the design procedures, as well as
demonstrate the potentials of the developed method for controlling very large dimensional truss structures.

Keywords. Decentralized control; Large-scale systems; Aerospace control

Introduction

Structural control has been an active area of R&D for the last decade in
response to an increasing demand in large space structure pointing and
control requirements. While there has been many innovations in *L*arge
*S*pace *S*tructure(LSS) control technology, one of the more difficult hur-
dles is the communication gap between structural engineers and control
engineers. The typical structural control problem begins with the structural engineer's finite element modeling and model reduction efforts,
from which a dynamic model of the structure is generated. The modeling phase is followed by a control design phase in which the structure
model is further reduced to a manageable dimension for control design.
While this level of interaction has been proven to be sufficient in the
past for structures with a few dominant flexible modes in the control
system bandwidth, it becomes inadequate when the complexity of the
structural control problem increases as the control system bandwidth
increases, the number of flexible modes inside this bandwidth increases
as the structural material becomes lighter, and the physical size of the
structure, as well as the number of actuators and sensors, grow by an
order of magnitude. In the last decade, the need emerged for a dis-
tinctive hybrid engineering discipline which has its origins in structural
engineering and control system engineering. This new interdisciplinary
field was designated as *C*ontrol *S*tructure *I*nteraction (CSI) Technology.

One of the key objectives in CSI Technology is to develop methods and
approaches in structural modeling and control while eliminating the
technical inconsistencies that are hidden in structural and control system engineering practices. Such development perhaps is most needed in
the design of distributed control for large space structures. Distributing or decentralizing LSS structural controllers means control action is
localized. Since the structure models are developed without any consideration of the controller's action, the opportunity to model the local
structural dynamic behavior in order to benefit the design of the distributed control laws is lost.

A connection between structural dynamics and large scale systems control was established by Young (1988a, 1988b) to benefit the design of
distributed structural controllers for LSS. The method developed in that
work detours from the conventional control system design path which
begins with a model of the open loop plant. Instead, the controlled
plant is assembled from *controlled components* in which the modeling'
phase and the control design phase are integrated at the component
level. The developed method is labeled as a *C*ontrolled *C*omponent
*S*ynthesis (*CCS*) method to reflect that it is motivated by the well developed *C*omponent *M*ode *S*ynthesis (*CMS*) methods - a collection of
structural analysis methods which has been demonstrated to be effective
for solving large complex structural analysis problems for almost three
decades. The design philosophy behind CCS is closely related to that of
the Subsystem Decomposition Approach in decentralized control. The
ideas behind CCS are stimulated by the subsystem decomposition viewpoint in large scale system theory on one hand, and by the component
mode synthesis methods in structural analysis.

In this paper, a review of CCS is first given, followed by an exposition
of a hierarchical CCS method in which the scope of CCS is expanded
to encompass a multi-level decentralized control system design of large
flexible structure control systems. Design of a multi-level decentralized
control system for a planar truss structure illustrates the key ideas of
hierarchical CCS.

Subsystem Decomposition

Decentralized control is intuitively appealing for structural control of
large flexible structures: It offers simplified control system implementations which only require the feedback of local measurements to close
the control loop for systems that may have a few hundred to thou-
sands of control loops. In *Subsystem Decomposition Approach* the
large structure model is first decomposed or partitioned into an interconnection of subsystem models. Local control designs are produced by
designing controllers for the local decoupled subsystem models which
are derived from the large structure model by discarding the interconnections. These local designs are solved independently, thereby reducing
the computational resource requirements for performing control designs
for very large structures. The behavior of the locally controlled structure
is analyzed using aggregation analysis methods: An aggregate variable
is defined to represent the subsystem's dynamic interactions with the
other subsystems. The order of the aggregate analysis problem is equal
to the number of subsystems which is determined by the decomposition
scheme. This approach is developed by Šiljak (1978) and his co-workers
. The work of Young (1983,1985), and Kida (1985) adopt this approach
for decentralized control designs.

Given a model of a large structure, a control system designer is faced
with the problem of decomposing the given model into an interconnection of subsystem models. One of the attractive features of the Subsystem Decomposition Approach is the use of local subsystem models,
rather than the model of the large structure, in designing decentralized
controllers. However, from the foregoing discussions, it is clear that
the large structure model must also be known to the designer in this
approach. Thus, the use of local models in the local control designs is
an option exercised by the designer to reduce the dimensionality of the
associated computational problems, and to improve the robustness of
the controlled structure with respect to parametric perturbations.

A more radical departure from the conventional large scale system design path calls for the development of the decoupled local subsystem
models without the use of the model of the large structure. Controlled
component synthesis adopts this new viewpoint on subsystem decomposition. Such a departure from large scale system theoretic perspectives
essentially coincides with the component mode synthesis viewpoints held
by structural analysts for quite some time.

Controlled Component Synthesis

The concept behind *C*omponent *M*ode *S*ynthesis (*CMS*) was first introduced by Hurty (1960). Craig (1977) provided a survey of CMS
methods developed between 1960 and 1976. More details on CMS can
be found in Meirovitch (1981) and Craig (1981). The fundamental idea
behind CMS methods is that a dynamic model of a large complex structure is to be built from models of its components. The components are
first modeled as individual structures; these component model data are
then processed to form a model of the large structure. The component
models are typically developed using a finite element method.

A critical design parameter in CMS methods is the choice of appropriate models or modes for the components. The selection of component
modes directly affects the accuracy to which the model of the coupled
structure that is synthesized from component model data, approximates
the finite element model of the coupled structure. The development of

CMS methods in the past two decades has been focused on the specification and the choice of component modes, and methods to handle the constraint equations.

$\underline{C}$ontrolled $\underline{C}$omponent $\underline{S}$ynthesis (CCS) is a framework for an integrated, component oriented, finite element modeling and structural control design. Similar to CMS methods, a CCS method is developed on the premise that a large complex controlled structure is to be built from *controlled components*: The finite element modeling and control design are performed for the individual components; the model of the large complex structure is assembled from the controlled components only for the purpose of performance evaluation and sensitivity studies.

The CCS method adopts the following modeling and control design considerations at the component level: Instead of using either the boundary loading, or the constraint modes approach in CMS as outlined before, we introduce a new approach for the development of component models based on boundary loading. For the design of controllers for the component, an interlocking control concept is developed to minimize the motion of the nodes that are adjacent to the boundary, thereby suppressing the transmission of mechanical disturbance from component to component in the coupled structure. A new synthesis procedure for deriving the coupled structure models from component models is also developed.

1. *Component Modeling for CCS*

A two component structure, as shown in Fig. 1, will be used to outlined the modeling and design procedure of CCS. Each of the structure component is composed of three finite elements. Identified in the figure by Roman numerals are the finite elements, and by solid circles are the element node points.

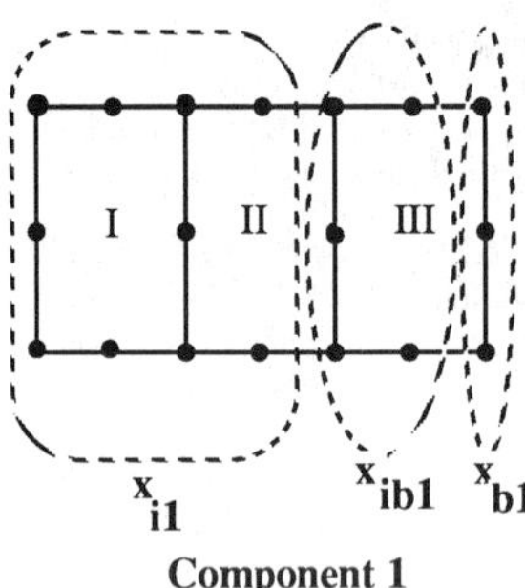
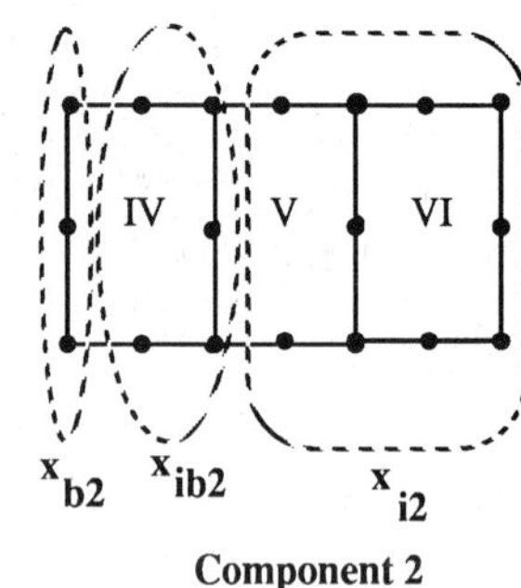

Fig. 1. A two component structure modeled with finite elements.

In CMS methods, the nodal coordinates of a component is partitioned into a set of internal(interior) coordinates, x_{is}, and a set of boundary coordinates, x_{bs}. In the CCS method, the internal coordinates are further subdivided into a group of internal boundary coordinates x_{ibs} and a group of internal coordinates x_{is}. Fig. 1 indicates how the three groups of coordinates are defined: The boundary coordinates are coordinates of the boundary element, such as element III of component 1, which are on the boundary. The remaining coordinates of the boundary element are designated the internal boundary coordinates. The remaining coordinates of the component are the internal coordinates.

Partitioning with respect to the three groups of coordinates, the finite element model for a component has the form,

$$\begin{bmatrix} M_{ii}^s & M_{iib}^s & 0 \\ M_{bii}^s & M_{ib}^s & M_{ibb}^s \\ 0 & M_{bib}^s & M_{bb}^s \end{bmatrix} \begin{bmatrix} \ddot{x}_{is} \\ \ddot{x}_{ibs} \\ \ddot{x}_{bs} \end{bmatrix} + \begin{bmatrix} K_{ii}^s & K_{iib}^s & 0 \\ K_{bii}^s & K_{ib}^s & K_{ibb}^s \\ 0 & K_{bib}^s & K_{bb}^s \end{bmatrix} \begin{bmatrix} x_{is} \\ x_{ibs} \\ x_{bs} \end{bmatrix} = \begin{bmatrix} f_i^s \\ f_{ib}^s \\ f_b^s \end{bmatrix} \tag{1}$$

The block bi-diagonality of the mass and stiffness matrices is a direct consequence of the chosen finite element mesh and the deliberate grouping of the coordinates for the component.

In developing a new approach to generate component models for CCS, we adopt the idea of boundary loading introduced by Benfield and

Hruda (1971) ; yet decide not to slave the IIB coordinates to the boundary coordinates. In boundary loading, the stiffness matrix K^s of a component is reduced by eliminating the internal coordinates x_{is}. Let the component coordinates and stiffness matrix be partitioned

$$x^s = \begin{bmatrix} x_{is} \\ x_{bs} \end{bmatrix} \quad , \quad K^s = \begin{bmatrix} K_{ii}^s & K_{ib}^s \\ K_{bi}^s & K_{bb}^s \end{bmatrix} \quad , \tag{2}$$

The reduced stiffness matrix with respect to the boundary coordinates is given by

$$K_b^s = T_b^{sT} K^s T_b^s \quad , \tag{3}$$

where

$$T_b^s = \begin{bmatrix} -K_{ii}^{s\,-1} K_{ib}^s \\ I \end{bmatrix} \quad . \tag{4}$$

The reduced mass matrix with respect to the boundary is derived following a similar process,

$$M_b^s = T_b^{sT} M^s T_b^s \quad . \tag{5}$$

In this new approach, instead of using the reduction matrix T_b^s, as given by Equation (4), a truncation matrix, which eliminates both the internal, and internal boundary degree of freedom, is used:

$$T_{bt}^s = \begin{bmatrix} 0 \\ 0 \\ I \end{bmatrix} . \tag{5}$$

The modifications to the mass and stiffness matrices of the component finite element model are computed using Equations (3) and (5) in the boundary loading approach, but with T_{bt}^s replaces T_b^s in these equations. The resulting modified mass and stiffness matrices for loading from the $j\,th$ to the $s\,th$ component are:

$$M_{bb,iL}^s = M_{bb}^s + M_{bb}^j , \tag{6}$$
$$K_{bb,iL}^s = K_{bb}^s + K_{bb}^j . \tag{7}$$

Note that these modifications are restricted to the submatrices corresponding to the boundary coordinates, *i.e.*, only the submatrices M_{bb}^s and K_{bb}^s are modified, the other submatrices in Equation (7) remain unchanged. We call this new approach to component modeling, Isolated Boundary Loading. The modified mass and stiffness matrices due to this approach can be obtained alternatively from the finite element modeling of an expanded component, *i.e.*, the original boundary of the component is extended one finite element into the adjacent component. The nodes of the expanded component consist of the original nodes of the component, and the internal boundary coordinates of the adjacent component. This alternative derivation of the component model, which is physically explicit, results from the interpretation that transforming with T_{bt}^s is equivalent to the removal of the internal, and internal boundary degree of freedom, *i.e.* $x_{is} = 0$ and $x_{ibs} = 0$. For isolated boundary loading from the $j\,th$ to the $s\,th$ component, the $s\,th$ component is expanded as shown in Fig. 2. The modified mass and stiffness matrices are obtained from the mass and stiffness matrices of the expanded component by deleting the rows and columns corresponding to the nodes in the expanded portion.

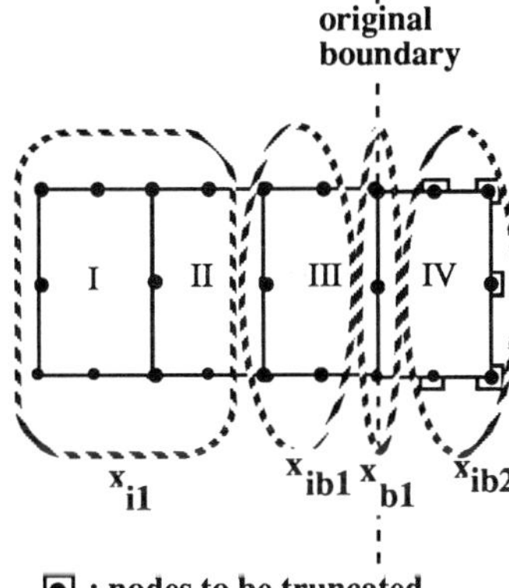

Fig. 2. Expanded component 1.

2. Connections to Overlapping Decompositions

As compared with the Subsystem Decomposition approach in which interconnections are discarded in local decoupled subsystem models, the developed component model captures the dynamic interactions with its neighboring components. It is therefore reasonable to expect that the component controllers designed with these component models would produce better decentralized control actions.

The component models developed using this new approach have a direct connection with the Subsystem Decomposition approach. These models are identical to the decoupled subsystem models if an overlapping decomposition is applied to the finite element model of the coupled structure. This is a key connection which allows the use of large scale system analysis tools developed in Ikeda and Šiljak (1980), Ikeda, and Šiljak, and White (1981), after the controlled component designs have been completed. In Young (1983,1985), decentralized control for truss structures has been investigated using decoupled subsystem models which are obtained by an overlapping decomposition of the structure. The performance analysis carried out in those earlier work can essentially be applied herein.

3. Interlocking Control Concept

The new insights gained from the newly developed component modeling approach in turn motivate a new component level control design concept which we call an *InterLocking Control* (ILC) Concept. Again, we use the two component structure to illustrate the idea. First, consider the model resulted from applying overlapping decomposition to the high fidelity model. It is an ISOS model which is given by

$$\begin{bmatrix} M_{ii}^p & M_{iib}^p & 0 \\ M_{bii}^p & M_{ib}^p & M_{ibb}^p \\ 0 & M_{bib}^p & M_{bb,iL}^s \end{bmatrix} \begin{bmatrix} \ddot{x}_{ip} \\ \ddot{x}_{ibp} \\ \ddot{x}_{bp} \end{bmatrix} + \begin{bmatrix} K_{ii}^p & K_{iib}^p & 0 \\ K_{bii}^p & K_{ib}^p & K_{ibb}^p \\ 0 & K_{bib}^p & K_{bb,iL}^s \end{bmatrix} \begin{bmatrix} x_{ip} \\ x_{ibp} \\ x_{bp} \end{bmatrix}$$
$$= \begin{bmatrix} f_i^p \\ f_{ib}^p \\ f_b^p \end{bmatrix} + \begin{bmatrix} 0 \\ 0 \\ M_{bib}^q \ddot{x}_{ibq} + K_{bib}^q x_{ibq} + f_b^q \end{bmatrix} \quad (8)$$

for all p and q in the index set $(p,q) = \{(s,j),(j,s)\}$. Note that the above equation with the second term on the RHS removed is identical to a component model for CCS. Interlocking Control refers to a component level control design concept in which collocated actuators and sensors are placed at the internal boundary degree of freedom, and the control law is designed, using the developed component model for CCS, to minimize the internal boundary coordinate motion. It is motivated by a closer examination of the above equation which shows that such minimization would localize the dynamic interactions of the coupled structure in the components. The component control action is designed to lock up its own internal boundary to realize a boundary condition which better approximates the one assumed in the component modeling of its adjacent components.

A convenient control design technique for this concept is the linear quadratic optimal regulator approach in which the internal boundary coordinates are considered as regulated outputs of the component to be weighted together with the component control inputs in the quadratic performance index. The resulting component control law minimizes this index. Using this approach, the ILC concept translates into a two step component control design process summarized below:

1. For the $s\,th$ component, use the component model

$$\begin{bmatrix} M_{ii}^s & M_{iib}^s & 0 \\ M_{bii}^s & M_{ib}^s & M_{ibb}^s \\ 0 & M_{bib}^s & M_{bb,iL}^s \end{bmatrix} \begin{bmatrix} \ddot{x}_{is} \\ \ddot{x}_{ibs} \\ \ddot{x}_{bs} \end{bmatrix} + \begin{bmatrix} K_{ii}^s & K_{iib}^s & 0 \\ K_{bii}^s & K_{ib}^s & K_{ibb}^s \\ 0 & K_{bib}^s & K_{bb,iL}^s \end{bmatrix} \begin{bmatrix} x_{is} \\ x_{ibs} \\ x_{bs} \end{bmatrix} = \begin{bmatrix} 0 \\ u^s \\ 0 \end{bmatrix}$$
$$y^s = x_{ibs}\,, \quad (9)$$

for control system design, where u^s and y^s denotes respectively the control force exerted by the actuators, and the sensor outputs, at the internal boundary coordinates.

2. Derive the component control law by minimizing the performance index,

$$J_c^s = \frac{1}{2} \int_0^\infty (y^{sT} y^s + u^{sT} R^s u^s) dt\,. \quad (10)$$

While the linear quadratic optimal regulator approach may require full component state feedback, other control design techniques which produce component controllers that use static and dynamic output feedback can also be used in the ILC concept, provided that the internal boundary coordinate motion is minimized.

At first glance, the component control law is a decentralized control law since it feeds back only measurements of the component. If we consider a state space whose state vector is composed of the components' coordinates and their time derivatives in the coupled structure, the component control law is indeed a decentralized one in this space. However, the state space of the coupled structure is a contraction (Ikeda and Šiljak, 1980) of this space, *i.e.*, it is the state space with all the duplicated boundary coordinates and their time derivatives removed. In the contracted state space, the boundary states are fed back to the adjacent component controllers.

Hierarchical CCS for Truss Structure

The hierarchical CCS method is illustrated with the design of a multi-level decentralized control law for a planar truss structure which is depicted in Fig. 3. This structure is constructed from three components: the top truss has six bays, the left support and the right support are two bay trusses. The top truss is a controlled structure for which a CCS design was carried out and documented in Young (1988a, 1988b). The left support and right support trusses are additional components, to be assembled with the top controlled truss to form a Π shape truss. Two levels of structural control are envisioned: in the lower level(Level 2) are the component controllers which realize the interlocking of the three components which form the top controlled truss, and in the higher level(Level 1), the component controllers which are to be designed to interlock the three components of the Π truss.

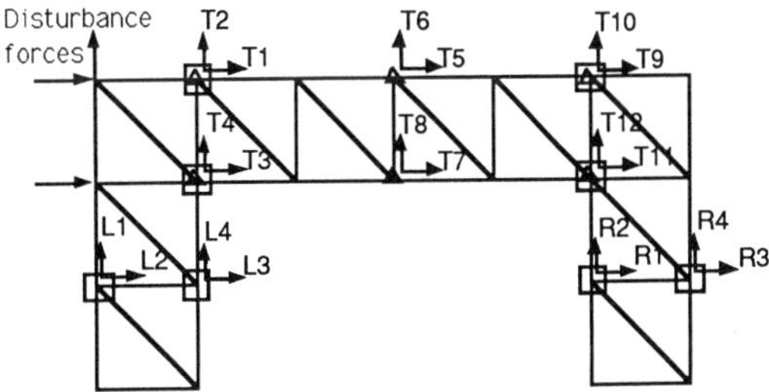

Fig. 3. A three component structure for hierarchical CCS design.

As in the CCS design of the top truss, the nodal coordinates are defined as the vertical and horizontal displacements at the joints. External forces applied at the nodes are decomposed into orthogonal components. The assumptions made are that the truss members are subjected to axial forces alone, and not bending moments; and the members are uniform rods of identical lengths L, mass per unit length m, cross-section area per unit length A and modulus of elasticity E.

The six bay top truss can be viewed as a structure consists of three identical components, namely the left component, the center component, and the right component, which are composed of the leftmost, the middle, and the rightmost two bays respectively. The six bay/three component truss structure is chosen to capture the essential characteristics of a truss consists of an arbitrary number of identical components, *i.e.*, a truss structure with an arbitrarily large number of two bay components is composed of the three same types of components identified in the six bay truss, with the center component duplicated as necessary. Thus, conclusions from the six bay/three component design apply equally well to the design of structural controls for a multiple bay truss.

For CCS, the component models are developed using the expanded component introduced in Isolated Boundary Loading. The mass and stiffness matrices of the expanded component are derived using a finite element method with the Ritz-Rayleigh approximation. The truss member mass and stiffness matrices expressed with respect to local coordinates and used in the assembly process are as derived in Meirovitch (1975):

$$K_{member} = \frac{EA}{L} \begin{bmatrix} 1 & -1 \\ -1 & 1 \end{bmatrix} \quad , M_{member} = \frac{mL}{6} \begin{bmatrix} 2 & 1 \\ 1 & 2 \end{bmatrix} \quad (11)$$

The component model is further scaled to remove the effects of the material properties: a new time variable $\tau = \left(\frac{m}{6EA}\right)^{\frac{1}{2}} Lt$ is introduced, and the nodal forces are scaled by $\frac{L}{EA}$.

In the hierarchical CCS design, we concentrate on the Level 1 controller designs in this paper, whereas the details of the Level 2 designs which produce the top controlled truss are referred to Young (1988b). For the Level 1 designs, we adopt an ILC scheme which elimniates the use of additional control actuators in the top truss. The nodes at which Level 2 collocated force actuators and displacement sensors are placed are marked by $\triangle$, whereas Level 1 control actuators are markerd by $\square$. Fig.3 shows the placements of the proposed Level 1 actuators, indicating the use of the same actuators in the top truss for both Level 1 and Level 2 control actions. As a result, it is necessary to generalize the notion of the ILC scheme - instead of restricting the minimization of the internal boundary node motion, we allow the ILC formulation to minimize an alternative set of nodes near the physical boundary of two adjacent components. We shall call these nodes which play a key role in ILC, interlocking nodes. We note that the internal boundary nodes of the top truss corresponding to the top/support boundary are not the interlocking nodes where Level 1 control actuators were placed.

The procedure to generate the component model from constraining the internal boundary modes of the adjacent component in the expanded component model is correspondingly modified to accomodate the introduction of intelocking nodes. The expanded component is defined as the physical component and the portion of the adjacent component which connects the boundary nodes to the interlocking nodes. The component model thus is derived from the deletion of the rows and columns of the expanded component's mass and stiffness matrices corresponding to the interlocking nodes of the adjacent components to form the component's mass and stiffness matrices.

For the choice of component models for the Level 1 top truss controllers, we exploit the existence of the Level 2 control actions which minimize the motions where the actuators are placed. The expanded components and the constrained nodes for the top truss Level 1 synthesis are shown in Fig.4. For the Level 1 left support controllers, as well as that for the right support, the corresponding exapnded components are shown in Fig.5.

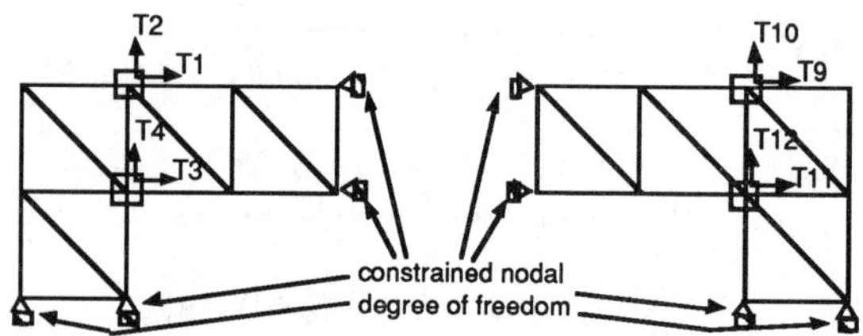

Fig. 4. Component models for hierarchical CCS design for the top component.

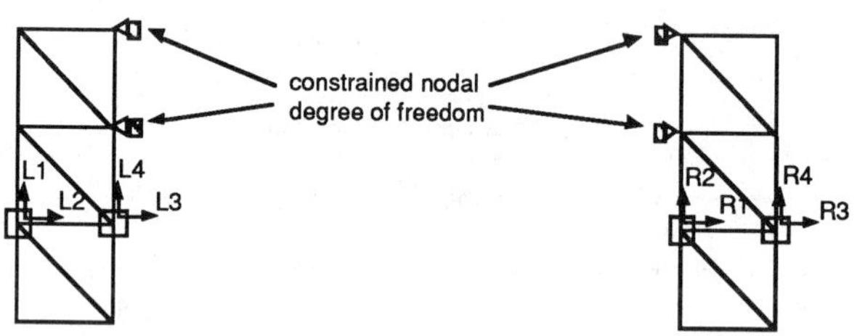

Fig. 5. Component models for hierarchical CCS design for the left and right support component.

No damping is assumed for the left and right support trusses , thus the open loop poles of these two components all lie on the imaginary axis. However, since the top truss is a controlled truss, the Level 2 controllers move the open loop (with respect to Level 1) poles of the Pi structure corresponding to the two support trusses slightly to the left and to the right of the imaginary axis, as shown in Fig.6. Note that although the Pi structure appears to be an unconstrained planar truss which implies there may be three rigid body modes, the Level 2 controllers imbedded in the top truss move the three double poles into the left half plane.

A linear quadratic regulator approach to ILC, as outlined in the last section, is adopted in the design of the Level 1 component controllers. There are four component control design problems in Level 1 - two for the top truss, one for interlocking the left support and another for the right support, and one each for the left support and for the right support. An identical control weighting matrix $R = .001I_{4\times4}$ for all four control designs is chosen. In each of these component control designs, an 24th order Riccati equation is solved to compute the optimal feedback gains, in contrast to an 88th order Riccati equation if an optimal centralized control approach is used.

The controlled components' poles, as well as the poles of the hierarchical controlled truss structure, are plotted in Fig.6a and Fig.6b. All the poles of the controlled structure have negative real parts, indicating that the closed loop system is asymptotically stable. That the pole locations of the controlled components are close to that of the controlled coupled structure indicates that the component models developed for CCS are effective for this structural control design.

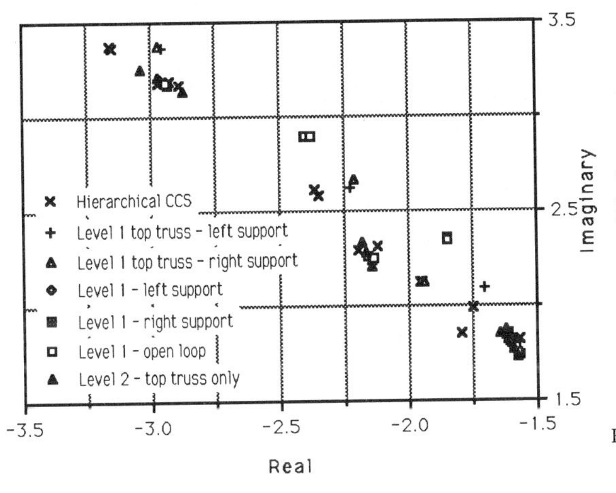

Fig. 6a. Pole locations of the controlled components and the controlled truss - large real parts.

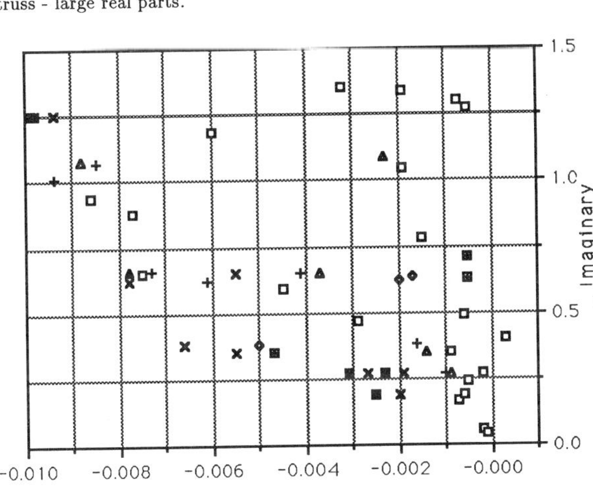

Fig. 6b. Pole locations of the controlled components and the controlled truss - small real parts.

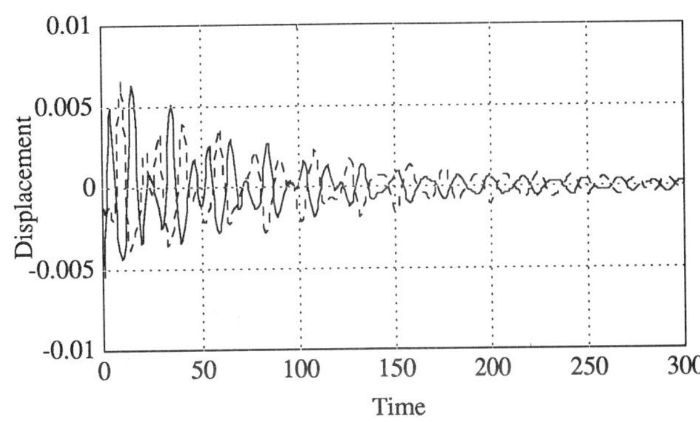

Fig. 7. Displacement time responses - Channel L3 (solid), channel T4 (dotted).

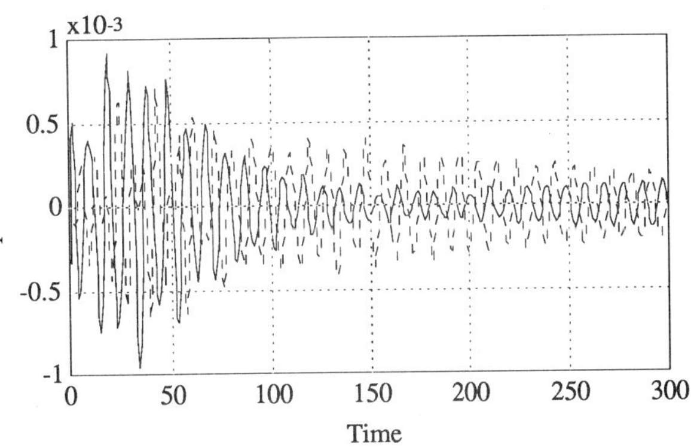

Fig. 8. Displacement time responses - Channel T5 (solid), channel T8 (dotted).

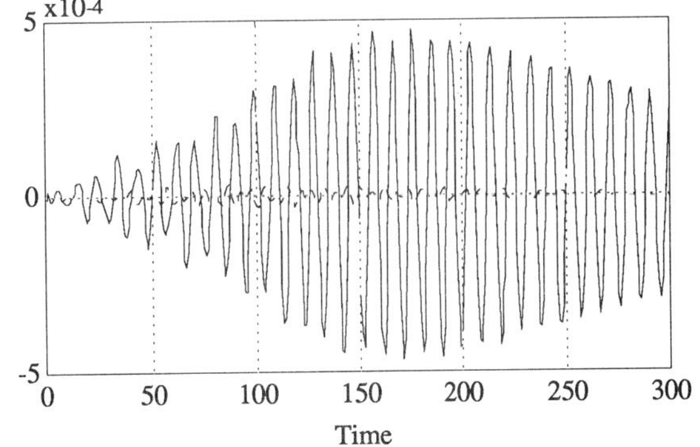

Fig. 9. Displacement time responses - Channel T9 (solid), channel R4 (dotted).

For transient response studies, we examine the response of the controlled structure to three disturbance force pulses of 0.5 seconds, simultaneously applied to the leftmost nodes of the top truss, as shown in Fig.3. The coupled structure is assumed to be in static equilibrium initially in the simulation. A sample of the sensor output time responses are shown in Figs.7-9. Two channels each are selected for the left, the center, and the right portions of the truss structure. The magnitudes of the displacement response drop by orders of magnitude for nodes that are further away from the disturbances. The delay effect of the force pulses on the displacements shown in Fig.9 is typical for the nodes in the right portion.

Conclusions

An extension of Controlled Component Synthesis to encompass a multi-level decentralized control framework for large space truss has been introduced in this paper. The design methodology is tested out in the design of a two level hierarchical CCS for a planar truss in which the key idea of assembly of controlled components in CCS has been effectively generalized to a multi-level assembly of controlled components.

References

Benfield, W. A., and R. F. Hruda (1971). Vibration Analysis of Structures by Component Mode Substitution. *AIAA Journal*, Vol. 9, No. 7, pp. 1255-1261.

Craig, R. R., Jr., and M. C. C. Bampton (1968). Coupling of Substructures for Dynamic Analyses. *AIAA Journal*, Vol. 6, No.7, pp. 1313-1319.

Craig, R. R., Jr. (1977). Methods of Component Mode Synthesis. *Vibration and Shock Digest*, Vol.9, pp. 3-10.

Craig, R.R., Jr. (1981). *Structural Dynamics*,Wiley.

Hurty, W. C. (1960). Vibration of Structural Systems by Component Mode Synthesis. *ASCE J. Engr. Mech. Div.*, Vol.85, pp.51-69.

Ikeda M. and D. D. Šiljak(1980). Overlapping Decompositions, Expansions, and Contractions of Dynamic Systems. *Large Scale Systems*, Vol. 1, pp. 29-38.

Ikeda M., D. D. Šiljak, and D. E. White(1981). Decentralized Control with Overlapping Information Set. *J. Opt. Theory Appl.*, Vol. 34, pp. 279-310.

Kida, T., I. Yamaguchi, and Y. Ohkami (1985). An Approach to Local Decentralized Control for Large Flexible Space Structures. AIAA Paper No. 85-1924-CP, *AIAA Guidance, Navigation and Control Conference, August 19-21, 1985, Snowmass, CO.*

Meirovitch L. (1975). *Elements of Vibration Analysis*, McGraw-Hill.

Meirovitch L. (1981). *Computational methods in Structural Dynamics*, Sijthoff & Noordoff.

Šiljak, D. D.(1978). *Large-Scale Dynamic Systems*, North-Holland.

Young, K. D. (1983). An Application of Decomposition Techniques to Control of Large Structures. *Proceedings of the Fourth VPI &SU /AIAA Symposium on Dynamics and Control of Large Structures*, Blacksburg, VA, June 6-8, 1983, pp.201-214.

Young, K. D. (1985). Approximate Finite Element Models for Structural Control. *Proceedings of the 24th IEEE Conference on Decision and Control*, Fort Lauderdale, FL, December 11-13, 1985, pp. 940-945.

Young, K.D. (1988a). Controlled Component Synthesis. *Computational Mechanics 1988*, S.N. Atluri and G. Yagawa (Editors), pp.44.iv.1-4, Springer-Verlag.

Young, K.D.(1988b). A Distributed Finite Element Modeling and Control Approach for Large Flexible Structures. AIAA Paper No. 88-4085-CP, *AIAA Guidance, Navigation and Control Conference, Minneapolis, MN, August 15-17, 1988*, to appear in AIAA Guidance, Control, and Dynamics, February, 1990.

A PARALLEL ALGORITHM FOR LARGE SCALE OPTIMAL CONTROL PROBLEMS USING SPATIAL DECOMPOSITION AND COORDINATION*

Xiaohong Guan and P. B. Luh

Department of Electrical & Systems Engineering, University of Connecticut, Storrs, CT 06269-3157, USA

Abstract. This paper presents a new parallel algorithm for solving large scale, discrete-time optimal control problems under parallel processing environments. The idea is to decompose an optimal control problem into a number of smaller subproblems by using the spatial decomposition and goal coordination scheme. The couplings among subsystems are relaxed by introducing coupling variables and Lagrange multipliers, and decomposed subproblems are solved in parallel by using the Differential Dynamic Programming (DDP) method. The Lagrange multipliers are selected as coordinating variables, and are updated at the high level. An analytical approach for estimating the computational complexity of decomposition/coordination algorithms is introduced, and relations between speed-up and major system parameters are established. It is shown that significant speed-ups are difficult to achieve by using conventional optimization techniques at the high level. The reduction in computational requirements for solving lower dimensional subproblems in parallel may be offset by a large number of high level iterations. The Parallel Variable Metric (PVM) is found to be a promising high level algorithm for loosely coupled systems. Numerical results show that comparing with one level DDP, the PVM/DDP algorithm can obtain significant speed-ups under a simulated parallel processing environment. Moreover, global variational feedback controls which are invaluable for on-line control systems are obtained.

Keywords. Large-scale systems; Parallel processing; Decomposition; Optimal control; Nonlinear programming.

1. Introduction

Large scale optimal control problems have long been an interesting research topic. The advent of parallel computers fosters a considerable amount of efforts to sovle these problems under parallel processing environments. One kind of approach is based on the idea of decomposition and coordination, where a large optimal control problem is decomposed into a number of subproblems and appropriate coordinating variables are introduced. This forms a two level structure, where the low level consists of many smaller optimal control subproblems, and the high level is a parameter optimization problem. With the coordinating varialbles given, low level subproblems are decoupled and solved in parallel. The global optimality is then obtained by iteratively updating coordinating varialbles at the high level. Successful methods for solving long horizon problems have been developed along the line of time decompsition, where the original problem is decomposed along the time axis ([CHA89], [TAN89]).

Problems with large state dimensions are generally

more difficult to sovle since the computational requirements increase drastically as the state dimension grows. Several spatial decomposition techniques have been developed, including goal coordination and mixed coordination ([PEA71], [SIN80], [JAM83], [PER84]). In both methods, coupling variables and Lagrange multipliers are introduced to relax the coupling among subproblems. In the mixed coordination, coupling variables and Lagrange multipliers are selected as coordinating variables; whileas in the goal coordination only Lagrange multipliers are selected as coordinating variables. In contrast to time decomposition, the major difficulties of spatial decomposition lie in the fact that the dimension of high level coordinating variables is generally proportional to the product of time horizon and the dimension of coupling variables. Since the number of high level iterations usually increases with the above product, the reduction in computational requirements for solving lower dimensional subproblems in parallel may be offset by a large number of high level iterations. Little work has been found in analyzing whether decomposition/coordination methods are better than undecomposed ones in terms of computational complexity and efficiency. Moreover, existing techniques generally do not provide global feedback controls.

This paper presents a new parallel algorithm for solving large scale, discrete-time optimal control problems by using the spatial decomposition and goal coordination scheme. The problem formulation and the decomposition/coordination framework are presented in Section 2. In Section 3, the Differential Dynamic Programming (DDP) of [YAK84] for solving low level subproblems and a few promising parameter optimization techniques are discussed. An analytical approach for estimating the computational complexity of decomposition and coordination algorithms is introduced, and a relationship between algorithm speed-up and problem structure/size is established. These are important, since one would like to have an idea about the speed-up of a parallel algorithm before actual implementation and testing. Our analysis shows that significant speed-ups are difficult to achieve by using conventional parameter optimization techniques at the high level.

To increase speed-up, parallel parameter optimization techniques have to be adopted at the high level. Our complexity analysis indicates that the Parallel Variable Metric (PVM) method reported in [STR73], [LAA85] and [LOO88] is the most promising one among several candidates for loosely coulped systems. The PVM/DDP algorithm developed in Section 4 is efficient, having quadratic convergence rate at both levels. Moreover, global variational feedback controls which are invaluable for on-line control systems are obtained.

The PVM/DDP algorithm is implemented in a simulated parallel processing environment. Numerical testing results presented in Section 5 indicate that, comparing with one level DDP, significant speed-ups are obtained. The results also validate the analytical relationship between speed-up and problem structure/size.

* The work was supported in part by the National Science Foundation under Grant ECS-8717167. The authors would like to thank Professor S. C. Chang of National Taiwan University for valuable comments.

2. Problem formulation

Consider the following dynamic system consisting of N interconnected subsystems:

$$x(t+1) = f_t(x(t), u(t)), \quad t=0,1,\ldots\ldots,T-1, \quad (2.1)$$

where

$$x(t) = [x_1^T(t), x_2^T(t),\ldots\ldots,x_N^T(t)]^T,$$
$$u(t) = [u_1^T(t), u_2^T(t),\ldots\ldots,u_N^T(t)]^T.$$

The variable $x_i(t) \in R^{n_i}$ is the state of the ith subsystem with $\sum_{i=1}^{N} n_i = n$, $u_i(t) \in R^{m_i}$ is the control of the ith subsystem with $\sum_{i=1}^{N} m_i = m$, and superscript T denotes transpose. The system dynamics is assumed to have the following structure:

$$f_t(x(t), u(t)) = \begin{bmatrix} f_{1t}(x_1(t), u_1(t)) + \sum_{j=2}^{N} A_{1j}(t)x_j(t) \\ \cdots\cdots \\ f_{it}(x_i(t), u_i(t)) + \sum_{\substack{j=1 \\ j\neq i}}^{N} A_{ij}(t)x_j(t) \\ \cdots\cdots \\ f_{Nt}(x_N(t), u_N(t)) + \sum_{j=1}^{N-1} A_{Nj}(t)x_j(t) \end{bmatrix} \quad (2.2)$$

Note that $f_{it}(\cdot)$ is a function of $x_i(t)$ and $u_i(t)$ only, and the coupling among subsystems is linear in state variables. The function $f_{it}(\cdot)$ is assumed to have continuous second order derivatives with respect to $x_i(t)$ and $u_i(t)$.

Let $y_i(\cdot)$ denote the coupling of the ith subsystem from other subsystems, i.e.,

$$y_i(t) \equiv \sum_{\substack{j=1 \\ j\neq i}}^{N} A_{ij}(t)x_j(t). \quad (2.3)$$

One can directly define $y_i(\cdot)$ as the "coupling variable". However, since the dimensions of coupling variables greatly affect computational complexity, a better definition is warranted. To do so, note that $A_{ij}(t)$ is an $n_i \times n_j$ matrix. It extracts information from $x_j(t)$ and feeds it to subsystem i. If this coupling information contains at most $r_i \; (\leq n_i)$ linearly independent combinations of $x_j(t)$, and this true for all $j\neq i$, then (2.3) can be rewritten as

$$y_i(t) = C_i(t)\sum_{\substack{j=1 \\ j\neq i}}^{N} M_{ij}(t)x_j(t), \quad (2.4)$$

where C_i is an $n_i \times r_i$ matrix, M_{ij} is a $r_i \times n_j$ matrix, and $\sum_{i=1}^{N} r_i = r$. For a loosely coupled system, r_i is usually smaller than n_i, therefore $r < n$. The coupling variables $z_i(\cdot)$ are then defined as

$$z_i(t) \equiv \sum_{\substack{j=1 \\ j\neq i}}^{N} M_{ij}(t)x_j(t), \quad t=0,\ldots,T-1, \quad i=1,\ldots,N, \quad (2.5)$$

with $z_i(t) \in R^{r_i}$. Equation (2.2) can then be rewritten as:

$$f_t(x(t), u(t)) = \begin{bmatrix} f_{1t}(x_1(t), u_1(t)) + C_1(t)z_1(t) \\ \cdots\cdots \\ f_{it}(x_i(t), u_i(t)) + C_i(t)z_i(t) \\ \cdots\cdots \\ f_{Nt}(x_N(t), u_N(t)) + C_N(t)z_N(t) \end{bmatrix}. \quad (2.6)$$

The cost function to be minimized is of the following additive form:

$$J \equiv \sum_{i=1}^{N} \left[\sum_{t=0}^{T-1} g_{it}(x_i(t), u_i(t), z_i(t)) + g_{iT}(x_i(T)) \right], \quad (2.7)$$

where $g_{it}(\cdot)$ and $g_{iT}(\cdot)$ are assumed to have continuous second order derivatives with respect to their arguments. Furthermore, $g_{it}(\cdot)$ is convex, has positive definite Hessian over $u_i(\cdot)$ and $z_i(\cdot)$, and $g_{iT}(\cdot)$ is convex over $x_i(T)$. The problem is then

(P):

$$\min_{u(t)} J, \quad (2.8)$$

subject to (2.6) with the initial condition x(0) given.

As metioned in the Introduction, the computational requirements in solving **P** increase drastically as the state dimension grows. The idea of spatial decomposition is to partition **P** into a number of smaller subproblems and solve them in parallel. To do so, couplings among subsystems are relaxed by using the Lagrange multiplier sequence

$$\lambda(t) = [\lambda_1^T(t), \lambda_2^T(t),\ldots\ldots,\lambda_N^T(t)]^T, \quad t=0,1\ldots,T-1, \quad (2.9)$$

where

$$\lambda_i(t) \in R^{r_i}, \; i=1,\ldots\ldots,N, \quad \text{and} \quad \lambda(t) \in R^r.$$

Then the Lagrangian can be written as

$$\begin{aligned} L &= \sum_{i=1}^{N} \left[\sum_{t=0}^{T-1} [\, g_{it}(x_i(t), u_i(t), z_i(t)) \right. \\ &\quad \left. + \lambda_i^T(t)(z_i(t) - \sum_{\substack{j=1 \\ j\neq i}}^{N} M_{ij}(t)x_j(t)) \,] + g_{iT}(x_i(T))\right], \\ &= \sum_{i=1}^{N} \left[\sum_{t=0}^{T-1} [\, g_{it}(x_i(t), u_i(t), z_i(t)) + \lambda_i^T(t)z_i(t) \right. \\ &\quad \left. - \sum_{\substack{j=1 \\ j\neq i}}^{N} \lambda_j^T M_{ji}(t)x_i(t) \,] + g_{iT}(x_i(T))\right]. \quad (2.10) \end{aligned}$$

For convenience, define a stack form of the Lagrange multipliers with dimension r·T as:

$$\lambda_s = [\lambda^T(0), \lambda^T(1),\ldots\ldots,\lambda^T(T-1)]^T. \quad (2.11)$$

By using the duality theorem ([LUE84]), problem **P** can be rewritten as:

$$\begin{aligned} \max_{\lambda_s} \min_{\substack{u(t) \\ z(t)}} L &= \max_{\lambda_s} \{\, \sum_{i=1}^{N} \min_{\substack{u_i(t) \\ z_i(t)}} [\, \sum_{t=0}^{T-1} [\, g_{it}(x_i(t), u_i(t), z_i(t)) \\ &\quad + \lambda_i^T(t)z_i(t) - \sum_{\substack{j=1 \\ j\neq i}}^{N} \lambda_j^T M_{ji}(t)x_i(t) \,] \\ &\quad + g_{iT}(x_i(T))]\,\}. \quad (2.12) \end{aligned}$$

By selecting Lagrange multipliers as coordinating variables, low level optimal control subproblems are formulated as follows:

(P-i), i=1,2,...,N:

$$\begin{aligned} \min_{\substack{u_i(t) \\ z_i(t)}} L_i, \quad \text{with } L_i &\equiv \sum_{t=0}^{T-1} [\, g_{it}(x_i(t), u_i(t), z_i(t)) \\ &\quad + \lambda_i^T(t)z_i(t) - \sum_{\substack{j=1 \\ j\neq i}}^{N} \lambda_j^T(t)M_{ji}(t)x_i(t) \,] \\ &\quad + g_{iT}(x_i(T)), \quad (2.13) \end{aligned}$$

subject to

$$x_i(t+1) = f_{it}(x_i(t), u_i(t)) + C_i(t)z_i(t). \quad (2.14)$$

Let $x_i^*(t)$, $u_i^*(t)$ and $z_i^*(t)$ denote, respectively, the optimal state, control and coupling variables of **P-i** for a given λ_s, and $L_i^*(\lambda_s)$ the optimal Lagrangian. Then the high level problem **P-H** is to select the best λ_s to maximize dual function:

(P-H):

$$\max_{\lambda_s} \Phi(\lambda_s), \quad \text{with} \quad \Phi(t) \equiv \sum_{i=1}^{n} L_i^*(\lambda_s). \quad (2.15)$$

Note that **P-H** is a parameter optimization problem where the dimention of λ_s is r·T.

3. The Low Level Differential Dynamic Programming Algorithm and Typical High Level Algorithms

3.1. The Low Level Differential Dynamic Programming Algorithm

There are many methods for solving low level subproblems (e.g., [BRY75], [SIN80], [JAM83]). However, several major disadvantages exist. Solving matrix Racatti equations for linear quadric (LQ) problems is efficient but the method can not be easily extended for problems with nonlinear system dynamics and/or nonquadratic cost functions. Gradient based algorithms are general but slow in convergence. Newton type algorithms usually require a lot more computations. Furthermore, these numerical methods generally do not provide feedback controls which are required for most on-line control systems. Dynamic programming, on the other hand, suffers from the curse of dimensionality.

To overcome the above difficulties, the Differential Dynamic Programming (DDP) of [YAK 84] is used to solve low level subproblems. DDP is a dynamic programming based successive approximation method, and has quadratic convergence rate. Suppose that the initial nominal state, control and coupling variables, $\{\bar{x}_i^0(t), \bar{u}_i^0(t), \bar{z}_i^0(t),$ t=0,1,......,T-1\}, are given. The method consists of a backward sweep deciding variational feedback control coefficients, and a forward sweep to update the nominal trajetory. Here coupling variables $\{z_i(\cdot)\}$ are treated as independent inputs to the ith subsystem. The DDP algorithm of [YAK84] is modified to contain coupling and coordinating variables, and to obtain global variational feedback controls.

The backward sweep is based on the quadratic approximations of the stage-wise cost and the optimal cost-to-go at each stage. By minimizing the sum of the above two costs, linear variational feedback controls are obtained:

$$\delta u_i(T) = \alpha_{1i}(T) + \beta_{1i}(T)\delta x_i(T), \qquad (3.1)$$

$$\delta z_i(T) = \alpha_{2i}(T) + \beta_{2i}(T)\delta x_i(T), \quad t=0,1,......,T-1. \quad (3.2)$$

This backward sweep starts from the terminal stage and works backwards in time, until the initial stage is reached.

In the forward sweep, variational feedback control coefficients are used to update nominal variables as follows:

$$\bar{u}_i^{l+1}(t) = \bar{u}_i^l(t) + \nu\alpha_{1i}(t) + \beta_{1i}(t)(\bar{x}_i^{l+1}(t) - \bar{x}_i^l(t)), \quad (3.3)$$

$$\bar{z}_i^{l+1}(t) = \bar{z}_i^l(t) + \nu\alpha_{2i}(t) + \beta_{2i}(t)(\bar{x}_i^{l+1}(t) - \bar{x}_i^l(t)), \quad (3.4)$$

$$\bar{x}_i^{l+1}(t+1) = f_i(\bar{x}_i^{l+1}(t), \bar{u}_i^{l+1}(t)) + C_i(t)\bar{z}_i^{l+1}(t), \quad (3.5)$$

where $\bar{x}_i^l(t)$ is nominal state at the lth iteration, etc. This forward sweep starts from the initial stage and works forward in time, until the terminal stage is reached. The value of ν is first set to one, and reduced by half if necessary until the total cost of new trajectory is lower than that of the original one. This ν reduction step is similar to the line search procedure in paramater optimization.

The backward sweep and forward sweep iterate until certain stopping criterion is satisfied. The final trajetory is optimal for the given $\{\lambda(t),$ t=0, 1,......,T-1\} if the system dynamics is linear and the cost function is convex. Otherwise, the result is locally optimal.

Note that $\delta z_i(t)$ is actually not an independent input to subsystem i. Rather, it is a linear combination of variational states of other subsystems. Thus in real operations, it would be desirable to have $\delta u_i(t)$ as a function of both $\delta x_i(t)$ and $\delta z_i(t)$. From the first order necessary conditions, it can be shown that $\delta u_i(t)$ can be expressed as follows:

$$\delta u_i(t) = \alpha_i(t) + \beta_i(t)\delta x_i(t) + \gamma_i(t)\delta z_i(t). \qquad (3.6)$$

Thus we have global variational feedback controls required for most on-line control systems. Derivations are provided in a detailed version of the paper.

All the computations involved in one iteration of DDP are matrix and vector manipulations. The complexity of matrix-matrix multiplication and matrix inversion is proportional to the cube of the matrix size for square matrices ([PRE88]), and is the dominating factor. The complexity of matrix-vector manipulation is proportional to the product of the dimensions of the matrix, and is negligible in this case. Thus the complexity of the ith subproblem is

$$C_{ddpi} = s_{ddp}N_{ddpi}Tn_i^3, \qquad (3.7)$$

where s_{ddp} is a scaling constant, N_{ddpi} is the product of total number of iterations and average number of ν reduction steps per iteration for the ith subproblem. Suppose that low level subproblems are solved in parallel with N processors. Then the complexity of the low level is

$$C_{ddp} = \max_i C_{ddpi}$$
$$= s_{ddp}N_{ddp}Tn_{ddp}^3, \qquad (3.8)$$

where N_{ddp} is the N_{ddpi} of the worst subproblem, and n_{ddp} is the corresponding state dimension.

If DDP is used to solve problem **P** directly without decomposition, the complexity of one level DDP is

$$C_0 = s_0N_0Tn^3, \qquad (3.9)$$

where s_0 is a scaling constant, N_0 is the product of total number of iterations and average number of ν reduction steps per iteration. It is important to note that the one level DDP algorithm can not make good use of the sparsity of the system dynamics. Even for an LQ problem, the sparsity of the system matrices generally does not lead to the sparsity of Hessian matrices of cost-to-go functions in the backward sweep. This can be seen from the computational procedure of the backward sweep of DDP.

The number of DDP iterations is generally problem dependent. However, for LQ problems, DDP can be completed in just one iteration (two if the computation for a given initial nominal trajectory is included). This is because DDP has the property of Newton's method for parameter optimization. Comparing (3.8) and (3.9), we can see potential savings in dealing with subproblems of smaller dimensions. The major difficulties of most spatial decomposition and coordination methods, however, lie in the fact that the number of high level iterations generally increases with the dimension of high level coordinating variables. The reduction in complexity for dealing with smaller subproblems may be offset by a large number of high level iterations. Selection of high level algorithm is thus of crucial importance to obtain significant speed-ups, and is discussed next.

3.2. Typical High Level Algorithms

To solve the high level parameter optimization problem, the Newton's method is efficient but requires the inverse of Hessian matrix of $\Phi(\lambda_s)$. This is difficult because $\{x_i^*(t), u_i^*(t), z_i^*(t)\}$ in (2.15) are not explicit functions of λ_s. Furthermore, inversing a big Hessian matrix of dimension $rT \times rT$ may result in numerical instability. Our concentration is therefore on the conjugate gradient and variable metric methods which are more attractive among others.

The conjugate gradient method updates λ_s as follows ([LUE84]):

$$\lambda_s^{k+1} = \lambda_s^k + \eta s^k, \qquad (3.10)$$

where

$$s^k = -\nabla\Phi(\lambda_s^k) + \frac{||\nabla\Phi(\lambda_s^k)||^2}{||\nabla\Phi(\lambda_s^{k-1})||^2}s^{k-1} \qquad (3.11)$$

is the conjugate gradient direction vector, k is the iteration index and η is the step size. From (2.10), $\nabla\Phi(\lambda_s)$ can be easily obtained once low level subproblems are solved:

$$\nabla\Phi(\lambda_i(t)) = z_i^*(t) - \sum_{\substack{i=1 \\ i \neq j}}^{N} M_{ij}(t)x_j^*(t). \qquad (3.12)$$

Thus the complexity of one high level iteration using the conjugate gradient algorithm is $O((rT)^2)$ resulting from matrix-vertor manipulations. The total complexity is therefore given by

$$C_{cg} = N_{cg}[s_{cg}(rT)^2 + C_{ddp}], \qquad (3.13)$$

where s_{cg} is a scaling constant, and N_{cg} is the product of total number of high level iterations and average number of function evaluations in a high level line search step.

The basic idea of the variable metric method is to approximate the inverse Hessian by using values of variables and gradients between two consecutive iterations ([LUE84]). This eliminates the major difficulty of the Newton's method for requiring to inverse high dimensional Hessian matrices. Suppose that the approximate inverse Hessian at the kth iteration is given by H^k, then the equation for updating H^k is

$$H^{k+1} = H^k + \frac{(\sigma^k - H^k y^k)(\sigma^k - H^k y^k)^T}{(\sigma^k - H^k y^k)^T y^k}, \qquad (3.14)$$

where

$$\sigma^k = \lambda_s^{k+1} - \lambda_s^k, \qquad (3.15)$$

$$y^k = \nabla\Phi(\lambda_s^{k+1}) - \nabla\Phi(\lambda_s^k), \qquad (3.16)$$

with $H^0 = I$, the identity matrix. Then following the Newton's method, coordinating variables are updated according to

$$\lambda_s^{k+1} = \lambda_s^k - \eta H^k \nabla\Phi(\lambda_s^k), \qquad (3.17)$$

where η is the step size. Similar to the analysis for the conjugate gradient method, the complexity of one high level iteration using the variable metric method is of order $O((rT)^2)$. The total complexity is

$$C_{vm} = N_{vm}[s_{vm}(rT_f)^2 + C_{ddp}], \qquad (3.18)$$

where s_{vm} is a scaling constant, N_{vm} is the product of total number of iterations and average number of function evaluations in a high level line search step.

The number of high level iterations is generally problem dependent. Comparison of algorithms, however, can be made based on the LQ assumption. It will be shown next that the number of high level iterations is deterministic in this case, and formulas for speed-ups can be explicitly derived.

Theorem 3.1. Suppose that the system dynamics (2.6) is linear, and stagewise cost functions (2.7) are of the following quadrtic form:

$$g_{it}(x_i(t), u_i(t), z_i(t)) = \frac{1}{2}[x_i^T(t)Q_i(t)x_i(t)$$

$$+ u_i^T(t)R_i(t)u_i(t) + z_i^T(t)S_i(t)z_i(t)], \qquad (3.19)$$

$$g_{iT}(x_i(T)) = \frac{1}{2}x_i^T(T)Q_i(T)x_i(T), \qquad (3.20)$$

where $Q_i(\cdot)$ are positive semidefinite, $R_i(\cdot)$ and $S_i(\cdot)$ are positive definite. Then the high level dual function $\Phi(\lambda_s)$ of (2.15) is of the following quadratic form:

$$\Phi(\lambda_s) = \lambda_s^T \Gamma \lambda_s + \zeta^T \lambda_s + constant, \qquad (3.21)$$

with Γ being negative definite.

The proof is provided in a detailed version of the paper.

It is known that for the quadratic function (3.21), the numbers of iterations using conjugate gradient and variable metric methods equal the dimension of λ_s i.e. rT ([LUE 84]). However, there is no formula for the average number of function evaluations per line search and scaling constants. From experience, it is asssumed that the average number of function evaluations per line search is 4. The scaling constants are determined based on the fact that the complexity of multiplying two square matrices is one times the cube of the matrix size, matrix inversion is $\frac{4}{3}$ times the cube of matrix size ([PRE88], p.38). Then by counting the number of dominating matrix manipulations and neglecting nondominating ones in each algorithm, the following scaling constants are obtained:

$$s_0 \approx 10, \quad s_{ddp} \approx 50, \quad s_{cg} \approx 1, \quad s_{vm} \approx 2. \qquad (3.22)$$

With above parameters we have:

$$C_0 = 20Tn^3, \qquad (3.23)$$

$$C_{cg} = 4rT[(rT)^2 + 100Tn_{ddp}^3], \qquad (3.24)$$

$$C_{vm} = 4rT[2(rT)^2 + 100Tn_{ddp}^3]. \qquad (3.25)$$

Note that speed-ups for two level algorithms are obtainable if n_{ddp}, r, and T are small. That is, the system is finely decomposed (n_{ddp} is much smaller than n), loosely coupled (r small), and has short time horizon. Even for such a system, speed-up may not be significant. Suppose $n_{ddp} = \frac{1}{15}n$, $r = \frac{1}{15}n$, then speed-ups are given by

$$S_{cg} = \frac{C_0}{C_{cg}} = \frac{15^4}{T[3T + 20n]}, \qquad (3.26)$$

$$S_{vm} = \frac{C_0}{C_{vm}} = \frac{15^4}{T[6T + 20n]}. \qquad (3.27)$$

Speed-ups are achievable only for small n and T, inconsistent with our original goal for solving problems with large dimensions. For example, with $n=150$, $n_{ddp} = 10$, $r = 10$, $T = 10$ and $N=10$, we have $S_{cg} = 1.67$, and $S_{vm} = 1.65$. The speed-ups are not significant for a ten processor system. This clearly says that the reduction in complexity by solving lower dimentional subproblems in parallel is offset by large numbers of high level iterations and function evaluations. Thus reducing numbers of high level iterations and function evaluations is crucial in obtaining significant speed-ups.

4. The High Level Parallel Variable Metric Algorithm

The key to improve our two level algorithm is to reduce the number of high level iterations and number of function evaluations in a high level line search step. This is possible by parallelizing the high level algorithm.

One version of parallel conjugate gradient methods is to concurrently evaluate several function values in a line search procedure ([LOO88]). The improvement, however, is not significant. Another parallel conjugate direction method performs K independent line searches along K conjugate directions, with K being the problem dimension ([MIA86]). Results from K line searches are then combined linearly to form a new point. Under the LQ assumption, the method requires only two parallel iterations to converge. This implies that low level subproblems have to be solved $2l_{pcg}$ times in parallel, where l_{pcg} is the number of parallel function evaluations per iteration. The number of processors needed is rTN since the dimension of the high level problem is rT and N processors are needed for each of the rT conjugate directions.

The parallel variable metric (PVM) method of [STR73], [LAA85] and [LAA88] is more attractive to us because it reduces the number of iterations but does not require many independent line searches. Under the LQ assumption, the algorithm has the property of Newton's method, i.e., it converges in just one iteration (two if solving low level subproblems for the given initial condition is included) and no line search is needed.

Let λ_s^k be the high level coordinating variable at the kth iteration and H^k the corresponding approximate inverse Hessian. The key idea of the PVM algorithm is to form rT linearly independent variables around λ_s^k and to compute their gradients in parallel. By using gradients corresponding to these rT linearly independent variables and that relating to λ_s^k, the approximate inverse Hessian is updated. Let

$$\delta^j = \varepsilon \cdot e^j \qquad j = 1, \ldots \ldots rT, \qquad (4.1)$$

where ε is a sufficiently small real number and e^j is the jth unit vector whose elements are zeros except the jth element being one. Let

$$\lambda_s^{kj} \equiv \lambda_s^k + \delta^j, \qquad j = 1,\ldots\ldots,rT, \qquad (4.2)$$

then λ_s^{kj}, j=1,......,rT, are rT linearly independent variables around λ_s. Define the partial updating matrix as follows:

$$\Psi^{kj} \equiv \Psi^{k(j-1)} + \tau^{kj} \, (\rho^{kj})(\rho^{kj})^T, \qquad j=1,\ldots\ldots rT, \quad (4.3)$$

where

$$\rho^{kj} = \delta^j - \Psi^{k(j-1)} y^{kj}, \qquad j=1,\ldots\ldots rT, \qquad (4.4)$$

$$y^{kj} = \nabla\Phi(\lambda_s^{kj}) - \nabla\Phi(\lambda_s^k), \qquad j = 1,\ldots\ldots rT, \qquad (4.5)$$

$$\tau^{kj} = ((\rho^{kj})^T(y^{kj}))^{-1}, \qquad j = 1,\ldots\ldots rT. \qquad (4.6)$$

The initial value of is $\Psi^{k0} = H^k$. Then the approximate inverse Hessian is obtained as a result of updating Ψ^{kj}:

$$H^{k+1} = \Psi^{k(rT)}. \qquad (4.7)$$

Finally, λ_s^k is updated according to (3.17). As mentioned in [LOO88], the PVM algorithm has the property of quadratic convergence.

The computaions involved in PVM excluding the calculation of $\nabla\Phi(\lambda_s^{kj})$ are rT sets matrix-vector manipulations of the dimension rT. This has the complexity of $O((rT)^3)$. Calculating $\nabla\Phi(\lambda_s^{kj})$ using (3.12) requires the optimal state and coupling variables of all low level subproblems. For each λ_s^{kj}, j=1,......,rT, N processor are needed to solve low level subproblems. Therefore NrT processors are required for the parallel computation of $\nabla\Phi(\lambda_s^{kj})$.

To reduce the number of processors needed, low level subproblems have to be grouped. The idea is to use one processor to solve problem **P-i** (rT+1) times for λ_s equal to λ_s^k and λ_s^{kj}, j=1,......,rT. Problem **P-i** is first solved for λ_s^k. The additional computations needed for λ_s^{kj}, j=1,......,rT are not much. The reason is that for LQ problems the Hessians of stage-wise cost functions are constant matrices. Therefore, computations of quadratic terms of cost-to-go and optimal cost-to-go at each stage need only to be done once for all the (rT+1) λ_s's. For non-LQ problems, Hessians obtained for λ_s^k can also be used as approximate ones for λ_s^{kj}, j=1,......,rT since ε is sufficiently small. Testing results presented in Section 5 justify this approximation. This leads to solving N subproblems by using N processors in parallel, while each subproblem is solved by one processor (rT+1) times with reduced effort. If more than N processors are available, further parallelization can also be done.

Now consider the complexity of the PVM/DDP algorithm. The addtional computations at the low level beyond (3.8) are rT sets of matrix-vector manipulations in the backward and forward sweeps with complexity $O(rTn_{ddp}^2)$. Therefore the complexity of the PVM/DDP algorithm is

$$C_{pvm} = N_{pvm}[s_{vh}(rT)^3 + N_{ddp}T(s_{ddp}n_{ddp}^3 + s_{vl}rTn_{ddp}^2)],(4.8)$$

where s_{vh} and s_{vl} are scaling constants, and N_{pvm} is the product of total number of iterations and average number of function evaluations in a high level line search step. Under the LQ assumption, $N_{pvm} = 2$. Selecting scaling constants as

$$s_{vh} \approx 8, \ s_{vl} \approx 200,$$

we have

$$C_{pvm} = 2 \, [8(rT)^3 + 2T(50n_{ddp}^3 + 200rTn_{ddp}^2)]. \qquad (4.9)$$

Then the speed-up for the PVM/DDP algorithm is given by

$$S_{pvm} = \frac{C_0}{C_{pvm}}$$

$$= \frac{5Tn^3}{4(rT)^3 + T(50n_{ddp}^3 + 200rTn_{ddp}^2)}. \qquad (4.10)$$

Suppose $n_{ddp} = \frac{1}{15}n$, $r = \frac{1}{15}n$, (4.10) becomes

$$S_{pvm} = \frac{5\times15^3}{4T^2 + 200T + 50}. \qquad (4.11)$$

Note that S_{pvm} is independent of n, but decreases as T increases. For the example of Section 3.2, one obtaines $S_{pvm} = 6.9$ with 10 processors. The speed-up is significant.

From (4.10) and (4.11), we conclude that the algorithm is particularly suitable for finely decomposed, loosely coupled systems with short horizon. As mentioned in Section 3.1, one level DDP can hardly make good use of these properties to reduce the computational complexity. If time horizon T is too large, the combination of time decomposition methods developed in [CHA89], [TAN89] and the spatial decomposition method presented here seems promising.

5. Numerical Testing Results

Testing of decomposition/coordination algorithms for large scale optimal control problems is not a easy job. There is no standard for system stucture, degree of coupling, etc. where the comparion of algorithms can be based on. Furthermore, dimensions of test problems should be large enough so that advantages of parallel decomposition and coordination algorithms can stand out. This, however, is very costly. In fact, few examples have been found in the literature that compare performance of decomposed algorithms versus undecomposed ones.

In this section, one problem with quadratic cost function and one problem with nonquadratic cost function are tested. The high level convergence criterion is

$$\nabla\Phi(\lambda_s) \le 0.0001. \qquad (5.1)$$

The convergence criterion for low level DDP is

$$\frac{|L_i^{l+1} - L_i^l|}{|L_i^l| + 1} \le 0.00001, \qquad (5.2)$$

with $L_i \equiv J$ for one level DDP. The testing is performed on an IBM 3090 mainframe computer under a simulated parallel processing environment in lack of a real parallel processing system. The number of processors is assumed to be the same as the number of low level subproblems. We also assume that the processing is synchronous and communication time is negligible. The low level CPU time at each iteration is obtained as the longest CPU time in solving individual subproblems for that iteration. The total CPU time is then taken as the high level time plus the the sum of low level CPU times for all iterations.

The first problem tested is an LQ problem. The system dynamics and cost function are provided in a detailed version of the paper. The main system parameters are: n=31, m=11, r=2, n_{ddp}=3 and N=11. Testing results are summarized in Table 1.

From Table 1, it can be seen that significant speed-ups are achieved by using the PVM/DDP algorithm for problems with short time horizons. Being consistent with the analysis of Section 4, estimated speed-ups are close to actual ones. Note that for T=10, 15 and 20 the number of one level DDP iterations is 3 instead of 2, the theoratical value for LQ problems. This may be caused by accumulate numerical errors since the sizes of matrices in one level DDP are large in these cases. The degree of sparsity also affects the speed-up. For example, if we reduce the dimensions of state and control variables but keep the dimension of coupling variables, a similar but relatively dense system is formed with n=20, m=10, r=2 and N=10. The resulting speed-up for T=6 is 1.6 in contrast to 6.83 in Table 1.

The second problem has the same system dynamics as the first one except that quartic terms with significant weightings are added to the quadratic cost function of the first problem. This cost function is provided in a detailed version of the paper. Results are summarized in Table 2.

The speed-ups in Table 2 are better than those in Table 1. This may be caused by the fact that quadratic approximations in DDP for small subproblems are more accurate than those for large undecomposed problems. Thus the one level DDP may require more iterations to converge. For example, with T=6, the number of DDP iterations is 23 for the undecomposed problem; while only 4 or 5 for each decomposed subproblem. Large number of one level DDP itera-

Table 1. Testing Results for Problem 1
with a Quadratic Cost Function

* number of one level DDP iterations
** number of high level iterations
\# Estimated speed-ups from (4.11)

T	IT_1^*	One-level DDP CPU time (s)	cost function	IT_h^{**}	PVM/DDP CPU time (s)	cost function	speed -up	$S_{pvm}^{\#}$
3	2	0.36	184.12	2	0.035	184.12	10.3	12.1
6	2	1.23	276.48	2	0.18	276.48	6.83	6.17
10	3	2.16	298.11	2	0.66	298.11	3.27	3.67
15	3	3.42	304.37	2	1.82	304.38	1.84	2.38
20	3	4.57	306.34	2	3.74	306.35	1.22	1.86

Table 2. Results for Problem 2
with a Nonquadratic Cost Function

* number of one level DDP iterations
** number of high level iterations

T	IT_1^*	One-level DDP CPU time (s)	cost function	IT_h^{**}	PVM/DDP CPU time (s)	cost function	speed -up
3	11	1.96	182.64	3	0.13	182.64	15.1
6	23	9.38	270.21	3	0.87	270.21	10.7
10	37	26.50	300.59	5	6.30	300.59	4.20
15	63	59.86	307.98	4	16.52	307.98	3.62

tions also indicate that weightings of nonquadratic terms are significant. Furthermore, the efficiency for T=3 (efficiency is defined as the ratio of speed-up versus the number of processors, and is a measure of how computation power is utilized) is $\frac{15.1}{11} = 1.37$, which is greater than one. This says that even by using one processor in this case, the PVM/DDP algorithm is better than the one level DDP in terms of computation time.

Another factor affecting speed-ups is the value of ε in (4.1). If ε is too small, the differences between high level variables λ_s^{kj} and λ_s^k and their corresponding gradients may be too small. This may result in numerical difficulties since the approximate inverse Hessian may not be accurate. On the other hand, ε can not be too large. As mentioned in Section 4, Hessians of stage-wise cost and cost-to-go functions corresponding to λ_s^k are used as approximations for those corresponding to λ_s^{kj}. Large ε could cause inaccuracy of these approximations. Testing for different values of ε is performed on problem 2 with T=6. The results are summerized in Table 3.

Table 3. Results for Different values of ε

ε	IT_h	cost function	CPU time (s)	speed-up
0.001	4	270.2087	1.14	8.23
0.002	3	270.2087	0.87	10.7
0.003	6	270.2092	1.78	5.27

Testing results validate our reasonig above. The value $\varepsilon = 0.002$ is used in generating Table 1 and Table 2.

6. Conclusions

This paper develops a new parallel algorithm for solving large scale, discrete-time optimal control problems by using the spatial decomposition and goal coordination scheme. In the process, an anlytical approach for evaluating computational complexity of decomposition/coordination algorithms is introduced and relations between speedups and major system parameters are established. The analysis sheds a new light on the issue of complexity for decomposition and coordination algorithms. It is found that reductions of high level iterations and function evaluations are crucial in obtaining significant speedups. By parallelizing the high level algorothm, the PVM/DDP algorithm developed achieves significant speedups comparing to one level DDP, and is particularly suitable for finely decomposed, loosely coupled systems with short time horizon under parallel processing environments. Furthermore, global variational feedback controls which are valuable for on-line control systems are obtained.

References

[BRY75] A. E. Bryson, Jr., Y. C. Ho, *Applied Optimal Control,* revised printing, Hemisphere Publishing Corporation, 1975.

[CHA89] S. C. Chang, T. S. Chang, P. B. Luh, "A Hierarchical Decomposition for Large Scale Optimal Control Problems with Parallel Processing Structure," *Automatica,* Vol. 25, No. 1, January 1989, pp. 77-86.

[JAM83] M. Jamshidi, *Large Scale Systems,* North Holland, 1983.

[LAA85] P. Van Laarhoven, "Parallel Variable Metric Algorithms for Unconstrained Optimization," *Mathematical Programming,* Vol. 33, 1985, pp. 68-81.

[LOO88] F. A. Loostma, K. M. Ragsdell, "State-of-the-Art in Parallel Nonlinear Optimization," *Parallel Computing,* Vol. 6, No. 2, 1988, pp. 133-155.

[LUE84] D. G. Luenberger, *Linear and Nonlinear Programming,* second edition, Addison Wesley, 1984.

[MIA86] X. Miao, P. B. Luh, S. C. Chang, "Parallel Conjugate Direction Method for Unconstrained Optimization," *Proceedings of the Second IASTED International Conference on Applied Control and Identification,* Los Angeles, CA, Dec. 1986, pp. 5-9.

[PEA71] J. D. Pearson, "Dynamic Decomposition Techniques," *Optimization Methods for Large Scale Systems,* D. A. Wismer (editor), McGraw-Hill, 1971.

[PER84] P. F. Perry, "Spatial and Time Decomposition Algorithms for Dynamic Nonlinear Network Optimization Using Duality," *Journal of Optimization Theory and Applications,* Vol. 42, No. 1, January 1984, pp. 77-101.

[PRE88] W. H. Press, et al, *Numerical Recipes in C,* Cambridge University Press, 1988.

[SIN80] M. G. Singh, *Dynamical Hierarchical Control,* North Holland, 1980.

[STR73] T. A. Straeter, "A Parallel Variable Metric Optimization Algorihm," *NASA Technical Note,* NASA TN D-7329, 1973.

[TAN89] J. Tang, P. B. Luh, T. S. Chang, "A Parallel Algorithm for Long Horizon Optimal control Problems Using the Mixed Coordination Method," *Proceedings of the 1989 American Control Conference,* Pittsburgh, PA, June, 1989, pp. 1783-1788.

[YAK84] S. Yakowits, B. Rutherford, "Computational Aspects of Discrete-Time Optimal Control," *Applied Mathematics and Computation,* Vol. 15, No. 1, 1984, pp. 29-45.

LARGE SCALE SYSTEMS WITH MULTIPLE OBJECTIVES: A NEGOTIATION PROCEDURE AND ITS APPLICATION IN POWER SYSTEMS

J. R. Cardarelli*, F. A. Gomide and K. Tarvainen*****

**Villares Indústria de Base S/A, 01516 São Paulo, SP, Brasil*
***UNICAMP/FEE/DCA, C.P. 6101, 13081 Campinas, SP, Brasil*
****Helsinki University of Technology, Otakaari 1, 02150 Espoo 15, Finland*

Abstract. An interactive negotiation procedure for large scale systems with multiple
objectives is proposed. It is assumed that there are multiple decision makers, who have
their own multiple objectives and who are dependent on each other via common resources
or physical connections. The negotiation procedure includes two repeated main steps:
the decision makers independent multicriteria optimization of their subsystems, and
a convenient step for trade-off between the decision makers. Assumptions guaranteeing
the convergence of the negotiation scheme are reviewed; among these, one assumption concerning independence of subsystems, is essential. The negotiation procedure is applied
in the operation planning of two coupled hydroelectrical power systems of the southeast
region of Brazil, and simulation results are included to show its usefulness in solving real world problems. The same operation planning problem is also solved by the
SEMOPS method. A comparison is made from the viewpoint of methodology, practice and
computational effort. The results obtained from the comparisons show that the proposed
scheme outperforms SEMOPS. Finally, conclusions and further work are addressed.

Keywords. Large-scale systems ; multiobjective optimization ; interactive methods ;
power management ; electric power systems.

INTRODUCTION

Real world problems, in greater number, are complex
and difficult to model and to solve. Most of the
time the classical, single objective optimization
theory is not enough to present more realistic solutions. The problems can involve more than one
decision maker, who needs to agree with one preferred solution. In addition, usually we have a
multiplicity of solutions that arise due to the
several objective functions inserted in the model
to search the best comprimise solution.

Considering that the reality does not have an ordinary and well known behaviour, new theories,
methods and tools have been developed to aid in
looking for better solutions. An example is the
multiobjective optimizatoin theory (Chankong, Haimes, 1978) which enables an analyst to take into
consideration more than one objective, including
the relationships among them (trade-offs). Decision
making theory (Keeney, Raiffa, 1976), with a group
utility function that possibly leads the decision
maker to use criteria explicitly from the model,
and the decomposition theory (Tarvainen, 1980; Gomide, 1982) which helps an analyst to transform
complex problems into a set of simpler problems
with special characteristics are further examples
of these developments intended to help analysts
and decision makers to search for the best compromise or preferred solution in complex, large scale
systems.

In this paper we present an interactive negotiation
procedure - INP - for large scale systems with multiple objectives as a method to obtain preferred
solutions for a class of problems characterized
by systems that have common resources or physical
interconnections. It is based on the exchange of
(trade-off) information between an analyst and the

decision makers aiming at a best compromise solution for a problem.

The INP assumes that there is a natural or induced
decomposition structure in the problem, where a
set of coupled (through resources) subproblems are
selected. Thus, problems involving multiple decision makers who have their own multiple objectives,
and that depend on each other via common resources
and where it is possible to decompose the overall
problem into a set of subproblems, are members
of the class of problems addressed by the negotiation procedure herein proposed.

The subproblems can be treated as a set of independent subproblems if one essential assumption
is satisfied. This assumption guarantees the autonomy of the decision makers, when the coupling
resources are fixed. It is assumed that the decision makers are not interested in each others
preference structure "per se", that is, when the
couplings are fixed, the decision makers are free
to trade-off their respective objectives in their
subproblems.

The INP proposed in this paper is based on trade-
off information, and there are no display difficulties. The decision makers have information on
their interdependence in the form of trade-offs.
A possible limitation of the proposed method is
the fact that it can be applied only in problems
that have common resources.

The procedure includes basically two repeated
steps: the decision makers' independent multiobjective optimization of their subproblems (subsystems) and a convenient step of trade-off between the decision makers.

The organization of the paper is as follows. Next

Section presents the mathematical problem formulation. The steps of INP are then detailed. Following, the method (procedure) analysis and its convergence properties established. An application of INP
in operation planning of a hydroelectrical power
system in southeast region of Brazil is addressed.
Next, the SEMOPS (Sequential Multiobjective Problem Solving Method) procedure is used to solve the
same problem. Comparisons of the two methods, from
the view point of methodology, practice and computational effort are presented. Finally, the
conclusions and further works are described.

MATHEMATICAL PROBLEM FORMULATION

Consider a system composed of a set of n coupled
subsystems where, associated with subsystem i, i=1,
..., n, we define x_i - the input, m_i - the decision, and y_i - the output, as variables belonging
to spaces of appropriate dimensions. Associated with
each subsystem there is a set of n_i objective functions, and a decision maker responsible for decision taking.

More precisely, we define the i-th subproblem as:

$$\min \begin{bmatrix} f_1^i (x_i, m_i, y_i) \\ f_2^i (x_i, m_i, y_i) \\ f_{n_i}^i (x_i, m_i, y_i) \end{bmatrix} \qquad (1)$$

$$DM_i : \qquad y_i = H_i(x_i, m_i), \qquad (2)$$

$$g_i(x_i, m_i, y_i) \le 0, \qquad (3)$$

$$x_i = \sum_{j=1}^{n} C_{ij}\, y_j \qquad (4)$$

where, n_i is the number of objectives of subproblem i, i=1, ..., n. The constraints model: the subsystem behaviour, the feasible region of the subproblem, and the couplings between subsystem, respectively.

The mathematical model above defines the characteristics of the problems to which the INP is addressed.

From now on, we use the words system and subsystems
to refer the problem and subproblems respectively.
The x_i's are, obviously, the coupling variables.

THE STEPS OF INP

The INP is basically composed by 5 steps, as follows:

REPEAT (UNTIL NO SIGNIFICANT CHANGE OCCURS)
. STEP 1 - The analyst fixes the coupling variables at feasible values;

. STEP 2 - The decision makers optimize their own
subsystems;

. STEP 3 - The analyst calculates the trade-off
among the subsystems. This is done
based on the objective functions values, and on the decision variables as
provided by the decision makers;

. STEP 4 - From the trade-off information, the
decision makers agree on direction and
amount of change in the objective
functions (using their own decision
rules);

. STEP 5 - The analyst calculates new values for
the coupling variables;

CONTINUE

In the first step, the analyst starts the procedure by fixing the coupling variables at feasible
values.

The optimization of the subsystems by the decision
makers, in the second step, can be done by any
appropriate technique, once the coupling variables
are respected.

The concept of trade-off (trade-offs are calculated
in the third step) is one of the most important
here. With it, the decision makers can negotiate to
find the best, overall compromise solution (the
fourth step). It is an important information for
the negotiation. In the last step, the analyst,
from the objective function values, and from the
decision variable values provided by decision makers, fixes coupling variables at new values.

ANALYSIS OF THE METHOD AND ITS CONVERGENCE

The INP is a method addressed for problems with
multiple decision makers who have their own multiple objectives. The decision for a subsystem is
normally taken by only one decision maker. However,
this method can be easily extended to solve subproblems that have more than one decision maker.
Other techniques of negotiation can also be used
without affecting the INP.

We say that INP converge to some limits when the
interactive process is continued by consistent decision makers, and if the limit values of the
objective functions correspond to the best compromise decisions, among all feasible decisions available to the decision makers. Now, we use the concept of utility function, that is, a function that
model the preference of the decision makers. Thus,
we say that INP converges if the utility function
converge to a limit value. Hence, the following two
assumptions are needed:

A1 - A best compromise solution exists;

A2 - An additive, differentiable group utility
function, not necessarily known, exists.

The existence of an additive utility function is an
important assumption to guarantee that, among the
solutions found by decision makers when they optimize their subsystems, is the best compromise solution for the system.

During the negotiation steps, that is, when the
decision makers are to choose one among all noninferior solutions, the decision makers use trade-
offs as provided by the analyst. With the trade-
offs, the decision makers can analyse the rate of
change among objectives. Hence, trade-off information is needed in the INP. Optimization of the subsystems can be done by any method, that does not
necessarily use the trade-off concept. A strong
condition, however, is the existence of trade-offs
among the subsystems. This means that the trade-
offs among the chosen objectives, representative
of the subsystems, need necessarily to exist. The
trade-offs guarantee the negotiation among the decision makers.

The third assumption is as follows:

A3 - All trade-offs between the objectives
exist, and all trade-offs in subsystems

optimization exist (Step 2)

As stated before, the most important point of this assumption is the existence of the trade-offs between the objectives chosen as representative of the subsystems. If we suppose that the negotiation will not be done through the trade-off, this assumption can be discarded.

Consider the group utility function, additive and differentiable. The indifferent trade-off is that one representing the opinion of the decision makers about one noninferior point where the partial objective trade-off is equal to the indifferent, subjective trade-off. In other words, we can state the fourth assumption as:

    A4 – The best compromise solution is the only point where the trade-offs are equal to the indifferent trade-offs of the group utility function (which ensures a local maximum to be a global one).

Now, at the limit values of the objective functions, we know that the objective, partial trade-offs between representative functions of subsystems are equal to the indifferent trade-off of the group utility function, and that the trade-offs in subsystem multiobjective optimization is also equal to the corresponding indifferent trade-offs. But note that, in subsystem optimizations, the couplings were kept constant. That is, trade-offs in the subsystem optimizations are not necessarily the trade-offs for the whole system. One simple assumption guaranteeing that trade-offs associated with the subsystems optimization are also trade-offs for the whole system, is the following:

    A5 – When determining any trade-off between objectives of the same subsystem, the fixed objectives fix the coupling variable associated with the particular subsystem.

If assumption A5 holds, then the trade-offs in the subsystem optimization (Step 2) are also trade-offs for the whole system. Thus, based on assumption A4 the method converges.

To see an interpretation of the most crucial assumptions, A5, let us put in a weaker form:

    A5' – Fixing the levels of objectives of n-1 subsystems, fixes the coupling variables of the remaining subsystem.

Assumption A5' is an independence condition for the subsystems, meaning that, when the levels of objectives of n-1 subsystems are fixed, the decision variables of n-1 subsystems do not affect the trade-offs of the remaining subsystem via changing the coupling variables of that particular subsystem.

One general class of systems where assumption A5' usually holds consists of these cases where couplings among subsystems are due to only common resources. In these cases, fixing objective functions levels for n-1 subsystems, usually fixes the amount of resources these n-1 subsystems require. Thus, the amount of resources for the remaining subsystem is fixed.

Above, we have suggested a simple interactive negotiation procedure, and considered its convergence requirements. It is clear that the assumption A5 (or A5') is the most crucial one. In fact, one can relax assumption A3: it is not necessary to have trade-offs between all objectives.

It should be noted that the suggested INP can be used also by a single decision maker.

In order to outline a practical way to work with the procedure just presented, a complementary remark is necessary. There are many techniques to solve multiobjective problems. Each of them have their own characteristics. Among the general techniques we have the ε-constraint method (Haimes and Chankong, 1983). With this method it is possible to transform the multiobjective case into a sequence of single objective problems. This is done by limiting the objective functions at values and including them in the restriction set, except one taken as a primary objective. If we use the ε-constraint, it is possible to obtain directly the trade-off between objectives (Gomide et alii, 1984, Gomide, 1981, Haimes and Chankong, 1983). Hence, the analyst have a way to obtain trade-offs, and use them as a negotiation procedure.

APPLICATION OF INP IN OPERATION PLANNING OF POWER SYSTEMS

This section addresses an application of INP in an operation planning of hydro power generation problem. Figure 1 shows the main reservoirs of the hydroelectrical power plants in the southeast region of Brazil. The reservoirs are arranged in two subsystems, representing two power companies that manage and operate them. The first subsystem comprises Marimbondo and Água Vermelha reservoirs, and the second São Simão and Ilha Solteira.

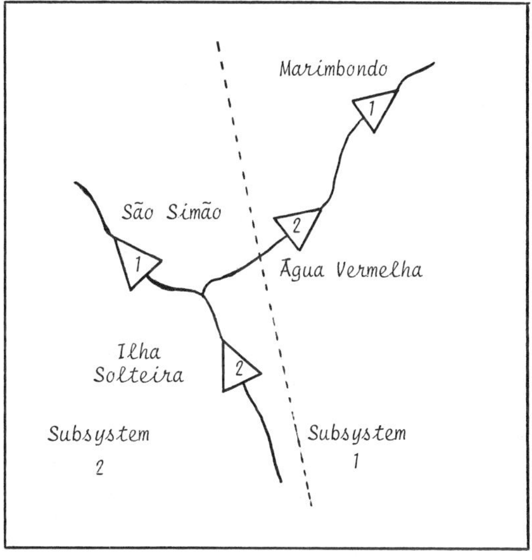

Fig. 1. Southeast Power System and Subsystems

We will denote by i the i-th subsystem, and by j the j-th reservoir of the i-th subsystem. Let's define the following:

$x_j^i(k)$: State variable. It is the capacity of the reservoir at the period k.

$u_j^i(k)$: Decision variable. The amount of turbinated water at the period k.

$y_j^i(k)$: Natural inflow at the period k.

The generation power (objective) function is given by:

$$f_j^i(x_j^i, u_j^i, k) = K_{T_j}^i\, x_j^i(T) + \sum_{k=0}^{T-1} \pi_j^i \cdot \rho \cdot u_j^i(k) \cdot g \cdot h_j^i(x_j^i(k))$$

where:

π_j^i : the efficiency in transforming potential falls into hydroelectrical power;

ρ : the specific weight of the water;

$u_j^i(k)$: the turbined water quantity in the period k;

g : the acceleration of the gravity;

$h_j^i(x_j^i(k))$: the difference between the water level, and the escape channel of the reservoirs, debited from the hydric energy lost;

i, j : 1,2.

The generation power function (differentiable, but non linear and non-convex) make the problem difficult to solve. In addition, the dimension of the problem normally increases computational difficulties.

The INP method is appropriate to this problem since it allows negotiations among the decision makers of the two power companies aiming a better use of hydric resources.

The scheme of negotiations, as applied to this problem, is depicted in Figure 2.

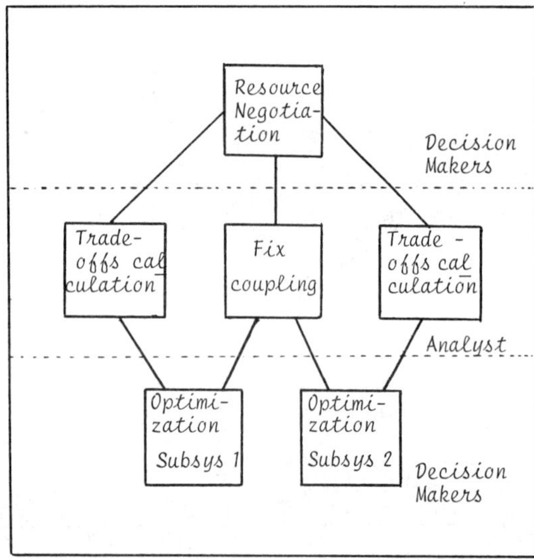

Fig. 2. Negotiation Scheme for the Power System

To simulate the negotiation among decision makers, the following utility function has been defined for testing and further comparison purposes:

$$U(f_1^1, f_1^2, f_2^1, f_2^2) = 0.2f_1^1 + 0.2f_2^1 + 0.4f_1^2 + 0.2f_2^2$$

The scheme of negotiation, as used in simulation, is shown in Figure 3.

From Figure 1, it is easily seen that the coupling variable is the decision variable $u_2^1(k)$, which is the physical connection between both subsystems. It is clear that assumption A5 holds in this case. Thus, the trade-offs of the subsystems are also the trade-offs of the overall system.

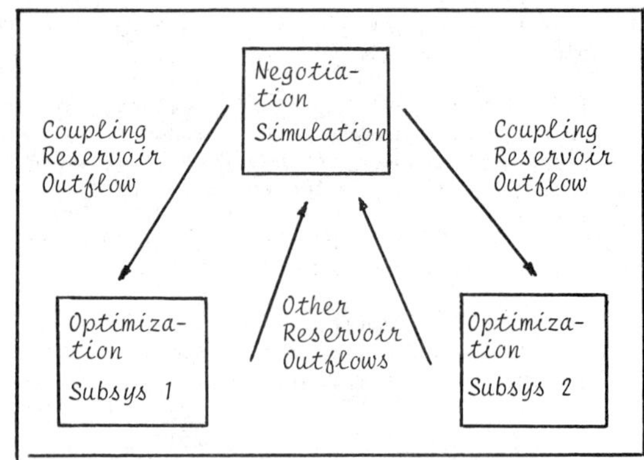

Fig. 3. Negotiation Simulation Scheme

From the utility function previously defined, we can now define the following subproblems:

Subproblem 1:

$$\max_{\text{s.t.}} U_1 \, [f_1^1 (x_1^1, u_1^1, k) , f_2^1 (x_2^1, u_2^1, k)]$$

$$u_2^1 (k) = c \; ; \; k = 0, \ldots, T-1$$

$$(x_j^i (k), u_j^i (k)) \in S_1 \; ; \; i = 1 \; ; \; k=0, \ldots, T \quad j = 1,2$$

Subproblem 2:

$$\max_{\text{s.t.}} U_2 \, [f_1^2 (x_1^2, u_1^2, k) , f_2^2 (x_2^2, u_2^2, k)]$$

$$u_2^1 (k) = c \; ; \; k = 0, \ldots, T-1$$

$$(x_j^i (k), u_j^i (k)) \in S_2 \; \begin{matrix} i = 2 \\ j = 1,2 \end{matrix} \; ; \; k=0,\ldots,T$$

Subproblem 3:

$$\max_{\text{s.t.}} U \, [f_1^1 (x_1^1, u_1^1, k) , f_2^1 (x_2^1, u_2^1, k) ,$$
$$f_1^2 (x_1^2, u_1^2, k) , f_2^2 (x_2^2, u_2^2, k)]$$

$$f_1^1 = c'$$

$$f_1^2 = c''$$

$$(x_j^i (k), u_j^i (k)) \in S_1 \cup S_2 \quad \begin{matrix} i = 1,2 \\ j = 1,2 \\ k = 0, \ldots, T \end{matrix}$$

$$x_j^i (k+1) = x_j^i (k) + y_j^i (k) - u_j^i (k) +$$
$$+ \sum_{(\ell,m) \in \Omega_j^i} u_m^\ell (k) \tag{5}$$

Where S1 and S2 include the equations of equilibrium of mass (5); the upper and lower bounds in state and decision variables of the respective subsystems, c,c' and c'' are constants fixed at appropriate values during the interactive process.

Subproblem 1 models the optimization problem faced by the decision makers of subsystem 1, with the coupling variable fixed at a value set by the analyst. Subproblem 2 is similar, but for decision maker of subsystem 2.

Subproblem 3 models the negotiation where all decision makers are involved. To proceed with negotiations, objective functions are selected to represent the subsystems, one for each subsystem. Objective functions f_2^1 and f_2^2 have been selected to represent subsystem 1 and 2 respectively. Due to its form, the objective functions values may be the same for several points. For instance, keeping the quantity of turbinated water constant instead of the objective function value, we are, in fact, selecting one among all possible points that corresponds to that objective function value. However, the important is that any of those trajectories will correspond to the same value of the utility function, reflecting the same preference of the decision maker. In this case, all trade-offs are partial trade-offs, since the levels of the objective, others than those used to represent each subsystem, are kept fixed during an interaction.

Table 1 presents the computational results obtained from the procedure simulation for several time horizons.

<u>TABLE 1 Computational Performance of INP</u>

PERIODS (MONTHS)	ITERATION NUMBER	UTILITY FUNCTION	CPU TIME (S)
6	3	7,95	15,56
7	3	9,13	29,86
8	3	10,14	43,10
9	3	11,14	65,31
10	3	12,04	119,11
11	3	12,92	118,24
12	3	13,95	162,24

The dependence of the system among the initial points, the penalty factor and the precision requested are the reason of the variations in the CPU time values. The INP, in all cases, has converged in a few interactions.

To solve the subproblems, an argumented Lagrangian method has been used and implemented in a VAX 1780 computer. The differences in CPU time are due to problem size (nº of periods).

APPLICATION OF THE SEMOPS METHOD

We now consider the same multiobjective problem as presented in the previous section, rewritten in an appropriated form to the SEMOPS method (Hwang and Masud, 1979):

$$\max\{f_1^1(x_1^1,\ u_1^1,\ k),\ f_2^1(x_2^1,\ u_2^1,\ k),\ f_1^2(x_1^2,\ u_1^2,\ k),$$

s.t.
$$f_2^2(x_2^2,\ u_2^2,\ k)\}$$

$$x_1^1(k+1) = x_1^1(k) + y_1^1(k) - u_1^1(k)$$

$$x_2^1(k+1) = x_2^1(k) + y_2^1(k) - u_2^1(k) + u_1^1(k)$$

$$x_1^2(k+1) = x_1^2(k) + y_1^2(k) - u_1^2(k)$$

MP:
$$x_2^2(k+1) = x_2^2(k) + y_2^2(k) - u_2^2(k) + u_2^1(k) +$$

$$+\ u_1^2(k)$$

$$\underline{x_j^i} \leq x_j^i(k) \leq \overline{x}_j^i \qquad k = 0,\ \dots,\ T$$
$$\underline{u_j^i} \leq u_j^i(k) \leq \overline{u}_j^i \qquad k = 1,\ \dots,\ T-1$$
$$\qquad\qquad\qquad\qquad i,j = 1,2$$

As goals and aspirations levels for the objectives functions, the values corresponding to the preferred solution obtained by INP are taken. This is because our goal now is to compare the methods, in achieving the same preferred solution.

The SEMOPS method is also interactive. After goals and aspirations levels are established, it defines surrogate functions, corresponding to decision makers aspirations.

As goals for the generation power functions, and aspirations levels we have:

$$f_1^1 > 9.7 \qquad f_2^1 > 9.0 \qquad f_1^2 > 14.8 \qquad f_2^2 > 21.2$$

$$AL_1^1 = 9.7 \qquad AL_2^1 = 9.0 \qquad AL_1^2 = 14.8 \qquad AL_2^2 = 21.2$$

which are the same values attained by the objective functions in the last section.

The objective functions limits as well as the corresponding dimensionless attainment indicator for the goals are:

$$f_1^1(x_1^1,\ u_1^1,\ k) \geq AL_1^1;\ d_1^1 = AL_1^1/f_1^1\ (x_1^1,\ u_1^1,\ k) \qquad (7)$$

$$f_2^1(x_2^1,\ u_2^1,\ k) \geq AL_2^1;\ d_2^1 = AL_2^1/f_2^1\ (x_2^1,\ u_2^1,\ k) \qquad (8)$$

$$f_1^2(x_1^2,\ u_1^2,\ k) \geq AL_1^2;\ d_1^2 = AL_1^2/f_1^2\ (x_1^2,\ u_1^2,\ k) \qquad (9)$$

$$f_2^2(x_2^2,\ u_2^2,\ k) \geq AL_2^2;\ d_2^2 = AL_2^2/f_2^2\ (x_2^2,\ u_2^2,\ k) \qquad (10)$$

The SEMOPS method is based on a sequence of optimization problems with the following surrogate objectives function:

$$S = \sum_{(i,j)\,\epsilon\,T'} d_j^i \qquad\qquad\qquad (11)$$

where (i,j) determine the objective functions that enter in function S definition.

In each iteration, a principal and several auxiliary problems are solved. The decision makers, according to some criteria, choose one of the objective functions to be deleted in S definitions, and include it into a constraint set for the next iteration.

Principal and auxiliary problem formulations for this application is detailed in Cardarelli (1987). For the SEMOPS method, see Monarchi (1980).

SEMOPS method application in operation planning problem has considered the problem structure. For instance, from the overall system topology, we see that reservoir 1 of subsystem 1 is more independent than the others. For this reason, the corresponding objective function has been incorporated in the set of constraints, keeping as goal the same values of the aspiration levels previously defined. From the results of the second iteration, and also due to the topology, f_2^1 has been included into the set of constraints, keeping the same aspiration level obtained. In the fourth iteration, the objective function f_2^2 has been fixed in the value obtained. The results of the last iteration is given in Table 2.

<u>TABLE 2 - SEMOPS Results - Last Iteration</u>

f^1	f^1	f^2	f^2	d_1	d_2	d_3	d^2	d_i^j	UTILITY FUNCTION	CPU TIME (S)
9,7	8,5	14,8	21,2	1,0	1,06	1,0	1,0	1,06	13,8	655,08
9,7	9,0	14,8	21,2	1,0	1,0	1,0	1,0	0,0	13,9	852,24

COMPARATIVE ANALYSES BETWEEN THE METHODS

The following results can be drawn after comparing
the main characteristics of the methods.

The INP preserves the form and properties of the
multiobjective optimization problem, while the
SEMOPS does not. For instance, SEMOP transforms a
linear optimization problem into a non-linear one.
Moreover, INP does not require any additional information other than trade-offs to be used. To
establish goals and aspiration levels, is a basic
premise for the SEMOPS method.

Furthermore, the INP naturally exploits a decomposition structure, while SEMOPS does not. This is
a simplification factor. For instance, in the power
system example each INP subproblem has 48 variables and 72 constraints. In SEMOPS, each principal
or secundary subproblem has 86 variables and 144
constraints. In addition, we have an increasing
constraint set due to objective functions inclu-
sions in each iteration.

From the point of view of computational performance
INP requires much less programming and CPU time
than SEMOPS. For instance, to generate a compromise
solution for a 12 months planning horizon, INP
took 162,24 seconds, SEMOPS 10,186.11 seconds.

CONCLUSIONS

There are some important INP characteristics that
we should mention:

. it is a natural method, since it does not require
any complementary information to be used;

. it is an interactive method, giving some freedom
to the decision maker on how to proceed with his
subsystem;

. it exploits decomposition structures to transform
a complex problem into a set of smaller and simpler problems;

. it converges fast;

. it does not introduce any additional complexity into the original problem.

Concerning INP application in the power system
planning example, it has been shown that it is more
efficient than previously developed methods, at
least for this class of problems.

Further work shall consider: extension of INP to
handle a broader class of problems, including those
with uncertainties and stochastic behaviour, as
well as to consider its combination with knowledge
based systems to elicidate preferences and to speed
compromise solution obtention.

REFERENCES

Cardarelli, J.R. (1987). _Interactive Decisions in
Large Scale Systems with Multiple Objective:
An Application in Power Systems_; MSc Thesis;
Universidade Estadual de Campinas - UNICAMP -
Campinas - Brasil (in Portuguese).

Chankong, V.; T. Haimes (1983). Multiobjective Decision Making: Theory and Methodology. North-
Holland, New York.

Gomide, F.A.C. (1982). Hierarchical Multistage,
Multiobjective Impact Analysis, PhD Disserta-
tion, Dep. of Systems Engineering, Case Western
Reserve University, Cleveland, Ohio, USA.

Gomide, F.A.C. (1980). A Multiobjective, Hierarchical Optimization Method for Multistage Systems;
Progress Report n. 2, Systems Engineering Department - Case Western Reserve University,
Cleveland, Ohio, USA.

Gomide, F.A.C.; K. Tarvainen; Y. Haimes (1984).
Iterative Negotiations on Multiple Objectives
in Large Scale Systems. 9th _World Congress of
IFAC_; Budapest, Hungary.

Haimes, Y.Y.; V. Chankong (1983). Optimization-
Based Methods for Multiobjective Decision-Making: Overview. _Large Scale Systems_; 1-33.

Hwang, C.L.; A.S.M. Masud (1979). Multiple Objective Decision Making Methods and Applications;
Springer-Verlag.

Kenney, R.L.; H. Raiffa (1976). Decisions with Multiple Objectives: Preferences and Value Trade-
offs. John Wiley, New York.

Tarvainen, K. (1980). Hierarchical Multiobjective
Optimization. Ph.D Dissertation, Dep. of Systems
Engineering, Case Western Reserve University,
Cleveland, Ohio, USA.

ACKNOWLEDGEMENT

This work was supported in part by CNPq, the Bra-
zilian National Research Council, under grant nº
300.729/86-3.

AN INTERACTIVE APPROACH TO THE MULTICRITERIA DECISION MAKING

G. Sotirov* and A. Voshchinin**

*Dept. of Automatics, Higher Inst. of Mechanical and Electrical Engineering, Sofia,
1156, Bulgaria
**Dept. of Automatics, Moscow Power Engineering Institute, Moscow, E250, USSR
(Joint Laboratory of Systems Optimization)

Abstract: The main purpose of this paper is to introduce a solution method for
multicriteria problems. The method offers a practical solution by combining judgement with
an automatic optimization technique in decision making. This is realised by using the
basic ideas behind the method of compromise programming. The "nearly equal" and the
"closest" sets of optimal weights of objective functions are established by using the
maximum entropy (MEP) and minimum squared deviation principle (MSDP) respectively. The
solutions in terms of deviation measure from the ideal point are presented to the decision-
maker (DM), and he is only asked to specify the amount by which specific alternatives
should exceed the "optimal" alternatives. Such interactive methods allow a re-adjustment
of optimal weights, displacement of the ideal point and new pre-decision conflict
formation, and leads to a decision in a finite number of iterations. A numerical example
is included to illustrate the proposed technique.

Keywords: Decision making; Multiple objective evaluation; maximum entropy principle;
minimum squared deviation principle.

INTRODUCTION

Most decision problems in real-world
situations have multiple objectives. It
is likely that these objectives are non-
commensurate and in conflict with one
another. In the mathematical sence, for
such prolems there is no unique optimum
solution. Therefore a compromise solution
with respect to the several objectives is
sought. In recent years, several methods
have been developed to solve multiobjective decision making problems [1],
[6].

Some of these methods, called interactive
methods, are based on progressive
exploration of the DM's preferences in
the objective space. When a decision
maker is unable to comminicate his
preference ordering explicitly, an
iterative technique is required to
solicit the preference information from
the DM in whatever form he is capable of
giving is (based largely on his
intuition). This information is used to
partially structure the problem (perhaps
producing a preliminary "best" solution).
The analyst then goes back to the
decision maker to extract more preference information and to allow him to
modify previously given information.
There are two types of interactive
methods. Those of the first type require
the DM to provide some trade-off amond
the attained values of objective
functions in order to determine the new
solution [4]. Interactive methods of the
second type require the DM to provide
some preference information by comparing
the various efficient solution in space
of objective functions or decision
variables [1].

The basic steps of many currently
available algorithms to solve multicriteria problems can be succintly stated
as follows:
1. Find an initial feasible solution;
2. Interact with the decision maker;
3. Obtain a new solution. If this or
    some previous solution is acceptab-
    le to the DM, stop; otherwise, go
    on to Step 2.

A common goal is to attempt to help a DM
make good (efficient) decisions that
would satisfy him. Since it is difficult
to provide an universal definition for
"satisfaction", we assume that a DM is
satisfied with a solution when he (at the
time of termination) feels that the
current solution is the best (most
preferable) that can be found using the
model. His concept of "best" here is
defined in terms of how much actual
achievement of aech objective deviates
from the desired goals or interactive
algorihms to solve multicriteria problems
follow the aabove basic structure, they
differ in terms of their philosophies/
approaches.

In this paper, an interactive linear
multiple objective method is introduced.
The method attempts to reduce the
complexity of information required from
the DM by combining the basic ideas
behing the methods of compromise
programming and MEP and MSDP in an
interactive way.

PROBLEM FORMULATION

When solving a multi-objective decision
problem, a Pareto optimal (or noninferior) solution is sought in most
cases [6]. Consider a general multicriteria linear programming (MCLP) of the
form:

$$\max \; [f_1(x), \ldots, f_m(x)] \qquad (1)$$
$$\text{s.t.} \; x \in X,$$

where $X = \{x \in R^n : x \geq 0, Ax \leq b, b \in R^k\}$, and
each $f_i(x)$, $i=1, \ldots, m$ is a linear
objective function.

Solving (1) for all noninferior extreme
points is equivalent to solving the
weighting problem P_w for all maximizing
extreme points [6].

$$\max \ \sum_{i=1}^{m} w_i f_i(x) \qquad\qquad (2)$$
$$\text{s.t. } x \in X, \ w \in W,$$

where $W = \{ w \in R^m : w \geq 0, \ \sum_{i=1}^{m} w_i = 1 \}$

There are different algoritms for solving the P_w problem. Unfortunately, the set of noninferior solutions is often quite large. Any attempts to reduce this set requires information about the DM's preferences, and thus an interactive approach is called for. In this paper the compromise programming and the method of displaced ideal, proposed by Zeleny [7], [8], are used to form an interactive strategy for reducing a solution set until the best-compromice solution can conveniently be selected.

Given a weight vector w, $x^w{}_p$ is a compromise solution of a multicriteria problem with respect to p if and only if it solves

$$\max \ L_p = \sum_{i=1}^{m} w_i (f_i(x) - f^o{}_i)^p \qquad (3)$$
$$\text{s.t. } x \in X, \ w \in W,$$

where $f^o{}_i = \max_{x \in x} f_i(x)$, $i=1, \ldots, m$ represents an ideal point.

In practical problems the ideal point is always unfeasible (otherwise there would be no conflicts), but it is conceivable that the nearest feasible solution could be an acceptable compromise for the DM.

The compromise set $X^c{}_w$, given the weight vector w, is defined as a set of all compromice solutions $x^p{}_w$, $1 \leq p \leq \infty$:

$$X^c{}_w = \{ \ x \in X : x \text{ solves (3) given w for}$$
$$\text{some } 1 \leq p \leq \infty \} \qquad (4)$$

If X is a polyhedron defined by a system of linear inequalities and each $f_i(x)$ is linear, then the ideal point f^o can be found by m simple linear programming problems. Furthermore, the compromise solutions f^1 and f^∞ can be found by a linear programms (the other compromise solution f^p, $1<p<$ can be found by convex programming).It should be noted that for the linear case, the compromise set $X^c{}_w$ is completely determined if $x^1{}_w$ and $x^\infty{}_w$ are known, since, as a function of p, f^p moves continuously and monotonously from f^1 to f^∞ [3]. Note that in a more general case when the set of possible outcomes $F=\{f(x):x \in X\}$, with f as its generic element, is convex and closed with respect to cyclic rotation, for $1<p<\infty$, $f^p{}_k = f^p{}_j$ for all $k, j=1, \ldots, m$. When $p = 1$, then there is at least one f^1 with $f^1{}_k = f^1{}_j$ for all $k, j=1, \ldots, m$. This property essentially shows that, if the stated assumptions are satisfied, then compromise solution can ensure an equal "utility" for eact criterion. Thus, an "equality principle" has been built into the compromise solution.

For the two-objective problem with $w>0$, $X^c{}_w$ is a set of points on the noninferior set f^o lying between $x^1{}_w$ and $x^\infty{}_w$. For problems with three or more objectives, Zeleny [7] suggests that the compromise set $X^c{}_w$ be approximated by the noninferior solutions of the following two-objective problem:

$$\max \ [L_1, \ L_\infty]$$
$$x \in X$$

If $X^c{}_w$ is still too large to select the best-compromise solution from it, obviously it should be reduced further by using, for instance, the method of so-called displaced ideal [8]. Based on this philisophy, it follows, that the ideal point with respect to the new $X^c{}_w$ should displase the previous point. The new ideal point $\tilde{f}^o$ can be calculated from:

$$\tilde{f}^o{}_i = f_i(x), \ i = 1, \ldots, m \qquad (5)$$
$$x \in X^c{}_w$$

Using this new ideal point, the new set $X^c{}_w$ can be constructed as follows:

$$\tilde{X}^c{}_w = \{ \ x:x \text{ solves}$$
$$\max \ \tilde{L}_p = \sum_{i=1}^{m} w_i (f_i(x) - \tilde{f}^o{}_i)^p,$$
$$x \in X^c{}_w$$
$$1 \leq p \leq \infty \} \qquad (6)$$

Once a reduced compromise set set $\tilde{X}^c{}_w$ is obtained, a search for the best-compromise solution can be resumed in an interactive way as before. This enables DM to make his choice more easily between the solutions.

It should be noted that the relative importance of each objective depends not only on how DM views these objectives in general (as reflected by the given w_is), but also on the context in which these objectives can assume values. This second component of relative importance is thus higly dependent on the set of alternatives we are operating with.

In this paper, the basic ideas behind the methods of compromise programming are combined with the maximum entropy and minimum squared deviation principles in order to modify the prefences in decision making and to select the best-compromise solution.

PROCEDURES FOR MODIFYING
THE PREFERENCES IN
DECISION MAKING

In this section we consider two simple computational procedures. For the first, or entropy-based, procedure, the point of departure is the concept of equal weights for all objectives [5]. For the second, or least squares procedure, the point of departure is a set of arbitrary weight.

In decision making problems MEP have been used to determine the probability of the input variables [2]. Here, a special case is considered where the outcomes of the variables are known and the probability ranges for various outcomes have been estimated by the DM. Thus, the probability ranges are the DM's subjective estimates of the probability of outcome occurrence. The probability function of the input variable is obtained by maximizing the entropy subject to the probability range constraints. This problem can be formulated mathematically as below:

$$\max H(h) = - \sum_{i=1}^{m} h_i \log h_i \qquad (7)$$

$$s.t. \sum_{i=1}^{m} h_i = 1$$

$$a_i \leq h_i \leq d_i, \quad i=1, \ldots, m$$

where $H(h)$ is the entropy, h_i is the probability of ith outcome and a_i, d_i are the lower and upper boundaries of h_i, respectively. Obviously (7) is a standart nonlinear programming problem and in fact the objective function $H(h)$ is separable.

Once (7) has been solved and a discrete probability function h is determined, the remaining problem in applying MEP is to consider these probabilities h_i as objective function weights w_i in the multicriteria problems. Thus, the optimal weights of objective functions in (3) would be automatically found for the DM in such a way that the weights are as "nearly equal" as possible.

Note that the motivation for maximizing entropy is pragmatic the maximum value of (7) without an interval constraint is $h_i = 1/m$ for all i's. That is, equal weights is the maximum entropy solution.

The least squares procedure computes the weights w_i, $i=1, \ldots, m$ for all objectives so that they are as "close" as possible to the given (arbitrary, previously determined) weights b_i. The concept of "close" is operationalized using the minimum squared deviation principle. Thus, the objective function in problem (7) takes the form of least squares-based function as follows:

$$\min \sum_{i=1}^{m} (w_i - b_i)^2 \qquad (8)$$

Obviously, the least squares procedure accommodates an arbitrary set of weights, including equal weights, and, therefore, is more general.

The purpose of the procedures proposed here of modifing the weight vector w is to compute, for two mutually non-dominated compromise solutions, the values of the new weights that make the weighted composites of the two solutions differ by an amount $r>0$ and to compute them in such a way that the weights are as "nearly equal" or as "closest to arbitrary weights" as possible. Thus, problem (7) assumes the forms:

$$\max H(w) = - \sum_{i=1}^{m} w_i \log w_i \qquad (9)$$

or

$$\min L(w) = \sum_{i=1}^{m} (w_i - b_i)^2 \qquad (10)$$

$$s.t. \sum_{i=1}^{m} w_i = 1$$

$$a_i \leq w_i \leq d_i, \quad i=1, \ldots, m$$

$$\sum_{i=1}^{m} [h_i f_i(x^p{}_w) - w_i f_i(x)] \leq r, \quad 1 \leq m$$

$$x \in X^c{}_w$$

An extension is to incorporate additional constraints, linear in the weights, that represent partial information provided by the DM. Additional constraints may include for instance, one of the following types of expressions:

(a) fixed weights $w_i = d_i$ for selected objectives;

(b) rank orders of weights: $w_1 \geq w_2 \geq \ldots \geq w_m$, as well as ranks of combined weights: $w_3 + w_4 + w_6 \geq w_1 + w_2$, and so forth;

(c) equality $f_i(x^p{}_w) = f_i(\tilde{x}^p{}_w)$, $\tilde{x}^p{}_w \in X^c{}_w$;

(d) inequality $f_i(x^p{}_w) \geq f_j(x^p{}_w)$;

(e) absolute difference $f_i(\tilde{x}^p{}_w) \geq f_j(x^p{}_w) + r_{ij}$ where $r_{ij} > 0$ is a specified difference;

(f) relative difference: $[f_i(\tilde{x}^p{}_w) - f_j(x^p{}_w)]/f_j(x^p{}_w) \geq r_{ij}$.

Therefore, the interactive procedure would proceed in a way determined by the DM's specific reactions.

THE PROPOSED APPROACH

One of the drawbacks of interactive methods is the difficulty in obtaining objective function weights by the DM even if the values of objective functions are presented to him on the same scale. In the proposed method, the optimal weight vector w is automatically generated in terms of maximum entropy or minimum squared deviation for the DM at each iteration.

The solution procedure starts by solving m simple LP problems to find the ideal point f^o. Then the weights w_i, $i=1, \ldots, m$ are determined by solving problems (9) or (10) and employed in (3) to find the compromise solutions for the specified values of f^o, w and p. All relevant information is presented to the DM. These incluse $X^c{}_w$, ideal points $f^o{}_i$, $i=1, \ldots, m$, the compromise solutions f^1, f^2, f^∞ with their weights and norms. If $X^c{}_w$ constains f^o or the DM is able to select the best-compromise solution from $X^c{}_w$, the process is terminated. Otherwise, through conversion, we can obtain new information from the DM on $X^c{}_w$, f^o, the weight vector w, and norm parameter p. The procedure is continued until the DM expressly states that he prefers one particular solution over all the others.

The algorithmic steps of interactive method can be summarised as follows:

0. Ask the DM to determine an arbitrary set of weights b_i, or to give bounded intervals $a_i \leq w_i \leq d_i$, where a_i, d_i are specified non-negative constants. Compute the values of the weights in

such a way that the weightsare as
"nearly equal", or as "closest to
the given weights b_i" as possible,
i.e. solve problem (9) or (10). Let
$w_i = w^o_i$, $i=1, \ldots, m$ are the optimal
weights.

1. Find the ideal point by solving m
   linear programming problems of the
   form: max $f_i(x)$ for each $i=1, \ldots, m$.
   $x \in X$

   Then the weights may be scaled, for
   instance, as follows: $w_i = w^o_i / f^o_i$, and
   let $p = 1, 2, \infty$ to begin with.

2. Construct the compromise set X^c_w by
   finding the set of noniferior
   solutions of

   $$\max \ [L_1, \ L_\infty]$$
   $$x \in X$$

   Note that
   $L_\infty = \max_{1 \le i \le m} [w_i(f_i(x) - f^o_i)]$.

3. If the DM is able to select the
   best compromise solution from X^c_w,
   or X^c_w contains f^o, stop: the best
   compromise solution has been found.
   If not, go on to Step 4 to obtain
   more information.

4. Modify weight vector w by asking DM
   to specify additional constraints to
   the problem of maximizing the
   entropy H(w), or minimizing the
   squared deviation L(w). Compute the
   modified weights values, i.e. solve
   problem (9) or (10). Let $\tilde{w}^o_i$, $i=1$,
   $\ldots, m$ are the new optimal weights.
   Set $X = X^c_w$ and go to Step 1.

It should be noted that, although no
mathematical convergence can be proved,
DM is expected to be rational and
consistent in providing the information
required from him, the procedure
termination taking place in a relatively
few iterations.

ILLUSTRATIVE EXAMPLE

Now the algorithmic steps of proposed
method for the case of "nearly equal"
weights are demonstrated and clarified by
means of the following two-criteria
problem [6]:

$$\max \ f_1(x) = 6x_1 + 4x_2$$
$$\max \ f_2(x) = x_1$$
$$x \in X,$$

where $X = \{x \in R^2 : x_1 + x_2 \le 100, \ 2x_1 + x_2 \le 150, \ x_1, x_2 \ge 0\}$.

Thought conversation, the bounded
intervals for the weights are specified
as follows: $0.4 \le w_1 \le 0.7$, $0.3 \le w_2 \le 0.6$. The
optimal weights are determined by solving

nonlinear problem (7): $w^o_1 = w^o_2 = 0.5$.

Ideal point f^o is (500, 75). Note that
this ideal solution is not feasible and,
therefore, inobtainable.

It should be noted that in the linear
case the compromise set X^c_w is completely
determined if x^1_w and x^∞_w are known. Now
we demonstrate how to find compromise
solutions f^1 and f^∞ :

(i) For the f^1 compromise solution we
    solve

    $$\min \ \{0.5[500-(6x_1+4x_2)]+0.5(75-x_1)\}$$
    $$x \in X$$

    We get $x^1_w = (50,50)$ and compromise
    solution $f^1 = (500, 50)$.

(ii) For the f^∞ compromise solution we
     solve

     $$\min \ y$$
     $$\text{s.t.} \ y \ge 0.5[500-(6x_1+4x_2)]$$
     $$y \ge 0.5(75-x_1)$$
     $$x \in X$$

     We get $x^\infty_w = (58.33, 33.33)$ and
     $f^\infty = (483.33, 58.33)$. Obviosly,
     x^1_w and x^∞_w lie on the same
     noninferior edge. Then,
     $X^c_w = \{x \in R^2 : 2x_1 + x_2 = 150, \ 50 \le x_1 \le 58.33, \ 33.33 \le x_2 \le 50\}$

Suppose the DM is unable to select the
best compromise solution from X^c_w. Then
throught conversion we specify some
additional constraints to the problem (9)
as follows:

$$\max \ (-w_1 \log w_1 - w_2 \log w_2)$$
$$\text{s.t.} \ w_1 + w_2 = 1$$
$$0.4 \le w_1 \le 0.7$$
$$0.3 \le w_2 \le 0.6$$
$$w_1 f_1(x) \le w^o_1 f_1(x^1_w) - 1$$
$$w_2 f_2(x) \le w^o_2 f_2(x^1_w) - 1$$
$$x \in X$$

We get optimal weights $\tilde{w}^o_1 = 0.498$,
$\tilde{w}^o_2 = 0.502$. To find the new ideal point,
we solve the following two linear
problems:

$$\max \ f_1(x)$$
$$x \in X^c_w$$
and
$$\max \ f_2(x)$$
$$x \in X^c_w$$

The ideal point $\tilde{f}^o$ is (500, 58.33). Now
we illustrate how to find the $\tilde{f}^1$ and $\tilde{f}^\infty$
compromise solutions:

(i) For the $\tilde{f}^1$ solution, we solve:

    $$\min \ \{0.498[500-(6x_1+4x_2)]+ +0.502(75-x_1)\}$$
    $$x \in X^c_w$$

    We get $\tilde{x}^1_w = x^1_w = (50,50)$ and
    compromise solution $\tilde{f}^1 = f^1 = (500,50)$.

(ii) For the $\tilde{f}^\infty$ solution, we solve

     $$\min \ y$$
     $$\text{s.t.} \ y \ge 0.498[500-(6x_1 + 4x_2)]$$
     $$y \ge 0.502(58.33 - x_1)$$
     $$x \in X^c_w$$

     We get $x^\infty_w = (52.79, 44.42)$ and
     compromise solution $\tilde{f}^\infty = (494.42, 52.79)$.

Suppose that the DM prefers the new
compromise solution $\tilde{f}^\infty$ when maximum

158

individual regret is emphasized. Thus,
the best-compromise solution of this
problem would be $x=(52.79, 44.42)$,
$f=(494.42, 52.79)$.

CONCLUSION

In this paper, an interactive method for
solving multicriteria problem is
introduced. The basic idea underlying the
proposed method is to combine, in an
iterative way, the maximum entropy or the
minimum squared deviation principles with
the concepts of compromise solutions and
compromise set in order to reduse a
solution set until the best-compromise
solution can be conveniently selected.
Noninferiority and feasibility of the
compromise solutions are guaranteed since
the decision-making area is contracted
after each iteration of the solution procedure. The advantage of solving an
optimization problem by the proposed
method is that the information required
is simple to provide and the most
probable solution is found in terms of
maximum entropy or minimum squared
deviation.

REFERENCES

[1] Chankong, V., and Y.Y. Haims (1983)
    _Multiobjective decision making:
    theory and methodology._ North-
    Holland Series in System Science and
    Engineering, 8, North-Holland, N.Y.
[2] Barron, H., and C.P. Schmidt (1988)
    _Sensitivity analysis of additive
    multiattribute value models._ Opns
    Res., 36, 122-127.
[3] Freimer, M., and P.L.Yu (1976) _Some
    new results on compromise solutions
    for group decision problems._ Mgmt
    Sci., 22, 688-693.
[4] Geoffrion,A.M. et al (1972) _An
    interactive approach for
    multicriteria optimization with an
    application to the operation of an
    academic department._ Mgmt Sci., 19,
    357-368.
[5] Sotirov G.R., _An interactive entropy
    based method for solving the multi-
    criteria problems._ IFAS Symp. on
    Large Scale Systems: Theory and
    Appl.,Berlin,1989
[6] Yu, P.L. (1985) _Multiple-criteria
    decision making: concepts,
    techniques, and extensions._ Plenum
    Press,N.Y.
[7] Zeleny,M. (1976) The theory of the
    displaced ideal. In _Multiple
    Criteria Decision Making:_ Kyoto 1975
    (M. Zeleny Ed.), Springer-Verlag,
    Berlin, 153-206.
[8] Zeleny, M. (1977) Adaptive
    displacement of preferences in
    decision making. _In Multiple
    Criteria Decision Making_ - TIMS
    Studies in Management Science
    (M.K.Starr and M.Zeleny, Eds),
    North-Holland, Amsterdam,147-158.

APPROXIMATIONS IN DECOMPOSITION OF LARGE-SCALE CONVEX PROGRAMS VIA A NONDIFFERENTIABLE OPTIMIZATION METHOD

K. C. Kiwiel

Systems Research Institute, Polish Academy of Sciences, Newelska 6, 01-447 Warsaw, Poland

Abstract. A proximal bundle method is presented for minimizing a nonsmooth convex function f. At each iteration it requires only one approximate evaluation of f and its ε-subgradient, and finds a search direction via quadratic programming. When applied to Lagrangian decomposition of convex programs, it allows for inexact solutions of decomposed subproblems; yet, increasing their required accuracy automatically, it asymptotically finds both primal and dual solutions. Some encouraging numerical experience is reported.

Keywords. Convex programming; decomposition; mathematical programming; nonlinear programming; nondifferentiable optimization; large-scale programming; proximal bundle methods.

INTRODUCTION

This paper presents a new decomposition method for solving separable convex programming problems. The motivation for this algorithm comes from the work of Kiwiel (1989b) on proximal bundle methods for nondifferentiable optimization.

The convex separable optimization problem is to

$$\text{maximize } \psi_0(z) := \sum_{i=1}^{n} \psi_{oi}(z_i)$$

$$\text{subject to } \psi_j(z):= \sum_{i=1}^{n} \psi_{ji}(z_i) \geq 0, \; j=1,\ldots,N,$$

$$z := (z_1,\ldots,z_n) \in Z := Z_1 \times \ldots \times Z_n, \tag{1}$$

where ψ_{ji} are (possibly nondifferentiable) closed proper concave functions on closed sets $Z_i \subset \text{dom } \psi_{ji} \subset \mathbb{R}^{n_i}$, $j=0,\ldots,N$, $i=1,\ldots,n$. Suppose that (1) is feasible. We assume that the dual function

$$f(x) = \sup\{ \psi_0(z) + \langle x,\psi(z)\rangle : z \in Z\} \tag{2}$$

is finite on an open neighborhood of $\mathbb{R}^N_+$, where $\psi(z) = (\psi_1(z),\ldots,\psi_n(z))$, that $X=\text{Argmin}\{f(x):x\geq0\}\neq\emptyset$, and that for each $x\geq0$ and $\varepsilon>0$ we can evaluate

$$f_i(x)=\sup\{\psi_{oi}(z_i)+\sum_{j=1}^{N}x_j\psi_{ji}(z_i): z_i\in Z_i\} \tag{3}$$

with accuracy ε/N by finding some $z_i(x,\varepsilon)$ in

$$Z_i(x,\varepsilon)=\{z_i\in Z_i: \psi_{oi}(z_i)+\sum_{j=1}^{N}x_j\psi_{ji}(z_i)\geq f_i(x)-\varepsilon/N\} \tag{4}$$

and $z(x,\varepsilon)=(z_1(x,\varepsilon),\ldots,z_n(x,\varepsilon))$, so that

$$f(x) = \sum_{i=1}^{n} f_i(x) \tag{5}$$

can be computed with accuracy ε

$$f(x) \geq \psi_0(z(x,\varepsilon)) + x^*\psi(z(x,\varepsilon)) \geq f(x)-\varepsilon. \tag{6}$$

The above assumptions are realistic in many applications (see, e.g. Demyanov and Vasiliev, 1985; Shor, 1985). For example, it may be impossible to find $z(x,0)$ in finite time, or calculating $z(x,\varepsilon)$ for a prescribed $\varepsilon>0$ may require much less work, e.g. when (3) involves solving a linear, quadratic or discrete problem by the methods of Gabasov and co-workers (1987).

Our extension of the proximal bundle method of Ki-

wiel (1989b) solves the dual problem to (1)

$$\text{minimize } f(x) \text{ over all } x \in S, \tag{7}$$

where $S=\mathbb{R}^N_+$. It is a feasible point method of "descent" in the sense of generating a sequence $\{x^k\} \subset S$ converging to some $\bar{x}\in X$, over-estimates $f^k_x\geq f(x^k)$ such that $f^k_x\downarrow f(\bar{x})$ and $f^{k+1}_x<f^k_x$ if $x^{k+1}\neq x^k$, tolerances $\varepsilon^k\downarrow 0$ and trial points $y^k\in S$ for evaluating approximate linearizations of f. More specifically, we say that $\bar{f}(.;y,\varepsilon)$ is an ε-linearization of f at y if it is an affine function of the form

$$\bar{f}(x;y,\varepsilon) = \bar{f}(y;y,\varepsilon) + \langle g_f(y,\varepsilon),x-y\rangle \; \forall x \tag{8a}$$

which supports the epigraph of f at y with tolerance ε in the sense that

$$f(x) \geq \bar{f}(x;y,\varepsilon) \; \forall x \in S, \tag{8b}$$

$$\bar{f}(y;y,\varepsilon) \geq f(y) - \varepsilon. \tag{8c}$$

Incidentally, $g_f(y,\varepsilon)$ is an ε-subgradient of f at y. Of course, by (2) and (6), we may use

$$\bar{f}(y;y,\varepsilon) = \psi_0(z(y,\varepsilon)) + \langle y,\psi(z(y,\varepsilon))\rangle, \tag{9a}$$

$$g_f(y,\varepsilon) = \psi(z(y,\varepsilon)). \tag{9b}$$

At the k-th iteration the method employs the following piecewise linear (polyhedral) lower approximation to f

$$\hat{f}^k(x) = \max\{ \bar{f}(x;y^j,\varepsilon^j): j\in J^k\} \tag{10}$$

with $J^k\subset\{1,\ldots,k\}$, $|J^k| \leq N+2$. The next trial point

$$y^{k+1} = \text{argmin}\{ \hat{f}^k(x) + u^k|x-x^k|^2/2: x \in S\}, \tag{11}$$

where $u^k>0$ is chosen by safeguarded quadratic interpolation to estimate the curvature of f between y^k and x^{k-1}. A *serious step* from x^k to $x^{k+1}=y^{k+1}$ occurs if y^{k+1} is significantly better than x^k in the sense that

$$f^{k+1}_y \leq f^k_x + m_L v^k, \tag{12}$$

where

$$f^{k+1}_y := \bar{f}(y^{k+1};y^{k+1},\varepsilon^{k+1}) + \varepsilon^{k+1} \tag{13}$$

overestimates $f(y^{k+1})$ by (8c),

$$\varepsilon^{k+1} = - (m_R-m_L)v^k \tag{14}$$

is the next tolerance, $0<m_L<m_R<1$ are fixed, and

$$v^k = \hat{f}^k(y^{k+1}) - f^k_x \qquad (15)$$

is the predicted descent (if $v^k \geq 0$ the algorithm may stop with an optimal x^k ; see (25)); then $f^{k+1}_x = f^{k+1}_y < f^k_x$. Otherwise, a *null step* $x^{k+1}=x^k$ improves the next model $\hat{f}^{k+1}$ with $\bar{f}(.;y^{k+1},\varepsilon^{k+1})$ and

$$\bar{f}(y^{k+1};y^{k+1},\varepsilon^{k+1})-f^k_x > m_L v^k - \varepsilon^{k+1} = m_R v^k > v^k \qquad (16)$$

so that a better next $y^{k+2} \neq y^{k+1}$ is found; cf. (15).

We show that $x^k \to \bar{x} \in X$ under no additional assumptions. Moreover, the method also produces (asymptotically) an optimal primal solution to (1) as follows. Solving (11) through quadratic programming (QP) yields Lagrange multipliers $\lambda^k_j \geq 0$, $j \in J^k$, summing up to 1, that produce the aggregate solution

$$z^k = \sum_{j \in J^k} \lambda^k_j z(y^j,\varepsilon^j) \qquad (17)$$

in Z (by convexity), such that

$$\psi_0(z^k) \geq \sup\{ \psi_0(z): \psi(z) \geq 0, z \in Z\} - \varepsilon^k_f, \qquad (18)$$

$$\psi_i(z^k) \geq \varepsilon^k_F, \quad i=1,\ldots,N, \qquad (19)$$

with easily computable tolerances $\varepsilon^k_f \to 0$ and $\varepsilon^k_F \to 0$.

Hence every accumulation point of $\{z^k\}$ (that exists e.g. for compact Z) solves (1); in practice one may stop if ε^k_f and ε^k_F are small enough. (We may add that the method will find a solution in a finite number of iterations if f is polyhedral, $\varepsilon^k \equiv 0$ and either $m_L =1$ in (12) or certain technical conditions are satisfied; see Kiwiel (1989c) for details).

There is extensive literature on decomposition in separable programming (see, e.g. the references of Sen and Sherali, 1986; Spingarn, 1985). Our method may be regarded as a regularized version of the classical cutting plane approach (Dantzig and Wolfe, 1960), that deletes the quadratic term in (11) and, hence, suffers from slow convergence and unbounded storage. The recently proposed augmented Lagrangian methods (see, e.g. Golshtein, 1987; Spingarn, 1985) are simpler and have strong convergence properties, but they involve additional nonlinearities in the objective of (3), and controlling their accuracy tolerances in not easy. Moreover, our method (just like the cutting plane ones) produces aggregate primal solutions (42) that have important "approximate discreteness" properties in the context of Lagrangian relaxation of discrete problems (see, e.g. Bertsekas, 1982; Lasdon, 1970).

In the context of nondifferentiable optimization methods that use approximate linearizations, we hope that our method is an improvement on those given in (Demyanov and Vasiliev, 1985; Kiwiel, 1985; Rzhevski and Kuncevich, 1985; Shor, 1985), since it generalizes one of the currently most efficient methods with exact linearizations (Kiwiel, 1989b).

The paper is organized as follows. The algorithm is derived in Section 2. Its global convergence is studied in Section 3. Implications for decomposition are studied in Section 4. Some modifications are described in Section 5. Our preliminary numerical experience is reported in Section 6. Finally, we have a conclusion section.

We use the following notation. We denote by $\langle .,. \rangle$ and $|.|$, respectively, the usual inner product and norm in $\mathbb{R}^N$. Both superscripts and subscripts are used to denote different vectors. For $\varepsilon \geq 0$, the ε-subdifferential of f at x is defined by

$$\partial_\varepsilon f(x) = \{p \in \mathbb{R}^N: f(y) \geq f(x)+\langle p,y-x \rangle - \varepsilon \ \forall y \in \mathbb{R}^N\}.$$

We denote by ∂f the ordinary subdifferential $\partial_0 f$. The mapping $\partial f(.)$ is locally bounded and f is locally Lipschitz continuous on S (Kiwiel, 1985a).

THE METHOD

We shall now describe our method for problem (7). We assume, for generality, that $S=\{x: h_i(x) \leq 0, i \in I\}$, where h_i are affine functions and I is finite, that f is convex and finite on an open neighborhood of S and that at each $y \in S$ we can find an ε-linearization of f for any $\varepsilon > 0$.

We start by specializing the results of Kiwiel (1989b) to subproblem (11). Let $g^j = g_f(y^j,\varepsilon^j)$, $f^k_j = \bar{f}(x^k;y^j,\varepsilon^j)$ and $\alpha^k_j = f^k_x - f^k_j \geq 0$ by (8b) and $f^k_x \geq f(x^k)$, $j \in J^k$. Then (11) can be solved by finding (d^k,v^k) to

$$\text{minimize} \quad v + u^k|d|^2/2,$$
$$\text{subject to} \quad -\alpha^k_j + \langle g^j,d \rangle \leq v \ \forall \ j \in J^k, \qquad (20)$$
$$h_i(x^k) + \langle \nabla h_i,d \rangle \leq 0 \ \forall \ i \in I,$$

since $y^{k+1}=x^k+d^k$. Denote the Lagrange multipliers of (20) by λ^k_j, $j \in J^k$, ν^k_i, $i \in I$.

Let δ_h denote the indicator function of S ($\delta_h(x)=0$ if $x \in S$, ∞ otherwise), and let $\hat{\phi}^k = \hat{f}^k + u^k|.-x^k|^2/2 + \delta_h$. As in (Kiwiel, 1989b), from (10) and the optimality condition $0 \in \partial \hat{\phi}^k(y^{k+1})$ for (11) we deduce the existence of $p^k_f \in \partial \hat{f}^k(y^{k+1})$ and $p^k_h \in \partial \delta_h(y^{k+1})$ such that the aggregate linearizations $\tilde{f}^k(x) = \hat{f}^k(y^{k+1})+ \langle p^k_f,x-y^{k+1} \rangle$ and $\tilde{\delta}^k_h(x) = \langle p^k_h,x-y^{k+1} \rangle$ minorize f and δ_h, respectively, and that

$$p^k_f + p^k_h + u^k d^k = 0, \qquad (21)$$

$$-v^k = u^k|d^k|^2 + \tilde{\alpha}^k_p = |p^k|^2/u^k + \tilde{\alpha}^k_p, \qquad (22)$$

$$\tilde{\alpha}^k_p = \tilde{\alpha}^k_{fp} + \tilde{\alpha}^k_h \geq 0, \qquad (23)$$

$$f(x) \geq f(x^k) + \langle p^k, x - x^k \rangle - \tilde{\alpha}^k_p \ \forall \ x \in S, \qquad (24)$$

$$f(x) \geq f(x^k)-|u^k v^k|^{1/2}|x-x^k|+v^k \ \forall \ x \in S, \qquad (25)$$

where $\tilde{\alpha}^k_{fp} = f^k_x - \tilde{f}^k(x^k) \geq 0$, $\tilde{\alpha}^k_h = -\tilde{\delta}^k_h(x^k) \geq 0$ and

$$p^k = p^k_f + p^k_h = -u^k d^k. \qquad (26)$$

(Hint: (24) can be derived by adding the inequalities $f \geq \tilde{f}^k$ and $\delta_h \geq \tilde{\delta}^k_h$ and using (21), (23) and (26); (22) follows from $v^k = \tilde{f}^k(y^{k+1}) - f^k_x = \tilde{f}^k(x^k) - f^k_x + p^k_f \times d^k = -\tilde{\alpha}^k_p - \tilde{\alpha}^k_h + p^k \times d^k$, whereas (25) results from (22)-(24) and the Cauchy-Schwartz inequality.) (21)-(26) also follow from the Karush-Kuhn-Tucker conditions for (20), with $p^k_f = \sum_j \lambda^k_j g^j$; cf. (Kiwiel, 1987).

The choice of weights u^k is crucial in practice, since "too large" u^k produce very short steps, whereas "too small" u^k may result in many null steps (Kiwiel, 1989b). Assuming temporarily that f behaves like a quadratic between x^k and y^{k+1} with the derivative $v^k = -u^k|d^k|^2$ at x^k along d^k, we have $f(y^{k+1})=f(x^k+d^k)=f(x^k)+v^k+\hat{u}|d^k|^2/2$, where $\hat{u}$ equals $2u^k(1-[f(y^{k+1})-f(x^k)]/v^k)$. To prevent drastic changes of u^k, we shall use

$$u^{k+1} = \min[\max\{u^{k+1}_{int}, u^k/10, u_{min}\}, 10u^k], \qquad (27)$$

$$u^{k+1}_{int} = 2u^k(1 - [f^{k+1}_y - f^k_x]/v^k),$$

where $0 < u_{min} \leq 1$ is fixed.

We may now state the method in detail.

Algorithm 1

Step 0 (*Initialization*). Select an initial point $x^1 \in S$, a final optimality tolerance $\varepsilon_s \geq 0$, improvement parameters $0 < m_L < m_R < 1$, an initial weight $u^1 > 0$, a lower bound for weights $u_{min} > 0$, the maximum number of stored subgradients $M \geq N+2$ and an initial accuracy tolerance $\varepsilon^1 > 0$. Set $y^1 = x^1$, $J^1 = \{1\}$, $f^1_1 = f(y^1; y^1, \varepsilon^1)$, $g^1 = g(y^1, \varepsilon^1)$ and $f^1_x = f^1_1 + \varepsilon^1$. Set the counters $k=1$, $l=0$ and $k(0)=1$.

Step 1 (*Direction finding*). Find the solution (d^k, v^k) of (20) and its multipliers λ^k_j such that the set $\hat{J}^k = \{j \in J^k: \lambda^k_j \neq 0\}$ satisfies $|\hat{J}^k| \leq M-1$. Compute $|p^k|$ and $\tilde{\alpha}^k_p$ from (22). Set ε^{k+1} by (14).

Step 2 (*Stopping criterion*). If $v^k \geq -\varepsilon_s$, terminate; otherwise, continue.

Step 3 (*Descent test*). Set $y^{k+1} = x^k + d^k$ and f^{k+1}_y by (13). If (12) holds, set $t^k_L = 1$, $f^{k+1}_x = f^{k+1}_y$, $k(l+1) = k+1$ and increase the counter of serious steps l by 1; otherwise, set $t^k_L = 0$ and $f^{k+1}_x = f^k_x$ (*null step*). Set $x^{k+1} = x^k + t^k_L d^k$.

Step 4 (*Linearization updating*). Select $\hat{J}^k_s$ such that $\hat{J}^k \subset J^k_s \subset J^k$ and $|J^k_s| \leq M-1$, and set $J^{k+1} = J^k_s \cup \{k+1\}$. Set $g^{k+1} = g_f(y^{k+1}, \varepsilon^{k+1})$, $f^{k+1}_{k+1} = \bar{f}(x^{k+1}; y^{k+1}, \varepsilon^{k+1})$ and $f^{k+1}_j = f^k_j + \langle g^j, x^{k+1} - x^k \rangle$ for $j \in J^k_s$.

Step 5 (*Weight updating*). If $x^{k+1} \neq x^k$, select $u^{k+1} \in [u_{min}, u^k]$ (e.g. by (27)); otherwise, either set $u^{k+1} = u^k$ or choose $u^{k+1} \in [u^k, 10u^k]$ (e.g. by (27)) if

$$\tilde{\alpha}^{k+1}_{k+1} > \max\{ |p^k| + \tilde{\alpha}^k_p, -10v^k \}. \qquad (28)$$

Step 6. Increase k by 1 and go to Step 1.

A few comments on the method are in order.

Step 1 may use the dual QP method of Kiwiel (1989a) which can solve efficiently sequences of related subproblems (20).

Step 2 is justified by the optimality estimate (25).

By the rules of Step 3,

$$x^k = x^{k(l)} \text{ if } k(1) \leq k < k(l+1), \qquad (29)$$

where we may let $k(l+1) = \infty$ if the number l of serious steps stays bounded.

At Step 4 one may let $J^{k+1} = J^k \cup \{k+1\}$ and then, if necessary, drop from J^{k+1} an index $j \in J^k \setminus \hat{J}^k$ with the largest error α^{k+1}_j.

Step 5 may use the weight updating procedure of Kiwiel (1989b), which has additional criteria for changing u^k.

With a suitable choice of u^1 and u_{min} (e.g. $u^1 = |g^1|$ and $u_{min} = 10^{-20} u^1$) and another stopping criterion,

the method can (at least in theory) be made invariant to the objective and constraint scaling; see (Kiwiel, 1989b).

CONVERGENCE

In this section we show that $\{x^k\} \to \bar{x} \in X = \text{Argmin}\{f \mid S\}$ if $X \neq \emptyset$. We assume, of course, that the tolerance $\varepsilon_s = 0$. Then (25) implies that upon termination $x^k \in X$. Hence we may suppose that the algorithm does not terminate. From lack of space, we shall only indicate how to modify the corresponding results of Kiwiel (1989b).

Consider the following condition

$$f(x^k) \geq f(\tilde{x}) \text{ for some fixed } \tilde{x} \in S_h \text{ and all } k, \qquad (30)$$

which holds if $X \neq \emptyset$ or $\tilde{x}$ is a cluster point of $\{x^k\}$, since $f^{k+1}_x \leq f^k_x$ and $f(x^k) \leq f^k_x$ for all k.

Lemma 1. If (30) holds then

$$\sum_{k=1}^{\infty} t^k_L |v^k| \leq [f^1_x - f(\tilde{x})]/m_L, \qquad (31)$$

and $v^k \xrightarrow{K} 0$ if $K = \{k: t^k_L = 1\}$ is infinite. Moreover, $x^k \to \bar{x}$ for some $\bar{x} \in S$.

Proof. Use (12) and (24) as in (Kiwiel, 1989b). $\square$

Let $\hat{f}^k_s = \max\{\bar{f}(.; y^j, \varepsilon^j): j \in J^k_s\}$, $\hat{\phi}^k_s = \hat{f}^k_s + u^k | . - x^k|^2/2 + \delta_h$ and

$$\eta^k = \min \hat{\phi}^k = \hat{f}^k(y^{k+1}) + u^k |y^{k+1} - x^k|^2/2. \qquad (32)$$

Note that $\eta^k \leq \hat{\phi}^k(x^k) \leq f(x^k)$. As in (Kiwiel, 1989b), we have $y^{k+1} = \text{argmin} \hat{\phi}_s$, $\hat{f}^k_s(y^{k+1}) = \hat{f}^k(y^{k+1})$, $\eta^k = \min \hat{\phi}_s$,

$$\hat{\phi}^k_s(x) \geq \eta^k + u^k |x - y^{k+1}|^2/2 \quad \forall \, x \in \mathbb{R}^N. \qquad (33)$$

Setting $x = x^k$ with $\hat{\phi}^k_s(x) \leq f(x)$ we get

$$u^k |y^{k+1} - x^k|^2/2 \leq f(x^k) - \eta^k. \qquad (34)$$

If $x^{k+1} = x^k$, then $\hat{f}^{k+1} \geq \hat{f}^k_s$, $u^{k+1} \geq u^k$, so $\hat{\phi}^{k+1} \geq \hat{\phi}^k_s$ and

$$\eta^k + u^k |y^{k+2} - y^{k+1}|^2/2 \leq \eta^{k+1} \leq f(x^k) \text{ if } x^{k+1} = x^k \qquad (35)$$

from (33). Letting $w^k = f^k_x - \eta^k$, we get from (15), (22)

$$w^k = u^k |d^k|^2/2 + \tilde{\alpha}^k_p = |p^k|^2/2u^k + \tilde{\alpha}^k_p, \qquad (36)$$

$$v^k \leq -w^k \leq v^k/2 \leq 0. \qquad (37)$$

Lemma 2. (i) If $k(l) \leq k \leq k' < k(l+1)$ then

$$w^k \leq w^{k'} \leq |g_f(x^{k(l)}, \varepsilon^{k(l)})|^2/2u^{k(l)} + \varepsilon^{k(l)}.$$

(ii) If (30) holds and $\{\varepsilon^k\}$ is bounded then $\{g^k\}$ is bounded.

(iii) If (30) holds and $\{\varepsilon^k\}$ is bounded then there exists $C < \infty$ such that

$$\alpha^{k+1}_{k+1} \leq C / \sqrt{u^k} + f^{k(l)}_x - f(x^{k(l)}) + \varepsilon^{k+1} \qquad (38)$$

if $k(1) \leq k < k(l+1)$ and $t^k_L = 0$.

Proof. (i) If $k = k(l)$ then $k \in J^k$ and

$$\eta^k \geq \min\{ \bar{f}(x; y^k, \varepsilon^k) + u^k |x - x^k|^2/2 : x \in S\}$$
$$\geq \min\{ f^k_k + \langle g^k, x - x^k \rangle + u^k |x - x^k|^2/2 : x \in \mathbb{R}^N\}$$
$$= f^k_k - |g^k|^2/2u^k,$$

so $w^k \leq f^k_x - f^k_k + |g^k|^2/2u^k = \varepsilon^k + |g^k|^2/2u^k$, and assertion (i) follows from (35).

(ii) Use Lemma 1, (14) with $m_L < m_R$ and (22) with

$u^k \geq u_{min}$ to obtain $y^{k+1} = x^k + d^k \to \bar{x}$. Then invoke the local boundedness of g_f.

(iii) We have $|d^k| \leq [(|g^{k(l)}|^2/2u_{min} + \varepsilon^{k(l)})/u^k]^{1/2}$ from (35), $u^k \geq u_{min}$, $y^{k+1} = x^{k(l)} + d^k$ and

$$\alpha_{k+1}^{k+1} = f_x^k - \bar{f}(x^k;y^{k+1},\varepsilon^{k+1}) \leq f(x^k) - f(y^{k+1}) + f_x^k - f(x^k) +$$
$$f(y^{k+1}) - \bar{f}(y^{k+1};y^{k+1},\varepsilon^{k+1}) + \langle g^{k+1},d^k \rangle$$
$$\leq |f(x^k) - f(y^{k+1})| + f_x^k - f(x^k) + \varepsilon^{k+1} + |g^{k+1}||d^k|,$$

so we may use (29) with $x^k \to \bar{x}$, the local boundedness of g_f and the local Lipschitz continuity of f to complete the proof. □

Lemma 3. If $x^k = x^{k(1)} = \bar{x}$ for some fixed 1 and all $k \geq k(1)$, then $w^k \downarrow 0$ and $v^k \to 0$.

Proof. By the rules of Step 5 and Lemma 2, $u^{k+1} \geq u^k$ and $w^{k+1} \leq w^k$ for all $k \geq k(1)$. By (35), $\eta^k \uparrow \bar{\eta} \leq f(\bar{x})$. Hence (34) shows that $\{y^k\}$ is bounded, while (35) yields $|y^{k+2} - y^{k+1}| \to 0$. Lemma 2(i), (14) and (37) imply that $\{g^k\}$ is bounded. Let $\bar{v} = \limsup v^k$ and choose $K' \subset \{1,2,\ldots\}$ such that $v^k \xrightarrow{K'} \bar{v}$. Let $k \geq k(1)$ and $\hat{\varepsilon}_f^k = \bar{f}(y^{k+1};y^{k+1},\varepsilon^{k+1}) - \hat{f}^k(y^{k+1})$. Then

$$\hat{\varepsilon}_f^k = \bar{f}(y^{k+2};y^{k+1},\varepsilon^{k+1}) - \hat{f}^k(y^{k+1}) - \langle g^{k+1}, y^{k+2} - y^{k+1} \rangle$$
$$\leq \hat{f}^{k+1}(y^{k+2}) - \hat{f}^k(y^{k+1}) + |g^{k+1}||y^{k+2} - y^{k+1}|$$
$$\leq v^{k+1} - v^k + |g^{k+1}||y^{k+2} - y^{k+1}|,$$

since $k+1 \in J^{k+1}$, so $\limsup\{\hat{\varepsilon}_f^k: k \in K'\} \leq 0$. But $f_y^{k+1} - f_x^k > m_L v^k$ for $k \geq k(1)$, so

$$\hat{\varepsilon}_f^k = f_y^{k+1} - f_x^k - v^k - \varepsilon^{k+1} > (m_L - 1 + m_R - m_L)v^k$$
$$\geq (1 - m_R)|v^k|$$

with $m_R \in (0,1)$ imply $\bar{v} = 0$. Then $w^k \downarrow 0$ and $v^k \to 0$ by (37). □

We may now state our principal result.

Theorem 1. Either $x^k \to \bar{x} \in X$ or $X = \emptyset$ and $|x^k| \to \infty$. In both cases $f_x^k \downarrow \inf\{f(x): x \in S\}$. If $\inf\{f(x): x \in S\} > -\infty$ then $v^k \to 0$.

Proof. If (30) holds, then the preceding results imply that $x^k \to \bar{x} \in X$ and $f_x^k \downarrow f(\bar{x}) = \tilde{f}(\tilde{x})$, so $\tilde{x} \in X$ and the definition of $\inf\{f(x): x \in S\}$ yields the desired conclusion. □

Thus the method must terminate if $\inf\{f(x): x \in S\} > -\infty$ and $\varepsilon_s > 0$.

APPLICATION TO DECOMPOSITION

Suppose now that algorithm 1 applied to problem (7) with $S = \mathbb{R}_+^N$, $X \neq \emptyset$ and ε-linearizations given by (9), calculates aggregate solutions z^k by (17).

Lemma 6. With $1 = (1,\ldots,1) \in \mathbb{R}^N$, we have

$$\psi_0(z^k) \geq f_x^k - \tilde{\alpha}_p^k - p^k \ast x^k$$
$$\geq f_x^k + v^k - |u^k v^k|^{1/2}|x^k|, \qquad (39)$$
$$\psi(z^k) \geq p_f^k \geq p^k \geq -|p^k|1$$
$$\geq -|u^k v^k|^{1/2}1, \qquad (40)$$
$$f_x^k \geq f(x^k) \geq \sup\{\psi_0(z): \psi(z) \geq 0, z \in Z\}. \qquad (41)$$

Proof. Since $y^{k+1} \in S = \mathbb{R}_+^N$, $p_h^k \in \partial\delta_h(y^{k+1})$ implies $p_h^k \leq 0$ and $p_h^k \ast y^{k+1} = 0$. Let $z^j = z(y^j,\varepsilon^j)$, $j \in J^k$. By (8a) and (9)-(11), we have $\hat{f}^k(y^{k+1}) = \sum_j \lambda_j^k \psi_0(z^j) + y^{k+1} \ast p_f^k$, and

$$\sum_j \lambda_j^k \psi_0(z^j) = \hat{f}^k(y^{k+1}) - \langle p_f^k + p_h^k, y^{k+1} \rangle = f_x^k - v^k - p^k \ast x^k + p^k \ast d^k,$$

so (39) follows from the concavity of ψ_0, (17), (22) and (26). Similarly,

$$\psi(z^k) \geq \sum_j \lambda_j^k \psi(z^j) = p_f^k = p^k - p_h^k \geq p^k$$

with (22) imply (40). (41) follows from the weak duality theorem. □

Comparing (18)-(19) with (39)-(41), we can identify the tolerances $\varepsilon_f^k = \tilde{\alpha}_p^k - p^k \ast x^k$ and $\varepsilon_F^k = |p^k|$. If $\{u^k\}$ is bounded, then $x^k \to \bar{x}$, $v^k \to 0$ and (22) imply that $\varepsilon_f^k \to 0$ and $\varepsilon_F^k \to 0$, so that $\{z^k\}$ is a generalized minimizing sequence for (1), and every accumulation point of $\{z^k\}$ (which exists, e.g. for compact Z) solves (1). On the other hand, if $\{u^k\}$ is unbounded then the proof of Lemma 5 shows that a subsequence of $\max\{\varepsilon_f^k, \varepsilon_F^k\}$ vanishes (cf. (28)). Of course, one may impose an upper bound on u^{k+1} at Step 5 to ensure the stronger convergence result.

MODIFICATIONS

To trade off storage and QP work per iteration for speed of convergence, one may replace subgradient selection with aggregation as in (Kiwiel, 1989b) (so that z^k is generated recusively and $M \geq 2$. The global convergence results are the same.

By exploiting the additive structure (5) of f one may increase the speed of convergence at the cost of more storage and work per iteration. To this end, use the approximations

$$\hat{f}^k(x) = \sum_{i=1}^n \hat{f}_i^k(x),$$
$$\hat{f}_i^k(x) = \max\{\bar{f}_i(x;y^j,\varepsilon^j/n): j \in J_i^k\}$$

constructed from the ε-linearizations of f_i defined via (8) with "f" replaced by "f_i", where the sets J_i^k satisfying $\sum_{i=1}^n |J_i^k| \leq M$ with $M \geq N+2n$ are selected by finding at most $N+n$ nonzero Lagrange multipliers λ_{ij}^k, $j \in J_i^k$, $i=1,\ldots,n$, of the corresponding extension of (20) (see Kiwiel, 1989b). In view of (3) and (4), we may use

$$\bar{f}(x;y^j,\varepsilon^j) = \psi_{0i}(z_i(y^j,\varepsilon^j)) + \sum_{j=1}^N x_j \psi_{ji}(z_i(y^j,\varepsilon^j))$$

and compute separate aggregate primal solutions

$$z_i^k = \sum_{j \in J_i^k} \lambda_{ij}^k z_i(y^j,\varepsilon^j), \quad i=1,\ldots,n, \qquad (42)$$

to form $z^k = (z_1^k,\ldots,z_n^k)$ for which the preceding convergence results hold. The fact that at most $N+n$ multipliers λ_{ij}^k are nonzero with

$$\sum_{j \in J_i^k} \lambda_{ij}^k = 1, \quad i=1,\ldots,n,$$

has important implications in Lagrangian relaxation of discrete problems (Bertsekas, 1982; Kiwiel and Toczyłowski, 1986).

NUMERICAL RESULTS

We shall now report on computational testing of the algorithm with a double precision Fortran code on an IBM PC/XT microcomputer with relative accuracy of 2.2×10^{-16} (=2.2E-16). The parameters had values m_L=0.1, m_R=0.2, ε^1=0.01, and ε_S=1E-6 in the stopping criterion $v^k \geq -\varepsilon_S(1+|f_x^k|)$.

We used the collection of fourteen nondifferentiable problems of the form (7) from (Kiwiel, 1989b). Table 1 in the Appendix contains results obtained for "maximally" inaccurate linearizations. This means that the problem subroutine evaluated an exact subgradient $g^{k+1} \in \partial f(y^{k+1})$ and set $\bar{f}(y^{k+1};y^{k+1}, \varepsilon^{k+1})=f(y^{k+1})-\varepsilon^{k+1}$, thus making g^{k+1} an ε^{k+1}-subgradient of f at y^{k+1}; the resulting linearizations satisfied (8) with equality in (8c). We also tested the opposite case, in which the subroutine returns exact function values $(\bar{f}(y^{k+1};y^{k+1},\varepsilon^{k+1})=f(y^{k+1}))$ but they are perturbed by the algorithm according to (13). In this case the results were similar to those in Table 1, in which k denotes the final iteration number (and the total number of function and subgradient evaluations), f($\bar{x}$) denotes the optimal value, and the left-hand columns contain results for exact linearizations $(\varepsilon^k \equiv 0)$ from (Kiwiel, 1989b). The largest growth of computational effort occurs on polyhedral functions (tests 3-6). This is not surprizing, since for such functions our method with exact linearizations finds minimizers in finitely many iterations (Kiwiel, 1989c), whereas ε-linearizations "soften" the edges of epigraphs of these functions, thus preventing finite convergence.

We may add, with regret, that we have found no comparable results in the literature.

Although our academic examples have no direct connection with decomposition, they suggest that our method is quite efficient, since some of them are considered to be difficult even for methods that use exact linearizations (Lemarechal, 1981; Shor, 1985). Our experience with decomposition of production scheduling problems (Kiwiel and Toczyłowski, 1986) will be reported elsewhere.

CONCLUSIONS

We have presented an extension of the proximal bundle method for convex nondifferentiable optimization to the case where only approximate objective linearizations are available. Our limited computational experience suggests that the method is promising. We have also exhibited important implications of this algorithm for decomposition of convex separable programs.

Acknowledgment. This research was supported by the Polish Academy of Sciences, under Project CPBP/0215.

REFERENCES

Bertsekas, D.P. (1982). *Constrained Optimization and Lagrange Multiplier Methods*. Academic Press, New York.

Dantzig, G.E., and P. Wolfe (1960). Decomposition principle for linear programs. *Operations Research*, 8, 101-111.

Demyanov, V.F., and L.V. Vasiliev (1985). *Nondifferentiable Optimization*. Optimization Software Inc., New York.

Gabasov, R., F.M. Kirilova, O.I. Kostyukova and V.M. Raketskii (1987). *Constructive Optimization Methods, Part 4, Convex Problems*. Izdatelstvo "Universitetskoye", Minsk (Russian).

Golshtein, E.G. (1987). A general approach to decomposition of optimizing systems. *Tekhnicheskaia Kibernetika*, No. 1, 59-69 (Russian).

Kiwiel, K.C. (1985a). *Methods of Descent for Nondifferentiable Optimization*. Lecture Notes in Mathematics 1133. Springer, Berlin.

Kiwiel, K.C. (1985b). An algorithm for nonsmooth convex minimization with errors. *Mathematics of Computation*, 45, 173-180.

Kiwiel, K.C. (1987). A constraint linearization method for nondifferentiable convex minimization. *Numerische Mathematik*, 51, 395-414.

Kiwiel, K.C. (1989a). A dual method for solving certain positive semi-definite quadratic programming problems. *SIAM Journal on Scientific and Statistical Computing*, 10.

Kiwiel, K.C. (1989b). Proximity control in bundle methods for convex nondifferentiable minimization. *Math. Programming* (to appear).

Kiwiel, K.C. (1989c). Exact penalty functions in proximal bundle methods for constrained convex nondifferentiable minimization. Technical Report, Systems Research Institute, Warsaw.

Kiwiel, K.C., and E. Toczłowski (1986). Aggregate subgradients in Lagrangian relaxations of discrete optimization problems. *Zeszyty Naukowe Politechniki Śląskiej, seria Automatyka z. 84*, (Wydawnictwo Naukowe Politechniki Śląskiej, Gliwice), 119-129 (Polish).

Lasdon, L.S. (1970). *Optimization Theory for Large Systems*. Macmillan, Toronto.

Lemarechal, C. (1982). Numerical experiments in nonsmooth optimization. In E.A. Nurminski (Ed.).*Progress in Nondifferentiable Optimization*. CP-82-S8. International Institute for Applied Systems Analysis, Laxenburg, Austria. pp.61-84.

Rzhevskii, S.V., and A.V. Kuncevich (1985). Application of an ε-subgradient method to the solution of the dual and primal problems of mathematical programming. *Kibernetika*, No. 5, 51-54 (Russian).

Sen, S., and H.D. Sherali (1986). A class of convergent primal-dual subgradient algorithms for decomposable convex programs. *Math. Programming*, 35, 279-297.

Shor, N.Z. (1985). *Minimization Methods for Nondifferentiable Functions*. Springer, Berlin.

Spingarn, J.E. (1985). Application of the method of partial inverses to convex programming decomposition. *Math. Programming*, 32, 199-223.

TABLE 1   Results for Exact and Approximate Linearizations

Test	N	Exact linearizations		Optimal value	Approximate linearizations	
		k	$f(x^k)$	$f(\bar{x})$	k	$f(x^k)$
1	5	29	22.600162	22.60016	35	22.60016
2	10	41	-0.8414074	-0.841408	45	-0.841407
3	50	52	6.0E-13	0	89	5.9E-7
4	48	180	-638565.00	-638565	435	-638564.51
5	50	16	3.6E-7	0	50	1.7E-6
6	30	7	3.5E-9	0	13	-2.3E-9
7	10	23	-0.3681664	-0.36811664	29	-0.3681658
8	4	14	0.7071074	0.7071068	25	0.7071069
9	4	9	1.0142141	1.0142136	14	1.0142136
10	6	47	0.0147064	0.0147063	61	0.0147066
11	5	10	-32.348679	-32.348679	14	-32.248679
12	4	20	-43.999961	-44	20	-43.999971
13	4	15	23.886767	23.886767	19	23.886767
14	6	23	68.829581	68.82956	26	68.82956

OPTIMAL REGULATOR FOR NEUTRAL SYSTEMS

N. Abe*, K. Uchida and E. Shimemura****

*Department of Precision Engineering, Meiji University, Higashimita,
Tama-ku Kawasaki, 214, Japan
**Department of Electrical Engineering, Waseda University, Okubo Shinjuku, 169,
Japan

Abstract: This paper is concerned with optimal regulator for neutral systems. We introduce Riccati type partial differential equations for neutral systems and derive existence conditions. Optimal regulator is constructed by using stationary solutions of the Riccati type partial differential equations. Some closed-loop properties of the optimal regulator are discussed.

Keywords: optimal regulator, neutral system, Riccati equation, circle condition

1.INTRODUCTION

Optimal regulator problem for retarded system was well-studied; it is now known that optimal regulator for retarded systems has good closed-loop properties the same as that for lumped parameter systems[1]. For neutral systems, a few researches about the optimal control are presented[2]-[5]. Datko[2] has demonstrated that the optimal control satisfied a linear feedback law. Sendaula[3] has suggested that the finite time interval optimal control law was characterized by the solutions of simultaneous differential equations. Ito and Tarn[4] have treated neutral system as an evolutional equations and discussed optimal regulator problem with operator Riccati equations. Pritchard and Salamon[5] have established a general semigroup framework for solving optimal regulator problem for distributed parameter systems with unbounded input and output operators; as one of these examples, they treat a neutral system.

In the previous researches stated above, it has been left some problems; for example, does the feedback control law have an integral kernel description ? If so, what form dose the kernel have ? In this paper we solve optimal regulator problem in a concrete form by using Riccati type partial differential equations and derive the optimal regulator of the integral kernel form. Some closed-loop properties of the optimal regulator similar to lumped parameter systems and retarded systems are studied.

In section 2 the finite time interval problem is formulated and solved. In section 3 we introduce Riccati type partial differential equations. The existence theorem of the Riccati type partial differential equations are established by applying the characteristic surface method[6]. In section 4 stationary Riccati type partial differntial equations are represented. In section 5 we define stabilizability and detectability for neutral systems and solve the optimal regulator problem. In section 6 closed-loop properties of the optimal regularor are mentioned.

2. OPTIMAL REGULATOR PROBLEM
IN FINITE TIME INTERVAL

We consider a system described by the linear neutral delay differential system

$$\dot{x}(t)=A_0 x(t)+A_1 x(t-h)+A_{-1}\dot{x}(t-h)+Bu(t) \tag{1}$$

where A_0, A_1, A_{-1} and B are nxn, nxn, nxn and nxm constant matricies, respectively. Initial condition $x(\beta)=\phi(\beta)$, $-h\leq\beta\leq0$ is C^1-function such that $\phi(0)=A_1\phi+A_{-1}\dot{\phi}$. The constraint guarantees C^1-solution $x(t)$. If ϕ is C^1-function without the constraint, then the solution $x(t)$ is discontinuous at points $t=nh$, $n=0,1,2,\cdots$.

Examples of linear neutral system are lossless transmission lines[8] or repetitive control systems[9].

We consider optimal regulator problem in a finite time interval for the linear neutral system described by (1). Let the finite time cost functional be given as

$$J(T)=\int_0^T [x'(t)C'Cx(t)+u'(t)Ru(t)]dt \tag{2}$$

where C is a constant nxr matrix and R is a positive definite constant mxm matrix. The prime denotes transpose. Consider the problem of finding a continuous control $u(t)$ in the finite time interval which minimizes the value of the finite time cost functional (2) subject to the system (1).

<u>Theorem 1</u>

Let $P_0(t)$, $P_1(t,\beta)$ and $P_2(t,\alpha,\beta)$, $0 \leq t \leq T$, $-h \leq \alpha \leq 0$, $-h \leq \beta \leq 0$ be a C^1-solution of the following Riccati type partial differential equations.

$$\frac{d}{dt}P_0(t) = -A_0'\Pi_0(t) - \Pi_0(t)'A_0$$
$$+ \Pi_0(t)'BR^{-1}B'\Pi_0(t)$$
$$- A_1'P_1'(t,0) - P_1(t,0)A_1 - C'C \qquad (3\text{-}a)$$

$$\frac{\partial}{\partial t}P_1(t,\beta) = \frac{\partial}{\partial \beta}P_1(t,\beta) - A_0'\Pi_1(t,\beta)$$
$$+ \Pi_0(t)'BR^{-1}B'\Pi_1(t,\beta)$$
$$- A_1'P_2(t,0,\beta) \qquad (3\text{-}b)$$

$$\frac{\partial}{\partial t}P_2(t,\alpha,\beta) = \frac{\partial}{\partial \alpha}P_2(t,\alpha,\beta)$$
$$- \frac{\partial}{\partial \beta}P_2(t,\alpha,\beta) - \Pi_1(t,\alpha)'$$
$$\cdot BR^{-1}B'\Pi_1(t,\beta) \qquad (3\text{-}c)$$

where $\Pi_0(t) = P_0(t) + A_{-1}'P_1'(t,0)$ and $\Pi_1(t,\beta) = P_1(t,\beta) + A_{-1}'P_2(t,0,\beta)$, with the terminal conditions,

$$P_0(T)=0, \ P_1(T,\beta)=0, \ P_2(T,\alpha,\beta)=0,$$
$$-h \leq \alpha \leq 0, \ -h \leq \beta \leq 0 \qquad (4)$$

the symmetric conditions

$$P_0'(t)=P_0(t), \ P_2'(t,\alpha,\beta)=P_2(t,\beta,\alpha) \qquad (5)$$

and the boundary conditions

$$\Pi_0(t) - P_1(t,-h) = 0,$$
$$\Pi_1(t,\beta) - P_2(t,-h,) = 0 \qquad (6)$$

Then the optimal control law $u^*(t)$ for cost functional (2) can be written as

$$u^*(t) = -R^{-1}B'\{\Pi_0(t)x(t)$$
$$+ \int_{-h}^{0} \Pi_1(t,\beta)(A_1 x(t+\beta) + A_{-1}\dot{x}(t+\beta))d\beta\} \qquad (7)$$

and the optimal cost $J^*(T)$ is represented as

$$J^*(T) = \phi'(0)P_0(0)\phi(0) + 2\phi'(0)\int_{-h}^{0}P_1(0,\beta)$$
$$\cdot (A_1\phi(\beta) + A_{-1}\dot{\phi}(\beta))d\beta$$
$$+ \int_{-h}^{0}\int_{-h}^{0}(A_1\phi(\alpha) + A_{-1}\dot{\phi}(\alpha))'P_2(0,\alpha,\beta)$$
$$\cdot (A_1\phi(\beta) + A_{-1}\dot{\phi}(\beta))d\alpha d\beta \qquad (8)$$

where ϕ is the initial function.

<u>Proof</u>

Let $x(t)$, $0 \leq t \leq T$ be a unique solution for the initial condition ϕ and control law $u(t)$ and $P_0(t)$, $P_1(t,\beta)$ and $P_2(t,\alpha,\beta)$ be the solution of (3)-(6). A functional $V(x)$ is defined by

$$V(x)=x'(t)P_0(t)x(t) + 2x'(t)\int_{-h}^{0}P_1(t,\beta)$$
$$\cdot (A_1 x(t+\beta) + A_{-1}\dot{x}(t+\beta))d\beta$$
$$+ \int_{-h}^{0}\int_{-h}^{0}(A_1 x(t+\alpha) + A_{-1}x(t+\alpha))'P_2(t,\alpha,\beta)$$
$$\cdot (A_1 x(t+\beta) + A_{-1}\dot{x}(t+\beta))d\alpha d\beta \qquad (9)$$

Now we consider the following identity.

$$V(x(T)) - V(x(0)) = \int_{-h}^{0}\dot{V}(x(t))dt \qquad (10)$$

Applying Riccati type partial differential equation (3), boundary condition (6) and the terminal conditions $V(x(T))=0$, to the right-hand side of the above identity, we obtain

$$J(T) = V(X(0)) + \int_{0}^{T}[u^*(t)-u(t)]'R$$
$$\cdot [u^*(t)-u(t)]dt \qquad (11)$$

where $u^*(t)$ is given by (7). It is clear that for any continuous control inputs the integration term of the right-hand side of (11) is nonnegative. It is also clear that only for $u^*(t)$ the integral term is zero. Since the first term of the right-hand side dose not depend on control input, $u^*(t)$ is the unique optimal control law and optimal cost is given by $V(x(0))$ that is (8). Δ

Now, we present global existence theorem of the Riccati type partial differential equations.

<u>Theorem 2</u>

If $||A_{-1}||_\infty < 1$ hold, then there exists a unique C^1-solution of Riccati type partial equations (3)-(6) in the region $R(T) = \{(t,\alpha,\beta) \ 0 \leq t \leq T, \ -h \leq \alpha \leq 0, \ -h \leq \beta \leq 0\}$, where $|| M ||_\infty = \max_i \sum_j |m_{ij}|$.

<u>Proof</u>

Denoting by $J^0(T)$ the cost functional (2) for the zero input ($u(t) \equiv 0$), we have that $J^*(T) \leq J^0(T)$. The boundness of the triplet $P_0(t)$, $P_1(t,\beta)$ and $P_2(t,\alpha,\beta)$ follows from this inequality and (8). Next, we consider a set of the characteristic curves which are solutions $d/dt\alpha(t) = -1$, $d/dt\beta(t) = -1$, that is,

$$C = \{(t,\alpha,\beta); \ \alpha=\alpha(t), \ \beta=\beta(t)\}.$$

Along each curve, we see that $P_1(t,\cdot)$ and $P_2(t,\cdot,\cdot)$ satisfy ordinary differential equations. The following statements are equivalent: the Riccati type partial differential equations (3) hold at all points (t,α,β) in the region $R(T)$ and the ordinary differential equations hold on all the characteristic curves in C on R(T). If $||A_{-1}||_\infty<1$ holds, then we can prove that the mapping F, defined by integrating (3-a) and ordinary differential equations along characteristic curves, is the contraction mapping. The fact means from the fixed point theorem and the boundness that there exists a unique solution in the region R(T). Δ

The difference between the existence theorem for neutral systems and retarded systems is that the neutral cases require an additional norm condition $||A_{-1}||_\infty< 1$. Note that the norm condition is a sufficient condition for existence of Riccati type partial differential equations.

4. STATIONARY SOLUTION OF RICCATI TYPE PARTIAL DIFFERENTIAL EQUATIONS

In this section we consider the existence theorem for stationary solution of the Riccati type partial differential equations. By using the solution, we construct the optimal regulator. The stationary solution is the triplet P_0, $P_1(\beta)$ and $P_2(\alpha,\beta)$, $-h\le\alpha\le 0$, $-h\le\beta\le 0$ that satisfies the following Riccati type partial differential equations.

$$A_0'\Pi_0+\Pi_0'A_0+A_1'P_1'(0)-P_1(0)A_1$$
$$-\Pi_0'BR^{-1}B'\Pi_0+C'C=0 \qquad (12\text{-a})$$

$$-\frac{\partial}{\partial\beta}P_1(\beta)-A_0'\Pi_1(\beta)$$
$$+A_1'P_2(0,\beta)-\Pi_0'BR^{-1}\,{}^3B'\Pi_1(\beta) = 0$$
$$(12\text{-b})$$

$$-\frac{\partial}{\partial\alpha}P_2(\alpha,\beta)-\frac{\partial}{\partial\beta}P_2(\alpha,\beta)$$
$$-\Pi_1(\alpha)'BR^{-1}B'\Pi_1(\beta) = 0 \qquad (12\text{-c})$$

where $\quad \Pi_0 = P_0+A_{-1}'P_1'(0)$

and $\quad \Pi_1(\beta) = P_1(\beta)+A_{-1}'P_2(0,\beta)$,

with the symmetric conditions
$$P'_0 = P_0, \quad P_2(\alpha,\beta) = P_2'(\beta,\alpha) \qquad (13)$$
and the boundary conditions
$$\Pi_0(0)-P_1(-h) = 0,$$
$$\Pi_1(\beta)-P_2(-h,\beta) = 0 \qquad (14)$$

Theorem 3

Suppose that $||A_{-1}||_\infty< 1$. If there exists a positive constant k such that
$$\max\{\max_i P_0^{ii}(t),$$
$$\max_{-h\le\beta\le 0}\max_i P_2^{ii}(t,\beta,\beta)\}\le k \qquad (15)$$
for t>0, where superscript ii denotes (i,i)-element of the matrix, then there is a C-solution of the stationary Riccati equations in the interval, $-h\le\alpha\le 0$, $-h\le\beta\le 0$.

Before turning to the proof of the theorem, we state a lemma for the properties of the nonstationary Riccati type partial differential equations.

Lemma 1

The solution of the Riccati type partial differential equations (3)-(6) has the following properties.

(i) $\quad P_0^{ii}(0)\ge 0$

(ii) $\quad P_2^{ii}(0,\beta,\beta)\ge 0$

(iii) $\quad P_0^{ij}(0)\le \frac{1}{2}(P_0^{ii}(0)+P_0^{jj}(0))$

(iv) $\quad P_2^{ij}(0,\alpha,\beta)\le \frac{1}{2}(P_2^{ii}(0,\beta,\beta)+P_2^{jj}(0,\alpha,\alpha))$

(v) $\quad P_1^{ij}(0,\beta)\le \frac{1}{2}(P_0^{ii}(0)+P_2^{jj}(0,\beta,\beta))$

This lemma is proved by using the relation J(T)≥0 and choosing suitably initial functions.

Proof of Theorem 3

Let u_1 and u_2 denote the optimal control for terminal condition T_1 and $T_2(T_1<T_2)$, respectively. Then the relation $J(u_2;\phi,T_2)\ge J(u_1;\phi,T_1)$ holds. From Lemma 1 (i), $P_0^{ii}(T_1)\le P_0^{ii}(T_2)$ and the boundness assumption (15), there exists the limit P_0^{ii} such that
$$\lim_{T\to\infty} P_0^{ii}(t) = P_0^{ii}. \qquad (16)$$
Furthermore, From Lemma 1 (iii), then
$$\lim_{T\to\infty} P_0^{ij}(t) = P_0^{ij}. \qquad (17)$$
If $T_2>T_1$, then $P_2^{ii}(T_1,\beta,\beta)\le P_2^{ii}(T_2\beta,\beta)$ and there exists the limit for fixed β such that
$$\lim_{T\to\infty} P_2^{ii}(t,\beta,\beta) = P_2^{ii}(\beta,\beta). \qquad (18)$$

We show that (18) is uniformly convergence with respect to β. Let c^{ii} denote the characteristic curve defined as follows.
$$c^{ii}=\{(\alpha,\beta,t)\quad \alpha=\beta, -h\le\beta\le 0,$$
$$t^i =t(\beta;\tau), \tau\ge 0\}$$
$$dt^i/d\beta = -1, -h\le\beta\le 0,$$
$$t^i(0;\tau)=\tau \qquad (19)$$
We consider $P_2^{ii}(t,\alpha,\beta)$ along this curve. $P_2^{ii}(t,\alpha,\beta)$ obeys the ordinary differential equation. Therefore, by using Ascoli-Arzela theorem, we obtain a sequence $\{\tau_k\}$ such that
$$\lim_{\tau_k\to\infty} P_2^{ii}(0,\beta,\beta;t_i(\beta;\tau_k))=\tilde{P}_2^{ii}(\beta,\beta) \qquad (20)$$
uniformly in β, $-h\le\beta\le 0$, for some continuous function $\tilde{P}_2^{ii}(\beta,\beta)$. From the uniqueness of the monotonicity conver-

gent sequence, we have $\tilde{P}_2^{\ ii}(\beta,\beta)$ $=P_2^{\ ii}(\beta,\beta)$. Furthermore, from Lemma 1 (iv), (v) and assumption (15), we show that non-diagonal terms of $|P_2(T_1,\alpha,\beta)-P_2(T_2,\alpha,\beta)|$ and $|P_1(T_1,\beta)-P_1(T_2,\beta)|$ are not greater than its diagonal terms. As mentioned above, the convergence

$$\lim_{T\to\infty} P_1^{\ ij}(t,\beta) = P_1^{\ ij}(\beta), \qquad (21)$$

$$\lim_{T\to\infty} P_2^{\ ij}(t,\alpha,\beta) = P_2^{\ ij}(\alpha,\beta) \qquad (22)$$

hold uniformly in (α,β) $-h\leq\alpha\leq0$, $-h\leq\beta\leq0$, where $P_1^{\ ij}(\beta)$ and $P_2^{\ ij}(\alpha,\beta)$ are continuous functions. $\triangle$

5. OPTIMAL REGULATOR

In this section we deal with optimal regulator for neutral systems under assumption of $||A_{-1}||_\infty < 1$ in the infinite time interval. We define stabilizability and detectability for neutral systems, and we show that if system (1) is stabilizable and detectable, then there exists the solution of the stationary Riccati equations. We show finally that the optimal closed-loop system is stable.

Definition 1 (stabilizability)

Suppose that control input is given by the following linear state feedback form

$$u(t)=K_0x(t)+\int_{-h}^{0} K_1(\beta)(A_1x(t+\beta)$$
$$+A_{-1}\dot{x}(t+\beta))d\beta \qquad (23)$$

where (K_0,K_1) $R^{nxm}xL^2([-h,0];R^{nxm})$. For the initial function ϕ, if the unique solution $x(t)$ satisfies the relation

$$\lim_{T\to\infty} \int_0^\tau x'(t)x(t)dt < \infty \qquad (24)$$

then the system is said to be L^2-stable. Furthermore, the system is stabilizable if there exists (K_0, K_1) for given initial function, such that the closed-loop system is L^2-stable.

Definition 2 (Detectable)

The system is detectable if the adjoint system is stabilizable.

$$\dot{y}(t) = A_0'y(t)+A_1'\dot{y}(t+h)$$
$$+A_{-1}'\dot{y}(t+h)+C'u(t) \qquad (25)$$

It is known that neutral root chains in the right half plane can not be shifted into the left half plane by linear state feedback control of the form (23)[10]. Since $\max|\lambda(M)|\leq||M||_\infty$, the assumption $||A_{-1}||_\infty<1$ (i.e. $\max|\lambda(A_{-1})|<1$) guarantees that neutral root chains lie in the left half plane, that is, there exist

only finite number of poles in the closed right half plane. Unless the assumption $\max|\lambda(A_{-1})|<1$ holds, the stabilizability problem has been left unsolved until now.

We now state an optimal regulator problem in the infinite time interval and derive the solution of the optimal regulator problem. Let the infinite time cost functional be given as

$$J(\infty) = \int_0^\infty [x'(t)C'Cx(t)+u'(t)Ru(t)]dt. \qquad (26)$$

Our problem is to find a control input which minimizes the cost functional (26) subject to the system (1).

Theorem 4

Suppose that $||A_{-1}||_\infty < 1$ and the system is stabilizable and detectable. There exists a unique optimal control law given as

$$u(t) = -R^{-1}B'\{\Pi_0x(t)$$
$$+\int_{-h}^{0}\Pi_1(\beta)(A_1x(t+\beta)+A_{-1}\dot{x}(t+\beta))d\beta\} \qquad (27)$$

where Π_0 and $\Pi_1(\beta)$ are defined by the solution of the stationary Riccati equations (12)-(14). The optimal cost $J(\infty)$ is represented as

$$J(\infty)=\phi'(0)P_0\phi(0)+2\phi'(0)\int_{-h}^{0}P_1(\beta)$$
$$\cdot(A_1\phi(\beta)+A_{-1}\dot{\phi}(\beta))d\beta$$
$$+\int_{-h}^{0}\int_{-h}^{0}(A_1\phi(\alpha)\quad A_{-1}\dot{\phi}(\alpha))'P_2(\alpha,\beta)$$
$$\cdot(A_1\phi(\beta)+A_{-1}\dot{\phi}(\beta))d\alpha d\beta. \qquad (28)$$

Proof

The optimality is proved in the same way as in the proof of Theorem 1. Since we assume that the system is stabilizable, there exists a feedback gain (K_0, K_1) such that the closed-loop system is L^2-stable. Let $x^+(t)$ and $u^+(t)$ denote the corresponding trajectory and input, respectively. Then we have

$$J(\tau)\leq\int_0^\tau [x^{+'}(t)C'Cx^+(t)$$
$$+u^{+'}(t)Ru^+(t)]dt. \qquad (29)$$

Since the closed-loop system is L^2-stable, $J(\tau)$ is bounded and we consider that (15) of Theorem 3 is satisfied for suitable choice of initial function. Therefore from Theorem 3 there exists the solution of the stationary Riccati equations . Zabczyk showed that the semigroup generated by the optimal closed-loop system is exponential stable by using an operator Lyapunov equation and the assumption of detectability[11]. We introduce the Lyapunov type partial differential equations. Applying similar procedure to that of Zabczyk to the Riccati type partial differential equations, we prove that the closed-loop system is L^2-stable under the assumption of detectability. Finally, we prove the uniqueness. Now suppose that there are two solutions of the stationary Riccati equations. We define the following functional

$$V^{12}(t)=x^{1\prime}(t)U_0 x^2(t)+x^{1\prime}(t)\int_{-h}^0 U_1(\beta)$$
$$\cdot(A_1 x^2(t+\beta)+A_{-1}\dot{x}^2(t+\beta))d\beta$$
$$+\int_{-h}^0 (A_1 x^1(t+\alpha)+A_{-1}\dot{x}^1(t+\alpha))^\prime U_1^\prime(\alpha)d\alpha$$
$$\cdot x^2(t)$$
$$+\int_{-h}^0 (A_1 x^1(t+\alpha)+A_{-1}\dot{x}^1(t+\alpha))^\prime U_2(\alpha,\beta)$$
$$\cdot(A_1 x^2(t+\beta)+A_{-1}\dot{x}^2(t+\beta))d\alpha d\beta \tag{30}$$

where U_i, $i=0,1,2$ are difference of two solutions and x^i, $i=1,2$ denote the trajectory of two closed-loop. Calculating $dV^{12}(t)/dt$, we have $dV^{12}(t)/dt = 0$. Since two closed-loop systems are L^2-stable, $V^{12}(t)\to 0$, $t\to\infty$ holds. This fact means that $U_0=0$, $U_1(\beta)=0$ and $U_2(\alpha,\beta)=0$, $-h\le\alpha\le 0$, $-h\le\beta\le 0$. Therefore, the uniqueness is proved. $\triangle$

6.CLOSED-LOOP PROPERTIES
We discuss in this section some properties of the optimal regulator analogous to that of the finite-dimensional optimal regulators, i,e, circle condition and the relation between hamiltonian matrix and closed-loop poles.

Multiplying (12-b) by $(A_1 e^{s\beta}+sA_{-1}e^{s\beta})$ from the right, integrating both sides with respect to β over the interval [-h, 0] by parts by using the boundary condition (14), we have

$$-P_1(0)(A_1+sA_{-1})+\Pi_0^\prime(A_1 e^{-sh}+sA_{-1}e^{-sh})$$
$$+\{sI+A_0^\prime-\Pi_0^\prime BR^{-1}B^\prime\}$$
$$\cdot\int_{-h}^0 \Pi_1(\beta)(A_1 e^{s\beta}+sA_{-1}e^{s\beta})d\beta$$
$$+(A_1^\prime-sA_{-1}^\prime)\int_{-h}^0 P_2(0,\beta)$$
$$\cdot(A_1 e^{s\beta}+sA_{-1}e^{s\beta})d\beta=0. \tag{31}$$

Multiplying transposed version of (12-b) in which the independent variable is denoted by α instead of β by $(A_1^\prime e^{-s\alpha}-sA_{-1}^\prime e^{-s\beta})$ from the left, and calculating the same integral operation as above, we have

$$-(A_1-sA_{-1})^\prime P_1(0)+(A_1 e^{sh}+sA_{-1}e^{sh})^\prime\Pi_0$$
$$-\int_{-h}^0 (A_1 e^{-s\alpha}-sA_{-1}e^{-s\alpha})^\prime\Pi_1^\prime(\alpha)$$
$$\cdot\{sI-A_0^\prime+BR^{-1}B^\prime\Pi_0\}d\alpha$$
$$+\int_{-h}^0 (A_1 e^{-s\alpha}-sA_{-1}e^{-s\alpha})^\prime P_2(\alpha,0)d\alpha$$
$$\cdot(A_1+sA_{-1})=0. \tag{32}$$

Furthermore, multiplying both sides of (12-c) by $(A_1 e^{s\beta}+sA_{-1}e^{s\beta})$ from the right and by $(A_1^\prime e^{-s\alpha}-sA_{-1}^\prime e^{-s\beta})$ from the left, integrating them with respect to α and β over the same interval [-h,0] by using the second boundary condition (14), we have

$$-2(A_1^\prime-sA_{-1}^\prime)\int_{-h}^0 P_2(0,\beta)$$
$$\cdot(A_1 e^{s\beta}+sA_{-1}e^{s\beta})d\beta$$
$$+2(A_1 e^{-sh}-sA_{-1}e^{-sh})^\prime\int_{-h}^0\Pi_1(\beta)$$
$$\cdot(A_1 e^{s\beta}+sA_{-1}e^{s\beta})d\beta$$
$$-\int_{-h}^0\int_{-h}^0 (A_1 e^{-s\alpha}-sA_{-1}e^{-s\alpha})^\prime\Pi_1(\alpha)^\prime$$
$$\cdot BR^{-1}B^\prime\Pi_1(\beta)(A_1 e^{s\beta}+sA_{-1}e^{s\beta})d\alpha d\beta = 0 \tag{33}$$

Now, adding together these equations and first Riccati equation (12-a) with sP_0-sP_0, that is (12-a)+(31)+(32)+(33), and rearranging terms, we obtain

$$F^\prime(-s)\Delta(s)+\Delta^\prime(-s)F(s)$$
$$+F^\prime(-s)BR^{-1}B^\prime F(s)-C^\prime C = 0 \tag{34}$$

where

$$F(s)=\Pi_0+\int_{-h}^0\Pi_1(\beta)(A_1 e^{s\beta}+sA_{-1}e^{s\beta})d\beta \tag{35}$$
$$\Delta(s)=sI-A_0-A_1 e^{-sh}-sA_{-1}e^{-sh}. \tag{36}$$

Then, by applying a standard successive manipulation for finite dimensional Riccati equation to (34), it can be rewritten as

$$[I+R^{-1}B^\prime F(-s)\Delta^{-1}(-s)B]^\prime R$$
$$\cdot[I+R^{-1}B^\prime F(-s)\Delta^{-1}(-s)B]$$
$$= R+B^\prime\Delta^{-T}(-s)C^\prime C\Delta^{-1}(s)B. \tag{36}$$

We can call it *Kalman equation*. Set $s=j\omega$ in the Kalman equation (36) and note that the matrix $B^\prime\Delta^{-T}(-j\omega)C^\prime C\Delta^{-1}(j\omega)B$ is positive semi-definite for all real ω. Thus it follows that

$$[I+R^{-1}B^\prime F(-j\omega)\Delta^{-1}(-j\omega)B]^* R$$
$$\cdot[I+R^{-1}B^\prime F(-j\omega)\Delta^{-1}(-j\omega)B] \ge R \tag{37}$$

for all real ω. we can call this inequality *circle condition*.

Circle condition plays an important role about robustness in finite dimensional systems and retarded systems[1]. Therefore, it may be true that optimal regulator for neutral systems has good closed-loop properties as same as the aboves.

Finally, we consider the relation between the hamiltonian matrix and the optimal closed-loop poles. The hamiltonian matrix is defined by
$$H(s) =$$

$$\begin{bmatrix} A_0 + e^{-sh}A_1 + se^{-sh}A_{-1} & -BR^{-1}B' \\ -C'C & -A_0' - e^{sh}A_1' - se^{sh}A_{-1}' \end{bmatrix}$$

$$(38)$$

Calculating $\det[sI-H(s)]$ with the Kalman equation (36), we have
$$\det[sI-H(s)]$$
$$= (-1)^n\det[\Delta'(-s)+F'(-s)BR^{-1}B']$$
$$\cdot\det[\Delta(s)+F(s)BR^{-1}B'] \quad (39)$$

Since closed-loop is L^2-stable, closed-loop poles lie in the left-half plane. The characteristic equation is given as $\det[\Delta(s)+BR^{-1}B'F(s)]$, the formura (39) ensures that the optimal closed-loop poles coincide with the eigenvalues associated with the hamiltonian matrix $H(s)$ in the left-half complex plane.

7.CONCLUSION

It has been shown that the infinite time interval optimal regulator for neutral systems is constructed by using the solution of the stationary Riccati equations on the assumption that no neutral chain lies in the closed-right half plane. It has been also shown that the optimal control law is given by the state feedback form and that the closed-loop system is L^2-stable and has the properties similar to those of finite dimensional systems and retarded time-delay systems. It is necessary furthermore to investigate whether optimal regulator problem is well-posed or not, when there exist infinite number of poles in the closed-right half plane.

It is known that there exist two different definition of state, one is $(\phi(0), A_1\phi+A_{-1}\dot{\phi})$ and the other is $(\phi(0)-A_{-1}\phi(-h), \phi)$. In this paper we have chosen the former as the definition of state. Even if we choose the latter, it is possible to discuss optimal regulator problem similaly.

REFERENCE
[1] K. Uchida, E. Shimemura and N. Abe: Circle condition and stability margin of the optimal regulator for systems with delays, Int. J. Control, Vol.46, No.4, 1203/1212, 1987
[2] R. Datko: Neutral autonomous functional equations with quadratic cost, SIAM J. Control, Vol.12, No.1, 70/82, 1974
[3] M. Sendaula: On the optimal control theory for neutral systems, IEEE Proc. ISCAS/76, 408/411, 1976
[4] K. Ito and T. J. Tarn: A linear quadratic optimal control for neutral systems, Nonlinear Analysis, Theory, Methods & Applications, Vol.9, No.7, 699/727, 1985
[5] A. J. Prichard and D. Salamon: The linear quadratic control problem for infinite dimensional systems with unbounded input and output operators, SIAM J. Control and Optimization, Vol.25, No.1, 121/144, 1987
[6] K. Uchida and E. Shimemura: Stationary solutions of matrix Riccati partial differential equations for linear systems with time-delay, SICE Symposium on Mathematical System Theory, 53/58, 1976
[7] R. Bellman and K. L. Cooke: "Differential-Difference Equations", Academic Press, 1963
[8] K. L. Cooke and D. W. Krumme: Differntial-difference equations and nonlinear initial-Boundary value problems for linear hyperbolic partial differential equations, J. Math. Anal. and Appl., 24, 372/387
[9] S. Hara, T. Omata and M. Nakano: Synthesis of Repetitive control systems and its applications, Proc. 24th IEEE Conf. Decision & Control, 1387/1392, 1985
[10] D. A. O'ccnnor and T. J. Tarn: On stabilization by state feedback for neutral differential difference equations, IEEE AC-28, No.5, 615/618, 1983
[11] J. Zabczyk: Remark on the algebraic Riccati equation in hilbert space, Appl. Math. & Optim., Vol.2, No.3, 251/258, 1976
[12] K. Uchida and E. Shimemura: closed-loop properties of the infinite-time linear-quadratic optimal regulator for systems with delays, Int. J. Control, Vol.43, No.3, 773/779, 1986

QUANTITATIVE DESIGN OF GAIN SCHEDULED CONTROLLERS FOR PLANTS WITH MEASURABLE TRANSPORT RATES

E. Boje

*Department of Electrical Engineering, University of Durban-Westville,
Private Bag X54001, Durban, 4000, South Africa*

Abstract. The Quantitative feedback theory (QFT) of Horowitz allows controller design for uncertain plants with transport delay to given specifications. The theory also provides the necessary insight to elaborate the design and include gain scheduling (adaptation) when the transport rate is measured. In industrial plants there may be high bandwidth requirements during normal (high transport rate, short delay) operation but lower transport rates during start–up may preclude the use of a fixed linear controller. The specifications during start–up need not necessarily be the same as during normal operation. This paper outlines a QFT approach to this problem and gives results measured on a physical device, built to demonstrate and test the design method.

Keywords. Feedback control, Time lag, Quantitative feedback theory, Gain scheduling.

INTRODUCTION

It is well known that dead time places fundamental limits on the bandwidth over which feedback benefits are achievable (Horowitz and Liao, 1984). Horowitz's Quantitative feedback theory allows the design of a fixed linear controller to achieve frequency–domain performance specifications on uncertainty reduction and disturbance rejection for a plant with dead time, given the plant (and transport rate) uncertainty and specifications compatible with the plant's non–minimum phase character. Using the Smith predictor or any other feedback structure in no way eases the limitation on feedback benefits resulting from the non–minimum–phase dynamics of the plant. In some problems there may be large transport delay in some phases of operation – an example being a process plant during start–up – but smaller transport delay, corresponding to high transport rates, during normal operation. If a fixed controller were designed for such a problem, the feedback benefits would be restricted during all phases of operation by the loop phase constraints during periods of low transport rates. In a number of industrial situations the "control" objective is however to achieve the fastest tracking and disturbance attenuation (i.e. highest bandwidth) possible without excessive overshoot and within the plant input limits. If the simple gain adaptation of multiplying the loop gain by the transport rate is allowed, this difficulty can be alleviated and the individual bandwidth limits associated with each value of delay can be approached with a controller which is linear (except for its gain).

This paper shows how to design a simple gain scheduled controller for the above problem. (The control algorithm is sufficiently simple that it could be coded on relatively cheap industrial programmable controllers.) The control design approach was applied to a hybrid simulator.

THEORY

Consider Figure 1 which shows an uncertain linear plant, with transfer function, $P(s) \in \mathcal{P}$, whose input arrives via some distributed process such as a pipe or conveyor belt of fixed length ℓ, with variable transport rate, $v(t)$. The transport behaviour is described by,

$$\frac{\partial x}{\partial t} + v(t) \frac{\partial x}{\partial z} = 0 \tag{1}$$

(x is the value of the controller output variable along the distributed process) with boundary condition,

$$x(z,t)\big|_{z=0} = y_c(t) \tag{2}$$

and the plant input is,

$$u_p(t) = x(z,t)\big|_{z=\ell} \tag{3}$$

(Alternatively,

$$u_p(t) = y_c(t-\tau) \tag{4}$$

where

$$\ell = \int_{t-\tau}^{t} v(\lambda)\, d\lambda \tag{5}$$

– the transport delay is determined implicitly via eq(5)).

The plant is imbedded in a two degree of freedom feedback structure where the feedback controller, $G(s)$, is designed for disturbance reduction and reduction in parameter sensitivity. The pre–filter, $F(s)$, is used to shape the input–output behaviour. The pre–filter design will not be addressed in this paper but note that it can be made to

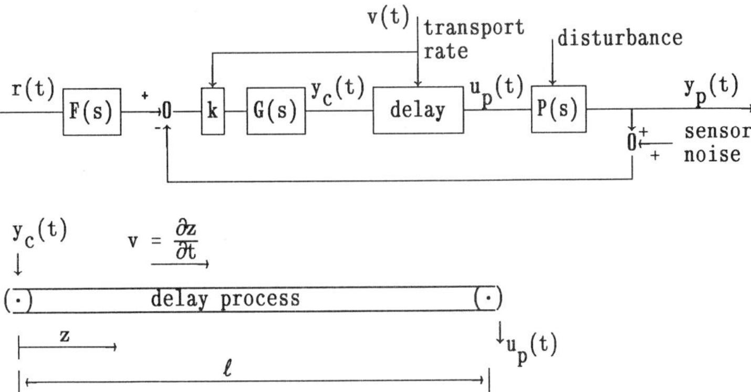

Figure 1 Plant with variable transport rate embedded in a two degree of freedom feedback structure with gain scheduling.

have a transport—rate dependent behaviour through the use of simple gain scheduling.

To start the design, it will be assumed that the transport rate, v(t), varies slowly enough to be taken as constant. The delay can be determined via,

$$\tau = \ell/v \tag{6}$$

The situation where this is not the case will be examined later.

PRINCIPAL IDEA OF GAIN SCHEDULING

<u>Example</u>

The origin of the design idea is best illustrated via a simple example with specifications,

i) $P(s) = \dfrac{1}{(s/0.1)+1}$ (with no uncertainty)

ii) $\ell = 1.0$; $v(t) \ \epsilon \ [1.0, \ 10.0]$ — low to high speeds are possible; speed varies slowly

iii) controller has only gain, $G(s) = k$

iv) achieve the highest closed loop bandwidth subject to $\left|\dfrac{1}{1+L(j\omega)}\right| \le 3dB \ \forall \ \omega$, (some gain and phase margins and disturbance rejection properties). $L(s) = P(s) \, e^{-\tau s} \, G(s)$ is the open loop transfer function

For the lowest speed which results in a delay of 1 second the gain is limited to $k = 4$. This value of controller gain gives a loop gain cross—over frequency of about 0.4 rad/s when the plant corresponding to the fastest transport rate is used. The maximum gain cross—over frequency achievable if this (fastest rate, shortest delay) case is considered individually is about 4.0 rad/s, for a controller gain of $k = 40$. The latter gain cross—over frequency approaches the "rule of thumb" limit given in Horowitz and Liao (1984) (and elsewhere) that a non minimum phase zero at ω_z constrains the open loop gain cross—over to about $\omega_c \le \omega_z/2$ (in this case, the 10 rad/s resulting from the "rule of thumb" cannot be achieved without violating specification (iv)). Figure 2 illustrates the maximum gain possible for each transport rate (assuming that the rate is constant).

With only time delay uncertainty, each plant template — i.e. the set of all possible plant gain and phase combinations, plotted on the logarithmic polar plot (Nichols chart) at a given frequency, ω_i (Horowitz and Sidi, 1972) — is a horizontal line of width $\omega_i(\tau_{max}-\tau_{min})$ radians and any closed loop specification of the form, $|T(j\omega)| \le \alpha$ will be at its limit at the longest time delay

around the gain cross—over point. Often the design task with a non—minimum phase system is to achieve the largest possible closed loop bandwidth without excessive overshoot of $|L(j\omega)/(1+L(j\omega))|$ or of $|1/(1+L(j\omega))|$. If the plant gain could be arranged (scheduled) so that the left hand (large phase) side of the template is turned down, the short time plants need not suffer the same bandwidth restrictions as the long dead—time plants. In other problems tried by the author, if the dynamic part of the controller design is first fixed by considering the shortest transport rate only (but the whole of the rest of the plant uncertainty set) and then the gain of the controller is modified to suit each possible transport rate, the gain versus transport rate curve can be approximated by a straight line through the origin, illustrated in Figure 2. Although not required for stability (unless the loop has one or more integrators), requiring the gain to go through zero turns the controller off when the transport rate is zero which in practical plants has the advantage of preventing input saturation under these conditions. Templates for $\left[P(j\omega) \, e^{-j\omega\tau} \right]$ and for $\left[v \, P(j\omega) \, e^{-j\omega\tau} \right]$ and bounds on the loop transmission from the above problem are shown in Figure 3 (for $\omega_i = 0.3$, 1.0 and 3.0 rad/s), to illustrate how the velocity can be used to achieve the above objective in a simple manner.

The above observations are explained as follows. If the minimum phase part of the open loop transfer function is rolling—off steadily at $-\alpha$ dB/decade around the gain cross—over point (at frequency, ω_c rad/s), the corresponding asymptotic phase is given by Bode's work (Horowitz, 1963, p.315) as, $-(\alpha/20)\times90°$. The transport delay contributes a further $-\omega_c\tau/\pi\times180°$ to the open loop phase. In this case, for two plants, P_1 and P_2, which are identical except for having different dead times, τ_1 and τ_2, to have the same phase margin requires that their gain cross over frequencies (ω_{c1} and ω_{c2}) are related as, $\omega_{c1} \tau_1 = \omega_{c2} \tau_2$. This in turn means that the gains must be such that $|L_1(j\omega_{c1})| = |L_2(j\omega_{c2})|$ or,

$$\frac{|L_1(j\omega)|}{|L_2(j\omega)|} = \left[\frac{\tau_2}{\tau_1}\right]^{(\alpha/20)} \tag{7}$$

around the cut—off frequency. Since in a non—minimum phase design it is reasonable for the roll off to be around -20 dB/decade at the gain cross—over point, the success of the design approach is explained.

In the low frequency part of the the design the additional gain uncertainty introduced by the gain scheduling restricts the possibility of designing conditionally stable loops. In many cases conditional stability is avoided completely but notice that the principal advantage of conditional stability is that a larger roll—off can be

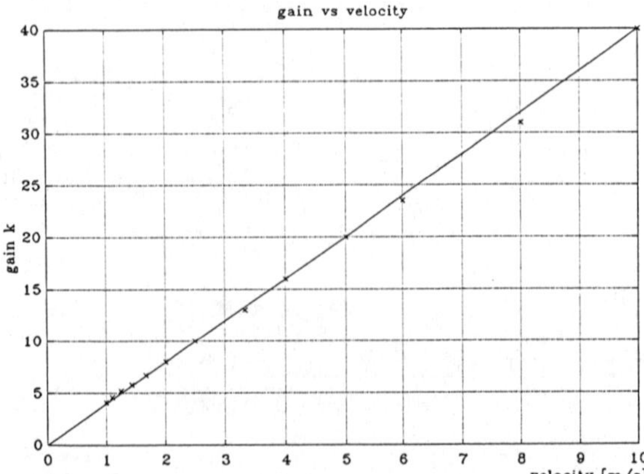

Figure 2 Maximum controller gain versus transport rate — Example

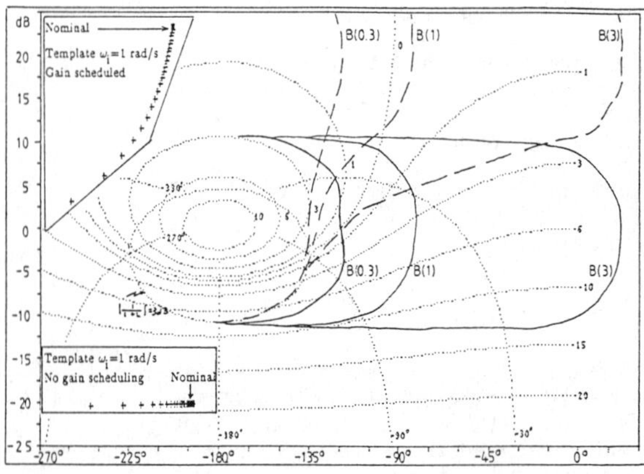

Figure 3 Templates and bounds — Example 1. Dashed lines are for plant modified by gain scheduling

achieved at low frequency. The reduction in the roll–off should be offset by the larger bandwidth for the fast–rate plants.

The additional gain uncertainty introduced by the gain scheduling does not necessarily affect the steady state behaviour of the closed loop in a low–passing type design problem as the loop gain can be made large at sufficiently low frequencies. This is of course provided that the plant uncertainty allows a gain cross–over point compatible with the specifications at some higher frequency.

THE EFFECT OF TIME VARYING TRANSPORT RATES ON SYSTEM PERFORMANCE

In the above, the transport rate was assumed to be fixed and measured. Under that assumption, the system's closed loop stability is not affected by the inclusion of the transport rate factor, v, in the controller gain. If the transport rate changes with time the resulting design is time varying but remains linear in its parameters (unless the velocity is determined in some way from the system's operating condition). The stability of this time varying system could for instance be guaranteed by imposing some differentiability conditions on $v(t)$ (the bounds of which are given in the problem statement) and some bounds $\dot{v}(t)$ so that Theorem 5.6.41 of Vidyasagar (1978, p.223 for linear systems with slowly varying parameters) could be applied.

The above approach will not be pursued here as in practical problems the interest is in performance as well as stability. In addition, while it may be possible to find a $v(t)$ to destabilise the design if no constraints are placed on the derivatives of $v(t)$, such a velocity profile is unlikely in practical situations. It will therefore be more instructive to examine the effect of step and/or sinusoidal changes in velocity on the behaviour of the system.

A rapid (step) change in velocity is a change from one (linear time invariant) closed–loop plant to another and the transient behaviour is the response of the second plant to the set of initial conditions left by the first plant. Good (high bandwidth, low overshoot) disturbance rejection properties of the control loop will ensure fast decay of these initial conditions (Horowitz, 1963, p. 441).

A LABORATORY EXPERIMENT

To test the design ideas presented here, the "transport delay simulator" shown in Figure 4, was constructed in our laboratory. It has a 2048 byte memory, addressed via a cyclic counter which is clocked by a voltage to frequency converter. The memory data is read from a bipolar 8 bit analog to digital (A/D) converter and is written (2048 cycles later) to a D/A converter. Linear plants and controllers were implemented using LM–741 type operational amplifiers. The rate dependent gain was obtained using an analog multiplier.

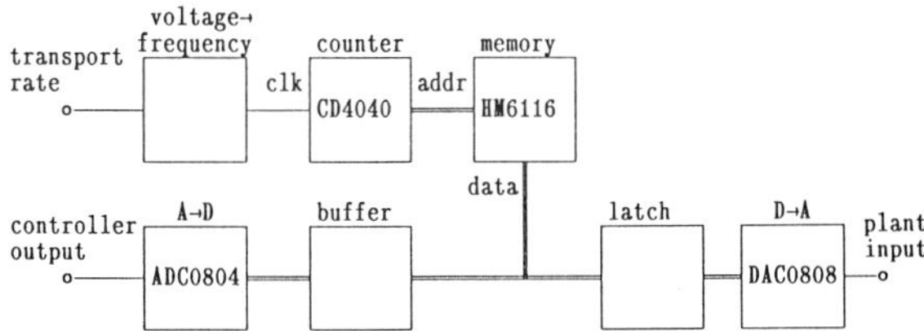

Figure 4 Adjustable transport rate simulator

The following plant was constructed:

$$P(s) = \frac{A}{(s/\omega_n)^2 + (3-A)(s/\omega_n) + 1} \tag{8}$$

with,

A ϵ [1.0, 2.5] (structured gain and damping factor uncertainty),

ω_n ϵ [0.09, 0.11] rad/s

and a transport delay of,

$$\tau \ \epsilon \ [1.0, \ 10.0] \ \text{seconds} \tag{9}$$

The plant step response for the two extreme values of A is shown in Figure 5.

The specifications were:

a) Achieve the highest closed loop bandwidth.

b) Zero steady–state output to step disturbances at the plant input.

c) $|L/(1+L)| \leq$ 3dB for all plants .

d) $|1/(1+L)| \leq$ 6dB for all plants.

The (arbitrarily chosen) nominal plant, P_0, had parameters, A=2.5, ω_n=0.1 and τ=1.0. Figure 6 shows templates (at ω_i = 0.1 and 0.3 rad/s) and bounds (at ω_i = 0.1, 0.3 and 1.0 rad/s) on the nominal loop transfer function, L_0, for the above plant, modified by the gain scheduled from the transport rate . Figure 6 also shows the nominal loop transfer function for the controller,

$$G(s) = \frac{0.1 \ ((s/0.1)^2 + 1.2 \ (s/0.1) + 1)}{s \ ((s/4.0)^2 + 2.0 \ (s/4.0) + 1)}$$

Figure 7 shows the closed loop system's response to step disturbances introduced at the plant output for the extreme parameter values, A=1.0, 2.5 and τ=1.0, 10.0.

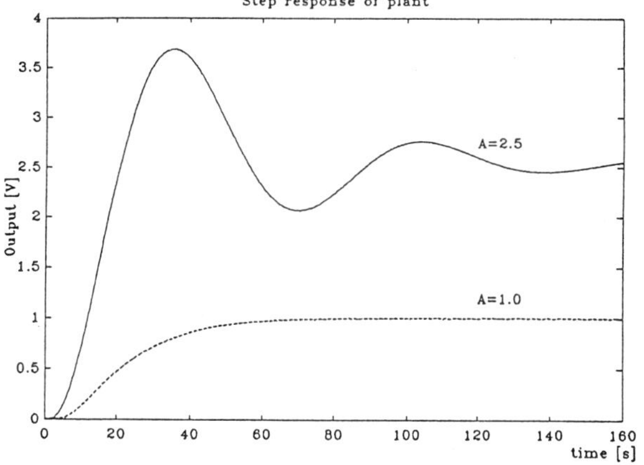

Figure 5 Step response of the uncompensated plant

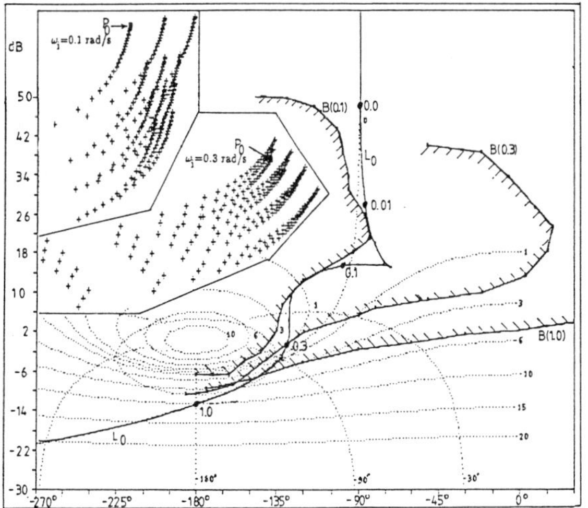

Figure 6 Templates, bounds on L_0 and an L_0 satisfying the bounds – Laboratory experiment

Figure 8 shows the effect of step changes (from one extreme value to the other) of the transport rate during transient behaviour.

<u>Remarks</u>

The analog components used were 10% tolerance. No special measures were taken to exclude noise from the circuits. Because of the high dynamic range of the controller the 8 bits of precision used in the digital portion of the circuit were only just sufficient for steady state accuracy. It is felt that this laboratory experiment has therefore some of the attributes found in industrial plants. The experiment has demonstrated the design method on a practical problem with very significant uncertainty, especially in the transport delay.

CONCLUSIONS

This paper has presented a very simple gain scheduling approach for plants with measurable transport rates. The design is executed using the quantitative feedback theory of Horowitz. The specifications during start—up may be different to those during normal operation. The simplicity of the approach of gain scheduling from a single measured variable with a fixed controller should be attractive in an industrial environment where additional feedback benefits can be obtained at the expense of a single additional measurement and a multiplication, well within the capabilities of modern programmable controllers. (See Eitelberg, 1987 for quantitative design for discrete controllers.)

The ideas of this paper can be extended to more complicated controller scheduling, whene there are a small number of measurable parameters which result in unacceptably large templates using the standard (fixed controller) quantitative feedback approach.

ACKNOWLEDGEMENTS

The comments of Prof E. Eitelberg are appreciated. The transport rate simulator was built by Mr I Haniff, a student at the University of Durban—Westville.

REFERENCES

EITELBERG, E. "Sampling rate design based on $(1-sT/2)$" *International Journal of Control*, 1988, Vol 48, No 4, pp 1423—1432.

HOROWITZ, I. *Synthesis of Feedback Systems*, Academic Press, New York, 1963, Library of Congress Number 63—12033

HOROWITZ, I. and LIAO, Y. "Limitations of non—minimum—phase feedback systems" *International Journal of Control*, 1984, Vol 40, No 5, pp 1003—1013.

HOROWITZ, I. and SIDI, M. "Synthesis of feedback systems with large plant ignorance for prescribed time—domain tolerances", *International Journal of Control*, Vol.16, No.2, pp.287—309, 1972.

VIDYASAGAR, M. *Nonlinear Systems Analysis*, Prentice—Hall, Englewood Cliffs, 1978, ISBN 0—13—623280—9

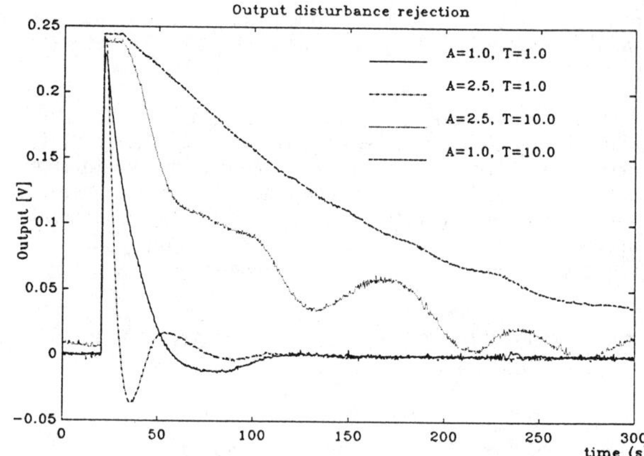

Figure 7 Rejection of step disturbances at the plant output

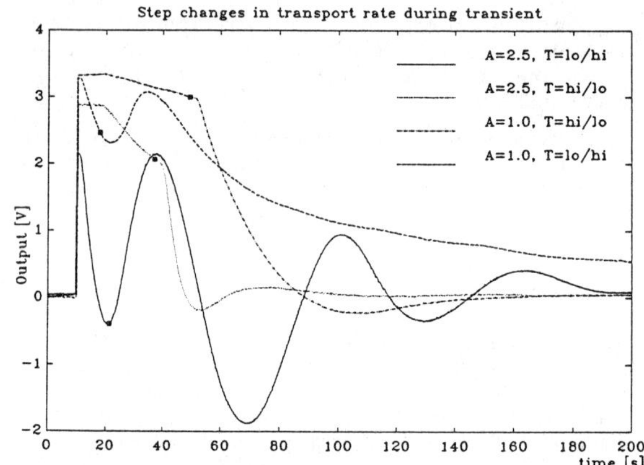

Figure 8 Effect of changing transport rates during transient response

SOME PROBLEMS OF QUALITATIVE CONTROL THEORY FOR TIME-DELAY SYSTEMS

V. M. Marchenko

Department of Mathematics, Byelorussian Institute of Technology, Minsk 220630, USSR

<u>Abstract</u>. The problem of modal control is extensively considered here for linear stationary time-delay systems of neutral type. To solve this problem a new type of regulators containing both time delays and state derivatives is introduced. Effective condition for the modal control by action of such regulators is given for the complete information case. In order to study the modal control problem for systems with incomplete information an analog of the well-known dynamic regulator proposed by Pearson for systems with no delay is used.

<u>Keywords</u>. Modal control; time-delay; neutral-type systems; difference regulators; incomplete information.

INTRODUCTION

One of the basic problems of qualitative theory of control for linear dynamic systems is the problem of modal control. This problem was considered in details for the linear stationary systems with no delay. General references to related works are the monographs of Porter and Grossley (1972) and Wonham (1979), the survey of Andreev (1977) and others.

During the last one or two decades a great deal of research works were concerned with the problem of modal control for linear stationary time-delay dynamic systems. The main part of them was related to retarded-type systems only. In this paper one can get acquainted with some references on this subject to learn the principal steps in the development of the modal control problem for time-delay systems, basic results and bibliography.

Early ivestigators (Krasovsky and Osipov, 1963; Osipov, 1965; Pandolfi, 1976) considered the problem of stabilization. Bulatov and al. (1974), Olbrot (1978)

studied a partial modal control which provides control by any finite part of spectrum by using a linear integral feedback. A new type of regulators, namely, the difference regulators were introduced by Morse (1976), Asmykovich and Marchenko (1976). In the paper of Asmykovich and Marchenko the problem was considered as a coefficient assignment problem for the system's characteristic equation. Later, Marchenko and Asmykovich (1980) extended this approach to investigation of multi-input retarded systems with several delays. They also used a time-delay dynamic regulator to solve the incomplete case modal control problem.

The conditions obtained for the modal controllability of retarded-type systems by action of difference regulators are very restrictive. To weaken these conditions linear integral regulators were introduced by Marchenko (1978). Other way of weakening the requirements to regulators was proposed by him (1987) by means of introduction of delays in control.

In this investigation an attempt is made

177

to extend the theory of modal control from retarded-type to neutral-type systems. Concerning other problems of the qualitative control theory such as controllability, canonical forms, reconstruction they are detailed by Gabasov and al. (1984).

MODAL CONTROL FOR COMPLETE INFORMATION CASE

Problem Formulation

Let us consider linear stationary objects with lags described by a differential-difference neutral-type equation of the form

$$\sum_{j=o}^{1} D_j \, \dot{x}(t - h_j) = \sum_{j=o}^{1} (A_j \, x(t - h_j)$$

$$+ B_j \, u(t - h_j)) \, , \qquad (1)$$

$$t > 0, \quad D_o = I_n,$$

where D_j, A_j are n by n constant matrices, B_j is a constant n-vector, $j = 0,1,\ldots,1$; $0 = h_o < h_1 < \ldots < h_1 \overset{\Delta}{=} h$ are constant delays.

For these systems the state concept (Marchenko, 1984) includes also an information on derivative $\dot{x}(t)$, $t > 0$. Therefore introduce a linear feedback as the linear difference regulator with derivatives in state

$$u(t) = \sum_{i=o}^{N} \sum_{j=o}^{M} Q_{ij}' \, \frac{d^i x(t - h_j^*)}{dt^i} \, , \qquad (2)$$

(the stroke (') denotes transposition)

where $h_j^* \in R$, $Q_{ij} \in R^n$, $(Q_{io} = 0$ for $i \geqslant 1)$, $i = 0,\ldots,N$; $j = 0,\ldots,M$; $0 = h_o^* < \ldots < h_M^*$.

The characteristic equation of the open system (1) is the following

$$\sum_{i=o}^{n} \sum_{j=o}^{J} r_{ij} \, p^i \, e^{-ps_j} = 0 \, , \qquad (3)$$

where $r_{no} = 1$; r_{ij}, s_j, $\ldots$ are real numbers; $0 = s_o < s_1 \ldots < s_J$; p is a complex variable.

The problem of modal control for the system (1), (2) will be regarded as the problem of control by coefficients r_{ij}, $\ldots$ of the equation (3). More precisely: the system (1) is called modally controllable by the regulator (2) if for any real numbers J^*, s_j^*, r_{ij}^*, $\ldots$; $0 = s_o^* < \ldots < s_{J^*}^*$; $r_{no}^* = 1$ there exists a linear feedback (2), i.e. there exist such numbers N, M, h_j^*, $\ldots$ and n-vectors Q_{ij}, $\ldots$, that

$$\det \, (\, p \, I_n + \sum_{j=1}^{1} p \, e^{-ph_j} D_j - \sum_{j=o}^{1} e^{-ph_j}$$

$$\times A_j - \sum_{j=o}^{1} e^{-ph_j} B_j \, (\sum_{j=o}^{M} Q_{oj}' \, e^{-ph_j^*}$$

$$+ \sum_{i=1}^{N} \sum_{j=1}^{M} p^i \, e^{-ph_j^*} Q_{ij}' \,) \,)$$

$$= \sum_{i=o}^{n} \sum_{j=o}^{J^*} r_{ij}^* \, p^i \, e^{-ps_j^*}$$

for all complex p .

Criteria for Modal Controllability

Let us denote

$$A(m,p) = p \sum_{j=1}^{1} m^{h_j} D_j - \sum_{j=o}^{1} m^{h_j} A_j \, ,$$

$$b(m) = \sum_{j=o}^{1} m^{h_j} B_j \, .$$

Then we have

<u>Theorem 1</u>. If the equality

$$\det \, (\, b(m), \, A(m,p) \, b(m), \, \ldots$$

$$\ldots, \ (A(m,p))^{n-1}\, b(m)\ ) \equiv \text{const} \neq 0 \tag{4}$$

is identically valid for all real $m \geqslant 0$
and for all complex p then the system (1)
is modally controllable by the regulator
(2).

For the proof of Theorem 1 see Appendix.

<u>Corollary 1</u>. If the condition (4) holds
then the system (1) is stabilized by the
regulator (2).

If we analyze the proof of Theorem 1 then
we can state

<u>Proposition 1</u>. The system (1) of retarded
type ($D_j = 0$, $j = 1,\ldots,l$) is modally
controllable by the regulator (2) if and
only if the identity (4) is valid for
$p = 0$.

MODAL CONTROL FOR INCOMPLETE
INFORMATION CASE

Consider the system (1) and suppose that
the output

$$y(t) = \sum_{i=o}^{K} \sum_{j=o}^{1} C_{ij}'\, \frac{d^{i}x(t - h_j)}{dt^{i}}\ , \tag{5}$$

$$t > 0\ ,$$

(where $C_{ij} \in R^{n}$, $\ldots$; $C_{1o} = 0$ for $i \geqslant 1$)

is possible to be measured only. In this
case it is impossible to make arbitrary
all coefficients of the characteristic
equation of the closed-loop system by
using the output feedback only. Therefore
let us consider a more general dynamic
regulator with delays

$$\frac{d^{n}u(t)}{dt^{n}} + \sum_{i=o}^{n-1} \sum_{j=o}^{M} a_{ij}\, \frac{d^{i}u(t - s_j^{*})}{dt^{i}}$$

$$+ \sum_{i=1}^{n-1} \sum_{j=1}^{M} \sum_{k=o}^{P} a_{ijk}\, \frac{d^{i+k}u(t - s_j^{*})}{dt^{i+k}}$$

$$= \sum_{i=o}^{n-1} \sum_{j=o}^{M} v_{ij}\, \frac{d^{i}y(t - s_j^{*})}{dt^{i}}$$

$$+ \sum_{i=1}^{n-1} \sum_{j=1}^{M} \sum_{k=o}^{P} v_{ijk}\, \frac{d^{i+k}y(t - s_j^{*})}{dt^{i+k}}\ . \tag{6}$$

Denoting

$$c(m,p) = \sum_{i=o}^{K} \sum_{j=o}^{1} C_{ij}\, p^{i}\, e^{-mh_j}$$

we have

<u>Theorem 2</u>. If the identity

$$\det (\ c(m,p),\ (A(m,p))'\,c(m,p),\ \ldots$$

$$\ldots,\ ((A(m,p)')^{n-1}\, c(m,p)\ ) \equiv \text{const} \neq 0$$

is valid along with the condition (4) then
the system (1),(5) is modally controllable
by regulator (6).

<u>Remark 1</u>. The definition of modal cont-
rollability of the system (1),(5) by regulator (6) is similar to the definition
for the system (1),(2): the system (1),
(5),(6) is modally controllable if there
exists a dynamic regulator for which the
characteristic equation of the corresponding closed-loop system (extended system)
has an arbitrary preassigned admissible
form.

The proof of Theorem 2 can be given using
the extended system in the similar way as
the proof of Theorem 1.

<u>Remark 2</u>. Several examples (Marchenko and
Asmykovich, 1980; Marchenko, 1987) of actual physical processes can be considered
to illustrate the result obtained.

APPENDIX

<u>Sketch of the Proof of Theorem 1</u>

By analogy with (Marchenko, 1987) we con-
stract a matrix $T(m,p) = (\ t_1(m,p),\ \ldots$

$$\ldots,\ t_n(m,p))\quad \text{where}$$

179

$$t_n(m,p) = b(m),$$

$$t_{n-1}(m,p) = A(m,p)\, b(m) + r_{n-1}(m,p) \times b(m),$$

$$\vdots$$

$$t_2(m,p) = (A(m,p))^{n-2}\, b(m) + \ldots + r_{n-i}(m,p)\, (A(m,p))^{n-2-i}\, b(m) + \ldots + r_2(m,p)\, b(m),$$

$$t_1(m,p) = (A(m,p))^{n-1}\, b(m) + \ldots + r_{n-i}(m,p)\, (A(m,p))^{n-1-i}\, b(m) + \ldots + r_1(m,p)\, b(m)$$

and the coefficients $r_i(m,p)$, ... are defined from the following

$$(A(m,p))^n\, b(m) = -\sum_{j=0}^{n-1} r_j(m,p) \times (A(m,p))^j\, b(m), \quad m \geqslant 0, \quad p \in \mathbf{C}.$$

Hence we have

$$\det\Big(\, p\, I_n + A(e^{-p},p) - b(e^{-p}) \times \sum_{i=o}^{N}\sum_{j=o}^{M} p^i\, e^{-ph_j^*}\, Q_{ij}'\, \Big) = \det\Big(\, T(e^{-p},p)^{-1}\,\big(\, p\, I_n + A(e^{-p},p) - b(e^{-p}) \times \sum_{i=o}^{N}\sum_{j=o}^{M} p^i\, e^{-ph_j^*}\, Q_{ij}'\,\big)\, T(e^{-p},p)\,\Big) =$$

$$= p^n + \sum_{j=o}^{n-1} p^j\, \Big(\, r_j(e^{-p},p) - \sum_{i=o}^{N}\sum_{j=o}^{M} p^i\, e^{-ph_j^*}\, Q_{ij}'\, T(e^{-p},p)\, (\underbrace{0,\ldots,0,1,0,\ldots,0}_{j+1})'\,\Big) = \sum_{i=o}^{n}\sum_{j=o}^{J} r_{ij}\, p^i\, e^{-ps_j} -$$

$$- \sum_{i=o}^{N}\sum_{j=o}^{M} p^i\, e^{-ph_j^*}\, Q_{ij}'\, T(e^{-p},p)\, (\, 1, p, \ldots, p^{n-1}\,)', \quad p \in R.$$

It is clear from here that the system (1) is modally controllable by the feedback (2) if there exist n-vectors Q_{ij}, ... such that the identity

$$\sum_{i=o}^{N}\sum_{j=o}^{M} Q_{ij}'\, p^i\, e^{-ph_j^*}\, T(e^{-p},p)\, (\, 1, p, \ldots, p^{n-1}\,)' = \Big(\, \sum_{j=o}^{J} r_{oj}\, e^{-ps_j} - \sum_{j=o}^{J^*} r_{oj}^*\, e^{-ps_j^*}, \ \sum_{j=o}^{J} r_{1j}\, e^{-ps_j}$$

$$- \sum_{j=o}^{J^*} r_{1j}^*\, e^{-ps_j^*}, \ \ldots, \ \sum_{j=o}^{J} r_{n-1j}\, e^{-ps_j} - \sum_{j=o}^{J^*} r_{n-1j}^*\, e^{-ps_j^*} + p \times \Big(\sum_{j=1}^{J} r_{nj}\, e^{-ps_j} - \sum_{j=1}^{J^*} r_{nj}^*\, e^{-ps_j^*}\Big) \times (\, 1, p, \ldots, p^{n-1}\,)', \quad p \in R,$$

holds. On the other hand, this identity is valid if the assumptions of Theorem 1 hold. This completes the proof of the theorem.

DISCUSSION

The regulator (2) is suitable for its technical realization since it requires the measurement of output signal at the present time moment and at several previous moments. However, the condition (4) of the existence of the regulator (2) is very limited. In particular, if the system (1),(2) is modally controllable then the system (1) is completely and pointwisely controllable (the definitions of complete and pointwise controllability are in (Gabasov and al., 1984)).

In order to weaken the condition (4) the linear feedback seems to be taken of the form

$$u(t) = \sum_{j=0}^{M} \int_{-N_j}^{0} [dQ_j(s)] x^{(j)}(t + s)$$

or more general

$$\sum_{j=0}^{K} \int_{-v_j}^{0} [dP_j(s)] u^{(j)}(t + s) =$$

$$= \sum_{j=0}^{M} \int_{-N_j}^{0} [dQ_j(s)] x^{(j)}(t + s)$$

where v_j, N_j, ..., are non-negative numbers; $Q_j(.)$, $P_j(.)$, ..., are matrix-functions of bounded variation elements;

$$x^{(j)}(t + s) = \frac{d^j x(t + s)}{dt^j} .$$

Note that the minimal state of the system (1) at time t is based on the information $(g_0, g(.))$ introduced by Marchenko (1984):

$$g_0 = x(t + 0),$$

$$g(t + s) = \sum_{h_j > s} (A_j x(t + s - h_j)$$

$$- D_j \dot{x}(t + s - h_j) + B_j u(t + s - h_j))$$

$$s \in [0, h].$$

Therefore it would be interesting to obtain a regulator as a linear feedback on the information $(g_0, g(.))$.

REFERENCES

Andreev, Yu.N. (1977). Algebraical state-variable techniques in the control theory for linear systems. _Automat. i Telemech._, _3_, 5-50.

Asmykovich, I.K., and V.M. Marchenko (1976). Spectrum control in systems with delay. _Automat. i Telemech._, _7_, 5-14.

Bulatov, V.I., T.S. Kalyuzhnaya and R.F. Naumovich (1974). Spectrum control of differential equations. _Differentsyal'nye Uravneniya_, _11_, 1946-1952.

Gabasov, R., F.M. Kirillova, V.M. Marchenko, and I.K. Asmykovich (1984). Problems of control and observation for infinitedimensional systems. _Inst. of Math. of Acad. Nauk BSSR_, Prepr. 1(186).

Krasovsky, N.N., and Yu.S. Osipov (1963). Stabilization of motions of the controlled object with delay in the control system. _Izves. Acad. Nauk SSSR. Techn. Kibern._, _6_, 3-15.

Marchenko, V.M. (1978). On the problem of modal control for linear systems with delay. _Dokl. Acad. Nauk BSSR_, _5_, 401-404.

Marchenko, V.M. (1984). Duality for the problems of control and observation of linear systems with after-effect. _Dokl. Acad. Nauk BSSR_, _4_, 297-300.

Marchenko, V.M. (1987). Modal control for linear time-delay retarded-feedback objects. In Y. Sunahara, S.G. Tzafestas, and T. Futagami (Eds.), _Modelling and Simulation of Distributed Parameter Systems_, Hiroshima Institute of Technology, Hiroshima, pp. 685-690.

Marchenko, V.M., and I.K. Asmykovich (1980). In S.G. Tzafestas (Ed.), _Simulation of Distributed-Parameter and Large-Scale Systems_, North-Holland Publ. Comp., Amsterdam-New York-Oxford, pp. 73-78.

Morse, A.S. (1976). Ring models for delay differential systems. _Atomatica_, _12_, 529-531.

Olbrot, A.W. (1978). Stabilizability, detectability and spectrum assignment for linear autonomous systems with general time delays. _IEEE Trans. Atomat. Control_, _23_, 887-890.

Osipov, Yu.S. (1965). On stabilization of controlled systems with delays. _Differentsyal'nye Uravneniya_, _5_, 606-618.

Pandolfi, L. (1976). Stabilization of neutral functional-differential equations. _J. Optimiz. Theory and Appl._, 20, 191-204.

Porter, B., and R. Grossley (1972). _Modal Control. Theory and Applications_. Taylor and Francis, London.

Wonham, W. (1979). _Linear Multivariable Control_, Springer-Verlag, 2nd ed.

LOW-ORDER RATIONAL ALL-PASS
APPROXIMATIONS TO e^{-s}

R. Piché

*Dept. of Mathematics, Tampere University of Technology, P.O. Box 527, SF-33101
Tampere 10, Finland*

Abstract

We compare, based on frequency response, three simple formulas (including the Padé formula) for
stable all-pass unit-magnitude transfer functions of order n which approximate the delay operator e^{-s}. We
show how a simple formula based on truncated product expansions of hyperbolic functions approximates
the delay over a larger bandwidth than the other two simple formulas. Formulas with optimum bandwidth
can be derived using numerical optimization; the coefficients for optimal formulas of order $n = 1$ and
$n = 2$ are presented and compared to the 'simple' formulas.

1 Introduction

Rational transfer function approximations to the delay operator
have a number of applications in control [4] and system identification [1]; they are often used to replace the delays in differential-
delay systems to yield a purely differential system, to which
standard techniques are then applied. In this paper we compare
a number of simple approximations using a frequency domain
criterion.

Functions which are especially well-suited to approximate
the delay operator e^{-s} are rational *inner* functions, that is, stable
all-pass transfer functions whose frequency response has unit
magnitude at all frequencies. A rational inner function of order
n has the form

$$T_n(s) := \prod_{k=1}^{n} \frac{\bar{c}_k + s}{c_k - s}$$

where the c_k are complex constants with negative real part. Since
the delay operator is itself a (nonrational) unit function, then, in
terms of the frequency response, the approximation error when
using unit functions is entirely in the phase. Let the frequency
response approximation error be defined as

$$E_n(\omega) := |T_n(i\omega) - e^{i\omega}|.$$

For small values of E_n, this error is approximately equal to the
phase error (expressed in radians), that is,

$$E_n(\omega) \simeq |\angle T_n(i\omega) - (-\omega)|.$$

The term 'phase error' will therefore also be used here to refer
to E_n.

A rational transfer function can of course only approximate
e^{-s} over a limited range of frequencies. A reasonable way of
evaluating an approximation is to use a frequency domain cost
function of the form

$$\sup_{\omega} f(\omega) E_n(\omega)$$

where f is a real even non-negative weight function whose form
depends on the eventual application. The specific evaluation criterion to be used in this work is based on a rectangular weight
function of width ω_0. The *maximum phase error* ϕ_0 of an approximation to e^{-s} is defined as the largest absolute frequency
response error (= phase error) at frequencies below ω_0, that is,

$$\phi_0 := \max_{0 \le \omega \le \omega_0} E_n(\omega).$$

Given a bandwidth ω_0, we seek an approximation which minimizes ϕ_0.

The evaluation criterion can also be stated from the dual
point of view, as follows. Let the *minimum bandwidth* ω_0 be
defined as the largest ω for which the phase error is less than
some fixed value ϕ_0, which represents the acceptable phase error for an application. The best approximation is then the one
with the largest ω_0.

In applications, one of the most popular approximations is
the Padé formula. Also used is a formula based on the identity
$e^z = \lim_{n\to\infty} (1 + z/n)^n$, which we shall term a Laguerre filter because, like the Laplace transform of Laguerre functions,
this formula has only one pole, of multiplicity n. A third simple
formula for a rational inner function approximation to the delay
(which appears to be original) can be constructed using a truncated product expansion of the hyperbolic functions cosh and
sinh. For brevity, this formula will be referred to as the Product
formula. All three formulas are identical in the case of first-order
rational approximation. Numerical results presented in the next
section indicate that for $n > 1$, the Product formula is better (in
terms of the bandwidth criterion) than the Padé formula, which
in turn is better than the Laguerre formula.

The three formulas mentioned above have the advantage of
being relatively simple to define. Approximations with wider
bandwidths can be found, however, using numerical optimization techniques. This is shown in the third section, where a nu-

merical method is used to construct rational inner approximations to e^{-s} of order 1 and 2.

2 Simple Formulas

2.1 Padé formula

The Padé approximation formula is defined as the rational function whose power series expansion agrees with that of the original function to as many terms as possible. The all-pass Padé formula for e^{-s} is

$$R_n(s) = \frac{\displaystyle\sum_{k=0}^{n} \frac{(n+k)!}{k!(n-k)!}(-s)^{n-k}}{\displaystyle\sum_{k=0}^{n} \frac{(n+k)!}{k!(n-k)!}(s)^{n-k}}.$$

The formulas for the first few orders are

$$R_1(s) = \frac{2-s}{2+s},$$

$$R_2(s) = \frac{12 - 6s + s^2}{12 + 6s + s^2},$$

$$R_3(s) = \frac{120 - 60s + 12s^2 - s^3}{120 + 60s + 12s^2 + s^3},$$

$$R_4(s) = \frac{1680 - 840s + 180s^2 - 20s^3 + s^4}{1680 + 840s + 180s^2 + 20s^3 + s^4}.$$

The transfer function R_n is stable for all n [1]. The power series expansion of $R_n(s)$ agrees with that of e^{-s} up to and including terms of order $2n$. Thus one expects the error to be very small at low frequencies, and this is confirmed by graphs of the phase error E_n, shown in Figures 1–4. Comparing these figures also indicates that, as one would expect, increasing the order improves the approximation, in the sense that the error remains small over a larger bandwidth.

2.2 Laguerre formula

A well-known convergent sequence of approximations to the exponential function is given by the formula

$$e^z \simeq (1 + z/n)^n,$$

which suggests the rational inner function approximation

$$e^{-s} = \frac{e^{-s/2}}{e^{s/2}} \simeq L_n(s) := \frac{(1 - s/2n)^n}{(1 + s/2n)^n}.$$

This rational function has only one pole, of multiplicity n, located at $-2n$. Since the Laplace transforms of Laguerre functions also have this property, we denote the above approximation a Laguerre formula. Figures 1–4 show the phase error of this formula for orders 1 to 4: it can be seen that the curve resembles the curve for the Padé formula, but the bandwidth is significantly smaller.

2.3 Product formula

Well-known formulas for the infinite product expansion of the hyperbolic functions are

$$\cosh(s/2) = \prod_{k=1}^{\infty} \left(1 + \frac{s^2}{(2k-1)^2 \pi^2}\right),$$

$$\sinh(s/2) = \frac{s}{2} \prod_{k=1}^{\infty} \left(1 + \frac{s^2}{4k^2 \pi^2}\right).$$

Making use of the identity

$$e^{-s} = \frac{\cosh(s/2) - \sinh(s/2)}{\cosh(s/2) + \sinh(s/2)}$$

and truncating the infinite products yields the approximation

$$P_n(s) := \frac{\displaystyle\prod_{k=1}^{\lfloor n/2 \rfloor} \left(1 + \frac{s^2}{(2k-1)^2\pi^2}\right) - \frac{s}{2}\prod_{k=1}^{\lfloor (n-1)/2 \rfloor}\left(1 + \frac{s^2}{4k^2\pi^2}\right)}{\displaystyle\prod_{k=1}^{\lfloor n/2 \rfloor} \left(1 + \frac{s^2}{(2k-1)^2\pi^2}\right) + \frac{s}{2}\prod_{k=1}^{\lfloor (n-1)/2 \rfloor}\left(1 + \frac{s^2}{4k^2\pi^2}\right)},$$

where the notation $\lfloor \cdot \rfloor$ means to take the integer part. For brevity, this approximation will be referred to as the Product formula. Note that $P_1 = L_1 = R_1$, that is, all three simple formulas considered in this section are identical in the case of an approximation of order 1. The transfer function $P_n(s)$ can be shown to be stable for all n by applying theorem 13 of chapter 15 of [2].

Figures 1–4 show graphs of the phase error of this formula at various orders: for $n > 1$, a number of small ripples are present at low frequencies, with the peaks occuring approximately at the values

$$\omega_k := (2k-1)\pi/2, \quad k = 2, 3 \ldots$$

The approximation is therefore not as good as the Padé formula at very low frequencies, but for sufficiently large phase error threshold ϕ_0, the Product formula gives a larger bandwidth. For instance, if formulas of order $n = 3$ are compared, then Figure 3 shows the following: if bandwidth is defined based on a phase error threshold $\phi_0 < .19$, then the Padé approximant R_3 is superior to the Product formula P_3; but if the bandwidth is judged on the basis of a phase error threshold larger than 0.19 radians, then the Product formula has the larger bandwidth. Figure 5 shows the bandwidths of the three simple formulas based on a phase error of $\pi/2$: it can be seen that the Product formula exhibits a significantly larger bandwidth for $n > 1$.

3 Equiripple Formula

The wider bandwidth exhibited by the Product formula suggests that an even better bandwidth may be achieved by an approximation of similar form as P_n, but whose first n error ripple peaks are all the same height. The formula for the rational inner function satisfying this constraint will be termed the Equiripple formula. The computation of the coefficients for the Equiripple formula requires numerical methods.

The Equiripple formula of order $n = 1$ is given by

$$B_1(s) = \frac{a_0 - s}{a_0 + s}.$$

By assigning values to the coefficient a_0 and plotting the corresponding error curve, it is a simple matter to find the first peak, which defines the maximum phase error for a certain bandwidth. These data are presented in figures 1 and 6. For example, if the approximation is required to be good up to a frequency of 2 radians per time unit, the Equiripple curve of Figure 1 indicates that a maximum phase error of 0.16 is achievable using a formula of order 1. The corresponding value for the coefficient is given by Figure 6 as $a_0 = 1.55$.

An approximation of order 2 has the form

$$B_2(s) = \frac{(a_1 - s)^2 + b_1^2}{(a_1 + s)^2 + b_1^2}.$$

The values for the coefficients can be found by solving the optimization problem described in the Introduction. We used the MATLAB toolbox routine NELDER, with the coefficients from the Product formula as starting values. The results are given in Table 1. Comparing the errors of the different formulas (Table 2), it can be seen that it is possible to find significantly better approximations than the simple formulas.

ω_0	a_1	b_1
2.8	2.5104	2.0209
3.0	2.2633	2.1504
3.5	2.4529	2.0450
4.0	2.1760	2.1185

Table 1: Equiripple formulas of order $n = 2$.

ω_0	Padé	Product	Equiripple
2.8	0.149	0.075	0.018
3.0	0.188	0.075	0.021
3.5	0.333	0.085	0.045
4.0	0.522	0.256	0.116

Table 2: Maximum phase error ϕ_0 for approximations of order $n = 2$.

4 Conclusion

The Product formula presented here has a significantly larger bandwidth than the Padé formula for approximations of order 2 and above. Even for approximations of order 1 and 2, one can find (using numerical optimization) formulas with significantly better bandwidth, but at the expense of losing the convenience of the simple formulas.

5 Acknowledgements

Thanks go to Frank Cameron and Pertti Mäkilä for fruitful discussions. This work was financially supported by the Academy of Finland.

References

[1] M. Agarwal and C. Canudas, "On-line estimation of time delay and continuous-time process parameters", American Control Conference, Seattle, 1986, 728–733.

[2] F. R. Gantmacher, *Théorie des matrices*, Tome 2, Dunod, Paris, 1966.

[3] J. R. Martinez, "Transfer functions of generalized Bessel polynomials," *IEEE Trans. Circuits and Systems*, **CAS-24**, 1977, 325–328.

[4] H. Stahl and P. Hippe, "Design of pole placing controllers for stable and unstable systems with pure time delay", *Int. J. Control* **45**(6) 1987, 2173–2182.

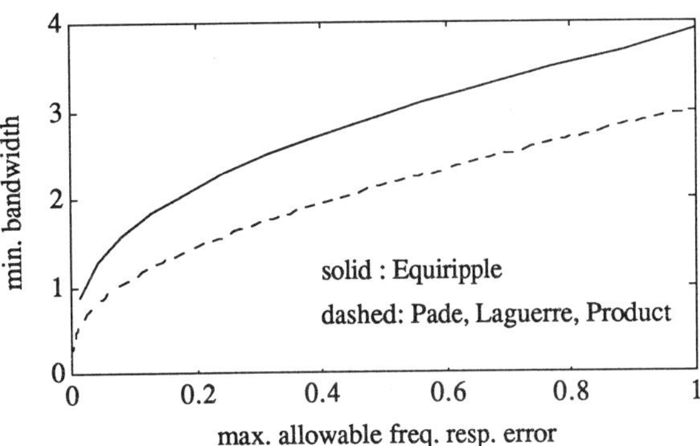

Figure 1: Bandwidth and error for formulas of order $n = 1$.

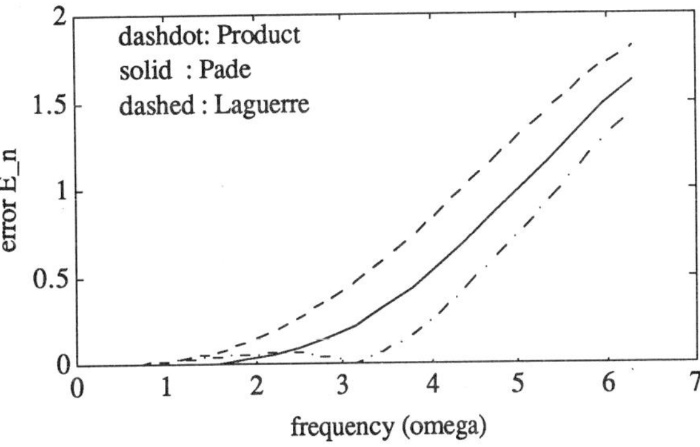

Figure 2: Frequency response error function for formulas of order $n = 2$.

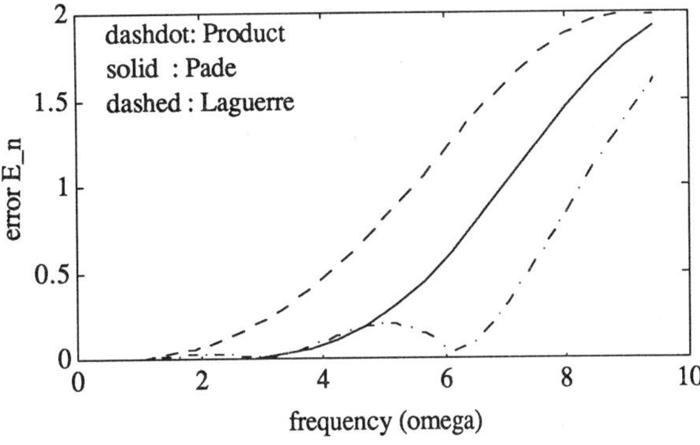

Figure 3: Frequency response error function for formulas of order $n = 3$.

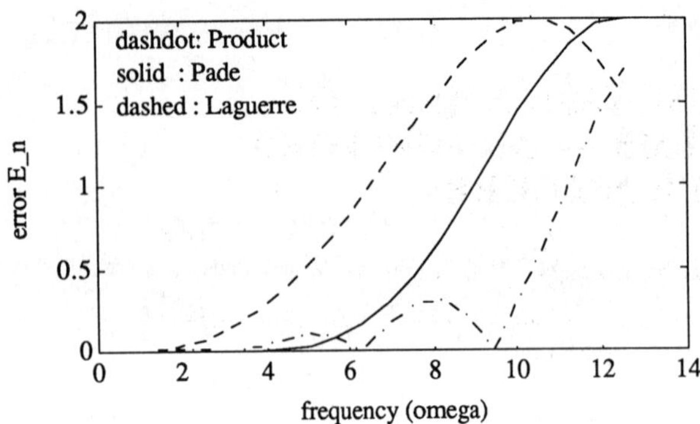

Figure 4: Frequency response error function for formulas of order $n = 4$.

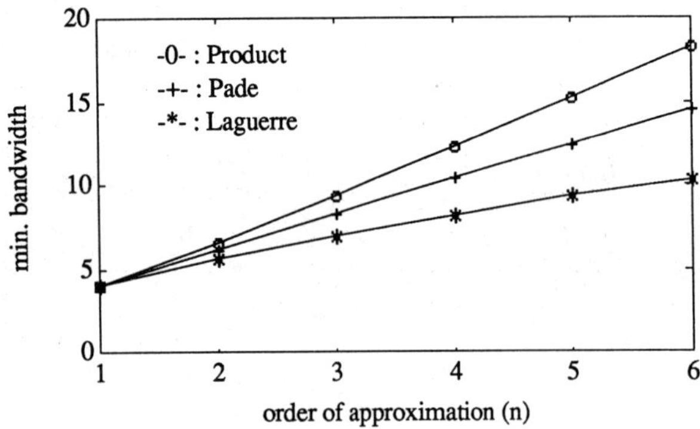

Figure 5: Bandwidth of approximation for error margin $\phi_0 = \pi/2$.

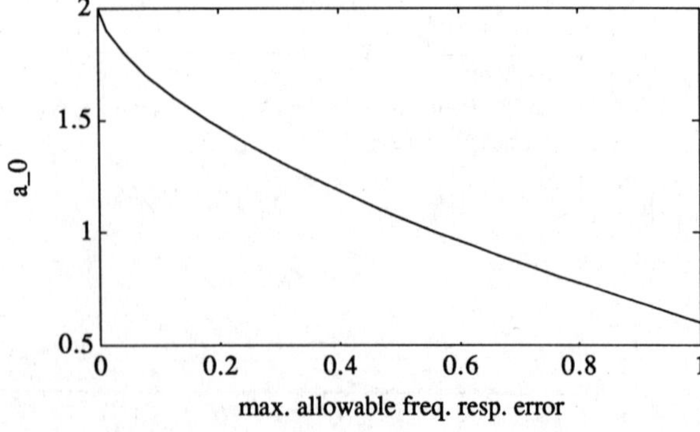

Figure 6: Values of first-order Equiripple formula coefficient (cf. Fig. 1).

FINITE SPECTRUM ASSIGNMENT OF TIME-DELAY SYSTEMS — A SIMPLIFIED DESIGN PROCEDURE

K. Watanabe, T. Ouchi and M. Nakatuyama

*Department of Electronic Engineering, Faculty of Engineering, Yamagata University,
Jonan, Yonezawa, 992, Japan*

Abstract   Finite spectrum assignment of time delay system is to locate n
poles at an arbitrarily preassigned set of points in the complex plane in
the same manner as delay-free systems, where n is the dimension of the
characteristic equation. The previous design method was complex. In this
paper, a simplified design procedure of the control law is presented.

Keywords   Time-delay systems; finite spectrum assignment; algorithm

INTRODUCTION

In general, a retarded differential system has an
infinite number of poles and controlling the
location of all poles is practically feasible.
Finite spectrum assignment is a linear static
feedback control such that the delay operator is
eliminated from the characteristic function of the
corresponding closed-loop system and n poles are
located at an arbitrarily preassigned set of points
in the complex plane in the same manner as
delay-free systems, where n is the dimension of the
differential equation describing the system.

The idea of finite spectrum assignment can be found
in papers by Maeda and Yamada(1975) and
Kamen(1978). If the system is reachable, algebraic
methods suggested by Morse(1976), Sontag(1976) and
Lee and Żak(1982) give simple feedback such that
the control matrix is over the ring of polynomials
in the delay operator. Reachability, however, is a
quite restrictive controllability condition. If the
system is not reachable, the control matrix over
the ring of polynomials can be easily enlarged to
the matrix with rational functions of the delay
operator.   The   feedback   control   is   usually
performed by a dynamical compensator.   The
closed-loop   system,   however,   contains   extra
dynamical modes which are not controllable. This is
not finite spectrum assignment.

To overcome the problem, Manitius and Olbrot(1979)
introduced the finite Laplace transforms in control
matrix and showed that the necessary condition for
finite spectrum assignment is for the system to be
spectral   controllable.   Watanabe   and
Ito(1983,1984a,1984b,1984c,1986) proved that the

spectral controllability is also the sufficient
condition   for   finite   spectrum   assignment   and
presented a general design procedure of finite
spectrum assignment. The algorithm was complex.
and further development was required.   Hyun, Shin
and Okubo(1987) presented a method which transforms
directly the control law with rational functions of
the delay operator to the one with polynomials and
the finite Laplace transforms . The control law
contains many extra finite Laplace transforms in
comparison of Watanabe's algorithm.

In this paper, a simplified design procedure of the
control matrix is presented by fusing the result by
Manitius   and   Olbrot(1979)   and   the   method   by
Watanabe (1983b,1984c).

FINITE SPECTRUM ASSIGNMENT

Consider the following system:
$$\dot{x}(t)=A(z)x(t)+b(z)u(t) \qquad (1)$$
where $x \in R^n$, $u(t) \in R$, and z is the right shift
(delay)operator which has the property such that
$zx(t)=x(t-h)$ for delay duration h>0. Furthermore,
$A(z) \in R^{n \times n}[z]$ and $b(z) \in R^{n \times 1}[z]$, where $R^{i \times j}[z]$
denotes the ixj matrix composed of real polynomials
in z. The control is given by

$$u(t)=f_x(z)x(t)+ \int_{-Nh}^{0} \phi(\tau)x(t+\tau)d\tau$$

$$+ \int_{-Nh}^{0} \psi(\tau)u(t+\tau)d\tau \qquad (2)$$

where   N   is   a   positive   integer, $\phi(\cdot) \in$
$L_2([-Nh,0],R^{1 \times n})$ and $\psi(\cdot) \in L_2([-Nh,0],R)$.

Taking the Laplace transform of (1) yields
$$sx(s)=A(z)x(s)+b(z)u(s) \qquad (3)$$
where $z=e^{-sh}$. The characteristic function of the

open-loop system is given by
$$\Delta(s,z) = |sI-A(z)|$$
$$= s^n + \alpha_1(z)s^{n-1} + \cdots + \alpha_n(z) \qquad (4)$$
where $\alpha_i(z)$ $(i=1,\cdots,n)$ are real polynomials in s. In general, the system has an infinite number of poles.

The control (2) is transformed to
$$u(s) = f_1(z,s)x(s) + f_2(z,s)u(s) \qquad (5)$$
where
$$f_1(z,s) = f_x(z)x(s) + \int_{-Nh}^{0} \phi(\tau)\exp(s\tau)d\tau \qquad (6.1)$$
$$f_2(z,s) = \int_{-Nh}^{0} \psi(\tau)\exp(s\tau)d\tau \qquad (6.2)$$
The characteristic function of the closed-loop system is given by
$$\Delta_T(s,z)$$
$$= \begin{vmatrix} sI-A(z) & -b(z) \\ -f_1(z,s) & 1-f_2(z,s) \end{vmatrix} \qquad (7)$$

If there exist $f_1(z,s)$ and $f_2(z,s)$ for arbitrary real numbers $\beta_1,\cdots,\beta_n$ such that
$$\Delta_T(s,z) = s^n + \beta_1 s^{n-1} + \cdots + \beta_n \qquad (8)$$
then the system (1) is said to be finite spectrum assignable. It is known that if and only if the system (1) is spectrally controllable, the system is finite spectrum assignable(Watanabe et. al 1984c). A simplified design procedure for computation of $f_1(z,s)$ and $f_2(z,s)$ is presented below.

SIMPLIFIED DESIGN PROCEDURE

It is supposed that the system (1) is spectrally controllable. The following equation holds.
$$\text{rank}[sI-A(e^{-sh}), b(e^{-sh})]=n, \quad \forall s \in C. \qquad (9)$$
This is equivalent to (Spong and Tran, 1981)
$$(a)\ \text{rank}_o[b(z),A(z)b(z),\cdots,A^{n-1}(z)b(z)]=n \qquad (10.1)$$
and
$$(b)\ \text{rank}[sI-A(e^{-sh}), b(e^{-sh})]=n, \quad \vee s \in \sigma(A) \cap \Lambda \qquad (10.2)$$
where (10.1) means that $\text{rank}[b(z),A(z)b(z),\cdots,A^{n-1}(z)b(z)]=n$ for all but finitely many $z \in C$, $\sigma(A)=\{s \in C \mid |sI-A(e^{-sh})|=0\}$ and $\Lambda=\{s \in C \mid \text{rank}[b(e^{-sh}),A(e^{-sh})b(e^{-sh}),\cdots,A(e^{-sh})^{n-1}b(e^{-sh})]<n\}$.

The characteristic function of the closed-loop system is expanded to
$$\Delta_T(s,z)$$
$$= |sI-A(z)| \; |1-f_2(z,s)-f_1(z,s)(sI-A(z))^{-1}b(z)|$$
$$= |sI-A(z)|$$
$$- [f_1(z,s),f_2(z,s)]\begin{bmatrix} \text{adj}(sI-A(z))b(z) \\ |sI-A(z)| \end{bmatrix}$$
$$= |sI-A(z)| - [f_1(z,s),f_2(z,s)]M(z)v(s) \qquad (11)$$
where
$$M(z) = \begin{bmatrix} \text{adj}(sI-A(z))b(z) \\ |sI-A(z)| \end{bmatrix} \qquad (12)$$
$$v(s) = [1,s,\cdots,s^n]^T \qquad (13)$$

By elementary row operations, $M(z)$ is transformed to
$$T_1(z)M(z)=M_1(z) \qquad (14)$$
where
$$M_1(z) = \begin{bmatrix} m_{11}(z) & & & 0 \\ \vdots & & \ddots & \\ m_{n1}(z) & & \cdots & m_{nn}(z) \\ m_{n+11}(z) & & m_{n+1n}(z) & 1 \end{bmatrix} \qquad (15)$$

The $m_{ij}(z)$ is the real polynomial in z, $m_{ii}(z) \neq 0$ $(i=1,2,\cdots,n)$ from (10.1) and the degree of $m_{ii}(z)$ is larger then that of $m_{ji}(z)$ $(j=i+1,\cdots,n+1)$. $T(z)$ is a unimodular matrix.

[lemma 1]
$$\frac{1}{m_{ii}(z)}\begin{bmatrix} m_{ii}(z) & & 0 \\ \vdots & \ddots & \\ m_{ii}(z) & \cdots & m_{ii}(z) \end{bmatrix} \in R^{i \times i}[z] \qquad (16)$$
The proof is given in Appendix 1.

Lemma 1 implies that
$$r(i)=(\text{the degree of } m_{i-1i-1}(z)) - (\text{the degree of } m_{ii}(z)) \geq 0 \qquad (17)$$

Suppose that
$$r(i)>1 \qquad (18.1)$$
for $i=c_1,c_2,\cdots,c_d$ $(c_1<c_2<\cdots<c_d)$ and
$$r(i)=0 \qquad (18.2)$$
for $i \neq c_1,c_2,\cdots,c_d$. Let
$$r=r(c_1)+r(c_2)+\cdots+r(c_d) \qquad (19)$$

By elementary row operations, we have $M_2(z)$ such that
$$T_2(z)M_1(z)=M_2(z) \qquad (20)$$
where
$$T_2(z) = \begin{bmatrix} T_{21}(z) \\ \vdots \\ T_{2d}(z) \end{bmatrix} \qquad (21)$$
$$T_{21}(z) = \begin{bmatrix} * & \cdots & * & z^{r(c_1)-1} & & 0 \\ \vdots & & \vdots & \vdots & & \\ * & \cdots & * & z & & \\ 0 & \cdots & 0 & 1 & & 0 \end{bmatrix} \qquad (22)$$
$$\begin{array}{c} \leftarrow \quad c_1 \quad \rightarrow \end{array}$$
$$M_2(z) = \begin{bmatrix} M_{21}(z) \\ \vdots \\ M_{2d}(z) \end{bmatrix} \qquad (23)$$
$$M_{21}(z) = \begin{bmatrix} * & \cdots & * & z^{r(c_1)-1} & m_{c_1c_1}(z) & & 0 \\ \vdots & & \vdots & & \vdots & & \\ * & \cdots & * & z & m_{c_1c_1}(z) & & \\ m_{c_11}(z) & \cdots & & & m_{c_1c_1}(z) & & 0 \end{bmatrix} \qquad (24)$$
asterisks denote polynomials in z and the degree of the asterisks in the j-th column of $M_{21}(z)$ is smaller than that of $m_{jj}(z)$ in $M_1(z)$.

$M_2(z)v(s)$ can be trarnsformed to

$$M_2(z)v(s)=P(s)v_\varpi(z) \tag{25}$$
where
$$v_\varpi(z)=[z^r,\cdots,z,1]^T \tag{26}$$

[lemma2] $P(s)$ is non-singular.
The proof is given by Appendix 2.

[lemma 3] Denote $P^{-1}(s)$ by
$$P^{-1}(s)=Q(s)=\begin{bmatrix}Q_1(s)\\ \vdots\\ Q_d(s)\end{bmatrix}\begin{matrix}r(c_1)\ \text{rows}\\ \\ r(c_d)\ \text{rows}\end{matrix} \tag{27}$$
The degree of the denominator of the non-zero elements in $Q_i(s)$ is at least by $r(c_i)-1$ larger than that of the numerator.
The proof is given by Appendix 3.

Let
$$R_i(s)=\begin{bmatrix}Q_i(s)\\ sQ_i(s)\\ \\ s^{c_i-2}Q_i(s)\end{bmatrix} \tag{28}$$

$$R_\varpi(s)=\begin{bmatrix}R_1(s)\\ \vdots\\ R_d(s)\end{bmatrix} \tag{29}$$

From lemma 3, the non-zero elements in $R_\varpi(s)$ are strictly proper. From Manitius and Olbrot (1979), there is a matrix $R_b(s)$ with rational functions of s such that $R(z,s)=R_\varpi(s)+R_b(s)z$ is the matrix with finite Laplace transforms.

[lemma 4]
$$R(z,s)M_2(z)v(s)=M_3(z)v(s) \tag{30}$$
where $M_3(z)$ is the matrix with polynomials in z and the degree of the elements in the i-th column of $M_3(z)$ is not larger than that of $m_{ii}(z)$ in $M_1(z)$.
The proof is gien by Appendix 4.

[lemma 5] If the system satisfies (10.2), then there exist the constant matrices Γ_1 and Γ_2 such that
$$[\ \Gamma_1,\quad \Gamma_2]\begin{bmatrix}M_1(z)\\ M_3(z)\end{bmatrix}=[I_{n\times n},0\ ] \tag{31}$$
The proof is given by Appendix 5.

Define $K(z,s)$ by
$$\begin{aligned}K(z,s)&=\{\ \Gamma_1+\Gamma_2(R_\varpi(s)+R_b(s)z)T_2(z)\}T_1(z)\\ &=[\ K_1(z,s),\quad K_2(z,s)\ ]\end{aligned} \tag{32}$$
$$\qquad\qquad n \qquad\qquad 1$$
Let
$$f_1(z,s)=[\ \alpha_n(z)-\beta_n,\ \cdots\cdots,\ \alpha_1(z)-\beta_1]K_1(z,s) \tag{33.1}$$
$$f_2(z,s)=[\ \alpha_n(z)-\beta_n,\ \cdots\cdots,\ \alpha_1(z)-\beta_1]K_2(z,s) \tag{33.2}$$
The $f_1(z,s)$ belongs to the class where the control consists of polynomials in z and the finite Laplace transforms. The $f_2(z,s)$ is the finite Laplace transforms because $f_2(z,s)$ is derived from

$$K_2(R_\varpi(s)+R_b(s)z)T_2(z)]T_1(z).$$

Substituting (32) into (7) yields
$$\begin{aligned}\triangle_f(s,z)&=\begin{vmatrix}sI-A(z)&-b(z)\\ -f_1(z,s)&1-f_2(z,s)\end{vmatrix}\\ &=|\ sI-A(z)\ |\ -\ [f_1(z,s),f_2(z,s)]M(z)v(s)\\ &=|\ sI-A(z)\ |\\ &\quad -[\ \alpha_n(z)-\beta_n,\ \cdots\cdots,\ \alpha_1(z)-\beta_1]\\ &\quad *[k_1(z,s),k_2(z,s]M(z)v(s)\\ &=|\ sI-A(z)\ |\\ &\quad -[\ \alpha_n(z)-\beta_n,\ \cdots\cdots,\ \alpha_1(z)-\beta_1]\begin{bmatrix}1\\ s\\ \vdots\\ s^{n-1}\end{bmatrix}\\ &=s^n+\beta_1 s^{n-1}+\cdots+\beta_n\end{aligned} \tag{34}$$
The system is finite spectrum assignable.

The algorithm is summarized as follows:

Step 1:Compute $M(z)$.
Step 2:Transform $M(z)$ to $M_1(z)$ as shown in (14).
Step 3:Construct $M_2(z)$ from $M_1(z)$ according to (20).
Step 4:Transform $M2(z)$ to $P(s)$ as shown in (25)
Step 5:Compute $Q(s)=P^{-1}(s)$ and $R_\varpi(s)$ given by (29). Construct the finite Laplace transform matrix $R(z,s)=R_\varpi(z,s)+R_b(z,s)z$.
Step 6:Compute $M_3(z)$ given by (30) from $R(z,s)$.
Step 7:Compute Γ_1 and Γ_2 which satisfy (31).
Step 8:Construct $f_1(z,s)$ and $f_2(z,s)$ given by (33).

The step 2,3 and 4 are refined on by fusing the results by Manitius and Olbrot(1979) and the one by Watanabe(1983). The algorithm is simplified in comparison with the previous one.

NUMERICAL EXAMPLE

Consider the system (1) with h=1,
$$A(z)=\begin{bmatrix}z&z^2\\ 0&0\end{bmatrix}\quad \text{and}\quad b(z)=\begin{bmatrix}0\\ z\end{bmatrix} \tag{35}$$
The characteristic function of the open-loop system is given by
$$|\ sI-A(z)\ |=s^2-zs \tag{36}$$

Step 1:$M(z)$ is given by
$$M(z)=\begin{bmatrix}z^3&0&0\\ z^2&z&0\\ 0&-z&1\end{bmatrix} \tag{37}$$
Step 2:The $T_1(z)$ and $M_1(z)$ are given by
$$T_1(z)=\begin{bmatrix}1&0&0\\ 0&1&0\\ 0&1&1\end{bmatrix} \tag{38}$$
$$M_1(z)=\begin{bmatrix}z^3&0&0\\ -z^2&z&0\\ -z^2&0&1\end{bmatrix} \tag{39}$$

Step 3:The $T_2(z)$ and $M_2(z)$ are computed as follows:

$$T_2(z) = \begin{bmatrix} 1 & z & 0 \\ 0 & 1 & 0 \\ 0 & 0 & 1 \end{bmatrix} \qquad (40)$$

$$M_2(z) = \begin{bmatrix} 0 & z^2 & 0 \\ -z^2 & z & 0 \\ -z^2 & 0 & 1 \end{bmatrix} \qquad (41)$$

Step 4:The $M_2(z)$ yields

$$P(s) = \left[\begin{array}{ccc} s & 0 & 0 \\ -1 & s & 0 \\ \hline -1 & 0 & s^2 \end{array}\right] \qquad (42)$$

Step 5:It follows from (42) that

$$Q(s) = \begin{bmatrix} Q_1(s) \\ Q_2(s) \end{bmatrix}$$

$$= \left[\begin{array}{ccc} \dfrac{1}{s} & 0 & 0 \\[2ex] \dfrac{1}{s^2} & \dfrac{1}{s} & 0 \\[2ex] \hline \dfrac{1}{s^3} & 0 & \dfrac{1}{s^2} \end{array}\right] \qquad (43)$$

$R_a(s)$ is given by

$$R_a(s) = \begin{bmatrix} \dfrac{1}{s} & 0 & 0 \\[2ex] \dfrac{1}{s^2} & \dfrac{1}{s} & 0 \\[2ex] \dfrac{1}{s^3} & 0 & \dfrac{1}{s^2} \\[2ex] \dfrac{1}{s^2} & 0 & \dfrac{1}{s} \end{bmatrix} \qquad (44)$$

The $R(z,s)$ is computed as follows:

$$R(z,s) = \begin{bmatrix} \dfrac{1-z}{s} & 0 & 0 \\[2ex] \dfrac{1-(s+1)z}{s^2} & \dfrac{1-z}{s} & 0 \\[2ex] \dfrac{1-(0.5s2+s+1)z}{s^3} & 0 & \dfrac{1-(s+1)z}{s^2} \\[2ex] \dfrac{1-(s+1)z}{s^2} & 0 & \dfrac{1-z}{s} \end{bmatrix} \qquad (45)$$

Step 6:The $M_3(z)$ is given by

$$M_3(z) = \begin{bmatrix} z^2-z^3 & 0 & 0 \\ z-z^2-z^3 & 0 & 0 \\ 1-z-0.5z^3 & -z & 0 \\ -z^3 & 1-z & 0 \end{bmatrix} \qquad (46)$$

Step 7:The matrices $\Gamma 1$ and $\Gamma 2$ which satisfy (32) are given by

$$\Gamma 1 = \begin{bmatrix} 3.5 & 1 & 0 \\ 2 & 1 & 0 \\ 1 & 0 & 1 \end{bmatrix} \qquad (47)$$

$$\Gamma 2 = \begin{bmatrix} 2 & 1 & 1 & 0 \\ 1 & 0 & 0 & 1 \\ 1 & 0 & 0 & 0 \end{bmatrix} \qquad (48)$$

Step 8:Computing $K_1(z,s)$ and $K_2(z,s)$ yields

$$K_1(z,s) = \begin{bmatrix} k_{11}(zs) & k_{12}(z,s) \\ K_{13}(z,s) & K_{14}(z,s) \end{bmatrix} \qquad (49)$$

$$K_2(z,s) = \begin{bmatrix} k_{21}(z,s) \\ K_{22}(z,s) \end{bmatrix} \qquad (50)$$

where

$$k_{11}(z,s) = 3.5 + 2\frac{1-z}{s} + \frac{1-(s+1)z}{s^2} + \frac{1-(0.5s^2+s+1)z}{s^3} \qquad (51)$$

$$K_{12}(z,s) = 1 + \frac{1-z}{s}(1+2z) + \frac{1-(s+1)z}{s^2}(1+z)$$
$$+ \frac{1-(0.5s^2+s+1)z}{s^3}z \qquad (52)$$

$$K_{13}(z,s) = 2 + \frac{1-z}{s} + \frac{1-(s+1)z}{s^2} \qquad (53)$$

$$K_{14}(z,s) = 1 + \frac{1-z}{s}(1+z) + \frac{1-(s+1)z}{s^2}z \qquad (54)$$

$$k_{21}(z,s) = \frac{1-(s+1)z}{s^2} \qquad (55)$$

$$K_{22}(z,s) = \frac{1-z}{s} \qquad (56)$$

Suppose that the characteristic function of the closed-loop system is required to be $(s+1)(s+2)$. The control matrices are given by

$$f_1(z,s) = [0-2,\ -z-3]K_1(z,s)$$
$$= \left[-13-2z- \frac{1-z}{s}(7+z) - \frac{1-(s+1)z}{s^2}(5+z) \right.$$
$$-2\frac{1-(0.5s^2+s+1)z}{s^3}, \quad -5-z-\frac{1-z}{s}(5+8z+z2)$$
$$\left.-\frac{1-(s+1)z}{s^2}(2+5z+z^2) - 2z\frac{1-(0.5s^2+s+1)z}{s^2} \right] \qquad (57)$$

$$f_2(z,s) = [0-2,\ -z-3]K_2(z,s)$$
$$= -2\frac{1-(s+1)z}{s^2} - \frac{1-z}{s}(3+z) \qquad (58)$$

CONCLUSION

A simplified design algorithm of control matrix for finite spectrum assignment is presented. This result is appliable to multivariable systems with delays via the results by Manitius (1985) and Watanabe (1986).

REFERENCES

Hyun Y.T., S.Shin and S. Okubo (1987). A Finite Spectrum Assignment Procedure for the Single-Input Linear Systems with Commensurate Time-Delays, Trans. SICE Japan, vol.23, no.4,pp.386-393.

Kamen, E.W. (1978). An Operator Theory of Linear Functional Differential Equations, J.Differential Equations, vol.27, pp.274-297.

Lee, E.B. and S.H.Zak (1982). On Spectrum Placement for LInear Time Invariant Delay Systems, IEEE Trans. on Automatic Control,

Maeda, H.,and H. Yamada (1975). On the Stabilizability of Linear Systems with Delays, Trans. SICE Japan, vol.11, no. 4, pp. 444-450.

Manitius,A.Z. and A.W.Olbrot (1979). Finite Spectrum Assignment Problem for Systems with Delays, IEEE Trans. on Automatic Control, vol.AC-21, no.4 pp.541-553.

Manitius, A.Z. and V.Manousiouthakis (1985). On Spectral Controllability of Multi-input time delay systems, Syst. Contr. Lett., vol.6, no. 3, pp. 199-205.

Morse, A. S. (1976). Ring Models for
   Delay-Differential Systems, Automatica, vol. 12,
   pp. 529-531.

Sontag, E. D. (1976). Linear Systems over
   Commutative Rings, A Survey, Richerche
   Automatica, vol. 7 no. 1 pp. 1-34.

Spong, M. W. and T. J. Tarn (1981). On the Spectral
   Controllability of Delay-Differential
   Equations, IEEE Trans. on Autoimatic Control,
   vol. AC-26, no. 2, pp. 527-528.

Watanabe, K., M. Ito, M. Kaneko and T. Ouchi
   (1983a). Finite Spectrum AssignmentPro blem for
   Systems with Delay in State Variables, IEEE
   Trans. on Automatic. Contolr, vol. AC-28, no. 4,
   pp. 506-508.

Watanabe, K. M. Ito and M. Kaneko (1983b). Finite
   Spectrum Assignment Problem for Systems with
   Multiple Commensurate Delays in State
   Variables, Int. J. Control, vol. 38, no. 5
   pp. 913-926.

Watanabe, K., M. Ito (1984a). A Necessary
   Condition for Spectral Controlability of Delay
   Systems, Int. J. Control, vol. 39, no. 2,
   pp. 363-374.

Watanabe, K. (1984b) Further Study of Spectral
   Controllability of Systems with Multiple
   Commensurate Delays in State Variables, Int. J.
   Control, vol. 39, no. 3 pp. 497-505.

Watanabe, k., M. Ito and M. Kaneko (1984c). Finite
   Spectrum Assignment Problem of Systems with
   Multiple Commensurate Delays in State and
   Control, Int. J. Control, vol. 39, no. 5,
   pp. 1073-1082.

Watanabe, K. and T. Ouchi (1985). An Observer of
   Systems with Delays in State Variables.
   Int. JH. Control, vol. 41, no. 1, pp. 217-229.

Watanabe, k. (1986). Finite Spectrum Assignment and
   Observer for Multivariable Systems with
   Commensurate Delays, IEEE Trans. on Automatic
   Control, vol. AC-31, no. 6, pp. 543-550.

APPENDIX 1 The proof of lemma 1: $T1(z)$ is denotd by

$$T_1(z) = \begin{bmatrix} T_{11}(z) & 0 \\ T_{12}(z)T_{11}(z) & 1 \end{bmatrix} \qquad (A.1)$$

Define $A(z)$ and $b(z)$ by
$$\bar{A}(z) = T_1(z)A(z)T_1^{-1}(z)$$
$$\bar{b}(z) = T_1(z)b(z) \qquad (A.2)$$

From (14) and (15), $\bar{A}(z)$ and $\bar{b}(z)$ are denoted by

$$\bar{A}(z) = \begin{bmatrix} a_{11}(z) & a_{12}(z) & & \\ & \vdots & & a_{n-1n}(z) \\ a_{n1}(z) & & --- & a_{nn}(z) \end{bmatrix} \qquad (A.3)$$

$$\bar{b}(z) = \begin{bmatrix} 0 \\ \vdots \\ 0 \\ a_{nn+1}(z) \end{bmatrix} \qquad (A.4)$$

This yields
$$T_1(z)M(z)v(s)$$

$$= \begin{bmatrix} a_{12}(z)a_{23}(z) \cdots a_{nn+1}(z) & & \\ a_{23}(z) \cdots a_{nn+1}(z) & | sI-\bar{A}(z) |_1 \\ \vdots & \\ a_{nn+1}(z) & | sI-\bar{A}(z) |_{n-1} \end{bmatrix} \qquad (A.5)$$

It follows from (A.5) that
$$m_{11}(z) = a_{1i+1}(z)a_{1i+12}(z) \cdots a_{nn+1}(z) \qquad (A.6)$$
This implies (16). $\square$

APPENDIX 2 The proof of lemma 2: $P(s)$ is the rxr squar matrix and is denoted by

$$P(s) = \begin{bmatrix} P_{11}(s) & \cdots & P_{1d}(s) \\ \vdots & & \vdots \\ P_{d1}(s) & \cdots & P_{dd}(s) \end{bmatrix} \qquad (A.7)$$

From (25), the degree of the diagonal elements of $Pii(s)$ is equal to c_i-1 and the degree of the elements in upper right triangular of $P_{ii}(s)$ except diagonal ones is c_i-1 at most. The degree of the elements in lower left triangular of $P_{ii}(s)$ except diagonal ones is smaller than c_i-1. Furthermor, the degree of the elements of $P_{ii+1}(s)$, $\cdots, P_{id}(s)$ is c_i-1 at most and that of $P_{i1}(s)$, $\cdots, P_{ii-1}(s)$ is smaller than c_i-1. These implies that
$$\det P_2(s) = s^q + (\text{polynomial of degree q-1 at most}) \qquad (A.8)$$
whrer
$$q = r(c_1)(c_1-1) + \cdots + r(c_d)(c_d-1) \qquad (A.9)$$
The matrix $P2(s)$ is non-singular. $\square$

APPENDIX 3 The proof of lemma 3: From the proof of lemma 2, the degree of the diagonal elements of $P_{ii}(s)$ is equal to c_i-1. Furthermore,

$$\begin{bmatrix} Q_1(s) \\ \vdots \\ Q_d(s) \end{bmatrix} = \begin{bmatrix} \text{adj}[P(s)] \\ | P(s) | \end{bmatrix} \qquad (A.10)$$

and the degree of $| P(s) |$ is q which is the sum of the degree of the diagonal elements in $P(s)$. These implies that the degree of the numerators of the non-zero elements in $Q_i(s)$ is at most $q-c_i+1$. The degree of the numerator of the non-zero elements in $Q_i(s)$ is at least by c_i-1 higher than that of the denominator. $\square$

APPENDIX 4 The proof of lemma 4: The following identity holds:
$$zR_b(s)M_2(z)v(s) = [R_a(s)+R_b(s)z]M_2(z)v(s)$$
$$-R_a(s)M_2(z)v(s) \qquad (A.11)$$
This yields
$$\| R_b(s)zM_2(e^{-sh})v(s) \|$$
$$\leq \| R_a(s)+R_b(s)e^{-sh} \| \, \| M_2(e^{-sh})v(s) \|$$
$$+ \| R_a(s)M_2(e^{-sh})v(s) \| \qquad (A.12)$$
where $\| \cdot \|$ is the matrix norm. Since $R_a(s)+R_b(s)z$ is the matrix with the finite Laplace transforms, the following inequality holds:
$$\| R_a(s)+R_b(s)e^{-sh} \| < \infty \qquad (A.13)$$
Since $M_2(z)$ consists of polynomials in z and $v(s)$ in s, we have
$$\| M_2(e^{-sh})v(s) \| < \infty \qquad (A.14)$$
It follows from (25) that the elements of $R_a(s)M_2(z)v(s)$ are the quasi-polynomials in s with

the polynomials in z as coeffceints. Then,

$$\| R_a(s) M_2(e^{-sh}) v(s) \| < \infty \tag{A.15}$$

Eqs. (A.12)–(A.14) imply that

$$\| R_b(s) z M_2(e^{-sh}) v(s) \| < \infty \tag{A.16}$$

$R_b(s) z M_2(e^{-sh}) v(s)$ consists of polynomials in s with polynomial coefficient of z. $[R_a(s)+R_b(s)z]*M_2(z)v(s)$ can be represented by (30). Since the degree of the column of $M_2(z)$ is less than that of the corresponding column of $M_1(z)$, the degree of the column of $M_3(z)$ does not exceed that of the corresponding column of M1(z). $\square$

APPENDIX 5 The proof of lemma 5: The (i,j)-th element of $M_1(z)$ is represented by

$$m_{ij}(z)=c_{ij0}z^P+c_{ij1}z^{P-1}+\cdots+c_{ijp} \tag{A.17}$$

where p is the degree of the j-th column of $M_1(z)$ and cijk are possibly non-zero real numbers. Let

$$C_{iij}=[c_{ij0},c_{ij1},\cdots,c_{ijp}] \tag{A.18}$$

and

$$C_1=\begin{bmatrix} C_{11} \\ : \\ C_{1d} \end{bmatrix}. \tag{A.19}$$

The matrix C_3 is constructed from $M_3(z)$ in the same manner as C_1. Let

$$C=\begin{bmatrix} C_1 \\ C_3 \end{bmatrix} \tag{A.20}$$

The C is the squar matric of size $n+r(c_1)(c_1-1)+\cdots+r(c_d)(c_d-1)$. If the C is singular, then there exist row vector k_1 and k_2 with real numbers such that

$$[k_1,k_2] \begin{bmatrix} M_1(z) \\ M_3(z) \end{bmatrix} = 0 \tag{A.21}$$

It follows that

$$[k_1+k_2\{R_a(s)+R_b(s)z\}T_2(z)]M_1(z)v(s)$$
$$=[k_1+k_2\{R_a(s)+R_b(s)z\}T_2(z)]$$
$$*\begin{bmatrix} I & 0 \\ T_{12}(z) & 1 \end{bmatrix}\begin{bmatrix} T_{11}(z)\,adj(sI-A(z))b(z) \\ |sI-A(z)| \end{bmatrix} =0 \tag{A.22}$$

This implies that there is a $i \in \{c_1,c_2,\cdots,c_d\}$ such taht

$$[* \cdots * \frac{q(z,s)}{p(s)} \quad 0\cdots 0]$$
$$\leftarrow i-1 \rightarrow$$
$$*\begin{bmatrix} a_{12}(z)a_{23}(z)\cdots a_{nn+1}(z) \\ a_{23}(z)\cdots a_{nn+1}(z)\,|sI-\bar A(z)|_1 \\ : \\ a_{ii+1}(z)\cdots a_{nn+1}(z)\,|sI-\bar A(z)|_{i-1} \\ : \\ a_{nn+1}(z)\,|sI-\bar A(z)|_{n-1} \\ |sI-A(z)| \end{bmatrix} =0 \tag{A.23}$$

It follows that

$$[* \cdots * \frac{q(z,s)}{p(s)} \quad]$$
$$\leftarrow i-1 \rightarrow$$
$$*\begin{bmatrix} a_{12}(z)a_{23}(z)\cdots a_{i-11}(z) \\ a_{23}(z)\cdots a_{i-11}(z)\,|sI-\bar A(z)|_1 \\ : \\ a_{i-11}(z)\,|sI-\bar A(z)|_{i-2} \\ |sI-\bar A(z)|_{i-1} \end{bmatrix} =0 \tag{A.24}$$

The $m_{i-1i-1}(z)/m_{ii}(z)=a_{i-1i}(z)$ is the polynomial of degree r(i)>1. Let the zerose of $a_{i-11}(z)=0$ be z_o. From (A.22), we have

$$\frac{q(z_o,s)}{p(s)}\,|sI-\bar A(z_o)|_{i-1}=0 \tag{A.25}$$

Since $|sI-\bar A(z_o)|_{i-1} \neq 0$, $q(z_o,s)=0$. Differentiating (A.24) with respect to z yields

$$\frac{\partial q(z,s)}{\partial z}\Big|_{z=z_o} = 0 \tag{A.26}$$

Continuing this procedure, we have

$$\frac{\partial^{r(i)-1}q(z,s)}{\partial^{r(i)-1}z}\Big|_{z=z_o} = 0 \tag{A.27}$$

The q(z,s) contains $a^{ii-1}(z)$ as a factor and the degree of z in q(z,s) dose not exceed r(i) because q(z,s) is given from $[Ra(s)+Rb(s)z]T2(z)$ and $T2(z)$ is of the form (22). The q(z,s) is represented by

$$q(z,s)=k(s)a_{i-11}(z) \tag{A.28}$$

where k(s) is the non-zero polynomial in s. The matrices p(s) and k(s) are co-prime.

Dividing (22) by $a_{i-11}(z)$ yields

$$[* \cdots * \frac{k(s)}{p(s)} \quad]$$
$$\leftarrow i-1 \rightarrow$$
$$*\begin{bmatrix} a_{12}(z)a_{23}(z)\cdots a_{i-2i-1}(z) \\ a_{23}(z)\cdots a_{i-2i-1}(z)\,|sI-\bar A(z)|_1 \\ : \\ a_{i-2i-1}(z)\,|sI-\bar A(z)|_{i-3} \\ |sI-\bar A(z)|_{i-2} \\ |sI-\bar A(z)|_{i-1} \end{bmatrix} =0 \tag{A.29}$$

It follows that

$$|\frac{k(s)}{p(s)}\,|sI-\bar A(z)|_{i-1}|$$
$$\leq |*a_{12}(z)a_{23}(z)\cdots a_{i-2i-1}(z)|$$
$$+|*a_{23}(z)\cdots a_{i-2i-1}(z)\,|sI-\bar A(z)|_1|$$
$$:$$
$$+|*a_{i-2i-1}(z)\,|sI-\bar A(z)|_{i-3}|$$
$$+|*\,|sI-\bar A(z)|_{i-2}| \tag{A.30}$$

Since asterisks denote polynomials in z and the finite Laplace transforms, they have limited values for $s=s_o$ which satisfies $p(s_o)=0$. Therefore,

$$|\frac{k(s_o)}{p(s_o)}\,|s_oI-\bar A(z)|_{i-1}| < \infty \tag{A.31}$$

Since $k(so) \neq 0$, it follows from (A.29) that

$$|s_oI-\bar A(z)|_{i-1}=0 \tag{A.32}$$

Since $q(z,s)/p(s)=k(s)a_{i-11}(z)/p(s)$ is the finite Laplace transform,

$$a_{i-11}(e^{-s_oh})=0 \tag{A.33}$$

Eqs. (A.32) and (A.33) yield that

$$rank[s_oI-\bar A(e^{-s_oh}), \bar b(e^{-s_oh})]< n \tag{A.34}$$

It follows that

$$rank[s_oI-A(e^{-s_oh}), b(e^{-s_oh})]< n \tag{A.35}$$

This contradicts to (10.2) and completes the proof. $\square$

DISCRETE-TIME LINEAR PERIODIC SYSTEMS:
STABILIZATION AND ZERO ERROR
REGULATION

P. Colaneri

*Centro di Teoria dei Sistemi CNR, Consiglio Nazionale delle Ricerche,
Piazza Leonardo da Vinci, 32, 20133 Milano, Italy*

Abstract. With reference to discrete-time linear periodic systems, two classical
problems of control theory are tackled in this paper: output stabilization and zero-
error regulation. The first problem is solved in two ways, involving a suitably
periodic pole placement technique and the theory of the difference periodic
Riccati equation. Moreover, a consistent definitions of zeros of periodic system and
of 'periodic integrators' allow to define a control structure capable of robustly
zeroing the error produced by periodic exogenous signals.

Keywords. Time-varying systems; Discrete-time systems; Stability criteria; Pole-
placement; Regulator theory.

INTRODUCTION

Research in periodic systems analysis and control
is motivated by two important, strongly connected,
issues: the use of periodic controllers for
enhancing the closed-loop behaviour of
time-invariant plants and the analysis of multirate
sampled-data control systems.
In the first line of reseach settle the gain margin
augmentation , the blocking zeros removal, the
decentralized fixed poles removal and the
simultaneous stabilization of a finite family of
linear time-invariant plants, to quote only a few.
In the recent paper by Francis and Georgiou (1988),
a number of the above arguments are analized in
detail.
On the other hand, the analysis of multirate
control, originated in the pioneer papers by Krank
(1957) and Jury et al. (1959), arises when the
inputs and the outputs of a time-invariant system
are subject to specified holding and/or sampling
mechanisms. In this way, a number of technological
constraints can be taken into account. For details
see e.g. the recent paper (Colaneri et al., 1989),
and the references quoted therein.
In this framework a further technique has been
introduced in Kabamba (1987), consisting of using
sampled-data hold functions for controlling
continuous-time linear systems.
On the theoretical side, a deep research in the
area of periodic systems has been recently carried
out. In particular, it is of interest here to
recall the survey paper by Bittanti (1986), where
the structural properties of linear periodic
systems, both continuous-time and discrete-time,
are summarized. Most results in this direction call
for strict relationships between discrete-time
periodic systems and time-invariant ones. In
particular, a state-sampled representation of a
T-periodic discrete-time system, introduced by
Meyer and Burrus (1975), has been proven to be
useful in the definition of zeros of periodic
systems (Bolzern et al., 1986 and Grasselli et al.,
1988) as well as in the analysis of the periodic
solutions of the periodic difference Riccati
equation, see (Bittanti et al., 1989). Further
research has been also carried out in order to
generalize the geometrical approach for solving
classical design problems, see (Grasselli et al.,
1987) and the references quoted there.
In the present paper, the plant at hand is
described from the very beginning by a
discrete-time linear periodic model.
In the first Section preliminary concepts of
periodic system theory are briefly recalled and the
concept of "periodic integrator" is introduced. Two
problems are then addressed: the output
stabilization and the zero-error regulation.
The first problem is tackled in the second Section
by using two different methods, involving the
state-sampling model of the periodic system and the
results on the periodic Riccati equation.
In the last Section the zero-error regulation
problem is finally solved by considering, as
exogenous signals acting on the plant, any periodic
signal. The cases when the plant is time-invariant
and/or the exogenous signals are constant are also
discussed. In so doing, important relationships
with the regulation of multirate sampled data
systems are clarified.

PRELIMINARIES AND PROBLEM FORMULATION

Some basic notions of periodic systems analysis are
recalled and two problems are finally formulated.

Preliminaries

Let the plant P be described by the linear
discrete-time T-periodic system:

$$x(t+1) = A(t)x(t) + B(t)u(t) \qquad (1.a)$$

$$y(t) = C(t)x(t) \qquad (1.b)$$

Here $A(t) \in \mathcal{R}^{n \times n}$, $B(t) \in \mathcal{R}^{n \times m}$, $C(t) \in \mathcal{R}^{p \times n}$ are
T-periodic function matrices, i.e. $A(t)=A(t+T)$,
$B(t)=B(t+T)$ and $C(t)=C(t+T)$, $\forall t \in Z$ (the set of
integers). $T \in Z^{+}$ (the set of positive integers) is
the system period. The transition matrix of system
(1) will be denoted by $\Phi_A(t,\tau)$. Matrix $\Phi_A(t_0+T,t_0)$
is called *monodromy matrix* at t_0. Its eigenvalues
are independent of t_0 and are named *characteristic
multipliers* of $A(.)$, see e.g. (Bittanti, 1986).
System (1), or, equivalently, matrix $A(.)$, is
asymptotically stable if and only if its
characteristic multipliers belong to the open unit
disk in the complex plane.

Time-invariant reformulation. The analysis of the
T-periodic system (1) can be carried out by
resorting to a time-invariant representation of
it. Precisely, following Meyer and Burrus (1975),
define the new input

$$\vartheta(k) = [\; u(kT+t_0)' \quad u(kt+1+t_0)'\ldots u(kT+T-1+t_0)' \;]'$$

and consider the system

$$\eta(k+1) = F\eta(k) + G\vartheta(k) \qquad (2.a)$$

$$\gamma(k) = H\eta(k) + E\vartheta(k), \qquad (2.b)$$

where $t_0 \in Z$,

$$F = \Phi_A(t_0+T,t_0) \in R^{nxn}, \quad G = [G_1\, G_2\, G_3\ldots G_T] \in R^{nxmT},$$

$$H = [H_1{'}\, H_2{'}\, H_3{'}\ldots H_T{'}]' \in R^{pTxn},$$

$$E = (\, E_{ij}\,) \in R^{pTxmT}, \quad \text{with}$$

$$G_i = \Phi_A(t_0+T,t_0+i)B(t_0+i-1),$$

$$H_i = C(t_0+i-1)\Phi_A(t_0+i-1,t_0)$$

$$E_{ij} = \begin{cases} C(t_0+i-1)\Phi_A(t_0+i-1,j+t_0)B(t_0+j-1) \;, & i < j \\[2mm] 0 & ,i \geq j \end{cases}$$

$$i,j = 1,2,\ldots T.$$

Given $\eta(0) = x(t_0)$ it results

$$\eta(k) = x(kT+t_0) \text{ and}$$

$$\gamma(k)=[\gamma(kT+t_0)'\; \gamma(kt+1+t_0)'\ldots\gamma(kT+T-1+t_0)']'.$$

In fact, system (2) can be thought as a state-sampled representation of system (1) at time t_0.

<u>Structural properties</u>. In view of the strong relationships between the input, state and output behaviour of systems (1) and (2), the structural properties, namely reachability, observability, stabilizability and detectability of the periodic system (1) at time t_0, can be equivalently analized by making reference to the time-invariant reformulation (2). Further equivalent notions of the structural properties of linear periodic systems have been also proposed in the literature. For more details the interested reader is referred to the survey paper by Bittanti (1986). For the proofs of Lemmas 1 and 2 below, see (Bittanti, Colaneri et al., 1986, Lemma 1) and (Bittanti, 1986, Theorem 5-c).

Lemma 1

The reachability [unobservability] subspace of $((A(.),B(.))\; [A(.),C(.)]$ at $t=t_0$ coincides with the reachability [unobservability] subspace of (F,G) $[F,H]$.

Lemma 2

The pair $(A(.),B(.))\; [A(.),C(.)]$ is stabilizable [detectable] if and only if the pair $(F,G)\; [F,H]$ is stabilizable [detectable]. ∎

<u>Zeros of periodic systems</u>. A definition of *zeros* of discrete-time linear periodic systems has been given in (Bolzern et al., 1986) by resorting to the time invariant reformulation (2). There it was shown that such definition allows to extend the well known blocking property of zeros. It will be shown that the definition below will play a central role in solving the error regulatotion problem. We restrict our attention to the square case, namely the case when the number of inputs is equal to the number of outputs, i.e. m=p.

Definition 1

A (complex) number λ is said to be a (trasmission) zero of (1) if it is a (trasmission) zero of (2), i.e. det $W(\lambda)=0$, where

$$W(\lambda) = \begin{bmatrix} \lambda I-F & -G \\ H & E \end{bmatrix} \in \mathcal{C}^{(n+mT)x(n+mT)}.$$
∎

A little thought on the meaning of the input, state and output vectors of the time-invariant reformulation (2) reveals that the nonzero zeros of (1), are independent of the initial sampling time t_0, see (Bolzern et al., 1986). On the contrary, see (Grasselli et al., 1987) the number of null zeros may vary with t_0, due to the fact that the dimensions of the reachable and observable parts of system (1) may vary with time.
For the definition of trasmission zeros of multivariable time-invariant systems, see. e.g. (Davison and Wang, 1974).

<u>T-Periodic Integrator</u>. In the following definition the concept of *T-periodic integrator* is introduced.

Definition 2

A T-periodic integrator is a T-order SISO system characterized by the tranfer function

$$I_T(z) = \frac{1}{z^T - 1} .$$
∎

In order to avoid any confusion, a 1-periodic integrator will be referred to as "simple integrator".

Problems Formulation

Two classical problems of systems and control theory will be addressed in the sequel: output stabilization and zero-error regulation.

<u>Output Stabilization</u>. The problem consists of constructing a T-periodic regulator R (see Fig.1), such that the overall closed-loop T-periodic system is asymptotically stable. This problem is solved in two different ways. In the first one a suitable (periodic) pole-assignment technique is undertaken, whereas, in the second one, the results on the theory of periodic Riccati equations are exploited.

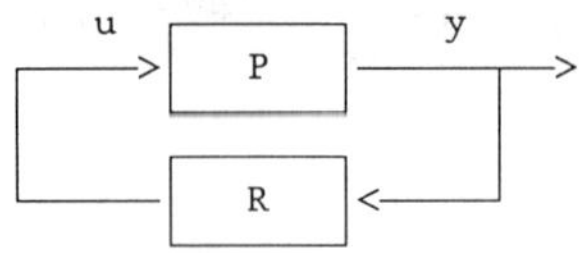

Fig.1

<u>Zero-error regulation</u>. With reference to the control-loop structure depicted in Fig.2, the problem consists of satisfying the following design specifications:
- Asymptotic stability
- Zero-error robust regulation, when $y^o(.)$ is any T-periodic signal.
More precisely, we aim at determining an internal structure of regulator R_1 such that, for any T-periodic stabilizing regulator R_2, the regulation error e(t) tends to zero as t tends to infinity for any T-periodic signal $y^o(.)$ and for any perturbation of the plant parameters which preserves closed-loop asymptotic stability.

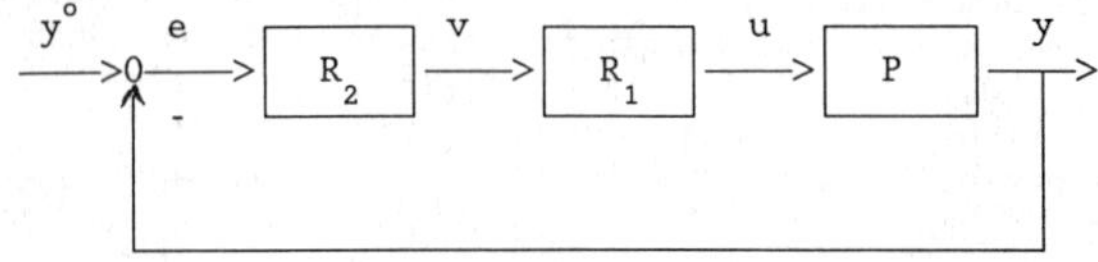

Fig.2

It will be shown that, in solving the problem above addressed, a main role will be plaied by the concept of zeros and the concept of periodic integrators of periodic systems (Definitions 1 and 2). The various cases when the exogenous signal $y^o(.)$ is constant and/or the plant is time-invariant will be finally dealt with. In this

194

respect, important relationships with the design of multirate sampled-data control systems will be clarified.

OUTPUT STABILIZATION

Pole-placement

As usual for this problem, we follow three steps. In the first, the state of system (1) is assumed to be measurable, whereas the recontruction of it is provided in the second step. The overall system is finally performed by substituting the reconstructed state to the 'true' one.

<u>State-feedback</u>. With reference to system (1), consider the sampled-data hold control law

$$u(t)=K(t)x(t_0+kT), \quad t \in [\ t_0+kT \ldots t_0+kT+T-1], \qquad (3)$$

where $K(.):Z \Rightarrow R^{m \times n}$ is a T-periodic matrix. Equation (3) corresponds to

$$\vartheta(k) = \bar{K} \ \eta(k) \qquad (4)$$

for the time-invariant system (2), with

$$\bar{K} = [\ \bar{K}_1' \quad \bar{K}_2' ..\bar{K}_T' \ ]' \ , \ \bar{K}_i \in R^{m \times n}, \ i = 1,2,\ldots,T, \quad (5.a)$$

$$\bar{K}_i = K(t_0+i-1), \ i=1,2,\ldots,T-1. \qquad (5.b)$$

Hence, the closed-loop system (2),(4) turns out to be characterized by matrix

$$F+G\bar{K} = \Phi_A(t_0+T,t_0)+ \sum_{i=0}^{T-1} \bar{G}_{i+1}\bar{K}_{i+1} \qquad (6)$$

Feedback (3) can be represented by the dynamic T-periodic control law acting on system (1):

$$z(t+1) = S(t)z(t) + Q(t)x(t) \qquad (7.a)$$

$$w(t) = S(t)z(t) + Q(t)x(t) \qquad (7.b)$$

$$u(t) = K(t)w(t). \qquad (7.c)$$

where $Q(.)$ and $S(.)$ are two nxn T-periodic matrices such that

$$Q(i) = I_n - S(i) = \begin{cases} 0 \ , & i \not= t_0 + kT \\ I_n, & i = t + kT \end{cases} , \ k \in Z \qquad (8)$$

The closed-loop T-periodic system (1),(7),(8) is characterized by matrix

$$A_1(t) = \begin{bmatrix} A(t)+B(t)K(t)Q(t) & B(t)K(t)S(t) \\ & \\ Q(t) & S(t) \end{bmatrix} \qquad (9)$$

Theorem 1

Matrix $A_1(.)$ possesses n characteristic multipliers equal to the eigenvalues of $F+G\bar{K}$ and n characteristic multipliers equal to zero.

Proof. The proof consists in a simple fact of matrix manipulations. Indeed, recalling (5) and (6), the monodromy matrix of $A_1(.)$ is

$$\Phi_{A_1}(t_0+T,t_0) = \begin{bmatrix} F+G\bar{K} & 0 \\ & \\ I_n & 0 \end{bmatrix} . \qquad \blacksquare$$

Notice that n characteristic multipliers of $A_1(.)$ are stable and equal to zero. This fact is due to the characteristic of the sampled-hold control law (3). A similar method for continuous-time linear periodic systems has been used by Kabamba (1986). As for the pole-placement problem, it is very well known that, by suitably selecting $\bar{K}$, it is possible

to arbitrarily assign n_r eigenvalues of $F+G\bar{K}$, n_r being the dimension of the reachability subspace of (F,G). From Lemmas 1 and 2, Theorem 1 and equations 5 the following result holds.

Corollary 1

(i) There exists a T-periodic matrices $K(.)$ such that $A_1(.)$ has n characteristic multipliers arbitrarily located, in complex conjugate pairs

(ii) There exists a T-periodic matrix $K(.)$: $Z \Rightarrow R^{m \times n}$ such that $A_1(.)$ is asymptotically stable if and only if $(A(.),B(.))$ is stabilizable. $\blacksquare$

<u>State-observer</u>. The dual version of the sampled-data hold control law (3) is now provided. Consider the following T-periodic system, which we candidate to yield the reconstruction $\hat{x}(t)$ of $x(t)$:

$$\hat{x}(t+1) = A(t)\hat{x}(t)+B(t)u(t)+Q(t+1)r(t+1) \qquad (10.a)$$

$$r(t+1) = S(t)r(t) + L(t)(C(t)\hat{x}(t)-y(t)), \qquad (10.b)$$

where $S(.)$ and $Q(.)$ are the T-periodic matrices defined in (8) and $L(.) : Z \Rightarrow R^{n \times p}$ is a T-periodic matrix. Equation (10.a) copies the system equation for $t \not= kT+t_0-1$, while at $t=kT+t_0-1$, a correcting term is added, consisting of the sum of the weighted output reconstruction errors $L(.)(C(.)\hat{x}(.)-y(.))$ over one period. Hence, it should be apparent that the sampled variable $\hat{\eta}(k) = \hat{x}(kT+t_0)$ satisfies the classical equation for a time-invariant observer:

$$\eta(k+1) = F\hat{\eta}(k) + G\vartheta(k) + \bar{L} \ (H \ \hat{\eta}(k) - \gamma(k)) \qquad (11)$$

with

$$\bar{L} = [\ \bar{L}_1 \ \bar{L}_2 \ \bar{L}_3 .. \ \bar{L}_T \ ], \quad \bar{L}_i \in R^{n \times p}, \ i= 1,\ldots,T \quad (12.a)$$

$$L(t_0+ i) = \bar{L}_{i+1} \ , \ i = 0, 1, \ldots , T-1. \qquad (12.b)$$

The dynamical matrix of system (2),(11) is

$$F+\bar{L}H = \Phi(t_0+T,t_0) + \sum_{i=0}^{T-1} \bar{L}_{i+1}\bar{H}_{i+1} \ , \qquad (13)$$

while, the T-periodic system (1),(10) is characterized by the dynamical matrix

$$A_2(t) = \begin{bmatrix} A(t)+Q(t+1)L(t)C(t) & Q(t+1) \\ & \\ L(t)C(t) & S(t) \end{bmatrix} \qquad (14)$$

Theorem 2

Matrix $A_2(.)$ possesses n characteristic multipliers equal to zero and n characteristic multipliers equal to the eigenvalues of $F+\bar{L}H$.

Proof. Recalling (12) and (13) and using some simple computations, it follows that the monodromy matrix of $A_2(.)$ takes the form:

$$\Phi_{A_2}(t_0+T,t_0) = \begin{bmatrix} F+\bar{L}H & 0 \\ & \\ * & 0 \end{bmatrix} ,$$

where $*$ denotes a block we don't consider. $\blacksquare$

Letting $q=[\hat{x}-x \quad r]$, the error dynamic satisfies $q(t+1)=A_2(t)q(t)$. Hence, if $A_2(.)$ is asymptotically stable, it results that $\hat{x}(t) \Rightarrow x(t)$ and system (10) can be considered as an asymptotic periodic observer for system (1).

Notice that matrix $A_2(.)$ has n characteristic multipliers equal to zero due to the summing-sampling mechanism given by (10). Define now as n_o the dimension of the unobservability subspace of (F,H). In view of Lemmas 1 and 2,

Theorem 2 and equations (12), the dual results of Corollaries 1 can now be stated.

Corollary 2

(i) There exists a T-periodic matrices $L(.)$ such that $A_2(.)$ has · n characteristic multipliers arbitrarily located, in complex conjugate pairs

(ii) There exists a T-periodic matrix $L(.)$: $Z \Rightarrow \mathcal{R}^{p \times n}$ such that $A_2(.)$ is asymptotically stable if and only if $(\hat{A}(.),C(.))$ is detectable. ∎

Output stabilization via pole-placement.

The regulator R of Fig.1 is obtained by substituting in the state-feedback control law (10) vector $\hat{x}(t)$ in place of $x(t)$, i.e. regulator R is described by

$$u(t) = K(t)w(t) \qquad (15.a)$$

$$w(t) = S(t)z(t) + Q(t)\hat{x}(t) \qquad (15.b)$$

$$\hat{x}(t+1) = A(t)\hat{x}(t)+B(t)u(t)+ Q(t+1)r(t+1) \qquad (15.c)$$

$$r(t+1) = S(t)r(t) + L(t)(C(t)\hat{x}(t)-y(t)) \qquad (15.d)$$

$$z(t+1) = S(t)z(t) + Q(t)\hat{x}(t) \qquad (15.e)$$

Letting the state-vector be $p = [\hat{x}'\ r'\ (\hat{x}-x)'\ z']'$ the T-periodic closed-loop system of Fig.1 can be written as

$$p(t+1) = A_3(t)p(t), \qquad (16.a)$$

$$A_3(.) = \begin{bmatrix} A_1(.) & * \\ 0 & A_2(.) \end{bmatrix}. \qquad (16.b)$$

From Corollaries 1,2, the following result holds.

Theorem 3

(i) There exist two T-periodic matrices, $K(.):Z \Rightarrow R^{m \times n}$ and $L(.):Z \Rightarrow R^{n \times p}$ such that $A_3(.)$ has n n- n multipliers arbitrarily assigned, in complex conjugate pairs.

(11) There exist T-periodic matrices $K(.)$: $Z \Rightarrow R^{n \times m}$ and $L(.)$: $Z \Rightarrow R^{p \times n}$ such that $A_3(.)$ is asymptotically stable if and only if $((A(.),B(.))$ is stabilizable and $((A(.),C(.))$ is detectable. ∎

In conclusion, note that the order of the regulator R obtained in this way is equal to 3n, since 2n characteristic multipliers are zero, due to the sampling, summing and holding mechanisms of (15).

Output stabilization via periodic Riccati equations

In this subsection we aim at determining a n-order T-periodic stabilizing regulator R (see Fig.1), directly described by:

$$u(t) = \tilde{K}(t)\tilde{x}(t) \qquad (17.a)$$

$$\tilde{x}(t+1)=A(t)\tilde{x}(t)+B(t)u(t)+\tilde{L}(t)(C(t)\tilde{x}(t)-y(t)) \qquad (17.b)$$

where $\tilde{L}(.):Z \Rightarrow \mathcal{R}^{n \times p}$ and $\tilde{K}(.):Z \Rightarrow \mathcal{R}^{m \times n}$ are T-periodic matrices such that $A(.)+B(.)\tilde{K}(.)$ and $A(.)+\tilde{L}(.)C(.)$ are asymptotically stable. As a matter of fact, if these requirements are achieved, the closed-loop T-periodic system (1),(17) of Fig.1 turns out to be asymptotically stable and described by:

$$\tilde{p}(t+1)=A_4(t)\tilde{p}(t) \qquad (18)$$

where $\tilde{p} = [x'\ (\tilde{x}-x)']'$ and

$$A_4(.) = \begin{bmatrix} A(.)+B(.)\tilde{K}(.) & * \\ 0 & A(.)+\tilde{L}(.)C(.) \end{bmatrix}.$$

The problem of finding conditions on the system matrices for the existence of stabilizing state-feedback control law and stable observer is an old and fashinating one, and goes back to the work of Wonham (1968) for the time-invariant case. Recently, the concepts of stabilizability and detectability of periodic systems have been completely clarified, see e.g Bittanti (1986). Hence the following result can be recalled.

Lemma 3

$(A(.),B(.))$ $[A(.),C(.)]$ is stabilizable [detectable] if and only if there exists a T-periodic matrix $\tilde{K}(.)$ $[\tilde{L}(.)]$ such that $A(.)+B(.)\tilde{K}(.)$ $[A(.)+\tilde{L}(.)C(.)]$ is asymptotically stable. ∎

Therefore, in agreement with Theorem 3, the output stabilization problem for the periodic system (1) is solvable if and only if the system is stabilizable and detectable. In this case the order of the regulator R of Fig.1 is equal to n. What is now missing is a method for computing the stabilizing T-periodic gains $\tilde{K}(.)$ and $\tilde{L}(.)$. We will analize this problem by exploiting the results available for the T-periodic difference Riccati equations (19) below, which can be associated with the pairs $((A(.),B(.))$ and $(A(.)),C(.))$, respectively.

$$\tilde{P}(t)=A'\tilde{P}(t+1)A-A'\tilde{P}(t+1)B(I_m +B'\tilde{P}(t+1)B)^{-1}B'\tilde{P}(t+1)A+\tilde{Q} \qquad (19.a)$$

$$\bar{P}(t+1) = A\bar{P}(t)A' - A\bar{P}(t)C'(I_p + C\bar{P}(t)C')^{-1}C'\bar{P}(t)A'+ \bar{Q} \qquad (19.b)$$

For simplicity, the dependence of $A(.),B(.)$ and $C(.)$ on time is not indicated in the equations above. Matrices $\tilde{Q}(.)$ and $\bar{Q}(.)$ are T-periodic symmetric and positive semidefinite $\forall t$. They will be specified in the sequel.

We try to determine two periodic stabilizing solutions of (19) , namely two T-periodic solutions $\tilde{P}(.)$ and $\bar{P}(.)$ such that $A(.)+B(.)\tilde{K}(.)$ and $A(.)+\tilde{L}(.)C(.)$ are both asymptotically stable, where

$$\tilde{K}(.)= -[I_m +B(t)'\tilde{P}(t+1)B(t)]^{-1}B(t)'\tilde{P}(t+1)A(t)' \qquad (20.a)$$

$$\tilde{L}(t)=-A(t)\bar{P}(t)C(t)'[I_p +C(t)\bar{P}(t)C(t)']^{-1} \qquad (20.b)$$

Recently, various issues regarding the discrete-time periodic Riccati equation have been given a satisfactory clarification, see e.g. (Bittanti, 1986). In (Bittanti et al., 1989)., a time-invariant reformulation approach has been taken. As a result, a bijective correspondence beetween the periodic solutions of the discrete-time periodic Riccati equation and the constant solutions of a suitable algebraic Riccati equation has been established. This gives an almost definitive answer to any questions arising in the periodic case.

As for equations (19), notice first that, if $\tilde{Q}(.)=B(.)'B(.)$ and $\bar{Q}(.)=C(.)'C(.)$, the duality relationships $A(t)<\longrightarrow A(-t)'$, $B(t)<\longrightarrow C(-t)'$ reflect into $\tilde{P}(t)<\longrightarrow \bar{P}(-t+1)$, $K(t)<\longrightarrow \tilde{L}(-t)'$. These relationships completely define the so-called duality principle for discrete-time systems, see (Bittanti, Colaneri et al. 1986). Moreover, system (1) enjoys the dual structural properties of the dual system characterized by the state, input and output matrices $A(-t)',C(-t)'$ and $B(-t)'$, respectively.

The computation of the periodic stabilizing gains $\tilde{K}(.)$ and $\tilde{L}(.)$ can be carried out in different ways. For example, the theorem below states an interesting convergence result, see (Bittanti et al., 1988).

Theorem 4

Let $\bar{Q}(.)=B(.)'B(.)$ and $\tilde{Q}(.)=C(.)'C(.)$. If system (1) is stabilizable and detectable, then $\forall t$

$$\lim_{t_1 \to +\infty} \tilde{\Pi}(t,t_1,U) = \tilde{P}(t), \quad \lim_{t_2 \to -\infty} \tilde{\Pi}(t,t_2,U) = \tilde{P}(t).$$

where $U \in \mathcal{R}^{n \times n}$ is any positive semidefinite matrix, $\tilde{P}(t)$ and $\tilde{P}(t)$ are the unique T-periodic solutions of (19) and $\tilde{\Pi}(t,\tau,U)$ $[\tilde{\Pi}(t,\tau,U)]$ is the solution at t of equation (19.a) [19.b] with terminal [initial] condition $\tilde{\Pi}(\tau,\tau,U)=U$ $[\tilde{\Pi}(\tau,\tau,U)=U]$. Moreover, $A(.)-B(.)\tilde{K}(.)$ and $A(.)-\tilde{L}(.)C(.)$ are asymptotically stable.

ZERO-ERROR REGULATION

In this Section, we consider the system depicted in Fig.2, where R_2 is any T-periodic stabilizing regulator. The structure of R_1 has to be determined in order to ensure zero-error[1] regulation when $y^o(.)$ belongs to a well specified class of functions. In particular, we consider two cases: in the first one, $y^o(.)$ is <u>any</u> T-periodic function whereas, in the second case, y^o is <u>any</u> constant signal. In both cases, the regulation problem is easily solved by considering the time-invariant reformulation (see Fig.3) associated with the periodic system of Fig.2. We here consider, without any loss of generality, the 'square' case, i.e. p=m.

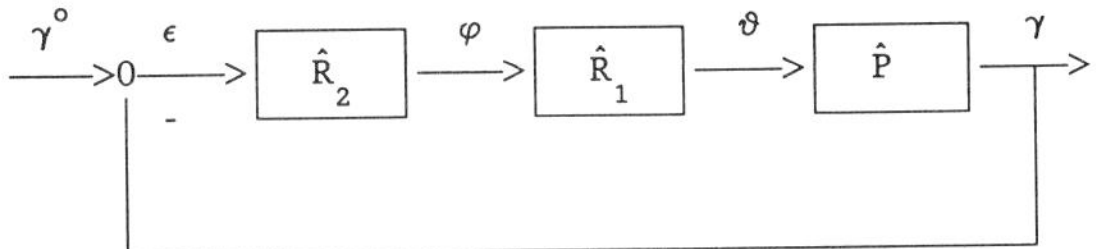

Fig.3

$\hat{P}$ is the time-invariant system (2),
$\hat{R}_1$ is a regulator plugged in the control system to ensure zero error regulation (the structure of which will be discussed in the sequel)
$\hat{R}_2$ is the time-invariant reformulation of any T-periodic stabilizing regulator $\hat{R}_2$.
By comparing Fig.2 and Fig.3, it is apparent that $\epsilon(k) \in R^{mT}$, $\gamma(k) \in R^{mT}$ and $\varphi(k)$ are given by

$$\epsilon(k)=[e(kT+t_0)' e(kT+t_0+1)' \ldots e(kT+t_0+T-1)']' \quad (21.a)$$

$$\gamma(k)=[y(kT+t_0)' y(kT+t_0+1)' \ldots y(kT+t_0+T-1)']' \quad (21.b)$$

$$\varphi(k)=[\varphi(kT+t_0)' \varphi(kT+t_0+1)' \ldots \varphi(kT+t_0+T-1)']' \quad (21.c)$$

Hence, $e(t) \Rightarrow 0$ if and only if $\epsilon(k) \Rightarrow 0$. Moreover, the time-invariant system of Fig.3 is asymptotically stable if and only if the T-periodic system of Fig.2 is asymptotically stable.

<u>T-periodic signals</u>. It is well known that the robust regulation problem for the time-invariant system of Fig.3 is solved when γ^o is constant if and only if $\hat{R}_1$ is constituted by mT noninteracting simple integrators. In Lemma 4 below the relationships beetween a T-periodic integrator (in the t-axis) and a simple integrator (in the k-axis) is stated.

Lemma 4

A T-periodic integrator (in the t-axis) is equivalent to T simple integrators (in the k-axis).

Proof. Recall Definition 2 and consider Figures 2 and 3. Let V(z) [U(z)] be the z-transform of $v(.)$ $[u(.)]$ and $\phi(z) = \{\phi_i(z), i=1,2,\ldots T\}$ $[\theta(z) = \{\theta_i(z), i=1,2,\ldots T\}]$ the z-transform of $\varphi(.)$ $[\vartheta(.)]$. The proof follows from well-known relationships beetween the z-transform of a signal (in the t-axis) and those of the associated sampled variables. Actually, if $\phi_i(z) = I_1(z)\theta_i(z)$, then

$$V(z)=\sum_{i=0}^{T-1} \phi_i(z^T) z^{-i} = I_T(z) \sum_{i=0}^{T-1} \theta_i(z^T) z^{-i} = I_T(z)U(z).$$

Conversely, if $V(z) = I_T(z) U(z)$, then, denoting by σ the T-th root of one, i.e. $\sigma = \exp(2\pi j/T)$, it follows that

$$\phi_i(z^T)= \frac{1}{T} z^{i-1} \sum_{i=0}^{T-1} V(z\sigma^i)\sigma^{(i-1)j} = I_T(z)\theta_i(z).$$

Hence $\phi_i(z) = I_1(z) \theta_i(z).$ ∎

From the Lemma above and the structure of $\hat{R}_1$ the following result is in order.

Theorem 5

Consider the T-periodic system depicted in Fig.2 and let

R_1 be constituted by m noninteracting T-periodic integrators
R_2 be any T-periodic stabilizing regulator.

The system error produced by any T-periodic signal $y^o(.)$ tends to zero as t tends to infinity in a robust way, i.e., for any perturbation of the plant parameters which preserves closed-loop asymptotic stability. ∎

The arguments presented in the previous Section can be now used for discussing the existence of a T-periodic regulator R_2. In order to apply Theorem 3, we have to consider the structural properties, namely stabilizability and detectability, of the 'augmented' plant (R_1,P), see Fig.2. Thanks to Lemma 2, these conditions can be analized by resorting to the time-invariant reformulation $(\hat{R}_1,\hat{P})$ associated with (R_1,P). Recall that $\hat{R}_1$ is constituted by mT noninteracting simple integrators. Hence, a standard result of linear systems theory ensures that $(\hat{R}_1,\hat{P})$ is stabilizable and detectable if and only if $\hat{P}$ is stabilizable and detectable and does not have transmission zeros equal to one. From the definitions of zeros of periodic systems and Lemma 2, the conclusion below holds true.

Lemma 5

The T-periodic system (R_1,P) is stabilizable and detectable if and only if P is stabilizable and detectable and does not possesses trasmission zeros equal to one. ∎

We are now in the position to state the main result of this Section.

Theorem 6

Consider Fig.2 and let R_1 be constituted by m noninteracting periodic integrators. If $((A(.),B(.))$ is stabilizable, $((A(.),C(.))$ is detectable and plant P doesn't have trasmission zeros equal to 1, then there exists a T-periodic regulator R_2 such that the closed-loop system of Fig.2 is asymptotically stable and zero-error regulation is achieved for any T-periodic reference signals and for any perturbation of the plant P which preserves asymptotic stability. ∎

Notice that a more complex structure of R_1 could be introduced containing the m parallel T periodic integrators as a part of it. In such a case, the same result holds true as well provided that cancellations do not occurr between unstables characteristic multipliers of P and zeros of R_1.

<u>Constant reference signal</u>. Consider now a constant reference signal y^o. Hence, $\gamma^o=[y^{o'} y^{o'} y^{o'} \ldots y^{o'}]'$ is a particular constant signal. One could try to simplify the order of regulator R_1 to get the zero-error regulation. This can be achieved by letting R_1 be composed by the cascade of m <u>simple</u> integrators with a suitable T-periodic gain.

Lemma 6

Consider Fig.2. Let R_2 be a (T-periodic) stabilizing regulator and R_1 be constited by the cascade connection of m noninteracting simple integrators with a periodic gain $N(.):Z \Rightarrow R^{m \times m}$ and let

$$N(t_0+i-1) = \sum_{j=1}^{T} \bar{W}_{ij} \, , \quad i=1,2,\ldots T \, ,$$

where $\bar{W}_{ij}$, $i=1,..,T$, $j=1,..,T$, are the mxm ordered matrices of the last mT rows and columns of matrix $W(1)^{-1}$, $W(1)$ being the 'system matrix' defined in Definition 1 and calculated in $\lambda=1$.
Hence, the unique asymptotic motion is constant and such that $y = y^0$, for any constant y^0.

Proof. Consider the system of Fig.3. The asymptotic stability implies that the unique equilibrium state $[x_R' \ x_P' \ x_I']'$ of P, R_1 and R_2 is such that

$$\begin{bmatrix} x_P \\ \bar{\vartheta} \end{bmatrix} W(1)^{-1} \begin{bmatrix} 0 \\ \gamma^0 \end{bmatrix}, \quad \bar{\vartheta}= \text{diag } \{N(i)\} \ [x_I' \ x_I' \ldots x_I' \ ]' \, .$$
$$i=t_0,\ldots,t_0+T-1$$

Hence, $x_I=y^0$ and $y=y^0$ yield the result. ∎

Notice however that, since R depends on the plant, the regulation is no longer robust unless only 'directional' perturbations of the plant are allowed or the plant is structurally such that

$$\sum_{j=1}^{T} \bar{W}_{ij} = \text{constant}, \ i = 1,2,\ldots,T \, . \tag{22}$$

This last condition could be exploited in order to get robustness results when the plant P is time-invariant, y^0 is constant and R_2 is any *T-periodic stabilizing* regulator constructed for enhancing the closed-loop performances. Notice that such a scheme naturally arise when dealing with *multirate sampled-data control systems*. Hence, in the sequel we will see that condition (22) is satisfied when the plant P is time-invariant. First notice that the time invariant reformulation $\hat{P}$ associated with the constant matrices A,B,C of P is

$$F=A^T, \ G=[A^{T-1}B \ A^{T-2}B \ \ldots B \ ], \ H'=[C' \ C'A' \ldots C'A'^{T-1}]$$

$$E = \{ E_{ij} \} \, , \quad E_{ij} = \begin{cases} 0 \, , & i \le j \\ \\ C \, A^{i-1-j}B \, , & i > j \end{cases} \, .$$

Moreover let the 'system matrix' be

$$J(\lambda) = \begin{bmatrix} \lambda I - A & - B \\ C & 0 \end{bmatrix} \in \mathcal{C}^{(n+m) \times (n+m)}$$

and suppose that P does not possess transmission zeros equal to one, i.e. det $J(1) \neq 0$.
Hence, it a simple but cumbersome fact of matrix manipulations to see that

$$\bar{W}_{ij} = \bar{Z}, \quad i=1,2,\ldots,T \ \ j=1,2,\ldots,T \tag{23}$$

where $\bar{Z}$ is the mxm matrix formed by the last m rows and columns of $J(1)^{-1}$. The following result holds.

Theorem 7

Consider Fig. 2. Let P be a time-invariant system, R_2 be a (T-periodic) stabilizing regulator and R_1 be constituted by m noninteracting integrators. Then, the closed-loop system error tends to zero for any constant reference signal y^0 robustly, i.e. for any perturbation of the system parameters which preserves closed-loop asymptotic stability.

Proof. The asymptotic stability implies that the unique asymptotic motion is constant and given by $x_R= 0$ and $\xi=[x_P \ x_I]$ such that

$$\xi = J(1)^{-1} \begin{bmatrix} 0 \\ y^0 \end{bmatrix},$$

where x_P , x_R and x_I are the states of the plant and regulators R_2 and R_1, respectively.

CONCLUSIONS

With reference to discrete-time linear periodic systems, two classical control problems have been defined and solved: stabilization and zero-error regulation. Among the various issues outlined herein, a first step toward a complete extension to such systems of the fundamental internal model principle has been provided.

REFERENCES

Bittanti S. (1986). Deterministic and stochastic linear periodic systems, in Time Series and Linear Systems, S.Bittanti ed., Springer Verlag, 141-182.

Bittanti S. and P.Bolzern (1986). On the structure theory of discrete-time linear systems, Int. J. Syst. Sci., 17,1, 33-47.

Bittanti S., P.Colaneri and G.De Nicolao (1986). Discrete-time linear periodic systems: a note on the reachability and controllability interval length", Systems and Control Letters, 8, 75-78.

Bittanti S., P.Colaneri and G.De Nicolao (1988). The difference periodic Riccati equation for the periodic prediction problem, IEEE Trans. AC, 33,8, 706-712.

Bittanti S., P.Colaneri and G.De Nicolao (1989). An algebraic Riccati equation for the discrete-time periodic prediction problem, Int.Rep. 89-002, Centro di Teoria dei Sistemi CNR, Dipartimento di Elettronica del Politecnico di Milano.

Bolzern P, P.Colaneri and R.Scattolini (1986). Zeros of discrete-time periodic linear systems, IEEE Trans. AC., 31, 1056-1058.

Colaneri P., R.Scattolini and N.Schiavoni (1989), Stabilization and regulation of multirate sampled-data linear systems, Automatica, to appear.

Davison E.J.and S.H.Wang (1974). Properties and calculation of trasmission zeros of linear multivariable systems", Automatica,10,643-658.

Francis B.A. and T.T.Georgiou (1988). Stability theory for linear time-invariant plants with periodic digital controllers, IEEE Trans. AC, 33, 9, 820-832.

Grasselli O. and S.Longhi (1987). Linear function dead-beat observers with disturbance localization for linear periodic discrete-time systems, Int. J. Contr., 45,5, 1603-1626.

Grasselli O. and S.Longhi (1988). Zeros and poles of linear periodic multivariable discrete-time systems", Circuits, Systems and Signal Proc.

Jury E.I., F.J.Mollin (1959). The analysis of sampled-data control systems with a periodically time-varying sampling rate, IRE Trans. AC., 24, 15-21.

Kabamba P.T.(1986). Monodromy eigenvalue assignment in linear periodic systems, IEEE Trans. AC., 31, 950-952.

Kabamba P.T.(1987). Control of linear systems using generalized sampled-data hold functions, IEEE Trans. AC. , 24, 772-783, 1987.

Krank G.M. (1957). Input-Output analysis of multi-rate feedback systems, IRE Trans. AC, 32,21-2.

Meyer R.A. and C.S.Burrus (1975).A unified analysis of multirate and periodically time-varying digital systems, IEEE Trans. CS., 22, 162-168.

Wonham W.M. (1968). On a matrix Riccati equation of stochastic control, SIAM J. Contr., 6,681-697.

OBSERVER SYNTHESIS BY EIGENSTRUCTURE ASSIGNMENT — AN APPLICATION TO "LOOP TRANSFER RECOVERY"*

J.-F. Magni

C.E.R.T.-D.E.R.A., B.P. 4025, F31055 Toulouse Cedex, France

Abstract. In this paper we give, in an eigenstructure assignment setting, a geometric approach to the synthesis of observers. This formulation allows to exhibit all the degrees of freedom and to give them a physical meaning. The problem of exact loop transfer recovery by mean of observers, more specifically by mean of minimal order observers, is then examined. The choice of the free parameters appearing for solving this problem reduces to a classical eigenstructure assignment problem.

Key words. Linear systems, multivariable control, minimal order observer, loop transfer recovery, pole assignment, eigenvector assignment.

1 Introduction.

Following the techniques relative to eigenstructure assignment introduced in Moore [7] for synthesizing control laws, simple ideas for synthesizing observers are given in this paper. The formulation of the observers used here is the one, based on the knowledge of a (C, A)-invariant subspace, introduced by Wonham [10]. Here, it is the construction of the involved (C, A)-invariant subspace which is concerned in the use of eigenstructure assignment techniques. The advantages of such an approach are the following. Firstly, let recall that the eigenstructure assignment relative to a control law allows to dibtribute the modes on the outputs to be controlled. Here the choice of the left eigenvectors and output directions allows to distribute the excitations (initial conditions, disturbances, inputs to be tracked) on the observer modes (see Comment 1 and [2]). Secondly, the choice of the left eigenvectors corresponds to the choice of the linear combinations of states which are observed. This fact allows to present in a simple unique framework the synthesis of observers of various kind : the exact or approximate observation of a single linear combination of states and the Luenberger observers of various order. Thirdly, all the degrees of freedom are clearly exhibited.

In this paper, we first consider the synthesis of observers and dual observers. It is shown how they can be simply synthetized by choosing vectors in some characteristic subspaces. More precisely we show that almost any choice of the vectors in these characteristic subspaces are valid. The loop transfer recovery problem is then examined from a modal point of view. The classical n-order observers are used to analyse the asymptotic solutions to this problem whilst the minimal order observers are used to solve it algebraically. The algebraic solution of the exact loop transfer recovery problem is explicitly formulated as an eigenstructure assignment problem. Finally some illustrative examples are given.

2 Observers and dual observers

2.1 Notations and preliminaries

We shall consider a linear time invariant control system :

$$\dot{x} = Ax + Bv \quad \text{and} \quad y = Cx \tag{1}$$

A, B and C, are linear mappings $A : \mathcal{X} \to \mathcal{X}$, $B : \mathcal{V} \to \mathcal{X}$ and $C : \mathcal{X} \to \mathcal{Y}$ where $\dim \mathcal{X} = n$, $\dim \mathcal{V} = m$ and $\dim \mathcal{Y} = p$. B and C are assumed to be of full rank. The vocabulary of the Geometric Approach will be used. The meaning of the following notations $\mathcal{V}^*_{\mathrm{Ker}\,C}$, $\mathcal{R}^*_{\mathrm{Ker}\,C}$, $\mathcal{V}^*_{b,\mathrm{Ker}\,C}$, $\mathcal{S}^*_{\mathrm{Im}\,B}$ etc ... is assumed to be known (see [6],[9], [11]).

Let λ be any complex number. The corresponding (A, B)-*characteristic subspace* is defined by :

$$\mathcal{S}(\lambda) = (A - \lambda I)^{-1}\mathrm{Im}B \tag{2}$$

Various properties of these subspaces are derived in Magni and Champetier [5]. If λ is not an uncontrollable eigenvalue, $\mathcal{S}(\lambda)$ is a m-dimensional subspace. Otherwise $\dim \mathcal{S}(\lambda) > m$. It is well known that $\mathcal{S}(\lambda)$ is the subspace containing all the vectors which can be assigned as right eigenvectors corresponding to the eigenvalue λ. Clearly, for any vector s in $\mathcal{S}(\lambda)$, there exists a unique m-dimensional vector w defined by

$$(A - \lambda I)s + Bw = 0 \tag{3}$$

This vector w will be called the *input direction* relative to (λ, s). Note that the (A, B)-characteristic subspaces are (A, B)-invariant and that they are invariant under state feedback.

By duality, for any complex number π, the (C, A)-*characteristic subspace* relative to π is defined by :

$$\mathcal{T}(\pi) = (A - \pi I)\mathrm{Ker}\,C \tag{4}$$

*This research was supported by the D.R.E.T

Clearly, for any vector u in $T(\pi)^{\perp}$, there exists a unique p-dimensional vector t defined by

$$u^*(A - \pi I) + t^* C = 0 \qquad (5)$$

This vector t will be called the *output direction* relative to (π, u). Finally note that the (C, A)-chracteristic subspaces are (C, A)-invariant and that they are invariant under output injection.

The key point for synthesizing observers is the following:

Lemma 2.1 - *The system defined by :*

$$\dot{z} = \pi z - t^* y + u^* B v \qquad (6)$$

where u, t, π satisfy :

$$u^*(A - \pi I) + t^* C = 0 \qquad (7)$$

is an observer of the variable $u^* x$. The dynamics of the observation is given by π.

Proof: Clearly from (1) and (6) we have :

$$\dot{z} - u^* \dot{x} = \pi z - t^* C x + u^* B v - u^* A x - u^* B v$$

hence from (7)

$$\dot{z} - u^* \dot{x} = \pi(z - u^* x) \quad \blacksquare$$

Comment 1 - Equations (6) and (7) make the advantages stated in the introduction conspicuous. Equation (6) gives the freedom relative to the choice of u and t when the observation dynamic is fixed. More precisely, choosing u^* is equivalent to choose which linear combination of states is to be observed and choosing respectively $u^* B$, $u^* x(0)$, t^* is equivalent to quantify the effect of inputs, initial states, measurements on the estimation error. The effect of state or measurement disturbances can be quantified similarly.

2.2 Observer synthesis

From Lemma 2.1 to synthesize a minimal order observer (i.e. of order equal to $n - p$) we have :

Procedure 2.2 - *Let $\{\pi_1, \ldots, \pi_{n-p}\}$ be a set of distinct complex conjugate numbers. Choose $n - p$ vectors satisfying the three following conditions*

$$u_i \in T(\pi_i)^{\perp} \qquad (8)$$

$$u_i = \bar{u}_j \ if \ \pi_i = \bar{\pi}_j \qquad (9)$$

$$rank[C^T, u_1, \ldots, u_{n-p}] = n \qquad (10)$$

Let U , T and Π denote respectively $[u_1, \ldots, u_{n-p}]^*$, $[t_1, \ldots, t_{n-p}]^*$ and $Diag\{\pi_1, \ldots, \pi_{n-p}\}$. Then the system illustrated in Figure 1 is a minimal order observer.

Comment 2 - Note that to synthesize a n-order observer it suffices to follow Procedure 2.2 but n vectors u_i satisfying (8) and (9) must be chosen instead of $n - p$. Condition (10) is to be replaced by $rank[u_1, \ldots, u_n] = n$ and the definition of β (see Figure 1) must be replaced by $\beta = U^{-1} z$ (note that β may as well be linear combinations of the states of

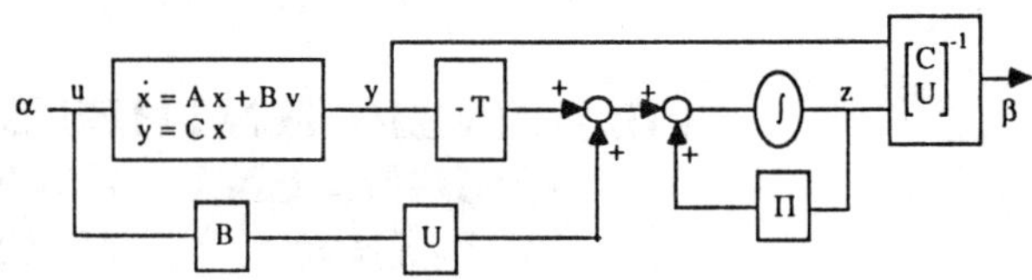

Figure 1 - $(n - p)$-order observer.

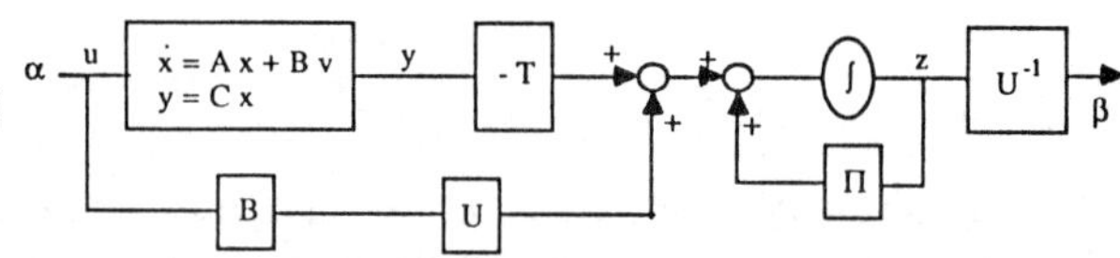

Figure 2 - n-order observer.

the observer and of the measurements). The resulting block-diagram is given Figure 2. The synthesis of Luenberger observers with order between $n - p$ and n is similar. The observation of a single linear function by mean of a $(\nu_0 - 1)$-order observer, where ν_0 denotes the observability index, proceeds easily from the fact that $\dim(\mathrm{Ker}\, C^{\perp} + T(\pi_1)^{\perp} + \ldots + T(\pi_{\nu_0-1})^{\perp}) = n$ (see [5]) which means that any vector u can be covered by a subspace spaned by $\nu_0 - 1$ vectors $u_i \in T(\pi_i)^{\perp}$ plus a vector in $\mathrm{Ker}\, C^{\perp}$. The observation of single linear functions may also be approximated by 1-dimensional observers by chosing in some subspace $T(\pi)^{\perp}$, the vector u which is the nearest to the desired one.

Now, let us state a well known result relative to observers:

Lemma 2.3 *The transfer matrix between α and β in Figures 1 and 2 is $(sI - A)^{-1} B$.*

2.3 Dual observer synthesis

To synthesize a $n - m$ order dual observer it suffices to proceed as follows :

Procedure 2.4 - *Let $\{\lambda_1, \ldots, \lambda_{n-m}\}$ be a set of distinct complex conjugate numbers. Choose $n - m$ vectors satisfying the three following conditions*

$$s_i \in \mathcal{S}(\lambda_i) \qquad (11)$$

$$s_i = \bar{s}_j \ if \ \lambda_i = \bar{\lambda}_j \qquad (12)$$

$$rank[B, s_1, \ldots, s_{n-m}] = n \qquad (13)$$

Let S , W and Λ denote respectively $[s_1, \ldots, s_{n-m}]$, $[w_1, \ldots, w_{n-m}]$ and $Diag\{\lambda_1, \ldots, \lambda_{n-m}\}$. Then the system illustrated in Figure 3 is a $(n - m)$-order dual observer.

Comment 2 is easily dualized. The corresponding n-order dual observer is given Figure 4 and Lemma 2.3 becomes:

Lemma 2.5 *The transfer matrix between α and β in Figures 3 and 4 is $C(sI - A)^{-1}$.*

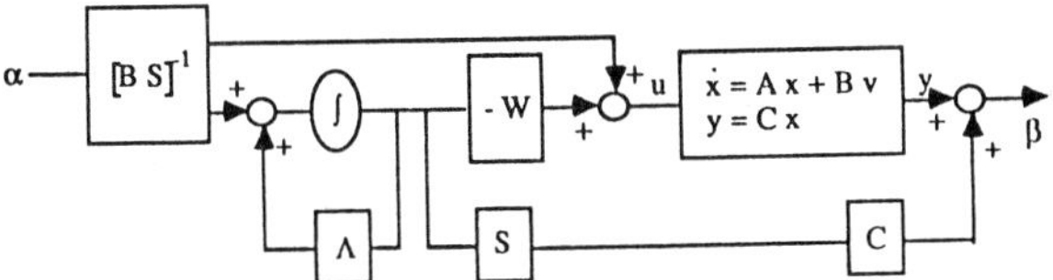

Figure 3 - $(n-m)$-order dual observer.

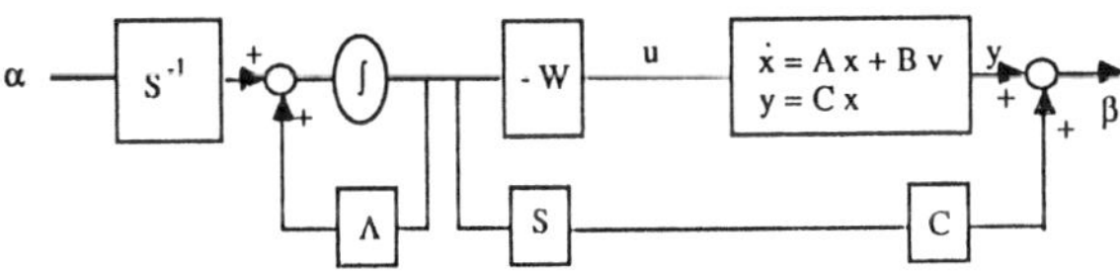

Figure 4 - n-order dual observer.

2.4 Genericity

In this section we give a technical result which is the counterpart of Lemma 3.5 in [11]. This result shows that the degrees of freedom appearing in Procedure 1 and 2 are true degrees of freedom because almost any choice of them is such that (10) or (13) holds. As already mentioned, only in order to avoid the used of the complicated notations relative to Jordan chains, we shall assume that if there exist uncontrollable or unobservable eigenvalues, they are distinct.

Proposition 2.6 - *Assume that if the pair (A,C) is unobservable, the unobservable eigenvalues are distinct. Let $\{\pi_1, \ldots, \pi_{n-p}\}$ be a set of distinct complex conjugate numbers. Then for almost any choice of the $n-p$ vectors u_i satisfying*

$$u_i \in \mathcal{T}(\pi_i)^{\perp} \text{ and } u_i = \bar{u}_j \text{ if } \pi_i = \bar{\pi}_j$$

we have

$$rank[C^T, u_1, \ldots, u_{n-p}] = n$$

if the set $\{\pi_1, \ldots, \pi_{n-p}\}$ contains all the unobservable eigenvalues of the pair (A,C). Conversely, if this set does not contain all the unobservable eigenvalues, the rank of the above matrix will always be less than n.

Proposition 2.7 - *Assume that if the pair (A,B) is uncontrollable, the uncontrollable eigenvalues are distinct. Let $\{\lambda_1, \ldots, \lambda_{n-m}\}$ be a set of distinct complex conjugate numbers. Then for almost any choice of the $n-m$ vectors s_i satisfying*

$$s_i \in \mathcal{S}(\lambda_i) \text{ and } s_i = \bar{s}_j \text{ if } \lambda_i = \bar{\lambda}_j$$

we have

$$rank[B, s_1, \ldots, s_{n-m}] = n$$

if the set $\{\lambda_1, \ldots, \lambda_{n-m}\}$ contains all the uncontrollable eigenvalues of the pair (A,B). Conversely, if this set does not contain all the uncontrollable eigenvalues, the rank of the above matrix will always be less than n.

Proof:We only prove the second Proposition.

Preliminaries. Let $\mathcal{R}^*$ denote the controllability subspace of the pair (A,B), $\iota : \mathcal{R}^* \to \mathcal{X}$ denote the canonical injection and (A_c, B_c) denote the restrictions to $\mathcal{R}^*$ of (A,B). $\mathcal{S}_c(\lambda)$ is the characteristic subspace corresponding to (A_c, B_c). Assume that the λ_i's are relabelled in such a way that $\{\lambda_{q+1}, \ldots, \lambda_{n-m}\}$ is the set of the uncontrollable eigenvalues of the pair (A,B) $(m+q = \dim \mathcal{R}^*)$. It is well known that the state space $\mathcal{X}$ can be decomposed as follows :

$$\mathcal{X} = \mathcal{R}^* \oplus < s_{\lambda_{q+1}} > \oplus \ldots \oplus < s_{\lambda_{n-m}} >$$

where s_{λ_i} denote the eigenvector of the matrix A corresponding to the uncontrollable eigenvalue λ_i. On this decomposition the mappings A and B have the following matrix representation :

$$A = \left[\begin{array}{cc} A_c & * \\ 0 & \text{Diag}\{\lambda_{q+1}, \ldots, \lambda_{n-m}\} \end{array} \right] \; ; \; B = \left[\begin{array}{c} B_c \\ 0 \end{array} \right]$$

Therefore, clearly :

$$\mathcal{S}(\lambda) = \left\{ \begin{array}{ll} \iota \mathcal{S}_c(\lambda) + s_\lambda & \text{if} \lambda \in \{\lambda_{q+1}, \ldots, \lambda_{n-m}\} \\ \iota \mathcal{S}_c(\lambda) & \text{otherwise} \end{array} \right. \tag{14}$$

The choice of the vectors s_i is assumed to be performed iteratively in the subspaces $\mathcal{S}(\lambda_i)$. If $i \leq q$, from (14), this is equivalent to choose iteratively the vectors $s_{c,i}$ $(s_i = \iota s_{c,i})$ in the subspaces $\mathcal{S}_c(\lambda_i)$. Let $\mathcal{S}_i$ denote $< s_1, \ldots, s_i >$ and for $i \leq q$, $\mathcal{S}_{c,i}$ denote $< s_{c,1}, \ldots, s_{c,i} > (\mathcal{S}_i = \iota \mathcal{S}_{c,i})$.

Proof of the "if part" : Firstly, assume that $i < q$ and that *for almost any vectors $s_{c,1} \in \mathcal{S}_c(\lambda_1), \ldots, s_{c,i} \in \mathcal{S}_c(\lambda_i)$ we have*

$$\dim(\mathcal{S}_{c,i} + \text{Im} B_c) = m + i$$

$\mathcal{S}_{c,i}$ is by definition a (A_c, B_c)-invariant subspace. As it does not intersec $\text{Im} B_c$, clearly : $\mathcal{R}^*_{\mathcal{S}_{c,i}} = 0$. Following Theorem 2.6 in [5], for all complex number λ which does not belong to $\{\lambda_1, \ldots, \lambda_i\}$, we have :

$$\dim(\mathcal{S}_c(\lambda) \cap \mathcal{S}_{c,i}) = \dim(\text{Im} B_c \cap \mathcal{S}_{c,i}) = 0$$

If $\mathcal{S}_c(\lambda_{i+1}) \subset \text{Im} B_c + \mathcal{S}_{c,i}$, from the above equality with $\lambda = \lambda_{i+1}$:

$$\mathcal{S}_c(\lambda_{i+1}) + \mathcal{S}_{c,i} = \text{Im} B_c + \mathcal{S}_{c,i}$$

As $\mathcal{S}_c(\lambda_{i+1})$ and $\mathcal{S}_{c,i}$ are (A_c, B_c)-invariant, so is $\text{Im} B_c + \mathcal{S}_{c,i}$. The pair (A_c, B_c) being controllable, it follows that this subspace must be equal to $\mathcal{R}^*$ because it contains $\text{Im} B_c$. This statement is in contradiction with the fact that by assumption $i < q$. Therefore $\mathcal{S}_c(\lambda_{i+1})$ cannot be a subset of $\text{Im} B_c + \mathcal{S}_{c,i}$ which means that *almost any vector $s_{c,i+1}$ in $\mathcal{S}_c(\lambda_{i+1})$ is such that* :

$$\dim(\mathcal{S}_{c,i+1} + \text{Im} B_c) = m + i + 1$$

Secondly, assume that $i \geq q$ and that $\dim(\mathcal{S}_i + \text{Im} B) = m + i$. In this case from (14) we still obviously have the fact that $\mathcal{S}(\lambda_{i+1})$ cannot be a subset of $\text{Im} B + \mathcal{S}_i$. We conclude as above. Finally Lemma 1 in Kimura [4] states that with the additional constraint $s_i = \bar{s}_j$ if $\lambda_i = \bar{\lambda}_j$, Proposition 2.7 still holds.

Proof of the "only if part" : As we have $\text{Im} B_c \subset \mathcal{R}^*$ and $\mathcal{S}_c(\lambda_i) \subset \mathcal{R}^*$, it is not possible to find $q+1$ vectors $s_{c,i} \in \mathcal{S}_c(\lambda_i)$ which satisfy $rank[B, s_{c,1}, \ldots, s_{c,q+1}] = m + q + 1$. ∎

3 Loop transfer recovery

In this section is analyzed the problem of loop transfer recovery defined by Doyle and Stein [3]. Our main purpopse is to derive a very simple approach for solving this problem by mean of a minimal order observer without introducing dynamics which must go to infinity. This problem will be called the *algebraic* loop transfer recovery problem by opposition to the classic *asymptotic* one.

Definition 3.1 *An observer as in Figure 1 or 2 is said to solve the loop transfer recovery problem if the transfer matrix between u and β is equal to $(sI - A)^{-1}B$.*

Definition 3.2 *A dual observer as in Figure 3 or 4 is said to solve the loop transfer recovery problem if the transfer matrix between α and y is equal to $C(sI - A)^{-1}$.*

In order to define the expectable limits in which the algebraic problem has a solution, let us first analyse the asymptotic one from a modal point of view.

3.1 The asymptotic loop transfer recovery problem

By simple inspection of Figure 2, from Lemma 2.3, it is clear that the loop transfer recovery problem can be solved asymptotically by mean of a n-order observer provided that the transfer matrix $U^{-1}(sI - \Pi)^{-1}UB$ can be made albitrarily close to the zero-transfer matrix. Similarly, by inspection of Figure 4 and from Lemma 2.5, the corresponding requirement relative to the n-order dual observers is that the transfer matrix $CS(sI - \Lambda)^{-1}S^{-1}$ can be made arbitrarily close to zero.

Let consider the second problem. We have :

$$CS(sI - \Lambda)^{-1}S^{-1} = C(sI - (A + BK))^{-1}$$

where $K = WS^{-1}$ and W is the matrix of the input directions (see (3) and Figure 4) which is defined by :

$$AS + BW = S\Lambda$$

Therefore the problem to solve is an *almost disturbance decoupling problem* which, following (Willems [9], Theorem 16), has a solution if and only if $\mathcal{X} \subset V_{b,\mathrm{Ker}\,C}^*$ where $V_{b,\mathrm{Ker}\,C}^*$ denotes the supremal almost invariant subspace "contained" in Ker C. It is shown in Malabre [6] that $V_{b,\mathrm{Ker}\,C}^* = S_{\mathrm{Im}\,B}^* + V_{\mathrm{Ker}\,C}^*$. Then the loop transfer recovery problem has an asymptotic solution if and only if $\dim(S_{\mathrm{Im}\,B}^* + V_{\mathrm{Ker}\,C}^*) = n$. Note that this is a right invertibility condition. By duality the same problem based on the use of a n-order observer has a solution if and only if $S_{\mathrm{Im}\,B}^* \cap V_{\mathrm{Ker}\,C}^* = 0$ which is a left invertibility condition. These invertibility conditions for solving the loop transfer recovery problem were as well derived in Soroka and Shaked [8] by using an alternative approach. These results are summarized in :

Proposition 3.3 *The loop transfer recovery problem can be solved* asymptotically *by mean of a n-order observer if and only if $S_{\mathrm{Im}\,B}^* \cap V_{\mathrm{Ker}\,C}^* = 0$*

Proposition 3.4 *The loop transfer recovery problem can be solved* asymptotically *by mean of a n-order dual observer if and only if $S_{\mathrm{Im}\,B}^* + V_{\mathrm{Ker}\,C}^* = \mathcal{X}$*

Comment 3 - The conditions of Proposition 3.3 and 3.4 are *generically* satisfied (see [9], Theorem 14) when respectively $m \leq p$ and $p \geq m$. See §3.3 (Example 1) for an example of a pathological systems.

Comment 4 - More generally, when we consider dual observers with order between $n - m$ and n, it is $C(sI - A - BK)^{-1}S$ which must be made arbitrarily close to zero. This is a problem of finding an almost (A, B)-invariant subspace $S = V_{b,\mathrm{Ker}\,C}$ which lies in the almost output nulling subspace of C. Note that $V_{b,\mathrm{Ker}\,C}$ must satisfy (13), i.e. Im $B + V_{b,\mathrm{Ker}\,C} = \mathcal{X}$ and that to minimize the number of poles which go to infinity the following condition is added : $V_{\mathrm{Ker}\,C}^* \subset V_{b,\mathrm{Ker}\,C}$. (See §3.3 and the appendix for an example of a construction of a subspace $V_{b,\mathrm{Ker}\,C}$.) We are now interested in solving the loop transfer recovery problem without introducing observer modes which approache infinity. From the above discussion, it appears that we must consider systems such that the subspace $V_{b,\mathrm{Ker}\,C}$ we are looking for, may reduce to $V_{\mathrm{Ker}\,C}^*$, i.e. systems satisfying Im $B + V_{\mathrm{Ker}\,C}^* = \mathcal{X}$. By duality, for finding $(n-p)$-order observers which solve the algebraic loop transfer recovery problem, we shall restrict our attention to systems such that Ker $C \cap S_{\mathrm{Im}\,B}^* = 0$.

3.2 The algebraic loop transfer recovery problem

By simple inspection of Figure 1, from Lemma 2.3, the algebraic loop transfer recovery problem has a solution by mean of a minimal order observer if and only if :

$$UB = 0 \tag{15}$$

Then we have :

Procedure 3.5 *(Apkarian et al [2]) To synthesize a $(n-p)$-order observer which solves the algebraic loop transfer recovery problem, follow Procedure 2.2 but replace (8) by :*

$$u_i \in T(\pi_i)^{\perp} \cap Im\,B^{\perp} \tag{16}$$

Clearly for the dual procedure (15) must be replaced by

$$CS = 0 \tag{17}$$

Then we have :

Procedure 3.6 *To synthesize a $(n-m)$-order dual observer, which solves the algebraic loop transfer recovery problem, follow Procedure 2.4 but replace (11) by :*

$$s_i \in \mathcal{S}(\lambda_i) \cap Ker\,C \tag{18}$$

In the previous subsection, we have justified that to apply Procedure 3.5, it was necessary to assume that Ker $C \cap S_{\mathrm{Im}\,B}^* = 0$ which is clearly equivalent to Ker $C \cap Im\,B = 0$. This property is assumed to be satisfied in the following proposition which justifies Procedure 3.5.

Proposition 3.7 *Assume that the triple (A, B, C) is such that $Ker\,C \cap Im\,B = 0$ and that its invariant zeros are distinct. Let $\{\pi_1, \ldots, \pi_{n-p}\}$ be a set of distinct complex conjugate numbers. Then for almost any choice of the $n - p$ vectors u_i satisfying*

$$u_i \in T(\pi_i)^{\perp} \cap Im\,B^{\perp} \quad and \quad u_i = \bar{u}_j \ if \ \pi_i = \bar{\pi}_j$$

we have

$$rank[C^T, u_1, \ldots, u_{n-p}] = n$$

if the set $\{\pi_1, \ldots, \pi_{n-p}\}$ contains all the invariant zeros of the triple (A, B, C). Conversely, if this condition is not satisfied, the rank of the above matrix will always be less than n.

Proposition 3.8 *Assume that the triple (A, B, C) is such that $Ker\,C + Im\,B = \mathcal{X}$ and that its invariant zeros are distinct. Let $\{\lambda_1, \ldots, \lambda_{n-m}\}$ be a set of distinct complex conjugate numbers. Then for almost any choice of the $n - m$ vectors s_i satisfying*

$$s_i \in \mathcal{S}(\lambda_i) \cap Ker\,C \ \text{ and } \ s_i = \bar{s}_j \ \text{ if } \ \lambda_i = \bar{\lambda}_j$$

we have

$$rank[B, s_1, \ldots, s_{n-m}] = n$$

if the set $\{\lambda_1, \ldots, \lambda_{n-m}\}$ contains all the invariant zeros of the triple (A, B, C). Conversely, if this condition is not satisfied, the rank of the above matrix will always be less than n.

Proof: The hypothesis $Ker\,C + Im\,B = \mathcal{X}$ has the obvious following consequence :

$$\mathcal{V}^*_{Ker\,C} = Ker\,C \tag{10}$$

therefore, there exists a state feedback K_0 such that

$$\hat{A}\,Ker\,C \subset Ker\,C$$

where $\hat{A}$ stands for $A + BK_0$. Let $(A_{Ker\,C}, B_{Ker\,C})$ denote the restriction to $Ker\,C$ of the pair $(\hat{A}, B)$ and ι' denote the canonical injection from $Ker\,C$ to $\mathcal{X}$. For some subspace Q, the state space $\mathcal{X}$ can be decomposed as follows :

$$\mathcal{X} = Ker\,C \oplus Q$$

On this decomposition, the mappings A and B have the following matrix representation :

$$A = \begin{bmatrix} A_{Ker\,C} & * \\ 0 & * \end{bmatrix} \ ; \ B = \begin{bmatrix} B_{Ker\,C} & * \\ 0 & * \end{bmatrix}$$

Then it is clear that the corresponding characteristic subspaces denoted $\mathcal{S}_{Ker\,C}(\lambda)$ satisfy:

$$\iota'\mathcal{S}_{Ker\,C}(\lambda) = \mathcal{S}(\lambda) \cap Ker\,C \tag{20}$$

From (19), the non controllable eigenvalues of the pair $(A_{Ker\,C}, B_{Ker\,C})$ are the invariant zeros of the triple (A, B, C) (see Anderson [1]). Therefore, from (20), Proposition 2.7 applied to $(A_{Ker\,C}, B_{Ker\,C})$ concludes the proof of Proposition 3.8. ■

Comment 5 - We have not dicussed the *stability* properties of the observers or dual observers considered all along this paper. Clearly stability in Propositions 2.6 and 2.7 is subject to the fact that respectively the pair (A, C) is detectable and the pair (A, B) is stabilizable. Similarly, stability in Propositions 3.7 and 3.8 is subject to the fact that (A, B, C) is minimum phase. Note that for non-minimum phase systems, the loop transfer recovery problem can be solved approximately by choosing stable eigenvalues instead of the zeros which lie in the right half plane. In this case, Equations (16) or (18) must not be solved exactly i.e. the vectors u_i or s_i must be close to $Im\,B^\perp$ or $Ker\,C$ but not inside these subspaces (otherwise the rank conditions of Propositions 3.7 and 3.8 cannot be satisfied) .

Comment 6 - The synthesis procedures proposed in this section can be used for squaring up the systems to be treated by the classical loop transfer recovery approach. For instance if $m < p$ and $Ker\,C \cap Im\,B = 0$, Procedure 3.5 can be used to compute a $(n - p \times n)$ matrix U. Then the augmented matrix "B" can be any matrix the columns of which span $Ker\,U$. In practice, the n-order observers obtained by the classical loop transfer recovery approaches are generally simplified by order reduction. The discussion given in Comment 4 gives a limit to this order reduction.

3.3 Numerical examples

Example 1 - This numerical example illustrates the problem which arises for systems which are not left invertible. Consider the following triple :

$$A = \begin{bmatrix} 0 & 1 & 0 & 0 \\ 0 & 0 & 1 & 0 \\ 0 & 0 & 0 & 1 \\ 0 & 0 & 0 & 0 \end{bmatrix} \ ; B = \begin{bmatrix} 0 & 0 \\ 0 & 1 \\ 0 & 0 \\ 1 & 0 \end{bmatrix} \ ; C = \begin{bmatrix} 1 & 0 & 0 & 0 \\ 0 & 1 & 0 & 0 \end{bmatrix}$$

It is straighforward to check that it does not satisfy the condition of Proposition 3.3. Then the asymptotic loop transfer recovery has no solution. This result can be checked by using the classical approach of [3] : when q approaches infinity, the gain $G(q) = \Sigma C^T$ of the Kalman filter corresponding to the following Riccati equation

$$A\Sigma + \Sigma A^T + q^2 B B^T - \Sigma C^T C \Sigma = 0$$

satisfies

$$\frac{G^T(q)}{q} \longrightarrow \begin{bmatrix} 0 & 0 & 0 & 0 \\ 0 & 1 & 1.414 & 1 \end{bmatrix}$$

which is not equal to B as claimed in [3]. (Note that this system is implicitly excluded in [3] because its "zero polynomial" is identically equal to zero.)

Example 2 - This second example illustrates Procedure 3.6. Consider the pair (A, C) of example 1 with the following input matrix B :

$$B^T = \begin{bmatrix} 1 & 1 & 2 & 0 \\ -3 & -1 & 0 & 4 \end{bmatrix}$$

This minimum phase system has two zeroes which are -1 and -2. Therefore to find the two-order dual observer which soves the algebraic loop transfer recovery problem it suffices to find s_i , $i = 1, 2$, which satisfy :

$$s_i \in \mathcal{S}(\lambda_i) \cap Ker\,C$$

for $\lambda_1 = -1$ and $\lambda_2 = -2$. This is equivalent to solve

$$\begin{bmatrix} A - \lambda_i I & B \\ C & 0 \end{bmatrix} \begin{bmatrix} s_i \\ w_i \end{bmatrix} = 0 \tag{21}$$

which gives simultaneously the vectors s_i and w_i. The following numerical values are obtained : $s_1 = [0 \ 0 \ -2 \ -4]^T$, $w_1 = [3 \ 1]^T$, $s_2 = [0 \ 0 \ -2 \ -2]^T$ $w_2 = [3 \ 1]^T$. Therefore the matrices S, W and Λ of Figure 3 must be :

$$S = \begin{bmatrix} s_1 & s_2 \end{bmatrix} \ ; \ W = \begin{bmatrix} w_1 & w_2 \end{bmatrix} \ ; \ \Lambda = Diag\{-1, -2\}$$

Example 3 - We consider the above example but with an actuator with dynamic equal to -3 on the first input. It is easily checked that

$$\mathcal{V}^*_{Ker\,C} = Im \begin{bmatrix} 0 & 0 \\ 0 & 0 \\ 1 & 0 \\ 0 & 1 \\ -1.5 & 0 \end{bmatrix} \ ; \ \mathcal{S}^*_{Im\,B} = Im \begin{bmatrix} 1 & -3 & 0 \\ 1 & -1 & 0 \\ 2 & 0 & 0 \\ 0 & 4 & 0 \\ 0 & 0 & 1 \end{bmatrix}$$

As we have $\mathcal{V}^*_{b, Ker\,C} = \mathcal{V}^*_{Ker\,C} + \mathcal{S}^*_{Im\,B} = \mathcal{X}$ there exists an asymptotic solution. Clearly :

$$\mathcal{V}^*_{b, Ker\,C} = \mathcal{V}^*_{Ker\,C} + Im\,B + Im\,ABw$$

where w is any two-dimensional vector with non-zero first element. Let $\mathcal{V}^*_{Ker\,C} + Im\,Bw + Im\,ABw$ be the almost (A, B)-invariant subspace "contained" in $Ker\,C$ denoted $\mathcal{V}_{b, Ker\,C}$ in Comment 4. To generate a family of (A, B)-invariant subspaces parametrized by ϵ which converges toward $\mathcal{V}_{b, Ker\,C}$ in the sense of the Grassmannian topology (see [9] for a definition) when ϵ approaches 0, it suffices to proceed as in example

2 for the two pairs of vectors (v_1, w_1) and (v_2, w_2) corresponding to the zeroes $\lambda_1 = -1$ and $\lambda_2 = -2$ by solving (21). For the third vector it suffices to consider $w_3(\epsilon) = w/\epsilon$ and $\lambda_3(\epsilon) = -1/\epsilon$ and to solve for $s_3(\epsilon)$ (see the appendix) :

$$\left[A - \lambda_3 I \quad B \right] \left[\begin{array}{c} s_3 \\ w_3 \end{array} \right] = 0$$

The following numerical values are obtained : $s_1 = [0 \ \ 0 \ - 2 - 4 \ \ 3]^T$, $w_1 = [2 \ \ 1]^T$, $s_2 = [0 \ \ 0 - 2 - 2 \ \ 3]^T$, $w_2 = [1 \ \ 1]^T$ and

$$s_3(\epsilon) = \frac{1}{3\epsilon - 1} \left[\begin{array}{c} -3\epsilon(1 - \epsilon + 2\epsilon^2) \\ -3\epsilon(1 - 2\epsilon) \\ -6\epsilon \\ 0 \\ 3 \end{array} \right] ; \ w_3(\epsilon) = \left[\begin{array}{c} 1/\epsilon \\ 0 \end{array} \right]$$

Therefore the matrices S, W and Λ of Figure 3 follow easily. By multiplying the matrix $[Bw \quad , s_3(\epsilon)]$ on the right hand side by an appropriate matrix it is easily checked that, as expected, the limit of $< Bw \quad , s_3(\epsilon) >$ is $< Bw, ABw >$ (in the sense recalled above).

APPENDIX

In this appendix we give a simple method for parametrizing families of (A, B)-invariant subspaces $\mathcal{V}(\epsilon)$ which tend toward a given *almost* (A, B)-invariant subspace $\mathcal{V}_b$ when ϵ approaches zero. This convergence should be understood in the Grassmannian sense (see [9]).

First, let us recall that any almost (A, B)-invariant subspace may be written as :

$$\mathcal{V}_b - \mathcal{V}_0 + \mathcal{B}_1 + \hat{A}\mathcal{B}_2 + \hat{A}^2\mathcal{B}_3 \ | \ ... \quad (22)$$

where $\hat{A} = A + BK$ for some state feedback K, $\{\mathcal{B}_i\}$ is a chain $(\text{Im } B \supset \mathcal{B}_1 \supset \mathcal{B}_2 \supset \mathcal{B}_3...)$ and $\mathcal{V}_0$ is some (A, B)-invariant subspace. For instance we have

$$\mathcal{V}^*_{b, \text{Ker } C} = \mathcal{V}^*_{\text{Ker } C} + \mathcal{S}^*_{\text{Im } B} = \mathcal{V}^*_{\text{Ker } C} + \text{Im } B + \hat{A}\mathcal{B}_2 + \hat{A}^2\mathcal{B}_3 + ...$$

For parametrizing a family of subspaces which converges toward $\mathcal{V}_b$ as in (22), it suffices to be able to parametrize a "subfamily" which converges toward :

$$\mathcal{V}'_b = \mathbf{b} + \hat{A}\mathbf{b} + ... + \hat{A}^{k-1}\mathbf{b} \quad (23)$$

where $\mathbf{b}$ is a one-dimensional subspace of $\text{Im } B$.

Proposition 3.9 *Assume that a set of k selfconjugate complex numbers $\{a_1, ..., a_k\}$ is chosen. Let w be a m-dimensional vector such that $\mathbf{b} = \text{Im } Bw$. Then the vectors $s_i(\epsilon)$ defined by :*

$$\left[\hat{A} - \frac{a_i}{\epsilon} I \quad B \right] \left[\begin{array}{c} s_i(\epsilon) \\ \frac{w}{\epsilon} \end{array} \right] = 0$$

are such that

$$\lim_{\epsilon \to 0} < s_1(\epsilon), ..., s_k(\epsilon) > = \mathcal{V}'_b$$

The corresponding state feedback is any matrix $K(\epsilon)$ continuous with respect to ϵ which satisfies :

$$K(\epsilon)[s_1(\epsilon), ..., s_k(\epsilon)] = \frac{1}{\epsilon}[w, ..., w]$$

To prove this result it suffices to consider the canonical form of the controllable part of the pair $(\hat{A}, \mathbf{b})$. On the corresponding basis, the vectors $s_i(\epsilon)$ have the following form ([5])

$$s_i(\epsilon) = [1, \frac{a_i}{\epsilon}, ..., \frac{a_i^{l-1}}{\epsilon^{l-1}}, 0, ..., 0]^T$$

where $l (\geq k)$ is the dimension of the controllable subspace. (Note that it is assumed that the complex numbers a_i/ϵ are not uncontrollable eigenvalues as this is always true for ϵ sufficiently small). On the same state space basis the subspace $\mathcal{V}'_b$ is given by

$$\mathcal{V}'_b = \text{Im } \left[0_{(l-k, k)} \quad I_k \quad 0_{(n-l, k)} \right]^T$$

Therefore the proof of Proposition 3.9 is a part of the proof of Theorem 6 in [9]. In Example 3, is used the following slightly modified version of this Proposition :

Corollary 3.10 *Assume that a set of $k - 1$ selfconjugate complex numbers $\{a_2, ..., a_k\}$ is chosen. Let w be a m-dimensional vector such that $\mathbf{b} = \text{Im } Bw$. Then the vectors $s_i(\epsilon)$ defined as in Proposition 3.9 are such that*

$$\lim_{\epsilon \to 0} \mathbf{b} + < s_2(\epsilon), ..., s_k(\epsilon) > = \mathcal{V}'_b$$

References

[1] B.D.O. Anderson. A note on transmission zeros of a transfer function matrix. *IEEE Transactions on Automatic Control*, AC-25:589–591, 1980.

[2] P. Apkarian, C. Champetier, and J.F. Magni. Design of a helicopter output feedback control law ... *to appear in Int. J. Control*, 1990.

[3] J.C. Doyle and G. Stein. Robustness with observers. *IEEE Transactions on Automatic Control*, AC-24:607–611, 1979.

[4] H. Kimura. Pole assignment by gain output feedback. *IEEE Transactions on Automatic Control*, AC-20:509–516, 1975.

[5] J.F. Magni and C. Champetier. A general framework for pole assignment algorithms. in *Proc. 27th I.E.E.E. conf. Decision Contr., Austin*, pages 2223–2229, December 1988.

[6] M. Malabre. Almost invariant subspaces, transmission and infinite zeros : a lattice interpretation. *Systems and Control Letters*, 1(6):347–355, 1982.

[7] B.C. Moore. On the flexibility offered by state feedback in multivariable system beyond ... *IEEE Transactions on Automatic Control*, AC-21:659–692, 1976.

[8] E. Soroka and U. Shaked. The LQG optimal regulation problem for systems with perfect measurements ... *IEEE Transactions on Automatic Control*, AC-33:941–944, 1988.

[9] J.C. Willems. Almost invariant subspaces : an approach to high gain feedback desing. Part1... *IEEE Transactions on Automatic Control*, AC-26:235–252, 1981.

[10] W.M. Wonham. Dynamic observers-geometric theory. *IEEE Transactions on Automatic Control*, AC-15:258–259, 1970.

[11] W.M. Wonham. *Linear Multivariable Control, a Geometric Approach*. Springer Verlag, New York Heidelberg Berlin, 1979.

A LYAPUNOV DESIGN APPROACH TO A STABLE POLE-PLACER FOR A MULTIVARIABLE NONMINIMUM-PHASE SYSTEM

N. Minamide*, J. Tsuboi* and P. N. Nikiforuk**

**Department of Electronics, Faculty of Engineering, Mie University, Kamihama-cho, Tsu-city, 514, Japan*
***Department of Mechanical Engineering, College of Engineering, University of Saskatchewan, Saskatoon, Saskatchewan, S7N 0W0, Canada*

Abstract. This paper proposes a new adaptive control strategy which is appliable to tho class of multivariable nonminimum phase systems described by proper full rank transfer matrices and may be implemented with a priori knowledge of the upper bound of the system observability indices. By representing the error system in state space form and introducing auxiliary signals superimposed in the error equation it is shown that there exist an appropriate parameter adjustment law and appropriate auxiliary signals so that the Lyapunov function to the error system can be constructed. A global asymptotic stability of a multi-input and multi-output nonminimum phase system is then established and the poles of the closed-loop system are shown to be adaptively assigned to the desired location.

Key words. Strictly positive real system; Lyapunov design method; matrix fraction description; global asymptotic stability; adaptive pole-placer

1. Introduction

The global stability of a single-input and single-output model reference adaptive control system has recently been established and has been extended to a multi-input, multi-output system. In order that the global stability can be guaranteed, the unknown system has to be minimum phase. This requirement renders the model reference approach as being impractical in many applications. This paper proposes a new adaptive control strategy which is applicable to the class of multivariable systems described by proper full rank transfer matrices and can be implemented with a priori knowledge of the upper bound of the system observability index.

A design methodology for adaptively assigning the closed-loop poles of a discrete time linear scalar system of known order has been recently proposed (Kreissermeier(1988)). Little has been done on extending these techniques to multi-input multi-output systems, although a globally stable indirect adaptive pole-placer for the multivariable nonminimum phase system of unknown order has recently been proposed using the idea of the dyadic controller (Minamide (1989)).

In this paper, the unknown system is modeled by the use of a column-reduced right coprime matrix fraction description of the system transfer matrix. One of the main contributions of the present paper is the development of a linear error equation for estimating the appropriate pole-placing feedback gains based on the sole assumption of an upper bound of the system observability index.

An important issue in the design of adaptive control systems is that of establishing the global stability of the overall system. Here, a Lyapunov design method is developed which uses the properties of the strictly positive real transfer function and the auxiliary signals superimposed in the error system. This method can adaptively assign the closed-loop poles of a discrete-time multivariable system and, at the same time, guarantee the global asymptotic stability of the overall system.

2. Problem Statement

Let the unknown system to be controlled be described by

$$A(q^{-1})z_t = u_t, \quad y_t = B(q^{-1})z_t \qquad (1)$$

where $u(t)$ is an $r \times 1$ input vector, $y(t)$ is an $m \times 1$ output vector, z_t is an $r \times 1$ partial state and $A(q^{-1})$, $B(q^{-1})$ are polynomial matrices in the delay operator q^{-1}.

The following a priori information on the system (1) is used:

A_1) The polynomial matrices $A(q^{-1})$ and $B(q^{-1})$ with $A(q^{-1})$ column-reduced constitute an irreducible right MFD (matrix fraction description) of the proper transfer function $G(q^{-1}) = B(q^{-1})A^{-1}(q^{-1})$.

A_2) Let $G(q^{-1}) = N^{-1}(q^{-1})M(q^{-1})$ be an irreducible left MFD of the transfer function $G(q^{-1})$ with $N(q^{-1})$ row-reduced. Then, the upper bound μ of the observability index $\mu_0 = \max_{1 \leq j \leq m} \partial_{rj}[N(q^{-1})]$ is known where $\partial_{rj}[N(q^{-1})]$ denotes the polynomial degree in the j-th row of $N(q^{-1})$.

For $j=1, 2, \cdots, r$, let $d_j^*(q^{-1})$ be a stable monic polynomial having at least one real root and of degree $\partial\{d_j^*(q^{-1})\} = \mu$ and let

$$D(q^{-1}) = \text{diag}[\, d_1^*(q^{-1}) \cdots d_r^*(q^{-1}) \,] \qquad (2)$$

The objective of control is to design an appropriate feedback control law which replaces the system poles by the desired ones determined by the zeros of $\det\{D(q^{-1})\}$.

<u>3. Basic Configuration</u>

Let it be assumed first that the system parameters are known and let a non-adaptive feedback control law be derived which assigns the poles of the systems (1) to the desired location.

For $j=1, 2, \cdots, r$, let $b_j^*(q^{-1})$ be a stable monic polynomial having at least one real root and of degree $\partial\{b_j^*(q^{-1})\} \leq \mu$ such that $a_j^*(q^{-1})$ and $b_j^*(q^{-1})$ are coprime and $\{b_j^*(q^{-1})/a_j^*(q^{-1})\}$ is strictly positive real.   Let

$$B^*(q^{-1}) = \text{diag}[\, b_1^*(q^{-1}) \cdots b_r^*(q^{-1}) \,]. \qquad (3)$$

[Lemma 1]

Consider the following set of the Diophantine equations.

$$- aR(q^{-1})A(q^{-1}) - aS(q^{-1})B(q^{-1})$$

$$- bP(q^{-1})A(q^{-1}) - bQ(q^{-1})B(q^{-1})$$

$$= D(q^{-1})A(0) - A(q^{-1}) \qquad (4)$$

$$cR(q^{-1})A(q^{-1}) + cS(q^{-1})B(q^{-1})$$

$$+ dP(q^{-1})A(q^{-1}) + dQ(q^{-1})B(q^{-1})$$

$$= B^*(q^{-1})A(0) - A(q^{-1}) \qquad (5)$$

where a, b, c and d are prespecified design parameters satisfying $ad-bc \neq 0$.   Then, (4) and (5) have a solution of the form

$$R(q^{-1}) = R_1 q^{-1} + \cdots + R_\mu q^{-\mu}, \quad R_j \epsilon R^{r \times r}$$

$$S(q^{-1}) = S_1 q^{-1} + \cdots + S_\mu q^{-\mu}, \quad S_j \epsilon R^{r \times m}$$

$$P(q^{-1}) = P_1 q^{-1} + \cdots + P_\mu q^{-\mu}, \quad P_j \epsilon R^{r \times r}$$

$$Q(q^{-1}) = Q_1 q^{-1} + \cdots + Q_\mu q^{-\mu}, \quad Q_j \epsilon R^{r \times m}$$

$$(j=1, 2, \cdots, \mu) \qquad (6)$$

<u>Proof</u>

The proof is omitted as it is relatively easy to produce.

To derive a pole-placing control law, multiplying (4) by z_t yields

$$z_t = A^{-1}(0)D^{-1}(q^{-1})\{\, u_t - aR(q^{-1})u_t$$

$$- aS(q^{-1})y_t - bP(q^{-1})u_t - bQ(q^{-1})y_t \,\} \qquad (7)$$

Similarly, multiplying (5) by z_t and using (7) yields

$$y_t^* \overset{\Delta}{=} u_t + cR(q^{-1})u_t + cS(q^{-1})y_t$$

$$+ dP(q^{-1})u_t + dQ(q^{-1})y_t$$

$$= B^*(q^{-1})A(0)z_t$$

$$= B^*(q^{-1})D^{-1}(q^{-1})\{\, u_t - aR(q^{-1})u_t$$

$$- aS(q^{-1})y_t - bP(q^{-1})u_t - bQ(q^{-1})y_t \,\} \qquad (8)$$

where y_t^* is a pseudo-output used as a target signal. Thus, the system poles will be replaced by the desired ones when the feedback control law

$$u_t = aR(q^{-1})u_t + aS(q^{-1})y_t$$

$$+ bP(q^{-1})u_t + bQ(q^{-1})y_t + r_t$$

$$\overset{\Delta}{=} aH^*\phi_t + bK^*\phi_t + r_t \qquad (9)$$

is used, where

$$H^* = [\, S_1 \cdots S_\mu \; R_1 \cdots R_\mu \,]$$

$$K^* = [\, Q_1 \cdots Q_\mu \; P_1 \cdots P_\mu \,] \qquad (10)$$

$$\phi_t' = [\, y_{t-1}' \cdots y_{t-\mu}' \; u_{t-1}' \cdots u_{t-\mu}' \,]$$

and r_t is a uniformly bounded external input. Using the control law (9), the system output y_t and the pseudo-output y_t^* have the following expressions, respectively.

$$y_t = B(q^{-1})z_t = B(q^{-1})A^{-1}(0)D^{-1}(q^{-1})r_t \qquad (11)$$

$$y_t^* = B^*(q^{-1})D^{-1}(q^{-1})r_t \qquad (12)$$

206

It is noted here that y_t^* is independent of the system parameters and can be generated as the target signal.

In adaptive control, the coefficients of the polynomial matrices $R(q^{-1})$, $S(q^{-1})$, $P(q^{-1})$ and $Q(q^{-1})$ in (8) and (9) are unknown and these are replaced by the estimated ones:

$$\hat{y}_t^* = u_t + c\hat{R}(q^{-1})u_t + c\hat{S}(q^{-1})y_t$$
$$+ d\hat{P}(q^{-1})u_t + d\hat{Q}(q^{-1})y_t + \eta_t$$
$$= u_t + c\hat{H}_t\phi_t + d\hat{K}_t\phi_t + \eta_t \qquad (13)$$

$$u_t = a\hat{R}_t(q^{-1})u_t + a\hat{S}_t(q^{-1})y_t$$
$$+ b\hat{P}_t(q^{-1})u_t + b\hat{Q}_t(q^{-1})y_t + r_t + \zeta_t$$
$$\overset{\Delta}{=} a\hat{H}_t\phi_t + b\hat{K}_t\phi_t + r_t + \zeta_t \qquad (14)$$

where the caret denotes the estimate of the corresponding quantity, and η_t and ζ_t are auxiliary signals to be specified later.

4. Representation of the error system

Nothing is said yet about how to estimate the appropriate pole-placing feedback gains. In order that this can be done, the state space representation of the error system which, in turn, leads to estimation of the feedback gains, will now be derived.

For this purpose, the feedback control law (14) is rewritten as

$$u_t = a\hat{H}_t\phi_t + b\hat{K}_t\phi_t + r_t + \zeta_t$$
$$= \{ A_0(q^{-1}) - D_0(q^{-1})A(0) \} z_t$$
$$+ a\overline{H}_t\phi_t + b\overline{K}_t\phi_t + r_t + \zeta_t \qquad (15)$$

where

$$A(q^{-1}) = A(0) + A_0(q^{-1})$$
$$D(q^{-1}) = I + D_0(q^{-1})$$

and the bar denotes the parameter mismatch. Similarly, using (1) and (15) yields

$$\hat{y}_t^* = B^*(q^{-1})A(0)z_t + c\overline{H}_t\phi_t + d\overline{K}_t\phi_t + \eta_t$$

$$= B_0^*(q^{-1})A(0)z_t - A_0(q^{-1})z_t + u_t + c\overline{H}_t\phi_t$$
$$+ d\overline{K}_t\phi_t + \eta_t$$
$$= \{ B_0^*(q^{-1}) - D_0(q^{-1}) \} A(0)z_t$$
$$+ (c+a)\overline{H}_t\phi_t + (d+b)\overline{K}_t\phi_t + r_t + \zeta_t + \eta_t \qquad (16)$$

where

$$B^*(q^{-1}) = I + B_0^*(q^{-1})$$

To derive a state space representation, it is to be noted first that from (1) and (15)

$$A(0)z_t = - A_0(q^{-1})z_t + u_t$$
$$= - D_0(q^{-1})A(0)z_t + a\overline{H}_t\phi_t + b\overline{K}_t\phi_t$$
$$+ r_t + \zeta_t \qquad (17)$$

Thus, taking as a state vector

$$x_t = \begin{bmatrix} A(0)z_{t-1} \\ A(0)z_{t-2} \\ \vdots \\ A(0)z_{t-\mu} \end{bmatrix} \in R^{r\mu}$$

gives

$$x_{t+1} = Ax_t + B(a\overline{H}_t\phi_t + b\overline{K}_t\phi_t + r_t + \zeta_t)$$
$$\hat{y}_t^* = Cx_t + D(a\overline{H}_t\phi_t + b\overline{H}_t\phi_t + r_t + \zeta_t)$$
$$+ c\overline{H}_t\phi_t + d\overline{K}_t\phi_t + \eta_t$$

where

$$A = \begin{bmatrix} -D_1 & \cdots & \cdots & -D_\mu \\ I & & & \\ & I & 0 & \\ & & \ddots & 0 \\ 0 & & I & \end{bmatrix}, \quad B = \begin{bmatrix} I \\ 0 \\ \vdots \\ 0 \end{bmatrix}$$

$$C = [B_1^*-D_1, \ B_2^*-D_2, \cdots, B_\mu^*-D_\mu], \quad D = I$$

and

$$B_0^*(q^{-1}) - D_0(q^{-1})$$
$$= (B_1^* - D_1)q^{-1} + \cdots + (B_\mu^* - D_\mu)q^{-\mu}$$

By direct calculation it is seen that the following identity holds.

$$B^*(q^{-1})D^{-1}(q^{-1}) = C(I - q^{-1}A)^{-1}B + D \qquad (18)$$

where, since $B^*(q^{-1})D^{-1}(q^{-1})$ is irreducible by assumption and $\partial\{\det D(q^{-1})\} = r\mu$, (18) is a minimal realization of $B^*(q^{-1})D^{-1}(q^{-1})$. Let

$$x_{t+1}^d = Ax_t^d + Br_t$$

$$y_t^d = Cx_t^d + Dr_t$$

be a state space representation of the reference output $y_t^* = B^*(q^{-1})D^{-1}(q^{-1})r_t$. Then, an error e_t, defined as the deviation of the actual target signal from the reference target signal

$$e_t = \hat{y}_t^* - y_t^* = \hat{y}_t^* - B^*(q^{-1})D^{-1}(q^{-1})r_t \qquad (19)$$

has the following state space representation

$$x_e(t+1) = Ax_e(t) + B(\ a\bar{H}_t\phi_t + b\bar{K}_t\phi_t + \zeta_t)$$
$$\qquad\qquad\qquad\qquad\qquad\qquad (20)$$
$$e_t = Cx_e(t) + D(\ a\bar{H}_t\phi_t + b\bar{K}_t\phi_t + \zeta_t)$$
$$+ c\bar{H}_t\phi_t + d\bar{K}_t\phi_t + \eta_t$$

where $x_e(t) = x_t^* - x_t^d$ is an error state vector. The error e_t also has the transfer function representation

$$e_t = B^*(q^{-1})D^{-1}(q^{-1})(\ a\alpha_t + b\beta_t + \zeta_t)$$
$$+ c\alpha_t + d\beta_t + \eta_t \qquad (21)$$

where the following short hand notations are introduced.

$$\alpha_t = \bar{H}_t\phi_t, \quad \beta_t = \bar{K}_t\phi_t$$

5. Some preliminary results

In the previous section, the representation of the error system in state space form has been developed. For establishing a global asymptotic stability of the error system, a construction of a suitable Lyapunov function is a key technical step. To help the calculation of the Lyapunov function, the following basic lemmas are first prepared.

[Lemma 2]

Given the error system

$$x_e(t+1) = Ax_e(t) + Bw_1(t)$$
$$\qquad\qquad\qquad\qquad\qquad (22)$$
$$e_t = Cx_e(t) + Dw_1(t) + w_2(t)$$

where $w_1(t) = a\alpha_t + b\beta_t + \zeta_t$ and $w_2(t) = c\alpha_t + d\beta_t + \eta_t$. Suppose that the transfer function $F(q^{-1}) = C(I - q^{-1}A)^{-1}B + D$ is strictly positive real, or equivalently that there exist a symmetric positive definite matrix P, a symmetric positive definite matrix Q, and matrices K and L such that

$$A'PA - P = -LL' - Q$$

$$B'PA + K'L' = C$$

$$K'K = D + D - B'PB$$

Let $V_1(t)$ be a positive scalar function defined by

$$V_1(t) = x_e'(t)Px_e(t) \qquad (23)$$

Then, the time difference $\Delta V_1(t) = V_1(t+1) - V_1(t)$ along the trajectory (22) takes the form:

$$\Delta V_1(t) = -x_e'(t)Qx_e(t) - \| L'x_e(t) - Ku_t \|^2$$
$$+ 2\{e_t - w_2(t)\}'w_1(t) \qquad (24)$$

The following lemma, which characterizes one of the useful properties of the adaptation scheme introduced below, will now be stated.

[Lemma 3]

Consider the following adaptation scheme:

$$\hat{H}_t = \hat{H}_{t-1} - \epsilon_1(t)\phi_t' \qquad (25.a)$$

$$\hat{K}_t = \hat{K}_{t-1} - \frac{\epsilon_2(t)\phi_t'}{1+\lambda\| \phi_t \|^2} \qquad (25.b)$$

Let

$$V_2(t) = \text{trace}\{\ \bar{H}_t'\bar{H}_t\ \} + \text{trace}\{\ \bar{K}_t'\bar{K}_t\ \} \qquad (26)$$

Then the time difference $\Delta V_2(t) = V_2(t) - V_2(t-1)$ is evaluated as

$$\Delta V_2(t) = -2\alpha_t'\epsilon_1(t) - 2\beta_t'\epsilon_2(t) - \lambda\beta_t'\beta_t$$
$$+ \lambda\beta_t'\beta_0(t-1) + \{\| \epsilon_1(t) \|^2$$

$$+ \frac{\| \varepsilon_2(t) \|^2}{1+\lambda| \phi_t |^2} \} \, \| \phi_t \|^2 \qquad (27)$$

where

$$\beta_0(t-1) = (\hat{K}_{t-1} - K^*)\phi_t = \overline{K}_{t-1}\phi_t \qquad (28)$$

6. Calculation of the Lyapunov function

One method will now be proposed that can guarantee the global asymptotic stability of the overall system by constructing the Lyapunov function to the error system (20) with the parameter adjustment law as in (25). In particular, consider the following set of equations:

$$e_t = F(q^{-1}) \{ \alpha_t + \beta_t + \zeta_t \} + \alpha_t + \eta_t \qquad (29)$$

$$\hat{H}_t = \hat{H}_{t-1} - \{ e_t - \xi_t \} \phi_t' \qquad (30.a)$$

$$\hat{K}_t = \hat{K}_{t-1} - \{ e_t - \zeta_t \} \phi_t'/(1+\sigma_t) \qquad (30.b)$$

where $\sigma_t=|\phi_t|^2$ and ζ_t, η_t and ξ_t are vector-valued functions to be determined appropriately. It is to be noted that (29) and (30) are obtained from (21) and (25) by letting

$$a=1, \quad b=1, \quad c=1, \quad d=0$$

$$\varepsilon_1(t) = e_t - \xi_t, \quad \varepsilon_2(t) = e_t - \zeta_t, \quad \lambda = 1$$

Now, as a possible candidate of the Lyapunov function, consider

$$V(t) = \{ V_1(t) + V_2(t) \}/2$$

where $V_1(t)$ and $V_2(t)$ are defined in (23) and (26), respectively. Using the lemma 2 and the lemma 3, the time defference $\Delta V(t)$ is calculated as

$$\Delta V(t) = n(t) + \beta_0'(t-1)\beta_0(t-1)/2$$
$$+ e_t'\zeta_t - \zeta_t'\eta_t + \| \zeta_t + \eta_t - \xi_t \|^2/2 \qquad (31)$$

where $n(t)$ represents the sum of the negative terms:

$$n(t) = - [x_e(t)Qx_e(t) + \| L'x_e(t) - Ku_t \|^2$$
$$+ \{\| \varepsilon_1(t) \|^2 + \| \varepsilon_2(t) \|^2/(1+\| \phi_t \|^2)\}$$

$$\cdot \| \phi_t \|^2 + \| \alpha_t + \beta_t \|^2 + \| \alpha_t + \zeta_t + \eta_t$$
$$- \xi_t \|^2 + \| \beta_t - \beta_0(t-1) \|^2]/2$$

Let the sum of the last three terms in (31) be designated as

$$W(t) = e_t'\zeta_t - \zeta_t'\eta_t + \| \zeta_t + \eta_t - \xi_t \|^2/2 \qquad (32)$$

To further execute the calculation, e_t will now be specified explicitly. Observe that from (19), (13) and (14) and using (30)

$$e_t = \hat{y}_t^* - y_t^* = u_t + \hat{H}_t\phi_t + \eta_t - F(q^{-1})r_t$$
$$= 2\hat{H}_t\phi_t + \hat{K}_t\phi_t + \zeta_t + \eta_t + \{ 1-F(q^{-1}) \} \, r_t$$
$$= m(t) + \zeta_t + \eta_t - 2\sigma(t)(e_t - \xi_t)$$
$$- \{ e_t - \zeta_t \} \sigma_t/(1 + \sigma_t) \qquad (33)$$

where

$$m(t) = 2\hat{H}_{t-1}\phi_t + \hat{K}_{t-1}\phi_t + \{ 1 - F(q^{-1}) \} r_t \qquad (34)$$

is an available signal at time t. Solving for e_t in (33) yields

$$e_t = \frac{(\nu_t+\sigma_t)\zeta_t + \nu_t \{ 2\sigma_t\xi_t+\eta_t+m(t) \}}{\omega_t} \qquad (35)$$

where

$$\omega_t = (2\sigma_t+1)(\sigma_t+1) + \sigma_t \qquad (36)$$

To determine appropriate ξ_t, η_t, and ζ_t, substitute (35) into (32) and find a vector ξ_t that minimizes the resulting equation with respect to ξ_t. The result is that $W(t)$ attains its minimum with respect to ξ_t

$$W(t) = - \frac{(\nu_t^2\sigma_t^2-\omega_t^2)\zeta_t'\zeta_t + \omega_t\sigma_t\zeta_t'\eta_t}{\omega_t^2}$$
$$- \frac{\nu_t m'(t)\zeta_t}{\omega_t} \qquad (37)$$

if and only if

$$\xi_t = \zeta_t + \eta_t - \nu_t\sigma_t\zeta_t/\omega_t \qquad (38)$$

Here, let

$$\zeta_t = - k(t)\mathrm{sgn}[m(t)]\| \phi_t \| \qquad (39)$$

$$\eta_t = (\frac{2\omega_t^2 - \nu_t^2 \sigma_t^2}{\sigma_t \omega_t})\zeta_t \qquad (40)$$

Then, (37) becomes

$$W(t) = - \zeta_t' \zeta_t - \frac{k(t)\nu_t}{\omega_t} \|m(t)\|_1 \| \phi_t \| \qquad (41)$$

where $\|m(t)\|_1$ is a ℓ_1-norm of $m(t)$. Substituting (41) into (31), we now have the following theorem which establishes the global stability of the overall system.

[Theorem 1]

Consider the error system (29) and the adaptation scheme (30) with e_t replaced by (35). Let the auxiliary variables ζ_t and η_t be specified as in (39) and (40), respectively, and ξ_t as in (38). Choose the transfer function $F(q^{-1})$ to be strictly positive real. Then there exists a positive decreasing gain $k(t)$ such that $V(t)$ becomes a Lyapunov function to the system comprised of (29) and (30) Consequently, the overall system is globally asymptotically stable and, in particular,

$$\lim_{t \to \infty} x_e(t) = 0$$

which implies that the system poles are asymptotically assigned to the desired location.

Proof

Note that by (28)

$$\| \beta_0(t-1)\|^2 \leq \| \overline{K}_{t-1} \|^2 \| \phi_t \|^2 \leq V(t-1)\| \phi_t \|^2$$

Therefore, if the magnitude of $k(t)$ is chosen so that

$$k^2(t) \geq V(t-1)/2$$

it is seen that

$$\Delta V(t) = n(t) - \frac{k(t)\nu_t}{\omega_t} \|m(t)\|_1 \| \phi_t \|$$

$$+ \beta_0'(t-1)\beta_0(t-1)/2 - k^2(t)\| \phi_t \|^2 \leq 0$$

which completes the proof.

Remark. Using the similar arguments, it can be shown that there exist an error system, a set of auxiliary signals and a parameter adjustment law with $\varepsilon_2(t)=0$ such that a Lyapunov function to the error system can be constructed. This sumplifies the adaptation scheme considerablly.

6. Conclusions

An adaptive pole-placer for a discrete-time multivariable nonminimum phase system has been proposed and a global asymptotic stability of the overall system has been established by the Lyapunov design method. In the course of constructing the Lyapunov function, uses are made of the strict positive realness of the transfer function associated with the error system and the auxiliary signals superimposed in the error equation. Although the global asymptotic stability of the overall system has been established using the particular error system with the auxiliary signals superimposed, further investigations need be carried out to obtain the simple error system with the improved auxiliary signals.

7. Reference

Kerisselmeier, G. and Smith, M. C. (1986) Stable adaptive regulatioin of arbitrary nth-order plants. IEEE Trans. Autmat. Contr. Vol. Ac-31, No. 4, pp. 299-305.

Minamide, N. and Kuroda, M. (1989). A Globally stable Adaptive Pole-Placer for a Discrete-Time Multivariable Nonminimum Phase Systems of Unknown Order. Trans. I. E. I. C., Vol. J72-A, pp. 302-308.

Goodwin, G. C. and Sin, K. S. (1981) Adaptive control of nonminimum phase systems. IEEE Trans. Automat. Contr. Vol. AC-26, pp. 478-483.

Elliot, H. (1982) Direct adaptive pole placement with application to nonminimum phase systems. IEEE Trans. Autmat. Contr. Vol. AC-27, pp720-722.

Egardt, B. and Samson, C. (1982) State adaptive control of nonminimum phase systems. Syst. Contr. Lett. Vol. 2, pp. 137-144.

Elliot, H., Cristi, R. and Das, M. (1985) Global stability of adaptive pole placement algorithms IEEE Trans. Autmat. Contr. Vol. AC-30, pp. 348-356.

Anderson, B. D. O. and Jonstone, R. M. (1985) Global adaptive pole positioning. IEEE Trans. automat. Contr. Vol. AC-30, pp. 11-22.

SOME ASPECTS ON POLE-PLACEMENT DESIGN

B. Wittenmark and K. J. Åström

*Department of Automatic Control, Lund Institute of Technology, Box 118,
S-221 00 Lund, Sweden*

Abstract. Pole-placement design for SISO systems are discussed in input-output formulation. It is shown how the different design parameters influence the performance of the closed loop system. For instance, the influences of sampling period, anti-aliasing filter, and desired closed loop performance are treated in detail. Rules of thumb for the choice of the different parameters are given.

Keywords: Pole-placement; controller design; sampled data systems

1. Introduction

In the pole-placement design method the dynamics of the process and the disturbances are specified with linear dynamical models. The specifications are given in terms of the desired closed loop poles. There are simple analytical procedures to determine if the problem can be solved and if so to give a solution.

Pole-placement control is simple. It is easy to understand and teach. It relates well to classical ideas like root locus and frequency domain methods. There are both state space and polynomial versions of the method. Hence it is a very good vehicle for bridging the gap between the ideas of internal (state space) and external (input-output) model approaches to linear systems. The method is almost identical for continuous and discrete time systems. Many other design methods can also be related to pole-placement. Typical examples are minimum variance, LQG, predictive control etc.

It thus appears that the pole-placement method is an almost ideal design method. It has, however, some severe drawbacks:

- It is difficult to relate engineering specifications to requirements on closed loop poles.

- All closed loop poles have to be specified. This is often too stringent. Bad choices of some poles may result in poor control

The controller consists of a feedback from the output of the process and a feedforward from the command signal. This results in a two-degree-of-freedom controller. The purpose of this paper is to suggest some refinements and design rules that make the pole-placement method a truly useful engineering design method.

The paper describes the pole-placement design for sampled data single-input-single-output systems.

Section 2 is a brief formulation of pole-placement design. The closed loop poles are given physical interpretation and grouped into different categories. This helps in selecting them. The notion of design variables is also introduced. These are the variables, which are at the disposal of the control designer. The control systems design problem is discussed in Section 3. Key issues like response to command signals, load disturbances, and measurement noise are investigated. Further robustness to process variations are discussed. In Section 4 we discuss how the criteria in Section 3 can be met by the pole-placement methods and we provide tools for investigating those aspects of the design problem that is not directly covered in the pole-placement design. Rules of thumb are given for the choice of the design parameters. Examples that illustrate the procedure are given in Section 5 and the paper ends with conclusions.

2. Pole-placement Design

The pole-placement design method for SISO sampled data systems is thoroughly discussed, for instance, in Åström and Wittenmark (1984). Let the process to be controlled be described by, see Figure 1,

$$y = x + e = \frac{B(q)}{A(q)}(u + v) + e \qquad (1)$$

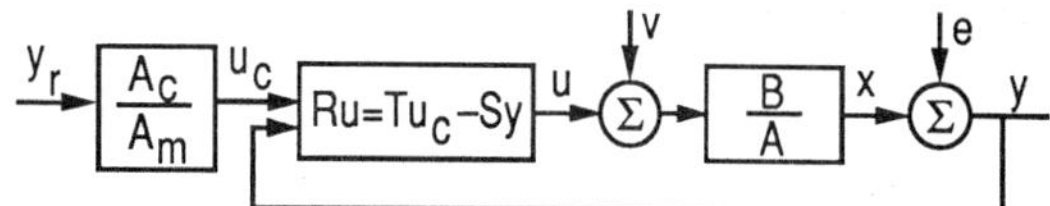

Figure 1. Process and controller configuration.

where u is the control variable, y the measured output, v load disturbance, and e measurement noise. A and B are polynomials in the forward shift operator q. A and B are assumed to be relatively prime. Further, it is assumed that A is monic. The pole-excess $d = \deg A - \deg B$ is the time delay of the process. Let the desired response from the reference signal y_r to the output be described by

$$A_m(q)y_m = B_m(q)y_r \tag{2}$$

Furthermore, let $A_o(q)$ be a predetermined stable polynomial specified by the designer. A_o is called the observer polynomial. The reason is discussed in the end of this section.

A general linear regulator can be described by

$$R(q)u = T(q)u_c - S(q)y \tag{3}$$

Elimination of u between (1) and (3) gives, when $e = 0$ and $v = 0$,

$$y = \frac{BT}{AR + BS}u_c$$

Introduce the polynomial A_c, that may be the same as A_m or an approximation of A_m. We want the closed loop system from u_c to y be governed by A_c. The signal u_c is generated from

$$u_c = \frac{A_c(q)}{A_m(q)}y_r \tag{4}$$

This gives the following condition for exact model following design

$$\frac{BT}{AR + BS} = \frac{B_m}{A_c} \tag{5}$$

where the denominator $AR + BS$ is the closed-loop characteristic polynomial. The denominator in (5) is usually of higher order than A_c. This implies that some poles and zeros are canceled. The polynomial B is factored as

$$B = B^+ B^-$$

where B^+ is a monic polynomial whose zeros are stable and so well damped that they can be canceled by the regulator. $B^+ = 1$ implies that there is no cancellation of any zeros. Since B^+ is canceled, it must also factor the closed-loop characteristic polynomial. The characteristic polynomial of the closed loop system will also contain the observer dynamics. The closed loop characteristic polynomial is thus given by

$$AR + BS = A_o A_c B^+$$

which is called the Diophantine equation. It follows that B^+ divides R. Hence

$$R = R_1 B^+ \tag{6}$$

$$AR_1 + B^- S = A_o A_c \tag{7}$$

Equation (7) has a unique solution if and only if A and B^- are relatively prime. Furthermore, it follows from (5) that B^- must divide B_m, i.e. $B_m = B^- B'_m$, and that

$$T = A_o B'_m \tag{8}$$

Pole-placement design is obtained by using the controller (3), where the polynomials R, S, and T are given by (6), (7), and (8). To get a causal controller, the model (2) must have the same or higher pole excess as the process (1). Further it is necessary that the observer polynomial is of sufficiently high order to guarantee that the Diophantine equation has a solution.

Disturbances with Known Dynamics

The pole-placement design can take into account disturbances with known dynamics. Let the disturbance v be generated from the dynamical system

$$A_d v = w$$

where w is a pulse, a set of widely spread pulses, or white noise. For example, a step disturbance is generated by

$$A_d(q) = q - 1$$

From the system model (1) and the regulator (3) we get

$$y = \frac{BT}{AR + BS}u_c + \frac{BR}{A_d(AR + BS)}w + \frac{AR}{AR + BS}e$$

$$u = \frac{AT}{AR + BS}u_c - \frac{BS}{A_d(AR + BS)}w - \frac{AS}{AR + BS}e \tag{0}$$

The closed-loop characteristic polynomial thus contains the disturbance dynamics as a factor. This polynomial is typically unstable. It follows from (9) that in order to maintain a finite output in case of these disturbances A_d must divide R. This would make y finite, but the controlled input u may be infinite to compensate for an infinite disturbance. The polynomial R must thus be of the form

$$R = R_1 A_d B^+ \tag{10}$$

This implies that a model for the disturbance dynamics is built into the regulator. The idea is called the internal model principle. See Francis and Wonham (1976). Notice that R will contain an integrator if the disturbance has step character.

Using (10), the design equation becomes

$$AA_d R_1 + B^- S = A_o A_c \tag{11}$$

The controller is now obtained from (8), (10), and (11).

State Space Discussion

State feedback and observer design can be interpreted in the polynomial formulation given above. In the state space formulation the closed loop characteristic polynomial from the reference value to the output is determined by the state feedback controller.

This corresponds to A_c in the polynomial method. The observer dynamics is in the state space formulation not controllable from the reference input. This thus corresponds to the Canceled dynamics specified by A_o in the polynomial design.

Design Parameters

The polynomial formulation of the pole-placement problem contains several design parameters. The designer must specify:

- Closed loop dynamics A_c
- Canceled zeros B^+
- Observer polynomial A_o
- Disturbance rejection polynomial A_d
- Sampling period h
- Anti-aliasing filter G_{aa}

These parameters are discussed in Section 4.

3. Key Design Issues

Control systems can be quite complicated because design is a compromise between many different factors. The following issues must typically be considered:

- Command signal following
- Load disturbances
- Measurement noise
- Model uncertainty
- Actuator saturation
- State constraint
- Controller complexity

We will now discuss how these factors are related to the design parameters given in Section 2. In a good design it is often necessary to grasp all factors. There are few design methods that consider all these factors. With pole-placement design the controller complexity is given directly by the model of the process and the disturbance. Responses to command signals, load disturbances and measurement noise are easy to assess. The effects of model uncertainty have to be investigated separately either by frequency response methods or by simulation. Actuator saturation is typically dealt with by extensions of classical anti-windup methods. See Åström and Wittenmark (1984). State constraints are normally dealt with by simulation. The separation of the closed loop poles into control poles (zeros of A_c) and observer poles (zeros of A_o) makes it easy to separate the response to command signals from the response to load disturbances and measurement noise. In the standard case we choose A_o to give appropriate response to load disturbances and measurement noise.

Command Signal Response

With standard pole-placement control the closed loop response is given by

$$y = \frac{B^- B_m'}{A_c} u_c = \frac{B^- B_m'}{A_m} y_r \tag{12}$$

The response is thus in essence governed by the plant zeros that are not Canceled and the polynomial A_c. The poles of A_m can not be chosen significantly faster than the zeros of B^-. The response to command signals does not depend on A_o. The load disturbance response does depend on A_c and A_o. It is possible to make a total separation of responses to command signals and load disturbances by using the two-degree-of-freedom structure shown in Figure 1. The disturbance dynamics is determined by $A_o A_c$ and the reference value dynamics is shaped through the feedforward given by (4)

Sensitivity to Disturbances

The transfer functions relating the plant output to measurement errors and load disturbances is given by

$$\begin{aligned}
x &= \frac{BT}{AR+BS} u_c + \frac{BR}{AR+BS} v - \frac{BS}{AR+BS} e \\
&= \frac{B_m}{A_c} u_c + \frac{B R_1 A_d}{A_o A_c} v - \frac{B^- S}{A_o A_c} e
\end{aligned} \tag{13}$$

The properties of the transfer functions can be assessed qualitatively from the location of the closed loop poles. The cancellation of disturbances with known dynamics is clearly seen from (13). For a more accurate assessment we recommend that to plot the Bode diagrams of the transfer functions.

Sensitivity to Model Errors

The sensitivity to unmodeled dynamics are well judged from the loop gain. In the sampled case this is given by

$$L = H_p H_c = \frac{BS}{AR}$$

where H_p and H_c are the pulse transfer functions of the process and the controller respectively. It is well known that the closed loop system is stable under plant variations ΔH_p provided that

$$\left| \frac{\Delta H_p}{H_p} \right| < \left| \frac{1+L}{L} \right| \tag{14}$$

A plot of the transfer function $(1+L)/L$ thus reveals the frequency ranges where high model precision is required. The relation is easy to apply when a design has been done. The right-hand side of (14) can be easily calculated and does not depend on the true transfer function. The conditions on the model precision can thus be expressed in terms of frequency domain conditions.

4. Choice of Design Parameters

In this section we will discuss the choice of design parameters that the control designer must do. The discussion results in practical rules of thumb that can be used.

Closed Loop Dynamics A_c

The polynomial A_c determines the closed loop behavior from the command signal u_c, but in the standard case it also influences the responses from the load disturbance and the measurement noise, see (13). The faster dynamics that are required the larger control signals are demanded. See (9). It is useful, also for sampled data systems, to specify the desired closed loop poles in continuous time form. Natural frequency and damping can then be used as design variables. A pair of control poles can be specified and the rest of the poles can be made a factor 1.5–2 faster than these poles.

On some occasions there are open loop poles that should not be changed, but kept in the closed loop system. One example is fast open loop poles. For discrete time systems it can also be sensible not to change the open loop poles introduced by the anti-aliasing filter.

The pole-placement design method has the drawback that all the poles have to be specified. There are methods, which instead place the poles within a specified region in the complex plane. See, for instance, Furuta and Kim (1987) and Wittenmark et al. (1987).

Canceled Zeros

The poles of the plant can be changed through feedback, but the zeros can be changed only through cancellation (B^+), and addition of the desired zeros (B'_m).

Sampled data systems frequently have zeros on the negative real axis ("sampling zeros") and outside the unit circle, see Åström et al. (1984). These zeros should not be Canceled even if they are inside the unit circle, since they will introduce ringing in the control signal. In the design it is therefore good practice not to change the open loop zeros but to keep them in the desired model.

Observer Polynomial

In the standard case the the observer polynomial is used to shape the responses for load disturbances and measurement noise, see (13). The control designer can choose A_o quite freely. As a rule of thumb the dynamics of A_o can be 1.5–2 times faster than the desired dynamics A_c.

Slow time constants in A_o will decrease the influence of measurement noise, but the reaction will be slow after a load disturbance. The influence of the measurement noise will increase when A_o is made faster. There is thus a compromise to be resolved depending on the nature of the disturbances acting on the system. When the properties of the disturbances are known, optimal filtering theory can be used to determine the optimal observer. For instance, if the system is described by

$$Ay = Bu + Ce$$

where e is white noise, then the optimal choice is $A_o = C$.

Sampled Data Systems

For sampled systems there are some additional considerations. The sampling period becomes a design parameter together with the anti-aliasing filter. They influence the response to measurement noise, load disturbances and the sensitivity to unmodeled dynamics. They have, however, little influence on the response to command signals. With a sampled system there will always be a delay in responding to load disturbances. A long sampling interval in connection with a properly anti-aliasing filter will make the system less sensitive to load disturbances.

Sampling Period

The sampling period should be chosen in relation to the closed loop bandwidth, which also includes the observer polynomial. A rule of thumb is to choose the sampling period h (in seconds) according to

$$\omega h \approx 0.2 - 0.6 \tag{15}$$

where ω is the natural frequency (in rad/s) of the corresponding continuous time control poles. This rule of thumb gives about 4–10 samples per rise time of the closed loop system or 15–45 samples per period.

It is also necessary to take the load disturbances into account when selecting the sampling period. A load disturbance occurring just after a sampling time will act on the system a full sampling period before it can be detected. It can thus be necessary to choose the sampling time according to the lower limit in (15). Too short a sampling period may introduce numerical difficulties in the calculation of the controller parameters and in the implementation of the controller.

Anti-aliasing Filter

Most analog sensors have filters to attenuate measurement noise in the signals. The filters are seldom chosen for a particular control application or for sampling the signals. It is very important in design of sampled data systems to incorporate an anti-aliasing filter. The filter should ideally remove all frequencies in the signals over the Nyquist frequency π/h rad/s. The high frequencies will otherwise be folded into low frequencies (aliasing) and will confuse the controller.

High order anti-aliasing filters are obtained by cascading filters of the form

$$G_{aa}(s) = \frac{\omega^2}{(s/\omega_B)^2 + 2\zeta\omega\,(s/\omega_B) + \omega^2} \tag{16}$$

where ω_B is the desired bandwidth of the filter. The parameters ω and ζ for different filters are given, for instance, in Åström and Wittenmark (1984).

Unless a quite high sampling rate is chosen it is necessary to include the anti-aliasing filter in the design of the controller. The necessary filter will otherwise deteriorate the performance of the closed loop system. The filter must be taken into account in the design of the regulator if the desired crossover

Table 1. Approximate time delay T_d of Bessel filters of different orders.

Order	$\omega_B T_d$
2	1.3
4	2.1
6	2.7

frequency is larger than about $\omega_B/10$, where ω_B is the bandwidth of the filter. The Bessel filter can, however, be approximated with a time delay, since the filter has linear phase for low frequencies. Table 1 shows the delay for different orders of the filter. This implies that the sampled data model including the anti-aliasing filter can be assumed to contain an additional time delay compared to the process.

5. Examples

The discussion of pole-placement design based on polynomial methods will be examplefied in this section. The influence of the different design parameters will be illustrated.

Nominal Design

Let the process be the harmonic oscillator with the transfer function

$$G(s) = \frac{\omega_0^2}{s^2 + \omega_0^2} \qquad \omega_0 = 1$$

The sampled pulse transfer function is

$$H(q) = \frac{(1-\beta)(q+1)}{q^2 - 2\beta q + 1} = \frac{B(q)}{A(q)} \qquad \beta = \cos(\omega_0 h)$$

The desired response is characterized by the continuous time characteristic equation

$$s^2 + 2\zeta\omega s + \omega^2 = 0$$

The sampled data form of this polynomial is $A_c(q) = A_m(q)$. Since the pulse transfer function has a zero at -1 no zero cancellation is allowed. This implies that $B^+ = 1$ and $B_m = B^- = B$. It is specified that the regulator should have integral action. This implies that the observer polynomial should be of second order.

$$s^2 + 2\zeta_{obs}\omega_{obs}s + \omega_{obs}^2$$

This polynomial is transferred to sampled data form $A_o(q)$.

The Diophantine equation is

$$(q^2 - 2\beta q + 1)R(q) + (1-\beta)(q+1)S(q) = A_o(q)A_m(q)$$

where R and S are of second order. The integrator is introduced by enforcing that $q - 1$ is a factor of R. The polynomial T is given by

$$T(q) = \frac{A_m(1)A_o(q)}{B(1)}$$

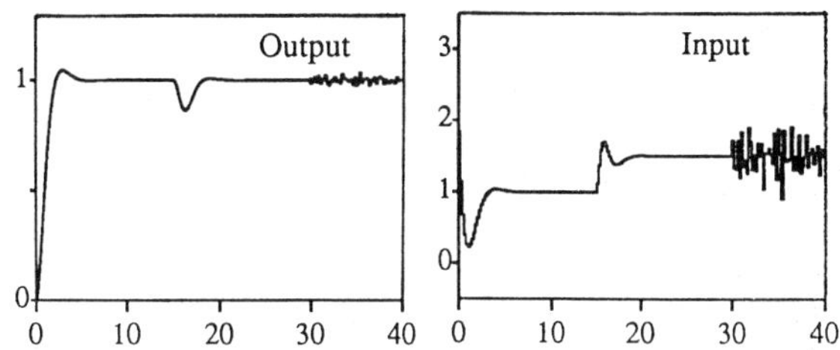

Figure 2. Simulation of the nominal design for the harmonic oscillator when $\omega = 1.5$, $\omega_{obs} = 3$, $\zeta = \zeta_{obs} = 0.7$ and $h = 0.2$, with integrator in the regulator.

The nominal parameter values are chosen as $\zeta = 0.7$, $\omega = 1.5$, $\zeta_{obs} = 0.7$, $\omega_{obs} = 3$ and $h = 0.2$. These specifications imply that significant damping is introduced and that the response speed is increased compared with the open loop system. Figure 2 shows the output and the input when the reference signal is a step at $t = 0$, a step disturbance at the input at $t = 15$ and discrete time white discrete time measurement noise with standard deviation 0.01 at $t = 30$.

Changing Observer Poles

Figure 3 shows the response when the observer poles are changed to $\omega_{obs} = 4$. The load disturbance is eliminated faster with a faster observer dynamics but the noise sensitivity also increases.

Changing the Sampling Period

The sampling period in the nominal design was chosen such that

$$\omega_{obs}h = 0.6$$

which is according to the upper limit of the rule of thumb. Figure 4 shows the responses when the sampling period is changed to $h = 0.1$ and 1. The controller will respond faster after load disturbances, when the sampling interval is decreased. A too long sampling interval will increase the deviation after the load disturbance. Also the aliasing effect is seen when there is measurement noise since no antialiasing filter is used.

Influence of Anti-aliasing Filter

The influence of the measurement noise in Figure 2 indicates that an anti-aliasing filter should be used to eliminate the effect of the disturbance. The filter

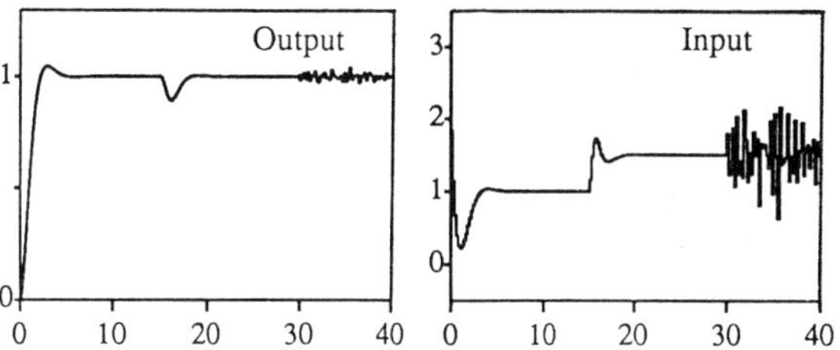

Figure 3. Response of the pole placement design for the harmonic oscillator for the observer dynamics $\omega_{obs} = 4$.

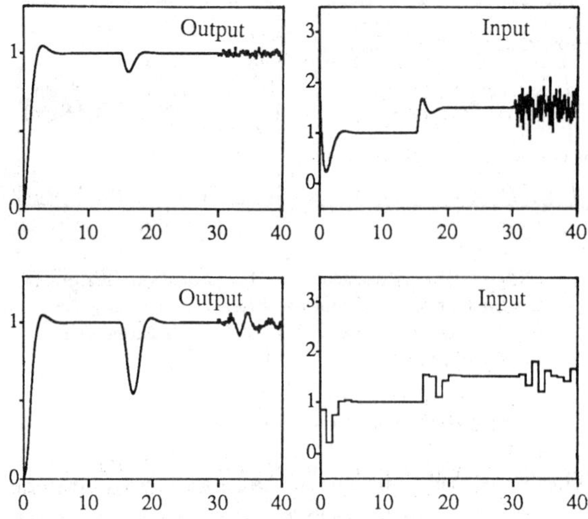

Figure 4. Response of the pole placement design for the harmonic oscillator for sampling intervals a) $h = 0.1$ b) $h = 1$.

will, however, also influence the closed loop response unless the bandwidth of the filter is an order of magnitude higher than the desired bandwidth of the closed loop system. The closed loop system will be unstable when the nominal controller is used on the system with a Bessel anti-aliasing filter, with $\omega_B = 2\pi$. The filter introduces too much phase lag at the cross over frequency. The filter dynamics should thus be considered when designing the controller. The Bessel filter can be well approximated by a delay. This simplifies the design and reduces the order of the controller. A sixth order Bessel filter is approximated as a delay $\tau = 2.7/\omega_B$. The filter bandwidth is chosen as $\omega_B = 2\pi$, which gives $\tau = 0.43$. To incorporate the delay it is necessary to increase the order of the regulator such that $\deg R = \deg S = \deg T = 5$. Figure 5 shows the response with anti-aliasing filter when the design is made by approximating the filter by a delay. The filter will significantly reduce the influence of the measurement noise. The filter will, however, also increase the deviation due to the load disturbance. The filter can be disregarded provided the filter bandwidth is more than ten times the desired closed loop bandwidth. This implies, however, also that the sampling frequency must be increased by an order of magnitude.

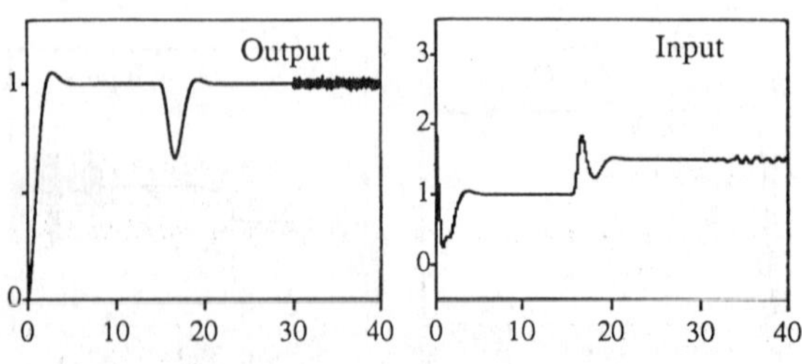

Figure 5. Response of pole-placement design when using anti-aliasing filter. The filter is approximated with a delay in the design.

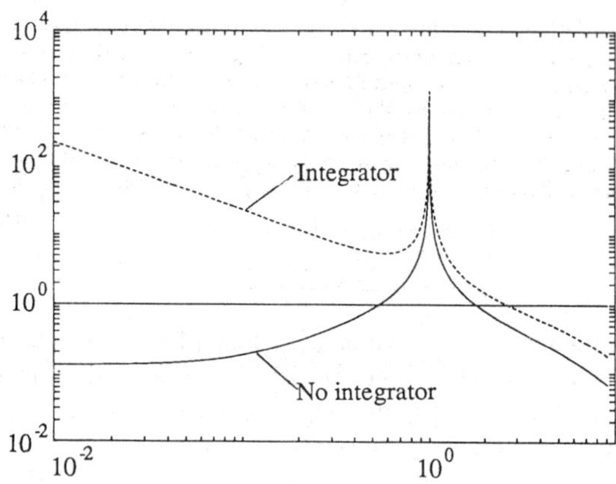

Figure 6. Bode diagram for the loop transfer function L in the nominal case when the controller is designed with and without integral action.

Robustness

The consequences of added unmodeled dynamics is illustrated next. Figure 6 shows the Bode diagram for the loop transfer function $L = H_p H_c$ when the pole placement controller is designed for the nominal case. For comparison the loop transfer function is shown for the case when no integral action is required in the controller. The enforced integral action implies that less model precision is needed for all frequencies.

6. Summary

The paper has given rules of thumb for selecting the design parameters in pole-placement design. With these rules it is quite easy to make a proper design of the closed loop system. The design must, however, be tested through analysis and simulations.

7. References

ÅSTRÖM, K.J., P. HAGANDER, and J. STERNBY (1984): "Zeros of sampled systems," *Automatica*, **20**, 31–38.

ÅSTRÖM, K.J. and B. WITTENMARK (1984): *Computer Controlled Systems*, Prentice Hall Inc, Englewood Cliffs, N.J.

FRANCIS, B.A. and W.M. WONHAM (1976): "The internal model principle of control theory," *Automatica*, **12**, 457–465.

FURUTA, K. and S.B. KIM (1987): "Pole assignment in a specified disk," *IEEE Trans. Aut. Control*, **AC-32**, 423–427.

WITTENMARK, B., R.J. EVANS, and Y.C. SOH (1987): "Constrained pole-placement using transformation and LQ-design," *Automatica*, **23**, 467–469.

CONTROLLER SYNTHESIS FOR MAXIMIZING DISTURBANCE BOUNDS IN LINEAR UNCERTAIN SYSTEMS

M. A. Franchek, S. Jayasuriya and M. J. Rabins

*Department of Mechanical Engineering, Texas A & M University, College Station,
TX 77843, USA*

ABSTRACT

A recently developed design methodology [1] [4] for maximizing the disturbance bound for SISO systems with no uncertainty is extended to SISO systems with structured uncertainty. In particular, we consider systems of the form

$$Y(s) = G_u(s, \alpha)U(s) + G_w(s, \alpha)W(s)$$

where $U(s)$ is the control input, $W(s)$ is the disturbance, $Y(s)$ is the output and α is a vector characterizing the bounded uncertain parameters. The objective is to synthesize a controller $G_c(s)$ to (i) maximize the disturbance bound (ii) simultaneously satisfy hard magnitude constraints on the control, output and intermediate states and (iii) satisfy bandwidth constraints. The methodology proposed is based on pointwise design in the frequency domain. The plant uncertainty is characterized by templates that depend on the frequency. Since the magnitude and phase vary with both frequency and the uncertain parameters, a Nichols chart is utilized to represent the plant templates. These templates are generated by bounding the uncertain plant by the so-called Kharitonov polynomials. Controller synthesis is accomplished by loop shaping so that the nominal loop transfer function stays within the acceptable boundaries in the Nichols chart corresponding to each frequency. In order to generate the allowable design regions, the state/control constraints and bandwidth limitations are incorporated into a set of target transfer functions. These target transfer functions are chosen on the basis of a unit step distrubance. Selecting the targets is particularly easy if only disturbance rejection is considered. The entire design process utilizes classical control theory and has much in common with QFT [5]. The design procedure is illustrated by an example.

I. INTRODUCTION

Two fundamental reasons for employing feedback are disturbance rejection and plant uncertainty. If a linear, certain, minimum phase plant with no external disturbances is given, a controller developed for reference tracking is the inverse of the plant dynamics with additional higher order poles added, making the controller realizable. Since this is never the case with real world systems, feedback is employed to achieve reference tracking despite the external disturbances and plant uncertainty.

Most design schemes tend to disregard the cost of feedback, namely bandwidth, thereby leading to large loop gains. As a result, excessive system bandwidths occured making the selected controller impractical for implementation. The amount of feedback required must be contingent on the amount of uncertainty in the system. Thus for small amounts of uncertainty, only a small amount of feedback should be specified. It should be clearly understood that controller selection is a series of design trade-offs between reference tracking, disturbance rejection, plant sensitivity and sensor noise rejection. These are generally difficult to embed in an optimization framework based on quadratic performance indices.

The need to incorporate the design trade-offs has led to numerous design techniques based on performance specifications in the frequency domain (i.e. H_∞, LQG/LTR, μ-synthesis). Yet these methods give an overdesigned system since generally a designer cannot specify desirable performance and bandwidths on individual states. If a hard constraint is required on only one state, then all the states are unjustly subjected to the same hard punishment. Also, the system designer never really develops an intuition for the natural plant attributes since controller selection is based on iterative optimization routines.

Only recently were other design methods developed making the bandwidth a concern [1]-[7]. One pointwise frequency design procedure based on Quantitative Feedback Theory was introduced providing a systematic approach to controller selection for a linear plant with structured uncertainty [1]-[4]. This method explicitly considers performance specifications and plant uncertainty while clearly showing the design trade-offs.

Proposed Frequency Design Methodology Reviewed

The proposed design metholdology [1]-[4] considers the following linear time invariant plant

$$\dot{x} = \mathbf{A}\,x + \mathbf{B}u + \mathbf{G}w \qquad (1)$$

$$y = [0 \ldots 0\ 0\ 1\ 0\ 0 \cdots 0]\,x \qquad (2)$$

where $x(t) \in \mathbf{R}^n$ is the states, $u(t) \in \mathbf{R}^1$ is the scalar control effort, $w(t) \in \mathbf{R}^1$ is the scalar disturbance and $y(t) \in \mathbf{R}^1$ is the scalar output. The system matrices are of appropriate dimensions.

This methodology is a simple, systematic approach to controller selection that maximizes the magnitude of a bounded input disturbance. Easily incorporated into the problem statement are prespecified control/state constraints, their desired bandwidths and the unknown magnitude of the persistent bounded disturbance. The key steps in controller synthesis are as follows :

- Control/state constraints and their desired bandwidths are embedded in the target transfer functions

- The closed loop transfer function for each state and the control effort are developed where the unknown controller transfer function is introduced as $G_c(s)$

- Substituting $s = j\omega$, a set of inequalities are expressed between the amplitudes of the target transfer function and the closed loop transfer functions.

- Classical loop shaping on the Nichols Chart is performed so that the compensated system remains within the allowable regions identified by the amplitude inequalities up to a frequency of one decade past the desired bandwith.

This design procedure gives a designer a feel for the plant dynamics while clearly allowing a decision among the design trade-offs. Also, because of the sharp controller designs, this method is expected to give a set of necessary and sufficient type theorems.

Closed Loop Transfer Function Development

The following transfer functions are written (Fig.(1))

$$\frac{X_i}{W} = \frac{G_{wi}}{1 + L} \tag{3}$$

$$\frac{X_j}{W} = G_{wj} + \frac{-G_{uj}G_{wi}}{G_{ui}} \frac{L}{1 + L} \tag{4}$$

$$\frac{U}{W} = \frac{-G_{wi}}{G_{ui}} \frac{L}{1 + L} \tag{5}$$

where

$$L = G_c G_{ui} \tag{6}$$

and i represents the feedback state and j represents the remaining states of the system.

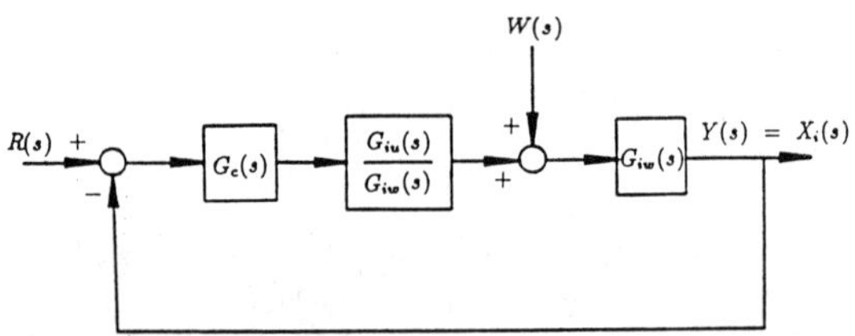

Figure 1: Disturbance Input Block Diagrams

Target Transfer Function Development

Consider a magnitude constraint on an arbitrary state, x_i

$$|x_i(t) - x_{io}(t)| \leq \beta \tag{7}$$

where x_{io} is the nominal state value due to a reference input. Also, impose bandwidth constraint, for example, of 10 rad/sec. The target transfer function representing this upper bound is

$$T_{x_i} = \frac{\beta}{\frac{s}{10} + 1} \tag{8}$$

Thus the target transfer function T_{x_i} represents the upper limit of the state x_i while explicity giving its desired bandwidth. Note that any other alterations or specifications can be incorporated into the target transfer functions. These are solely customized by the designer.

Introducing Plant Uncertainty

From (6) it is evident that for a linear certain plant, $L(j\omega)$ is a single point at each particular frequency. If $G_{ui}(j\omega)$ is uncertain, then the single point for $L(j\omega)$ is replaced by an area (template) outlining the family of points produced by parameter variations.

Note that these templates must be generated at each frequency of interest. One current method of area (template) generation is incremental increasing of the transfer function coefficients over their uncertainty range and then drawing a boundary around the points. Although this method is effective, it is time consuming and tedious and could obscure

possible design insight. Therefore, it is desirable to replace this brute force approach with a quicker analytical scheme.

A concise, systematic procedure greatly reducing the family of plants needed to analyze during controller synthesis is sought. The ultimate goal is to reduce the infinite number of transfer functions to four; in particular, those that represent the maximum and minimum amplitudes and phase angles of the plant at each frequency. Note that a coefficient combination used at one frequency is usually not the same at another frequency. The uncertainty addressed is structured uncertainty; i.e., the number of equations describing the system is known, but their coefficients vary within bounded sets. Neglecting higher order dynamics is valid since controller synthesis in the frequency domain allows appropriate amplitude roll-offs at high frequencies.

Section 2, develops the problem statement. Section 3 gives an example demonstrating the ability of the method. Section 4 offers conclusions and continuing research directions being explored.

II. PROBLEM FORMULATION

Consider a linear time invariant (LTI) dynamic plant represented as

$$G_p(s, \alpha) = \frac{b_m s^m + b_{m-1} s^{m-1} + \cdots + b_1 s + b_0}{s^n + a_{n-1} s^{n-1} + \cdots + a_1 s + a_0} \tag{9}$$

where α denotes a parameter indexing the uncertainty, and b_j, $j = 0, 1, 2, \cdots, m$ and a_i, $i = 0, 1, 2, \cdots, (n-1)$ are real numbers constrained within bounded sets. One intention of this paper is to develop the minimal number of transfer functions that describe the plant uncertainty.

Designing the controller based on the four extreme points corresponding to maximum and minimum amplitudes and phase angles will satisfy the entire family of plants produced by the uncertainty. Thus, the maximum and minimum amplitude of $G_p(j\omega, \alpha)$ is required over its phase range. The magnitude of (9) is

$$|G_p(j\omega, \alpha)| = \frac{\sqrt{Re_N^2 + Im_N^2}}{\sqrt{Re_D^2 + Im_D^2}} \tag{10}$$

Maximizing (10) would require maximizing the numerator while minimizing the denominator. Thus the problem statement is altered to contain a subtask that maximizes and minimizes the magnitude of a two-dimensional vector obtained from a polynomial whose coefficients vary within bounded sets.

Consider a general, fixed order polynomial written as

$$P(s) = p_n s^n + p_{n-1} s^{n-1} + \cdots + p_1 s + p_0 \tag{11}$$

where

$$p_{i,\alpha} \in [p_i^{min}, p_i^{max}], \quad i = 0, 1, 2, \cdots, n \tag{12}$$

The extremes for (11) are formed by applying Kharitonov polynomials which give coefficient combinations determining if an interval polynomial is Hurwitz [8]. These same coefficient combinations are employed to determine the four extremes of the uncertain plant.

The four Kharitonov polynominals associated with (11) are

$$\delta^1(s) = p_0^{max} + p_1^{min} s + p_2^{min} s^2 + p_3^{max} s^3 + p_4^{max} s^4 + \cdots \tag{13}$$

$$\delta^2(s) = p_0^{max} + p_1^{max} s + p_2^{min} s^2 + p_3^{min} s^3 + p_4^{max} s^4 + \cdots \tag{14}$$

$$\delta^3(s) = p_0^{min} + p_1^{min} s + p_2^{max} s^2 + p_3^{max} s^3 + p_4^{min} s^4 + \cdots \tag{15}$$

$$\delta^4(s) = p_0^{min} + p_1^{max} s + p_2^{max} s^2 + p_3^{min} s^3 + p_4^{min} s^4 + \cdots \tag{16}$$

Substituting $s = j\omega$ into the four spanning polynomials and grouping real and imaginary parts, the same as grouping even and odd powers, gives

$$\delta^1(j\omega) = p_e^{max}(\omega) + j\,\omega\,p_o^{min}(\omega) \tag{17}$$

$$\delta^2(j\omega) = p_e^{max}(\omega) + j\,\omega\,p_o^{max}(\omega) \tag{18}$$

$$\delta^3(j\omega) = p_e^{min}(\omega) + j\,\omega\,p_o^{min}(\omega) \tag{19}$$

$$\delta^4(j\omega) = p_e^{min}(\omega) + j\,\omega\,p_o^{max}(\omega) \tag{20}$$

where

$$p_e^{max}(\omega) = p_0^{max} - p_2^{min}\omega^2 + p_4^{max}\omega^4 - p_6^{min}\omega^6 + \cdots \tag{21}$$

$$p_e^{min}(\omega) = p_0^{min} - p_2^{max}\omega^2 + p_4^{min}\omega^4 - p_6^{max}\omega^6 + \cdots \tag{22}$$

$$p_o^{max}(\omega) = p_1^{max} - p_3^{min}\omega^2 + p_5^{max}\omega^4 - p_7^{min}\omega^6 + \cdots \tag{23}$$

$$p_o^{min}(\omega) = p_1^{min} - p_3^{max}\omega^2 + p_5^{min}\omega^4 - p_7^{max}\omega^6 + \cdots \tag{24}$$

The subscript e represents the even terms and the subscript o represents the odd terms. The magnitude for each polynomial is written as

$$\left| \delta^i(j\omega) \right| = \sqrt{[p_e(j\omega)]^2 + [p_o(j\omega)]^2} \tag{25}$$

Plotting the odd and even polynomials of δ^i for $i = 1, 2, 3, 4$ identifies the zero crossings of the pair. Correcting these polynomials at the frequencies for which the curves cross the frequency axis, yields the maxima and minima for the even and odd polynomials, thereby giving the desired coefficient combinations. These corrections are illustrated by the dashed and dotted lines in Fig 2 for p_e.

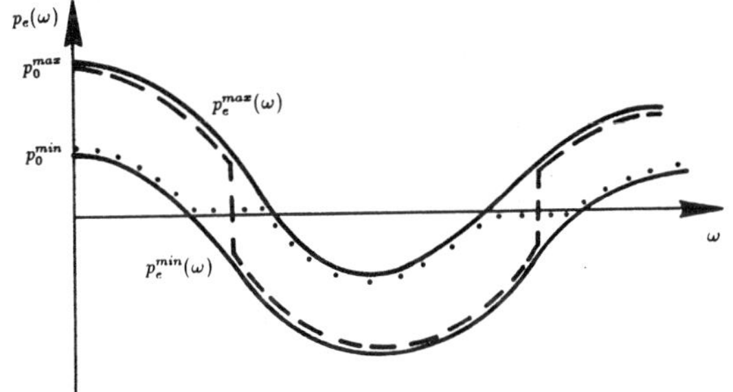

Figure 2: Characteristics of the Kharitonov Polynomials

Thus, at a particular frequency, the maximum and minimum value for the real and imaginary (even and odd) components can be easily determined. It is then possible to bound the interval polynomial parts, real and imaginary, in the complex plane (Fig. 3). Resulting is a box containing all possible real and imaginary combinations due to plant parameter variations; namely, identifying the extreme amplitude and phase angle of an interval polynomial.

Extending these results, it is possible to bound a linear plant with structured uncertainty. This is achieved by constructing a box in the complex plane for both the numerator and denominator of (10). Then identifying the information shown in Fig. 3, the extremes for gain amplitude and phase angle are determined. Using these four extremes, a set of boundaries for each can be developed outlining a region on the Nichols chart where the compensated plant must remain.

III. EXAMPLE

Consider the following second order system [6]

$$\dot{\mathbf{x}} = \begin{bmatrix} 0 & 1 \\ 0 & -a \end{bmatrix} \mathbf{x} + \begin{bmatrix} 0 \\ k \end{bmatrix} u + \begin{bmatrix} 0 \\ k \end{bmatrix} w \tag{26}$$

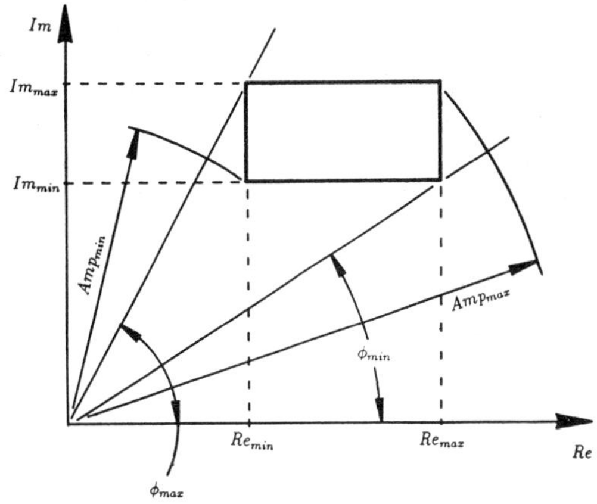

Figure 3: Interval Polynomial Box

where

$$a \in [3, 5] \tag{27}$$

and

$$k \in [1, 100] \tag{28}$$

For purposes of demonstrating the design process, the goal is to select a controller such that when the system is subjected to a unit step input disturbance

$$|x_1 - x_{1o}| \le 0.04 \tag{29}$$

$$|x_2 - x_{2o}| \le 1.0 \tag{30}$$

$$|u - u_o| \le 2.0 \tag{31}$$

where the subscript 'o' represents the operating value of the state/controller due to a reference input.

Through simple algebraic manipulations, the following transfer functions can be written

$$\frac{Y}{U} = \mathbf{G_u} = \begin{bmatrix} G_{u1} \\ G_{u2} \end{bmatrix} = \frac{1}{\Delta} \begin{bmatrix} k \\ ks \end{bmatrix} \tag{32}$$

$$\frac{Y}{W} = \mathbf{G_w} = \begin{bmatrix} G_{w1} \\ G_{w2} \end{bmatrix} = \frac{1}{\Delta} \begin{bmatrix} k \\ ks \end{bmatrix} \tag{33}$$

where

$$\Delta = s(s + a) \tag{34}$$

Next the following transfer functions can be written as

$$\frac{X_1}{W} = \frac{G_{w1}}{1 + L} \tag{35}$$

$$\frac{X_2}{W} = G_{w2} + \frac{-G_{u2}G_{w1}}{G_{u1}} \frac{L}{1 + L} \tag{36}$$

$$\frac{U}{W} = \frac{-G_{w1}}{G_{u1}} \frac{L}{1 + L} \tag{37}$$

where

$$L = G_c G_{u1} \tag{38}$$

The target transfer functions based on the state/control constraints are

$$\frac{X_1}{W} = \frac{0.04}{\frac{s}{20} + 1} \tag{39}$$

$$\frac{X_2}{W} = \frac{1.0}{\frac{s}{20} + 1} \tag{40}$$

$$\frac{U}{W} = 2.0 \tag{41}$$

The dynamics of $\frac{1}{\frac{s}{20}+1}$ was included in the state targets so that each would have a finite bandwidth. Note that if the targets were strictly constants, a controller would be sought providing the compensated system with an infinite bandwidth for the targets. Note that a bandwidth was not specified in the control target so as to reduce the complexity of graph interpretation. Selecting the target transfer function bandwidth allows the designer to specify explicitly the bandwidth of each state and the controller.

Since each target represents an upper bound, the following inequalities can be written:

$$\frac{X_1}{W} \; : \; \left|\frac{0.04}{\frac{s}{20}+1}\right| \geq \left|\frac{G_{w1}}{1+L}\right| \tag{42}$$

$$\frac{X_2}{W} \; : \; \left|\frac{1.0}{\frac{s}{20}+1}\right| \geq \left|G_{w2} + \frac{-G_{u2}G_{w1}}{G_{u1}}\frac{L}{1+L}\right| \tag{43}$$

$$\frac{U}{W} \; : \; |2.0| \geq \left|\frac{-G_{w1}}{G_{u1}}\frac{L}{1+L}\right| \tag{44}$$

Remark : There is a direct coupling among the transfer functions in the inequalities (42)-(44). Consider inequality (43). If G_{w2} is chosen so that the gain amplitude is maximized, then by (32) and (33) the coefficients of G_{u2} have been chosen. Regardless, G_{w1} always equals G_{u1} since they are the same. Although this is a special case specific to this example, these types of observations are required when developing the boundaries.

The boundaries shown in Fig. 4 - 7 were developed from (42)-(44). Each figure deserves special attention so that the meaning of each boundary is clear. Consider Fig. 4. The solid line template labeled G_{u1} is the current, uncompensated location of the linear uncertain plant. Each corner of the template has a number 1–4 that represents maximum amplitude, minimum amplitude, maximum phase and minimum phase, respectively. For each extreme point a set of boundaries is developed. Note, in the general case, there is a set of boundaries for each of the four extreme points, where each set contains one boundary for the controller, one for the feedback state and one for each remaining state. Thus if the order of the system is n, then a total of $4\,(n+1)$ boundaries exist on a single Nichols chart at a particular frequency. Of course, $n+1$ boundaries are used per extreme point.

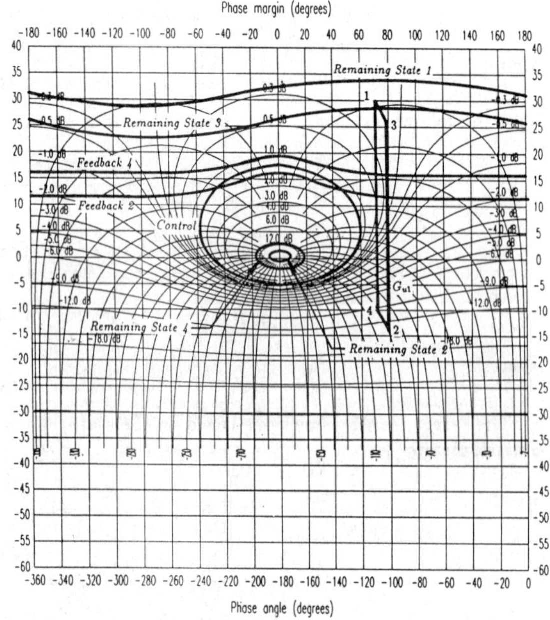

Figure 4: Constraint Boundaries at $\omega \; = \; 1$ rad/sec

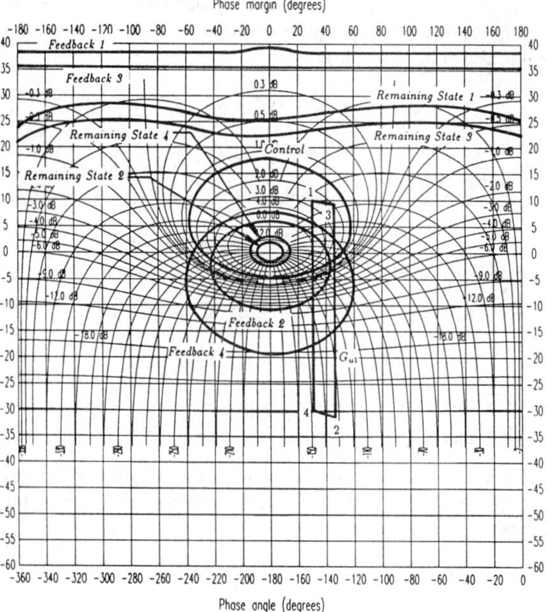

Figure 5: Constraint Boundaries at $\omega \; = \; 5$ rad/sec

Figure 6: Constraint Boundaries at $\omega \; = \; 10$ rad/sec

From (44), (32) and (33), it is clear that the target inequality for the controller reduces to

$$\left|\frac{1.0}{\frac{s}{20}+1}\right| \geq \left|\frac{L}{1+L}\right| \tag{45}$$

Therefore, only one controller boundary is required for all four extreme points.

For (42) and (43) a set of boundaries can be developed for each point on the template. So by applying our bounding technique, the coefficients are identified that maximize and minimize the amplitude and phase for the linear uncertain system at a particular frequency.

Reviewing Fig. 4 for the feedback state boundaries only, it is clear from (42) that each represents a lower **open loop** bound. Therefore the compensated plant must have each of its extreme points above its respective boundary. Note that *Feedback 1* is the boundary developed for the feedback

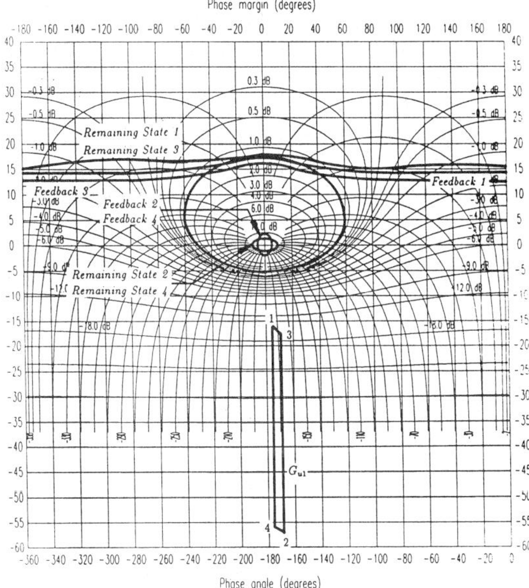

Figure 7: Constraint Boundaries at $\omega = 25$ rad/sec

state corresponding to the coefficient combination used to develop point 1, the maximum amplitude. Also note that the coefficient combination used at one frequency determining an extreme point is not necessarily the same at another frequency.

Continuing to review Fig. 4 for the remaining state, it is clear that from (43) each represents an upper **closed loop** bound. Thus the compensated plant must have each of its extreme points below their respective boundary. Before the boundaries are interpreted, consider the intersection for the closed loop amplitude of 2.0 db and a closed loop phase of $-2°$ (this corresponds to an open loop amplitude and phase of about 14 db and 172°, respectively). For a constant closed loop phase of $-2°$, the closed loop amplitude is decreased if the point is moved on the $-2°$ closed loop phase curve toward the negative closed loop amplitude region. Yet from boundary of *Remaining State 1* in Fig. 4, two possible amplitude solutions exist per one specific closed loop phase. This seems to be an interference in boundaries. Actually this implies that the G_{w1} point lies outside the circle developed by the target bound $\frac{1.0}{\frac{s}{20}+1}$. Since the controller must force the system within the circle (review (43)) the result is an upper and lower bound at a specific set of phases, represented by Fig. 8. For a complete explanation of inequality interpretation review [1]-[4]. Thus the sections of closed loop phase that lie above *Remaining State 1* and *Remaining State 3* are possible locations for the compensated plants for points 1 and 3, respectively.

The last two boundaries based on the remaining state have a target radius greater than $|G_{w2}|$. Thus at each particular closed loop phase, there is only one amplitude condition that must be satisfied.

With all the boundaries developed, the desired characteristics of the controller are easily identified; i.e., at each frequency the required change in amplitude and the required phase shift are obtainable. So at the frequency range of interest, a set of bounds on amplitude and phase are given. Thus a controller with frequency information bounded by the given constraints meets the prespecified state/control constraints along with the specified performance represented by the target transfer functions for a step disturbance. One controller satisfying all the boundaries for the system is

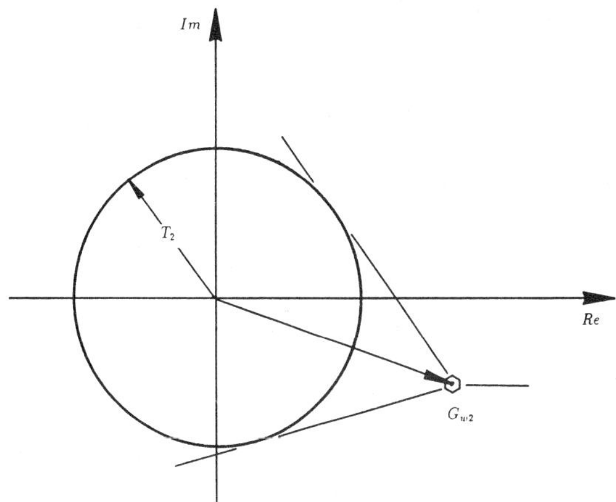

Figure 8: Graphic Representation for $|T_2| < |G_{w2}|$

$$G_c(s) = \frac{1.3\ 10^6(s+1)(s+5)}{(s+100)(s+200)}. \tag{46}$$

Remark : When choosing the bounds on the controller, a sense of gain and phase margin must be included on the loop transfer function. Note that guaranteeing performance implies that stability is guaranteed. Hence if the boundaries allow arbitrary phase and gain margins, then the controller bounds at that frequency should be based on trying to obtain large phase and gain margins.

Figs. 9 - 11 give the transient response based on a unit step input disturbance for the selected controller (46).

Notice that the control exhibits a high bandwidth. If it is desirable to limit its bandwidth, appropriate amplitude roll-off is required during loop shaping.

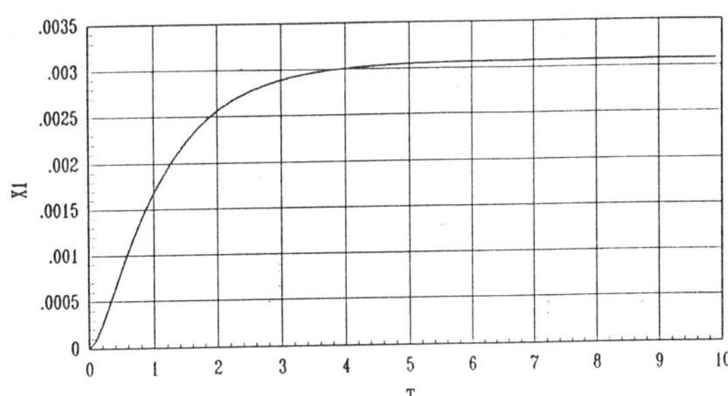

Figure 9: Transient Response of x_1 due to a Unit Input Disturbance

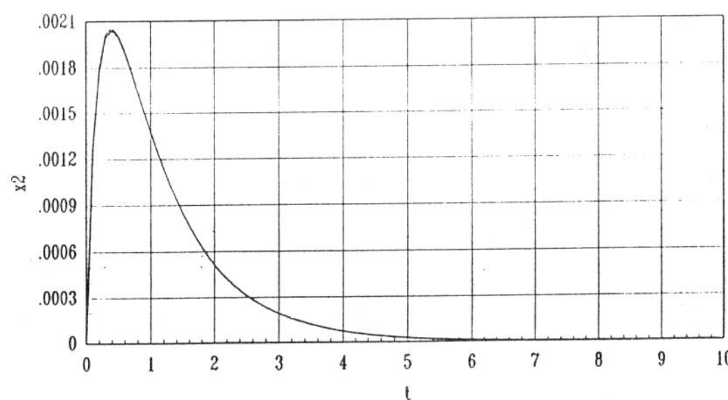

Figure 10: Transient Response of x_2 due to a Unit Input Disturbance

IV. CONCLUSIONS

The problem presented is simple in that only two coefficients have uncertainty, which facilitates understanding of the concept of the proposed method without concern over detail work. Currently, this method has introduced more

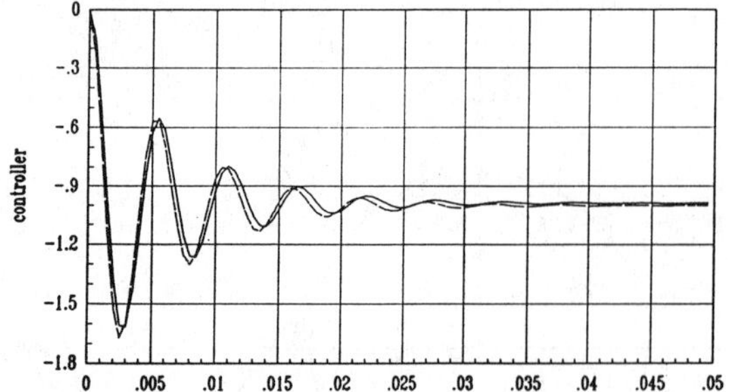

Figure 11: Transient Response of u due to a Unit Input Disturbance

questions within robust control. For instance, the shape of the template associated with the given system was concave. Would it matter if the shape of the plant were convex? How could this characteristic of the plant be identified by the four extreme points? Could the number of boundaries required be reduced by considering the boundaries for only the state and/or control nearest saturation? Of course, the identification of the state(s)/control nearest saturation would require a measure or an index to be developed. If the system were non-minimum phase, could this method find a controller that both stabilizes the system and satisfies the prespecifed constraints and performances?

New results concerning the mapping between the time and frequency domain has been developed that provides a method to determine if the augmented system remains bounded when it is mapped from the frequency domain to the time domain. Also, it has been shown that the maximum magnitude of a persistent bounded disturbance a given uncertain system can reject is not contingent on the feedback state. Therefore the feedback state is chosen based on the control effort required to reject the input disturbance and the ability of representing the controller with a realizable transfer function whose coefficients are real. These results are presented in a separate paper [9].

REFERENCES

1. Jayasuriya, S. and Franchek, M.A., "Frequency Domain Design for Prespecified State and Control Constraints under Persistent Bounded Disturbances," *Proceedings of the IEEE Conference in Decision and Control*, Austin, Texas, 1988.

2. Franchek, M.A., "Synthesis of Controllers for Prespecified Performance in Linear Uncertain Systems", M.S. Thesis, Texas A&M University, Dec. 1988.

3. Jayasuriya, S. and Franchek, M.A., "Loop Shaping for Robust Performance in Systems with Structured Perturbations," *Proceedings of the ASME Winter Annual Meeting*, Chicago, Illinois, 1988.

4. Jayasuriya, S. and Franchek, M.A., "Maximal Persistent Bounded Disturbance Rejection for a Class of SISO Systems with Structured Perturbations," *Proceedings of the IEEE International Conference on Control and Application*, Tel Aviv, Israel, April 1989.

5. Horowitz, I.M., "Quantitative Synthesis of Uncertain MIMO Feedback Systems," International J. of Control, Vol. 30, 1979.

6. Krishnan, K.R. and Horowitz, I.M., "Synthesis of a Non-linear Feedback System with Significant Plant-Ignorance for Prespecified System Tolerances," *International Journal Control*, Vol. 19, No. 4, 1974

7. Horowitz, I.M., and Sidi, M., "Synthesis of Feedback Systems with Large Plant Ignorance for Prescribed Time Domain Tolerances," *International Journal Control*, Vol. 16, No. 2, 1972.

8. Bhattacharyya, S.P., *Robust Stabilization Against Structured Perturbations*, Lecture Notes in Control and Information Sciences, Springer-Verlag, Vol. 99, 1987.

9. Zentgraf, P. and Jayasuriya, S., "On QFT-Type Design for Maximizing the Size of a Persistent Disturbance," ASME Winter Annual Meeting, Special Publication, San Francisco, CA, Dec. 1989 (to appear)

A FREQUENCY DOMAIN METHOD FOR LOW ORDER CONTROLLER DESIGN

M. Lilja

*Department of Automatic Control, Lund Institute of Technology, P.O. Box 118,
S-221 00 Lund, Sweden*

Abstract. The Ziegler-Nichols "self oscillation" method has been used in some
decades as a quick and easy way to compute parameters for PI and PID controllers
in process control. This method uses only information about one point on the
Nyquist curve of the process. This paper describes a design method, which uses
process frequency response data at several frequencies. A desired closed loop model
is specified and controller parameters are computed using a weighted least squares
criterion in the frequency domain. The technique can also be interpreted as an
approximate pole placement design thereby establishing a connection between
frequency domain methods and pole placement.

Keywords: Frequency domain, low order controllers, least squares method, approximation, pole placement

1. Introduction

A linear, time invariant system can be described
by

$$y = G(p)(u + d) \qquad (1)$$

where y is the process output, u the control
signal, d a load disturbance and $p = d/dt$ is
the differential operator. The transfer function G
need not be rational, i.e. the model could be of
infinite dimension. A general linear two-degree-
of-freedom controller can be written as

$$u = H(p)r - F(p)y \qquad (2)$$

where r is the reference input and H and F
are transfer functions. The Ziegler-Nichols self
oscillation method often gives a closed loop
step response with better damping than the
Ziegler-Nichols step response method. The self
oscillation method corresponds to identifying
one point on the Nyquist curve of G, namely
one intersection with the negative real axis.
The controller parameters are then computed
from the corresponding frequency ω_1 and the
magnitude $|G(i\omega_1)|$. It is rather obvious that

the closed loop response will be improved by
considering more than just one frequency. In
the "dominant pole design" described in [Åström
and Hägglund (1985)] the frequency response is
given at two frequencies and the order of the
controller is at most two.

The design method described in this paper
uses arbitrarily many frequencies. It differs from
other methods for design of low order controllers
in the frequency domain in the respect that no
"intermediate" low order process model is used.
Instead the least squares method is used to fit
the controller parameters directly to a specified
closed loop transfer function. This gives a larger
freedom in the choice of the controller order relatively to the order of the closed loop model specification. The "intermediate process model" approach is found e.g. in [Goodwin and Salgado
(1989)] which also considers uncertainties formulated within a stochastic framework and robust
design based on the envelope of the uncertain
Nyquist curve. The formulation in this paper
contains no stochastic considerations or uncertainty assumptions.

2. Description of the method

Only finite dimensional controllers will be considered. The transfer functions H and F in Eq. 2 can then be chosen as

$$H(s) = \frac{T(s)}{R(s)}$$

$$F(s) = \frac{S(s)}{R(s)}$$

where the polynomials R, S and T are given by

$$R(s) = s^{n_R} + r_1 s^{n_R-1} + \cdots + r_{n_R}$$
$$S(s) = s_0 s^{n_S} + s_1 s^{n_S-1} + \cdots + s_{n_S}$$
$$T(s) = t_0 A_o(s)$$

The polynomial $A_o(s)$, which can be interpreted as an observer polynomial, is specified by the user. Let the desired closed loop transfer function be given by $G_m(s)$. The actual closed loop transfer function is given by

$$G_{cl}(s) = \frac{G(s)T(s)}{R(s) + G(s)S(s)} \qquad (3)$$

Introduce the relative closed loop model error

$$E(s) = \frac{G_{cl}(s) - G_m(s)}{G_{cl}(s)}$$

A simple calculation gives that

$$E(s) = -1 + F_R(s)\frac{R(s)}{t_0} + F_S(s)\frac{S(s)}{t_0} \qquad (4)$$

where

$$F_R(s) = \frac{G_m(s)}{G(s)A_o(s)}R(s)$$

$$F_S(s) = \frac{G_m(s)}{A_o(s)}S(s)$$

From Eq. 4 it follows that the controller parametrization

$$\theta = \frac{1}{t_0}(1, r_1, \ldots, r_{n_R}, s_0, \ldots, s_{n_S})^T$$

gives linearity in $E(s)$. Choose a set of complex numbers $\mathcal{Z} = \{z_i\}_{i=1}^N$ and define the loss function

$$J = \sum_{i=1}^N w_i^2 |E(z_i)|^2$$

where w_i^2 are weightings. Let $n_R = \deg R$ and $n_S = \deg S$ and introduce the notation

$$\phi_R(s) = F_R(s)\begin{pmatrix} s^{n_R} & s^{n_R-1} & \ldots & 1 \end{pmatrix}$$

$$\phi_S(s) = F_S(s)\begin{pmatrix} s^{n_S} & s^{n_S-1} & \ldots & 1 \end{pmatrix}$$

$$\phi(s) = \begin{pmatrix} \phi_R(s) & \phi_S(s) \end{pmatrix}$$

$$\Phi = \begin{pmatrix} w_1\phi(z_1) \\ w_2\phi(z_2) \\ \vdots \\ w_N\phi(z_N) \end{pmatrix}, \quad \psi = \begin{pmatrix} w_1 \\ w_2 \\ \vdots \\ w_N \end{pmatrix}$$

The loss function can then be written as

$$J(\theta) = (\Phi\theta - \psi)^*(\Phi\theta - \psi)$$

where $*$ denotes conjugate transpose. Since J is a non-negative definite and quadratic it has a unique global minimum precisely when the matrix Φ has full column rank. Column rank deficiency occurs e.g. if

$$N < n_R + n_S + 2 \qquad (5)$$

This can be avoided by choosing sufficiently many complex frequencies z_j. Another case, for which Φ lacks full column rank, is when there are solutions of lower order.

In continuous time design the finite point set $\mathcal{Z}$ is typically chosen as a subset of the imaginary axis but other choices are possible such as points near the poles of the desired closed loop model (compare [Åström (1988)]). The design method is also applicable to discrete time systems in which case $\mathcal{Z}$ is chosen as a subset of the unit circle. In general the set $\mathcal{Z}$ is chosen to be closed under conjugation. Together with the conditions $\overline{F_R(s)} = F_R(\overline{s})$ and $\overline{F_S(s)} = F_S(\overline{s})$ this implies that $\overline{\Phi} = \Pi\Phi$ and $\overline{\psi} = \Pi\psi$ for some $N \times N$ permutation matrix Π. This guarantees that the least squares solution

$$\hat{\theta} = (\Phi^*\Phi)^{-1}\Phi^*\psi$$

is real. This is true since

$$\overline{\hat{\theta}} = (\overline{\Phi}^*\overline{\Phi})^{-1}\overline{\Phi}^*\overline{\psi} = (\Phi^*\Pi^T\Pi\Phi)^{-1}\Phi^*\Pi^T\Pi\psi =$$
$$= (\Phi^*\Phi)^{-1}\Phi^*\psi = \hat{\theta}$$

where the fact that every permutation matrix is orthogonal has been used. The property $\overline{G(s)} = G(\overline{s})$ is equivalent to $G(s)$ having a real inverse Laplace transform. In the following, the notation $\Omega = \{\omega_1, \ldots, \omega_m\}$ refers to the approximation set $\mathcal{Z} = \{i\omega_1, \ldots, i\omega_m, -i\omega_1, \ldots, -i\omega_m\}$.

By a simple modification of the functions $F_R(s)$ and $F_S(s)$ it is possible to incorporate pre-specified factors in the polynomials $R(s)$ and $S(s)$ by introducing $R(s) = R_1(s)R_2(s)$ and $S(s) = S_1(s)S_2(s)$ with R_1 and S_1 being the fixed parts of the respective polynomials. The condition Eq. 5 is modified to $N < n_{R_2} + n_{S_2} + 2$. For example, if integration is to be included in the controller, the pre-specified part of R is chosen as $R_1(s) = s$ and the function F_R is modified to $R_1(s)F_R(s) = sF_R(s)$.

The least squares method is implemented as a function in the matrix manipulation language PRO-MATLAB [Pro-Matlab] using the built in QR-factorization with pivoting. This MATLAB function has been used in the example presented in the next section.

3. Example

Consider the system

$$G(s) = \frac{1}{(s+1)^8}$$

Some low order controllers will be computed for this system. Starting with the choice of approximation frequencies, it is important to know for which frequencies the Nyquist curve intersects the real and imaginary axes. These frequencies will be denoted by $\omega_{-90°}$, $\omega_{-180°}$, $\ldots$ with obvious notation. For this specific case we have that $\omega_{-n\cdot90°} = \tan n\frac{\pi}{16}$, $n = 1, 2, \ldots$. The numerical values for the first frequencies are $\omega_{-90°} \approx 0.199$, $\omega_{-180°} \approx 0.414$ and $\omega_{-270°} \approx 0.668$.

Let the expression $P(n, \omega, \varphi)$ denote a polynomial of n'th order with poles evenly spread out on a circular arc with radius ω and bounded by the angles $\pm\varphi$ with the negative real axis. This choice of pole pattern is rather arbitrary but is motivated from the observation that it gives a more well-damped step response than e.g. the commonly used Butterworth configuration, especially for high order systems.

First order controller

To begin with, a first order controller with integration (a PI-controller) will be designed. The phase of the controller varies between $-90°$ and $0°$. This means that the bandwidth of the closed loop system should not exceed the bandwidth of the open loop system. Taking the values of the "axis frequencies" into account the approximation frequencies are chosen as $\Omega = \{0, 0.01, 0.1, 0.2\}$. The desired closed loop model is somewhat arbitrarily chosen to be of third order with relative degree one. Before specifying this, a third order process model $\hat{G}(s)$ is fitted to $G(s)$ at the chosen frequencies. This gives

$$\hat{G}(s) = \frac{0.1229s^2 - 0.1607s + 0.0807}{s^3 + 1.0955s^2 + 0.4847s + 0.0807}$$

The Nyquist curves for $G(s)$ and $\hat{G}(s)$ fits very well for frequencies below 0.2 rad/s as Figure 1

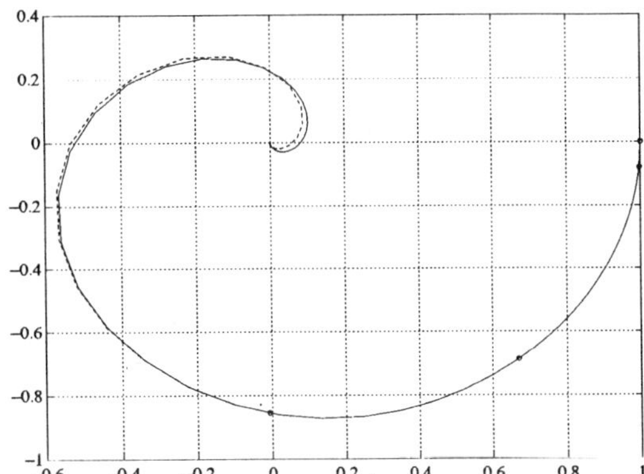

Figure 1. Nyquist curves of $G(s) = 1/(s+1)^8$ (dashed curve) and a third order approximation $\hat{G}(s)$ (solid curve). The approximation frequencies $\Omega = \{0, 0.01, 0.1, 0.2\}$ are marked with o's.

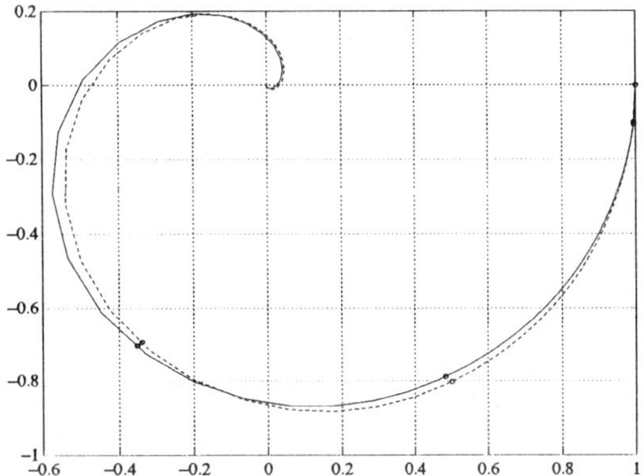

Figure 2. Nyquist curves of desired closed loop system (dashed curve) and actual closed loop system (solid curve) when using a first order controller with integration. The fitting frequencies $\Omega = \{0, 0.01, 0.1, 0.2\}$ are marked with o's.

indicates. Let $\hat{B}(s)$ and $\hat{A}(s)$ denote the numerator and denominator polynomials respectively in $\hat{G}(s)$. The zeros of $\hat{B}(s)$, $0.653 \pm i0.479$, are kept in the model $G_m(s)$. The polynomials A_m and A_o are chosen as $A_m(s) = P(3, 0.3, 45°)$ and $A_o(s) = s + 0.3$. A first order controller with integration obtained by fitting at the frequencies $\Omega = \{0, 0.01, 0.1, 0.2\}$ results in the closed loop Nyquist curves shown in Figure 2. The zero frequency is included in order to get a small error in the closed loop static gain. By choosing a sufficiently large weighting for this frequency the error can be made arbitrarily small. The controller output u and the process output y of the closed loop are shown in Figure 3 for the case when the reference signal r is a unit step at $t = 0$ and the load disturbance d is a negative unit step at $t = 50$. The step response of the desired model from r to y is also given in the plot. The step responses follows very close. If the desired closed

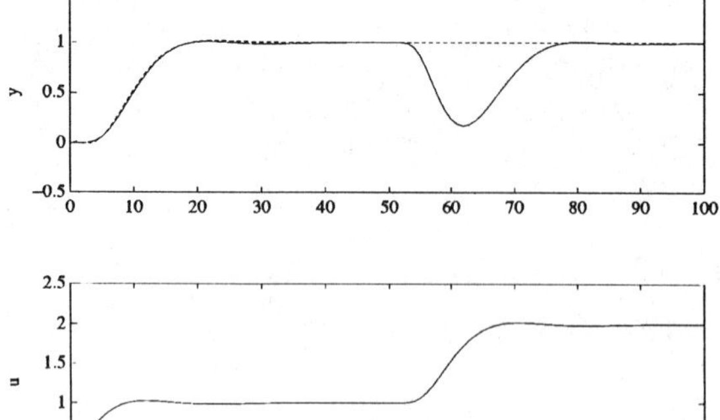

Figure 3. Step responses of desired closed loop system (dashed curve) and actual closed loop system (solid curve) when using a first order controller with integration. The fitting frequencies are given by $\Omega = \{0, 0.01, 0.1, 0.2\}$.

loop poles are moved further out from the origin the fitting will deteriorate and when the polynomials A_m and A_o are chosen as $P(3, \omega_m, 45°)$ with $\omega_m = 1.1$ the actual closed loop system is unstable.

Second order controller

By increasing the order of the controller, it is possible to get more phase lead in the loop which means that the bandwidth can be further increased. A second order controller with integration (PID-controller) is computed using the approximation frequencies $\Omega = \{0, 0.01, 0.1, 0.2, 0.4\}$. The polynomials A_m and $A_o(s)$ are this time chosen as $A_m(s) = P(3, 0.4, 45°)$ and $A_o(s) = P(2, 0.4, 45°)$. The closed loop Nyquist curves are shown in Figure 4.

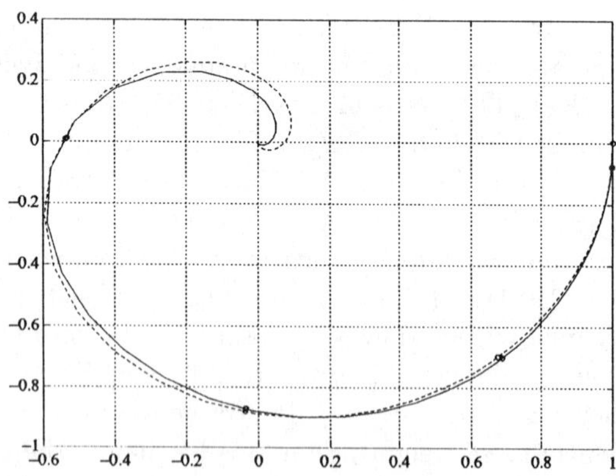

Figure 4. Nyquist curves of desired closed loop system (dashed curve) and actual closed loop system (solid curve) when using a second order controller with integration. The fitting frequencies $\Omega = \{0, 0.01, 0.1, 0.2, 0.4\}$ are marked by o's.

The step response of the closed loop system

agrees well with the specified model as Figure 5 shows.

When the specified poles are moved further out from the origin, the design procedure gives an unstable controller. The reason for this is intuitively that more phase lead is required. The only way a fixed order controller can give such a phase advance is to make it unstable. The resulting closed loop system will also be unstable. This is due to the fact that the instability of the controller implies that the Nyquist curve should encircle the point -1. This is not captured by the optimization procedure, since the approximation frequencies are much lower than the frequencies for which the loop Nyquist curve is close to -1.

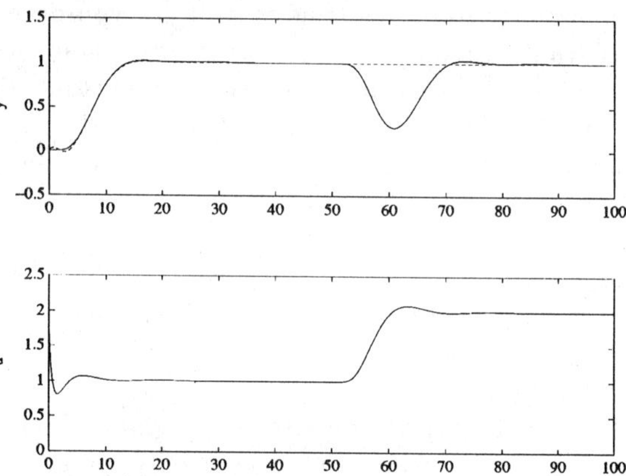

Figure 5. Step responses of desired closed loop system (dashed curve) and actual closed loop system (solid curve) when using a second order controller with integration. The fitting frequencies are given by $\Omega = \{0, 0.01, 0.1, 0.2, 0.4\}$.

In fact, the instability occurs for much smaller values of ω_m (≈ 0.47) than in the first order case. The reason for this is that in the first order case the single controller pole is fixed to zero, so that there is no possibility to have any controller poles in the open right half plane. This clearly indicates that unstable controllers are not desirable.

Third order controller

The bandwidth of the closed loop system can be made larger if the controller order is increased. In the third order case the polynomials $A_m(s) = A_o(s) = P(3, 0.5, 45°)$ were chosen and the approximation frequencies were the same as in the second order case, namely $\Omega = \{0, 0.01, 0.1, 0.2, 0.4\}$. The Nyquist curves of the closed loop system and the reference model fits well together below $\omega = 0.4$ (Fig.6). In this case the controller poles are relatively well damped but the controller will eventually become unstable as the specified bandwidth is increased. The

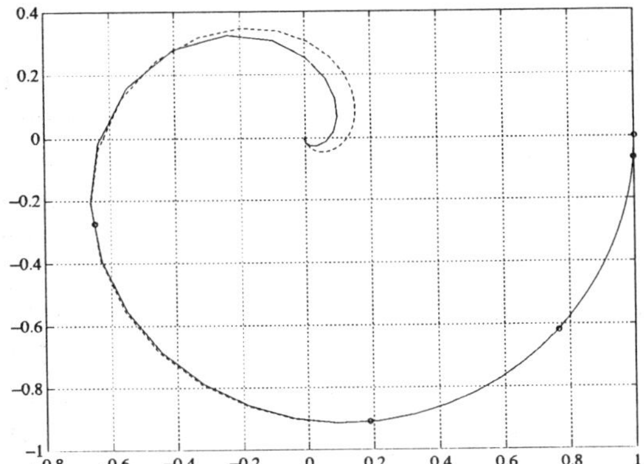

Figure 6. Nyquist curves of desired closed loop system (dashed curve) and actual closed loop system (solid curve) when using a third order controller with integration. The fitting frequencies $\Omega = \{0, 0.01, 0.1, 0.2, 0.4\}$ are marked by o's.

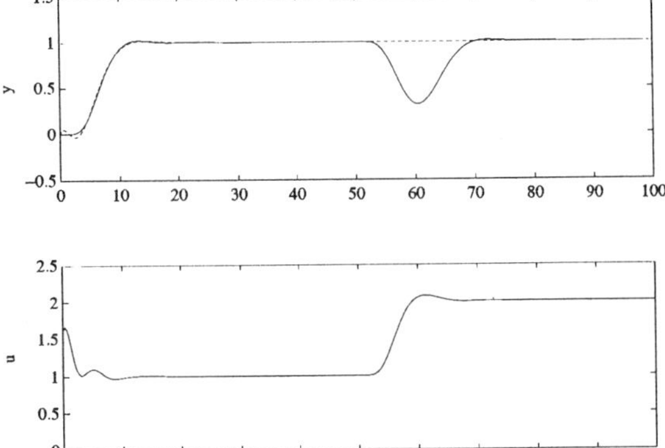

Figure 7. Step responses of desired closed loop system (dashed curve) and actual closed loop system (solid curve) when using a third order controller with integration. The fitting frequencies are given by $\Omega = \{0, 0.01, 0.1, 0.2, 0.4\}$.

actual and desired closed loop step responses are shown in Figure 7.

Discussion

The benefits of increasing controller complexity can be judged from Fig. 3, 5 and 7. When going from a PI controller (Fig. 3) to a PID controller (Fig. 5) the 50% time is decreased from 10 seconds to 8. The maximum error due to the load disturbance is decreased from 0.8 to 0.7. When increasing the controller order from two to three the enhancement is less significant but still noticeable.

As the order of the controller is increased, some caution has to be taken. The number of approximation points (frequencies) must be sufficiently large. Also, the order of the desired closed loop model may have to be increased to

get a stable controller. Another possibility is to force the controller to be stable by keeping the polynomial R fixed to a stable polynomial R_1. The order of the controller may then have to be increased to get a good Nyquist curve fitting.

4. Comparison with pole placement

When the process transfer function $G(s)$ is rational ($G(s) = B(s)/A(s)$) an ordinary pole placement design can be made (see Chapter 10 [Åström and Wittenmark (1984)]). The order of the controller is then determined by the order of $G(s)$ and the degree of the observer polynomial $A_o(s)$. Let the closed loop model be given by $G_m(s) = B_m(s)/A_m(s)$ where B_m and A_m are polynomials and let $B_m(s)$ be a constant multiple of $B(s)$ (no cancellation of process zeros). The relation in Eq. 3 can then be written as

$$A(s)R(s) + B(s)S(s) = A_m(s)A_o(s)$$

which is the well-known Diophantine-Aryabhatta-Bezout (DAB) equation. Pole placement design is thus a special case of the proposed method. This gives a connection between frequency domain methods and pole placement.

5. Conclusions

A new method for design of low order controllers for stable processes has been proposed. The method uses a least squares method to find a linear controller of specified order from a number of points on the Nyquist curve of the process. Since the design is made in the frequency domain the process model may be of infinite order. By repeating the optimization for controllers having different complexity it is also possible to clearly determine the benefits of increasing the order of the controller. The design method is illustrated by an example where the process model is of eighth order. The low order controllers thus obtained are shown to give good accuracy with respect to the desired closed loop step response.

6. Acknowledgement

I would like to thank my thesis supervisor Professor Karl Johan Åström for many useful discussions and for his enthusiastic and encouraging support. This project was supported by the National Swedish Board of Technical Development (STU) under contract 89-00403P.

7. References

ÅSTRÖM AND WITTENMARK (1984): *Computer Controlled Systems*, Prentice-Hall.

ÅSTRÖM AND HÄGGLUND (1985): "Dominant pole design," Technical report TRFT-7282, Department of Automatic Control, Lund Institute of Technology, Lund, Sweden.

ÅSTRÖM (1988): "Dominant pole placement design of PI regulators," Technical report TRFT-7381, Department of Automatic Control, Lund Institute of Technology, Lund, Sweden.

GOODWIN AND SALGADO (1989): "Quantification of uncertainty using an embedding principle," Proc. American Control Conference, June 1989, Pittsburg USA.

PRO-MATLAB (1987): *PRO-MATLAB User's guide*, The MathWorks, Inc., Sherborn, Mass., USA.

PARAMETRIZATION OF ALL STATE FEEDBACK H$^\infty$ CONTROLLERS

K. Z. Liu and T. Mita

*Department of Electrical and Electronic Engineering, Chiba University, 1-33,
Yayoi-cho, Chiba-city 260, Japan*

Abstract . The state feedback H^∞ control problem is considered in this paper. It is shown that under mild
conditions this problem can be solved simply by using the conjugation approach proposed by Kimura(1988,
1989) with some extended treatment to the case that $T_3(s)$ is tall and strictly proper. The parametrization of
all state feedback controllers achieving the prescribed norm bound is obtained. This is an eminent difference
from previous works on state feedback H^∞ control problem where only a static feedback control law(the central
solution of our result) is provided. Both four block and one block problems are solved in this paper. The
results of this paper embrace those of Doyle, et al.(1988) and Petersen(1988).

Keywords. H^∞ control; state feedback; parametrization of controllers; conjugation theory; model-matching
problem; one block problem; four block problem.

1. INTRODUCTION

H^∞ control problems with state feedback have been studied by
several researchers(Petersen, 1987, 1988, Khargonekar, et al.,
1988, Doyle, et al., 1988) in the past few years. All these works
are based on arguments around some untraditional Riccati equations. Particularly, in these works, only a static feedback controller is derived. On the other hand, the conjugation approach
of Kimura(1988, 1989) deals with H^∞ control problems based
on a simple notion of J-lossless conjugation, it provides us with
a unified framework of H^∞ control theory. This theory has
successfully solved the output feedback standard H^∞ problem
(Kawatani, et al., 1989),(Kimura, et al., 1988). The purpose of
this paper is trying to attack the state feedback H^∞ problem
with conjugation theory, and to give the parametrization of state
feedback controllers achieving the prescribed norm bound.

However, in applying the conjugation theory for model-matching
problem to the state feedback H^∞ problem, there arises such difficulty as $T_3(s)$ is strictly proper and tall so that the conjugation
theory can not be applied directly. We shall show that some pre-
treatment can get over this difficulty and that the state feedback
H^∞ control problem can be solved by conjugation theory, under
mild conditions. The complete parametrization of H^∞ controllers
is obtained. It is made clear that the static controller given in
(Doyle, et al., 1988) or (Petersen, 1988) is just the central solution of state feedback H^∞ controllers. The underlying idea is
to factor $T_3(s)$ as $T_{3o}(s)T_{3i}(s)$ where $T_{3i}(s)$ is fat, proper and
contains all interpolation constraints to H^∞ optimization(That
is, unstable zeros and rank deficit). Then we can absorb $T_{3o}(s)$
into $Q(s)$ and apply the conjugation theory for normal four block
problem(Kimura, et al., 1988).

Furthermore, because the solution of four block problem does
not cover that of one block problem, the solution of one block
problem is also given.

In this paper, let RH^∞ denote the set of rational proper stable
transfer function matrices, BH^∞ be its subset with H^∞-norm
less than 1. We use the notations

$$\left[\begin{array}{c|c} A & B \\ \hline C & D \end{array}\right] := D + C(sI - A)^{-1}B$$

$$\|U\|_\infty := \sup_\omega[\sigma_{max}(U(j\omega))]$$

where $\sigma_{max}(U)$ denotes the maximal singular value of U.

2. PROBLEM STATEMENT

Consider the feedback configuration(Fig.1) of the linear time-
invariant system.

$$G(s) = \left[\begin{array}{cc} G_{11} & G_{12} \\ G_{21} & G_{22} \end{array}\right] = \left[\begin{array}{c|cc} A & B_1 & B_2 \\ \hline C_1 & D_{11} & D_{12} \\ I & 0 & 0 \end{array}\right]$$

$$\left[\begin{array}{c} z \\ y \end{array}\right] = G \left[\begin{array}{c} w \\ u \end{array}\right] \tag{1}$$

In Fig.1 $K(s)$ is the stabilizing controller. The state space equations associated with (1) is described by

$$\begin{aligned}
\dot{x}(t) &= Ax(t) + B_1 w(t) + B_2 u(t) \\
z(t) &= C_1 x(t) + D_{11} w(t) + D_{12} u(t) \\
y(t) &= x(t)
\end{aligned} \tag{2}$$

where $x \in R^n$ is the state, $w \in R^r$ is the disturbance, $u \in R^p$ is
the control input, $z \in R^m$ is the controlled output and $y \in R^n$ is
the measured output(i.e., state feedback).

We make the following assumptions :

A1. (A, B_2) *is stabilizable.*
A2. (C_1, A) *is observable.*
A3. *The matrix D_{12} is column fullrank and $m > p$.*
A4. $D_{11} = 0$.

Remark 1. *Assumptions A1, A2 are the conditions used in
(Kimura, 1988). A1 is assumed for the stabilizability of $G(s)$, A2
is assumed to assure that Riccati equation (11) may have positive
definite solution.*

*For simplicity, assumption $D_{11} = 0$ is made here. The case $D_{11} \neq
0$ can be solved without essential difficulty in the same way, or
can be derived from the result for $D_{11} = 0$ via loop shifting
(Safanov, et al., 1988). Note that we assume $m > p$ so that the
problem is a four block one. As for the one block problem, see
section 4. The latter problem is identical to that considered by
Petersen(1988).*

The problem here is to attenuate the response $z(t)$ to the disturbance $w(t)$. That is, to minimize

$$\alpha := \sup_{\|w\|_2 \leq 1} \|z\|_2 \tag{3}$$

under the condition of internal stability. After suitable normal-
ization, the problem is simplified to finding a stabilizing controller

$K(s)$ such that $\alpha < 1$. A controller that achieves $\alpha < 1$ and stabilizes the closed loop system is called a desired controller hereafter. As is well-known, this boils down to finding $Q(s) \in RH^\infty$ such that(Francis, 1987)

$$\|T_1(s) - T_2(s)Q(s)T_3(s)\|_\infty < 1, \ Q(s) \in RH^\infty \qquad (4)$$

$$: \begin{bmatrix} T_1(s) & T_2(s) \\ T_3(s) & 0 \end{bmatrix} = \left[\begin{array}{cc|cc} A_F & -B_2F & B_1 & B_2 \\ 0 & A_H & B_1 & 0 \\ \hline C_1 + D_{12}F & -D_{12}F & 0 & D_{12} \\ 0 & I & 0 & 0 \end{array} \right] \qquad (5)$$

where F and H are chosen such that $A_F := A + B_2F, A_H := A + H$ are stable matrices. The class of stabilizing controllers is represented as

$$K(s) = (Y_2 - M_2Q)(X_2 - N_2Q)^{-1}$$

$$= (\tilde{X}_2 - Q\tilde{N}_2)^{-1}(\tilde{Y}_2 - Q\tilde{M}_2) \qquad (6)$$

$$: \begin{bmatrix} \tilde{X}_2 & -\tilde{Y}_2 \\ -\tilde{N}_2 & \tilde{M}_2 \end{bmatrix} \begin{bmatrix} M_2 & Y_2 \\ N_2 & X_2 \end{bmatrix} = I$$

where $\tilde{X}_2, \tilde{Y}_2, \tilde{M}_2, \tilde{N}_2$ are given by

$$\begin{bmatrix} \tilde{X}_2(s) & \tilde{Y}_2(s) \\ \tilde{N}_2(s) & \tilde{M}_2(s) \end{bmatrix} = \left[\begin{array}{c|cc} A_H & B_2 & H \\ \hline -F & I & 0 \\ I & 0 & I \end{array} \right] \qquad (7)$$

We note that although $T_3(s) = (sI - A_H)^{-1}B_1$ is strictly proper and tall, $T_3(s)$ has no finite unstable zeros except an infinite one. On the other hand, B_1 may not be of fullrank. If we assume $rankB_1 = q \leq r$, then by suitable column transformation, B_1 can be written as

$$B_1 = \tilde{B}_1 D_3 \quad , \quad D_3 D_3^T = I_q \qquad (8)$$

$$: \ \tilde{B}_1 \in R^{n \times q} \quad , \quad D_3 \in R^{q \times r}$$

in which $\tilde{B}_1$ is of fullrank(A column transformation yields $B_1 = \tilde{B}_1 D_3$, and $D_3 D_3^T$ can be factored as $D_3 D_3^T = UU^T$ with U invertible, then (8) is deduced by putting $D_3 = U^{-1}D_3, \tilde{B}_1 = \tilde{B}_1 U$). Apparently, $(sI - A_H)^{-1}\tilde{B}_1$ will not impose any constraint to H^∞ optimization because it has neither finite unstable zeros nor rank deficit(As can be seen later, the infinite zero does not affect H^∞ optimization). Therefore, we can absorb it into Q(s).

Now, We redefine

$$Q'(s) = Q(s)(sI - A_H)^{-1}\tilde{B}_1, \ T_3(s) = D_3 \qquad (9)$$

Then $T_3(s)$ becomes right invertible. Also, assumption A3 ensures that $T_2(s)$ is left invertible. Hence, our problem described by (4) becomes

$$\|T_1(s) - T_2(s)Q'(s)T_3(s)\|_\infty < 1 \qquad (10)$$

which falls right into the catalogue of normal four block problem. This transformation of free parameter Q(s) enables us to apply the theory of (Kimura, et al., 1988) to solve problem (10) for $Q'(s)$ in the first, then derive Q(s) that satisfies (4) from $Q'(s)$ via (9).

3. DERIVATION OF SOLUTIONS

This section is divided into four steps. The first step is concerned with the derivations of parameters defined in (Kimura, et al., 1988), the second step is the parametrization of all $Q'(s) \in RH^\infty$ that solve problem (10), the third step derives Q(s) that satisfies (4) from $Q'(s)$ and the last step characterizes all desired H^∞ controllers.

Step 1.

Due to assumption A3, there exist $D_{12}^+, D_{12}^\perp$ satisfying

$$\begin{bmatrix} D_{12}^+ \\ D_{12}^\perp \end{bmatrix} [D_{12}, (D_{12}^\perp)^T] = \begin{bmatrix} I_p & 0 \\ 0 & I_{m-p} \end{bmatrix}$$

Let

$$\hat{A}_2 = A - B_2D_{12}^+C_1 \ , \ C_l = D_{12}^\perp C_1 \ , \ K_l = -X_lC_1^T(D_{12}^\perp)^T \ ,$$

$$L_l = B_2D_{12}^+ - X_lC_1^T(D_{12}^\perp)^T D_{12}^\perp \ , \ M_l = -B_1$$

in which X_l is the solution of the Riccati equation

$$\hat{A}_2X_l + X_l\hat{A}_2^T + X_lC_1^T(D_{12}^\perp)^T D_{12}^\perp C_1X_l +$$

$$B_1B_1^T - B_2D_{12}^+(D_{12}^+)^T B_2^T = 0 \qquad (11)$$

that makes $-(\hat{A}_2 - K_lC_l)$ stable.

Moreover, due to the definition of D_3, there exist $D_3^+, D_3^\perp$ satisfying

$$\begin{bmatrix} D_3 \\ (D_3^\perp)^T \end{bmatrix} [D_3^+, D_3^\perp] = \begin{bmatrix} I_q & 0 \\ 0 & I_{r-q} \end{bmatrix}$$

We get from (8) that $D_3^+ = D_3^T$. Since $B_1 = \tilde{B}_1D_3$, it follows that

$$\begin{bmatrix} -T_3 \\ (D_3^\perp)^T(I - \tilde{T}_1T_1) \end{bmatrix} [-D_3^+, D_3^\perp] = \begin{bmatrix} I_q & 0 \\ 0 & I_{r-q} \end{bmatrix}.$$

This means $V_r(s)$ defined in (Kimura, et al., 1988) is $[-D_3^+, D_3^\perp]$. Therefore, in applying the conjugation theory, those parts that are concerned with $V_r(s)$ should be omitted.

It is easy to verify that when $WD_{12}^+(D_{12}^+)^TW^T = I_p$ holds for some matrix W, then

$$D_{co} = \begin{bmatrix} (D_{12}^+)^TW^T & 0 \\ 0 & D_3^T \end{bmatrix}$$

becomes (J, J')-unitary(i.e., $D_{co}^T J D_{co} = J'$) where $J = I_m \oplus -I_r, J' = I_p \oplus -I_q$.

Step 2.

With these preceding preparations, we are now ready to apply theorem 3 of (Kimura, et al., 1988) to our problem described in (10). In the following discussion, we will omit the variable s for simplicity when it is obvious from context that a coefficient matrix or a vector is a function of s.

Theorem 1. *There exists a $Q'(s) \in RH^\infty$ satisfying (10) if and only if the solution X_l of (11) exists, which is positive definite and stabilizes $-(\hat{A}_2 - K_lC_l)$. In that case, $Q'(s) \in RH^\infty$ is parametrized as*

$$Q'(s) = (\Pi_{11}S + \Pi_{12})(\Pi_{21}S + \Pi_{22})^{-1}, \ S(s) \in BH^\infty, \qquad (12)$$

where

$$\Pi(s) = \begin{bmatrix} \Pi_{11} & \Pi_{12} \\ \Pi_{21} & \Pi_{22} \end{bmatrix}$$

$$= \left[\begin{array}{cc|cc} A_H & -B_1B_1^T & 0 & \tilde{B}_1 \\ 0 & -(\hat{A}_2 - K_lC_l)^T & -X_l^{-1}B_2W^{-1} & -X_l^{-1}\tilde{B}_1 \\ \hline -F & -(F - F_\infty)X_l & -W^{-1} & 0 \\ 0 & -\tilde{B}_1^T & 0 & I_q \end{array} \right] \qquad (13)$$

with

$$F_\infty = -[D_{12}^+C_1 + D_{12}^+(D_{12}^+)^T B_2^T X_l^{-1}] \qquad (14) \qquad \square$$

(See the Appendix)

However, the following representation of $Q'(s)$ is more suitable for deriving Q(s) from $Q'(s)$.

$$Q'(s) = \Delta_{11}' + \Delta_{12}'S(I - \Delta_{22}'S)^{-1}\Delta_{21}', \ S(s) \in BH^\infty \qquad (15)$$

$$: \ \Delta'(s) = \begin{bmatrix} \Delta_{11}' & \Delta_{12}' \\ \Delta_{21}' & \Delta_{22}' \end{bmatrix} = \left[\begin{array}{c|c} A_{\Delta'} & B_{\Delta'} \\ \hline C_{\Delta'} & D_{\Delta'} \end{array} \right]$$

The parameters of $\Delta'(s)$ are calculated from $\Pi(s)$ as(refer to Genin, et al.(1983) or Kimura (1989) for formulae)

$$A_{\Delta'} = \begin{bmatrix} A_H & 0 \\ 0 & -(\hat{A}_2 - K_lC_l)^T - X_l^{-1}B_1B_1^T \end{bmatrix} ,$$

$$B_{\Delta'} = \left[\begin{array}{cc} -\tilde{B}_1 & 0 \\ X_l^{-1}\tilde{B}_1 & X_l^{-1}B_2 W^{-1} \end{array} \right] ,$$

$$C_{\Delta'} = \left[\begin{array}{cc} F & (F - F_\infty)X_l \\ 0 & -\tilde{B}_1^T \end{array} \right] ,$$

$$D_{\Delta'} = \left[\begin{array}{cc} 0 & -W^{-1} \\ I_q & 0 \end{array} \right] .$$

Here we have utilized $B_1 B_1^T = \tilde{B}_1 \tilde{B}_1^T$. From (11), we see that

$$\begin{aligned} & X_l[-(\hat{A}_2 - K_l C_l)^T - X_l^{-1}B_1 B_1^T]X_l^{-1} \\ =\ & [-X_l \hat{A}_2^T - X_l C_1^T (D_{12}^\perp)^T D_{12}^\perp C_l X_l - B_1 B_1^T]X_l^{-1} \\ =\ & [\hat{A}_2 X_l - B_2 D_{12}^+(D_{12}^+)^T B_2^T]X_l^{-1} \\ =\ & A + B_2 F_\infty \end{aligned}$$

Using this relation, $\Delta'(s)$ finally can be written as

$$\Delta'(s) = \left[\begin{array}{cc|cc} \Lambda_H & 0 & -\tilde{B}_1 & 0 \\ 0 & A + B_2 F_\infty & \tilde{B}_1 & B_2 W^{-1} \\ \hline F & F - F_\infty & 0 & -W^{-1} \\ 0 & -\tilde{B}_1^T X_l^{-1} & I_q & 0 \end{array} \right] \quad (16)$$

by taking a similarity transformation in $\Delta'(s)$ with $T = \left[\begin{array}{cc} I & 0 \\ 0 & X_l^{-1} \end{array} \right]$. Further, (11) can be rewritten as

$$\begin{aligned} & (A + B_2 F_\infty)X_l + X_l(A + B_2 F_\infty)^T \\ =\ & -B_2 D_{12}^+(D_{12}^+)^T B_2^T - B_1 B_1^T - K_l K_l^T \end{aligned} \quad (17)$$

Therefore, $(A + B_2 F_\infty)$ must be stable because $X_l > 0$ and $(A + B_2 F_\infty, B_2)$ is stabilizable. So, $\Delta'(s)$ is stable.

Step 3.

From (15) and the definition of $Q'(s)$, we have

$$Q(s)(sI - A_H)^{-1}\tilde{B}_1 = \Delta'_{11} + \Delta'_{12}S(I - \Delta'_{22}S)^{-1}\Delta'_{21} \quad (18)$$

That is to say if $Q(s) \in RH^\infty$ satisfies this equation, then $\alpha < 1$. Note that $(sI - A_H)^{-1}\tilde{B}_1$ is not right invertible so that Q(s) is not uniquely determined from $Q'(s)$.

Now we go to extract $(sI - A_H)^{-1}\tilde{B}_1$ from Δ'_{11} and Δ'_{21}. Making use of the identity $s(sI - A)^{-1} = I + (sI - A)^{-1}A$, we have

$$\begin{aligned} \Delta'_{11} =\ & -F(sI - A_H)^{-1}\tilde{B}_1 + (F - F_\infty)(sI - A - B_2 F_\infty)^{-1} \\ & \bullet(sI - A_H)(sI - A_H)^{-1}\tilde{B}_1 \\ =\ & \{-F + (F - F_\infty)[I + (sI - A - B_2 F_\infty)^{-1} \\ & \bullet(A + B_2 F_\infty - A_H)]\}(sI - A_H)^{-1}\tilde{B}_1 \\ =\ & [-F_\infty + (F - F_\infty)(sI - A - B_2 F_\infty)^{-1}(B_2 F_\infty - H)] \\ & \bullet(sI - A_H)^{-1}\tilde{B}_1 \end{aligned}$$

and

$$\begin{aligned} \Delta'_{21} =\ & I_q - \tilde{B}_1^T X_l^{-1}(sI - A - B_2 F_\infty)^{-1}\tilde{B}_1 \\ =\ & [\tilde{B}_1^+ - \tilde{B}_1^T X_l^{-1}(sI - A - B_2 F_\infty)^{-1}](sI - A_H) \\ & \bullet(sI - A_H)^{-1}\tilde{B}_1 \end{aligned}$$

where $\tilde{B}_1^+ \tilde{B}_1 = I_q$. Therefore, from (18) and Theorem 1, we obtain the following corollary.

Corollary 1. $\alpha < 1$ if and only if the solution X_l of (11) exists, which is positive definite and stabilizes $-(\hat{A}_2 - K_l C_l)$. In that case, all solutions $Q(s) \in RH^\infty$ are parametrized as

$$Q(s) = \Delta_{11} + \Delta_{12}S(I - \Delta_{22}S)^{-1}\Delta_{21}, \ S(s) \in BH^\infty \quad (19)$$

with

$$\begin{aligned} \Delta_{11} &= (F - F_\infty)(sI - A - B_2 F_\infty)^{-1}(B_2 F_\infty - H) - F_\infty + U(s) \\ \Delta_{12} &= [-I_p + (F - F_\infty)(sI - A - B_2 F_\infty)^{-1}B_2]W^{-1} \\ \Delta_{21} &= [\tilde{B}_1^+ - \tilde{B}_1^T X_l^{-1}(sI - A - B_2 F_\infty)^{-1}](sI - A_H) \\ \Delta_{22} &= -\tilde{B}_1^T X_l^{-1}(sI - A - B_2 F_\infty)^{-1}B_2 W^{-1} \end{aligned} \quad (20)$$

in which $U(s) \in RH^\infty \bigcap Null[(sI - A_H)^{-1}\tilde{B}_1]$. $\quad\square$

Apparently Δ_{21} is improper and Δ_{12} is proper. Hence, to ensure the causality of Q(s), $S(s) \in BH^\infty$ must be strictly proper. In the derivation of desired controllers, we only consider the case $U(s) = 0$, the general case may be rather involved.

Step 4.

The last step is to show that all desired controllers are of a special structure, and that the final result has nothing to do with the coprime factorization of the plant $G(s)$.

For the linear fractionally represented controller $K(s) = (\Phi_{11}Q + \Phi_{12})(\Phi_{21}Q + \Phi_{22})^{-1}$ (Φ_{ij} is the (i,j) block of Φ, Δ_{ij} is the (i,j) block of Δ, etc..), K(s) has a scattering structure (see Fig.2) where $\left[\begin{array}{c} b_1 \\ a_1 \end{array} \right] = \Phi \left[\begin{array}{c} a_2 \\ b_2 \end{array} \right]$, $b_1 = K(s)a_1$, $a_2 = Q(s)b_2$. Fig.2 can be transformed into Fig.3 where $\left[\begin{array}{c} b_1 \\ h_2 \end{array} \right] = J \left[\begin{array}{c} a_1 \\ a_2 \end{array} \right]$. Meanwhile Q(s) of (19) has a similar structure as Fig.3 for Δ, therefore K(s) finally possesses the structure of Fig.4 with $\left[\begin{array}{c} a_2 \\ b_3 \end{array} \right] = \Delta \left[\begin{array}{c} b_2 \\ a_3 \end{array} \right]$, $a_3 = S(s)b_3$. Let $\Psi := \Phi^{-1}$, we have $\left[\begin{array}{c} a_2 \\ b_2 \end{array} \right] = \Psi \left[\begin{array}{c} b_1 \\ a_1 \end{array} \right]$. Then it is a straightforward calculation to obtain

$$\left[\begin{array}{c} b_1 \\ b_3 \end{array} \right] = \left[\begin{array}{cc} \Sigma_{11} & \Sigma_{12} \\ \Sigma_{21} & \Sigma_{22} \end{array} \right] \left[\begin{array}{c} a_1 \\ a_3 \end{array} \right] \quad (21)$$

where

$$\begin{aligned} \Sigma_{11} &= (\Psi_{11} - \Delta_{11}\Psi_{21})^{-1}(\Delta_{11}\Psi_{22} - \Psi_{12}) \\ \Sigma_{12} &= (\Psi_{11} - \Delta_{11}\Psi_{21})^{-1}\Delta_{11} \\ \Sigma_{21} &= \Delta_{21}\Psi_{22} + \Delta_{21}\Psi_{21}\Sigma_{11} \\ \Sigma_{22} &= \Delta_{22} + \Delta_{21}\Psi_{21}\Sigma_{12} \end{aligned} \quad (22)$$

Therefore, the controllers K(s) can be represented by

$$K(s) = \Sigma_{11} + \Sigma_{12}S(I - \Sigma_{22}S)^{-1}\Sigma_{21}, \ S(s) \in BH^\infty \quad (23)$$

In our problem, K(s) is given by (6), so $\Phi = \left[\begin{array}{cc} M_2 & -Y_2 \\ N_2 & -X_2 \end{array} \right]$. Furthermore(use (6))

$$\begin{aligned} \Psi &= \left[\begin{array}{cc} I & 0 \\ 0 & -I \end{array} \right]^{-1} \left[\begin{array}{cc} M_2 & Y_2 \\ N_2 & X_2 \end{array} \right]^{-1} \\ &= \left[\begin{array}{cc} I & 0 \\ 0 & -I \end{array} \right] \left[\begin{array}{cc} \tilde{X}_2 & -\tilde{Y}_2 \\ -\tilde{N}_2 & \tilde{M}_2 \end{array} \right] = \left[\begin{array}{cc} \tilde{X}_2 & -\tilde{Y}_2 \\ \tilde{N}_2 & -\tilde{M}_2 \end{array} \right] \end{aligned}$$

Thus, we have

$$\begin{aligned} \Psi_{11} - \Delta_{11}\Psi_{21} &= \tilde{X}_2 - \Delta_{11}\tilde{N}_2 \\ &= I - (F - F_\infty)(sI - A - B_2 F_\infty)^{-1}B_2 \\ \Delta_{11}\Psi_{22} - \Psi_{12} &= \tilde{Y}_2 - \Delta_{11}\tilde{M}_2 \\ &= [I - (F - F_\infty)(sI - A - B_2 F_\infty)^{-1}B_2]F_\infty \end{aligned}$$

Substituting them into (22), we obtain the class of H^∞ controllers achieving $\alpha < 1$.

Theorem 2. *The class of desired controllers under the condition* $U(s) = 0$ *is given by (23) with*

$$\begin{aligned} \Sigma_{11} &= F_\infty \\ \Sigma_{12} &= -W^{-1} \\ \Sigma_{21} &= -s\tilde{B}_1^+ + \tilde{B}_1^+(A + B_2 F_\infty) + \tilde{B}_1^T X_l^{-1} \\ \Sigma_{22} &= -\tilde{B}_1^+ B_2 W^{-1} \quad\quad\quad\quad\quad\square \end{aligned} \quad (24)$$

In (23), S(s) can only be taken strictly proper in order to ensure the causality of H^∞ controllers. When S(s)=0, the corresponding

H^∞ controller is called the central solution.

Before proving Theorem 2, we give the following lemma.

Lemma 1. Let $G(s) = C(sI - A)^{-1}B$ be any arbitrary system that has no poles on the imaginary axis. If there exist matrices P and K (include 0) satisfying

$$PA + A^T P + PBB^T P + C^T C + K^T K = 0$$

then $\|G\|_\infty \lesssim 1$.

Proof. This is obvious because

$$
\begin{aligned}
& I - B^T(-j\omega I - A^T)^{-1}C^T C(j\omega I - A)^{-1}B \\
= \ & I - B^T(-j\omega I - A^T)^{-1}[P(j\omega I - A) + (-j\omega I - A^T)P \\
& -PBB^T P - K^T K](j\omega I - A)^{-1}B \\
= \ & I - B^T(-j\omega I - A^T)^{-1}PB - B^T P(j\omega I - A)^{-1}B \\
& +B^T(-j\omega I - A^T)^{-1}PBB^T P(j\omega I - A)^{-1}B \\
& +B^T(-j\omega I - A^T)^{-1}K^T K(j\omega I - A)^{-1}B \\
= \ & [I - B^T P(j\omega I - A)^{-1}B]^*[I - B^T P(j\omega I - A)^{-1}B] \\
& +[K(j\omega I - A)^{-1}B]^*[K(j\omega I - A)^{-1}B] \\
\geq \ & 0 \qquad\qquad \square
\end{aligned}
$$

Proof of Theorem 2 :

From the preceding arguments, we see that controllers given in Theorem 2 achieve $\alpha < 1$. Therefore, to prove K(s) is really the desired controller, we need only to show that the closed loop system is stable. Noting that the closed loop system is of the configuration of Fig.5, it is routine to obtain

$$z = [\Gamma_{11} + \Gamma_{12}S(I - \Gamma_{22}S)^{-1}\Gamma_{21}]w \tag{25}$$

$$
\begin{aligned}
\Gamma_{11} &= G_{11} + G_{12}\Sigma_{11}(I - G_{22}\Sigma_{11})^{-1}G_{21} \\
\Gamma_{12} &= G_{12}[I + \Sigma_{11}(I - G_{22}\Sigma_{11})^{-1}G_{22}]\Sigma_{12} \\
\Gamma_{21} &= \Sigma_{21}(I - G_{22}\Sigma_{11})^{-1}G_{21} \\
\Gamma_{22} &= \Sigma_{22} + \Sigma_{21}(I - G_{22}\Sigma_{11})^{-1}G_{22}\Sigma_{12}
\end{aligned}
\tag{26}
$$

Substituting (1) and (24) into $\Gamma(s)$, we find

$$\Gamma(s) = \left[
\begin{array}{c|cc}
A + B_2 F_\infty & B_1 & -B_2 W^{-1} \\
\hline
C_1 + D_{12}F_\infty & 0 & -D_{12}W^{-1} \\
\tilde{B}_1^T X_l^{-1} & -D_3 & 0
\end{array}
\right] \tag{27}$$

which is obviously stable. In view of (25), to prove the stability of the closed loop system, we need only to verify that $(I - \Gamma_{22}S)^{-1}$ is stable. Since $D_{12}^+(D_{12}^+)^T = W^{-1}W^{-T}$, $B_1 B_1^T = \tilde{B}_1 \tilde{B}_1^T$, (17) is equivalent to

$$
X_l^{-1}(A + B_2 F_\infty) + (A + B_2 F_\infty)^T X_l^{-1} = \\
-X_l^{-1}B_2 W^{-1}W^{-T}B_2^T X_l^{-1} - X_l^{-1}\tilde{B}_1\tilde{B}_1^T X_l^{-1} - X_l^{-1}K_l K_l^T X_l^{-1}
$$

So we conclude by Lemma 1 that $\|\Gamma_{22}\|_\infty \leq 1$. Since $S(s) \in BH^\infty$, $\Gamma_{22}S$ is stable and $\|\Gamma_{22}S\|_\infty \leq \|\Gamma_{22}\|_\infty\|S\|_\infty < 1$. Then, according to small gain theorem, $(I - \Gamma_{22}S)^{-1}$ must be stable. That is, the closed loop system is stable. $\square$

Remark 2. *When we impose the restriction $D_{12}^T[C_1, D_{12}] = [0, I]$, then $D_{12}^+ = D_{12}^T$. $D_{12}^+ C_1 = 0, D_{12}^+(D_{12}^\perp)^T = 0$ implies there must be a matrix C such that $C_1 = (D_{12}^\perp)^T C$, so $C_1^T(D_{12}^\perp)^T D_{12}^\perp C_1 = C^T D_{12}^\perp(D_{12}^\perp)^T \bullet D_{12}^\perp(D_{12}^\perp)^T C = C^T C$ and $C_1^T C_1 = C^T D_{12}^\perp(D_{12}^\perp)^T C = C^T C$ hold. Therefore, we have $C_1^T(D_{12}^\perp)^T D_{12}^\perp C_1 = C_1^T C_1$. Hence, (11) reduces to*

$$AX_l + X_l A^T + X_l C_1^T C_1 X_l + B_1 B_1^T - B_2 B_2^T = 0 \tag{28}$$

$X_l > 0$ implies $P = X_l^{-1} > 0$. Furthermore, as (Doyle, et al., 1988) we define a Hamiltonian as

$$H := \left[
\begin{array}{cc}
A & B_1 B_1^T - B_2 B_2^T \\
-C_1^T C_1 & -A^T
\end{array}
\right] \tag{29}$$

A similarity transformation with $T = \left[\begin{array}{cc} I & X_l \\ 0 & I \end{array}\right]$ and the substitution of (28) show that H is similar to

$$\left[
\begin{array}{cc}
A + X_l C_1^T C_1 & 0 \\
-C_1^T C_1 & -(A + X_l C_1^T C_1)^T
\end{array}
\right] \tag{30}$$

Because $\hat{A}_2 - K_l C_l = A + X_l C_1^T C_1$, we see that the requirement that $-(\hat{A}_2 - K_l C_l)$ be stable means that H has no eigenvalues on the imaginary axis and $\chi_-(H)$ is complementary to $Im\left[\begin{array}{c} I \\ 0 \end{array}\right]$. Therefore, the solvability condition here is the same as that of Doyle, et al.(1988). Their (sub)optimal control law is just the central solution $K(s) = F_\infty$ of (23).

4. SOLUTION OF ONE BLOCK PROBLEM

For the case that D_{12} is square and nonsingular, the problem turns out to be a one block problem(Petersen, 1988) whose solution is not contained in that of four block problem for reasons below.

Making a control input transformation $v = C_1 x + D_{12}u$ in (2), (2) becomes

$$
\begin{aligned}
\dot{x}(t) &= (A - B_2 D_{12}^{-1}C_1)x(t) + B_1 w(t) + B_2 D_{12}^{-1}v(t) \\
z(t) &= v(t)
\end{aligned}
\tag{31}
$$

Because the C-matrix in $z(t)$ is 0, for this system, $\hat{A}_2 - K_l C_l = A - B_2 D_{12}^{-1}C_1$ corresponds to the zeros of the transfer function of $u(t)$ to $z(t)$ which in general contain both stable and antistable ones, so the requirement that $-(\hat{A}_2 - K_l C_l)$ be stable is violated.

However, this problem can be equally easily solved by using the conjugation theory for one block model-matching problem. For simplicity of notation, we formulate this problem as

$$
\begin{aligned}
\dot{x}(t) &= Ax(t) + Dw(t) + Bv(t) \\
z(t) &= v(t) \\
y(t) &= x(t)
\end{aligned}
\tag{32}
$$

Since when A is stable, a trivial solution is $v(t) = 0$, we assume

A1. The pair (A,B) is stabilizable.
A2. A is not stable but has no eigenvalues on the imaginary axis.
A3. D matrix is column fullrank.

Without loss of generality, we can write

$$A = \left[\begin{array}{cc} A_1 & A_2 \\ 0 & A_3 \end{array}\right], \ B = \left[\begin{array}{c} B_1 \\ B_2 \end{array}\right], \ D = \left[\begin{array}{c} D_1 \\ D_2 \end{array}\right] \tag{33}$$

in which A_1 is stable, A_3 is antistable. By the stabilizability of (A, B),the pair (A_3, B_2) must be controllable. Thus in the following two Lyapunov equations associated with system (32)

$$A_3 S + S A_3^T = B_2 B_2^T \tag{34}$$

$$A_3 T + T A_3^T = D_2 D_2^T \tag{35}$$

we have unique solutions $S > 0$, $T \geq 0$.

Due to A1, F and H can be chosen such that $A_F := A + BF, A_H := A + H$ are stable matrices. Then, the corresponding $T_1(s), T_2(s), T_3(s)$ and other parameters of (4), (6) are(Francis, 1987)

$$\left[\begin{array}{cc} T_1(s) & T_2(s) \\ T_3(s) & 0 \end{array}\right] = \left[
\begin{array}{cc|cc}
A_F & -H & 0 & B \\
0 & A_H & D & 0 \\
\hline
F & 0 & 0 & I \\
0 & I & 0 & 0
\end{array}
\right],$$

$$\left[\begin{array}{cc} \tilde{X}_2(s) & \tilde{Y}_2(s) \\ \tilde{N}_2(s) & \tilde{M}_2(s) \end{array}\right] = \left[
\begin{array}{c|cc}
A_H & B & H \\
\hline
-F & I & 0 \\
I & 0 & I
\end{array}
\right]$$

Similar to the four block case, although $T_3(s)$ is strictly proper and tall, $T_3(s)$ has no finite unstable zeros so that we can absorb it into Q(s). Secondly, the zeros of $T_2(s)$ are the eigenvalues of A in which both stable ones and antistable ones exist. In order to apply conjugation theory, we factor $T_2(s)$ as the product of an inner matrix and an outer matrix. It is not difficult to see that

$$T_2(s) = \Theta(s)[T_2^{-1}(s)\Theta(s)]^{-1} \qquad (36)$$

$$: \Theta(s) = \left[\begin{array}{c|c} -A_3^T & -S^{-1}B_2 \\ \hline B_2^T & I \end{array}\right],$$

$$T_2^{-1}(s)\Theta(s) = \left[\begin{array}{c|c} A - BB_2^T S^{-1}(0,I) & -B \\ \hline F + B_2^T S^{-1}(0,I) & I \end{array}\right]$$

is such a factorization(This is actually obtained by the conjugation of A_3 mode in $T_2(s)^{-1}$).

Absorbing $T_3(s), [T_2^{-1}(s)\Theta(s)]^{-1}$ into Q(s), the problem becomes

$$\|T_1(s) - \Theta(s)Q'(s)\|_\infty < 1 \qquad (37)$$

$$: Q'(s) = [T_2^{-1}(s)\Theta(s)]^{-1}Q(s)T_3(s)$$

Write

$$T_1(s) = \left[\begin{array}{c|c} A_1 & B_1 \\ \hline C_1 & D_1 \end{array}\right], \quad \Theta(s) = \left[\begin{array}{c|c} A_2 & B_2 \\ \hline C_2 & D_2 \end{array}\right]$$

and define

$$\hat{A}_2 := A_2 - B_2 D_2^{-1} C_2 = S^{-1}A_3 S, \quad L := B_2 D_2^{-1} = -S^{-1}B_2$$

then the Sylvester equation $\hat{A}_2 R - R A_1 = L C_1$ has a unique solution $R = S^{-1}[0, I, 0, I]$. So

$$M := L D_1 + R B_1 = S^{-1}D_2$$

Applying theorem 7.1 b of (Kimura, 1989), we obtain

Theorem 3. $\alpha < 1$ if and only if $P > 0$ in

$$\hat{A}_2 P + P \hat{A}_2^T = LL^T - MM^T \qquad (38)$$

In that case, $Q'(s)$ is parametrized as

$$Q'(s) = \Delta'_{11} + \Delta'_{12}V(I - \Delta'_{22}V)^{-1}\Delta'_{21}, \ V(s) \in BH^\infty \qquad (39)$$

$$: \left[\begin{array}{cc} \Delta'_{11} & \Delta'_{12} \\ \Delta'_{21} & \Delta'_{22} \end{array}\right] =$$

$$\left[\begin{array}{cc|cc} A_1 & 0 & B_1 & 0 \\ 0 & \hat{A}_2 - LL^T P^{-1} & M & -L \\ \hline D_2^{-1}(C_1 + C_2 R) & -D_2^{-1}(C_2 + L^T P^{-1}) & 0 & -D_2^{-1} \\ 0 & -M^T P^{-1} & I & 0 \end{array}\right] \qquad \Box$$

However, (38) is equivalent to

$$A_3(SPS) + (SPS)A_3^T = B_2 B_2^T - D_2 D_2^T \qquad (40)$$

Comparing this with (34), (35), we get uniquely

$$SPS = S - T := Q_0$$

by the stability of $-A_3$. Therefore, $P > 0 \iff S > T$. That is, the solvability condition here is the same as that of (Petersen, 1988). Furthermore, (40) can be written as

$$(A_3 - B_2 B_2^T Q_0^{-1})Q_0 + Q_0(A_3 - B_2 B_2^T Q_0^{-1})^T = -B_2 B_2^T - D_2 D_2^T \qquad (41)$$

Then, $(A_3 - B_2 B_2^T Q_0^{-1})$ must be stable because $Q_0 > 0$ and $(A_3 - B_2 B_2^T Q_0^{-1}, B_2)$ is controllable.

It is readily verified by suitable similarity transformations that A_F mode is unobservable and $(A_3 - B_2 B_2^T S^{-1})$ mode is uncontrollable in $[T_2^{-1}(s)\Theta(s)]\Delta'_{11}$. Eliminating them out, we obtain

$$[T_2^{-1}(s)\Theta(s)]\Delta'_{11} = (F - F_\infty)(sI - A - BF_\infty)^{-1}D - FT_3(s)$$

$$= [-F_\infty + (F - F_\infty)(sI - A - BF_\infty)^{-1}(BF_\infty - H)]T_3(s)$$

in which

$$F_\infty = -B_2^T Q_0^{-1}[0, I] \qquad (42)$$

Similarly, $(A_3 - B_2 B_2^T S^{-1})$ mode can be shown to be uncontrollable in $[T_2^{-1}(s)\Theta(s)]\Delta'_{12}$. Eliminating it out,

$$[T_2^{-1}(s)\Theta(s)]\Delta'_{12} = -I + (F - F_\infty)(sI - A - BF_\infty)^{-1}B$$

Moreover, Δ'_{21} can be rewritten as

$$\Delta'_{21} = I - D_2^T Q_0^{-1}(sI - A_3 - B_2 B_2^T Q_0^{-1})^{-1}D_2$$
$$= I - D_2^T Q_0^{-1}(sI - A_3 - B_2 B_2^T Q_0^{-1})^{-1}[0, I]D$$
$$= I - D_2^T Q_0^{-1}[0, I](sI - A - BF_\infty)^{-1}D$$
$$= \{D^+ - D_2^T Q_0^{-1}[0, I](sI - A - BF_\infty)^{-1}\}(sI - A_H)T_3(s)$$

where $D^+ D = I$. Therefore Q(s) is parametrized as

$$Q(s) = \Delta_{11} + \Delta_{12}V(I - \Delta_{22}V)^{-1}\Delta_{21}, \ V(s) \in BH^\infty \qquad (43)$$

with

$$\Delta_{11} = -F_\infty + (F - F_\infty)(sI - A - BF_\infty)^{-1}(BF_\infty - H) + U(s)$$
$$\Delta_{12} = -I + (F - F_\infty)(sI - A - BF_\infty)^{-1}B$$
$$\Delta_{21} = \{D^+ - D_2^T Q_0^{-1}[0, I](sI - A - BF_\infty)^{-1}\}(sI - A_H)$$
$$\Delta_{22} = -D_2^T Q_0^{-1}[0, I](sI - A - BF_\infty)^{-1}B \qquad (44)$$

in which $U(s) \in RH^\infty \bigcap Null[T_3(s)]$. To ensure the causality of Q(s), $V(s) \in BH^\infty$ must be strictly proper.

Similar argument as that of step 4 of section 3 yields the class of H^∞ controllers achieving $\alpha < 1$. We only give the conclusion here, details are omitted since they are redundant.

Theorem 4. *The class of desired controllers under the condition $U(s) = 0$ is given by*

$$K(s) = \Sigma_{11} + \Sigma_{12}V(I - \Sigma_{22}V)^{-1}\Sigma_{21}, \ V(s) \in BH^\infty \qquad (45)$$

where $V(s)$ is strictly proper and

$$\begin{aligned}
\Sigma_{11} &= F_\infty \\
\Sigma_{12} &= -I \\
\Sigma_{21} &= -sD^+ + D^+(A + BF_\infty) + D_2^T Q_0^{-1}[0, I] \\
\Sigma_{22} &= -D^+ B
\end{aligned} \qquad (46)$$

Proof. As in the four block case, we have to prove the stability of the closed loop system. Applying (25),(26), we have

$$z = [\Gamma_{11} + \Gamma_{12}V(I - \Gamma_{22}V)^{-1}\Gamma_{21}]w$$

$$: \ \Gamma(s) = \left[\begin{array}{c|cc} A_3 - B_2 B_2^T Q_0^{-1} & D_2 & -B_2 \\ \hline -B_2^T Q_0^{-1} & 0 & -I \\ D_2^T Q_0^{-1} & -I & 0 \end{array}\right].$$

Obviously, $\Gamma(s)$ is stable. Therefore, we need only to prove the stability of $(I - \Gamma_{22}V)^{-1}$. Since $V(s) \in BH^\infty$, it is sufficient to show that $\|\Gamma_{22}\|_\infty \leq 1$. This is clear from Lemma 1 because (41) can be rewritten as

$$Q_0^{-1}(A_3 - B_2 B_2^T Q_0^{-1}) + (A_3 - B_2 B_2^T Q_0^{-1})^T Q_0^{-1}$$

$$= -Q_0^{-1}B_2 B_2^T Q_0^{-1} - Q_0^{-1}D_2 D_2^T Q_0^{-1} \qquad \Box$$

Let $V(s) = 0$ in (45), we see that the controller $K(s) = F_\infty$ given by Petersen(1988) is just our central solution.

5. CONCLUSION

The state feedback H^∞ control problem is solved by using the conjugation approach(Kimura, 1988, 1989). The parametrization of all state feedback H^∞ controllers is obtained, which differs from previous works on state feedback H^∞ control problem where only the central solution is given. Both four block and one block problems are solved in this paper. Our results embrace those of (Doyle, et al., 1988), (Petersen, 1988) as special cases.

REFERENCES

Doyle, J.C., Glover, K., Khargonekar, P.P., and Francis, B.A. (1988). State space solutions to standard H_2 and H_∞ control problems. *Proc. of Amer. Control Conf.*, 1691-1696.

Francis, B.A. (1987). *A course in H_∞ control theory*. Lecture Notes in Control and Information Sciences (Berlin: Springer-Verlag).

Genin, Y., Van Dooren, P., and Kailath, K. (1983). On Σ-lossless transfer functions and related questions. *Linear Algebra and Appl.*, **50**, 251-275.

Kawatani, R., and Kimura, H. (1989). Synthesis of reduced order H^∞ controllers based on conjugation. To appear in Int. J. Control.

Khargonekar, P.P., Petersen, I.R., and Rotea, M.A. (1988). H^∞ optimal control with state feedback. *I.E.E.E. Trans. Autom. Control*, **33**, 786-788.

Kimura, H. (1989). Conjugation, interpolation and model-matching in H^∞. *Int. J. Control*, **49**, 269-307.

Kimura, H., and Kawatani, R. (1988). Synthesis of H^∞ controllers based on conjugation. *Proc. of 27th I.E.E.E. Conf. on Decision and Control*, 7-13.

Petersen, I.R. (1987). Disturbance attenuation and H^∞ optimization: A design method based on the algebraic Riccati equation. *I.E.E.E. Trans. Autom. Control*, **32**, 427-429.

Petersen, I.R. (1988). Complete result for a class of state feedback disturbance attenuation problems. *Proc. of 27th I.E.E.E. Conf. on Decision and Control*, 1349-1353.

Safanov, M.G., and Limbeer, D.J.N. (1988). Simplifying the H^∞ theory via loop shifting. *Proc. of 27th I.E.E.E. Conf. on Decision and Control*, 1399-1404.

Appendix: The calculation of $\Pi(s)$

Let $\Pi(s) = D_\Pi + C_\Pi(sI - A_\Pi)^{-1}B_\Pi$, $T_i(s) = \underline{D_i} + \underline{C_i}(sI - \underline{A_i})^{-1}\underline{B_i}$. Direct application of Theorem 3 of (Kimura, et al., 1988) yields (Note we have cancelled those parts that are associated with $V_r(s)$ and the dynamics of $T_3(s) = D_3$.)

$$A_\Pi = \begin{bmatrix} A_F & -B_2F & -B_1B_1^T \\ 0 & A_H & -B_1B_1^T \\ 0 & 0 & -(\dot{A}_2 - K_lC_l)^T \end{bmatrix},$$

$$B_\Pi = \left\{ \begin{bmatrix} \begin{pmatrix} I & 0 & 0 \\ 0 & I & 0 \end{pmatrix} \begin{pmatrix} 0 & B_1 \\ \underline{B_2D_2^+} & -\underline{B_2D_2^+D_1} \\ 0 & \end{pmatrix} \end{bmatrix} + \begin{bmatrix} 0 \\ 0 \\ I \end{bmatrix} X_l^{-1}[-L_l, M_l] \right\} D_{co}$$

$$= \begin{bmatrix} 0 & \tilde{B}_1 \\ 0 & \tilde{B}_1 \\ -X_l^{-1}B_2W^{-1} & -X_l^{-1}\tilde{B}_1 \end{bmatrix}$$

$$C_\Pi = \begin{bmatrix} \underline{D_2^+}(\underline{C_1}, \underline{C_2}) \begin{pmatrix} I & 0 \\ 0 & I \\ -I & 0 \end{pmatrix} & C_{13} \\ 0 & D_3M_l^T \end{bmatrix}$$

$$= \begin{bmatrix} 0 & -F & -(F - F_\infty)X_l \\ 0 & 0 & -\tilde{B}_1^T \end{bmatrix}$$

in which $D_3M_l^T = -D_3B_1^T = -D_3D_3^T\tilde{B}_1^T = -\tilde{B}_1^T$ is used. Also, we have

$$D_\Pi = \begin{bmatrix} -W^{-1} & 0 \\ 0 & I_q \end{bmatrix}.$$

Apparently, A_F mode is unobservable. Hence, (13) is finally obtained by the elimination of A_F mode. $\quad\square$

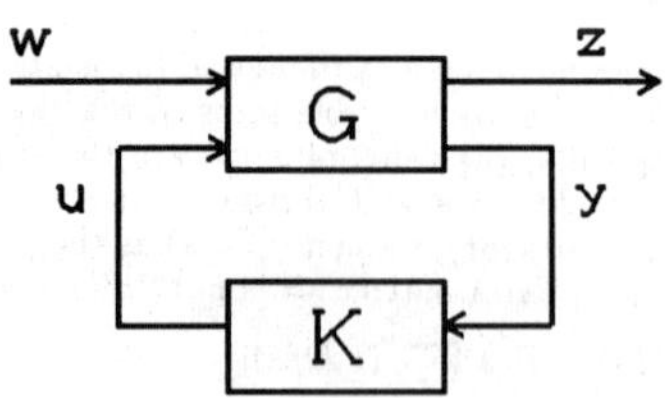

Fig.1　The feedback system

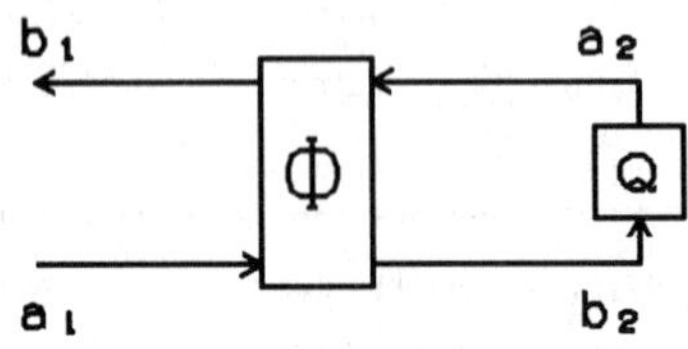

Fig.2　The scattering structure of K(s)

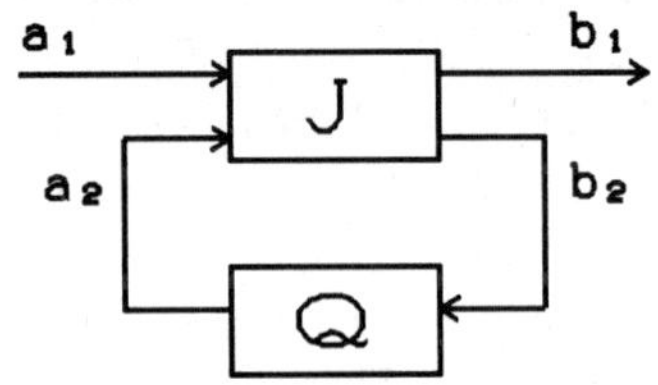

Fig.3　Another structure of K(s)

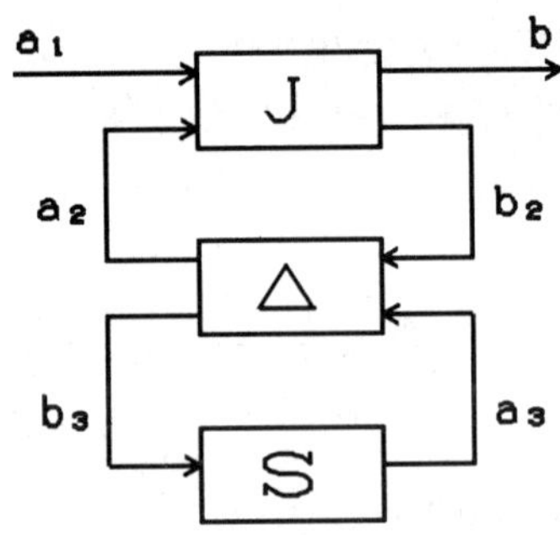

Fig.4　The structure of desired H^∞ controllers

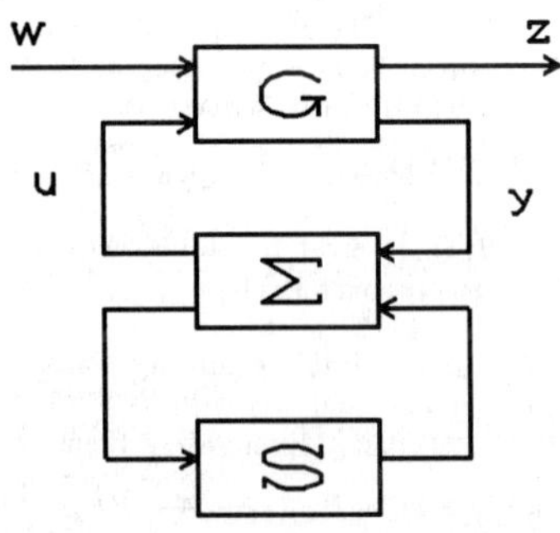

Fig.5　The configuration of desired H^∞ closed loop system

ROBUST DISTURBANCE REJECTION CONSTRAINT IN THE DESIGN OF FEEDBACK LINEAR SYSTEMS

A. Osorio-Cordero

CIEA del I.P.N. Depto. de Ing. Eléctrica, A.P. 14-740 07000, México, D.F., México

Abstract - The aim of this paper is the study and quantification of the conditions required to attain robust disturbance rejection. By robust disturbance rejection it is understood the rejection to disturbances at the output of a linear feedback system in the presence of plant nominal model uncertainties. Intuitively disturbance rejection is more difficult to achieve in the presence of plant model uncertainties than in their absence. In this paper we make this idea clear and quantify it for square plants in terms of the size of the plant. The norm employed throughout the paper is the Euclidean norm, which turns to be the largest singular value of the matrix in question.

Key Words - Feedback control, disturbance rejection, robust control.

NOMENCLATURE

R, C fields of real and complex numbers

C_+ ($Z \in C$ Re $(Z) \geq 0$), closed right-half complex plane.

C_- ($Z \in C$ Re $(Z) < 0$), left-half complex plane.

$R[S]$ ring of polynomials in S with coefficients in R.

$R(S)$ field of rational functions in S with coefficientes in R.

$Rp(S)$ ring of proper rational functions in S with coefficientes in R.

$Rpo(S)$ ring of strictly proper rational functions in S with coefficients in R.

$Rp(S)$ euclidean ring of proper rational functions in S and analytic in C_+.

$Rpo(S)$ euclidean ring of strictly proper rational functions in S and analytic in C_+.

$A \in M(B)$ matrix A with elements in the set B.

η_A^+ number of C_+ poles of $A \in M(R(S))$, counted according to its Mac Millan degree.

$\bar{\sigma}(A), \underline{\sigma}(A)$ Largest and smallest singular value respectively of $A \in M(B)$.

1. INTRODUCTION

It is well known that one of the main purposes of feedback is the reduction of the sensitivity of the system to plant uncertainties. Another advantage of feedback is the possibility of achieving output disturbance rejection. However the question arises: under which conditions is it possible to achieve both?

This problem was addressed by Chen and Desoer (1983) and also a solution was provided.

Consider the linear time invariant lumped multivariable system S(P,C,F) shown in figure 1. consisting of a plant, precompensator and feedback compensator.

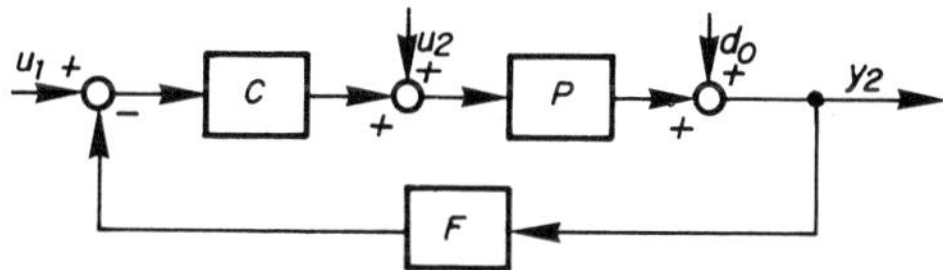

Fig. 1 The nominal system S(P,C,F).

Suppose P is subject to additive perturbations $\Delta P \in D$.

Denote by S (P + ΔP, C, F) the perturbed feedback system.

ASSUMPTIONS

A.1. $P \in R_{po}^{n_o \times n_i}(S), C \in R_p^{n_i \times n_o}(S)$ and $F \in R_p^{n_o \times n_o}(S)$ (1)

A.2. $D \stackrel{\triangle}{=} \{\Delta P :$

       a)$\Delta P \in R_{po}^{n_o \times n_i}(S)$ (2)

       b) $\| \Delta P(jw) \| \leq \ell(w), \quad \forall w \in R_+$ (3)

       c)$\eta_{p+\Delta p}^+ = \eta_p^+\}$ (4)

Where $w \to \ell(w)$ is a given "tolerance" function satisfying.

A.3. $\ell : R_+ \to R_+ \backslash \{0\}$ is continuos, and

A.4. $\exists$ a positive integer k s.t. $\ell(w)w^k > 1$ for all w sufficiently large.

                                        ⊙

<u>DEFINITION</u> (Robust Disturbance Rejection)

We say that the system S (P, C, F) achieves robust disturbance rejection in the frequency band of interest Ω over the class of additive perturbations D if and only if:

i) The system S $(P + \Delta P, C, F)$ is exponentially stable $\forall \Delta P \epsilon D$ and

ii) $\| H_{y_2 d_0}(\Delta P)(jw) \| << 1 \quad \forall w \epsilon \Omega, \forall \Delta P \epsilon D$ (5)

Where

$$H_{y_2 d_0}(\Delta P) = [I_{n_0} + (P + \Delta P)CF]^{-1} \stackrel{\triangle}{=} (T')^{-1} \quad (6)$$

and T' is the return difference of the perturbed system.

∅

The following lemma due to Chen and Desoer (1982) gives necessary and sufficient conditions for Robust stability:

LEMMA 1. Necessary and sufficient condition for Robust stability.

Consider the system S $(P + \Delta P, C, F)$ satisfying assumptions A.1 - A.4 u.t.c. if the system S (P, C, F) is exponentially stable, then the system S $(P + \Delta P, C, F)$ is exponentially stable, $\forall \Delta P \epsilon D$.

$$\Longleftrightarrow$$

$$\| CF(I_{n_0} + PCF)^{-1}(jw) \| \leq 1/\ell(w), \quad \forall w \epsilon \mathbf{R}_+ \quad (7)$$

Which in the case of a square plant can be written as

$$\| P^{-1}H(jw) \| \leq 1/\ell(w), \quad \forall w \epsilon \mathbf{R}_+ \quad (8)$$
and where, $H \stackrel{\triangle}{=} H_{y_2 u_1}$ from now on.

∅

PROOF. See Chen and Desoer (1982)

⊗

The following theorem is the main result of Chen and Desoer (1983)

THEOREM (Necessary and sufficient condition for robust disturbance rejection).

Let assumptions A.1 - A.4 hold and let $n_o \leq n_i$. Suppose that

$$\underline{\sigma}[P(jw)] \gg \ell(w), \quad \forall w \epsilon \Omega \quad (9)$$

u.t.c., if:

i) the system S (P, C, F) is exponentially stable; and

ii) $\underline{\sigma}[I_{no} + PCF(jw)] \gg 1, \quad \forall w \epsilon \Omega;$

then

The system S (P, C, F) achieves robust disturbance rejection in the band Ω over the class of additive perturbations D.

$$\Longleftrightarrow$$

i) $\bar{\sigma}[CF(I_{no} + PCF)^{-1}(jw)] \leq 1/\ell(w), \forall w \epsilon \mathbf{R}_+ \backslash \Omega \quad (10)$

ii) $\bar{\sigma}[CF(I_{n0} + PCF)^{-1}(jw)] \ll 1/\ell(w), \forall w \epsilon \Omega \quad (11)$

∅

PROOF. See Chen and Desoer (1983)

In the following we will make equation (11) more explicit in such a way that it will be able to be used as a design constraint in a mathematical programming problem.

2. PRELIMINARY RESULTS.

Let

$$T = I + PCF \quad \text{and} \quad (12)$$

$$T' = I + (P + \Delta P)CF \quad (13)$$

be the nominal and perturbed return difference functions respectively of the system of figure 1; they are related by:

$$T' = [I + \Delta P P^{-1} H]T \quad (14)$$

Where we have assumed that $n_i = n_o$

We impose on the nominal system two extra assumptions and we will see its consequences in the following assertion:

ASSERTION 1.

Let the system S (P, C, F) satisfy assumptions A.1 - A.4 and

A.5. $\quad \| T^{-1} \| \leq \epsilon \quad (0 < \epsilon \leq 1) \quad (15)$

A.6. $\quad \| P^{-1} \| \leq a/\ell(w) \quad 0 < a < 1 \quad (16)$

Where $\epsilon > 0$ and ℓ are given and a is determined from the size of P^{-1}. Define $\bar{a}$ by

$$\bar{a} \stackrel{\triangle}{=} a(1 + \epsilon) < 1 \quad (17)$$

Then

$$\| H \| \leq 1 + \epsilon \quad (18)$$

$$\| \Delta P P^{-1} H \| \leq \bar{a} \quad (19)$$

$$\| P^{-1} H \| \leq \frac{\bar{a}}{\ell} < \frac{1}{\ell} \quad (20)$$

∅

PROOF. By properties of norms and simple algebra

⊗

Note that A.6 i.e.

$$\| P^{-1} \| \le a/\ell, \, o < a < 1 \implies \| P \| > \ell > \| \Delta P \|$$

Observe that A.5 implies nominal disturbance rejection.

ASSERTION 2.

Suppose the system S (P, C, F) satisfies assumptions A.1 - A.6 and suppose it is exponentially stable then the perturbed system S (P + Δ P, C, F) is exponentially stable $\forall \Delta P \epsilon D$.

$$\oslash$$

PROOF.

By assertion 1, $\| P^{-1}H \| \le \bar{a}/\ell < 1/\ell$. Then by Lemma 1, S (P + Δ P, C, F) is exponentially stable $\forall \Delta P \epsilon D$.

$$\otimes$$

ASSERTION 3.

Let the system satisfy assumptions A1 - A5 and $n_i = n_o$ then

$$\| P^{-1} \| \le a/\ell$$

$$\implies$$

$$\| (T')^{-1} \| \le b \| T^{-1} \| \qquad \forall \Delta P \epsilon D \qquad (21)$$

Where

$$b = \frac{1}{1 - \bar{a}} > 1 \qquad (22)$$

$$\oslash$$

PROOF.

For $A \epsilon R_p^{m \times m}(S)$ and $\| A \| < 1$ the following holds

$$\| (I + A)^{-1} \| \le \frac{1}{1 - \| A \|} \qquad (23)$$

then

$$\| (I + \Delta P P^{-1} H)^{-1} \| \le \frac{1}{1 - \| \Delta P P^{-1} H \|} \le \frac{1}{1 - \bar{a}} \qquad (24)$$

Where use of (19) was made. On the other hand, taking inverse of (13).

$$(T')^{-1} = T^{-1}[I + \Delta P P^{-1} H]^{-1} \qquad (25)$$

Then, taking norms in (25), making use of the multiplicative property of norms and making use of (24), (21) results where b is given by (22).

$$\otimes$$

A.5 implies that the system S (P, C, F) has the disturbance rejection property. Equation (21) of assertion 3 implies that

$$\| (T')^{-1} \| \le b \| T^{-1} \| \le b\epsilon \qquad \forall \Delta P \epsilon D \qquad (26)$$

then if

$$b\epsilon < 1 \qquad (27)$$

the perturbed system S (P + ΔP, C, F) will also have the disturbance rejection property.

From (27) we get the condition the nominal design needs to satisfy in terms of "the size" of the plant (namely in terms of a) in order to achieve robust disturbance rejection.

$$b\epsilon = \frac{1}{1 - \bar{a}}\epsilon < 1 \qquad (28)$$

$$\Rightarrow$$

$$\epsilon < \frac{1 - a}{1 + a} \qquad (29)$$

or

$$a < \frac{1 - \epsilon}{1 + \epsilon}$$

in other words, in order to achieve robust disturbance rejection, the nominal design must be such that.

$$\| T^{-1} \| \le \epsilon < \frac{1 - a}{1 + a} \qquad (30)$$

3. MAIN RESULT

The previous results can be summarized in the following assertion.

ASSERTION 4.

Let the system S (P, C, F) satisfy the set of assumptions A.1 - A.6. u.t.c. if

i) The nominal system S (P, C, F) is exponentially stable and

ii) $\epsilon < \frac{1-a}{a+a}$ $\quad (\epsilon < 1 - \bar{a})$

then

$$\| (T')^{-1} \| \le b\epsilon < 1, \qquad \forall \Delta P \epsilon D \qquad (31)$$

i.e.

The system S (P, C, F) achieves robust disturbance rejection over the class of additive perturbations D.

$$\Longleftrightarrow$$

$$\parallel P^{-1}H \parallel \leq \frac{\bar{a}}{\ell} < \frac{1-\epsilon}{\ell} \tag{32}$$

$$\oslash$$

PROOF.

As far as robust stability is concerned, the proof of this assertion is direct from Lemma 1 because $1 - \epsilon < 1$. For this reason, we concentrate in the robust disturbance rejection issue.

We show first that (32) $\Longrightarrow$ (31).

By equations (14), (24) and A.5 the following can be written

$$\parallel (T')^{-1} \parallel = \parallel T^{-1}(I + \Delta P P^{-1}H)^{-1} \parallel$$
$$\leq \parallel T^{-1} \parallel \parallel (I + \Delta P P^{-1}H)^{-1} \parallel \leq \frac{\epsilon}{1-\bar{a}} = b\epsilon \tag{33}$$

but $\epsilon < 1 - \bar{a}$ in ii) of assertion 4, so

$$\parallel (T')^{-1} \parallel \leq b\epsilon < 1$$

Now we show (31) $\Longrightarrow$ (32) making use of the results given by Chen and Desoer (1983) in Lemma 3.4. This is done by contradiction.

Assume the negation of (32) as true:

$$\exists w_1 \epsilon R + \quad \parallel P_1^{-1}H_1 \parallel > \frac{\bar{a}}{\ell_1} \geq \frac{1-\epsilon}{\ell_1} \tag{34}$$

Subindex 1 meaning the function in question is evaluated at frequency w_1. Substituting $\parallel A \parallel$ by $\bar{\sigma}$ [A] and from the properties of singular values, (31) can be written as:

$$\underline{\sigma}[T]' > 1 \geq \frac{1}{b\epsilon} \tag{35}$$

from (14) and the fact that $\underline{\sigma}$ [A B] $\leq \underline{\sigma}$ [A] $\bar{\sigma}$ [B] the following can be written

$$\underline{\sigma}[T'] \leq \underline{\sigma}[I + \Delta P P^{-1}H]\bar{\sigma}[T] \tag{36}$$

Chen and Desoer in Lemma 3.4 (1983) constructed a $\Delta P_1 = \Delta P(jw_1) \ni$

$$\underline{\sigma}[I + \Delta P_1 P_1^{-1}H] = 1 - \ell_1\bar{\sigma}[P_1^{-1}H_1] + \epsilon' = \eta \tag{37}$$

Where $\epsilon' > 0$ is arbitrarily small.

Substitution of (34) in (37) gives:

$$\eta = 1 - \ell_1\bar{\sigma}[P_1^{-1}H_1] + \epsilon' < 1 - \ell_1\frac{\bar{a}}{\ell_1} + \epsilon' = 1 - \bar{a} + \epsilon'$$

$$= \underline{\sigma}[I + \Delta P_1 P_1^{-1}H_1] < 1 - \bar{a} + \epsilon' \tag{38}$$

However, A.5: $\bar{\sigma}[T^{-1}] \leq \epsilon$

$$\Longrightarrow \bar{\sigma}(T) \geq \frac{1}{\epsilon} \tag{39}$$

Evaluating (36) and (39) at jw_1 the following can be written:

$$\underline{\sigma}[T'_1] \leq \eta\bar{\sigma}[T_1] < (1 - \bar{a} + \epsilon')\frac{1}{\epsilon} \leq (1 - \bar{a} + \epsilon')\bar{\sigma}[T]$$

or

$$\underline{\sigma}(T'_1) < \frac{1 - \bar{a} + \epsilon'}{\epsilon} = \frac{1 - \bar{a}}{\epsilon} + \frac{\epsilon'}{\epsilon} \approx \frac{1 - \bar{a}}{\epsilon} = \frac{1}{b\epsilon} \tag{40}$$

but (40) contradicts (31), so (31) $\Longrightarrow$ (32) must be true.

$$\otimes$$

The results presented up to here are valid at a given frequency. In fact assumptions such as

$$\parallel T^{-1} \parallel \leq \epsilon < 1$$

are known to be valid in a range of frequencies, not only at one particular frequency neither for all $w \in R +$.

Bode showed that sensitivity can not arbitrarily be decreased over all $w \in R +$. If sensitivity is reduced in a band of frequencies it increases in its complement.

Also assumption A.6: $\parallel P^{-1} \parallel \leq a/\ell$ is not satisfied at all frequencies but in a band of frequencies.

In general the size of the perturbations increases at high frequencies and eventually it might be larger than the size of the plant itself at those frequencies.

These comments imply that the assumptions that have been made are sensible and valid at low and medium frequencies that are the operation range frequencies of a system.

The next proposition is the main result of this paper. It takes into account the remarks we have made and is a direct consequence of Lemma 1 and the results presented here.

PROPOSITION (Main Result).

Let the system satisfy assumptions A.1 - A.4 and also.

A.5 $\parallel T^{-1} \parallel \leq \epsilon \quad \forall w \epsilon \Omega, \epsilon > 0 \tag{41}$

A.6 $\parallel P^{-1} \parallel \leq a/\ell \leq \frac{a}{\parallel \Delta P \parallel}, \quad a < \frac{1}{1+\epsilon} \quad \forall w \epsilon \Omega \tag{42}$

u.t.c. if

i) The system S (P, C, F) is exponentially stable

ii) $\epsilon < \frac{1-a}{1+a}, \forall w \epsilon \Omega \tag{43}$

(i.e. the nominal system has the disturbance rejection property in the band of frequencies Ω).

Then,

the system S (P, C, F) achieves robust disturbance rejection
in the band Ω over the class of additive perturbations.

$$\Longleftrightarrow$$

i) $\| P^{-1}H \| \leq 1/\ell(w), \qquad \forall w \epsilon R_+ \backslash \Omega$ (44)

ii) $\| P^{-1}H \| \leq \frac{\bar{a}}{\ell} < \frac{1-\epsilon}{\ell}, \qquad \forall w \epsilon \Omega$ (45)

$$\oslash$$

PROOF.

The proof. follows directly from lemma 1 and from Assertion
4.

$$\otimes$$

CONCLUSIONS

From the proposition that is the main result, it can be seen
that the condition to assure robust disturbance rejection
presented by Chen and Desoer (1983) in their paper mentioned earlier i.e.

$$\| P^{-1}H \| \ll 1/\ell(w) \qquad \forall w \epsilon \Omega \qquad (46)$$

is made more specific by eliminating the rather vague "$\ll$"
sign.

On the other hand condition (46) tells that depending on the
value of "a" (indirectly depending on the size of the plant),
the condition to achieve robust disturbance rejection may
not be as restrictive as Chen and Desoer (1983) suggest
in (46), but anyway it is more restrictive than the condition
to achieve robust stability, mamely.

$$\| P^{-1}H \| \leq 1/\ell(w) \qquad \forall w \epsilon R_+ \qquad (47)$$

The importance of the quantification we have obtained is
that the results presented can be employed in the design
of linear systems because they have been expressed as
inequalities. These inequalities can be used as inequality
design constraints in a mathematical programming problem.

REFERENCES

Chen, M.J. and Desoer C.A. (1982). Necessary and sufficient condition for robust stability of linear distributed feedback systems. Int. J. Control, Vol, 35, No. 2, 255 - 267.

Chen, M.J. and Desoer, C. A. (1983). The problem of guaranteeing disturbance rejection in linear multivariable feedback
systems. Int. J. Control, Vol. 37, No. 2, 305 - 313.

DESIGN OF OBSERVER FOR LINEAR DESCRIPTOR SYSTEMS

S. Kawaji

*Department of Electrical Engineering & Computer Science, Kumamoto University,
Kumamoto 860, Japan*

Abstract. This paper is concerned with the problem of designing an observer for linear
descriptor systems $E\dot{x} = Ax + Bu$, $y = Cx$. At first, an observer is presupposed to have
the same structure as Luenberger Observer, and the fundamental equations that the plant
and the observer must satify are derived. Based on these equations, it is shown that
an observer exists if the system is observable in the sense of Rosenbrock. And the
design methods of an identity observer and a minimal-order observer are presented by
utilizing the generalized matrix inverse. Secondly, the realizability condition is relaxed by eliminationg the purely static modes of the system, and it is shown that the
observer can be realized if the system is observable in the sense of Verghese.

Keywords. descriptor system; observability; observers; purely static modes; system
order reduction

INTRODUCTION

Linear descriptor system is a mixture of dynamic
and algebraic equations, and can be served as a
useful method in analysis of many real and practical situations. Due to its extensive applications,
the descriptor system has attracted much attention
in recent years (See the survey paper by Lewis, 198
6).

For descriptor systems, the basic design method is
the feedback of the descriptor variables, and an
"observer" is indispensable for the implementation
of the control law via incomplete state observations. A considerable amount of research has been
devoted to the design of observers (El-Tohami,1983;
Shafai, 1987; Kobayasi, 1987; Kawaji, 1988). The
traditional approach to such an observer is to separate the dynamic equations from the algebraic equations, and to estimate the unobservable parts from
the measured outputs.

In this paper, contrary to this, an observer is
presupposed to have the same structure as Luenberger
Observer (Luenberger, 1971). Firstly, the fundamental equations that the system and the observer
must satisfy are derived. Based on these equations,
sufficient conditions are given for the existence
of observer, and methods of designing an identity
observer and a minimal order observer are presented
by utilizing the generalized matrix inverse. Secondly, the realizability condition is relaxed by
eliminating the purely static parts of the system.
Finally, an example is offered to illustrate the
procedure.

DEFINITIONS AND FUNDAMENTAL EQUATIONS

Consider the linear descriptor system described by

$$E\dot{x} = Ax + Bu$$
$$y = Cx \tag{1}$$

where $x \in R^n$ is the descriptor variables, $u \in R^m$ is the
input vector, $y \in R^p$ is the output vector. E, A, B, C
are constant maticies of appropriate dimensions,
and rank $C = p$. E is not necessarily nonsingular
and rank $E = r$ $(\leq n)$. The invertibility of $\lambda E - A$ is
assumed for the solution of (1) to be unique.

In system (1), there exist exponential modes, impulsive modes, and the purely static modes. Consequently, two kinds of observability are defined as
follows (Cobb 1984)
(i) System (1) is *R-observable* (observable in the
sense of Rosenbrock) iff

$$\text{rank} \begin{bmatrix} E \\ C \end{bmatrix} = n \tag{2}$$

$$\text{rank} \begin{bmatrix} \lambda E - A \\ C \end{bmatrix} = n, \quad \forall \lambda \in \mathbb{C} \tag{3}$$

(ii) Use an allowed transformation to bring the
modal observability matrix of system (1) to the form

$$\begin{bmatrix} \lambda E_1 - A_1 & A_2 \\ C_1 & C_2 \end{bmatrix} \text{ with } E_1 \text{ of full column rank}$$

Then system (1) is *V-observable* (observable in the
sense of Verghese) iff

$$\text{rank} \begin{bmatrix} E_1 & A_2 \\ 0 & C_2 \end{bmatrix} = n \tag{4}$$

$$\text{rank} \begin{bmatrix} \lambda E - A \\ C \end{bmatrix} = n, \quad \forall \lambda \in \mathbb{C} \tag{3}$$

Next consider the related system driven by $u(t)$ and
$y(t)$ of system (1) and giving the output $w(t)$

$$\dot{z} = \hat{A}z + \hat{B}y + \hat{J}u$$
$$w = \hat{C}z + \hat{D}y \tag{5}$$

where $z \in R^q$, and $\hat{A}$, $\hat{B}$, $\hat{J}$, $\hat{C}$, $\hat{D}$ are matrices of proper
dimensions.

<u>Definition</u>   If, for a specified matrix K,

$$\lim_{t \to \infty} [w(t) - Kx(t)] = 0, \quad \forall x(0-), z(0), u(\cdot) \tag{6}$$

holds, then system (5) is said to be a Kx-observer
for system (1). If $K = I_n$, system (5) is called a
state observer.

The following theorem containes the fundamental results of observer theory

<u>Theorem 1</u>   System (5) is an observer for system (1)

241

if

$$\text{Re } \lambda_i(\hat{A}) < 0, \quad i = 1, \cdots, q \tag{7a}$$

and if there exists a matrix $U \varepsilon R^{q \times n}$ such that

$$\hat{A}UE + \hat{B}C = UA \tag{7b}$$

$$\hat{J} = UB \tag{7c}$$

$$\hat{C}UE + \hat{D}C = K \tag{7d}$$

(Proof) Let

$$\xi \overset{\Delta}{=} z - UEx$$

then ξ is governed by

$$\dot{\xi} = \hat{A}\xi + (\hat{A}UE + \hat{B}C - UA)x + (\hat{J} - UB)u$$

which by involking (7b), (7c) reduces to

$$\dot{\xi} = \hat{A}\xi$$

It follows from (7a) that $\xi(t) \to 0$ as $t \to \infty$. Then

$$w = \hat{C}\xi + (\hat{C}UE + \hat{D}C)x$$

and hence, from (7d), $w(t) \to Kx(t)$ as $t \to \infty$.

Q.E.D.

Remark: Note that Theorem 1 provides a sufficient condition. While, if E is nonsingular, (7b)-(7d) reduce to

$$\begin{aligned} \hat{A}\bar{U} + \hat{B}C &= \bar{U}E^{-1}A \\ \hat{J} &= \bar{U}E^{-1}B \\ \hat{C}\bar{U} + \hat{D}C &= K \end{aligned} \tag{8}$$

where $\bar{U} \overset{\Delta}{=} UE$. It is well known in linear system that (8) is the necessary and sufficient condition.

In the following, only the case $K = I_n$, that is, a state observer is investigated.

DESIGN OF OBSERVERS

In the preceding section, a sufficient condition that a Luenberger-type system (5) can qualify as an observer for the descriptor system (1) was derived. The observer design problem is to find a matrix U such that the conditions (7a)-(7d) are satisfied when E, A, B, C, and observer's order q are given. In this section, an identity observer and a minimal order observer will be constructed.

Firstly, it is required from the condition (7d) that

$$\text{rank} \begin{bmatrix} E \\ C \end{bmatrix} = n \tag{9}$$

which implies that the impulse modes and the purely static modes of system (1) must be observable (Cobb, 1984). It is noted from (9) that $r \geq n - p$. When the condition (9) is satisfied, there exist two matricies $E^{\#} \varepsilon R^{n \times n}$ and $C^{\#} \varepsilon R^{n \times p}$ such that

$$[E^{\#} \quad C^{\#}] \begin{bmatrix} E \\ C \end{bmatrix} = I_n \tag{10}$$

i.e., $[E^{\#} \quad C^{\#}]$ is a generalized matrix inverse of $[E^T \quad C^T]^T$. Here we define a class

$$\begin{aligned} \Gamma \overset{\Delta}{=} \{ (E^{\#}, \ C^{\#}) : \det E^{\#} &\neq 0, \\ E^{\#}E + C^{\#}C &= I_n \} \end{aligned} \tag{11}$$

Then we have the following (Kawaji, 1988).

<u>Lemma</u> Γ is not empty.

(The proof is omitted). An algorithm to find an element of Γ is given in Appendix.

Also the condition (7d) indicates

$$\text{Rank} \begin{bmatrix} UE \\ C \end{bmatrix} = n \tag{12}$$

from which we have $n-p \leq \text{rank } U \leq n$. It follows that minimal order of Luenberger-type observer is $n-p$. In the sequel, an identity observer and a minimal order observer constructed by utilizing eqn. (10).

[I] Identity observer

For system (1), choosing an any $(E^{\#}, \ C^{\#}) \varepsilon \Gamma$ and letting

$$\hat{C} = I_n, \quad U = E^{\#}, \quad \hat{D} = C^{\#} \tag{13}$$

we get

$$\hat{C}UE + \hat{D}C = E^{\#}E + C^{\#}C = I_n$$

Condition (7d) is thus satisfied, and rank $U = n$. Substituting (13) into (7b) and use of (10) yields

$$\hat{A} + (\hat{B} - \hat{A}C^{\#})C = E^{\#}A \tag{14}$$

Letting $K = \hat{B} - \hat{A}C^{\#}$, we get

$$\hat{A} = E^{\#}A - KC \tag{15}$$

$$\hat{B} = \hat{A}C^{\#} + K = E^{\#}AC^{\#} + K - KCC^{\#} \tag{16}$$

Furthermore,

$$\hat{J} = E^{\#}B \tag{17}$$

from (7c). Therefore, we obtained an identity observer as

$$\begin{aligned} \dot{z} &= (E^{\#}A - KC)z + (E^{\#}AC^{\#} + K - KCC^{\#})y + E^{\#}Bu \\ \hat{x} &= z + C^{\#}y \end{aligned} \tag{18}$$

The remaining problem is the determination of the parameter K in such a way that a stable observer results.

For this, in (15), $(C, E^{\#}A)$ must be observable (the detectability is sufficient, strictly speaking. But an observer with freedom to choose all eigenvalues is frequently important.). That is,

$$\text{rank} \begin{bmatrix} \lambda I_n - E^{\#}A \\ C \end{bmatrix} = n, \quad \forall \lambda \ \varepsilon \ \mathbb{C} \tag{19}$$

is necessary. Noticing $(E^{\#}, \ C^{\#}) \ \varepsilon \ \Gamma$

$$\begin{bmatrix} \lambda I_n - E^{\#}A \\ C \end{bmatrix} = \begin{bmatrix} E^{\#} & \lambda C^{\#} \\ 0 & I_p \end{bmatrix} \begin{bmatrix} \lambda E - A \\ C \end{bmatrix} \tag{20}$$

and $\det E^{\#} = n$. Therefore we have

$$\text{rank} \begin{bmatrix} \lambda E - A \\ C \end{bmatrix} = n, \quad \forall \lambda \ \varepsilon \ \mathbb{C} \tag{21}$$

which implies that the exponential modes of system (1) must be observable (Cobb, 1984).

Thus we come to the theorem giving sufficient condition for the existence of an identity observer.

<u>Theorem 2</u>   An identity observer for system (1) can be realized if (9) and (21) holds, i.e. system (1) is R-observable.

[II] Minimal order observer

As is well known, the utilization of all the output information results a reduced-order observer. We will construct a minimal order observer since rank $C = p$.

There exists a matrix $U^{\#} \varepsilon R^{(n-p) \times n}$ such that

$$T^{-1} = \begin{bmatrix} U^{\#} \\ C \end{bmatrix} \begin{matrix} \}n-p \\ \}p \end{matrix} \tag{22}$$

is nonsingular. Denote

$$T = [H \quad D] \tag{23}$$

then we have

$$HU^{\#} + DC = I_n \tag{24}$$

242

Using any $(E^\#, C^\#) \in \Gamma$ and (24), we get

$$HU^\# E^\# E = (I_n - DC)(I_n - C^\# C)$$

$$= I_n - \{C^\# + D(I_p - CC^\#)\}C$$

or

$$[HU^\# E^\#\ \ C^\# + D(I_p - CC^\#)]\begin{bmatrix} E \\ C \end{bmatrix} = I_n \qquad (25)$$

So, taking account of rank $U = n-p$, set

$$\hat{C} = H$$
$$\hat{D} = C^\# + D(I_p - CC^\#) \qquad (26)$$
$$U = U^\# E^\#$$

then condition (7d) is satisfied. Substituting (26) into (7b) and rearranging them, we get

$$\hat{A}U^\# + (\hat{B} - \hat{A}U^\# C^\#)C = U^\# E^\# A$$

or

$$[\hat{A}\ \ \hat{B} - \hat{A}U^\# C^\#] = U^\# E^\# A \begin{bmatrix} U^\# \\ C \end{bmatrix}^{-1} \qquad (27)$$

from which

$$\hat{A} = U^\# E^\# A \begin{bmatrix} U^\# \\ C \end{bmatrix}^{-1} \begin{bmatrix} I_{n-p} \\ 0 \end{bmatrix} \qquad (28)$$

$$\hat{B} = U^\# E^\# A \begin{bmatrix} U^\# \\ C \end{bmatrix}^{-1} \begin{bmatrix} 0 \\ I_p \end{bmatrix} + \hat{A}U^\# C^\# \qquad (29)$$

Also condition (7c) gives

$$\hat{J} = U^\# E^\# B \qquad (30)$$

The stability of $\hat{A}$ still remains: it is necessary to find a matrix $U^\#$ rendering $\hat{A}$ in (28) stable. For this, $\tilde{C}$ is firstly constructed such that

$$\begin{bmatrix} \tilde{C} \\ C \end{bmatrix} (\overset{\Delta}{=} S^{-1})$$

is nonsingular. $U^\#$ satifying (22) is chosen as

$$U^\# = [I_{n-p}\ \ L]S^{-1} \qquad (31)$$

From (28) we have

$$\hat{A} = [I_{n-p}\ \ L]S^{-1}E^\# AS \begin{bmatrix} I_{n-p} \\ 0 \end{bmatrix}$$

$$= A_{11} + LA_{21} \qquad (32)$$

where

$$S^{-1}E^\# AS = \begin{bmatrix} A_{11} & A_{12} \\ A_{21} & A_{22} \end{bmatrix} \begin{matrix} \}n-p \\ \}p \end{matrix}$$
$$\underbrace{\phantom{A_{11}}}_{n-p}\ \underbrace{\phantom{A_{12}}}_{p}$$

It follows that if (A_{21}, A_{11}) is observable, we can determine the parameter L so that $\hat{A}$ has arbitrarily assignable poles. But we have

$$\begin{bmatrix} S^{-1} & 0 \\ 0 & I_p \end{bmatrix}\begin{bmatrix} \lambda I_n - E^\# A \\ C \end{bmatrix}S$$

$$= \begin{bmatrix} \lambda I_{n-p} - A_{11} & -A_{12} \\ -A_{21} & \lambda I_p - A_{22} \\ 0 & I_p \end{bmatrix}$$

since $(E^\#, C^\#) \in \Gamma$. Therefore, the observability of (A_{21}, A_{11})

$$\text{rank} \begin{bmatrix} \lambda I_{n-p} - A_{11} \\ A_{21} \end{bmatrix} = n-p, \ \forall \lambda \in C \qquad (38)$$

is equivalent to (19), hence to (21).

We have thus obtained a minimal order observer as

$$\dot{z} = (A_{11} + LA_{21})z + \{A_{12} + LA_{22}$$
$$- (A_{11} + LA_{21})(L - [I_{n-p}\ \ L]S^{-1}C^\#)\}y$$
$$+ [I_{n-p}\ \ L]\,S^{-1}E^\# Bu \qquad (39)$$

$$\hat{x} = S\begin{bmatrix} I_{n-p} \\ 0 \end{bmatrix}z + \{C^\# + S\begin{bmatrix} -L \\ I_p \end{bmatrix}(I_p - CC^\#)\}y$$

<u>Theorem 3</u> A minimal order observer for system (1) can be realized if (9) and (21) holds, i.e. system (1) is R-observable.

Remark: For the case $E = I_n$, $(I_n, 0) \in \Gamma$. It follows that the realizability condition (9) is obvious, and (21) reduces to the observability condition of (C, A). And an identity observer (18) leads to

$$\dot{z} = (A - KC)z + Ky + Bu$$
$$\hat{x} = z \qquad (40)$$

which was proposed by Luenberger (1971). Also, a minimal order observer (39) is written as

$$\dot{z} = (A_{11} + LA_{21})z + \{A_{12} + LA_{22}$$
$$- (A_{11} + LA_{21})L\}y + [I_{n-p}\ \ L]S^{-1}Bu \qquad (41)$$

$$\hat{x} = S\begin{bmatrix} I_{n-p} \\ 0 \end{bmatrix}z + S\begin{bmatrix} -L \\ I_p \end{bmatrix}y$$

which was proposed by Gopinath (1971).

RELAXATION OF RELAIZABILITY CONDITION

The condition in Theorem 1 or Theorem 2 is severe. In this section, the realizability condition for observer is relaxed by using the elimination of the purely static modes of the system.

Under the regularity assumption of system (1), two nonsingular matrices P and Q exists such that

$$PEQ = \text{diag}\{I_{n1},\ J,\ 0\}$$

$$PAQ = \text{diag}\{A_1,\ I_{n2},\ I_{n3}\}$$

where $J = \text{diag}\{J_{r1}, \cdots, J_{r\bar{s}}\}$, $J_{ri} \in R^{ri \times ri}$ is nilpotent $(ri \geq 2)$, $n1 = \deg \det(\lambda E - A)$, $n2 = r1 + \cdots + r\bar{s}$, $n1 + n2 + n3 = n$ (Gantmacher 1959). Then system (1) is r.s.e. to

$$\begin{bmatrix} I_{n1} & 0 & 0 \\ 0 & J & 0 \\ 0 & 0 & 0 \end{bmatrix}\begin{bmatrix} \dot{z}_1 \\ \dot{z}_2 \\ \dot{z}_3 \end{bmatrix} = \begin{bmatrix} A_1 & 0 & 0 \\ 0 & I_{n2} & 0 \\ 0 & 0 & I_{n3} \end{bmatrix}\begin{bmatrix} z_1 \\ z_2 \\ z_3 \end{bmatrix} + \begin{bmatrix} B_1 \\ B_2 \\ B_3 \end{bmatrix}u$$

$$y = [C_1\ \ C_2\ \ C_3]\begin{bmatrix} z_1 \\ z_2 \\ z_3 \end{bmatrix} \qquad (42)$$

where $z = Q^{-1}x$, $PB = [B_1^T\ B_2^T\ B_3^T]^T$, $CQ = [C_1\ C_2\ C_3]$.

It is noted that $z_3 = -B_3 u$, i.e. z_3 is directly connected with the input and has no dynamics. We call it the purely static modes.

By eliminatin this mode, we obtain a reduced system as

$$\begin{bmatrix} I_{n1} & 0 \\ 0 & J \end{bmatrix}\begin{bmatrix} \dot{z}_1 \\ \dot{z}_2 \end{bmatrix} = \begin{bmatrix} A_1 & 0 \\ 0 & I_{n2} \end{bmatrix}\begin{bmatrix} z_1 \\ z_2 \end{bmatrix} + \begin{bmatrix} B_1 \\ B_2 \end{bmatrix}u$$

$$y = [C_1\ \ C_2]\begin{bmatrix} z_1 \\ z_2 \end{bmatrix} - C_3 B_3 u \qquad (43)$$

Next we obtain is the relation between the observability of system (42) and that of system (43).

Proposition   If system (1) is V-observable, then the reduced system obtained by eliminating the purely static modes is R-observable.

(Proof)   Assume that system (1) is V-observable. By applying condition (3) to system (42),

$$\text{rank} \begin{bmatrix} \lambda I_{n1}-A_1 & 0 & 0 \\ 0 & \lambda J-I_{n2} & 0 \\ 0 & 0 & -I_{n3} \\ C_1 & C_2 & C_3 \end{bmatrix} = n, \quad \forall \lambda \in \mathbb{C}$$

Therefore

$$\text{rank} \begin{bmatrix} \lambda I_{n1}-A_1 & 0 \\ 0 & \lambda J-I_{n2} \\ C_1 & C_2 \end{bmatrix} = n_1+n_2, \quad \forall \lambda \in \mathbb{C} \quad (44)$$

Next, there exist two nonsingular matrices S and T for nilpotent matrix J such that

$$SJT = \begin{bmatrix} I_{n2-\bar{s}} & 0 \\ 0 & 0 \end{bmatrix}, \quad ST = \begin{bmatrix} T_{11} & T_{12} \\ T_{21} & 0 \end{bmatrix}$$

(the proof is simple and omitted). Denote

$$C_2 T = [\underbrace{C_{21}}_{n2-\bar{s}} \quad \underbrace{C_{22}}_{\bar{s}}]$$

Then condition (4) is equivalent to

$$\text{rank} \begin{bmatrix} I_{n1} & 0 & 0 & 0 \\ 0 & I_{n2-\bar{s}} & T_{12} & 0 \\ 0 & 0 & 0 & 0 \\ 0 & 0 & 0 & I_{n3} \\ 0 & 0 & C_{22} & C_3 \end{bmatrix} = n$$

which shows that

$$\text{rank } C_{22} = \bar{s} \quad (45)$$

We will now check the R-observability of the system obtained by eliminating the purely static modes of system (42) (i.e., system (43)). Condition (3) is satified by (44). Also condition (4) is shown to be satified by the relation

$$\text{rank} \begin{bmatrix} I_{n1} & 0 \\ 0 & J \\ C_1 & C_2 \end{bmatrix} = \text{rank} \begin{bmatrix} I_{n1} & 0 & 0 \\ 0 & S & 0 \\ 0 & 0 & I_p \end{bmatrix} \begin{bmatrix} I_{n1} & 0 \\ 0 & J \\ C_1 & C_2 \end{bmatrix} \begin{bmatrix} I_{n1} & 0 \\ 0 & T \end{bmatrix}$$

$$\begin{bmatrix} I_{n1} & 0 & 0 \\ 0 & I_{n2-s} & 0 \\ 0 & 0 & 0 \\ C_1 & C_{21} & C_{22} \end{bmatrix}$$

and use of (45).   Q.E.D.

Remark: Armentano (1986) showed the same results by using a geometric approach. Our derivation is based on the structure of regular pencile and is simple.

Finnaly, an observer for the V-observable system is constructed as follows: Denote $y_* \stackrel{\Delta}{=} y + C_3 B_3 u$ and $k = \text{rank } [C_1 \; C_2]$. Let R be the matrix which picks up k independent rows from y_*, and let $\bar{y}$ be new output. Then system (43) is written as

$$\bar{E}\dot{\bar{z}} = \bar{A}\bar{z} + \bar{B}u \quad (46)$$
$$\bar{y} = \bar{C}\bar{z}$$

where

$$\bar{z} = \begin{bmatrix} z_1 \\ z_2 \end{bmatrix}, \quad \bar{E} = \begin{bmatrix} I_{n1} & 0 \\ 0 & J \end{bmatrix}, \quad \bar{A} = \begin{bmatrix} A_1 & 0 \\ 0 & I_{n2} \end{bmatrix},$$

$$\bar{B} = \begin{bmatrix} B_1 \\ B_2 \end{bmatrix}, \quad \bar{C} = R[C_1 \; C_2]$$

We know

$$\text{rank} \begin{bmatrix} \bar{E} \\ \bar{C} \end{bmatrix} = n_1 + n_2 \quad (47)$$

Then an obsever for system (46) is constructed by using our procedure as

$$\dot{\hat{\xi}} = F\omega + \bar{G}\bar{y} + J_2 u$$
$$\hat{\bar{z}} = H\omega + \bar{D}\bar{y} \quad (48)$$

where

$$\bar{E}^{\#}\bar{E} + \bar{C}^{\#}\bar{C} = I_{n1+n2}, \quad W^{-1} = \begin{bmatrix} \tilde{\bar{C}} \\ \bar{C} \end{bmatrix}$$

$$[F \quad G] = [I_{n1+n2-k} \quad \bar{L}]W^{-1}\bar{E}^{\#}\bar{A}W$$

$$[H \quad D] = W\begin{bmatrix} I_{n1+n2-k} & -\bar{L} \\ 0 & I_k \end{bmatrix} \quad (49)$$

$$[J_1 \; J_2] = [I_{n1+n2-k} \quad L]W^{-1}[\bar{E}^{\#}\bar{A}\bar{C}^{\#} \quad E^{\#}\bar{B}]$$

$$\bar{G} = J_1 + G(I_k - \bar{C}\bar{C}^{\#})$$

$$\bar{D} = \bar{C}^{\#} + D(I_k - \bar{C}\bar{C}\#)$$

It follows that a minimal order observer for system (1) is obtained as

$$\dot{\omega} = F\omega + \bar{G}Ry + (J_2 + \bar{G}RC_3 B_3)u$$
$$\hat{x} = Q\begin{bmatrix} H\omega + \bar{D}Ry + \bar{D}RC_3 B_3 u \\ -B_3 u \end{bmatrix} \quad (50)$$

whose order is $n_1 + n_2 - k$.

EXAMPLE

Consider a following descriptor system (n = 3, p = 2, r = 2)

$$\begin{bmatrix} 1 & 0 & 0 \\ 0 & 0 & 1 \\ 0 & 0 & 0 \end{bmatrix}\dot{x} = \begin{bmatrix} -5 & 0 & 0 \\ 0 & 1 & 0 \\ 0 & 0 & 1 \end{bmatrix}x + \begin{bmatrix} 1 \\ 0 \\ -1 \end{bmatrix}u$$

$$y = \begin{bmatrix} 1 & 0 & 0 \\ 0 & 1 & 0 \end{bmatrix}x$$

We see that conditions (9), (21) are satisfied, and hence this system is R-observable. It follows from Theorem 2 or Theorem 3 that an observer with arbitrarily assignable poles can be realized.

First we find $(E^{\#}, C^{\#}) \in \Gamma$ as

$$E^{\#} = \begin{bmatrix} 1 & 0 & 0 \\ 0 & 0 & 1 \\ 0 & 1 & 0 \end{bmatrix}, \quad C^{\#} = \begin{bmatrix} 0 & 0 \\ 0 & 1 \\ 0 & 0 \end{bmatrix}$$

(i) identity observer

$$E^{\#}A = \begin{bmatrix} -5 & 0 & 0 \\ 0 & 0 & 1 \\ 0 & 1 & 0 \end{bmatrix}$$

Letting $\{-\alpha, -\beta, -\gamma\}$ be the set of poles of $\hat{A}$, we have, by simple calculation,

$$K = \begin{bmatrix} -5 & k_4 \\ 0 & k_5 \\ -1 & k_6 \end{bmatrix}$$

where $k_4 = \alpha\beta\gamma$, $k_5 = \alpha+\beta+\gamma$, $k_6 = 1+\alpha\beta+\beta\gamma+\gamma\alpha$. Eqn.(18) gives the identity observer:

$$\dot{z} = \begin{bmatrix} 0 & -k_4 & 0 \\ 0 & -k_5 & 0 \\ 1 & 1-k_6 & 1 \end{bmatrix}z + \begin{bmatrix} -5 & 0 \\ 0 & 0 \\ -1 & 1 \end{bmatrix}y + \begin{bmatrix} 1 \\ -1 \\ 0 \end{bmatrix}u$$

$$\hat{x} = z + \begin{bmatrix} 0 & 0 \\ 0 & 1 \\ 0 & 0 \end{bmatrix} y$$

(ii) minimal order observer

Choosing as $\tilde{C} = [0 \ \ 0 \ \ 1]$ so that

$$S^{-1} = \begin{bmatrix} 0 & 0 & 1 \\ 1 & 0 & 0 \\ 0 & 1 & 0 \end{bmatrix}$$

is nonsingular, then we have

$$S^{-1} E^{\#} AS = \begin{bmatrix} 0 & 0 & 1 \\ 0 & -5 & 0 \\ 0 & 1 & 0 \end{bmatrix}$$

In this case $A_{11} = 0$, $A_{21} = [0 \ \ 1]^T$. So letting $-\alpha$ be the pole of observer, we get $L = [l_1 \ \ -\alpha]$, where l_1 is arbitrary. Eqn.(39) gives the minimal order observer:

$$\dot{z} = -\alpha z + [l_1 \alpha - 5 l_1 \ \ 1] y + (l_1 + \alpha) u$$

$$\hat{x} = \begin{bmatrix} 0 \\ 0 \\ 1 \end{bmatrix} z + \begin{bmatrix} 1 & 0 \\ 0 & 1 \\ -l_1 & 0 \end{bmatrix} y$$

CONCLUSION

In this paper, an observer for descriptor systems is presupposed to have the same structure as one of Luenberger, and the fundamental observer equations have been derived. Based on these equations, the realizability conditions have been studied. As a result, an observer can be constructed if the system is observable in the sense of Rosenbrock. And the design method has been presented. The results are extension of those for regular systems.

Further it has been shown that the realizability conditions are relaxed by elimination of the purely static modes, and an observer can be constructed if the system is observable in the sense of Verghese.

It should be mentioned that other observer problems such as linear function observer, unknown-input observer, etc. could be discussed based on the fundamental equations.

ACKNOWLEDGEMENT

The author would like to express his appreciation to Prof. Katsuhisa Furuta of Tokyo Institute of Technology.

REFERENCES

Armentano, V.A. (1986). The pencil (sE-A) and controllability- observability for generalized linear systems; A geometric approach, SIAM J. Control and Optimizization, 24, 616-638

Cobb, J.D. (1984). Controllability, observability and duality in singular systems, IEEE Trans. Autom. Control, 29, 1076-1082

Dai, L. (1988). Observers for discrete singular systems, IEEE Trans. Autom. Control, 33, 187-191

El-Tohami, M., Nagy, V.L. and Mukundan, R. (1983). On the design of observers for generalized state space systems using singular value decomposition, Int. J. Control, 38, 673-683

Gantmacher, F.R. (1974). The Theory of Matrices, Chelsea, New York

Kawaji, S. and Narahashi, S. (1988). Design of observer for descriptor systems (in Japanese), Trans. of SICE, 24, 141-148

Kawaji, S., Iwai, Z. and Inoue, A. (1988). Observers (in Japanese), Koronasha, Tokyo

Kobayashi, N. and Nakamizo, T. (1987). On observers for linear descriptor systems (in Japanese), Trans. of SICE, 23, 1342-1345

Lewis, F.L. (1986) A survey of linear singular systems, Circuit Syst. & Signal Process, 5, 3-36

Luenberger, D.G. (1971). An introduction to observers, IEEE Trans. Autom. Control, 16, 596-602

Rosenbrock, H.H. (1974). Structural properties of linear dynamical systems, Int. J. Control, 20, 191-202

Shafai, B. and Carroll, R.L. (1987). Design of minimal order observer for singular systems, Int. J. Control, 45, 1075-1081

Verghese, G.C., Levy, B.C. and Kailath, T. (1981). A generalized state-space for singular systems, IEEE Trans. Autom. Control, 26, 811-831

APPENDIX

Algorithm for $(E^{\#}, C^{\#}) \in \Gamma$

step 1. Find two nonsingular matrices P_1 and P_2 such that

$$\begin{bmatrix} P_1 & 0 \\ 0 & P_2 \end{bmatrix} \begin{bmatrix} E \\ C \end{bmatrix} = \begin{bmatrix} E_1 \\ E_2 \\ C_1 \\ C_2 \end{bmatrix} \begin{matrix} \}r \\ \}n-r \\ \}n-r \\ \}p-n+r \end{matrix}$$

where

$$\text{rank } E_1 = r, \quad \text{rank } \begin{bmatrix} E_1 \\ C_1 \end{bmatrix} = n$$

step 2. Let

$$W = \begin{bmatrix} E_1 \\ E_2 + C_1 \end{bmatrix}$$

$$M = \begin{bmatrix} 0 & 0 \\ I_{n-r} & 0 \end{bmatrix} \begin{matrix} \}r \\ \}n-r \end{matrix}$$

$$\underbrace{\phantom{I_{n-r}}}_{n-r} \ \underbrace{}_{p-n+r}$$

step 3. $E^{\#}$ and $C^{\#}$ are given as

$$E^{\#} = W^{-1} P_1$$

$$C^{\#} = W^{-1} M P_2$$

REALIZING THE ACTION OF A CASCADE COMPENSATOR BY STATE FEEDBACK

V. Kučera

*Institute of Information Theory and Automation, Czechoslovak Academy of Sciences,
182 08 Prague 8, Czechoslovakia*

Abstract. The class of dynamic cascade compensators is characterized whose effect on a given linear generalized state-space system can be represented by a linear static state feedback. Internal properness and stability of the resulting closed-loop system are investigated. A simple procedure is proposed for the calculation of any and all feedback gains given the system and the compensator.

Keywords. Linear systems; Generalized state-space systems; State feedback; Cascade compensation; System theory.

INTRODUCTION

Consider a *generalized state-space system* governed by the equation

$$E\dot{x} = Fx + Gu \tag{1}$$

where $x \in \mathbf{R}^n$, $u \in \mathbf{R}^m$ and E, F and G are constant matrices with entries in $\mathbf{R}$, the field of real numbers.

When $sE - F$ is square and non-singular, the system (1) is said to be *regular* and gives rise to the transfer function

$$T(s) = (sE - F)^{-1}G, \tag{2}$$

which is a rational $n \times m$ matrix. For background, motivation and elements of the theory of such systems the reader is referred to Lewis (1986).

The application of state feedback

$$u = Lx + Mv, \tag{3}$$

where $v \in \mathbf{R}^m$ and L, M are constant matrices with entries in $\mathbf{R}$, results in the closed-loop system

$$E\dot{x} = (F + GL)x + GMv. \tag{4}$$

We say the feedback (3) is *admissible* if both $I_m - LT(s)$ and M are non-singular matrices. The resulting system is then regular and gives rise to the transfer function

$$T_{L,M}(s) = T(s)\left[I_m - LT(s)\right]^{-1}M. \tag{5}$$

Obviously, the effect of admissible state feedback on the given system can be represented by a dynamic compensator cascaded with the system and having the transfer function

$$C(s) = \left[I_m - LT(s)\right]^{-1}M. \tag{6}$$

The aim of the paper is to study the converse problem: Given a compensator with rational transfer function $C(s)$, can its action on the given system be realized by applying an admissible state feedback?

The answer is a qualified affirmative. We shall give conditions which single out precisely those $C(s)$ whose action can be so realized. Furthermore, it will be shown which additional conditions imposed on $C(s)$ correspond to internal properness and/or stability of the feedback system. The proofs are constructive and lead to a simple calculation of any and all feedback gains L, M given $T(s)$ and $C(s)$.

These results generalize those obtained by Hautus and Heymann (1978) for controllable state-space systems

$$\dot{x} = Fx + Gu, \tag{7}$$

namely that the action of $C(s)$ on (7) can be implemented by an admissible state feedback (3) if and only if $C(s)$ is biproper and for every polynomial vector $u(s)$ such that $T(s)u(s)$ is polynomial, $C^{-1}(s)u(s)$ is also polynomial. Results like these provide a vehicle for the study of state feedback control problems by means of algebraic properties of transfer functions.

ADMISSIBLE FEEDBACK

The application of state feedback (3) may result in a non-regular closed-loop system (4), no matter whether the original system (1) is regular or not. An admissible state feedback (3), however, preserves the regularity.

Lemma 1. Let $sE - F$ be non-singular and $T(s)$ defined by (2). Then, for any compatible matrix L with entries in $\mathbf{R}$, $sE - F - GL$ is non-singular if and only if $I_m - LT(s)$ is non-singular.

Proof: We write (2) as

$$(sE - F)T(s) = G$$

and compose it with $GLT(s)$ to obtain

$$(sE - F - GL)T(s) = G[I_m - LT(s)].$$

If $I_m - LT(s)$ is non-singular, then

$$X(s) = T(s)[I_m - LT(s)]^{-1}$$

is a rational solution of the equation

$$(sE - F - GL)X(s) = G.$$

Hence, over the field $\mathbf{R}(s)$ of rational functions,

$$
\begin{aligned}
\mathrm{rank}(sE - F - GL) &= \mathrm{rank}[sE - F - GL \ \ G] \\
&= \mathrm{rank}[sE - F \ \ G]\begin{bmatrix} I_n & 0 \\ -L & I_m \end{bmatrix} \\
&= \mathrm{rank}[sE - F \ \ G] \\
&= n
\end{aligned}
$$

so that $sE - F - GL$ is non-singular.

On the other hand, let $sE - F - GL$ be non-singular. Then

$$Y(s) = (sE - F - GL)^{-1}G$$

is a rational solution of the equation

$$Y(s)[I_m - LT(s)] = T(s).$$

Hence, over $\mathbf{R}(s)$,

$$
\begin{aligned}
\mathrm{rank}[I_m - LT(s)] &= \mathrm{rank}\begin{bmatrix} I_m - LT(s) \\ T(s) \end{bmatrix} \\
&= \mathrm{rank}\begin{bmatrix} I_m & -L \\ 0 & I_n \end{bmatrix}\begin{bmatrix} I_m \\ T(s) \end{bmatrix} \\
&= \mathrm{rank}\begin{bmatrix} I_m \\ T(s) \end{bmatrix} \\
&= m
\end{aligned}
$$

so that $I_m - LT(s)$ is non-singular. $\square$

We are now in a position to interpret the notion of admissible state feedback. The non-singularity of $I_m - LT(s)$ preserves the regularity of the system while the non-singularity of M preserves the control abilities of the input. Therefore, no loss of these crucial properties occurs through the application of an admissible state feedback.

FEEDBACK IMPLEMENTATION

We have seen that the action of an admissible state feedback (3) on a regular system (1) can be represented by a dynamic precompensator with transfer function (6). To study the converse problem, we start by writing the transfer function (2) of system (1) in the matrix fraction form

$$T(s) = B(s)A^{-1}(s) \tag{8}$$

where $A(s)$ and $B(s)$ are relatively right prime polynomial matrices such that the composite matrix

$$\begin{bmatrix} A(s) \\ B(s) \end{bmatrix} \tag{9}$$

is column reduced.

In order to prove our main result, we shall need the following lemma, adapted from Kučera and Zagalak (1989), which provides a condition for a polynomial equation to have a constant solution.

Lemma 2. The equation

$$XP(s) + YQ(s) = R(s) \tag{10}$$

for matrices $P(s), Q(s)$ and $R(s)$ of size $m \times m$, $n \times m$ and $m \times m$, respectively, with entries in $\mathbf{R}[s]$, the ring of polynomials in the indeterminate s over $\mathbf{R}$, has a solution pair X, Y with entries in $\mathbf{R}$ such that X is non-singular if and only if the rows of the matrix $\begin{bmatrix} P(s) \\ Q(s) \end{bmatrix}$ span the same $\mathbf{R}$-linear space as those of $\begin{bmatrix} R(s) \\ Q(s) \end{bmatrix}$.

Proof: For necessity, let X, Y be a solution pair of (10) with entries in $\mathbf{R}$. Then

$$\begin{bmatrix} X & Y \\ 0 & I_n \end{bmatrix}\begin{bmatrix} P(s) \\ Q(s) \end{bmatrix} = \begin{bmatrix} R(s) \\ Q(s) \end{bmatrix}.$$

If X is non-singular, the constant transformation matrix is also non-singular and the rows of the matrices

$$\begin{bmatrix} P(s) \\ Q(s) \end{bmatrix}, \begin{bmatrix} R(s) \\ Q(s) \end{bmatrix}$$

span the same $\mathbf{R}$-linear space.

To prove sufficiency, let T_1 and T_2 be non-singular matrices with entries in $\mathbf{R}$ such that

$$T_1 P(s) = \begin{bmatrix} P_1(s) \\ P_2(s) \\ 0 \end{bmatrix}, \quad T_2 R(s) = \begin{bmatrix} R_1(s) \\ R_2(s) \\ 0 \end{bmatrix}$$

where the rows of $\begin{bmatrix} P_1(s) \\ P_2(s) \end{bmatrix}$ as well as those of $\begin{bmatrix} R_1(s) \\ R_2(s) \end{bmatrix}$ are $\mathbf{R}$-linearly independent, and where the rows of $P_1(s)$ and $R_1(s)$ do not belong to the $\mathbf{R}$-linear row space of $Q(s)$ while those of $P_2(s)$ and $R_2(s)$ do. If the rows of $\begin{bmatrix} P(s) \\ Q(s) \end{bmatrix}$ and $\begin{bmatrix} R(s) \\ Q(s) \end{bmatrix}$ generate the same $\mathbf{R}$-linear space, then so do the rows of $P_1(s)$ and $R_1(s)$. Let the number of these rows be p. Then there exists a non-singular $p \times p$ matrix X_1 having entries in $\mathbf{R}$ such that

$$X_1 P_1(s) = R_1(s)$$

and matrices X_2, Y_2 with entries in $\mathbf{R}$ such that

$$X_2 Q(s) = \begin{bmatrix} R_2(s) \\ 0 \end{bmatrix}, \quad Y_2 Q(s) = \begin{bmatrix} P_2(s) \\ 0 \end{bmatrix}.$$

So we have

$$\begin{bmatrix} X_1 & 0 & 0 \\ 0 & I_{m-p} & X_2 - Y_2 \\ 0 & 0 & I_m \end{bmatrix} \begin{bmatrix} P_1(s) \\ P_2(s) \\ 0 \\ Q(s) \end{bmatrix} = \begin{bmatrix} R_1(s) \\ R_2(s) \\ 0 \\ Q(s) \end{bmatrix}.$$

It follows that the pair

$$X = T_2^{-1} \begin{bmatrix} X_1(s) & 0 \\ 0 & I_{m-p} \end{bmatrix} T_1, \quad Y = T_2^{-1} \begin{bmatrix} 0 \\ X_2 - Y_2 \end{bmatrix}$$

solves the equation (10) and X is non-singular. $\qquad\square$

Theorem 1. Let (1) be a regular system, $T(s)$ in (8) its transfer function and $C(s)$ a rational $m \times m$ matrix with entries in $\mathbf{R}(s)$. Then there exists an admissible state feedback (3) such that

$$C(s) = [I_m - LT(s)]^{-1}M$$

if and only it the following conditions all hold:

(a) $C(s)$ is non-singular;

(b) $C^{-1}(s)A(s)$ is polynomial;

(c) the rows of $\begin{bmatrix} A(s) \\ B(s) \end{bmatrix}$ span the same $\mathbf{R}$-linear space as those of $\begin{bmatrix} C^{-1}(s)A(s) \\ B(s) \end{bmatrix}$.

Proof: To prove necessity, let constant matrices L, M defining an admissible feedback (3) exist and satisfy

$$\begin{aligned} C(s) &= [I_m - LT(s)]^{-1}M \\ &= A(s)[A(s) - LB(s)]^{-1}M. \end{aligned}$$

The condition (a) is an immediate consequence of admissibility. Write

$$C^{-1}A(s) = M^{-1}[A(s) - LB(s)],$$

which implies (b). We then have

$$\begin{bmatrix} C^{-1}(s)A(s) \\ B(s) \end{bmatrix} = \begin{bmatrix} M^{-1} & -ML^{-1} \\ 0 & I_n \end{bmatrix} \begin{bmatrix} A(s) \\ B(s) \end{bmatrix}$$

which shows that the polynomial matrices

$$\begin{bmatrix} A(s) \\ B(s) \end{bmatrix}, \begin{bmatrix} C^{-1}(s)A(s) \\ B(s) \end{bmatrix}$$

are related by a contant non-singular matrix. This proves (c).

For sufficiency, let (a), (b) hold and consider the equation

$$XA(s) + YB(s) = C^{-1}(s)A(s) \tag{11}$$

in polynomial matrices. In view of (c), the hypotheses of Lemma 1 are satisfied. It follows that (11) has a constant soluton pair X, Y such that X is non-singular. Put

$$L = -X^{-1}Y, \quad M = X^{-1}.$$

Then (11) yields

$$M^{-1}[A(s) - LB(s)] = C^{-1}(s)A(s) \tag{12}$$

which is equivalent to

$$C(s) = [I_m - LT(s)]^{-1}M. \qquad\square$$

In case the transfer function (8) of system (1) is *strictly proper*, the column degrees of $A(s)$ dominate in (9),

$$\deg \mathrm{col}_i\, A(s) = 1 + \deg \mathrm{col}_i\, B(s), \quad i = 1, 2, ..., m$$

so that the highest-column-degree coefficient matrices satisfy

$$\begin{bmatrix} A(s) \\ B(s) \end{bmatrix}_{hc} = \begin{bmatrix} A(s)_{hc} \\ 0 \end{bmatrix}.$$

Hence the condition (c) of Theorem 1 holds if and only if

$$\deg \mathrm{col}_i\, A(s) = \deg \mathrm{col}_i\, C^{-1}(s)A(s), \quad i = 1, 2, ..., m$$

and

$$A(s)_{hc} = U[C^{-1}(s)A(s)]_{hc}$$

for a non-singular matrix U with entries in $\mathbf{R}$. This is the case if and only if $C(s)$ is a biproper matrix. Furthermore, the condition (b) is equivalent to $C^{-1}(s)u(s)$ being polynomial for every polynomial $u(s)$ such that $T(s)u(s)$ is also polynomial. We have thus obtained the result of Hautus and Heymann (1978) for ordinary state-space systems.

PROPERNESS AND STABILITY

The feedback implementation of a cascade compensator has numerous practical advantages over the feedforward scheme. These include the possibility of designing a system which is internally proper and/or stable. It is therefore natural to ask the following question: Which additional properties of $C(s)$ will result in an internally proper and/or stable feedback system (4)?

We say with Kučera (1986) that a regular, generalized state-space system like (1) is *internally proper* if it has no infinite eigenvalue, i.e., if its free response $x(t)$ is devoid of impulses at $t = 0$ for every initial state $x(0-)$.

We say that a regular, generalized-state space system like (1) is *internally stable* if it has no finite unstable eigenvalues, i.e., if its free response $x(t)$ tends to zero when $t \to \infty$ for every initial state $x(-0)$.

Only *controllable* eigenvalues can be altered by feedback. A finite eigenvalue s_o of (1) is said to be controllable if

$$\mathrm{rank}[sE - F \quad G] = n$$

for $s = s_o$ while the infinite eigenvalue of (1) is said to be controllable if

$$\mathrm{rank}[(1/s)E - F \quad G] = n$$

for $s = 0$, see Verghese, Lévy, and Kailath (1981) and Cobb (1984).

Let k denote the total column degree of the polynomial

matrix (9), that is,

$$k = \sum_{i=1}^{m} \deg \operatorname{col}_i \begin{bmatrix} A(s) \\ B(s) \end{bmatrix}.$$

Then we have the following result.

Theorem 2. Let (1) be a regular system whose infinite eigenvalue is controllable, let $T(s)$ in (8) be its transfer function and $C(s)$ a rational $m \times m$ matrix. Suppose that there exists an admissible state feedback (3) such that

$$C(s) = [I_m - LT(s)]^{-1}M.$$

Then the resulting closed-loop system (4) is internally proper if and only if

$$\deg \det C^{-1}(s)A(s) = k. \tag{13}$$

Proof: It follows from (4) and (5) that

$$(sE - F - GL)^{-1}G = B(s)[A(s) - LB(s)]^{-1}.$$

Therefore the number of controllable eigenvalues of (4), counted with multiplicities, is equal to

$$\begin{aligned}
\sum_{i=1}^{m} \deg \operatorname{col}_i \begin{bmatrix} A(s) - LB(s) \\ B(s) \end{bmatrix} &= \\
= \sum_{i=1}^{m} \deg \operatorname{col}_i \begin{bmatrix} I_m & -L \\ 0 & I_n \end{bmatrix} \begin{bmatrix} A(s) \\ B(s) \end{bmatrix} \\
= \sum_{i=1}^{m} \deg \operatorname{col}_i \begin{bmatrix} A(s) \\ B(s) \end{bmatrix} \\
= k.
\end{aligned}$$

Theorem 1 implies that $C^{-1}(s)A(s)$ is a polynomial matrix. The number of finite controllable eigenvalues of (4), multiplicities included, is given by (12) as

$$\deg \det[A(s) - LB(s)] = \deg \det C^{-1}(s)A(s).$$

Hence (4) has no infinite controllable eigenvalue if and only if (13) holds.

Now the non-controllable eigenvalues of (4), if any, are those of (1). By assumption they are all finite. Thus (4) has no impulsive behaviour at all if and only if (13) holds. $\square$

Theorem 3. Let (1) be a regular system whose finite unstable eigenvalues are all controllable, let $T(s)$ in (8) be its transfer function and $C(s)$ a rational $m \times m$ matrix. Suppose that there exists an admissible state feedback (3) such that

$$C(s) = [I_m - LT(s)]^{-1}M.$$

Then the resulting closed-loop system (4) is internally stable if and only if $\det C^{-1}(s)A(s)$ is a strictly Hurwitz polynomial.

Proof: It follows from (4) and (5) that

$$(sE - F - GL)^{-1}G = B(s)[A(s) - LB(s)]^{-1}.$$

Hence, the finite controllable eigenvalues of (4) are given by $\det[A(s) - LB(s)]$. Theorem 1 implies that $C^{-1}(s)A(s)$ is a polynomial matrix. Using (12), $\det C^{-1}(s)A(s)$ is equal to $\det[A(s) - LB(s)]$ up to a non-zero constant multiplier.

Thus (4) has no finite unstable controllable eigenvalues if and only if $\det C^{-1}(s)A(s)$ is a strictly Hurwitz polynomial.

Now the finite non-controllable eigenvalues of (4), if any, are those of (1). By assumption, they are all stable. Thus (4) yields no unstable behaviour at all if and only if $\det C^{-1}(s)A(s)$ is a strictly Hurwitz polynomial. $\square$

CONSTRUCTION

The sufficiency proof of Theorem 1 is constructive and provides a procedure for the calculation of L and M given $T(s)$ and $C(s)$. The procedure can be summarized as follows.

(a) Factorize $T(s)$ as in (8),

$$T(s) = B(s)A^{-1}(s),$$

where $A(s)$ and $B(s)$ are relatively right prime polynomial matrices such that
$$\begin{bmatrix} A(s) \\ B(s) \end{bmatrix}$$
is column reduced.

(b) Calculate $C^{-1}(s)A(s)$; if it fails to be polynomial, stop.

(c) Solve the equation (11),

$$XA(s) + YB(s) = C^{-1}(s)A(s),$$

for a constant solution pair X, Y with X non-singular; if it fails to exist, stop.

(d) Put $L = -X^{-1}Y$, $\quad M = X^{-1}$.

In general, there are more that one pair of matrices L, M that implement the action of the given $C(s)$ upon $T(s)$. Any and all such matrices are obtained by this procedure.

EXAMPLES

Two illustrative examples are included.

Example 1. Consider a system (1) h⸺ing the coefficient matrices

$$E = \begin{bmatrix} 0 & 1 & 0 & 0 \\ 0 & 0 & 0 & 0 \\ 0 & 0 & 1 & 0 \\ 0 & 0 & 0 & 1 \end{bmatrix}, \quad F = \begin{bmatrix} 1 & 0 & 0 & 0 \\ 0 & 1 & 0 & 0 \\ 0 & 0 & 0 & 0 \\ 0 & 0 & 0 & -1 \end{bmatrix}, \quad G = \begin{bmatrix} 0 \\ -1 \\ 1 \\ 0 \end{bmatrix}$$

and the transfer function

$$T(s) = \begin{bmatrix} s^2 \\ s \\ 1 \\ 0 \end{bmatrix} \begin{bmatrix} s \end{bmatrix}^{-1}.$$

Then the compensator

$$C(s) = \frac{s}{s^2 + 2s + 1}$$

satisfies the conditions (a) through (c) of Theorem 1 and its effect upon (1) can be realized by any state feedback (3) having the gains

$$L = \begin{bmatrix} -\alpha & 1 - 2\alpha & -\alpha & \beta \end{bmatrix}, \quad M = \alpha$$

with arbitrary $\alpha \neq 0$ and β.

Applying Theorems 2 and 3, the resulting closed-loop system is seen to be internally proper and stable; it has one non-controllable eigenvalue equal to -1. A direct implementation of the feedforward compensator would result in an unstable system. $\square$

Example 2. Consider a single-input regular system (1) over $\mathbf{R}$ with transfer function

$$T(s) = \frac{B(s)}{a(s)}$$

where $a(s)$ is a polynomial that is relatively right prime with the polynomial vector $B(s)$. We wish to characterize, in an explicit manner, all cascade compensator transfer functions $C(s)$ whose effect on (1) is realizable by an admissible state feedback (3).

The conditions (a) and (b) of Theorem 1 imply that

$$C(s) = \frac{a(s)}{p(s)}$$

for a polynomial $p(s) \neq 0$. By (c), the rows of the matrices

$$\begin{bmatrix} a(s) \\ B(s) \end{bmatrix}, \quad \begin{bmatrix} p(s) \\ B(s) \end{bmatrix}$$

span the same $\mathbf{R}$-linear space. Hence if we let k denote the highest degree occurring in $a(s)$ and $B(s)$, then

$$\deg p(s) \leq k$$

with equality holding when $T(s)$ is strictly proper. The integer k is known as the controllability index of (1), see Kučera and Zagalak (1989). $\square$

CONCLUSIONS

Given a regular, generalized state-space system (1), the transfer function $C(s)$ of all dynamic cascade compensators whose effect on (1) can be realized by an admissible state feedback (3) has been described in Theorem 1. The properties of $C(s)$ which correspond to internal properness and stability of the closed-loop system (4) have been given in Theorems 2 and 3, respectively.

A major role is played by the matrix $C^{-1}(s)A(s)$. It must be a polynomial matrix for a feedback implementation of $C(s)$ to exist, and its determinant is the characteristic polynomial of the controllable part of the closed-loop system.

Theorem 1 extends Theorem 5.10 of Hautus and Heymann (1978) to the realm of generalized, not necessarily controllable, state-space systems. The result is constructive and provides a useful tool for the study of state feedback control problems by means of algebraic properties of transfer function.

REFERENCES

Cobb, J. D. (1984). Controllability, observability, and duality in singular systems. *IEEE Trans. Automat. Contr.*, AC-29, 1076-1082.

Hautus, M. L. J. and M. Heymann (1978). Linear feedback - an algebraic approach. *SIAM J. Contr. Optimiz.*, 16, 83-105.

Kučera, V. (1986). Internal properness and stability in linear systems. *Kybernetika*, 22, 1-18.

Kučera, V. and P. Zagalak (1989). Constant solutions of polynomial equations. *Preprints IFAC Workshop on System Structure and Control*, Prague, Czechoslovakia. pp. 33-36.

Lewis, F. L. (1986). A survey of linear singular systems. *J. Circuits, Systems, Signal Proc.*, Special Issue on Singular Systems, 5, 3-36.

Verghese, G. C., B. C. Lévy and T. Kailath (1981). A generalized state-space for singular systems. *IEEE Trans. Automat. Contr.*, AC-26, 811-831.

SUBOPTIMAL REGULATION OF SINGULARLY PERTURBED SYSTEMS BY DESCRIPTOR VARIABLE APPROACH

Yue-yun Wang and Dan-jie Pan

*Department of Automatic Control, Shanghai Jiao Tong University,
Shanghai 200030, PRC*

Abstract. This paper first treats the optimal regulator problem for descriptor systems. Using the Hamiltonian minimization. Bender and Laub (1987) once solved the problem. We present the conditions for the existence of a unique solution to such a problem, which removes the conservativeness left in Bender and Laub's paper. We show differently how to compute the optimal feedback gains. Following the above, the paper next investigates the connections between the two finite regulator problems for both descriptor and singularly perturbed systems. Then, it poses a new approach to designing near-optimum regulators for singularly perturbed systems whose reduced model may exhibit impulses. Finally, a design example is presented for illustrating our theorems.

Keywords. Singular perturbation; suboptimal control; decomposition; optimal regulators; descriptor systems.

I. INTRODUCTION

We consider the finite linear optimal regulator problem

$$J=\frac{1}{2}z'(t_f)Gz(t_f)+\frac{1}{2}\int_0^{t_f}(z'Qz+v'Rv)dt \tag{1}$$

$$\dot{z}_1=A_1z_1+A_2z_2+B_1v \quad z_1(0-)=z_1^0$$
$$\mu\dot{z}_2=A_3z_1+A_4z_2+B_2v \quad z_2(0-)=z_2^0 \tag{2}$$

Where

$$G=\begin{bmatrix} G_1 & \mu G_{12} \\ \mu G'_{12} & \mu G_2 \end{bmatrix}\geqslant 0 \qquad Q=\begin{bmatrix} Q_1 & Q_{12} \\ Q'_{12} & Q_2 \end{bmatrix}\geqslant 0$$

$$R>0$$

z_1, z_2 and v are $k-$, $l-$ and $m-$ dimensional vectors, $z'=[z_1' \ z_2']$, and μ is a small positive parameter. If $\mu=0$, the above problem is reduced to the problem

$$J=\frac{1}{2}x_1'(t_f)G_1x_1(t_f)+\frac{1}{2}\int_0^{t_f}(x'Qx+u'Ru)dt \tag{3}$$

$$E\dot{x}=Ax+Bu$$
$$y=Cx \qquad x(0-)=\begin{bmatrix} z_1^0 \\ z_2^0 \end{bmatrix} \tag{4}$$

Where

$$E=\begin{bmatrix} I & 0 \\ 0 & 0 \end{bmatrix} \qquad A=\begin{bmatrix} A_1 & A_2 \\ A_3 & A_4 \end{bmatrix}$$

$$B=[B_1' \ B_2']' \qquad C=[C_1 \ C_2]$$

with $C_1'C_1=Q_1$, $C_1'C_2=Q_{12}$ and $C_2'C_2=Q_2$, and the descriptor vector x is partitioned into $x'=[x_1' \ x_2']$. (3),(4) is a standard LQ problem of descriptor systems, which has been recently treated in (Cobb, 1983; Cheng, 1987; Bender and Laub, 1987; Lewis, 1985 etc.) Different from the approaches taken by others, Bender and Laub (1987) used the calculas of variations to solve this optimization problem. They posed somewhat conservative conditions on the existence of the impulse-free optimal solution. They computed the optimal feedback gains from a state transition matrix for the finite horizon problem, or from solving a generalized eigenvalue problem for the infinite horizon problem.

In this paper, we first contribute to refinements and improvements of Bender's techniques. We present the conditions for the existence of a unique solution to the problem (3). These conditions, in contrast to (Bender,1987), are the most useful in characterization of the impulse-free optimization. In order to relate the optimal control to the descriptor trajectory, we show how to compute the optimal feedback regulator, which can be decomposed into two independent subregulators. One subregulator corresponds to the solution of a reduced Riccati equation P(t), and another corresponds to a constant pole-shifting gain F.

The paper latter investigates the relationships between the optimal feedback regulators, trajectories and costs for both problems(1) and (3). When the fast modes in(2) can be neglected (or A_4^{-1} exists), the topic was partly answered by Haddad and Kokotovic (1971), where the two optimal controls were related.

If however, neglecting fast modes causes impulses (or A_4 singular), the relationship becomes unclear. In this case, even the composite control (Chow,1976) may not be applicable for near-optimum decomposition of the problem(1)(2).

Fortunately in either case (A_4^{-1} exists or not), we show that the optimal regulator and trajectory of the problem (1)(2) will tend to those of (3)(4). Thus, the near-optimum regulator of the singularly perturbed systems corresponds to the optimal one for the descriptor systems. Design of the near-optimum regulator, as shown in the first part, involves separately solving two lower order subregulators. This new approach has improved the two-stage method (Kokotovic,1972) and the composite control (Chow,1976) where in the impulsive case the two subregulators can't be designed separately.

Because of limited space, we shall state many of results without proof, the complete version of this paper is available at the author's hands(Wang, 1987)

II. LINEAR REGULATOR FOR DESCRIPTOR SYSTEMS

It will prove necessary to summarize some well-known results as follows. The descriptor system (4) is assumed to be regular, i.e. $|sE-A| \not\equiv 0$. Denote $r=\deg|sE-A|$, then the system (4) has numbers of r finite dynamic modes, $(k-r)$ impulsive modes and 1 infinite nondynamic modes.

The definitions of mode controllability and observability are referred to (Verghese, 1981; Bender, 1987). We here only state the testing lemmas without proof.

Lemma 1: If A_4 is invertible, then the system (4) has no impulsive modes and is regular.

Lemma 2: The $(k-r)$ impulsive modes of the system (4) are controllable and observable, if and only if $[A_4\ B_2]$ and $[A_4'\ C_2']$ are of full row rank, respectively.

Lemma 3: The r finite modes of the system (4) are controllable and observable, if and only if

$$\mathrm{rank}[sE-A\ B]=k+1$$

and

$$\mathrm{rank}[sE'-A'\ C']=k+1, \forall\ \text{finite } s$$

We now solve the above finite horizon problem. In terms of the Hamilton-lagrange theory, Bender and Laub(1987) has derived the two point boundary value problem, the necessary conditions for (3) to be minimized.

$$\begin{bmatrix} E & O \\ O & E' \end{bmatrix}\begin{bmatrix} \dot{x} \\ \dot{\lambda} \end{bmatrix}=\begin{bmatrix} A & -BR^{-1}B' \\ -Q & -A' \end{bmatrix}\begin{bmatrix} x \\ \lambda \end{bmatrix}$$

$$u=-R^{-1}B'\lambda$$

with the boundary condition

$$Ex(0-)=\begin{bmatrix} z_1^0 \\ 0 \end{bmatrix} \qquad E'\lambda(t_f)=E'GEx(t_f)$$

We rewrite this necessary conditions as

$$\begin{bmatrix} \dot{x}_1 \\ \dot{\lambda}_1 \end{bmatrix}=\begin{bmatrix} A_1 & -S_1 \\ -Q_1 & -A_1' \end{bmatrix}\begin{bmatrix} x_1 \\ \lambda_1 \end{bmatrix}+\begin{bmatrix} A_2 & -S_{12} \\ -Q_{12} & -A_3' \end{bmatrix}\begin{bmatrix} x_2 \\ \lambda_2 \end{bmatrix} \tag{5a}$$

$$0=\begin{bmatrix} A_3 & -S_{12}' \\ -Q_{12}' & -A_2' \end{bmatrix}\begin{bmatrix} x_1 \\ \lambda_1 \end{bmatrix}+\begin{bmatrix} A_4 & -S_2 \\ -Q_2 & -A_4' \end{bmatrix}\begin{bmatrix} x_2 \\ \lambda_2 \end{bmatrix} \tag{5b}$$

$$u=-R^{-1}(B_1'\lambda_1+B_2'\lambda_2) \tag{5c}$$

with the boundary conditions

$$x_1(0-)=z_1^0 \qquad \lambda_1(t_f)=G_1x_1(t_f)$$

Where $\lambda'=[\lambda_1'\ \lambda_2']$ is the codescriptor vector, and $S_1=B_1R^{-1}B_1'$, $S_{12}=B_1R^{-1}B_2'$ and $S_2=B_2R^{-1}B_2'$. To simplify the equation, we define:

$$T_1=\begin{bmatrix} A_1 & -S_1 \\ -Q_1 & -A_1' \end{bmatrix} \qquad T_2=\begin{bmatrix} A_2 & -S_{12} \\ -Q_{12} & -A_3' \end{bmatrix}$$

$$T_3=\begin{bmatrix} A_3 & -S_{12}' \\ -Q_{12}' & -A_2' \end{bmatrix} \qquad T_4=\begin{bmatrix} A_4 & -S_2 \\ -Q_2 & -A_4' \end{bmatrix} \tag{6}$$

For (3) to be finite and minimized, the optimal control and trajectory should correspond to the unique, impulse-free solution of (5) (Bender,1987). By Lemma 1, this requires the invertibility of T_4. Bender (1987) proved that T_4 is invertible if $[A_4\ B_2]$ has full row rank and $Q_2=C_2'C_2$ is positive definite. Thus, they required that the number of outputs in (4) be at least as large as the number of the infinite nondynamic modes of (4). This conservative constraint will be removed by our following lemma.

Lemma 4: T_4 is invertible if and only if $[A_4\ B_2]$ and $[A_4'\ C_2']$ are of full row rank.

Proof: Let $\alpha=[\alpha_1\ \alpha_2]$ be a row vector, such that

$$\alpha T_4=0 \tag{7}$$

or equivalently

$$\alpha_1A_4-\alpha_2C_2'C_2=0, \quad -\alpha_1B_2R^{-1}B_2'-\alpha_2A_4'=0 \tag{8}$$

Post multiplying the first equation in (8) by α_2', and the second equation by α_1', then adding the two equations yields

$$\alpha_1B_2R^{-1}B_2'\alpha_1'+\alpha_2C_2'C_2\alpha_2'=0$$

So $\quad \alpha_1B_2=0, \qquad \alpha_2C_2'=0 \tag{9a}$

The (9a) together with (8) means that

$$\alpha_1 A_4 = 0, \qquad \alpha_2 A_4' = 0 \qquad (9b)$$

Since $[A_4 \ B_2]$ and $[A_4' \ C_2']$ are of full row rank, there must be $\alpha = [\alpha_1 \ \alpha_2] = 0$. This proves the sufficiency.

The proof of necessity:

Assume $|T_4| \neq 0$, where T_4 is a 21×21 matrix. Since

$$T_4 = \begin{bmatrix} A_4 & -S_2 \\ -Q_2 & -A_4' \end{bmatrix} = \begin{bmatrix} A_4 & B_2 & 0 & 0 \\ 0 & 0 & -C_2' & -A_4' \end{bmatrix} \begin{bmatrix} I & 0 \\ 0 & -R^{-1}B_2' \\ C_2 & 0 \\ 0 & I \end{bmatrix}$$

one has

$$21 = \text{rank } T_4 \leqslant \text{rank} \begin{bmatrix} A_4 & B_2 & 0 & 0 \\ 0 & 0 & -C'_2 & -A'_4 \end{bmatrix} \leqslant 21$$

Thus

$$\text{rank} \begin{bmatrix} A_4 & B_2 & 0 & 0 \\ 0 & 0 & -C'_2 & -A'_4 \end{bmatrix} = 21$$

or $[A_4 \ B_2]$ and $[A'_4 \ C'_2]$ are of full row rank ∎

According to Lemma 2, Lemma 4 has a system-theoretic interpretation. For T_4 to be invertible, the zero eigenvalues of A_4 must be the controllable and observable modes of the triple (A_4, B_2, C_2) or equivalently, the impulsive modes of (4) must be controllable and observable.

Suppose the hypotheses of Lemma 4 are met, we now derive a Riccati differential equation. Using (5b) to eliminate $[x_2' \ \lambda_2']'$ from (5a) yields

$$\begin{bmatrix} \dot{x}_1 \\ \lambda_1 \end{bmatrix} = \begin{bmatrix} A_0 & -S_0 \\ -Q_0 & -A'_0 \end{bmatrix} \begin{bmatrix} x_1 \\ \lambda_1 \end{bmatrix} \qquad (10)$$

with the boundary condition

$$x_1(0\text{-}) = z_1^0$$

$$\lambda_1(t_f) = G_1 x_1(t_f)$$

Where

$$\begin{bmatrix} A_0 & -S_0 \\ -Q_0 & -A_0' \end{bmatrix} = T_1 - T_2 T_4^{-1} T_3 \qquad (11)$$

Among the blocks in (11), it can be checked that $S_0 = S_0'$ $Q_0 = Q_0'$, and the block $(2,2)$ is the transpose of the $k \times k$ matrix A_0, multiplied by minus one. Equation (10) is just like the usual two-point boundary value problem of linear-guadratic optimal control. Thus λ_1 and x_1 are related by

$$\lambda_1 = P(t) x_1$$

Where $P(t)$ is the unique solution of the Riccati differential equation

$$-\frac{dP(t)}{dt} = P(t) A_0 + A_0' P(t) - P(t) S_0 P(t) + Q_0$$

$$P(t_f) = G_1. \qquad (12)$$

Remark 1: It can further be shown that Q_0 and S_0 are positive semidefinite. The proof involves lenthy calculations which is found in (Wang, 1987). Wang (1987) also shows that if both impulsive modes and finite modes of (4) are controllable and observable, as $t_f \to \infty$, the solution $P(t)$ becomes a constant positive definite matrix, which satisfies the corresponding algebraic Riccati equation.

Next, substituting λ_1 back into (5b) and (10), one gets

$$\begin{matrix} x_2 - L_1 x_1 \\ \lambda_2 = L_2 x_1 \end{matrix} \qquad \begin{bmatrix} L_1 \\ L_2 \end{bmatrix} = -T_4^{-1} T_3 \begin{bmatrix} I \\ P(t) \end{bmatrix} \qquad (13)$$

and the unique optimal trajectory x^* satisfying

$$\dot{x}_1 = (A_0 - S_0 P(t)) x_1 \qquad x_1(0\text{-}) = z_1^0 \qquad (14)$$

$$0 = L_1 x_1 - x_2$$

Implementation of the Optimal Control: To relate the optimal control $u^*(t)$ to the descriptor vector $x(t)$, we substitute λ_1 and λ_2 into (5c) to get

$$u^*(t) = -R^{-1}(B_1' P(t) + B_2' L_2) x_1 \qquad (15)$$

then, we use $x_2 - L_1 x_1 = 0$ rewriting (15) into

$$u^*(t) = -R^{-1}(B_1' P(t) + B_2' L_2) x_1 - R^{-1} B_2' F(x_2 - L_1 x_1)$$
$$= -R^{-1} B' K x \qquad (16)$$

Where $K = \begin{bmatrix} P & 0 \\ L_2 - FL_1, F \end{bmatrix} \qquad (16')$

P, L_1 and L_2 are time dependent, F is an 1×1 constant matrix to be determined.

Lemma 5: The lower trianqular block matrix K will be the optimal feedback gain if F makes $(A_4 - S_2 F)$ invertible.

Proof: Applying (16) into (4) yields the closed-loop system

$$\dot{x}_1 = \bar{A}_1 x_1 + \bar{A}_2 x_2 \qquad x_1(0\text{-}) = z_1^0 \qquad (17)$$

$$0 = \bar{A}_3 x_1 + \bar{A}_4 x_2$$

Where $\bar{A}_1$, $\bar{A}_2$, $\bar{A}_3$ and $\bar{A}_4$ are $k \times k$, $k \times 1$, $1 \times k$ and 1×1 respectively. Note that $\bar{A}_4 = A_4 - S_2 F$. Since $[A_4 \ B_2]$ is full row rank, so is $[A_4 \ S_2]$. Thus, there exist pole-shifting gain F that makes $\bar{A}_4$ invertible. If $|\bar{A}_4| \neq 0$, for the same initial data $x_1(0\text{-}) = z_1^0$, the optimal trajectory x^* of (14) must be an impulse-free solution of (17). Moreover, since $|\bar{A}_4| \neq 0$, the solution of (17) is unique, which therefore is just equivalent to the optimal trajectory x^*. Hence, one can implement the optimal control by (16).

■■

To this step, one knows how to compute the optimal feedback gain. Note that although $u^*(t)$ is unique, the gain matrix K is not for F can be an arbitrary matrix making $\bar{A}_4$ invertible. It will be seen latter the special structure of K will play an important role in singular perturbation of linear regulators.

III. DESIGN OF NEAR-OPTIMUM REGULATOR

In this section, we present results establishing the relations between the two regulator problems (1) and (3). Haddad (1971), and Chow (1976) investigated this topic under the condition of A_4 in (4) being invertible. In contrast to theirs, we only make the following assumption without requiring invertibility of A_4.

Assumption 1: The triple (A_4, B_2, C_2) of the system (4) is controllable and observable.

Note that the hypotheses of Lemma 4 holds under this assumption. The optimal control for the problem (1) (2) is

$$v^*(t) = -R^{-1}[B_1' \ \frac{1}{\mu}B_2']\bar{K}z \qquad (18)$$

Where

$$\bar{K} = \begin{bmatrix} \bar{K}_1 & \mu\bar{K}_2 \\ \mu\bar{K}_2' & \mu\bar{K}_3 \end{bmatrix} \qquad (18')$$

is the solution of an $(k+1)\times(k+1)$ Riccati equation.

To connect the two optimal gains K of (16') and $\bar{K}$ of (18'), we state the following theorem.

Theorem 1: Under the Assumption 1, if one chooses F in K to be the positive definte solution of

$$FA_4 + A_4'F - FB_2R^{-1}B_2'F + Q_2 = 0 \qquad (19)$$

then as the small parameter μ in (2) tends to zero, the following limits exist

$$\bar{K}_1 \to P, \quad \bar{K}_2 \to (L_2 - FL_1)', \quad \bar{K}_3 \to F \qquad (20)$$
$$t\in[0, t_f], \quad t\in[0, t_f), \quad t\in[0, t_f)$$

Proof: Let F satisfy (19). Operating the similar transformations on T_4, one gets

$$\begin{bmatrix} I & 0 \\ -F & I \end{bmatrix}\begin{bmatrix} A_4 & -S_2 \\ -Q_2 & -A'_4 \end{bmatrix}\begin{bmatrix} I & 0 \\ F & I \end{bmatrix} = \begin{bmatrix} \bar{A}_4 & -S_2 \\ 0 & -\bar{A}_4' \end{bmatrix} \qquad (21)$$

Where $\bar{A}_4 = A_4 - S_2F$ is stable, Thus from (21)

$$T_4^{-1} = \begin{bmatrix} I & 0 \\ F & I \end{bmatrix}\begin{bmatrix} \bar{A}_4^{-1} & -\bar{A}_4^{-1}S_2\bar{A}_4^{-T} \\ 0 & -\bar{A}_4^{-T} \end{bmatrix}\begin{bmatrix} I & 0 \\ -F & I \end{bmatrix} \qquad (22)$$

Next, define

$$N_1 = -(A_2 - S_{12}F)(A_4 - S_2F)^{-1} \qquad (23a)$$

$$N_2 = (Q_{12} + A_3'F)(A_4 - S_2F)^{-1} \qquad (23b)$$

Now substitute T_4^{-1} of (22) and T_1, T_2 and T_3 of (6) into (11). By calculation, the bolcks A_0, S_0 and Q_0 can be arranged into

$$A_0 = A_1 + N_1A_3 + S_{12}N_2' + N_1S_2N_2' \qquad (24a)$$

$$S_0 = B_0R^{-1}B_0', \qquad B_0 = B_1 + N_1B_2 \qquad (24b)$$

$$Q_0 = -N_2A_3 - A_3'N_2' - N_2S_2N_2' + Q_1 \qquad (24c)$$

The triple (A_0, S_0, Q_0) defines the coefficient matrices of the Riccati equation (12).

Substituting T_4^{-1} of (22) into (13), then premultiplying both sides of (13) by $\begin{bmatrix} I & 0 \\ -F & I \end{bmatrix}$

we have

$$\begin{bmatrix} I & 0 \\ -F & I \end{bmatrix}\begin{bmatrix} L_1 \\ L_2 \end{bmatrix} = -\begin{bmatrix} * & * \\ N_2' & -N_1' \end{bmatrix}\begin{bmatrix} I \\ P \end{bmatrix}$$

Hence

$$L_2 - FL_1 = N_1'P - N_2' \qquad (25)$$

Kokotovic and Yackel (1972) once in their paper found the limits of the optimal gain $\bar{K}$ as $\mu \to 0$. Denote these limits to be (K_1, K_2, K_3). In comparison of our (19)(24)(25) with (11c)(15b)(16)(17) of (Kokotovic and Yackel, 1972), it is seen that $K_1 = P$, $K_2 = (L_2 - FL_1)'$ and $K_3 = F$. Thus, as $\mu \to 0$, (20) follows directly from (Kokotovic and Yackel, 1972, Theorem 1.)

■■

As a result of our Theorem 1:

$$\lim_{\mu\to 0} v^*(t) = \bar{v}(t) = -R^{-1}B'Kz \qquad t\in[0, t_f) \qquad (26)$$

Concerning the relations between the trajectories and costs, we state the following lemmas without proof.

Lemma 6: If the near-optimum control $\bar{v}(t) = -R^{-1}B'Kz$, instead of $v^*(t)$, is applied to the system (2), as $\mu \to 0^+$. the resulting trajectory z will tend to the optimal trajectory x^* of the closed-loop system (17)

i.e. $z_1 \to x_1^* \quad \forall[0\ t_f]$

$z_2 \to x_2^* \quad \forall[t_1\ t_f]$ Where $t_1 > 0$ is arbitrarily close to zero.

The proof of this Lemma is based on the fact that $\bar{A}_4$ is stable for F is the positive definite solution of (19). Thus the trajectory z will tend to its reduced one as $\mu \to 0^+$, which corresponds to the trajectory x^* of (17).

Lemma 7: Let $J(v^*)$ denote the optimal cost (1) and $J(u^*)$ denote the optimal cost (3), then

$$\lim_{\mu \to 0} J(v^*) = J(u^*) \qquad (27)$$

The proof of Lemma 7 is based on (26) and Lemma 6. Finally we summarize the above results. The design of the near-optimum regulators for the problem (1)(2) is developed into the following algorithm:

Step 1) Compute the blocks A_0, S_0 and Q_0 from

$$\begin{bmatrix} A_0 & -S_0 \\ -Q_0 & -A_0' \end{bmatrix} = T_1 - T_2 T_4^{-1} T_3$$

Step 2) Solve the Riccati equation

$$-\frac{dP(t)}{dt} = P(t)A_0 + A_0'P(t) - P(t)S_0P(t) + Q_0$$

$$P(t_f) = G_1$$

Step 3) Compute $L_1(t)$ and $L_2(t)$

Step 4) Solve the positive definite F from (19), and in terms of (16'), form the near-optimum gain $-R^{-1}B'K$

Remark 2: Whenever A_4 is invertible or not, the fast gain F and the slow gain $P(t)$ can always be solved separately. In addition, under the conditions of Remark 1, Theorem 1 is also true when the terminal time t_f approaches infinity (see Wang,1987).

IV. DESIGN EXAMPLE

We consider the familar speed control problem for an electrical DC motor (Kokotovic and Yackel,1972)

$$\frac{d\omega}{dt} = (D/G)i \qquad (28a)$$

$$\lambda L\frac{di}{dt} = -C\omega - Ri + v \qquad (28b)$$

Where ω, i and v are speed, current and voltage deviations from their respective nominal values $\omega_0 = 400\text{rad/s}$. The motor constants $R = 7.9\Omega$, $L = 0.0136H$, $C = 0.0246\text{v·s/rad}$, $G = 1.32 \times 10^{-6}\text{kg·m}^2$. In (28) the armature inductance, being a small parameter, is multiplied by a factor λ. Let the cost functional be

$$J = \frac{0.01}{2}\omega^2(t_f) + \frac{1}{2}\int_0^{t_f}(\omega^2 + 4600i^2 + 30v^2)dt \qquad (29)$$

with $t_f = 10\text{ms}$. The optimization problem to find the optimal regulator $\bar{K}$ is solved on an IBM/PC computer. Meantime, the algorithm for design of suboptimal regulators is applied to compute $P(t)$ and F separately. The figures below simulate the design results, which checks with our Theorem 1.

The solid curves represent the optimal solution $\bar{K}_1$, $\bar{K}_2$ and $\bar{K}_3$. The dotted curves represent the near-optimum solution. Note that the convergences of $\bar{K}_2$ and $\bar{K}_3$ are not uniform on the whole interval $[0 \ t_f]$ due to the boundary phenomenon near t_f.

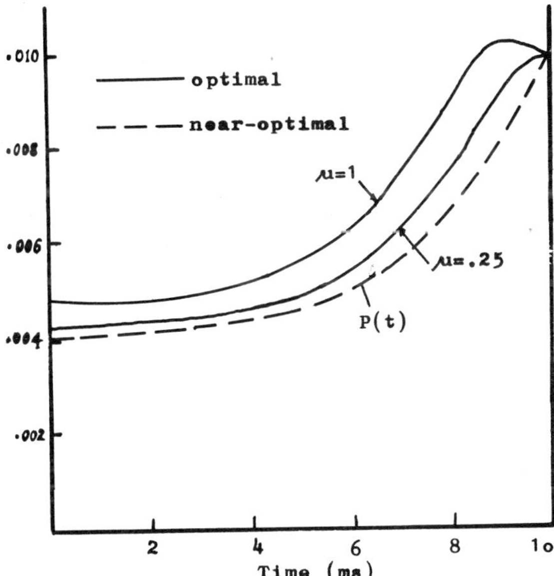

Fig. 1. Optimal and near-optimal gains $\bar{K}_1$ and P

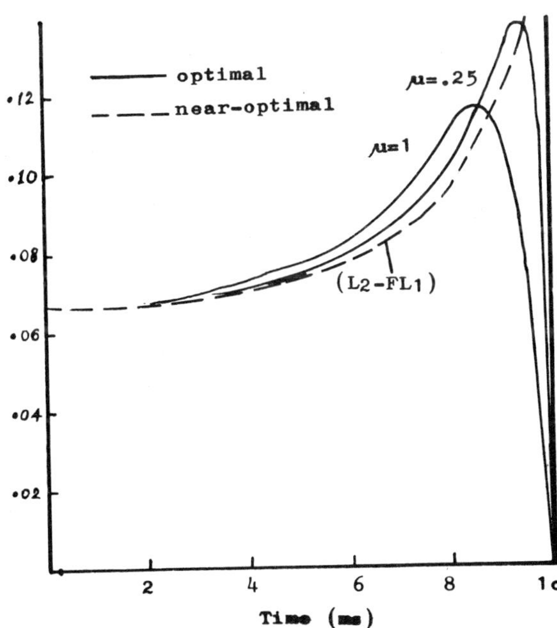

Fig. 2. Optimal and near-optimal gains $\bar{K}_2$ and $(L_2 - FL_1)'$

257

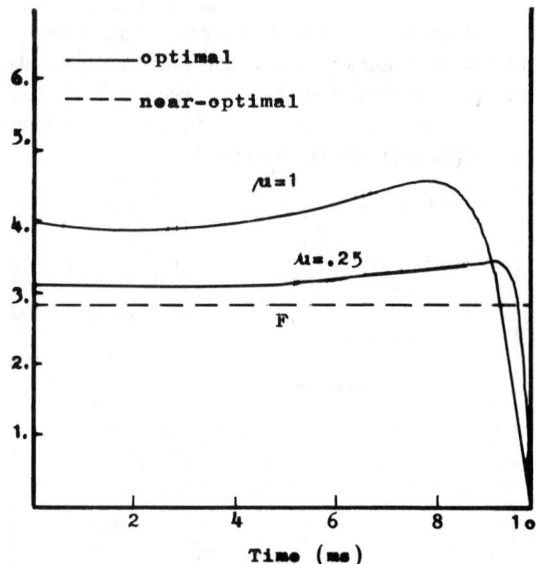

Fig. 3. Optimal and near- optimal gains
$\bar{K}_3$ and F.

V. CONCLUSION

By relating the optimal feedback controls, trajectories and costs between the optimization problem (1) and (3), this paper helps clarify the structural properties for both descriptor and singularly perturbed linear regulators. Moreover, a new approach is developed to design near-optimum regulators for singularly perturbed systems. Such design is decomposed into separate solutions of two regulator problems of lower order. The decomposition is applicable whenever the reduced system (4) exhibits impulses or not. Hence, the approach advances over the previous ones which were only confined to the non-impulsive case, (Haddad and Kokotovic, 1971; Chow and Kokotovic, 1976) or could not be decomposable (Kokotovic and Yackel, 1972).

Acknowledgment

The authors would like to thank Dr. D.J. Bender for his helpful discussions on this investigation.

REFERENCE

Bender, D.J., and A.J. Laub (1987). The linear quadratic optimal regulator for descriptor systems. IEEE Trans. Automat. Contr., 32. 672—688

Cobb, D. (1983). Descriptor variable systems and optimal state regulation. IEEE Trans. Automat. Contr., 28, 601—611.

Chow, J.H., and P.V. Kokotovic (1976). A decomposition of near-optimum regulators for systems with slow and fast modes. IEEE Trans. Automat. Contr., 21, 701—705.

Cheng, Z.L., H.M. Hong, and J.F. Zhang (1987). The optimal state regulation of generalized dynamic systems with quadratic cost function. 10th IFAC World Congress,, July 27—31, Munich.

Haddad, A.H., and P.V. Kokotovic (1971). Note on singular perturbation of linear state regulators. IEEE Trans. Automat. Contr., 16, 279—281.

Kokotovic, P.V., and R.A. Yackel (1972). Singular perturbation of linear regulators: Basic theorem. IEEE Trans. Automat. Contr., 17, 29—37.

Lewis, F.L (1985). Preliminary notes on optimal control for singular systems.Proc. IEEE CDC Conf., 266—272.

Verghese, G.C., B.C. Levy and T. Kailath (1981). A generalized state-space for singular systems. IEEE Trans. Automat. Contr., 26, 811—831.

Wang, Y.Y. (1987). Feedback control of descriptor systems. Ph.D. Dissertation Shanghai Jiao Tong Univ., Jan., Shanghai.

SOME NEW ASPECTS ON DESCRIPTOR FORM
REPRESENTATION OF SYSTEMS

R. Ylinen and H. Blomberg

Helsinki University of Technology, Otakaari 5 A, SF-02150 Espoo, Finland

Abstract. Properties of the descriptor form representation of linear time-invariant multivariable systems have been studied using polynomial system methodology. A new normal form for the input-output equivalence on descriptor form representations has been presented. This normal form makes the complicated and partly misleading controllability and observability considerations presented in the literature clear and easy to carry out.

Keywords. multivariable systems; descriptor form; controllability; observability

INTRODUCTION

A general representation of a finite dimensional linear input-output (i/o) system is a regular generator

$$[A(z)\,\vdots\,-B(z)] \begin{matrix} (x,y) \\ u \end{matrix} \quad \text{with } \det A(z) \neq 0, \qquad (1)$$

where $A(z)$ and $B(z)$ are polynomial matrices in an operator z on a signal space χ. The i/o relation, or more specifically, (u,y) relation generated by (1) is the set $S\{u,y\}$ of pairs (u,y) such that $((x,y),u)$ belongs to the kernel of $[A(z)\,\vdots\,-B(z)]$. Using the polynomial methods the internal output variable x can be eliminated and an i/o generator for $S\{u,y\}$ be obtained. The controllability and observability properties of the system have been shown to be related to the existence or nonexistence of certain common divisors in (1) (Blomberg and Ylinen,1983).

The representations of the same i/o relation $S\{u,y\}$ are said to be i/o or (u,y) equivalent to each other. There are many special forms of representation for i/o relations. The state-space representation is probably the most advanced one but for many purposes the more elementary i/o generator representation or the so-called descriptor form representation can replace it. In the descriptor form (1) is of the special form

$$\left[\begin{array}{c|c|c} Ez-F & 0 & -G \\ \hline -H & I & -K(z) \end{array} \right] \begin{matrix} x \quad y \quad u \end{matrix} \quad \text{with } \det(Ez-F) \neq 0. \qquad (2)$$

The general controllability and observability concepts apply, of course, to the descriptor form representation, so that new special definitions and methods for the descriptor form are not needed. Nevertheless, a number of papers and studies considering the subject has been published during the last ten years. Unfortunately, these studies have led to an unnecessarily complicated and partly even misleading theory. The reason for this is that the possibility to construct $(u,(x,y))$ equivalent representations to (2) by applying elementary row operations represented by polynomial matrices has not been recognized - only elementary row operations represented by constant matrices have been used.

The controllability and observability studies in the literature are based on a so-called normal form, which is obtained using constant row and column operations. In this paper we will show that this form can be reduced further by elementary polynomial row operations to a remarkably simpler form with respect to the interesting controllability and observability properties.

The theory for descriptor form representations as presented in the literature, cf. Campbell, 1980, Helmke and Shayman, 1989, Luenberger, 1977, 1978, Murota, 1987, Rosenbrock, 1974, Shayman, 1988, Shin and Kabamba, 1988, Verghese et al., 1981, Yamada and Luenberger 1958a, 1958b, Yip and Sinovec, 1981, and many others, is in principle based on well-known results concerning nonsingular and singular pencils of matrices, see for instance Gantmacher, 1959, volume II, chapter XII.

DESCRIPTOR FORM REPRESENTATION OF LINEAR CONSTANT SYSTEMS

Let z be either the differentiation operator $p:\chi\rightarrow\chi$, or the unit prediction operator $q:\chi\rightarrow\chi$, with χ a suitable signal space. In the case $z=p$, χ can be interpreted as the space of all complex-valued (alternatively real-valued) infinitely continuously differentiable functions defined on $\mathbf{R}$, denoted by C^∞ - more generally, χ can be the space of all complex-valued (alternatively real-valued) piecewise infinitely regularly differentiable generalized functions on $\mathbf{R}$, denoted by $\underline{D}$. In the case $z=q$, χ can be taken as the space of all sequences of complex (alternatively real) numbers $(x_0,x_1,x_2,\ldots)$, i.e.

as $\mathbf{C}^{\mathbf{N}_0}$ $(\mathbf{R}^{\mathbf{N}_0})$ with $\mathbf{N}_0 \underline{\Delta} \{0,1,2,\ldots\}$. These spaces along with the polynomials in z ($z=p$ or q), also called operator polynomials, have been discussed in detail in Blomberg and Ylinen (1983). Essentially, the operator polynomials in z are treated as ordinary polynomials in an indeterminate. They form a ring and an integral domain, generally denoted by $\mathbf{C}[z]$, with respect to the usual operations for polynomials, appropriately interpreted. This ring also allows the construction of an associated field of quotients, denoted by $\mathbf{C}(z)$. χ is considered as a (left) $\mathbf{C}[z]$-module.

Consider a given system modelled by the set of input-output (i/o) pairs (u,y) satisfying the module equation

$$A(z)y=B(z)u \tag{3}$$

with $\det A(z)\neq 0$, i.e. represented by the regular i/o generator

$$\begin{array}{cc} y & u \\ [A(z)\,\vert\,-B(z)]. \end{array} \tag{4}$$

(4) may also appear in a little more general form with y above replaced by a pair (y_1,y):

$$\begin{array}{cc} (y_1,y) & u \\ [A(z)\,\vert\,-B(z)], \end{array} \tag{5}$$

where y_1 has q_1 components and y q components; $A(z)$ in (5) is of size (q_1+q) by (q_1+q). The i/o relation, or more specifically, (u,y) relation $S\{u,y\}$ generated by (5) is as usual interpreted as the set of all (u,y) that can be formed from the set of all $(u,(y_1,y))$ generated by (5).

Suppose now that constant matrices E,F,G,H with E and F square, as well as a polynomial matrix K(z) of proper sizes can be found such that $\det(Ez-F)\neq 0$, and such that the i/o generator

$$\begin{array}{ccc} x & y & u \end{array}$$
$$\begin{bmatrix} Ez-F & 0 & -G \\ \hline -H & I & -K(z) \end{bmatrix} \tag{6}$$

is (u,y) equivalent to (4) or (5) as the case may be. In (5) and (6) y_1 and x are internal output signal vectors, x having, say, ℓ components, and the (u,y) relation generated by (6) is interpreted in the usual way. y is also called the overall output signal vector, and (y_1,y) in (5) as well as (x,y) in (6) are called internal-overall output signal vectors. Note that the i/o generator appearing in (6) is regular.

Conversely, the original system equations may occasionally be given in the form of an i/o generator (6), and then a (u,y) equivalent i/o generator (4) can be obtained by eliminating the internal signal vector x from (6) using the standard elimination procedure (see Blomberg and Ylinen,1983, sections 1.4 and 7.1).

Now (6) is called a <u>descriptor form representation</u> of the given system, or alternatively an i/o generator in <u>descriptor form</u> for that system. The descriptor form representation have certain advantages from a structural point of view compared with a state-space representation. However, it is found that (6) as given above is not general enough for the present purpose, a <u>modified</u> descriptor form representation will occasionally be needed with G in (6) replaced by a corresponding polynomial matrix G(z). For E=I, (6) represents a standard state-space representation. Referring to this, and to the fact that E most often is assumed singular, (6) is sometimes also called a generalized state-space representation, or a singular representation (cf. Verghese et al., 1981, Campbell, 1980).

An i/o generator in descriptor form, (u,y) equivalent to (4) or (5), can be found as follows. The first step would be to replace the i/o generator in question by a (u,y) equivalent i/o generator such that the new A(z) would contain polynomials in z of at most the first degree. If necessary, the usual simple trick of introducing suitable new variables will do the job. For a more detailed exposition of the procedure, see Schumacher, 1988. A simple rewriting of the i/o generator so obtained then yields an i/o generator of the form (6), possibly with the constant matrix G replaced by a polynomial matrix G(z).

<u>Note 1</u>
(i) It is readily concluded that generally it holds that rank $E\geq\deg\det(Ez-F)$, equality holding if (Ez-F) is row proper. In many cases it may also be possible, if desired, to increase the rank of E by means of elementary row operations applied to $[Ez-F\,\vert\,-G(z)]$. Rank E is therefore not a characteristic system quantity, but rank $E=\deg\det(Ez-F)$ turns out to be an important condition which may be imposed on an i/o generator in descriptor form. This condition can thus always be made to be fulfilled - without loss of generality.

(ii) It may well happen that the first result obtained is an i/o generator in modified descriptor form with G in (6) replaced by a polynomial matrix G(z). In such a case it is always possible to construct a new i/o generator in descriptor form (u,y) equivalent (but possibly not (u,(x,y)) equivalent) to the original one with G(z) replaced by a constant matrix G. The construction goes as follows.

First a unimodular $Q_1(z)$ is determined so that $(Ez-F)Q_1(z)$ becomes column proper. Note that $(Ez-F)Q_1(z)$ can then again be written as Ez-F for some new E and F. Next suitable combinations of the columns of $(Ez-F)Q_1(z)$, possibly multiplied by polynomials in z, are added to the columns of -G(z) so that in the end a constant matrix appears in the place of G(z). These operations determine the matrix $Q_3(z)$. The procedure described above resembles the algorithm presented in Blomberg and Ylinen, 1983, section 1.5 for constructing a feasible feedback system from a first candidate.

The final i/o generator in descriptor form is thus:

$$\begin{array}{ccc} \tilde{x} & y & u \end{array}$$
$$\begin{bmatrix} (Ez-F)Q_1(z) & 0 & (Ez-F)Q_1(z)Q_3(z)-G(z) \\ \hline -HQ_1(z) & I & -HQ_1(z)Q_3(z)-K(z) \end{bmatrix} \tag{7}$$

with

$$\tilde{x}=Q_1^{-1}(z)x-Q_3(z)u \tag{8}$$

and with $(Ez-F)Q_1(z)Q_3(z)-G(z)$ a constant matrix. ∎

CONTROLLABILITY AND OBSERVABILITY ASPECTS

An i/o generator in descriptor form (6), possibly in the modified form with G replaced by a polynomial matrix G(z), is of course subject to standard controllability and observability concepts as defined for system representations by means of polynomial matrices. As discussed in detail in Blomberg and Ylinen, 1983, controllability and observability properties are related to the existence or nonexistence of certain common divisors and their properties. A brief presentation of the basic facts follows.(see also Blomberg and Ylinen, 1983).

Consider an i/o generator in modified descriptor form (6) with G replaced by G(z) and let

$$\begin{array}{ccc} x & y & u \end{array}$$
$$\begin{bmatrix} M(z) & A_2(z) & -B_1(z) \\ \hline 0 & A_4(z) & -B_2(z) \end{bmatrix} \tag{9}$$

be an i/o generator in upper block triangular form, (u,(x,y)) equivalent to (6). Note that M(z) is a greatest common right divisor (GCRD) of Ez-F and H, or equivalently, of the rows of Ez-F and H. Let further $S\{u,(x,y)\}$, $S\{u,x\}$ and $S\{u,y\}$ denote the various i/o relations generated by (6) and (9). Then the fundamental controllability and observability properties of the system under consideration are as follows:

260

<u>Note 2</u>

(i) S{u,(x,y)} is controllable if and only if the columns of (6), or equivalently, the columns of (9) are left coprime.

(ii) S{u,x} is controllable if and only if (the columns of) $Ez-F$ and $G(z)$ are left coprime.

(iii) S{u,y} is controllable if and only if (the columns of) $A_4(z)$ and $B_2(z)$ are left coprime.

(iv) S{u,(x,y)} is x(u,y) observable if and only if (the rows of) $Ez-F$ and H are right coprime, or equivalently, if and only if $M(z)$ is unimodular. ∎

Greatest common left or right divisors can be constructed, and left or right coprimeness tested as described in Blomberg and Ylinen, 1983, section 1.3, cf. also (9) and Note 2 (iv) above. Let the method be used to test the controllability of S{u,(x,y)}, S{u,x} and S{u,y} according to Note 2 (i)...(iii). Thus consider (6) with G replaced by G(z). By means of elementary column operations it can be transformed first to

$$\begin{bmatrix} Ez-F & | & 0 & | & -G(z) \\ \hline 0 & | & I & | & 0 \end{bmatrix} \qquad (10)$$

and then to

$$\begin{bmatrix} L(z) & | & 0 & | & 0 \\ \hline 0 & | & I & | & 0 \end{bmatrix}, \qquad (11)$$

where $L(z)$ is a GCLD of $Ez-F$ and $G(z)$. It follows from (11) that $\mathrm{diag}\{L(z),I\}$ is a GCLD of the columns of (6). Accordingly, for $L(z)$ unimodular $Ez-F$ and $G(z)$ are left coprime, and the columns of (6) as well as the columns of (9) are also left coprime. From (9) it can be concluded that then also $A_4(z)$ and $B_2(z)$ must be left coprime. On comparing the above results with Note 2 (i)...(iii) it follows that:

<u>Note 3</u>
(i)   If S{u,(x,y)} is controllable, then so are S{u,x} and S{u,y}.

(ii) S{u,(x,y)} is controllable if and only if S{u,x} is controllable.

(iii) If $M(z)$ is unimodular, then S{u,(x,y)} is controllable if and only if S{u,y} is controllable.

(iv) For $M(z)$ unimodular, the controllability properties of S{u,(x,y)}, S{u,x} and S{u,y} according to Note 2 (i)...(iii) are all mutually equivalent, i.e. controllability of any one of S{u,(x,y)}, S{u,x} and S{u,y} implies controllability of all the others. ∎

The results cited above are as such valid for differential systems with z interpreted as the differentiation operator p. For difference systems with z interpreted as the unit prediction operator q there are some additional aspects, see Kalman et al, 1969, sections 2.3 and 2.6, and Blomberg and Ylinen, 1983, sections 12.4 and 13.1.

NORMAL FORMS FOR SYSTEM EQUATIONS IN DESCRIPTOR FORM

Consider a set of system equations given in modified descriptor form (cf.(6); det(Ez-F)≠0):

$$\begin{array}{cc} & \begin{array}{ccc} \ell & q & r \end{array} \\ \begin{array}{c} \ell \\ q \end{array} & \begin{bmatrix} Ez-F & | & 0 & | & -G(z) \\ \hline -H & | & I & | & -K(z) \end{bmatrix} \end{array} \begin{array}{c} \ell \\ q \\ \\ r \end{array} \begin{bmatrix} x \\ - \\ y \\ = \\ u \end{bmatrix} =0, \qquad (12)$$

where the matrix sizes are as indicated. Let deg det(Ez-F) be equal to ℓ_1 (≥0) with Ez-F not necessarily row proper, i.e. with $\ell_1\le$rank $E\le\ell$, a row proper Ez-F implying rank E=deg det(Ez-F)=ℓ_1.

Below it will be shown that there are matrices $P(z)$ and Q of proper sizes with $P(z)$ unimodular and Q constant and nonsingular such that

$$\begin{array}{cc} & \begin{array}{ccc} \ell & q & r \end{array} \\ \begin{array}{c} \ell \\ q \end{array} & \begin{bmatrix} P(z)(Ez-F)Q & | & 0 & | & -P(z)G(z) \\ \hline -HQ & | & I & | & -K(z) \end{bmatrix} \end{array} \begin{array}{c} \ell \\ q \\ \\ r \end{array} \begin{bmatrix} \tilde{x} \\ - \\ y \\ = \\ u \end{bmatrix} =0 \qquad (13)$$

with $\tilde{x}=Q^{-1}x$ takes the form ($\ell_1+\ell_2=\ell$, $\tilde{x}=(\tilde{x}_1,\tilde{x}_2)$):

$$\begin{array}{cc} & \begin{array}{cccc} \ell_1 & \ell_2 & q & r \end{array} \\ \begin{array}{c} \ell_1 \\ \ell_2 \\ q \end{array} & \begin{bmatrix} E_1z-F_1 & | & 0 & | & 0 & | & -G_1(z) \\ \hline 0 & | & I & | & 0 & | & -G_2(z) \\ \hline -H_1 & | & -H_2 & | & I & | & -K(z) \end{bmatrix} \end{array} \begin{array}{c} \ell_1 \\ \ell_2 \\ q \\ \\ r \end{array} \begin{bmatrix} \tilde{x}_1 \\ -- \\ \tilde{x}_2 \\ = \\ y \\ == \\ u \end{bmatrix} =0, \qquad (14)$$

where E_1 is nonsingular for $\ell_1>0$. Of course, if ℓ_1 or $\ell_2=0$, then the corresponding rows and columns do not appear in (14). In what follows it is assumed that ℓ_1 and ℓ_2 are nonzero. If this does not hold, the results can be modified in an obvious way.

Now (14) could be even further simplified in a number of ways. In particular, the second block row in (14) multiplied from the left by H_2 can be added to the third block row to yield:

$$\begin{array}{cc} & \begin{array}{cccc} \ell_1 & \ell_2 & q & r \end{array} \\ \begin{array}{c} \ell_1 \\ \ell_2 \\ q \end{array} & \begin{bmatrix} E_1z-F_1 & | & 0 & | & 0 & | & -G_1(z) \\ \hline 0 & | & I & | & 0 & | & -G_2(z) \\ \hline -H_1 & | & 0 & | & I & | & -K(z)-H_2G_2(z) \end{bmatrix} \end{array} \begin{array}{c} \ell_1 \\ \ell_2 \\ q \\ \\ r \end{array} \begin{bmatrix} \tilde{x}_1 \\ -- \\ \tilde{x}_2 \\ -- \\ y \\ == \\ u \end{bmatrix} =0. \qquad (15)$$

Of course, (14),(15) generate, by construction, the same (u,y) relation as (12).

(14),(15) are loosely speaking called "normal forms" corresponding to (12) with respect to the type of transformations applied. It should be stressed that the normal forms used in this context do not generally qualify as true canonical forms for the transformations considered, for details, see Helmke and Shayman, 1989.

P(z) and Q above can be found in a number of steps. The procedure is not unique, and the final results are unique only with respect to the general structure of the normal forms.

<u>Step 1</u> Determine a unimodular $P_1(z)$ so that $P_1(z)(Ez-F)$ becomes row proper with the ℓ_1 uppermost rows being of first degree, and with the ℓ_2 bottom rows being of zero degree. Thus $P_1(z)(Ez-F)$ reads in partitioned form:

$$\begin{array}{cc} & \begin{array}{cc} \ell_1 & \ell_2 \end{array} \\ \begin{array}{c} \ell_1 \\ \ell_2 \end{array} & \begin{bmatrix} E_1z-F_1 & | & E_2z-F_2 \\ \hline -F_3 & | & -F_4 \end{bmatrix} \end{array} \qquad (16)$$

for some E_1, E_2, F_1, F_2, F_3 and F_4 of some suitable form. From the row properness of (16) it follows that $[E_1 \mid E_2]$ and $[-F_3 \mid -F_4]$ are of full rank ℓ_1 and ℓ_2 respectively. If $Ez-F$ in (12) happens to be row proper, then $P_1(z)$ can be chosen constant and nonsingular.

Step 2 As the next step it is in general worthwhile to reorder the columns of (16), i.e. to multiply (16) from the right by a permutation matrix Q_1, so that the resulting $P_1(z)(Ez-F)Q_1$ becomes

$$
\begin{array}{cc} \ell_1 & \ell_2 \end{array}
\begin{array}{c} \ell_1 \\ \ell_2 \end{array}
\left[
\begin{array}{c|c}
E_1 z - F_1 & E_2 z - F_2 \\
\hline
-F_3 & -F_4
\end{array}
\right]
\qquad (17)
$$

for some nonsingular E_1, and some E_2, F_1, F_2, F_3 and F_4, not necessarily the same as in (16). E_1 in (17) can, for instance, be formed from ℓ_1 linearly independent columns picked from $[E_1 \mid E_2]$ in (16).

This step turns out to be crucial with respect to how $\tilde{x}_1$ and $\tilde{x}_2$ in (14),(15) are related to the original variables appearing in x. The variables associated with the first ℓ_1 columns in (17) either form the components of the final signal vector $\tilde{x}_1$ - this happens if E_2 in (17) is the zero matrix - or else appear as additive parts of those components.

Step 3 Next (17) is multiplied from the right by the constant nonsingular matrix

$$
Q_2 =
\begin{array}{c} \ell_1 \\ \ell_2 \end{array}
\left[
\begin{array}{c|c}
I & -E_1^{-1}E_2 \\
\hline
0 & I
\end{array}
\right]
\qquad (18)
$$

resulting in (E_1, F_1 and F_3 as in (17))

$$
\begin{array}{c} \ell_1 \\ \ell_2 \end{array}
\left[
\begin{array}{c|c}
E_1 z - F_1 & -\tilde{F}_2 \\
\hline
-F_3 & -\tilde{F}_4
\end{array}
\right]
\qquad (19)
$$

with

$$
\begin{cases}
\tilde{F}_2 = F_2 - F_1 E_1^{-1} E_2, \\
\tilde{F}_4 = F_4 - F_3 E_1^{-1} E_2.
\end{cases}
\qquad (20)
$$

(19) is, of course, still row proper implying that $\tilde{F}_4$ is nonsingular.

Step 4 Finally, (19) is multiplied from the right by the constant nonsingular matrix

$$
Q_3 =
\begin{array}{c} \ell_1 \\ \ell_2 \end{array}
\left[
\begin{array}{c|c}
I & 0 \\
\hline
-\tilde{F}_4^{-1} F_3 & I
\end{array}
\right],
\qquad (21)
$$

and from the left by the constant nonsingular matrix

$$
P_2 =
\begin{array}{c} \ell_1 \\ \ell_2 \end{array}
\left[
\begin{array}{c|c}
I & -\tilde{F}_2 \tilde{F}_4^{-1} \\
\hline
0 & -\tilde{F}_4^{-1}
\end{array}
\right]
\qquad (22)
$$

resulting in $P_2 P_1(z)(Ez-F)Q_1 Q_2 Q_3$ given by

$$
\begin{array}{c} \ell_1 \\ \ell_2 \end{array}
\left[
\begin{array}{c|c}
E_1 z - F_1 + \tilde{F}_2 \tilde{F}_4^{-1} F_3 & 0 \\
\hline
0 & I
\end{array}
\right],
\qquad (23)
$$

which is of the required form as appearing in (14),(15). Thus $P_2 P_1(z)$ can be taken as $P(z)$ and $Q_1 Q_2 Q_3$ as Q in (13). Further, if $Ez-F$ in (12) happens to be row proper, then (12) can be transformed to the form (13) using only constant nonsingular transformation matrices $P(z)=P$ and Q.

RELATIONSHIP BETWEEN THE NEW SIGNAL VECTOR $\tilde{x}$ AND THE ORIGINAL SIGNAL VECTOR x

Suppose that (16) satisfies the conditions imposed on (17), i.e. suppose that $Q_1 = I$. Then the new signal vector $\tilde{x} = (\tilde{x}_1, \tilde{x}_2)$ in (13),(14) and (15) is obtained from the original signal vector x in (12) - partitioned as (x_1, x_2) in the same way as $\tilde{x}$ - according to

$$
\begin{array}{c} \ell_1 \\ \ell_2 \end{array}
\left[
\begin{array}{c}
\tilde{x}_1 \\
\hline
\tilde{x}_2
\end{array}
\right]
=
\begin{array}{c} \ell_1 \\ \ell_2 \end{array}
\left[
\begin{array}{c|c}
I & E_1^{-1}E_2 \\
\hline
\tilde{F}_4^{-1}F_3 & I + \tilde{F}_4^{-1}F_3 E_1^{-1}E_2
\end{array}
\right]
\begin{array}{c} \ell_1 \\ \ell_2 \end{array}
\left[
\begin{array}{c}
x_1 \\
\hline
x_2
\end{array}
\right]
\qquad (24)
$$

From this it can be seen that a particular component of x_1 also qualifies as the corresponding component of $\tilde{x}_1$ only if the corresponding row of $E_1^{-1}E_2$ happens to be a zero row. This is an important result, because $\tilde{x}_1$ will be the state vector in the final state-space representation of the system. Looking back at the expressions (17),...,(24) it seems clear that in order to obtain simple expressions, the reordering contained in (17) should be performed so that E_1 is close to I and E_2 as sparse as possible.

CONTROLLABILITY AND OBSERVABILITY
ASPECTS CONCERNING THE NORMAL FORMS

Consider (12) and (14),(15). It is readily concluded, that $Ez-F$ and $G(z)$ are left coprime if and only if $E_1 z - F_1$ and $G_1(z)$ are left coprime, and $Ez-F$ and H are right coprime if and only if $E_1 z - F_1$ and H_1 are right coprime.

A number of i/o relations can be associated with (12) and (14),(15). There are the i/o relations $S\{u,(x,y)\}$, $S\{u,x\}$ and $S\{u,y\}$ generated by (12), and the i/o relations $\tilde{S}\{u,(\tilde{x},y)\}$, $\tilde{S}\{u,\tilde{x}\}$ and $\tilde{S}\{u,y\}$ with $\tilde{S}\{u,y\}=S\{u,y\}$ generated by (14),(15). In addition, (14),(15) also generate i/o relations of the form $\tilde{S}\{u,(\tilde{x}_1,y)\}$, $\tilde{S}\{u,\tilde{x}_1\}$, $\tilde{S}\{u,(\tilde{x}_2,y)\}$ and $\tilde{S}\{u,\tilde{x}_2\}$.

On the basis of the results cited, it is then readily concluded that there is a close correspondence between the controllability and observability properties of the i/o relations mentioned above. For instance:

Note 4

(i)
$\begin{cases} S\{u,(x,y)\} & \text{is controllable,} \\ \tilde{S}\{u,(\tilde{x},y)\} & -"- \\ \tilde{S}\{u,(\tilde{x}_1,y)\} & -"- \end{cases}$

are mutually equivalent statements.

(ii)
$\begin{cases} S\{u,x\} & \text{is controllable,} \\ \tilde{S}\{u,\tilde{x}\} & -"- \\ \tilde{S}\{u,\tilde{x}_1\} & -"- \end{cases}$

are mutually equivalent statements.

(iii)
$\begin{cases} S\{u,(x,y)\} & \text{is } x(u,y) \text{ observable,} \\ \tilde{S}\{u,(\tilde{x},y)\} & \text{is } \tilde{x}(u,y) \text{ observable,} \\ \tilde{S}\{u,(\tilde{x}_1,y)\} & \text{is } \tilde{x}_1(u,y) \text{ observable} \end{cases}$

are mutually equivalent statements.

(iv) $\tilde{S}\{u,\tilde{x}_2\}$ is always controllable as well as $\tilde{x}_2(u)$ observable. ∎

Thus the interesting controllability and observability properties associated with (14),(15) are related only to the variable $\tilde{x}_1$, there are no dynamic modes associated with $\tilde{x}_2$.

DISCUSSION

Note 5
(i) In the literature (see for instance Yip and Sinovec, 1981, Yamada and Luenberger, 1985a, Murota, 1987, chapter 3) transformations of the original system equations in descriptor form as given above by (12) (in the literature $G(z)$ is assumed to be a constant matrix - the modified form used here does not influence the discussion to follow) according to (13) are considered only for $P(z)$ and Q constant and nonsingular. Referring to the results obtained it then follows that if the original $Ez-F$ in (12) happens to be row proper, then the normal forms (14),(15) can be achieved in this way. But if $Ez-F$ is not row proper with
$\deg \det(Ez-F)=\ell_1$, $\ell_1<\text{rank } E\leq\ell=\ell_1+\ell_2$, then (14) has to be replaced by the normal form (cf. Yip and Sinovec, 1981, Murota, 1987, chapter 3, Helmke and Shayman, 1989, see also Gantmacher, 1959, volume II, chapter XII):

$$
\begin{array}{cccc}
\ell_1 & \ell_2 & q & r
\end{array}
$$
$$
\begin{array}{c}
\ell_1 \\ \ell_2 \\ q
\end{array}
\left[
\begin{array}{c|c|c|c}
E_1z-F_1 & 0 & 0 & -G_1(z) \\
\hline
0 & E_4z-I & 0 & -G_2(z) \\
\hline
-H_1 & -H_2 & I & -K(z)
\end{array}
\right]
\begin{array}{c}
\ell_1 \\ \ell_2 \\ q
\end{array}
\left[
\begin{array}{c}
\tilde{x}_1 \\ \tilde{x}_2 \\ y \\ \hline u
\end{array}
\right]=0, \quad (25)
$$

with E_1 nonsingular, rank $E_4=\text{rank } E-\ell_1$ and E_4 nilpotent, i.e. $E_4{}^m=0$, $E_4{}^{m-1}\neq0$ for some positive integer m. Of course, the submatrices appearing in (25) are not necessarily the same as those in (14). Note that the nilpotency of E_4 implies that (can be verified by direct calculation)

$$(E_4z-I)^{-1}=-I-E_4z-E_4{}^2z^2-\ldots-E_4{}^{m-1}z^{m-1}. \quad (26)$$

Consequently, the inverse of E_4z-I exists as a polynomial matrix, implying that E_4z-I is unimodular. Accordingly, (25) can again by means of elementary row operations be transformed first to the form (14), and thereafter to the form (15).

(ii) The controllability and observability concepts used in this paper are "dynamic" in the sense that they are closely related to the possibility of pole placements in feedback (control) systems by means of dynamic feedback. This problem has been discussed in detail in Blomberg and Ylinen, 1983, sections 1.5 and 7.12, see also Blomberg and Ylinen, 1983, section 7.11.

In addition, other than dynamic control properties may occasionally be of interest, for instance the possibility to keep some of the system variables (or their rate of change) at some prescribed constant values for an arbitrary long period of time in a steady state situation, or to force some system variables to take some prescribed values at some prescribed time instant. Requirements of the latter kind concerning the variable $\tilde{x}_2$ in (14),(15) and (25) have given rise to a particular controllability concept called "C-controllability" associated with system equations in descriptor form, see Yip and Sinovec, 1981. The dynamic controllability as appearing above is then called "R-controllability", and it is implied by the C-controllability.

(iii) As was stated above, original system equations given in (modified) descriptor form according to (12) are in the literature studied under transformations described by (13) with $P(z)$ and Q constant and nonsingular, i.e. $\tilde{x}$ transformations by more general unimodular polynomial matrices $P(z)$ are not recognized. The corresponding normal form may then be as given in (25) containing a nonzero E_4. The correct interpretation of this normal form - in particular with respect to the variable $\tilde{x}_2$ - has caused much trouble and confusion.

To begin with recall that (25), within the present algebraic framework, always can be transformed to the normal forms (14),(15) by means of elementary row operations. Consequently, there is in this case no mysticism regarding $\tilde{x}_2$ - according to (14),(15), $\tilde{x}_2$ is simply given by $\tilde{x}_2=G_2(z)u$, or equivalently, using the notations appearing in (25), by $\tilde{x}_2=(E_4z-I)^{-1}G_2(z)u$. Transfer properness then requires that $G_2(z)$ and $(E_4z-I)^{-1}G_2(z)$ respectively are constant matrices, and C-controllability implies that they are of rank ℓ_2. The crucial question arises: under what conditions is the present algebraic framework valid, in particular, under what conditions are the elementary row operations allowed? Note that transformations represented by multiplications by constant nonsingular matrices are always allowed - they simply represent one-to-one transformations of equations and variables.

The validity of the polynomial framework as used in this book was studied in great detail in Blomberg and Ylinen, 1983, see in particular part III of the book. The matter is by no means trivial, and only a brief and rather loosely formulated summary of the main results can be given here. A complete presentation would require complicated concepts like generalized functions and projection mappings.

Two main cases will be distinguished.

Case 1 The basic system equations as given by (3) with $\det A(z)\neq0$, are in this case valid "all the time", meaning that the signal values $y(t),(zy)(t),\ldots,u(t),(zu)(t),\ldots$ at any particular time instant t belonging to the time interval on which the system is defined satisfy the same basic system equations. In this case the fundamental theory presented in Blomberg and Ylinen, 1983, chapters 6 and 12 is applicable, implying that the i/o relation $S\{u,y\}$ generated by (3) is invariant with respect to elementary row operations applied to $[A(z)|-B(z)]$. The algebraic framework utilized above is thus valid, and all the results presented hold true.

In particular, every set of system equations (12) given in modified descriptor form is in this case (u,y) equivalent to a set of equations in the normal forms (14),(15). Accordingly, only ℓ_1 mutually independent initial conditions, i.e. those associated with $\tilde{x}_1$, can be chosen freely at any particular initial time. The normal form (25) presented in the literature is here highly misleading. It suggests that ℓ_2 free initial conditions could be associated with $\tilde{x}_2$ in addition to the ℓ_1 free initial conditions associated with $\tilde{x}_1$. The initial conditions associated with $\tilde{x}_2$ would then produce a set of impulses (for $z=p$) or pulse sequences (for $z=q$) spanning a rank E_4-dimensional vector space of such functions (see Murota, 1987, chapter 3, theorem 13.3; cf. also the Laplace transform and Z-transform calculi applied to (25)). But this is an erroneous conclusion, because $\tilde{x}_2(t)$ is at any time t uniquely determined by $u(t),(zu)(t),\ldots$ as $\tilde{x}_2(t)=(G_2(z)u)(t)$ as can be seen from (14),(15). Recall also that rank E_4 by no means is characteristic for the underlying system - rank E_4 is merely a property of the actual representation.
Case 2 To begin with suppose that there are two different systems defined on a common time interval of definition in the following way.

System 1 is given by system equations in some suitable form, say in the form (3), and they are valid all the time.

System 2 is given by the system equations for system 1 to which a set of "interconnection constraints" is added. These interconnection constraints force some of the input and output signals to coincide. Again the system equations for system 2 are valid all the time.

From these systems a system 3 is formed so that system 3 is governed by the system equations for system 1 up to a time instant t_0 belonging to the time interval of definition; at t_0 a "switching" occurs in the system meaning that from t_0 on system 3 is governed by the system equations for system 2 with initial conditions at t_0 determined by system 1. System 3 is thus a linear time-varying system to which the present polynomial algebra is not as such applicable. Transformations represented by multiplications by constant nonsingular matrices are still allowed.

The main result obtained in Blomberg and Ylinen, 1983, can now be presented briefly in the following form.

If system 1 is strictly transfer proper, then the behaviour of system 2 and system 3 on the time interval from t_0 on coincide, i.e. system 2 and system 3 generate the same i/o signal segments on this time interval. Consequently, as far as the behaviour on this time interval is concerned, all the results presented in case 1 above are also valid for system 2 as well as for system 3. The present polynomial algebraic framework is applicable, and the normal forms (14),(15) can be obtained.

If system 1 is transfer proper but not strictly transfer proper, then the above results hold almost always, except perhaps for some particular parameter values.

Because difference systems that are not transfer proper are noncausal, they do not appear as realistic models of physical systems, and the conditions for the above results to be true can thus be considered to hold quite generally for difference systems.

In the case of differential systems it may be sensible to consider also the possibility of a nonproper system 1. If so then there may appear differences in the behaviour of system 2 and system 3 on the time interval from t_0 on. These differences - if they occur - consist of additional impulse transients at t_0 appearing in the behaviour of system 3. In this case the present polynomial algebraic framework is not applicable to system 3 on the time interval from t_0 on, and the normal forms (14),(15) can not be obtained from (25). Accordingly, this is the only case in which the normal form (25) is of some relevance.

(iv) Some details in the above presentation have been inspired by a dissertation by Eklund, 1971. This holds in particular for the transformation procedure contained in Step 1,...,Step 4 for transforming (12) to the normal forms (14),(15). It is interesting to note that the construction given by Eklund works only under certain conditions, these conditions being equivalent to the condition that Ez-F in (12) is row proper. ∎

CONCLUSION

Application of polynomial methods to the descriptor form representation has appeared useful for analysis of linear time-invariant systems. In particular, the new so-called normal form for the input-output equivalence on descriptor form representations constructed using polynomial row operations and constant column operations simplifies the observability and controllability considerations presented in the literature.

REFERENCES

Blomberg,H. and Ylinen,R. (1983). Algebraic theory for multivariable linear systems. Academic Press.

Campbell,S.L. (1980). Singular systems of differential equations. Pitman.

Eklund,K. (1971). Linear drum boiler-turbine models. Report 7117, Division of Automatic Control, Lund Institute of Technology.

Gantmacher,F.R. (1959). The theory of matrices, volumes I and II. Chelsea.

Helmke,U. and Shayman,M.A. (1989)."A canonical form for controllable singular systems." Systems & Control Letters 12, 111-122.

Kalman,R.E., Falb,P.L. and Arbib,M.A. (1969). Topics in mathematical system theory. McGraw-Hill.

Luenberger,D.G. (1977)."Dynamic equations in descriptor forms." IEEE Trans. Automat. Contr.,AC-22, 312-321.

Luenberger,D.G. (1978)."Time-invariant descriptor systems." Automatica, 14, 473-480.

Murota,K. (1987). Systems analysis by graphs and matroids; structural solvability and controllability. Springer.

Rosenbrock,H.H. (1974)."Structural properties of linear dynamical systems." Int. J. Contr., 20, 191-202.

Rosenbrock,H.H. and Pugh,A.C. (1974)."Contributions to a hierarchical theory of systems.". Int. J. Contr. 19, 845-867.

Shayman,M.A. (1988)."Homogeneous indices, feedback invariants and control structure theorem for generalized linear systems." Siam J. Contr. Optim., 26, 387-400.

Schumacher,J.M. (1988)."Transformations of linear systems under external equivalence." Linear Algebra Appl., 102, 1-33.

Shin,K-C. and Kabamba,P. (1988)."Observation and estimation in linear descriptor systems with constrained dynamical systems." J. Dynamic Systems, Measurement, Contr.,110, 255-265.

Verghese,G.C., Lévy,B.C. and Kailath,T. (1981)."A generalized state-space for singular systems." IEEE Trans. Automat. Contr., AC-26, 811-831.

Yamada,T. and Luenberger,D.G. (1985a)."Generic controllability theorems for descriptor systems." IEEE Trans. Automat. Contr., AC-30, 144-152.

Yamada,T. and Luenberger,D.G. (1985b)."Algorithms to verify generic causality and controllability of descriptor systems." IEEE Trans. Automat. Contr., AC-30, 874-880.

Yip,E.L. and Sinovec,R.F. (1981)."Solvability, controllability and observability of continuous descriptor systems." IEEE Trans. Automat. Contr., AC-26, 702-707.

THE LINEAR-QUADRATIC OPTIMAL REGULATOR OF DESCRIPTOR SYSTEMS WITH THE TERMINAL STATE CONSTRAINED[+]

Cheng Zhaolin*, Yan Jiuxi and Meng Zhaobo***

**Mathematics Department, Shandong University, Jinan, PRC*
***Shandong Arch. and Civil Eng. Inst., Jinan, PRC*

Abstract. In this paper, the linear-quadratic optimal regulator of descriptor systems with the terminal state constrained is investigated and the explicit expression for the optimal control is .obtained.

Keywords. Optimal control; regulator theory; controllability; observability; descriptor systems; Riccati equations

INTRODUCTION

The system

$$E\dot{x} = Ax + Bu \tag{1}$$

is called a descriptor system where $x \in R^n$, $u \in R^r$, $E, A \in R^{n \times n}$, $B \in R^{n \times r}$, E, A, B are constant, E is singular and E, A satisfy $\det[sE-A] \neq 0$. The linear-quadratic optimal regulator problem of descriptor systems, i.e. the LQ problem , was discussed by many authors (Pandolfi, 1981; Cobb, 1983; Bender and Laub, 1987). It is sure that for the finite horizon LQ problem , the terminal state of the optimal trajectory of system (1) can not equal zero in general. However , in most industrial and economical problems the dynamic processes of descriptor systems are required to reach its equilibrium position. Hence , it is necessary to further investigate the LQ problem of system (1) with the terminal state constrained. Let LQC represent the above problem in which the optimal control is required to drive the dynamic process of system (1) to its equilibrium position at the terminal time T , i.e. the optimal trajectory x(t) must satisfy Ex(T)=0. In this paper, subject to the strong controllability and the pulse-observability of the system , the LQC problem is solved by transforming into that of a regular state-space system, and an explicit expression of the optimal control is also obtained .

MAIN RESULTS

Definition 1. System (1) is called controllable if

+ Research supported by the National Science Fundation

$$\text{rank } [sE-A \; B]=n, \qquad \text{for any } s \in C. \tag{2}$$

Defination 2. System (1) is called pulse-controllable if there exists a constant matrix $K \in R^{r \times n}$, such that

$$\text{degree } \{\det[sE-(A+BK)]\}=\text{rank } E. \tag{3}$$

Defination 3. System (1) is called strongly controllable if it is both controllable and pulse-controllable.

Lemma 1 (Cobb, 1981). If system (1) is pulse-controllable, then for any complex s there exist constant matrices $F, G \in R^{n \times n}$, $K \in R^{r \times n}$, and F, G are nonsingular, such that

$$F[sE-(A+BK) \; B]\begin{bmatrix} G & 0 \\ 0 & I_r \end{bmatrix} = \begin{bmatrix} sI_{n_E}-A_1 & 0 & B_1 \\ 0 & -I_{n-n_E} & B_2 \end{bmatrix} \tag{4}$$

where $n_E=\text{rank } E$, $A_1 \in R^{n_E \times n_E}$, $B_1 \in R^{n_E \times r}$, $B_2 \in R^{(n-n_E) \times r}$

Definition 4. The following system

$$\begin{cases} E\dot{x}=Ax+Bu \\ y=Cx \end{cases} \tag{5}$$

is called pulse-observable if the system

$$E^T \dot{x} = A^T x + C^T v \tag{6}$$

is pulse-controllable.

Lemma 2 (Verghese, 1978). System (5) is pulse-observable iff there do not exist constant vectors $\alpha, \beta \in R^n$, and $\beta \neq 0$ such that

$$\begin{bmatrix} sE-A \\ C \end{bmatrix} \beta = \begin{bmatrix} E\,\alpha \\ 0 \end{bmatrix}, \qquad \text{for any} \quad s \in C. \tag{7}$$

Consider the LQ problem of system (1) with the cost

$$J(u, Ex(0)) = \int_0^T (x^{\tau} Qx + u^{\tau} Ru)\,dt \tag{8}$$

the initial constraint

$$Ex(0) = \xi \tag{9}$$

and the terminal constraint

$$Ex(T) \triangleq Ex(T-0) = 0 \tag{10}$$

where ξ in formula (9) is given and $\xi \in$ Range E, Q, R in cost (8) are constant matrices and $Q \in R^{n \times n}$, $R \in R^{\tau \times \tau}$, $Q \geqslant 0$, $R > 0$. Let $\mathcal{U}$ be the set of admissible controls in which any admissible control $u(t)$ is sufficiently smooth, such that $J(u, Ex(0)) < \infty$, and the trajectory of system (1) satisfies constraints (9)-(10). Regard this problem as problem LQC.

Notice that all initial states satisfing constraint (9) compose a linear manifold

$$\mathcal{X} = \{ x(0) \mid x(0) \in R^n, \ Ex(0) = \xi \}. \tag{11}$$

Hence, solving problem LQC is actually to search for optimal control $u^* \in \mathcal{U}$ and $x^*(0) \in \mathcal{X}$, such that

$$\begin{array}{c} J(u^*, Ex^*(0)) = \min\ J(u, Ex(0)) . \\ u \in \mathcal{U} \\ x(0) \in \mathcal{X} \end{array} \tag{12}$$

Decompose the weighting matrix Q in cost (8) as $Q = C^{\tau} C$. and construct the system

$$\begin{cases} E\dot{x} = Ax + Bu \\ y = Cx \end{cases} \tag{1)c}$$

where A, B is the parameter matrices of system (1), and C is defined by $Q = C^{\tau} C$. It is easy to prove that the pulse-observability of system (1)c only depends on matrices E, A and Q, but not on the decomposition manner of Q.

<u>Theorem 1</u>. Suppose that syatem (1) is strongly controllable, and system (1)c is pulse-observable, then there exists the solution of the problem LQC , and the optimal control can be expressed as

$$u^*(t) = Hx(t) + f(t) \tag{13}$$

where H is a constant matrix depending on parameter matrices E, A, B and weighting matrices Q, R, and $f(t)$ is a nonlinear control function depending on matrices E, A, B, Q, R and the initial constraint.

Proof. The proof is constructive , includes:

1. Reforming cost (8). In view of the strony controllability of system (1) and lemma 1, we can find constant matrix $K \in R^{\tau \times n}$ and nonsingular matrices $F, G \in R^{n \times n}$, such that formula (4) holds, i.e. for any admissible control $u \in \mathcal{U}$, and the trajectory x of system (1) which is driven by u and satisfies constraints (9)-(10), the equations

$$v = u - Kx \tag{14}$$

$$x = G \cdot \tilde{x} = G[x_1^{\tau} \quad x_2^{\tau}]^{\tau} \tag{15}$$

$$\dot{x}_1 = A_1 x_1 + B_1 v \tag{16}$$

$$0 = x_2 + B_2 v \tag{17}$$

$$x_1(0) = [I_{n_E} \quad 0]F\,\xi \tag{18}$$

$$x_1(T) \triangleq x_1(T-0) = 0 \tag{19}$$

become identical, where $v \in R^{\tau}$, $x_1 \in R^{n_E}$, $x_2 \in R^{n-n_E}$, $A_1 \in R^{n_E \times n_E}$, $B_1 \in R^{n_E \times \tau}$, $B_2 \in R^{(n-n_E) \times \tau}$, and $n_E = $ rank E.

Substituting u, x in cost (8) by v, x_1, x_2 in (14), (15) and (17) , we obtain

$$J(u, Ex(0)) = \int_0^T \begin{bmatrix} x_1 \\ v \end{bmatrix}^{\tau} M \begin{bmatrix} x_1 \\ v \end{bmatrix} dt \tag{20}$$

where the weighting matrix

$$M = \begin{bmatrix} I_{n_E} & 0 \\ 0 & -B_2 \\ 0 & I_{\tau} \end{bmatrix}^{\tau} \begin{bmatrix} G^{\tau} & 0 \\ 0 & I_{\tau} \end{bmatrix} \begin{bmatrix} I_n & K^{\tau} \\ 0 & I_{\tau} \end{bmatrix} \begin{bmatrix} Q & 0 \\ 0 & R \end{bmatrix} \begin{bmatrix} I_n & 0 \\ K & I_{\tau} \end{bmatrix} \begin{bmatrix} G & 0 \\ 0 & I_{\tau} \end{bmatrix} \begin{bmatrix} I_{n_E} & 0 \\ 0 & -B_2 \\ 0 & I_{\tau} \end{bmatrix}$$

$$= \begin{bmatrix} M_{11} & M_{12} \\ M_{12}^{\tau} & M_{22} \end{bmatrix} \tag{21}$$

and $M_{11} \in R^{n_E \times n_E}$, $M_{12} \in R^{n_E \times \tau}$, $M_{22} \in R^{\tau \times \tau}$. It is proved in Appendix 1 that $M_{22} > 0$. Note

$$N = M_{11} - M_{12} M_{22}^{-1} M_{12}^{\tau} \tag{22}$$

$$\omega = M_{22}^{-1} M_{12}^{\tau} x_1 + v. \tag{23}$$

Thus cost (20), equivalently cost (8), can be rewritten as

$$J(u, Ex(0)) = \int_0^T (x_1^{\tau} Nx_1 + \omega^{\tau} M_{22}\, \omega)\,dt \triangleq J(\omega, [I_{n_E} \quad 0]F\,\xi) \tag{24}$$

where $N \geqslant 0$. By (16), (23), (18)-(19) we know yet that x_1 in cost (24) is the trajectory of system

$$\dot{x}_1 = (A_1 - B_1 M_{22}^{-1} M_{12}^{\tau})x_1 + B_1\, \omega \tag{25}$$

with initial value

$$x_1(0) = [I_{n_E} \; 0] F \xi \qquad (26)$$

and terminal constraint

$$x_1(T) \overset{\Delta}{=} x_1(T-0) = 0. \qquad (27)$$

2. Equivalence of the LQC problems. Regard the optimal control problem of the regular state-space system (25) with cost (24) and constraints (26)-(27) as problem LQCR. Let Ω be the set of admissible controls in which any admissible control $\omega(t)$ is sufficiently smooth, such that $J(\omega, [I_{n_E} \; 0] F \xi) < \infty$, and the trajectory of system (25) satisfies constraints (26)-(27).

Now we prove that problem LQC is equivalent to problem LQCR. In the section 1 of the proof, it is showed that if (u,x) is any pair of admissible control-trajectory of problem LQC, then the x_1 and ω which are determined by u and x according to (14), (15) and (23) satisfy equations (25) - (27), i.e. (ω, x_1) compose a pair of admissible control-trajectory of problem LQCR. Hence we have

$$\inf_{\substack{\omega \in \Omega}} J(\omega, [I_{n_E} \; 0] F \xi) \leqslant \inf_{\substack{u \in \mathcal{U} \\ x(0) \in \mathcal{X}}} J(u, Ex(0)). \qquad (28)$$

Otherwise, it is also evident that if (ω, x_1) is any pair of admissible control-trajectory of problem LQCR, then the x and u which are determined by ω and x_1 according to (23), (17) and (14)-(15) satisfy eqation (1) and constraints (9)--(10) i.e. (u,x) compose a pair of admissible control-trajectory of problem LQC. Thus

$$\inf_{\substack{u \in \mathcal{U} \\ x(0) \in \mathcal{X}}} J(u, Ex(0)) \leqslant \inf_{\substack{\omega \in \Omega}} J(\omega, [I_{n_E} \; 0] F \xi). \qquad (29)$$

Associating (28) with (29) there gives

$$\inf_{\substack{u \in \mathcal{U} \\ x(0) \in \mathcal{X}}} J(u, Ex(0)) = \inf_{\substack{\omega \in \Omega}} J(\omega, [I_{n_E} \; 0] F \xi). \qquad (30)$$

This shows that problem LQC is equvalent to problem LQCR, and the optimal cost values are the same. Moreover, the optimal solutions of problem LQC can be determined by that of problem LQCR according to formulas (23),(17),(14)-(15), and otherwise is also true.

3. Solving problem LQCR. It is known that problem LQCR is only relative to the regular state-space systems. We solve the problem by solving a sequence of the free terminal problems. Regard the following free terminal problem of system (25) with initial value (26) and the cost

$$J(\omega, [I_{n_E} \; 0] F \xi, S_k) = x_1^{\tau}(T) S_k x_1(T)$$

$$+ \int_0^T (x_1^{\tau} N x_1 + \omega^{\tau} M_{22} \omega)\,dt, \quad \text{for } k=1,2,\ldots \quad (31)$$

as problem LQNCR , where $S_k \in R^{n_E \times n_E}$, $S_{k+1} > S_k > 0$, $S_k \to \infty$ as $k \to \infty$. It is well known that the optimal solution of the problem LQNCR is (Anderson, 1971):

$$\omega_k^*(t) = -M_{22}^{-1} B_1^{\tau} P_k(t) x_1(t) \qquad (32)$$

and the optimal cost value is

$$J(\omega, [I_{n_E} \; 0] F \xi, S_k)$$

$$= \xi^{\tau} F^{\tau} [I_{n_E} \; 0]^{\tau} P_k(0) [I_{n_E} \; 0] F \xi . \qquad (33)$$

where $P_k(t)$ is the solution of the following Riccati differential equation

$$\dot{P} = -P(A_1 - B_1 M_{22}^{-1} M_{12}^{\tau}) - (A_1 - B_1 M_{22}^{-1} M_{12}^{\tau})^{\tau} P$$

$$+ P B_1 M_{22}^{-1} B_1^{\tau} P - N \qquad (34)$$

with the terminal constraint

$$P(T) = S_k . \qquad (35)$$

Note

$$D = \begin{bmatrix} A_1 - B_1 M_{22}^{-1} M_{12}^{\tau} & -B_1 M_{22}^{-1} B_1^{\tau} \\ -N & -(A_1 - B_1 M_{22}^{-1} M_{12}^{\tau})^{\tau} \end{bmatrix} \qquad (36)$$

$$\exp(D(t-T)) = \begin{bmatrix} L_{11}(t-T) & L_{12}(t-T) \\ L_{21}(t-T) & L_{22}(t-T) \end{bmatrix} \qquad (37)$$

where $L_{ij}(t-T) \in R^{n_E \times n_E}$, $i,j = 1,2$. By the characteristics of Riccati equations (refer to Appendix 2, Theorem 2.2) we have

$$P_k(t) = (L_{21}(t-T) + L_{22}(t-T) S_k)$$

$$\cdot (L_{11}(t-T) + L_{12}(t-T) S_k)^{-1}. \qquad (38)$$

Consider the limit of $P_k(t)$ as $k \to \infty$. In terms of the strong controllability of system (1), we know that system (25) is controllable. This guarantees that the matrix $L_{12}(t-T)$ is nonsingular as $t \in [0,T)$ (refer to Appendix 2, Theorem 2.3). Associating with $S_k^{-1} \to 0$ as $k \to \infty$, so

$$P(t) \overset{\Delta}{=} \lim_{k \to \infty} P_k(t) = L_{22}(t-T) L_{12}^{-1}(t-T),$$

$$\text{for } t \in [0,T). \qquad (39)$$

It is easy to know that $P(t)$ satisfies Riccati equation (34) and $P(t) > 0$ as $t \in [0,T)$.

Now we construct the optimal solution of problem

LQCR. Let the feedback control law

$$\omega^*(t)=-M_{22}^{-1}B_1^{\tau}P(t)x_1(t),\quad \text{for } t\in[0,T]. \quad (40)$$

Then the trajectory of system (25) which is driven by $\omega^*(t)$ and satisfies constraints (26)-(27) is

$$x_1^*(t)=L_{12}(t-T)L_{12}^{-1}(-T)[I_{n_E}\ 0]F\xi,\quad \text{for } t\in[0,T]. \quad (41)$$

Integrate

$$\frac{d}{dt}(x_1^{*\tau}(t)P(t)x_1^*(t))$$

$$=-(x_1^{*\tau}(t)Nx_1^*(t)+\omega^{*\tau}(t)M_{22}\omega^*(t)) \quad (42)$$

from 0 to T, by (39)-(41) it follows that

$$J(\omega^*,[I_{n_E}\ 0]F\xi)$$

$$=\xi^{\tau}F^{\tau}[I_{n_E}\ 0]^{\tau}L_{22}(-T)L_{12}^{-1}(-T)[I_{n_E}\ 0]F\xi. \quad (43)$$

Next we point out that $\omega^*(t)$ is the optimal control of problem LQCR. In fact, if $\omega(t)$ is any admissible control of problem LQCR, then $\omega(t)$ is also any admissible control of problem LQNCR, hence, it follows by (33) that

$$J(\omega,[I_{n_E}\ 0]F\xi)=J(\omega,[I_{n_E}\ 0]F\xi,S_k)$$

$$\geqslant \xi^{\tau}F^{\tau}[I_{n_E}\ 0]^{\tau}P_k(0)[I_{n_E}\ 0]F\xi. \quad (44)$$

Let $k\to\infty$, associating (39) with (43)-(44) we obtain

$$J(\omega,[I_{n_E}\ 0]F\xi)\geqslant \xi^{\tau}F^{\tau}[I_{n_E}\ 0]^{\tau}P(0)[I_{n_E}\ 0]F\xi$$

$$=J(\omega^*,[I_{n_E}\ 0]F\xi) \quad (45)$$

i.e. $\omega^*(t)$ is the optimal control of problem LQCR, and the optimal cost value of problem LQCR is given by (43).

4. Solving problem LQC. By the equavalence of problem LQC with problem LQCR and formulas (14),(15), (17),(23),(39)-(41) we know the optimal control of problem LQC is

$$u^*(t)=Hx(t)+f(t),\quad \text{for } t\in[0,T] \quad (46)$$

where

$$H=K-M_{22}^{-1}M_{12}^{\tau}[I_{n_E}\ 0]G^{-1} \quad (47)$$

$$f(t)=-M_{22}^{-1}B_1^{\tau}L_{22}(t-T)L_{12}^{-1}(-T)[I_{n_E}\ 0]F\xi,$$

$$\text{for } t\in[0,T] \quad (48)$$

H is constant determined by matrices E, A, B, Q, R, and f(t) is a nonlinear function determined by E, A, B, Q, R and the initial constraint (9).

The optimal trajectory of problem LQC is

$$x^*(t)=G\begin{bmatrix}L_{12}(t-T)\\ B_2M_{22}^{-1}(B_1^{\tau}L_{22}(t-T)+M_{12}^{\tau}L_{12}(t-T))\end{bmatrix}$$

$$\cdot L_{12}^{-1}(-T)[I_{n_E}\ 0]F\xi,\quad \text{for } t\in[0,T] \quad (49)$$

and the optimal cost value is

$$J(u^*,Ex^*(0))$$

$$=\xi^{\tau}F^{\tau}[I_{n_E}\ 0]^{\tau}L_{22}(-T)L_{12}^{-1}(-T)[I_{n_E}\ 0]F\xi. \quad (50)$$

This concludes the proof. $\square$

Example. Give a system

$$\begin{bmatrix}0 & 1 & 0\\ 0 & 0 & 0\\ 0 & 0 & 0\end{bmatrix}\dot{x}=\begin{bmatrix}1 & 0 & 0\\ 0 & 1 & 0\\ 0 & 0 & 1\end{bmatrix}x+\begin{bmatrix}0 & 0\\ 1 & 0\\ 0 & 1\end{bmatrix}u. \quad (51)$$

with the initial constraint and the terminal constraint

$$Ex(0)=[1\ 0\ 0]^{\tau},\qquad EX(1)=0 \quad (52)$$

and the cost

$$J(u,Ex(0))=\int_0^1(x^{\tau}\begin{bmatrix}1 & 0 & 0\\ 0 & 0 & 0\\ 0 & 0 & 0\end{bmatrix}x+u^{\tau}\begin{bmatrix}1 & 0\\ 0 & 1\end{bmatrix}u)dt. \quad (53)$$

Consider the optimal regulator problem .Of course, the above system satisfies all that Theorem 1 requires . Hence, the optimal control is

$$u^*(t)=\begin{bmatrix}1 & -1 & 0\\ 0 & 0 & 0\end{bmatrix}x(t)+\frac{1}{sh1}\begin{bmatrix}ch(t-1)\\ 0\end{bmatrix},$$

$$\text{for } t\in[0,1] \quad \cdot(54)$$

the optimal trajectory is

$$x^*(t)=\frac{1}{sh1}\begin{bmatrix}-ch(t-1) & -sh(t-1) & 0\end{bmatrix}^{\tau},$$

$$\text{for } t\in[0,1] \quad (55)$$

and the optimal cost value is

$$J(u^*,Ex^*(0))=cth1. \quad (56)$$

CONCLUSION

The main results in this paper includes the following two aspects: 1. Subject to the strong controllability of system (1) and the pulse-observability of system (1)c, the equivalence of problem LQC with problem LQCR is given, hence, solving problem LQC can be reduced to solving problem LQCR ,and the latter is only relative to the regular state-space system; 2. The explicit expression of the optimal control of problem LQC is given which shows that the optimal control can be represented as the sum of the constant gain state feedback and the nonlinear control function, thus the regulation of the control precision can be realized by regulating the feedback gain and the nonlinear input.

REFERENCES

Anderson, B.D.O. and J.B. Moore (1971), <u>Linear Optimal Control</u>, Prentice-Hall.

Bender, D.J.and A.J. Laub (1987), The linear-quadratic optimal regulator for descriptor systems, <u>IEEE, Trans. Aut. Control</u>, AC-32, 8, 672-688.

Cobb, D. (1983), Descriptor variable systems and optimal state regulation, <u>IEEE, Trans. Aut. Control</u>, AC-28, 5, 601-611.

Cobb, D. (1981), Feedback and pole placement in descriptor varibles systems, <u>Int. J. Contr.</u>, 35, 6, 1135-1146.

Pandolfi, L. (1981). On the regulator problem for the linear degenerate control systems, <u>J. Opti. Theorey Appli.</u>, 33, 241-254.

Verghese, G. (1978), Infinite-frequency behavior in generalized dynamical systems, <u>Ph. D. Dissertation, Dep. Electrical Engineering, Stanford University.</u>

APPENDIX 1.

Proof of the matrix $M_{22} > 0$.

Assume the contrary, i.e. that M_{22} degenerate, then there exists the vector $\alpha_1 \in R^r$, and $\alpha_1 \neq 0$, such that $\alpha_1^\tau M_{22} \alpha_1 = 0$. Noticing that $R > 0$, $Q = C^\tau C$, and

$$M_{22} = \begin{bmatrix} 0 \\ B_2 \end{bmatrix}^\tau G^\tau Q G \begin{bmatrix} 0 \\ B_2 \end{bmatrix} + (I_\gamma - KG \begin{bmatrix} 0 \\ B_2 \end{bmatrix})^\tau R (I_\gamma - KG \begin{bmatrix} 0 \\ B_2 \end{bmatrix}) \quad (1.1)$$

we have

$$CG \begin{bmatrix} 0 \\ B_2 \end{bmatrix} \alpha_1 = 0 \quad (1.2)$$

$$\alpha_1 = KG \begin{bmatrix} 0 \\ B_2 \end{bmatrix} \alpha_1 . \quad (1.3)$$

Note

$$\alpha = G \begin{bmatrix} B_1 \\ 0 \end{bmatrix} \alpha_1 , \quad \beta = G \begin{bmatrix} 0 \\ B_2 \end{bmatrix} \alpha_1 \quad (1.4)$$

evidently, $C\beta = 0$ by (1.2), and by (4), (1.3)-(1.4) it follows that

$$[sE-A] \beta = F^{-1} F [sE-(A+BK)] G \begin{bmatrix} 0 \\ B_2 \end{bmatrix} \alpha_1 + F^{-1} FBKG \begin{bmatrix} 0 \\ B_2 \end{bmatrix} \alpha_1$$

$$= F^{-1} \begin{bmatrix} sI_{n_E} - A_1 & 0 \\ 0 & -I_{n-n_E} \end{bmatrix} \begin{bmatrix} 0 \\ B_2 \end{bmatrix} \alpha_1 + F^{-1} \begin{bmatrix} B_1 \\ B_2 \end{bmatrix} \alpha_1$$

$$= F^{-1} FEG \begin{bmatrix} B_1 \\ 0 \end{bmatrix} \alpha_1 = E\alpha \quad , \text{ for any } s \in C. \quad (1.5)$$

Associating (1.5) with $C\beta = 0$, we obtain

$$\begin{bmatrix} sE-A \\ C \end{bmatrix} \beta = \begin{bmatrix} E\alpha \\ 0 \end{bmatrix} , \quad \text{ for any } s \in C. \quad (1.6)$$

In view of that the system (1)c is pulse-observable, applying lemma 2 to (1.6) we obtain $\beta = 0$. Further, $\alpha_1 = 0$ by (1.3). This contradicts the hypothesis, and the proof is completed. $\quad\square$

APPENDIX 2.

The solution of Riccati differential equations.

Consider following equations

$$\dot{P}(t) = -P(t)A - A^\tau P(t) + P(t)BR^{-1} B^\tau P(t) - C^\tau C \quad (2.1)$$

$$P(T) = S \quad (2.2)$$

$$\begin{bmatrix} \dot{\Phi}(t) \\ \dot{\Psi}(t) \end{bmatrix} = \begin{bmatrix} A & -BR^{-1} B^\tau \\ -C^\tau C & -A^\tau \end{bmatrix} \begin{bmatrix} \Phi(t) \\ \Psi(t) \end{bmatrix} \quad (2.3)$$

$$\begin{bmatrix} \Phi(T) \\ \Psi(T) \end{bmatrix} = \begin{bmatrix} I \\ S \end{bmatrix} \quad (2.4)$$

where matrices A, B, C, R, S are constant, I is the unit matrix, and $R > 0$, $S \geqslant 0$. It is well known that solving of Riccati equation (2.1)-(2.2) can be reduced to solving of linear equation (2.3)-(2.4). Now we give out some important properties of Riccati equations relating to this paper.

Theorem 2.1 (Anderson, 1971). Let $P(t)$ and $[\Phi^\tau \ \Psi^\tau]^\tau$ be solutions of equations (2.1)-(2.2) and equations (2.3)-(2.4) respectively. Then the following properties

$$P(t) \geqslant 0 \quad (2.5)$$

$$\Phi(t) \text{ is nonsingular} \quad (2.6)$$

$$P(t) = \Psi(t) \Phi^{-1}(t) \quad (2.7)$$

hold, for $t \in [0, T]$.

Theorem 2.2. Let

$$D=\begin{bmatrix} A & -BR^{-1}B^{\tau} \\ -C^{\tau}C & -A^{\tau} \end{bmatrix} \qquad (2.8)$$

$$\exp(D(t-T))=\begin{bmatrix} L_{11}(t-T) & L_{12}(t-T) \\ L_{21}(t-T) & L_{22}(t-T) \end{bmatrix} \qquad (2.9)$$

where $L_{ij}(t-T)$, $i,j=1,2$, are square matrices with same order, then for $t\in[0,T]$ the following properties

$$L_{11}(t-T) \text{ and } L_{22}(t-T) \text{ are nonsingular} \qquad (2.10)$$

$$\Phi(t)=L_{11}(t-T)+L_{12}(t-T)S \qquad (2.11)$$

$$\Psi(t)=L_{21}(t-T)+L_{22}(t-T)S \qquad (2.12)$$

$$P(t)=(L_{21}(t-T)+L_{22}(t-T)S)$$

$$\cdot (L_{11}(t-T)+L_{12}(t-T)S)^{-1} \qquad (2.13)$$

hold.

Proof. Obviously, equations (2.11)-(2.13) hold. Let $S=0$, by (2.11) and (2.6), we know that $L_{11}(t-T)$ is nonsingular. In the same way investigating the following equations

$$\dot{P}(t)=P(t)A^{\tau}+AP(t)+P(t)C^{\tau}CP(t)-BR^{-1}B^{\tau} \qquad (2.14)$$

$$P(T)=0 \qquad (2.15)$$

we obtain that $L_{22}(t-T)$ is nonsingular . Hence properties in (2.10) hold. $\qquad\square$

Theorem 2.3. As $t\in[0,T)$, $L_{12}(t-T)$ is nonsingular if the matrix pair $[A,B]$ is controllable.

Proof. It is known (Anderson, 1971) that as $t\in[0,T)$, the solution $P(t)$ of equations (2.1)-(2.2) is positive definite if the matrix pair $[A,C]$ is observable. Applying this result and Theorem 2.2 to equations (2.14)-(2.15), the theorem is obtained. $\qquad\square$

CONVERSION OF THE POLYNOMIAL CONTINUOUS-TIME MODEL TO THE DELTA DISCRETE-TIME AND VICE VERSA

J. Ježek

*Institute of Information Theory and Automation, Czechoslovak Academy of Sciences,
182 08 Prague 8, Czechoslovakia*

Abstract. The paper describes an algorithm for converting the continuous-time polynomial model $B(p)/A(p)$ of a linear system to the discrete-time $b(\Delta)/a(\Delta)$ and vice versa. Here p is the derivative operator and Δ the difference one. The delta-models are more convenient than the commonly used z-models, especially for small sampling periods. The algorithm is based on computing the exponential, logarithm and related functions in the algebra of congruences of polynomials.

Keywords. Discrete time systems; continuous time systems; polynomials; algebraic system theory; difference equations; numerical methods.

1.INTRODUCTION

There are two ways of control of continuous-time processes: the continuous-time control and the discrete-time one. Each of them has a different basic principle of work: the continuous-time control works unintermittently and is implemented by analogue elements, the discrete-time one works in periodic instants of time and is implemented by digital elements.

Each of these two approaches has its mathematical models and the theory based on it: the differential equations or the discrete-time ones. For a concrete system, the model can be obtained either by analyzing the structure of the system or by measurement and identification. Often it is necessary to convert the continuous-time model to the discrete-time one or vice versa.

The paper is devoted to relation between the continuous-time and the discrete-time models and also to algorithms for conversion of the models. It presents the conversion algorithms for a new type of discrete-time models.

2.SIGNALS AND OPERATORS

The continuous-time signals $x(t)$ are functions of $t \geq 0$. Their Laplace transform $\bar{x}(p)$ is

$$\bar{x}(p) = \int_0^\infty x(t)\mathrm{e}^{-pt}\mathrm{d}t. \tag{1}$$

For rational signals of order N, it is

$$\bar{x}(p) = \frac{B(p)}{A(p)} = \frac{B_0 p^N + ... + B_N}{A_0 p^N + ... + A_N}$$

with $\bar{x}(\infty) = 0$, i.e. deg $B <$ deg A; we do not allow Dirac impulses. The fraction can be normalized by $A_0 = 1$.

The complex variable p can be also interpreted as the derivative operator. The transforms can be also expressed in complex variable $s = 1/p$, interpreted as the integration operator:

$$\bar{x}(s) = \frac{\hat{B}(s)}{\hat{A}(s)} = \frac{B_0 + ... + B_N s^N}{A_0 + ... + A_N s^N}.$$

The using of $\hat{B}(s), \hat{A}(s)$ is in some situations more suitable than using $B(p), A(p)$. The coefficients are the same but in reverse ordering. The properties are $\bar{x}(0) = 0$, i.e. $B(s)$ but not $A(s)$ divisible by s. It also means $A_0 \neq 0, B_0 = 0$.

The signals can be discretized with sampling period $T > 0$ and phase ε, where

$$t = (n + \varepsilon)T, \quad n = 0, 1, 2..., \quad 0 \leq \varepsilon < 1. \tag{2}$$

We denote it

$$x_n(\varepsilon) = D\{x(t)\} \tag{3}$$

For the discrete-time signals, the z-transform is commonly used where z can be interpreted as the advance operator. However, there is an unpleasant property of z-transforms: they do not converge for $T \to 0$ to anything useful. This difficulty can be overcome, see (Nagy and Ježek, 1986), by using, instead of z, the complex variable

$$\Delta = (z - 1)/T, \quad z = T\Delta + 1$$

interpreted as the difference operator:

$$\Delta x_n = (x_{n+1} - x_n)/T.$$

The summation operator $\sigma = 1/\Delta$

$$\sigma x_n = T \sum_{k=0}^{n-1} x_k$$

can be also used.

The Δ-transform $\bar{x}(\Delta, \varepsilon)$ is

$$\bar{x}(\Delta, \varepsilon) = \sum_{n=0}^{\infty} x_n(\varepsilon)(T\Delta + 1)^{-n} \qquad (4)$$

We use the symbol D from (3) also for transforms

$$\bar{x}(\Delta, \varepsilon) = D\{\bar{x}(p)\}.$$

For rational signals, the transform is

$$\bar{x}(\Delta, \varepsilon) = \frac{b(\Delta, \varepsilon)}{a(\Delta)} = \frac{b_0(\varepsilon)\Delta^N + \ldots + b_N(\varepsilon)}{a_0 \Delta^N + \ldots + a_N} \qquad (5)$$

with $\bar{x}(\infty, \varepsilon) \neq \infty$, or

$$\bar{x}(\sigma, \varepsilon) = \frac{\hat{b}(\sigma, \varepsilon)}{\hat{a}(\sigma)} = \frac{b_0(\varepsilon) + \ldots + b_N(\varepsilon)\sigma^N}{a_0 + \ldots + a_N \sigma^N} \qquad (6)$$

with $\bar{x}(0, \varepsilon) \neq \infty$. It can be normalized by $a_0 = 1$.

The connection between operators is

$$\Delta = (e^{Tp} - 1)/T, \qquad (7)$$

$$p = (\ln(T\Delta + 1))/T. \qquad (8)$$

For $T \to 0$, it is $\Delta \to p$, $\sigma \to s$; it can be seen in (7), (8). By comparing (1) and (4) we see that the transform converges

$$T\bar{x}(\Delta, \varepsilon) \to \bar{x}(p).$$

For the coefficients (normalized):

$$Tb_i(\varepsilon) \to B_i, \qquad (9)$$

$$a_i \to A_i.$$

E.g. for

$$x(t) = e^{At} \qquad (10)$$

$$x_n(\varepsilon) = e^{AT(n+\varepsilon)} \qquad (11)$$

the transforms are

$$\bar{x}(p) = \frac{1}{p - A}, \qquad (12)$$

$$\bar{x}(s) = \frac{s}{1 - As}, \qquad$$

$$\bar{x}(\Delta, \varepsilon) = \frac{T\Delta + 1}{T\Delta + 1 - e^{AT}} e^{AT\varepsilon}, \qquad (13)$$

$$\bar{x}(\sigma, \varepsilon) = \frac{T + \sigma}{T + (1 - e^{AT})\sigma} e^{AT\varepsilon}.$$

The convergence is evident.

The rational continuous-time signal can be decomposed into partial fractions. With simple roots, they are (10),(12), in the general case

$$x(t) = \frac{t^k}{k!} e^{At}, \qquad (14)$$

$$\bar{x}(p) = \frac{1}{(p - A)^{k+1}}. \qquad (15)$$

For the discrete-time signal:

$$x_n = \frac{n^{(k)}}{k!} e^{ATn}, \qquad (16)$$

$$\bar{x}(\Delta) = \frac{(T\Delta + 1)e^{ATk}}{(T\Delta + 1 - e^{AT})^{k+1}} \qquad (17)$$

where $n^{(k)} = n(n-1)\ldots(n-k+1)$ is the "factorial power", a function more suitable than n^k. The $\bar{x}(\Delta, \varepsilon)$ has a more complicated formula; it can be expressed as a sum of (17) terms.

We can derive some properties of general $\bar{x}(\Delta)$ from the fractions (17). The poles $p_0 = A$, $\Delta_0 = (e^{AT} - 1)/T$ fulfil the same mapping (7) as the operators p, Δ. It maps the strip

$$-2\pi/T < \text{Im } p_0 < 2\pi/T \qquad (18)$$

in the p-plane to the whole Δ-plane with the exception of

$$\Delta_0 \text{ real}, \quad -\infty < \Delta_0 \leq -1/T. \qquad (19)$$

With condition (18) for all poles in the signal, the conversion (3) for signals of N-th order is bijective and we can define

$$x(t) = D^{-1}\{x_n(\varepsilon)\},$$

$$\bar{x}(p) = D^{-1}\{\bar{x}(\Delta, \varepsilon)\}.$$

The poles (19) are excluded; they are possible in the discrete-time signals but they have no continuous-time counterpart of the same order. The pole $\Delta_0 = -1/T$ represent a mode of finite duration, it has no continuous-time counterpart of finite order at all.

Furthermore, we see from (17) that the general $\bar{x}(\Delta, \varepsilon)$ has a zero point $\Delta_0 = -1/T$, i.e. $b(\Delta, \varepsilon)$ but not $a(\Delta)$ is divisible by $(T\Delta + 1)$. In (6), $\bar{x}(\sigma, \varepsilon)$ has a zero point $\sigma_0 = -T$, the factor in numerator is $(T + \sigma)$.

All these properties of Δ-transforms are analogical to those of z-transforms but now it may be seen how they approach for $T \to 0$ the properties of p-transforms.

3.CONVERSION OF SIGNALS

The general problem is now the algorithm for conversion of rational signals. With respect to (9), we define it as follows: given $N, B_i, A_i (i = 1 \ldots N), T, \varepsilon$, compute Tb_i, a_i. The backward conversion proceeds from Tb_i, a_i to B_i, A_i of the same N.

Several algorithms were designed for the conversion problems. They are based on the same principles as the known algorithms for z-transform. The standard procedure uses the partial fraction expansion: we compute the roots of $A(p)$, convert every fraction according to (15), (17) and finally we sum the converted fractions. A bit of complication consists in the fact that (14) deals with t^k but (16) with $n^{(k)}$; the discretization of (15) is not (17) but a combination of such terms. A further complication consists in complex roots. Nevertheless, it is possible to construct a general algorithm. But this way was not followed because of numerical experience from the z-transform algorithm which says that the precision reached is not high due to computation of roots, fraction decomposition and composition.

It is natural to look for an algorithm where the root computation is not required. In the state-space theory, the continuous-time and the discrete-time equations are

$$dx/dt = Ax, \qquad \Delta x_n = Fx_n$$

and the conversion is given by matrix function

$$F = (e^{AT} - I)/T$$

which can be computed by power series without computation of eigenvalues. For our problem of polynomial conversion, it is of course possible to go from the given polynomials to matrices, to convert these and to go back from them to polynomials. It was performed (for z-transforms), see (Ježek 1968), and it worked numerically well. But this way is too complicated; a more direct way is described in the next sections.

4.CONGRUENCES OF POLYNOMIALS

In this section, we develop a mathematical device for the polynomial algebra which is analogical to matrix functions, see also (Vostrý 1977). Recall that for matrix A with eigenvalues α_i, the result of function $\phi = f(A)$ has eigenvalues $\varphi_i = f(\alpha_i)$. The polynomial analogy of this is: for polynomial A with roots α_i, how to compute polynomial ϕ whose roots are $\varphi_i = f(\alpha_i)$.

To begin our construction, consider operators $F(q)$, polynomial in some operator q but applied only to signals y which satisfy equation $M(q)y = 0$. For such signals, we can replace $M(q)$ by zero; specially any $F(q)$ can be replaced by its remainder $F_R(q)$ from the division $F(q) = M(q)Q(q) + F_R(q)$.

Speaking more exactly: given a fixed $M(q)$ of degree N, going from the algebra of polynomials $R[q]$, we define the algebra $R[q]/M$ of classes of polynomials congruent modulo M. Its dimension is N, its base is $1, q, \ldots, q^{N-1}$. Multiplication is performed by polynomial multiplication followed by taking the remainder mod M.

Let us mention some properties of this algebra. The inverse $G = 1/F$ is defined by $FG = 1 \bmod M$, which leads to polynomial equation $FG + MQ = 1$ in $R[q]$ for unknown G, Q with condition $\deg G < \deg M$. So, the inverse exists if F is coprime with M.

Every F in $R[q]/M$ has its annihilating polynomial $H = \mathrm{ann}(F)$; it is the polynomial of least degree for which $H(F) \bmod M = 0$. Its degree is at most N; $H(q)$ can be computed by solving a set of linear equations for coefficients H_i. It is clear that the annihilating polynomial of q, i.e. of the generating element of the algebra, is $M(q)$. This is something like the Cayley-Hamilton theorem from the matrix algebra.

In $R[q]$, the norm of a polynomial may be defined $\|F(q)\| = \sum_i |F_i|$. In $R[q]/M$, we can define a norm

$$\|F(q)\|_M = \max_{j=0}^{N-1} \|q^j F(q) \bmod M\|.$$

It satisfies all axioms of norm in an algebra, specially

$$\|F + G\|_M \leq \|F\|_M + \|G\|_M,$$

$$\|FG\|_M \leq \|F\|_M \|G\|_M.$$

Our algebra is now equipped with norm and topology, we can use e.g. power series, iterative processes. We can construct functions similarly to functions of matrices. Analogically to the spectrum of a matrix, we define the spectrum of A by roots α_i of $\mathrm{ann}(A)$. For any function $f(\lambda)$, the spectrum of $\phi = f(A)$ is $\varphi_i = f(\alpha_i)$.

The problem of computing functions of polynomials is now solved. Given $A(q)$ with roots α_i, polynomial $\phi = f(A)$ with roots $\varphi_i = f(\alpha_i)$ can be obtained by computing $f(q)$ in algebra $R[q]/A$ and taking the annihilating polynomial.

This procedure can be used for numerical computations. Compared with the matrix algebra, the computational complexity is lower: $\sim N^2$ operations (e.g. for multiplication, inverse), $\sim N$ storage in comparison with $\sim N^3$ operations, $\sim N^2$ storage. The only process requiring $\sim N^3$, $\sim N^2$ is computation of annihilating polynomial.

5.ALGORITHM FOR SIGNAL CONVERSION

In our problem of signal conversion, we follow the way described for z-transform by (Vostrý 1978). We begin with separate conversion $A(p)$ to $a(\Delta)$. The roots of these polynomials are related by (7), (8). According to Section 4, the conversion consists in $R[p]/A$ - algebra computation of the function $pf(pT)$ where

$$f(\lambda) = (e^\lambda - 1)/\lambda.$$

It is performed by power series

$$f(\lambda) = 1 + \frac{\lambda}{2!} + \frac{\lambda^2}{3!} + \ldots \tag{20}$$

convergent for all λ. For practical computations, (20) works well for $\|\lambda\| < 1$. This condition is reached by repeated using the formula

$$f(\lambda) = f(\lambda/2)[1 + (\lambda/2)f(\lambda/2)/2].$$

The backward conversion $a(\Delta)$ to $A(p)$ consists in $R[\Delta]/a$ - algebra computation of the function $\Delta g(1 + T\Delta)$ where

$$g(\lambda) = (\ln \lambda)/(\lambda - 1).$$

It is performed by power series

$$g(\lambda) = \frac{2}{\lambda + 1} \left[1 + \frac{1}{3}\left(\frac{\lambda - 1}{\lambda + 1}\right)^2 + \frac{1}{5}\left(\frac{\lambda - 1}{\lambda + 1}\right)^4 + \ldots \right]$$

convergent for $\|\lambda - 1\| < 1$, suitable for computation for $\|\lambda - 1\| < 1/2$. This condition is reached by repeated using of the formula

$$g(\lambda) = 2g(\sqrt{\lambda})/(\sqrt{\lambda} + 1).$$

The square root $\sqrt{\lambda}$ is computed by the iterative process

$$x_0 = 1, \quad x_{i+1} = (x_i + \lambda/x_i)/2,$$

$$i \to \infty, \quad x_i \to \sqrt{\lambda}.$$

The next step in our algorithm of conversion $\bar{x}(p)$ to $\bar{x}(\Delta, \varepsilon)$ is computation of $b(\Delta, \varepsilon)$. First, we compute numbers $\Delta^k x_0(\varepsilon)$ by the inverse Laplace transform

$$x_n(\varepsilon) = \frac{1}{2\pi i} \int \frac{B(p)}{A(p)} e^{pT(n+\varepsilon)} \mathrm{d}p.$$

Taking the k-th difference and setting $n = 0$ we have

$$\Delta^k x_0(\varepsilon) = \frac{1}{2\pi i} \int \frac{B(p)}{A(p)} \left(\frac{e^{pT} - 1}{T}\right)^k e^{pT\varepsilon} \mathrm{d}p.$$

Such expressions of the form

$$I = \frac{1}{2\pi i} \int \frac{\phi(p)}{A(p)} \mathrm{d}p$$

where $\phi(p)$ is analytic on the spectrum of $A(p)$, can be computed by congruences of polynomials. Taking the remainder $\phi_R(p)$

$$\phi(p) = Q(p)A(p) + \phi_R(p)$$

we have

$$I = \frac{1}{2\pi i}\int \left[Q(p) + \frac{\phi_R(p)}{A(p)}\right]dp.$$

The integration path can be taken as a closed curve: the imaginary axis plus the semicircle in the right half-plane (the radius approaches infinity). The first term in the integral is zero as the integrand is analytic. The second term is computed easily as the ratio of leading coefficients in the numerator and in the denominator.

For $\varepsilon \neq 0$, we must also compute $e^{pT\varepsilon} \bmod A(p)$. It is performed by power series

$$e^\lambda = 1 + \lambda + \frac{\lambda^2}{2!} + \ldots$$

combined with repeated using of the formula

$$e^\lambda = (e^{\lambda/2})^2.$$

Now back to our problem of computing $Tb(\Delta,\varepsilon)$. From the Newton interpolation formula

$$x_n(\varepsilon) = \sum_{k=0}^{\infty}(\Delta^k x_0(\varepsilon))\frac{n^{(k)}}{k!} \tag{21}$$

by taking the transform, using (16),(17) we obtain

$$T\bar{x}(\sigma,\varepsilon) = (T+\sigma)\sum_{k=0}^{\infty}(\Delta^k x_0(\varepsilon))\sigma^k. \tag{22}$$

So, having computed the numbers $\Delta^k x_0(\varepsilon)$ we can compute coefficients of σ-series for $T\bar{x}(\sigma,\varepsilon)$. Finally, we have

$$T\hat{b}(\sigma,\varepsilon) = \hat{a}(\sigma)T\bar{x}(\sigma,\varepsilon)$$

by polynomial multiplication. It suffices to take $k = 0\ldots N$.

For the backward conversion, $B(p)$ from $Tb(\Delta,\varepsilon)$, $a(\Delta)$, $A(p)$ we use the inverse transform

$$x_n(\varepsilon) = \frac{1}{2\pi i}\int \frac{b(\Delta,\varepsilon)}{a(\Delta)}(T\Delta+1)^n\frac{Td\Delta}{T\Delta+1} \tag{23}$$

For rational signals, (23) gives the signal also for noninteger times (2):

$$x(t) = \frac{1}{2\pi i}\int \frac{b(\Delta,\varepsilon)}{a(\Delta)}(T\Delta+1)^{t/T-\varepsilon}\frac{Td\Delta}{T\Delta+1}.$$

Taking the k-th derivative and setting $t = 0$, we have

$$x^{(k)}(0) = \frac{1}{2\pi i}\int \frac{b(\Delta,\varepsilon)}{a(\Delta)}\left(\frac{\ln(T\Delta+1)}{T}\right)^k(T\Delta+1)^{-\varepsilon}\frac{Td\Delta}{T\Delta+1}.$$

The numbers $x^{(k)}(0)$ can be computed by congruences of polynomials. Note that $b(\Delta,\varepsilon)$ is divisible by $(T\Delta+1)$.

For $\varepsilon \neq 0$, we must also compute $(T\Delta+1)^{-\varepsilon} \bmod a(\Delta)$. The general power (in any algebra) can be computed by exponential and logarithm but a better way was designed by using binary code, square and square roots, see (Ježek 1988).

Now in our problem of computing $B(p)$, from the Taylor interpolation formula

$$x(t) = \sum_{k=0}^{\infty} x^{(k)}(0)\frac{t^k}{k!} \tag{24}$$

by taking the transform

$$\bar{x}(s) = s\sum_{k=0}^{\infty} x^{(k)}(0)s^k \tag{25}$$

we see that $\hat{B}(s)$ can be computed by polynomial multiplication

$$\hat{B}(s) = \hat{A}(s)\bar{X}(s)$$

with $k = 0\ldots N$. Note that the structure is in both conversions the same, for $T \to 0$ (21) converges to (24), (22) to (25).

6. CONVERSION OF TRANSFER FUNCTION

Up to now, we have dealt with conversion of signals or of their transforms. The conversion of transfer functions needs something more. A linear system is supposed in the form

$$y = \frac{b}{a}u + \frac{c}{a}e$$

with output y, input u and white noise e. This scheme may be continuous-time or discrete-time, polynomials a, b, c in p or in Δ. Each of transfer functions b/a and c/a is converted in a different way.

The continuous-time transfer function $H(p) = B(p)/A(p)$ is Laplace transform of the impulse response $h(t)$, response to $u(t) = \delta(t)$. In digital control, the input is not the Dirac pulse but is held constant for $0 \leq t < T$ and then reset. In z-transform, the discrete transfer function is

$$H(z,\varepsilon) = (1 - z^{-1})D\left\{\frac{H(p)}{p}\right\}.$$

In our transform, it is

$$H(\Delta,\varepsilon) = \frac{\Delta}{T\Delta+1}D\left\{\frac{H(p)}{p}\right\}.$$

So, when the continuous-time transfer function is converted to the discrete-time one, it is first divided by p, then converted TD and finally the numerator divided by $T\Delta+1$ and the denominator by Δ (they are divisible by these). For $T \to 0$, it is $H(\Delta,\varepsilon) \to H(p)$. The backward conversion goes

$$H(p) = p\frac{D^{-1}}{T}\left\{\frac{T\Delta+1}{\Delta}H(\Delta,\varepsilon)\right\}.$$

When looking for a discrete-time equivalent of the noise filter $C(p)/A(p)$, we want the discretely generated noise to have the same statistical properties as the continuously generated one. We require that their autocorrelation functions coincide for $t = nT$. In transforms, the spectral densities are converted

$$\frac{c^*c}{a^*a} = D\left\{\frac{C^*C}{A^*A}\right\}.$$

The asterisk denotes the adjoint operator

$$p^* + p = 0, \quad s^* + s = 0,$$

$$\Delta^* + \Delta = T\Delta^*\Delta, \quad \sigma^* + \sigma = T.$$

The autocorrelation function is two-sided symmetric: $R(-\tau) = R(\tau)$. Our discretization D is defined for one-sided functions ($\tau \geq 0$); so we discretize one side of it only. The algorithm is as follows:

1) Given C, A, the continuous-time spectral density is decomposed into D/A

$$\frac{C^*C}{A^*A} = \frac{D}{A} + \frac{D^*}{A^*}.$$

It is performed by solving a symmetric polynomial equation.

2) D/A is converted to discrete-time d/a. A special care must be taken for $t = 0$; this point is common to D/A and D^*/A^*. We must take only a half of it:

$$\frac{d(\Delta)}{a(\Delta)} = D\left\{\frac{D(p)}{A(p)}\right\} - \frac{1}{2}\left[\frac{D}{A}\right]_{t=0}.$$

3) The resulting discrete-time spectral density is factorized into c/a

$$\frac{d}{a} + \frac{d^*}{a^*} = \frac{c^*c}{a^*a}.$$

It is performed by solving the polynomial equation for spectral factorization.

In the same way, an algorithm for backward conversion is constructed. For more details on adjoint, symmetry and polynomial equations, see (Ježek 1983, Nagy and Ježek 1986).

When investigating the convergence $T \to 0$, we must realize that the autocorrelation function $R(\tau)$ of $e(t)$ is $\delta(t)$ but R_n of e_n is 1 for $n = 0$ and 0 for $n \neq 0$. So R_n does not converge to $R(\tau)$ but R_n/T does. It implies

$$\sqrt{T}\frac{c(\Delta)}{a(\Delta)} \to \frac{C(p)}{A(p)}.$$

7.CONCLUSION

The discrete time Δ-models have interesting properties. For any finite value of T, they are equivalent to the commonly used z-models but for $T \to 0$ their behaviour differ dramatically. The z-models do not converge to anything useful but Δ-models converge to p-models. In this way, a unified theory of discrete-time and continuous-time control emerges. All equations for discrete-time control synthesis depend on the parameter T; the continuous-time control is given simply by $T \to 0$. This is true even numerically: when carefully programmed, the algorithms work even for the case $T = 0$ without difficulties. The only difference consists in the implementation of the controller.

The conversion algorithms were implemented on the computer and tested in the same way as the old algorithms for z-models. For reasonable data, they work well. The difficulties appear mainly when some poles are close to the forbidden values (19). It happens for T too great compared with time constants or oscillation periods of the system. This behavior is the same as for z-models. On the other side, the difficulties with T too small disappeared, the algorithms work for T arbitrarily small, even for $T = 0$.

REFERENCES

Ježek,J. (1968). Algorithm for computing the discrete time transfer function of a linear dynamic system (in Czech). Kybernetika 4, 3, 246-259.

Ježek,J. (1983). Conjugated and symmetric polynomial equations I, II. Kybernetika 19, 2, 121-130 and 3, 196-211.

Ježek,J. (1988). An efficient algorithm for computing real power of a matrix and a related matrix function. Aplikace matematiky, 33, 1, 22-32.

Nagy,I. and Ježek,J. (1986). Polynomial LQ control synthesis for delta-operator models. 2-nd IFAC workshop on adaptive systems, Lund, Sweden, pp.323-327.

Vostrý,Z. (1977). Congruence of analytic functions modulo a polynomial. Kybernetika, 13, 2, 116-137.

Vostrý,Z. (1978). Polynomial approach to conversion between Laplace and Z transforms. Kybernetika 14, 4, 292-306.

FEEDBACK SYSTEM DESIGN AND STRUCTURAL PROPERTIES: THE ROLE AND USE OF RATIONAL MATRIX GCD's

V. X. Le and M. G. Safonov

*Dept. of Electrical Engineering-Systems, University of Southern California,
Los Angeles, CA 90089-0781, USA*

ABSTRACT. A state-space construction for rational matrix greatest common divisors (GCD's) is given. It is shown how the GCD results can be used to solve the problem of designing stable minimum-phase squaring-down compensators for multivariable plants, leading to a direct state-space construction for such compensators and a state-space solution to "fat-plant" H-infinity control problems. The theory also provides a state-space construction for minimal bases for rational vector spaces in the sense of Forney. The results make use of the concepts of strongly observable systems and maximally unobservable systems initiated in the respective works of Silverman and of Wonham and build upon the concepts introduced in the state-space GCD extraction results for polynomial matrices of Silverman and Van Dooren.

KEYWORDS. Algebraic system theory, feedback control, inverse systems, linear systems, matrix fraction descriptions, matrix polynomial equations, multivariable control systems, state-space methods, synthesis methods, zeros.

1. INTRODUCTION

In the early 1980's the matrix fraction description (MFD) viewpoint on feedback system theory led significant simplifications in controller design and has broadened the scope of achievable design objectives. An important milestone was the paper of Desoer and co-workers (1980) in which, via a basic ring-theoretic setup, the key parametrization result of Youla, Jabr and Bongiorno (1976) was rederived and simplified. This development was followed by a flurry of research results on coprime MFD's of matrix plant transfer functions and the related Bezout-Diophantine equality, notably Vidyasagar, Schneider and Francis (1982), and Nett, Jacobson and Balas (1984), which paved the way for recent advances in H∞ control theory (see Doyle (1984), Francis (1987), and Safonov and co-workers (1987a)).

Although the concept and the many implications of coprime MFD's in linear system theory are by now well exploited and well understood, the closely related subject of matrix greatest common divisors (GCD's) has received relatively little attention. A probable reason for the oversight is the lack of a reliable algorithm for computing rational matrix GCD's. One of the main results in this paper is a state-space algorithm to fill this gap (Theorem 2.1). Further, we show that the state-space rational matrix GCD results can be directly applied to solve such problems as minimal basis construction for rational vector spaces (section 2), squaring-down non-square matrix transfer functions (section 3), and H∞ optimal control for fat plants (section 4), thus establishing the central role of rational matrix GCD's in feedback and system theory. In the remainder of this section we present a brief overview on the matrix GCD extraction problem and its relations to the aforementioned topics, deferring the technical details to the subsequent sections. But, first, we introduce the required notation.

It is assumed that the reader is already familiar with the basic concept of an algebraic ring (see, e.g., Jacobson (1953), Vidyasagar (1985)). We consider a ring $\mathbf{G}$. An element $u \in \mathbf{G}$ is called a unit in $\mathbf{G}$ if its inverse u^{-1} is also an element of $\mathbf{G}$. Let $\mathbf{M(G)}$ denote the set of matrices whose elements belong to $\mathbf{G}$ and let $\mathbf{U(G)}$ be the subset of $\mathbf{M(G)}$ which consists of all square, invertible matrices whose inverses also belong to $\mathbf{M(G)}$. A matrix $\mathbf{U} \in \mathbf{U(G)}$ is said to be G-unimodular. It is assumed that the reader is also familiar with the standard definitions for greatest common right divisor (GCRD), coprimeness, and matrix fraction description (MFD) for matrices in $\mathbf{M(G)}$. The rational matrix GCD results in the paper are developed strictly for the case of GCRD at no loss of generality since the complementary results for the case of GCLD follow by duality.

We next specify $\mathbf{R}$ to be the ring of all rational functions of s and $\mathbf{R_p}$ to be the subring of all proper rational functions of s. Let $\mathbf{R_{p\Omega}}$ be the subring of $\mathbf{R_p}$ which consists of all proper rational functions of s that are analytic in the region Ω of the complex s-plane. Then, as in the case of the ring $\mathbf{G}$, $\mathbf{M(R_{p\Omega})}$ denotes the set of matrices whose elements belong to $\mathbf{R_{p\Omega}}$ and $\mathbf{U(R_{p\Omega})}$ denotes the subset of $\mathbf{M(R_{p\Omega})}$ consisting of all $\mathbf{R_{p\Omega}}$-unimodular matrices. A matrix in $\mathbf{M(R_{p\Omega})}$ is said to be Ω-stable; if additionally it does not have any zero in Ω then it is said to be Ω-miniphase. We now state the following lemma on matrix GCRD which is basic to the subsequent results (for proof, see Vidyasagar (1985, Chapt. 4, Appendix B)).

Lemma 1.1: Let $G_1(s)$, $G_2(s) \in \mathbf{M(R_{p\Omega})}$ with $G_1(s)$ an $r_1 \times m$ matrix and $G_2(s)$ an $r_2 \times m$ matrix and assume that $r = r_1 + r_2 > m$. Then, there exists an $\mathbf{R_{p\Omega}}$-unimodular matrix $\tilde{U}(s) \in \mathbf{U(R_{p\Omega})}$ such that

$$\tilde{U}(s) \cdot \begin{bmatrix} G_1(s) \\ G_2(s) \end{bmatrix} = \begin{bmatrix} \tilde{R}(s) \\ 0 \end{bmatrix} \begin{array}{l} m \\ r-m \end{array}$$

$$\underbrace{\qquad}_{m}$$

where $\tilde{R}(s) \in \mathbf{M(R_{p\Omega})}$ is a square matrix. Further, $\tilde{R}(s)$ is a GCRD of $G_1(s)$ and $G_2(s)$. □

The lemma implies, in essence, that a GCRD of a matrix in $\mathbf{M(R_{p\Omega})}$ may be obtained by left-multiplying it with a suitable $\mathbf{R_{p\Omega}}$-unimodular matrix.

With the notation and the basic lemma thus stated, we now formulate the GCRD extraction problem and briefly describe its relations to the problems of state-space construction of minimal bases, of squaring-down non-square plants, and of state-space synthesis of H∞ optimal controllers for fat plants. We call a matrix tall when it has more rows than columns and wide when it has more columns than rows. The problem we consider is the following: Given a tall matrix $N(s) \in \mathbf{M(R_{p\Omega})}$, i.e., $N(s)$ is Ω-stable, which has full column-rank and the following state-space realization

$$N(s) \stackrel{s}{=} \left[\begin{array}{c|c} A & B \\ \hline C & D \end{array} \right],$$

how does one construct an $\mathbf{R_{p\Omega}}$-unimodular matrix

$$\tilde{U}(s) = \begin{bmatrix} U(s) \\ U_\perp(s) \end{bmatrix}$$

such that

$$\tilde{U}(s)N(s) = \begin{bmatrix} \tilde{R}(s) \\ 0 \end{bmatrix}, \tag{1.1}$$

where $\tilde{R}(s) \in \mathbf{M(R_{p\Omega})}$ is square and is a GCRD of $N(s)$? Existence of such a matrix $\tilde{U}(s)$ is ensured by Lemma 1.1, and we will show in section 2 its state-space construction. Once $\tilde{U}(s)$ is determined, computing

$$\tilde{U}^{-1}(s) = \tilde{V}(s) = \left[V(s) \mid V_\perp(s) \right],$$

where $\tilde{V}(s)$ is $\mathbf{R_{p\Omega}}$-unimodular since $\tilde{U}(s)$ is $\mathbf{R_{p\Omega}}$-unimodular, readily yields the following factorization for $N(s)$

$$N(s) = \left[V(s) \mid V_\perp(s) \right] \cdot \begin{bmatrix} \tilde{R}(s) \\ 0 \end{bmatrix} = V(s)\tilde{R}(s). \tag{1.2}$$

In Forney (1975), $V(s)$ and $V_\perp(s)$ of (1.2) form the so-called basis and dual basis for $N(s)$, whose state-space constructions are also given in section 2.

The equivalence between the solution to the GCRD problem and that of the squaring down problem may be seen as follows. For a tall plant $N(s) \in \mathbf{M(R_{p\Omega})}$, via (1.1), $U(s)N(s) = \tilde{R}(s)$ which, by specifying Ω to be the right half of the complex s-plane, $U(s)$ becomes a stable minimum-phase squaring-down (SD) post compensator having the key property that it does not induce any new zero in Ω in the squared-down plant $\tilde{R}(s)$. The properties of stability and minimum-phase are the key properties required of dynamic SD compensators as has been implied in Saberi and Sannuti (1988). In section 3 we will discuss the squaring down problem in further detail and give a state-space construction for U(s) for the general case $N(s) \notin \mathbf{M}(\mathbf{R}_{p\Omega})$.

We next sketch the role of the GCD results in solving the $H\infty$ control synthesis for fat plants. The term <u>fat plant</u> denotes a plant-feedback controller interconnection as in Figure 1.1, where the matrix $P_{12}(s)$ is wide and/or $P_{21}(s)$ is tall. "Fat" here signifies a situation where there are more sensor outputs than disturbance signals and/or more actuator inputs than penalized outputs, the former being a common occurence in aircraft flight control systems. The $H\infty$ control problem is the following: Given

$$P(s) = \begin{bmatrix} P_{11} & P_{12} \\ P_{21} & P_{22} \end{bmatrix}(s) \stackrel{s}{=} \left[\begin{array}{c|cc} A & B_1 & B_2 \\ \hline C_1 & D_{11} & D_{12} \\ C_2 & D_{21} & D_{22} \end{array} \right].$$

Find all stabilizing K's that satisfy $\left\| T_{y_1 u_1} \right\|_\infty \le 1$.

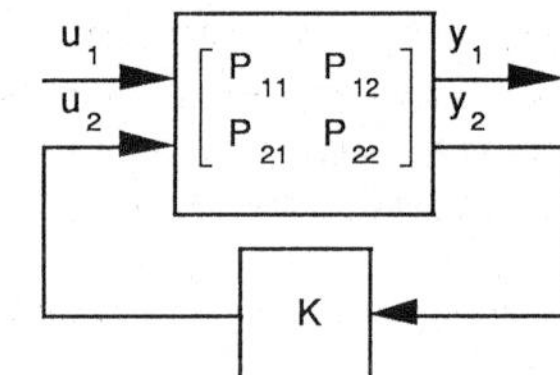

Figure 1.1. Plant-feedback controller system

Note, first of all, that the existing $H\infty$ design methods for "non-fat" plants, i.e., those plants such that $P_{12}(s)$ is not wide and $P_{21}(s)$ is not tall (e.g., Doyle (1984), Francis (1987), Safonov and co-workers (1987), Doyle and co-workers (1988), Glover and Doyle (1988), Safonov and Limebeer (1988) and Glover and co-workers (1989)) do not apply to the case of fat plants. But if, following the Youla parametrization to transform the problem to an open-loop stable $H\infty$ control synthesis as in Fig. 1.2, we insert $U_{12}(s)$ and $U_{21}(s)$, which are the stable minimum-phase SD pre- and post-compensators for $T_{12}(s)$ and $T_{21}(s)$, the problem reduces to the non-fat $H\infty$ control synthesis, for which the techniques of Doyle and co-workers (1988), Glover and Doyle (1988), Safonov and Limebeer (1988) and Glover and co-workers (1989) may be applied. This is possible because, it turns out that, the stable minimum-phase SD compensators $U_{12}(s)$ and $U_{21}(s)$

do not affect the set of closed-loop transfer functions $T_{y_1 u_1}$ realizable via stabilizing feedbacks.

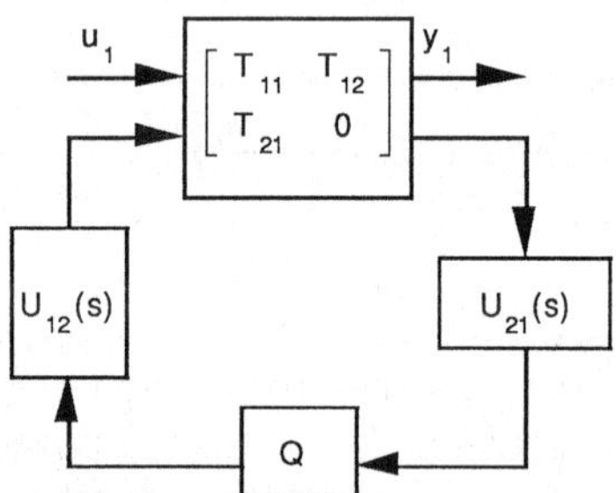

Figure 1.2. System with SD compensators

As a closing remark for this introductory section, we present a cursory review of the existing literature on the matrix GCD extraction problem. Most standard textbooks (e.g., Rosenbrock (1970), Wolovich (1974), Kailath (1980)) restrict attention to the case where N(s) is a polynomial matrix, i.e., $\Omega = \mathbb{C}$. These references suggest algorithms for constructing the GCRD $\tilde{R}(s)$ and the unimodular polynomial matrix $\tilde{U}(s) \in \mathbf{M}(\mathbf{R}_\mathbb{C})$ of Lemma 1.1 via elementary row operations on N(s). A state-space technique to construct the GCRD $\tilde{R}(s)$ of a polynomial matrix $N(s) \in \mathbf{M}(\mathbf{R}_\mathbb{C})$, but not the corresponding unimodular $\tilde{U}(s)$ was developed by Silverman and Van Dooren (1978). In this paper, we build upon the work of Silverman and Van Dooren to develop state-space algorithm for computing not only $\tilde{R}(s)$ but also $\tilde{U}(s)$, $\tilde{V}(s)$ for the case $N(s) \in \mathbf{M}(\mathbf{R}_{p\Omega})$.

2. MAIN RESULTS

In this section we develop the main theorem of the paper, Theorem 2.1, which, given a state-space realization for a tall matrix $N(s) \in \mathbf{M}(\mathbf{R}_{p\Omega})$, provides state-space realizations for an $\tilde{R}(s) \in \mathbf{M}(\mathbf{R}_{p\Omega})$ and an $\mathbf{R}_{p\Omega}$-unimodular matrix $\tilde{U}(s)$ that satisfies

$$\tilde{U}(s)N(s) = \begin{bmatrix} \tilde{R}(s) \\ 0 \end{bmatrix},$$

where $\tilde{R}(s)$ is square and is a GCRD of N(s). For the case of wide matrices in $\mathbf{M}(\mathbf{R}_{p\Omega})$, the dual of Theorem 2.1 may be readily derived to yield GCLD results.

We begin by considering a full-rank, tall matrix $N(s) \in \mathbf{M}(\mathbf{R}_{p\Omega})$ having a not-necessarily-minimal state-space realization

$$N(s) \stackrel{s}{=} \left[\begin{array}{c|c} A & B \\ \hline C & D \end{array} \right],$$

where $A \in \mathbb{C}^{n \times n}$, $B \in \mathbb{C}^{n \times m}$, $C \in \mathbb{C}^{r \times n}$, $D \in \mathbb{C}^{r \times m}$, and $r \ge m$. We note that the full-rank condition on N(s) is necessary for the GCRD extraction problem to be well posed (see Kailath (1980, p. 378)). Further, for simplicity of notation, the following two assumptions on D are taken:

<u>Assumption 1:</u> D has full column rank, i.e., N(s) has no zero at $s = \infty$ (see Kailath (1980, p. 449)). $\square$

<u>Assumption 2:</u> D has orthonormal columns, i.e., $D^* D = I_{m \times m}$. $\square$

The foregoing two assumptions incur no loss of generality as other problems can readily be made to satisfy these assumptions via simple transformations. Satisfaction of Assumption 1 may be accomplished by a number of transformations, and for the reader's convenience, we give the details of one such transformation in Appendix B. Once Assumption 1 is satisfied, right-multiplying N(s) by a suitably chosen invertible constant matrix causes Assumption 2 to be satisfied. Such a constant matrix may be constructed from the singular-value decomposition (SVD) of D as follows. Let the SVD of D be given by

$$D = U_D \begin{pmatrix} \Sigma_D \\ 0 \end{pmatrix} V_D^*$$

$$= \begin{bmatrix} U_{D_1} & | & U_{D_2} \end{bmatrix} \begin{pmatrix} \Sigma_D \\ 0 \end{pmatrix} V_D^*,$$

where $\Sigma_D = \mathrm{diag}\,(d_1, d_2, ..., d_m) \in \mathbb{R}^{m \times m}$, $U_D \in \mathbb{C}^{r \times r}$, $V_D \in \mathbb{C}^{m \times m}$ and U_D, V_D are unitary matrices. If we set $S = V_D \Sigma_D^{-1}$, then

$$N(s)S = N'(s) \stackrel{s}{=} \left[\begin{array}{c|c} A & B' \\ \hline C & D' \end{array} \right]$$

is such that $D' = U_{D_1}$ has orthonormal columns.

In the following, we define:

$$\left[\begin{array}{c|c} \tilde{A} & \tilde{B} \\ \hline \tilde{C} & \tilde{D} \end{array} \right] \equiv \left[\begin{array}{c|c} \tilde{T}(A - BD^* C)\tilde{T}^{-1} & \tilde{T}B \\ \hline D_\perp D_\perp^* C \tilde{T}^{-1} & D \end{array} \right]$$

$$= \left[\begin{array}{cccc|c|c} \tilde{A}_{00} & 0 & 0 & 0 & 0 & \tilde{n}_0 \\ \tilde{A}_{10} & \tilde{A}_{11} & 0 & 0 & \tilde{B}_1 & \tilde{n}_1 \\ \tilde{A}_{20} & 0 & \tilde{A}_{22} & 0 & 0 & \tilde{n}_2 \\ \tilde{A}_{30} & \tilde{A}_{31} & \tilde{A}_{32} & \tilde{A}_{33} & \tilde{B}_3 & \tilde{n}_3 \\ \hline \tilde{C}_0 & \tilde{C}_1 & 0 & 0 & D & \end{array} \right] \quad (2.1)$$

where $D_\perp \in \mathbb{C}^{r \times (r-m)}$ is such that $[D \mid D_\perp]$ is unitary; e.g., $D_\perp = U_{D_2}$, and $\tilde{T}$ is an observable-controllable-canonical transformation matrix (see Kalman (1963)) chosen so that $(\tilde{A}_{11}, \tilde{B}_1, \tilde{C}_1, D)$ is a minimal realization of $(\tilde{A}, \tilde{B}, \tilde{C}, \tilde{D})$. As shown in Fig. 2.1, $\tilde{N}(s) = \tilde{C}(Is - \tilde{A})^{-1}\tilde{B} + D$ arises from the system $N(s) = C(Is - A)^{-1}B + D$ via the output feedback D^*. The key property of the feedback matrix D^*, as was first noted by Silverman (1976), is that it makes $\left(D_\perp D_\perp^* C, A - BD^* C \right)$ "maximally unobservable"; that is, each zero of $\tilde{N}(s)$ is cancelled by a pole. In other words, every transmission zero of N(s) becomes an eigenvalue of the controllable, but unobservable, part of the $\tilde{A}$-

matrix, viz. $\tilde{A}_{33}$; this, incidentally, provides a simple method for computing the zeros of any not-necessarily-square transfer function matrix N(s). To maintain continuity of the main development and for the reader's convenience, the concept of maximally unobservable systems and the related concept of strongly observable systems are discussed further in Appendix A. For notational convenience, the matrices $\tilde{T}$ and $\tilde{T}^{-1}$ are partitioned conformably with $\tilde{A}$ to give

$$\tilde{T} = \left[\tilde{T}_0^* \; \tilde{T}_1^* \; \tilde{T}_2^* \; \tilde{T}_3^* \right]^*, \quad \tilde{S} = \left[\tilde{S}_0 \; \tilde{S}_1 \; \tilde{S}_2 \; \tilde{S}_3 \right] \stackrel{\Delta}{=} \tilde{T}^{-1}. \qquad (2.2)$$

We note that Chiang and Safonov (1988) contains numerically robust model reduction algorithms based on the methods of Safonov, Chiang and Limebeer (1987b) and Safonov and Chiang (1988) which may be used to reliably compute the matrices $\tilde{T}_1, \tilde{S}_1, \tilde{A}_{11}, \tilde{B}_1, \tilde{C}_1$ that determine the controllable-observable part of the system $(\tilde{A}, \tilde{B}, \tilde{C}, \tilde{D})$ given in (2.1).

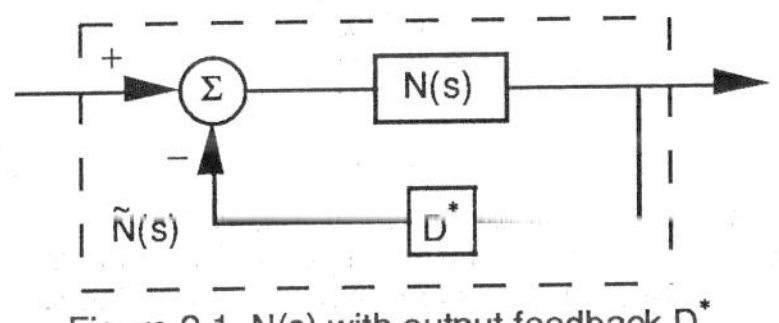

Figure 2.1. N(s) with output feedback D^*

The following lemma is immediate.

<u>Lemma 2.1:</u> There exist an $m \times \tilde{n}_1$ state-feedback $\tilde{F}_1$ and an $\tilde{n}_1 \times r$ output-injection $\tilde{H}_1$ such that

$$(Is - \tilde{A}_{11} - \tilde{B}_1 \tilde{F}_1)^{-1} \text{ and } (Is - \tilde{A}_{11} - \tilde{H}_1 \tilde{C}_1)^{-1}$$

are Ω-stable. $\quad\square$

<u>Proof:</u> The fact that $(\tilde{C}_1, \tilde{A}_{11})$ is, by construction, observable ensures the existence of an Ω-stabilizing $\tilde{H}_1$ such that $(Is - \tilde{A}_{11} - \tilde{H}_1 \tilde{C}_1)^{-1}$ is Ω-stable. Furthermore, $(\tilde{A}, \tilde{B}) = (\tilde{T} [A - BD^*C] \tilde{T}^{-1}, \tilde{T}B)$ is Ω-stabilizable since, indeed, the state-feedback $D^*C\tilde{T}^{-1}$ results in the Ω-stable system $(\tilde{T}A \tilde{T}^{-1}, \tilde{T}B)$. It follows that $(\tilde{A}_{11}, \tilde{B}_1)$ is Ω-stabilizable, and hence there exists an Ω-stabilizing state-feedback $\tilde{F}_1$. $\quad\square$

The main theorem of the paper follows.

<u>Theorem 2.1:</u> Let

$$\tilde{U}(s) = \begin{bmatrix} U(s) \\ U_\perp(s) \end{bmatrix} \stackrel{s}{=} \left[\begin{array}{c|cc} \tilde{A}_{11} + \tilde{H}_1 \tilde{C}_1 & \tilde{B}_1 D^* - \tilde{H}_1 D^* D_\perp^* \\ \hline -\tilde{F}_1 & D^* \\ -D_\perp^* C & D_\perp^* \end{array} \right], \qquad (2.3)$$

where $\tilde{F}_1$ and $\tilde{H}_1$ are, respectively, any $m \times \tilde{n}_1$ state-feedback and $\tilde{n}_1 \times r$ output-injection that make $(Is - \tilde{A}_{11} - \tilde{B}_1 \tilde{F}_1)^{-1}$ and $(Is - \tilde{A}_{11} - \tilde{H}_1 \tilde{C}_1)^{-1}$ Ω-stable. Then, $\tilde{U}(s)$ is a <u>minimal</u> realization of an $\mathbf{R}_{p\Omega}$-unimodular matrix whose inverse $\tilde{V}(s)$ has the following <u>minimal</u> realization

$$\tilde{V}(s) = \tilde{U}^{-1}(s) = \left[V(s) \mid V_\perp(s) \right]$$

$$\stackrel{s}{=} \left[\begin{array}{c|cc} \tilde{A}_{11} + \tilde{B}_1 \tilde{F}_1 & \tilde{B}_1 & -\tilde{H}_1 D_\perp \\ \hline \tilde{C}_1 + D\tilde{F}_1 & D & D_\perp \end{array} \right], \qquad (2.4)$$

and

$$\tilde{U}(s) \cdot N(s) = \begin{bmatrix} \tilde{R}(s) \\ 0 \end{bmatrix},$$

where

$$\tilde{R}(s) \stackrel{s}{=} \left[\begin{array}{c|c} A & B \\ \hline -\tilde{F}_1 \tilde{T}_1 + D^*C & I \end{array} \right] \in \mathbf{M}(\mathbf{R}_{p\Omega})$$

is a square matrix and is a GCRD of N(s). $\quad\square$

<u>Remark 2.1:</u> (2.3) and (2.4) are the reduced (with non-minimal non-Ω-stable modes removed) representations of, respectively, the following non-minimal realizations:

$$\left[\begin{array}{c|c} A - BD^*C + \tilde{S}_1 \tilde{H}_1 D_\perp D_\perp^* C & BD^* - \tilde{S}_1 \tilde{H}_1 D_\perp D_\perp^* \\ \hline -\tilde{F}_1 \tilde{T}_1 & D^* \\ -D_\perp^* C & D_\perp^* \end{array} \right] \qquad (2.5)$$

and

$$\left[\begin{array}{c|cc} A - BD^*C + B\tilde{F}_1 \tilde{T}_1 & B & -\tilde{S}_1 \tilde{H}_1 D_\perp \\ \hline D_\perp D_\perp^* C + D\tilde{F}_1 \tilde{T}_1 & D & D_\perp \end{array} \right]. \qquad (2.6)$$

This follows by applying the state-coordinate transformation (2.2) to (2.5) and (2.6) and truncating the uncontrollable/unobservable parts. $\quad\square$

<u>Proof of Theorem 2.1:</u> Existence of the stabilizing $\tilde{F}_1$ and $\tilde{H}_1$ follows directly from Lemma 2.1. What remains to be proved are: (i) $\tilde{U}(s)$ is $\mathbf{R}_{p\Omega}$-unimodular, (ii) $\tilde{U}(s)$ and $\tilde{V}(s)$ are minimal, and (iii) that $\tilde{R}(s)$ is a square matrix, $\tilde{R}(s) \in \mathbf{M}(\mathbf{R}_{p\Omega})$, and $\tilde{R}(s)$ is a GCRD of N(s).

(i) To prove that $\tilde{U}(s)$ is $\mathbf{R}_{p\Omega}$-unimodular, it suffices to show that $\tilde{U}^{-1}(s)$ is Ω-stable. Noting that for any square and invertible G(s) having realization

$$G(s) \stackrel{s}{=} \left[\begin{array}{c|c} A' & B' \\ \hline C' & D' \end{array} \right]$$

$G^{-1}(s)$ has the realization

$$G^{-1}(s) \stackrel{s}{=} \left[\begin{array}{c|c} A' - B'D'^{-1}C' & B'D'^{-1} \\ \hline -D'^{-1}C' & D'^{-1} \end{array} \right],$$

it follows immediately that the realization for $\tilde{V}(s) = \tilde{U}^{-1}(s)$ is given by (2.4). That $\tilde{U}^{-1}(s)$ is Ω-stable is clear since $\tilde{F}_1$ is, by hypothesis, an Ω-stabilizing state feedback.

(ii) To establish minimality of $\tilde{V}(s)$, it suffices to establish minimality of $V(s) \equiv \tilde{V}(s)\begin{bmatrix} I \\ 0 \end{bmatrix}$. By construction, $(\tilde{A}_{11}, \tilde{B}_1)$ is controllable; further, by invariance of controllability under state feedback, it results that $(\tilde{A}_{11} + \tilde{B}_1 \tilde{F}_1, \tilde{B}_1)$ is controllable. Also, by construction, the system $(\tilde{A}_{11}, \tilde{B}_1, \tilde{C}_1, D)$ is strongly observable (see Appendix A), and therefore remains observable for any state-feedback $\tilde{F}_1$. Thus, the realization (2.4) of $\tilde{V}(s)$ is minimal for all $\tilde{F}_1, \tilde{H}_1$ and, obviously, minimality of $\tilde{U}(s) = \tilde{V}^{-1}(s)$ follows from the minimality of $\tilde{V}(s)$.

(iii) The proof that $\tilde{R}(s)$ is a square matrix, $\tilde{R}(s) \in \mathbf{M}(\mathbf{R}_{p\Omega})$, and $\tilde{R}(s)$ is a GCRD of N(s) is accomplished as follows. Computing the realization for $\tilde{R}(s)$ via left-multiplying N(s) with (2.5) to get

$$\tilde{R}(s) = \tilde{U}(s) \cdot N(s)$$

$$\stackrel{s}{=} \left[\begin{array}{c|c} A - BD^*C + \tilde{S}_1 \tilde{H}_1 D^* D_\perp^* C & BD^* - \tilde{S}_1 \tilde{H}_1 D^* D_\perp^* \\ \hline -\tilde{F}_1 \tilde{T}_1 & D^* \\ -D_\perp^* C & D_\perp^* \end{array} \right] \times \left[\begin{array}{c|c} A & B \\ \hline C & D \end{array} \right]$$

$$\stackrel{s}{=} \left[\begin{array}{cc|c} A - BD^*C + \tilde{S}_1 \tilde{H}_1 D^* D_\perp^* C & BD^*C - \tilde{S}_1 \tilde{H}_1 D^* D_\perp^* C & B \\ 0 & A & B \\ \hline -\tilde{F}_1 \tilde{T}_1 & D^*C & I \\ -D_\perp^* C & D_\perp^* C & 0 \end{array} \right]$$

$$\stackrel{s}{=} \left[\begin{array}{cc|c} A - BD^*C + \tilde{S}_1 \tilde{H}_1 D_\perp D_\perp^* C & 0 & 0 \\ 0 & A & B \\ \hline -\tilde{F}_1 \tilde{T}_1 & -\tilde{F}_1 \tilde{T}_1 + D^*C & I \\ -D_\perp^* C & 0 & 0 \end{array} \right] \qquad (2.7)$$

where the latter equality (2.7) follows by applying the state-coordinate transformation

$$\begin{pmatrix} x_1 \\ x_2 \end{pmatrix} \leftarrow \begin{bmatrix} I & I \\ 0 & I \end{bmatrix} \begin{pmatrix} x_1 \\ x_2 \end{pmatrix}.$$

Deleting the first n states (which are evidently uncontrollable) yields

$$\tilde{R}(s) \stackrel{s}{=} \left[\begin{array}{c|c} A & B \\ \hline -\tilde{F}_1 \tilde{T}_1 + D^*C & I \end{array} \right]$$

which shows that $\tilde{R}(s)$ is a square matrix, $\tilde{R}(s) \in \mathbf{M}(\mathbf{R}_{p\Omega})$, and $\tilde{R}(s)$ is a GCRD of N(s). $\quad\square$

Remark 2.2: In the case that N(s) is a polynomial matrix, $\tilde{U}(s)$ and $\tilde{V}(s)$ become the usual unimodular polynomial matrices if, following Silverman and Van Dooren (1978), we first perform the s-plane change of variable $\tilde{s} = 1/s$ to map the point $s = \infty$ to the origin, then taking Ω to be $\mathbb{C}/\{0\}$ we use $\tilde{F}_1$ and $\tilde{H}_1$ to place poles of $\tilde{U}'(\tilde{s})$, $\tilde{V}'(\tilde{s})$ at the origin, and finally reverse the transformation $s \leftarrow 1/\tilde{s}$ to obtain polynomial matrices $\tilde{U}(s)$, $\tilde{V}(s)$. In Silverman and Van Dooren (1978), the authors work with the following direct factorization of $N'(\tilde{s})$

$$N'(\tilde{s}) = V'(\tilde{s})\tilde{R}'(\tilde{s}) = \begin{bmatrix} V'_1(\tilde{s}) \\ V'_2(\tilde{s}) \end{bmatrix} \cdot \tilde{R}'(\tilde{s}) \, ,$$

where the system $V'(\tilde{s})$ is maximally unobservable (see Appendix A) and has poles at $\tilde{s} = 0$ to yield $\tilde{R}'(\tilde{s})$, from which the desired polynomial GCRD $\tilde{R}(s)$ of N(s) is obtained using the change of variable $s \leftarrow 1/\tilde{s}$. Our main theorem, Theorem 2.1, is more general in that besides including the polynomial matrix GCD extraction results of Silverman and Van Dooren as a special case, by noting that $V'(\tilde{s})$ being maximally unobservable implies that $(\tilde{V}_1'(\tilde{s}), \tilde{V}_2'(\tilde{s}))$ and, hence, $(V_1(s), V_2(s))$ are coprime which, by partitioning $\tilde{U}(s)$ conformably with V(s) as

$$\tilde{U}(s) = \begin{bmatrix} U_1(s) & U_2(s) \\ U_{1\perp}(s) & U_{2\perp}(s) \end{bmatrix} ,$$

also gives a state-space solution to the following polynomial matrix Bezout identity

$$U_1(s)V_1(s) + U_2(s)V_2(s) = I \, .$$

Remark 2.3: Forney (1975, Sect. 4) gives an algorithm to obtain a minimal basis from a non-minimal basis for a given rational vector space (i.e., a vector space of rational functions of s) and suggests that the simplest alternate approach is via GCD extraction of the basis.

A state-space realization for the bases $\tilde{V}(s) = [V(s) \mid V_\perp(s)]$ is given by (2.4) whose MacMillan degree, i.e., the dimension of $(\tilde{A}_{11} + \tilde{B}_1 \tilde{F}_1)$, is the invariant dynamical order of N(s). Rational minimal bases and their duals are important in the problem of determining minimal-order inverses and in obtaining rational solution X(s) to matrix equalities of the type A(s)X(s) = B(s) where A(s) and B(s) are given rational matrices as has been described in Forney (1975) and Kailath (1980).
Remark 2.4: The constructions of the $R_{p\Omega}$-unimodular matrices in the GCD extraction problem make use of the concepts of strongly observable systems and maximally unobservable systems from the respective works of Silverman (1976) and Wonham (1979). A detailed discussion of these concepts is given in Appendix A.

3. THE SQUARING DOWN PROBLEM

In this section we discuss the squaring down (SD) problem and present a direct state-space solution to the synthesis of dynamic SD compensators for non-square plants. The discussion begins with a brief review of the literature.

The term "squaring down" was coined in the work of MacFarlane and Karcanias (1976) who investigated the problem of designing a full-rank constant matrix K in cascade with a non-square plant G(s) such that the overall transfer matrix KG(s) (for G(s) tall) or G(s)K (for G(s) wide) is square and has the key property that it does not have any additional RHP zero. The main objective of squaring down is to suit the squared-down plants to certain classical feedback design methods which require a square plant, e.g., characteristic-gain loci of MacFarlane and Kouvaritakis (1977), diagonal dominance of Rosenbrock (1974), etc. It is known that some modern control methods, notably state-space H∞ synthesis, also require the use of SD compensators under certain circumstances, e.g., in the case of fat plants as described in Vidyasagar (1985, Sect. 6.5) in which there are more control actuator inputs than penalized outputs and/or more sensor outputs than disturbance inputs (see section 4). The SD problem was later pursued by other researchers, including Kouvaritakis and MacFarlane (1976), Vardulakis (1980), Davison (1983), Sannuti and Saberi (1987), etc. However, the solution to the synthesis of constant SD compensators proved elusive.

By relaxing the constraint that the SD compensator be constant to the weaker requirement that it be stable and minimum phase, Aplevich (1981) and Saberi and Sannuti (1988) obtain concrete constructions for dynamic SD compensators. The constructive algorithm described in Aplevich (1981) is iterative and involves computing the inverse of a system of the same order as that of the given plant at each iteration. In Saberi and Sannuti (1988) the key step in the synthesis of a dynamic SD compensator is the construction of a special coordinate basis for the plant, which involves a certain regrouping of the input-output variables and defining new state variables.

The contribution of this paper to the SD problem is in the formula for the state-space realization of an Ω-stable Ω-miniphase SD compensator. Unlike the approach of Saberi and Sannuti (1988), our realization solution is given directly in terms of the state-space data of the plant. The formula builds on Theorem 2.1 to describe, for a tall not-necessarily Ω-stable rxm plant G(s) whose not-necessarily-minimal realization

$$G(s) \stackrel{s}{=} \left[\begin{array}{c|c} A & B \\ \hline C & D \end{array} \right] \tag{3.1}$$

where $A \in \mathbb{C}^{n \times n}$, $B \in \mathbb{C}^{n \times m}$, $C \in \mathbb{C}^{r \times n}$, $D \in \mathbb{C}^{r \times m}$, $r \geq m$, is Ω-stabilizable, i.e., there exists a matrix F such that $(Is - A + BF)^{-1}$ is Ω-stable, a state-space realization for a dynamic SD post-compensator U(s) that has the following key properties:
(i) U(s) is Ω-stable and Ω-miniphase.
(ii) The squared-down plant $\tilde{G}(s) = U(s)G(s)$ has the same "A-matrix" as the plant G(s), i.e., the same characteristic equation.
(iii) The squared-down plant $\tilde{G}(s) = U(s)G(s)$ has no new zero in Ω.

In the following, it is assumed that G(s) also satisfies Assumptions 1 and 2 of section 2, i.e., D has orthonormal columns. Adopting the notation (2.1) and (2.2), the formula for U(s) is given in the theorem.
Theorem 3.1: Let

$$\tilde{U}(s) = \begin{bmatrix} U(s) \\ U_\perp(s) \end{bmatrix} \stackrel{s}{=} \left[\begin{array}{c|cc} \tilde{A}_{11} + \tilde{H}_1\tilde{C}_1 & \tilde{B}_1 D^* - \tilde{H}_1 D^* D_\perp^* \\ \hline -\tilde{F}_1 & D^* \\ -D_\perp^* C & D_\perp^* \end{array} \right] , \tag{3.2}$$

where $\tilde{F}_1$ and $\tilde{H}_1$ are, respectively, any $m \times \tilde{n}_1$ state-feedback and $\tilde{n}_1 \times r$ output-injection that make $(Is - \tilde{A}_{11} - \tilde{B}_1\tilde{F}_1)^{-1}$ and $(Is - \tilde{A}_{11} - \tilde{H}_1\tilde{C}_1)^{-1}$ Ω-stable. Then, U(s) satisfies the properties (i), (ii) and (iii) and the squared-down plant has the realization

$$\tilde{G}(s) = U(s)G(s) \stackrel{s}{=} \left[\begin{array}{c|c} A & B \\ \hline -\tilde{F}_1\tilde{T}_1 + D^*C & D \end{array} \right] . \tag{3.3}$$

$\square$

Proof: Existence of an Ω-stabilizing $\tilde{F}_1$ is ensured by the hypothesis that (A,B) is Ω-stabilizable while existence of an Ω-stabilizing output-injection $\tilde{H}_1$ follows from $(\tilde{C}_1, \tilde{A}_{11})$ being observable by construction— cf. (2.1). That properties (i) and (ii) are satisfied follows from the constructed realization for U(s) and the resultant realization for $\tilde{G}(s)$, see the proof of Theorem 2.1. To show that property (iii) is satisfied we note that the zeros of $\tilde{G}(s)$ are the poles of $\tilde{G}^{-1}(s)$, which are the characteristic roots of $(Is - A + BD^*C - B\tilde{F}_1\tilde{T}_1)^{-1}$. Making use of (2.1), the zeros of $\tilde{G}(s)$ are thus the eigenvalues of

$$\begin{pmatrix} \tilde{A}_{11} + \tilde{B}_1\tilde{F}_1 & 0 \\ \tilde{A}_{31} + \tilde{B}_3\tilde{F}_1 & \tilde{A}_{33} \end{pmatrix} .$$

The zeros of $\tilde{G}(s)$ corresponding to the eigenvalues of $(\tilde{A}_{11} + \tilde{B}_1\tilde{F}_1)$ are the Ω-miniphase transmission zeros induced by the dynamic SD compensator U(s), whereas the zeros of $\tilde{G}(s)$ corresponding to the eigenvalues of $\tilde{A}_{33}$ are identical to the zeros of G(s). $\square$

Remark 3.1: If the realization (3.1) for G(s) is Ω-detectable, i.e., there exists an output-injection H such that $(Is - A + HC)^{-1}$ is Ω-stable, then $\tilde{G}(s) = U(s)G(s)$ is also Ω-detectable. This and the property of Ω-stabilizability of $\tilde{G}(s)$ by (3.3) imply that detectability and stabilizability are preserved via Ω-miniphase squaring-down compensators, a result previously proved in Saberi and Sannuti (1988). $\square$
Remark 3.2: The dynamical order of the squaring-down compensator U(s) is precisely the invariant dynamical order of the plant G(s) (see Remark 2.3). $\square$

4. H-INFINITY OPTIMAL CONTROL FOR FAT PLANTS

In this section we briefly describe the application of the state-space GCD theory to yield a solution to the problem of H∞-optimal control synthesis for fat plants. The fat-plant problem has previously been treated by Vidyasagar (1985, Sect. 6.5) who, by way of the Youla parametrization, considers an H∞ optimization with blocking-RHP-zeros constraint and gives a solution via the matrix Nevanlinna-Pick interpolation theory of Delsarte, Genin and Kamp (1979). In the following, we treat a more general fat-plant problem without the blocking-RHP-zeros restriction and give a synthesis solution via application of the state-space GCD theory. The key step in the synthesis, subsequent to the Youla parametrization step, is the insertion of stable minimum-phase squaring-down compensators (Lemma 4.1) to transform a fat open-loop stable H∞ synthesis to a non-fat open-loop stable H∞ synthesis, for which the techniques of Doyle and co-workers (1988), Glover and Doyle (1988), Safonov and Limebeer (1988), and Glover and co-workers (1989) may be applied.

Recall from Doyle (1984), Francis (1987) and Safonov and co-workers (1987a) that the Youla parametrization transforms the set of closed-loop transfer functions

$$T_{y_1 u_1} = P_{11} + P_{12} K (I - P_{22} K)^{-1} P_{21}$$

into the set

$$T_{y_1 u_1} = T_{11} - T_{12} Q T_{21}$$

where T_{11}, T_{12}, T_{21} are stable and may be realized via stabilizing state-feedback and output-injection matrices, and Q is any stable matrix. The H∞ control problem thus becomes: Find all stable Q 's such that

$$\left\| T_{11} - T_{12} Q T_{21} \right\|_\infty \leq 1. \tag{4.1}$$

As has been remarked earlier, existing techniques of Doyle (1984), Francis (1987), Safonov and co-workers (1987a), Doyle and co-workers (1988), Glover and Doyle (1988), Safonov and Limebeer (1988), and Glover and co-workers (1989) do not apply in the case that T_{12} is a wide matrix and/or T_{21} is a tall matrix, which are characteristic of fat plants. The following lemma, however, provides the needed machinery to circumvent those difficulties.

Lemma 4.1: Let $\hat{U}_{12}(s) = [U_{12}(s) \mid U_{12\perp}(s)], \hat{V}_{12}(s) = \begin{bmatrix} V_{12}(s) \\ V_{12\perp}(s) \end{bmatrix}$ and

$\hat{R}_{12}(s)$ solve the GCLD problem for $T_{12}(s)$, i.e.,

$$T_{12}(s)\hat{U}_{12}(s) = \left[\hat{R}_{12}(s) \mid 0 \right]; T_{12}(s) = \left[\hat{R}_{12}(s) \mid 0 \right] \hat{V}_{12}(s),$$

and let $\tilde{U}_{21}(s) = \begin{bmatrix} U_{21}(s) \\ U_{21\perp}(s) \end{bmatrix}, \tilde{V}_{21}(s) = [V_{21}(s) \mid V_{21\perp}(s)]$, and $\tilde{R}_{21}(s)$

solve the GCRD problem for $T_{21}(s)$, i.e.,

$$\tilde{U}_{21}(s)T_{21}(s) = \begin{bmatrix} \tilde{R}_{21}(s) \\ 0 \end{bmatrix}; T_{21}(s) = \tilde{V}_{21}(s) \begin{bmatrix} \tilde{R}_{21}(s) \\ 0 \end{bmatrix}.$$

Then, Q solves the problem (4.1) if and only if Q has the form

$$Q = \hat{U}_{12} \begin{bmatrix} \bar{Q}_{11} & \bar{Q}_{12} \\ \bar{Q}_{21} & \bar{Q}_{22} \end{bmatrix} \tilde{U}_{21}$$

where $\bar{Q}_{12}, \bar{Q}_{21}, \bar{Q}_{22}$ are any stable matrices and where $\bar{Q}_{11}$ is stable and is a solution to the non-fat problem

$$\left\| T_{11} - \hat{R}_{12} \bar{Q}_{11} \tilde{R}_{21} \right\|_\infty \leq 1 . \qquad \square$$

Proof: The proof follows immediately from the observation that the two sets {Q | Q stable} and

$$\left\{ \bar{Q} \mid \bar{Q} = \begin{bmatrix} \bar{Q}_{11} & \bar{Q}_{12} \\ \bar{Q}_{21} & \bar{Q}_{22} \end{bmatrix}, \bar{Q} \text{ stable} \right\}$$

are equivalent due to the properties of stability and minimum-phase of $\hat{U}_{12}(s)$ and $\tilde{U}_{21}(s)$. $\square$

Remark 4.1: Solving for $\bar{Q}_{11}$'s may be efficiently accomplished via the techniques of Doyle and co-workers (1988), Glover and Doyle (1988), Safonov and Limebeer (1988), and Glover and co-workers (1989). $\square$

Remark 4.2: Three fat-plant cases are distinguished: (i) Fat Actuator : dim $(u_2) >$ dim (y_1) and dim $(y_2) \leq$ dim (u_1), (ii) Fat Sensor : dim $(y_2) >$ dim (u_1) and dim $(u_2) \leq$ dim (y_1), and (iii) Doubly Fat : dim $(u_2) >$ dim (y_1) and dim $(y_2) >$ dim (u_1). For the fat-sensor case, $\hat{U}_{12}(s) = \hat{V}_{12}(s) = I$.

Likewise, in the fat-actuator case, $\tilde{U}_{21}(s) = \tilde{V}_{21}(s) = I.$ $\square$

Remark 4.3: By section 3, the matrices $U_{12}(s)$ and $U_{21}(s)$ in Lemma 4.1 are the squaring-down compensators for the matrices $T_{12}(s)$ and $T_{21}(s)$, respectively, as in Fig. 1.2. State-space realizations for $U_{12}(s)$ and $U_{21}(s)$ may be obtained via Theorem 2.1. $\square$

Remark 4.4: The RHP zeros of $\hat{R}_{12}(s)$ and $\tilde{R}_{21}(s)$ are identical to those of $T_{12}(s)$ and $T_{21}(s)$, respectively. $\square$

Motivated by the need for a state-space theory for the design of stable minimum-phase squaring-down compensators for multivariable feedback control systems, we have developed state-space formulas for rational matrix GCD computation and pointed out the role and use of rational matrix GCD's in addressing several control synthesis issues. We introduced the concept of an $R_{p\Omega}$-unimodular matrix , defined as an invertible, square rational transfer-function matrix that has the property that both it and its inverse are proper and analytic in Ω. Then, building upon and extending the work of Silverman and Van Dooren (1978), we developed state-space constructions for $R_{p\Omega}$-unimodular matrices $\tilde{U}(s)$ and $\tilde{V}(s)$ which, for a given tall matrix $N(s) \in M(R_{p\Omega})$, generate the GCRD $\tilde{R}(s)$ of $N(s)$, i.e.,

$$\tilde{U}(s)N(s) = \begin{bmatrix} U(s) \\ U_\perp(s) \end{bmatrix} N(s) = \begin{bmatrix} \tilde{R}(s) \\ 0 \end{bmatrix}$$

$$N(s) = \tilde{V}(s) \begin{bmatrix} \tilde{R}(s) \\ 0 \end{bmatrix} = \left[V(s) \mid V_\perp(s) \right] \begin{bmatrix} \tilde{R}(s) \\ 0 \end{bmatrix} .$$

The problem of matrix GCD computation has thus been reduced to a standard state-feedback stabilization or pole-placement problem where the region of analyticity Ω may be specified as desired. Further, our results also provide state-space realizations for bases and dual bases (in the sense of Forney (1975)) for rational vector spaces. Although the main developments in the paper center on the ring $R_{p\Omega}$ of proper rational Ω-stable functions, i.e., of lumped continuous-time or discrete-time systems, the algebraic nature of the results indicates that the technique may be similarly applied to other rings as well, e.g., the Hermite ring of distributed systems (see Vidyasagar (1985, Chpt. 8)).

In addressing the role and use of rational matrix GCD's in control synthesis, we have described the problem of designing dynamic squaring-down compensators for multivariable plants and have shown in section 3 that the synthesis of such compensators is equivalent to solving for U(s). A state-space realization of a stable minimum-phase squaring-down compensator U(s) is given in Theorem 3.1. Finally, in section 4, we have shown how the squaring-down results enable one to solve the problem of state-space H-infinity control synthesis for fat plants.

In this appendix, we briefly discuss the key results of Silverman (1976) and Wonham (1979) on, respectively, strongly observable systems and maximally unobservable systems. These results are central to the development of the state-space formulas for GCD computation in section 2. First, we state the following definitions.

Definition A.1: A state-feedback compensated system

$$G_F(s) \stackrel{s}{=} \left[\begin{array}{c|c} A + BF & B \\ \hline C + DF & D \end{array} \right]$$

is said to be maximally unobservable if F minimizes the rank of the observability matrix of (C+DF, A+BF). $\square$

Definition A.2: An observable system G(s) having dynamical equations

$$\begin{cases} \dot{x} = Ax + Bu \\ y = Cx + Du \end{cases}$$

is said to be strongly observable if it remains observable for every state feedback u = − Fx. $\square$

We observe that central to these defined notions are the use of state feedback for compensation and its effects on the system's observability. It is well known that the use of state feedback is mainly to achieve pole placement, i.e., to move the system's poles to desired location, which, in general, alters the system's observability. For the case of Def. A.1, the feedback F causes the system's observable subspace to have the smallest dimension whereas Def. A.2 gives the case of those systems whose observability, i.e., the dimension of the observable subspace, is invariant for any state feedback. It turns out that there is a key connection between the concepts of maximally unobservable systems and strongly observable systems, that is, any observable realization of a maximally unobservable system is strongly observable. This connection was exploited in Silverman (1976) to give the following results.

Consider a tall system G(s) having a minimal state-space realization

$$G(s) \stackrel{s}{=} \left[\begin{array}{c|c} A & B \\ \hline C & D \end{array} \right]$$

where $A \in \mathbb{C}^{n \times n}, B \in \mathbb{C}^{n \times m}, C \in \mathbb{C}^{r \times n}, D \in \mathbb{C}^{r \times m}$, and assume that D has orthonormal columns. Then,

<u>Lemma A.1:</u> (Silverman (1976)) G(s) is strongly observable if and only if it has no transmission zeros. □

<u>Lemma A.2:</u> (Silverman (1976)) $F = -D^*C$ renders $G_F(s)$ maximally unobservable. □

Adopting the notation (2.1) and (2.2), it is clear that the rows of the matrix $\tilde{T}_1$ span the rows of the observability matrix of $(\tilde{C}, \tilde{A})$ where

$$\tilde{A} \overset{\Delta}{=} A - BD^*C, \quad \tilde{C} \overset{\Delta}{=} D_\perp D_\perp^* C.$$

Then the following theorem by Silverman (1976) gives a complete characterization of state feedbacks $\tilde{F}$ which cause $G_F(s)$ to be maximally unobservable.

<u>Theorem A.1:</u> (Silverman (1976)) $G_F(s)$ is maximally unobservable if and only if $\tilde{F} = -D^*C + \tilde{F}_1\tilde{T}_1$ for some $\tilde{F}_1$. □

APPENDIX B

In this appendix we describe a transformation that accounts for the zeros at infinity of N(s), i.e., the case when the D-matrix of N(s) does not have full column rank at s = ∞, in the GCRD extraction problem. The algorithm uses the bilinear mapping $\tilde{s} = (\alpha s + \beta)/(\gamma s + \delta)$ to relocate troublesome zeros at s = ∞ to the finite point $\tilde{s} = \alpha/\gamma$. Let N(s) have the state-space realization

$$N(s) \overset{s}{=} \left[\begin{array}{c|c} A & B \\ \hline C & D \end{array} \right],$$

where $A \in \mathbb{C}^{n \times n}$, $B \in \mathbb{C}^{n \times m}$, $C \in \mathbb{C}^{r \times n}$, $D \in \mathbb{C}^{r \times m}$, and $r \geq m$. It is assumed a priori that N(s) have full normal rank m.

<u>Bilinear Mapping of the s-Plane:</u> (Kailath (1980), Safonov (1987)) For $\gamma \neq 0$ and $\alpha\delta - \beta\gamma \neq 0$, the transformation $\tilde{s} = (\alpha s + \beta)/(\gamma s + \delta)$ maps the zeros at s = ∞ of N(s) to the point $\tilde{s} = \alpha/\gamma$ to give

$$N'(\tilde{s}) = C'(I\tilde{s} - A')^{-1}B' + D'$$

$$\overset{s}{=} \left[\begin{array}{c|c} (\beta I + \alpha A)(\delta I + \gamma A)^{-1} & (\alpha\delta - \beta\gamma)C(\delta I + \gamma A)^{-1}B \\ \hline C(\delta I + \gamma A)^{-1} & D - \gamma C(\delta I + \gamma A)^{-1}B \end{array} \right]$$

such that D' has full column rank. Solving the GCRD problem for $N'(\tilde{s})$ via Theorem 2.1, one obtains

$$\tilde{U}'(\tilde{s}) \cdot N'(\tilde{s}) = \left[\begin{array}{c} \tilde{R}'(\tilde{s}) \\ 0 \end{array} \right]$$

which, via the inverse conformal mapping $s = (\delta\tilde{s} - \beta)/(\alpha - \gamma\tilde{s})$, readily yields the desired solutions $\tilde{U}(s)$ and $\tilde{R}(s)$. □

REFERENCES

Aplevich, J.D. (1981). On the squaring problem in feedback design. <u>Proc. Joint Automatic Control Conf.</u>, Charlottesville, VA.

Chiang, R.Y. and Safonov, M.G. (1988). <u>Robust Control Toolbox</u>, South Natick, MA: Mathworks, 1988.

Davison, E.J. (1983). Some properties of minimum phase systems and 'squared-down' systems. <u>IEEE Trans. Automat. Contr.</u>, vol. AC-28, no. 2, pp. 221-222.

Delsarte, P.R., Genin, Y. and Kamp, Y (1979). The Nevanlinna-Pick problem for matrix-valued functions. <u>SIAM J. Appl. Math</u>, 36, pp. 47-61.

Desoer, C.A., Liu, R.W., Murray, J. and Saeks, R. (1980). Feedback system design: The fractional approach to analysis and synthesis. <u>IEEE Trans. Automat. Contr.</u>, vol. AC-25, pp. 399-412.

Doyle, J.C. (1984). <u>Lecture Notes in Advances in Multivariable Control</u>, ONR/Honeywell Workshop, Minneapolis.

Doyle, J.C., Glover, K., Khagonekar, P. and Francis, B.A. (1988). State-space solutions to standard H_2 and H_∞ control problems. <u>Proc. of American Control Conf.</u>, Atlanta, GA.

Forney, Jr., G.D.(1975). Minimal bases of rational vector spaces, with applications to multivariable linear systems. <u>SIAM J. Control</u>, vol. 13, no. 3, pp. 493-520.

Francis, B.A. (1987). A course in H∞ control. <u>Lecture Notes in Control and Information Sciences</u>, vol. 88, Berlin: Springer-Verlag.

Glover, K. and Doyle, J. (1988). State-space formulae for all stabilizing controllers that satisfy an H∞ norm bound and relations to risk sensitivity. <u>Systems and Control Letters</u>, vol. 11, pp. 167-172.

Glover, K., Limebeer, D.J.N., Doyle, J.C., Kasenally, E.M., Safonov, M.G. (1989). A characterization of all solutions to the four block general distance problem. Preprint.

Jacobson, N. (1953). <u>Lectures in Abstract Algebra</u>, vol. 1, New York: Van Nostrand.

Kailath, T. (1980). <u>Linear Systems</u>, Englewood Cliffs, N.J.: Prentice Hall.

Kalman, R.E. (1963). Mathematical description of linear dynamical systems. <u>SIAM J. Control</u>, vol. 1, pp. 152-192.

Kouvaritakis, B. and MacFarlane, A.G.J. (1976). Geometric approach to analysis and synthesis of system zeros: Part 2, nonsquare systems. <u>Int. J. Contr.</u>, vol. 23, no. 2, 167-181.

MacFarlane, A.G.J. and Karcanias, N. (1976). Poles and zeros of linear multivariable systems: a survey of the algebraic, geometric and complex-variable theory. <u>Int. J. Contr.</u>, vol. 24, no. 1, pp. 33-74.

MacFarlane, A.G.J. and Kouvaritakis, B. (1977). A design technique for linear multivariable feedback systems. <u>Int. J. Contr.</u>, vol. 25, pp. 837-879.

Moore, B.C. and Silverman, L.M. (1974). A time domain characterization of the invariant factors of a system transfer function. <u>Proc. of Joint Automatic Control Conf.</u>, Austin, TX.

Nett, C.N., Jacobson, C.A.and Balas, M.J. (1984). "A connection between state-space and doubly coprime fractional representations," <u>IEEE Trans. Automat. Contr.</u>, vol. AC-29, pp. 831-832.

Rosenbrock, H.H. (1970). <u>State Space and Multivariable Theory</u>, London: Nelson.

Rosenbrock, H.H. (1974). <u>Computer-Aided Control System Design</u>, New York: Academic.

Saberi, A. and Sannuti, P. (1988). "Squaring down by static and dynamic compensators," <u>IEEE Trans. Automat. Contr.</u>, vol. 33, no. 4, pp. 358-365.

Safonov, M.G. (1987). "Imaginary-axis zeros in multivariable H∞ optimal control," <u>Modeling, Robustness and Sensitivity Reduction in Control Systems</u>, R.F. Curtain, Ed., Berlin: Springer Verlag.

Safonov, M.G., Jonckheere, E.A., Verma, M. and Limebeer, D.J.N. (1987a). "Synthesis of positive real multivariable feedback systems," <u>Int. J. Contr.</u>, vol. 45, no. 3, pp. 817-842.

Safonov, M.G., Chiang, R.Y. and Limebeer, D.J.N. (1987b). "Hankel model reduction without balancing: A descriptor approach," <u>Proc. IEEE Conf. on Decision and Control</u>, Los Angeles, CA.

Safonov, M.G. and Chiang, R.Y. (1988). "A Schur method for balanced model reduction," <u>Proc. American Control Conf.</u>, Atlanta, GA.

Safonov, M.G. and Limebeer, D.J.N. (1988). "Simplifying H∞ Control Theory Via Loop Shifting," <u>Proc. IEEE Conf. on Decision and Control</u>, Austin, TX.

Sannuti, P. and Saberi, A. (1987). "A special coordinate basis of multivariable linear systems: finite and infinite zero structure, squaring down, and decoupling," <u>Int. J. Contr.</u>, vol. 45, no. 5, pp. 1655-1704.

Silverman, L.M. (1969). "Inversion of multivariable systems," <u>IEEE Trans. on Automat. Contr.</u>, vol. AC-14, pp. 270-276.

Silverman, L.M. (1976). "Discrete Riccati equations: alternative algorithm, asymptotic properties, and system theory interpretations," <u>Control and Dynamic Systems</u>, C.T. Leondes, Ed., New York: Academic Press.

Silverman, L.M. and Van Dooren, P. (1978). "A system theoretic approach for GCD extraction," <u>Proc. IEEE Conf. on Decision and Control</u>, San Diego, CA.

Vardulakis, A.I.G. (1980). "Zero placement and the 'squaring down' problem: a polynomial matrix approach," <u>Int. J. Contr.</u>, vol. 31, no. 5, pp. 821- 832.

Vidyasagar, M., Schneider, H. and Francis, B.A. (1982). "Algebraic and topological aspects of feedback stabilization," <u>IEEE Trans. Automat. Contr.</u>, vol. AC-27, pp. 880-894.

Vidyasagar, M. (1985). <u>Control System Synthesis: A Factorization Approach</u>, Cambridge, MA: M.I.T. Press.

Wolovich, W. (1974). <u>Linear Multivariable Control</u>, New York: Springer-Verlag.

Wonham, W.M. (1979). <u>Linear Multivariable Control: A Geometric Approach</u>, New York: Springer-Verlag.

Youla, D.C., Jabr, H.A. and Bongiorno, Jr., J.J. (1976). "Modern Wiener-Hopf design of optimal controllers, Part 2: The multivariable case," <u>IEEE Trans. Automat. Contr.</u>, vol. AC-21, pp. 319-338.

INVERTIBILITY AND INVERSION OF LINEAR PERIODIC SYSTEMS

A. M. Perdon*, G. Conte and S. Longhi*****

*Dipartimento di Metodi e Modelli Matematici per le Scienze Applicate, Università di
Padova, Padova, Italy
**Dipartimento di Matematica, Università di Genova, Genova, Italy
***Dipartimento di Elettronica e Automatica, Università di Ancona, Ancona, Italy*

Abstract. In this paper the invertibility problem is solved for periodic discrete-time systems. Both existence
conditions and synthesis procedures of a left or a right inverse system are given.

Keywords. Inverse systems, Periodic systems, Discrete time systems, Multivariable systems, Time-varying
systems, System theory.

INTRODUCTION

Linear discrete-time periodic systems are useful in modelling
several processes as well as in designing controllers for other
classes of systems and their study has received an increasing
attention in the last years (see e.g. Araki and Yamamoto, 1986;
Bittanti, 1986; Francis and Georgiou, 1988; Grasselli et al.,
1979,1980; Grasselli and Longhi, 1983; Kaczorek, 1985;
Khargonecar et al., 1985; Liu and Kuo, 1979; Meyer and
Burrus, 1975; Olbrot, 1987; Richards, 1983; Willems et al.,
1984). In particular, algebraic and, more recently, geometric
characterizations of structural properties have been anlyzed and
applied to solving the state estimation problem and various
control problem, such as the eigenvalue assignment problem,
the disturbance decoupling problem, the output regulator
problem, the model matching problem (see e.g. Bittanti and
Bolzern, 1985; Bolzern et al., 1986; Conte et al., 1988, 1989;
Grasselli, 1984; Grasselli and Lampariello, 1981; Grasselli and
Longhi, 1986, 1987, 1988a,b, 1989; Kono, 1980, 1983;
Longhi et al., 1989; Verriest, 1988).

The aim of this paper is to study the left and the right
invertibility of linear discrete-time periodic systems and to give
a characterization of these properties in terms of the structural
indices described in (Longhi et al., 1989). Roughly speaking,
we obtain an extension to the periodic framework of the results
holding in the time-invariant case which relate the left or right
invertibility to the presence of as many zeros at infinity as the
number of inputs or, respectively, as the number of outputs.
Furthermore, using the Periodic Structure Algorithm
introduced in (Longhi et al, 1989), we obtain a synthesis
procedure for constructing the left or the right inverse, if any,
of a given periodic system.

The paper is organized as follows. In Section 1 we recall the
notion of Periodic Structure at infinity introduced in (Longhi et
al, 1989) and the steps of Periodic Structure Algorithm. In
Section 2 we describe the notions of left and right invertibility,
we are dealing with and, in the Main Theorem, we characterize
them in terms of the indices $q_r(k)$ obtained by the Periodic
Structure Algorithm. Then, we provide the construction of a
left or a right inverse in the case in which the necessary and
sufficient conditions for its existence are satisfied. Finally, in
Section 3 we analyze a numerical example.

PRELIMINARIES AND NOTATIONS

We consider linear discrete time periodic systems Σ of the form

$$x(k+1) = A(k)\, x(k) + B(k)\, u(k) \qquad (1a)$$

$$y(k) = C(k)\, x(k) \qquad (1b)$$

where $k \in Z$, $x(.) \in X := \mathbb{R}^n$ is the state, $u(.) \in U := \mathbb{R}^m$ is the
input, $y(.) \in Y := \mathbb{R}^p$ is the output and A(.), B(.) and C(.) are
periodic matrices of suitable dimensions, with period ω and real
coefficients.

Generalizing the classical Silverman's Algorithm (Silverman,
1969), a structure algorithm for systems of the above kind,
called the Periodic Structure Algorithm, has been developed in
(Longhi et al., 1989). The algorithm reads as follows:

Periodic Structure Algorithm .
Given a system Σ of the form (1) set, preliminarily, $y_0(k) =
y(k)$, $C_0(k) = C(k)$, $q_0(k) = 0$ and $d_0(k) = p$ (= dim Y) for all k.

Step 1)
Write $M_0(k) = [\, 0 \quad I_{d_0(k+1)} \,]$ and compute

$$M_0(k) \begin{bmatrix} y_0(k) \\ y_0(k+1) \end{bmatrix} = y_0(k+1) = C_0(k+1)x(k+1)$$
$$= C_0(k+1)A(k)x(k)+C_0(k+1)B(k)u(k).$$

Define $q_1(k) := \operatorname{rank} C_0(k+1)B(k)$.
If $S_1(k)$ denotes a nonsingular matrix such that

$$D_1(k) = S_1(k)C(k+1)B(k) = \begin{bmatrix} D_1(k) \\ 0 \end{bmatrix} \begin{matrix} \} q_1(k)\ \text{indp.rows} \\ \} d_0(k+1)+q_0(k)-q_1(k) \end{matrix}$$

set

$$y_1(k)=S_1(k)M_0(k)\begin{bmatrix} y_0(k) \\ y_0(k+1) \end{bmatrix}=\begin{bmatrix} C_1(k) \\ \tilde{C}_1(k) \end{bmatrix}x(k) +\begin{bmatrix} D_1(k) \\ 0 \end{bmatrix}u(k).$$

Step i)
Let $\quad y_{i-1}(k) = \begin{bmatrix} C_{i-1}(k) \\ \tilde{C}_{i-1}(k) \end{bmatrix}x(k) + \begin{bmatrix} D_{i-1}(k) \\ 0 \end{bmatrix}u(k)\quad$ and set

$$M_{i-1}(k) = \begin{bmatrix} I_{q_{i-1}(k)} & 0 & 0 & 0 \\ 0 & & 0 & 0 & I_{d_{i-1}(k+1)} \end{bmatrix},$$

where $d_{i-1}(k) = d_{i-2}(k+1)+q_{i-2}(k)-q_{i-1}(k)$. Then compute

$$M_{i-1}(k)\begin{bmatrix} y_{i-1}(k) \\ y_{i-1}(k+1) \end{bmatrix} = \begin{bmatrix} C_{i-1}(k)x(k) \\ \tilde{C}_{i-1}(k+1)x(k+1) \end{bmatrix} + \begin{bmatrix} D_{i-1}(k)u(k) \\ 0 \end{bmatrix}$$
$$= \begin{bmatrix} C_{i-1}(k) \\ \tilde{C}_{i-1}(k+1)A(k) \end{bmatrix}x(k) + \begin{bmatrix} D_{i-1}(k) \\ \tilde{C}_{i-1}(k+1)B(k) \end{bmatrix}u(k)$$

and define $q_i(k) := \operatorname{rank} \begin{bmatrix} D_{i-1}(k) \\ \tilde{C}_{i-1}(k+1)B(k) \end{bmatrix}$.

If $S_i(k)$ denotes a nonsingular matrix such that

283

$$D_i(k) = S_i(k) \begin{bmatrix} D_{i-1}(k) \\ \tilde{C}_{i-1}(k+1)B(k) \end{bmatrix} = \begin{bmatrix} D_i(k) \\ 0 \end{bmatrix} \begin{matrix} \} \; q_i(k) \; \text{indp.rows} \\ \} \; d_i(k) \end{matrix} \; ,$$

define

$$y_i(k) = S_i(k)M_{i-1}(k) \begin{bmatrix} y_{i-1}(k) \\ y_{i-1}(k+1) \end{bmatrix} = \begin{bmatrix} C_i(k) \\ \tilde{C}_i(k) \end{bmatrix} x(k) + \begin{bmatrix} D_i(k) \\ 0 \end{bmatrix} u(k).$$

(Remark that dim $y_i(k)$ is not constant but, as the dimensions of $M_{i-1}(k)$, it depends on k.) ❑

The above algorithm stops, say at step $r \le n\omega$, when no more independent rows in the matrix which multiplies u(k) can be obtained. This implies in particular that $q_{r+i}(k) = q_r(k)$ for all k and all $i \ge 1$, and the r-tuple of integer valued periodic functions $(q_1(k),...,q_r(k))$ turns out to contain, as described in (Longhi et al, 1989; Conte et al., 1989), a fundamental structural information on Σ. More precisely, assuming that the columns of the input matrix B(k) in (1) are independent for all k, the difference $q_r(k) - q_i(k)$ represents the analoguous, with respect to the time invariant case, of the number of zeros at infinity of order greater than i and $q_r(k)$ represents the analogous of the number of zeros at infinity. If we set

$$p_{i+1}(k+1) := q_r(k) - q_i(k) \qquad \text{for } i \ge 0 \qquad (2)$$

we have:

Definition 1. (Longhi et al., 1989, Def. 1.5) *The r-uple of integer valued ω-periodic functions $(p_1(k),...,p_r(k))$, defined by means of (2), is called the periodic structure at infinity of the periodic system Σ.*

Remark 1. The $p_i(k)$ can be also defined directly, as actually has been done in (Longhi et al. 1989), in terms of the Periodic Invariant Subspace Algorithm (Grasselli and Longhi, 1986). More precisely, given a system Σ of the form (1), if $\{X = V^0(k), \; \text{Ker } C(k) = V^1(k), \; V^2(k), \; ....., \; V^r(k) = V^*(k)\}$ is the sequence of subspaces generated by such algorithm which converges to the maximum periodic controlled subspace $V^*(k)$ contained in Ker C(k), we have

$$p_{i+1}(k) = \dim \frac{\text{Im } B(k-1) \cap V^i(k)}{\text{Im } B(k-1) \cap V^*(k)} \qquad \forall k \in Z \qquad (3)$$

(see (Longhi et al., 1989) for details). ❑

A different characterization of the $q_i(k)$'s which will be used in the following is given in terms of Toeplitz matrices associated to Σ. Let us consider the periodic matrices $T_i(k)$ defined recursively as

$T_0(k)$= empty matrix (for consistency in formulas)

$$T_i(k) = \begin{bmatrix} C(k+1)B(k) & \vline & 0 \\ C(k+2)A(k+1)B(k) & \vline & \\ " & \vline & T_{i-1}(k+1) \\ C(k+i)A(k+i-1)...A(k+1)B(k) & \vline & \end{bmatrix} \; \text{for } i \ge 1$$

then we have:

Proposition 1. (Longhi et al. 1989, Prop. 2.5) *The following relation holds for all $i \ge 1$ and all $k \in Z$:*

$$q_i(k) = \text{rank } T_i(k) - \text{rank } T_{i-1}(k+1) \; . \qquad (4)$$

LEFT AND RIGHT INVERTIBILY

The left and right invertibility of periodic systems of the form (1) have been considered by Kono (1981, 1983). In this section, after stating suitable definitions which generalize slightly those of Kono (1981, 1983), we characterize the left and right invertibility in terms of the structural indices $q_i(k)$'s.

Moreover, we describe a conctruction of a left or right inverse based on the Periodic Structure Algorithm. In the following the symbol Z^+ denotes the set of non-negative integers.

Definition 2. *System Σ is said to left invertible at time k, if for every $h \in Z^+$ there exist an integer $\sigma \in Z^+$ such that the input u(.) is uniquely determined over the interval [k-h-1,k] by the knowledge of the initial state x(k-h-1) and of the output y(.) over the interval [k-h,k+σ].*

Definition 3. *System Σ is said to right invertible at time k, if for every $h \in Z^+$, $x_0 \in X$ and reference function $w(.) \in Y$ defined over the interval [k,k+h+1] there exist an integer $\sigma \in Z^+$ and an input u(.) defined over the interval [k-σ,k+h] such that for the initial state x(k-σ)=x_0 the output y(j) = w(j) for all $j \in [k,k+h+1]$.*

Remark 2. The above definitions coincide with those given by Kono (1981, 1983) when the initial state is assumed to be zero. The notion of right and left invertibility here introduced coincide, respectively with the notion of output function reproducibility (Brockett and Mesarovic`, 1965) and input function observability (Wolovich, 1974). ❑

Remark 3. Since the left (right) invertibility at time k is easily seen to imply the left (right) invertibility at time k-j (k+j) for any $j \in Z^+$ and since Σ is periodic,the notion of left (right) invertibility is actually independent on k (compare with (Kono, 1983)). In the following we will therefore speak simply of left (right) invertibility without any time specification. ❑

The main result of the section is the following.

Theorem 1. *A system Σ of the form (1) is*
(i) *left invertible if and only if $q_r(k) = m$ for all integer k,*
(ii) *right invertible if and only if $q_r(k) = p$ for all integer k.*

Remark 4. In the linear time-invariant case, the left and the right invertibility of a system Σ are characterized by the fact that its transfer matrix has full column or full row rank. This is equivalent to having m zeros at infinity, or p zeros at infinity, respectively. The above conditions appear therefore to be the extension to the periodic framerwork of those holding in the time-invariant case (Conte et al., 1989). ❑

Proof of Theorem 1. (i) The necessary and sufficient condition given in (Kono, 1983) for left invertibility with zero initial state becames in our notation

$$\text{rank} T_{(n+1)\omega}(k) - \text{rank} T_{n\omega}(k) = m\omega \qquad (5)$$

for any arbitrary integer k. Since

$$\text{rank} T_{(n+1)\omega}(k) - \text{rank} T_{n\omega}(k) = q_{(n+1)\omega-1}(k) \\ + q_{(n+1)\omega-2}(k+1) + ... + q_{n\omega}(k+\omega-1) \qquad (6)$$

and $n\omega \ge r$, the equality (5) is equivalent to $q_r(k)=m$ for all k. Furthmore, the assumption of zero initial state is easily seen to be unnecessary.
(ii) The necessary and sufficient condition given in (Kono, 1983) for right invertibility with zero initial state becames in our notation

$$\text{rank} T_{(n+1)\omega}(k) - \text{rank} T_{n\omega}(k) = p\omega \qquad (7)$$

for any arbitrary integer k. Therefore, by relation (6) the equality (7) is equivalent to $q_r(k)=p$ for all k. Also in this case, the assumption of zero initial state is easily seen to be unnecessary. ❑

Construction of a left inverse.
Let a system Σ of the form (1) be given and assume that $q_r(k)$ equals m, the number of inputs, for all k. Then, the final step of the Periodic Structure Algorithm yields

$$y_r(k) = \begin{bmatrix} C_r(k) \\ \tilde{C}_r(k) \end{bmatrix} x(k) + \begin{bmatrix} D_r(k) \\ 0 \end{bmatrix} u(k)$$

where $D_r(k)$ is a square non singular matrix for all k. Denoting by $y_{r,m}(k)$ the vector consisting of the first m components of $y_r(k)$, we can write

$$u(k) = D_r^{-1}(k)[\, y_{r,m}(k) - C_r(k)x(k)\,]$$

Remark that $y_{r,m}(k)$ depends on the outputs $y(k), \ldots, y(k+r)$ of Σ. More precisely, $y_{r,m}(k)$ can be obtained making use of the following relation

$$\begin{bmatrix} y_r(k+h\omega) \\ y_r(k+1+h\omega) \\ \ldots \\ y_r(k+\omega-1+h\omega) \end{bmatrix} = \prod_{i=1}^{r} K_{i,k}H_{i,k} \begin{bmatrix} y(k+h\omega) \\ y(k+1+h\omega) \\ \ldots \\ y(k+\omega-1+h\omega) \end{bmatrix} \quad \forall k \in \mathbf{Z}^+ \quad (8)$$

where the matices $K_{i,k}$ and $H_{i,k}$ (i=1,...,r) are defined as follows

$$K_{i,k} = \text{diag}\{\, S_i(k),\, S_i(k+1),\, \ldots ,\, S_i(k+\omega-1)\, \}$$

$S_i(k)$ being the matrix introduced at the i-th step of the Periodic Structure Algorithm,

$$H_{i,k} = \begin{bmatrix} M^\circ_{i-1}(k) & & 0 \\ 0 & \text{diag}\left\{\begin{bmatrix} M''_{i-1}(k+1) \\ M^\circ_{i-1}(k+1) \end{bmatrix}; \cdots ; \begin{bmatrix} M''_{i-1}(k+\omega-1) \\ M^\circ_{i-1}(k+\omega-1) \end{bmatrix}\right\} \\ M''_{i-1}(k)\delta^\omega & & 0 \end{bmatrix}$$

where δ is the forward shift operator, and for all integer k:

$$M^\circ_0(k) := \text{empty matrix} , \quad M''_0(k) := I_{d_0(k)}$$

$$M^\circ_j(k) := [\, I_{q_j(k)}\ 0\,], \quad M''_j(k) := [\, 0\ I_{d_j(k)}\,] \quad j=1,\ldots, r-1$$

with $q_{j-1}(k) + d_{j-1}(k+1)$ columns in $M^\circ_j(k)$ and $M''_j(k)$. The dimensions of $S_i(.)$, $M^\circ_{i-1}(.)$ and $M''_{i-1}(.)$ are in general time-varying, while the dimensions of $K_{i,k}$ and $H_{i,k}$ are time-invariant. In particular, $\dim K_{i,k} = \dim H_{i,k} = \omega p \times \omega p$.

Now, the system Σ_L

$$z(k+1) = (A(k)-B(k)D_r^{-1}(k)C_r(k))z(k)+B(k)D_r^{-1}(k)y_{r,m}(k) \quad (9a)$$

$$u(k) = -D_r^{-1}(k)C_r(k)z(k) + D_r^{-1}(k)y_{r,m}(k) \quad (9b)$$

is a left inverse of Σ which, initialized at $z(k-h-1) = x(k-h-1)$ with input $y_{r,m}(j)$ for $j \in [k-h-1,k]$, produces as output $u(j)$ for $j \in [k-h-1,k]$. $\quad\Box$

<u>Construction of a right inverse.</u>
Let a system Σ of the form (1) be given and assume that $q_r(k)$ equals p, the number of outputs, for all k. Then, the final step of the Periodic Structure Algorithm yields

$$y_r(k) = C_r(k)x(k) + D_r(k)u(k) \quad \forall k \in \mathbf{Z} \quad (10)$$

where the rank of $D_r(k)$ is p for all integer k and the dimension of $y_r(.)$ are time-invariant and equals to p. Define a ω-periodic subspace of U:

$$U_2(k) := \text{Ker } D_r(k) \quad \forall k \in \mathbf{Z}$$

and decompose U as $U = U_1(k) \oplus U_2(k)$, with $U_1(.)$ ω-periodic. Then, assuming as basis of U at time k the union of ω-periodic basis of the $U_1(k)$ and $U_2(k)$, we have

$$D_r(k) = [\, \bar{D}_r(k)\ \ 0\,] \quad \forall k \in \mathbf{Z}$$

where $\bar{D}_r(k)$ is non-singular for all integer k. Hence, the relation (10) takes the form

$$y_r(k) = C_r(k)x(k) + \bar{D}_r(k)u_1(k) \quad \forall k \in \mathbf{Z}$$

where $u_1(k) \in U_1(k)$. Denote by $w_r(k)$ the function defined by the following relation

$$\begin{bmatrix} w_r(k+h\omega) \\ w_r(k+1+h\omega) \\ \ldots \\ w_r(k+\omega-1+h\omega) \end{bmatrix} = \prod_{i=1}^{r} K_{i,k}H_{i,k} \begin{bmatrix} w(k+h\omega) \\ w(k+1+h\omega) \\ \ldots \\ w(k+\omega-1+h\omega) \end{bmatrix} \quad \forall k \in \mathbf{Z}^+$$

where $w(.) \in Y$ is the reference function. Remark that $w_r(k)$ depends on the reference function values $w(k), \ldots, w(k+r)$.

Then, the system Σ_R

$$z(k+1) = (A(k)-B(k)\bar{D}_r^{-1}(k)C_r(k))z(k)+B(k)\bar{D}_r^{-1}(k)w_r(k)$$

$$\bar{u}(k) = -\bar{D}_r^{-1}(k)C_r(k)z(k) + \bar{D}_r^{-1}(k)w_r(k)$$

is a right inverse of Σ. In fact, system Σ with initial state $x(k-r) = x_0$ and input $\bar{u}(j)$ for $j \in [k-r,k+h]$ produces an output $y(i) = w(i)$ for $i \in [k,k+h+1]$, where $\bar{u}(.)$ is the output of Σ_R initialized at $z(k-r) = x_0$. $\quad\Box$

A NUMERICAL EXAMPLE

Consider as a numerical example the system $\Sigma = (A(k), B(k), C(k))$ proposed by Kono (1983), where $n=\omega=2$, $p=m=1$ and

$$A(0) = \begin{bmatrix} 1 & 0 \\ 0 & 1 \end{bmatrix}, \ A(1) = \begin{bmatrix} 0 & 1 \\ 1 & 0 \end{bmatrix}; \ B(k) = \begin{bmatrix} 1 \\ 0 \end{bmatrix} \ k=0,1$$

$$C(0) = [\, 1\ 0\,], \ C(1) = [\, 0\ 1\,]$$

The system Σ is not regular in the sense of (Silverman, 1969), since the rank of $C(k+1)B(k)$ is time-varying. Applying the Periodic Structure Algorithm, at the first step one has:

$$y_1(0) = [\, 0\ 1\,]x(0), \qquad y_1(1) = [\, 0\ 1\,]x(1) + u(1)$$

$q_1(0) = 0$, $q_1(1) = 1$ and $S_1(k) = 1$.

At the second step, from $d_1(k) = d_0(k+1)+q_0(k)-q_1(k)$ it follows that $d_1(0) = 1$, $d_1(1) = 0$,

$$M_1(0) = \text{empty matrix}, \quad M_1(1) = \begin{bmatrix} 1 & 0 \\ 0 & 1 \end{bmatrix}$$

$$M_1(0) \begin{bmatrix} y_1(0) \\ y_1(1) \end{bmatrix} = y_2(0) = \text{empty vector}$$

$$M_1(1) \begin{bmatrix} y_1(1) \\ y_1(2) \end{bmatrix} = \begin{bmatrix} 0 & 1 \\ [0\ 1]A(1) \end{bmatrix}x(1) + \begin{bmatrix} 1 \\ [0\ 1]B(1) \end{bmatrix}u(1)$$
$$= \begin{bmatrix} 0 & 1 \\ 1 & 0 \end{bmatrix}x(1) + \begin{bmatrix} 1 \\ 0 \end{bmatrix}u(1) = y_2(1)$$

$q_2(0) = 0$, $q_1(1) = 1$, $S_2(0) = $ empty matrix and $S_2(1) = I_2$.

At the third step, one has: $d_2(0) = 0$, $d_2(1) = 1$,

$$M_2(0) = [\, 0\ 1\,], \quad M_2(1) = [\, 1\ 0\,]$$

$$M_2(0) \begin{bmatrix} y_2(0) \\ y_2(1) \end{bmatrix} = [\, 1\ 0\,]A(0)x(0) + [\, 1\ 0\,]B(0)u(0)$$
$$= [\, 1\ 0\,]x(0) + u(0) = y_3(0)$$

$$M_2(1) \begin{bmatrix} y_2(1) \\ y_2(2) \end{bmatrix} = [\, 0\ 1\,]x(1) + u(1) = y_3(1)$$

$q_3(k) = 1$ and $S_3(k) = 1$ for $k=0,1$.

Then, by Theorem 1 with $r = 3$, the system Σ is left and right invertible, as stated already in (Kono, 1983). Now, by the procedures introduced above, it is possible to construct a left and right inverse of Σ.

For $r = 3$, one has $C_r(0) = [\ 1\quad 0\ ]$, $C_r(1) = [\ 0\quad 1\ ]$, $D_r(k) = 1$ for all $k \in Z$ and the system Σ_L defined by (9) has the following form:

$$z(k+1) = A^L(k)z(k) + B^L(k)y_3(k)$$

$$\bar{u}(k) = C^L(k)z(k) + D^L(k)y_3(k)$$

where

$$A^L(0) = \begin{bmatrix} 0 & 0 \\ 0 & 1 \end{bmatrix}, \quad A^L(1) = \begin{bmatrix} 0 & 0 \\ 1 & 0 \end{bmatrix}; \qquad B^L(k) = B(k) \quad k=0,1$$

$$C^L(0) = [\text{-}1\ 0\ ], \quad C^L(1) = [\ 0\ \text{-}1\ ].$$

Moreover, by relation (8)

$$\begin{bmatrix} y_3(0) \\ y_3(1) \end{bmatrix} = \begin{bmatrix} 0 & \delta^2 \\ \delta^2 & 0 \end{bmatrix} \begin{bmatrix} y(0) \\ y(1) \end{bmatrix} = \begin{bmatrix} y(3) \\ y_3(2) \end{bmatrix} \tag{11a}$$

or, equivalently,

$$\begin{bmatrix} y_3(1) \\ y_3(2) \end{bmatrix} = \begin{bmatrix} 0 & 1 \\ \delta^4 & 0 \end{bmatrix} \begin{bmatrix} y(1) \\ y(2) \end{bmatrix} = \begin{bmatrix} y(2) \\ y_3(5) \end{bmatrix} \tag{11b}$$

The time-varying bank of forward shifters defining (11) has the form:

$$y_3(k) = \begin{cases} y(k+2) & \text{for all even integer } k \\ y(k+1) & \text{for all odd integer } k \end{cases}$$

For example, with initial state $x(0) = [\ 1\quad 2\ ]'$ a possible input/output sequence of Σ has the following form:

k	0	1	2	3	4	5	6	7	8	9	10	11
u(k)	1	2	0	-1	-1	-1	-2	0	1	1	1	-
y(k)	1	2	4	2	1	4	3	0	0	1	2	1

From this sequence of $y(.)$, the system Σ_L with initial state $z(0) = [\ 1\quad 2\ ]$ and the bank of forward shifter defined by (11) produces the following sequences

k	0	1	2	3	4	5	6	7	8	9	10	11
$y_3(k)$	2	4	4	1	0	3	1	0	1	2	-	-
$\bar{u}(k)$	1	2	0	-1	-1	-1	-2	0	1	1	-	-

where $\bar{u}(i) = u(i)$ for $i \in [0,9]$.

CONCLUSIONS

Necessary and sufficient conditions have been given for the left or right invertibility of a linear periodic discrete-time system, together with synthesis procedures of a left or a right inverse. These results allow to solve the output function reproducibility and the input function observability of a periodic system.

REFERENCES

Araki, M and K. Yamamoto (1986). Multivariable multirate sampled-data systems: state space description, transfer characteristics, and Nyquist criterion, IEEE Trans. Aut. Control, 31, 145-154.

Bittanti, S. (1986). Deterministic and stochastic linear periodic systems, in Bittanti, S. (Ed.), Time Series and Linear Systems, Springer-Verlag, Berlin, pp. 141-182.

Bittanti, S. and P. Bolzern (1985). Reachability and controllability of discrete time linear periodic systems, IEEE Trans. Autom. Control, 30.

Bolzern, P, P. Colaneri and R. Scattolini (1986). Zeros of discrete-time linear periodic systems, IEEE Trans. Autom. Control, 31, 1057-1058.

Brockett, R.W. and M.D. Mesarovic (1965). The reproducibility of multivariable systems, J. Math. Analysis and Applications, 11, 548-563.

Conte, G., S. Longhi and A. Perdon (1988). Structure at infinity for discrete time linear periodic systemss, 27th IEEE Conf. on Decision and Control, Austin (Texas), 902-903.

Conte, G., A. Perdon and S. Longhi (1989). Zero structures at infinity of linear periodic systems, (submitted).

Francis, B.A. and T.T. Georgiou (1988). Stability theory for linear time-invariant plants with periodic digital controllers, IEEE Trans. Autom. Control, 33, (9), 820-832.

Grasselli, O.M. (1984). A canonical decomposition of linear periodic discrete- time systems , Int. J. Control, 40, (1), 201-214.

Grasselli, O.M., A. Isidori and F. Nicolo'(1979). Output regulation of a class of bilinear systems under constant disturbances, Automatica, 15, 189-195.

Grasselli, O.M., A. Isidori and F. Nicolo' (1980). Dead-beat control of discrete-time bilinear systems, Int. J. Control, 12, (1), 31-39.

Grasselli, O.M. and F. Lampariello (1981). Dead-beat control of linear periodic discrete-time systems , Int. J. Control, 33, (6), 1091-1106.

Grasselli, O.M. and S. Longhi (1983). On the stabilization of a class of bilinear systems , Int. J. Control, 37, (2), 413-420.

Grasselli, O.M. and S. Longhi (1986). Disturbance localization with dead-beat control for linear periodic discrete-time systems, Int. J. Control, 44, (5), 1319-1347.

Grasselli, O.M. and S. Longhi (1987). Linear function dead-beat observers with disturbance localization for linear periodic discrete-time systems, Int. J. Control, 45, (5), 1603-1627.

Grasselli, O.M. and S. Longhi (1988a). Disturbance localization with dead-beat control by measurement feedback for linear periodic discrete-time systems, Automatica, 24, (3), 375-385.

Grasselli, O.M. and S. Longhi (1988b). Zeros and poles of linear periodic multivariable discrete-time systems, Circuits Systems Signal Processing, 7, (3), 361-380.

Grasselli, O.M. and S. Longhi (1989). Eigenvalue assignment for linear periodic discrete-time non-reachable systems. Proc. IEEE Inter. Conf. on Control and Appl., Jerusalem (Israel), paper no. RA-2-4.

Kaczorek, T. (1985). Pole placement for linear discrete-time systems by periodic output-feedback, Syst. Control Lett., 6, 267-269.

Khargonekar, P.P., K. Polla and A. Tannenbaum (1985). Robust control of linear time-invariant plants using periodic compensation, IEEE Trans. Autom. Control, 30, 1088-1096.

Kono, M. (1980). Eigenvalue assignment in linear periodic discrete-time systems, Int. J. Control, Vol. 32, No. 1, 149-158.

Kono, M. (1981). Invertibility of linear time-varying discrete-time systems, Syst. Control Lett., 1, 148-153.

Kono, M. (1983). Left invertibility in linear periodic discrete time systems, Trans. Soc. Instrum. and Control Eng. (Japan), 21, (12), pp.1255-1260.

Liou, M.L. and Y.L. Kuo (1979). Exact analysis of switched capacitor circuits with arbitrary inputs, IEEE Trans. Ccts Syst., 26, 213-223.

Longhi, S, A. Perdon and G.Conte (1989). Geometric and algebraic structure at infinity of discrete-time linear periodic systems, Linear Algebra and Applications, Special Issue, (to appear).

Meyer, R.A. and C.S. Burrus (1975). A unifed analysis of multirate and periodically time-varying digital filters, IEEE Trans. Circuits Syst., 22, (3), 162-168.

Olbrot, A.V. (1987). Robust stabilization of uncertain systems by periodic feedback, Int. J. ontrol, 45, 747-758.

Richards, J.A.(1983). Analysis of Periodically Time-Varying Systems, Springer, Berlin.

Silverman, L. (1969). Inversion of multivariable linear systems, IEEE Trans. Autom. Control, 14, (3), 270-276.

Verriest, E.I. (1988). The operational transfer function and parametrization of N-periodic systems. Proceedings of the 27th IEEE Conference on Decision and Control, Austin (Texas), 1994-1999.

Willems,J.L., V. Kucèra and P. Brunovsky (1984). On the Assignment of invariant factors by time-varying feedback strategies, Syst. Control Lett., 5,75-80.

Wolovich, W.A. (1974). Linear Multivariable Systems, Springer-Verlag, New york.

A CHARACTERIZATION OF ALL REALIZATIONS WITH VECTOR SECOND ORDER DYNAMICS

R. Schrama

*Laboratory for Measurement and Control, Department of Mechanical Engineering,
Delft University of Technology, Mekelweg 2, 2628 CD Delft, The Netherlands*

Abstract. Related to the equation $M\ddot{x}+L\dot{x}+Kx=f$, which is often used in modeling mechanical systems, there exists a state space realization with a typical structure. This representation will be called a vector second order (vso) realization and all realare said to be vso compatible. In this paper we derive sets of necessary and of sufficient conditions on a realization (A,B,C,D) to be vso compatible. By means of a new canonical form we reduce these conditions to one set, which is both necessary and sufficient for a controllable realization to be vso compatible. Finally we show that the vso compatibility is a generic property for a large class of realizations.

Keywords. Canonical forms; mechanical systems; modeling; model structure; linear systems; state-space methods; system theory.

INTRODUCTION

Motivated by our interest in approximate identification of complex mechanical systems, we investigate systems described by the *vector second order* (vso) differential equation

$$M\ddot{q} + L_q\dot{q} + K_q q = N_q u \tag{1a}$$

(Skelton,1988,p.85) and the *measurement equation*

$$y = C_{q1}q + C_{q2}\dot{q}. \tag{1b}$$

The variables $u{\in}\mathbb{R}^m$, $y{\in}\mathbb{R}^p$ are called the input and output respectively, and $q{\in}\mathbb{R}^n$ may be considered as a second order state. All matrices are real and of appropriate dimensions. M is assumed to be non-singular as in all references concerning Eq.(1a) and clearly the state space representation

$$\begin{bmatrix} \dot{x} \\ \ddot{x} \end{bmatrix} = \begin{bmatrix} 0 & I \\ -K & -L \end{bmatrix}\begin{bmatrix} x \\ \dot{x} \end{bmatrix} + \begin{bmatrix} 0 \\ N \end{bmatrix}u \tag{2a}$$

$$y = [C_1 \; C_2]\begin{bmatrix} x \\ \dot{x} \end{bmatrix} \tag{2b}$$

is input-output (i/o) equivalent to Eq.(1).

In literature these sets of equations have been subject of interest in several areas of systems theory. Information on the structure of the state space and input matrices of Eq.(2) has been used in several areas: control design (Williams and Montgomery, 1982), controllability and observability analysis (Hughes and Skelton, 1980; Laub and Arnold, 1984), model reduction (Blelloch, 1987), and stability conditions (Shieh et al., 1987). In identification however structural information is used only under the additional assumption, that all elements of the physical displacement x (Hendricks et al., 1984) are measured independently: $y{=}x$ as in Goyder (1980), Natke (1983) and Yeh and Yang (1987) or at least C_1 is square and invertible and $C_2{=}0$ as in Roether (1987). In the more general case, i.e. Eq.(2b) pur sang, identification is performed in a first order context using no structural information (Juang, 1987; Kano, 1986; Sridar and Aubrun, 1986; Van den Bossche et al., 1986).

This observation has led to the research presented in this paper: we investigate the structure of the set $\mathcal{S}_{2n}$ consisting of all state space realizations of order 2n with m inputs and p outputs; n, m and p are fixed within a set. (A,B,C,D) denotes a realization in $\mathcal{S}_{2n}$, and by the state space isomorphism it induces a class of i/o equivalent realizations. In formulating exactly the problem investigated, we need the following two definitions.

Definition 1.
A *vso realization* (A_v,B_v,C_v,D) is a state space realization with structured system matrices as in

$$\begin{bmatrix} \dot{x} \\ \ddot{x} \end{bmatrix} = \begin{bmatrix} 0 & I \\ A_{v21} & A_{v22} \end{bmatrix}\begin{bmatrix} x \\ \dot{x} \end{bmatrix} + \begin{bmatrix} 0 \\ B_2 \end{bmatrix}u \tag{3a}$$

$$y = [C_1 \; C_2]\begin{bmatrix} x \\ \dot{x} \end{bmatrix} + Du \tag{3b}$$

with $A_{v21},A_{v22}{\in}\mathbb{R}^{n\times n}$, $B_2{\in}\mathbb{R}^{n\times m}$, $C_1,C_2{\in}\mathbb{R}^{p\times n}$ and $D{\in}\mathbb{R}^{p\times m}$. The corresponding state variable is called a *vso state* and $\mathcal{S}_{vso}$ denotes the set of all vso realizations of order 2n.

Definition 2.
A *vso compatible realization* is any realization (A,B,C,D), that admits an i/o equivalent vso realization; i.e. $A{\in}\mathbb{R}^{2n\times 2n}$, $B{\in}\mathbb{R}^{2n\times m}$ and there exists a matrix $T{\in}\mathbb{R}^{2n\times 2n}$, $|T|{\neq}0$ such that (TAT^{-1},TB,CT^{-1},D) is a vso realization. Any such T is said to be *vso-reaching* and the set of 2n dimensional vso compatible realizations is indicated as $\mathcal{S}_{vsoc}$.

In contrast with Eq.(2) we admit an algebraic coupling D between u and y. For mechanical systems this implies the possibility of acceleration measurements, but it will be of no further relevance here. In the literature the vso state is often regarded as physical displacement and velocity. However from our point of view, i.e. approximate identification, this interpretation is not an issue. Accordingly we don't impose any assumptions on the output matrix $[C_1 \; C_2]$.

Now we state precisely the problem studied here. Given the inclusion $\mathcal{S}_{vsoc} \subset \mathcal{S}_{2n}$, we examine to what extend a set $\mathcal{S}_{2n}$ is covered by its adherent set $\mathcal{S}_{vsoc}$; i.e. we investigate which realizations $(A,B,C,D) \in \mathcal{S}_{2n}$ are vso compatible. Our approach follows from the next outline of the paper.

In the next section we first show, that we may restrict our research to quadruples (A,B,C,D) with a full rank input matrix B. Then for purposes of convenience a slightly different structured representation but with equally partitioned state space matrix is introduced. Subsequently two necessary conditions on a realization (A,B,C,D) to be vso compatible are presented. By assuming these conditions to hold in the sequel of the paper, we restrict this research to an important class of realizations. The third section starts with an investigation of the relation between an arbitrary element in $\mathcal{S}_{vsoc}$ and an i/o equivalent element in $\mathcal{S}_{vso}$. This enables us to derive sets of necessary and of sufficient conditions on (A,B,C,D) to be vso compatible. In the following section a new canonical form is introduced, by which the two sets of conditions collapse into a pair of simple necessary and sufficient conditions on a controllable realization to be an element of $\mathcal{S}_{vsoc}$. Further in section 5 we show that under the assumptions of section 2, the set $\mathcal{S}_{vsoc}$ is generic within $\mathcal{S}_{2n}$ in a sense, that will be clarified later. The final section deals with the conclusions and future research.

PRELIMINARY RESULTS

Our first lemma shows, that for the characterization of all vso compatible realizations we may restrict our research to quadruples (A,B,C,D) with a full rank input matrix B, i.e. realizations with independent inputs only.

Lemma 1.
Let the input matrix $B \in \mathbb{R}^{2n \times k}$ with rank $\rho(B)=m$ of a realization (A,B,C,D) be factorized as $B = B_m \cdot L$ with $B_m \in \mathbb{R}^{2n \times m}$ and $L \in \mathbb{R}^{m \times k}$, then (A,B,C,D) is vso compatible if and only if (A,B_m,C,D) is vso compatible.

Proof. See appendix.

For purposes of convenience we transform the vso realizations of Eq.(3) into the different but intimately related realization (A_w,B_v,C_w,D):

$$T_s := \begin{bmatrix} I & 0 \\ -A_{v21} & I \end{bmatrix}$$

$$A_w := T_s A_v T_s^{-1} = \begin{bmatrix} A_{v22} & I \\ A_{v21} & 0 \end{bmatrix}, \quad C_w := C_v T_s^{-1}. \qquad (4)$$

With a slight abuse of notion we will call this realization a vso realization also.
Next assuming the full rankness of B and B_v to hold, two essential properties of vso realizations can be derived from Eq.(3a). First since $\rho(B)=m$, the matrix $[B_v \,|\, A_w B_v]$ has full column rank. Hence all controllability indices $\eta_i, i=1,.,m$ are larger than two. Secondly by the structure of B_v, given in Eq.(3a), we have that $\rho(B_v) \leq n$, which implies that the number of indenpendent inputs is smaller than half the state space dimension.

Lemma 2.
A realization (A,B,C,D) with a full rank input matrix $B \in \mathbb{R}^{2n \times m}$ is vso compatible only if $m \leq n$ and all controllability indices $\eta_i, i=1,.,m$ are larger than or equal to 2.

Proof.
By invariance under i/o equivalence these properties are necessary for vso compatibility.

In the sequel a realization (A,B,C,D) is assumed to have a full rank input matrix. Similarly we also suppose the conditions in lemma 2 to hold throughout the paper. We like to remark, that considering our objective of approximate identification of complex mechanical systems, these conditions do not impose severe restrictions on the set of realizations investigated. For in practice not only the number of independent inputs of these systems is very often smaller than half the the dimension of the state space, but also the condition on the controllability indices is almost always met within $\mathcal{S}_{2n}$ as we will shown later on in section 5.

CONDITIONS FOR VSO COMPATIBILITY

In this section we will present sets of necessary and of sufficient conditions on a realization $(A,B,C,D) \in \mathcal{S}_{2n}$ to be vso compatible. As can be seen from definitions 1 and 2 this property depends just on the pair (A,B). By the following two notions we can examine these matrices one at a time.

Definition 4.
An i/o equivalence transformation matrix $T \in \mathbb{R}^{2n \times 2n}$, $|T| \neq 0$ is called A_w-reaching for $A \in \mathbb{R}^{2n \times 2n}$ if the matrix TAT^{-1} is structured as A_w of Eq.(4).
Analogously T is called B_v-reaching for $B \in \mathbb{R}^{2n \times m}$ if TB is as B_v in Eq.(3).

The next lemma considers the existence of an A_w-reaching matrix T. Moreover it shows an essential property of any such transformation matrix.

Lemma 3.
There exists an A_w-reaching transformation for a matrix $A \in \mathbb{R}^{2n \times 2n}$ if and only if A admits a matrix $E \in \mathbb{R}^{2n \times n}$, such that the composed matrix $[AE \,|\, E]$ is invertible. Moreover matrix T is A_w-reaching if and only if it satisfies $T^{-1}=[AE \,|\, E]$ for the appropriate matrix E.

Proof. See appendix.

Using this result the existence of an A_w-reaching T for A can be related to the geometric multiplicities (Lancaster and Tismenetsky, 1985, p.159) of the eigenvalues of A. We state the next lemma without proving it here.

Lemma 4 (Schrama, 1989).
A matrix $A \in \mathbb{R}^{2n \times 2n}$ admits an A_w-reaching T if and only if the geometric multiplicity of each eigenvalue of A is smaller than or equal to n.

Now a natural question is, given a pair (A,B) what T is both A_w-reaching for A and B_v-reaching for B? For in this case, the realization (A,B,C,D) would be vso compatible. However since this implies a combination of conditions on T as well as on it's inverse, the problem seems to be untractable. On the other hand if (A,B,C,D) is vso compatible, then there always exists a T such that $TAT^{-1}=A_w$ and

$$TB = B_c := \begin{bmatrix} 0 \\ B_{c2} \end{bmatrix}, \quad B_{c2} = \begin{bmatrix} I_m \\ 0 \end{bmatrix} \in \mathbb{R}^{n \times m}, \qquad (5)$$

where in anticipation of the next section the index c stands for 'canonical'. This observation enables us to derive the following sets of necessary and of sufficient conditions on a realization (A,B,C,D) to be vso compatible. Note that the assumptions of section 2 have been overruled.

Theorem 1.
A 2n dimensional state space realization (A,B,C,D) with k inputs and p outputs is vso compatible only if the next conditions are satisfied.
1. No controllability index η_i, i=1,.,k equals 1.
2. Each eigenvalue of A has a geometric multiplicity smaller than or equal to n.
3. $\dim(\mathcal{B} \cap \mathcal{A}) \leq \dim(\mathcal{A}) - n$, where $\mathcal{A}=\mathrm{Im}(A)$ and $\mathcal{B}=\mathrm{Im}(B)$.

Proof.
We consider only realizations with a full rank input matrix (extension to a rank deficient B is trivial from lemma 1). Conditions 1 and 2 make up lemma's 2 and 4, since the property $\eta_i > 1$ implies that $\rho(B)=m \leq n$. The last condition states: looking for $[AE \,|\, E]$ with B as the first columns of E in order to arrive at Eq.(5) not too many columns of B may be spanned by the columns of A if A is rank deficient, since then $[AE \,|\, A]$ cannot be invertible. For more details see Schrama (1989).

Theorem 2.
A 2n dimensional state space realization (A,B,C,D) with k inputs and p outputs is vso compatible if it satisfies conditions 1 and 2 of theorem 1 and
4. If (A,B,C,D) has n_u uncontrollable modes, then each corresponding eigenvalue has a geometric multiplicity smaller than $n_u/2$ or $(n_u+1)/2$ if n_u is even or odd.

Proof.
As in the proof of theorem 1 we consider only realizations with B of full rank. Let (A,B,C,D) be given in Kalman decomposed form (Van Dooren, 1981) and let the number of uncontrollable modes be even, then apply lemma 4 and theorem 3 (next section) to the uncontrollable respectively controllable part. Again we refer to Schrama (1989) for more details.

THE SET OF ALL CONTROLLABLE
VSO COMPATIBLE REALIZATIONS

Now we show, that the condition on the controllability indices, as presented in theorem 1, is adequate for describing the set of all controllable vso compatible realizations as a subset of all controllable realizations in $\mathcal{S}_{2n}$. This characterization is stated precisely in the next theorem; again the assumptions of section 2 have been overruled.

Theorem 3.
A 2n dimensional controllable state space realization (A,B,C,D) with k inputs and p outputs is vso compatible if and only if no controllability index η_i, i=1,..,k equals 1.

Proof.
The *only if* part of this theorem is immediate from theorem 1 (condition *1*). In the sequel of this section a canonical form with the structure of a vso realization is defined on the set of realizations, that satisfies the condition mentioned in the theorem. This proves the *if* part.

We outline the remainder of this section, since it is quite involved. First we introduce some necessary concepts. While we add some notions in proposition 1, we construct a selection rule for any controllable (A,B,C,D), which satisfies the condition $\eta_i \neq 1$. Based on this selection rule we formalize the new canonical vso realization in lemma 5. Finally we end this section with some elaboration on this canonical form.

Definition 5.
– $\mathcal{S}_{2n}{}^c$ and $\mathcal{S}_{2n}{}^s$ denote respectively the subset of all controllable $(A,B,C,D)\in\mathcal{S}_{2n}$ and the subset of all $(A,B,C,D)\in\mathcal{S}_{2n}{}^c$ with $\eta_i\neq 1$. Here $\mathcal{S}_{2n}{}^s$ is the set of interest.
– Let $\mathcal{V}_1,\mathcal{V}_2$ be two sets containing a possibly different number of equally dimensioned vectors. Then $\ell[\mathcal{V}_1,\mathcal{V}_2]$ determines the number of all vectors in the compound set $\mathcal{V}_1 \cup \mathcal{V}_2$ minus the rank of the matrix generated by putting all vectors side by side. If the result is zero, then all vectors in the compound set are linearly independent.
– $R(k)$ denotes the k-th m-block of the controllability matrix R: $R(k) = A^{k-1}B\in\mathbb{R}^{2n\times m}$, k=1,..,2n. Moreover $R(k)^i$, i=1,..,m is the i-th column in $R(k)$.
– A set $\mathcal{I}_{ps}$ of pairwise selection indices ν_i is a set of even integers, that like the controllability indices, correspond to a selection of linearly independent columns of the controllability matrix R. The associated vectors, called the *pairwise selected vectors*, constitute the set $\mathcal{V}_{ps}$ with N_{ps} elements.

It is well known in the literature, that the interdependence of the columns in the finite controllability matrix R, $R:=[B\ AB\ \cdots\ A^{2n-1}B]$ enables the definition of a selection rule, that generates a non-singular transformation matrix for any realization (A,B,C,D) in $\mathcal{S}_{2n}{}^c$. Since any such selection rule is based on invariants under i/o equivalence, it gives rise to a set of canonical forms (Denham 1974, Glover and Willems 1974, Luenberger 1967).

Proposition 1.
Let (A,B,C,D) be any realization in $\mathcal{S}_{2n}{}^s$. Then the following four step selection rule gives rise to a non-singular matrix T_c.
1. Generate the set $\mathcal{I}_{ps}$ of ν_i, i=1,..,m as $\nu_i=\eta_i$ for η_i is even or zero and $\nu_i=\eta_i-1$ for η_i is odd. Also generate the corresponding set $\mathcal{V}_{ps}$ of N_{ps} vectors. If all indices η_i are even then go to step *4*.
2. Next arrange all odd controllability indices η_i in increasing order of magnitude. If two odd controllability indices η_p and η_q are equal, then select them in increasing order of their indices: p<q $\rightarrow$ ..,η_p,η_q,.. Then extend the set $\mathcal{V}_{ps}$ by way of performing the next linear independency test for all odd controllability indices η_i corresponding to the sequence just arranged:
If $\ell[\{R(\eta_i)^i, R(\eta_{i+1})^i\}, \mathcal{V}_{ps}] = 0$, then
- Extend $\mathcal{V}_{ps}$ with the two vectors $R(\eta_i)^i$, $R(\eta_{i+1})^i$. Inherently increase ν_i and N_{ps} by 2 (see step *1*).
- If $N_{ps}=2n$ then go to step *4*.
3. For pairs (A,B), that not yet resulted in $N_{ps}=2n$, generate two new sequences of integers: first let the *index set of remaining vectors* $\mathcal{I}_{rv}$ consist of all i corresponding to an odd η_i, for which ν_i was not increased in step *2*. Secondly create *the index set of available vectors* $\mathcal{I}_{av}$, consisting of all i,i=1,..,m, that do not belong to $\mathcal{I}_{rv}$ nor correspond to $\eta_i=0$. Then $\mathcal{V}_{ps}$ is extendend by means of the next test. Take the smallest pairwise selection index ν_j in the set $\{\nu_i | i\in\mathcal{I}_{av}\}$. If this minimum is not unique, take the one with the smallest index j.

If $\ell[\{R(\nu_j+1)^j, R(\nu_j+2)^j\}, \mathcal{V}_{ps}] = 0$, then
- Extend $\mathcal{V}_{ps}$ by the vectors $R(\nu_j+1)^j$, $R(\nu_j+2)^j$. Inherently increase ν_i and N_{ps} by 2 (see step *1*).
- If $N_{ps}=2n$ then go to step *4*.
else, eliminate j from the index set $\mathcal{I}_{av}$.
Modify the set $\{\nu_i | i\in\mathcal{I}_{av}\}$ and perform this test till $\mathcal{I}_{av}=\emptyset$.
4 Use pairwise selection indices ν_i to generate the matrix T_c as follows. Let ν_{max} be the possibly non unique maximum of all ν_i, then arrange all elements of $\mathcal{V}_{ps}$ into

$$E_1:= [R_r(2)\ R_r(4)\ ...\ R_r(\nu_{max})]$$

$$E_2:= [R_r(1)\ R_r(3)\ ...\ R_r(\nu_{max}-1)] \tag{6}$$

where $R_r(k)$ is the result of omitting those columns in $R(k)$, that are not selected by the indices ν_i. If step *4* has been reached by a go-to-statement in step *1*, *2* or *3*, i.e. $\Sigma\nu_i=2n$, then T_c^{-1} can be composed as $T_c^{-1}:= [E_1|E_2] = [AE_2|E_2]$. Otherwise $E_2, E_1\in\mathbb{R}^{2n\times(n-j)}$ and another 2j independent columns have to be added using the index set $\mathcal{I}_{rv}$: Order all odd η_i with $i\in\mathcal{I}_{rv}$ first with respect to their value and secondly with respect to their index analogously to step *2*. Initialize the *set $\mathcal{V}_{ss}$ of simply selected vectors* as an empty set. Repeat the next test for the sequence of η_i's as just ordered until $\mathcal{V}_{ss}$ consists of 2j columns:
If $\ell[\{R(\eta_i)^i\}, \{\mathcal{V}_{ps} \cup \mathcal{V}_{ss}\}] = 0$, then extend $\mathcal{V}_{ss}$ by the vector $R(\eta_i)^i$; else omit η_i from the sequence.
This results in an unique sequence of 2j indices η_i ordered corresponding to the columns $R(\eta_i)^i$ consecutively incorporated into $\mathcal{V}_{ss}$. Say η_g, η_h are the first and second elements in this sequence then modify E_1, E_2 as

$$E_1:= [E_1\ R(\eta_g+1)^g + R(\eta_h)^h]$$

$$E_2:= [E_2\ R(\eta_g)^g + R(\eta_h-1)^h] \tag{7}$$

Repeat this modification for η_g, η_h being respectively the third and fourth, the fifth and sixth, etc. elements in the sequence. Finally create T_c through $T_c^{-1}:=[E_1|E_2]$.

Remark 1.
This selection rule is not applicable to all realizations in $\mathcal{S}_{2n}{}^c$, since η_h-1 in Eq.(7) might be zero for some pair (A,B) and $R(0)$ has not been defined.

We admit that this selection rule is rather a hideous one, but it just serves as a means to prove theorem 2. In the next example the selection rule is illustrated.

Example 1.
Let the controllability indices of a pair (A_w,B_v) be $(\eta_1,..,\eta_6) = (6,4,3,5,3,5)$. All columns in the controllability matrix R, that are in $\mathcal{V}_{ps}$ after step *1*, are indicated by p and likewise s denotes a remaining column, which gets selected through an index η_i. Note that $(\nu_1,..,\nu_6)= (6,4,2,4,2,4)$ and

$$R = [pppppp\,|\,pppppp\,|\,ppspsp\,|\,pp\text{-}p\text{-}p\,|\,p\text{-}\text{-}s\text{-}s\,|\,p\text{-}\text{-}\text{-}\text{-}\text{-}]$$

In step *2* the odd controllability indices are arranged as 3, 3, 5, 5 for i respectively 2, 5, 4, 6. We will denote this as 3_2, 3_5, 5_4, 5_6. Then check whether or not $(R(3)^3, R(4)^3)$ may be incorporated in $\mathcal{V}_{ps}$. This is possible only if $R(4)^3$ depends on one of the columns indicated with s except for $R(3)^3$. But since $R(4)^3$ depends on it's preceding columns, the only possible candidate is $R(3)^5$. Say the test indeed satisfies $\ell[.]=0$, then $(\nu_1,..,\nu_6)$ becomes $(6,4,4,4,2,4)$ and

$$R = [pppppp\,|\,pppppp\,|\,pppp\underline{s}sp\,|\,pppp\text{-}p\,|\,p\text{-}\text{-}s\text{-}s\,|\,p\text{-}\text{-}\text{-}\text{-}\text{-}]$$

Say neither $(R(5)^4, R(6)^4)$ or $(R(5)^6, R(6)^6)$ may be incorporated in $\mathcal{V}_{ps}$. Then in step *3* create $\mathcal{I}_{rv}=\{4,5,6\}$, $\mathcal{I}_{av}=\{1,2,3\}$ and order the minima of $\{\nu_1,\nu_2,\nu_3\}$ as (ν_2,ν_3). Check consecutively whether the pair $(R(5)^2, R(6)^2)$ or $(R(5)^3, R(6)^3)$ satisfies the linearity test. If one of both pairs of columns would be independent, then we have 2n vectors. However say this is not the case. Then $\mathcal{I}_{av}$ becomes $\{1\}$ and we check $(R(7)^1, R(8)^1)$ as a candidate pair for $\mathcal{V}_{ps}$. Again let this does not work out. For a discussion on the peculiarity of such an example we refer to the end of this section. Now we arrive at step *4* and we seek for the set $\mathcal{V}_{ss}$. Since we just assumed $(R(3)^3, R(4)^3)$ to be selected $R(3)^5$ cannot be independent (see earlier in this example) and therefore 5 can be eliminated from $\mathcal{I}_{rv}$. The remaining set $\{R_4(5), R_6(5), \mathcal{V}_{ps}\}$ consists of· 2n independent columns.

Finally create T_c from E_1 and E_2:

$$E_1=[\ R(2)\ |\ R(4)^{(1\ 2\ 3\ 4\ 5)}\ |\ R(6)^1\ |\ R(5)^6+R(6)^4\ ]$$

$$E_2=[\ R(1)\ |\ R(3)^{(1\ 2\ 3\ 4\ 6)}\ |\ R(5)^1\ |\ R(4)^6+R(5)^4\ ]$$

which is the same as

$$E_1=[\ AB\ |\ A^3[b_1\ b_2\ b_3\ b_4\ b_6]\ |\ A^5b_1\ |\ A^4b_6+A^5b_4\ ]$$

$$E_2=[\ B\ |\ A^2[b_1\ b_2\ b_3\ b_4\ b_6]\ |\ A^4b_1\ |\ A^3b_6+A^4b_4\ ]$$

with b_i the i-th column of B. $\qquad\qquad\square$

<u>Lemma 5.</u>
Let the transformation matrix T_c be determined for any realization $(A,B,C,D)\in\mathcal{S}_{2n}{}^s$ by the selection rule of proposition 1. Then the induced set of all realizations $(A_c,B_c,C_c,D):=(T_cAT_c{}^{-1},T_cB,CT_c{}^{-1},D)$ constitutes a set of controllable canonical vso realizations for the i/o equivalence on the set $\mathcal{S}_{2n}{}^s$.

<u>Proof.</u> See appendix.

As mentioned previously by virtue of these canonical vso realizations proving the *if* part of theorem 2 now has become self-evident.

The pairwise selection indices ν_i are based on the interdependence of the columns in the controllability matrix, and therefore they are invariants (MacLane and Birkhof 1967, Denham 1974) for the i/o equivalence. Furthermore for any pair (A,B) with $N_{ps}=\Sigma\nu_i<2n$, the sequence of $2n-N_{ps}$ odd controllability indices, corresponding to the vectors in $\mathcal{V}_{ss}$, is also invariant for i/o equivalence.
By the invariance of the ν_i's some genericity properties of these indices may be examined through the first Luenberger controllable canonical form for i/o equivalence (Luenberger, 1967). – For a definition of genericity we refer to the next section. – It can be shown, that within $\mathcal{S}_{2n}{}^c$ generically $2n$ vectors are selected in step *2*. And similarly within the subset of $\mathcal{S}_{2n}{}^s$ for which step *3* has to be performed N_{ps} will generically be increased to $2n$ at that stage. Instead of a formal proof we give the next illustrative example.

<u>Example 2.</u>
Let a pair (A,B) with controllability indices $(3,3,4)$ have the first Luenberger controllable canonical form. Then the controllability matrix looks like

$$R=\left[\begin{array}{ccc|ccc|ccc|ccccccc}
1&0&0&0&0&0&0&0&0&p_1&q_1&0&\cdot&\cdot&r_1&\cdot&\cdot&s_1\\
0&0&0&1&0&0&0&0&0&p_2&q_2&0&\cdot&\cdot&r_2&\cdot&\cdot&s_2\\
0&0&0&0&0&0&1&0&0&p_3&q_3&0&\cdot&\cdot&r_3&\cdot&\cdot&s_3\\
0&1&0&0&0&0&0&0&0&p_4&q_4&0&\cdot&\cdot&r_4&\cdot&\cdot&s_4\\
0&0&0&0&1&0&0&0&0&p_5&q_5&0&\cdot&\cdot&r_5&\cdot&\cdot&s_5\\
0&0&0&0&0&0&0&1&0&p_6&q_6&0&\cdot&\cdot&r_6&\cdot&\cdot&s_6\\
0&0&1&0&0&0&0&0&0&p_7&q_7&0&\cdot&\cdot&r_7&\cdot&\cdot&s_7\\
0&0&0&0&0&1&0&0&0&p_8&q_8&0&\cdot&\cdot&r_8&\cdot&\cdot&s_8\\
0&0&0&0&0&0&0&0&1&p_9&q_9&0&\cdot&\cdot&r_9&\cdot&\cdot&s_9\\
0&0&0&0&0&0&0&0&0&0&0&1&0&0&r_{10}&0&0&s_{10}
\end{array}\right]\cdots$$

The columns with elements p_i, q_i and r_i are respectively the third, sixth and tenth column in the matrix A. These 28 parameters constitute a set of pairs (A,B) with given controllability indices. Step *2* of the selection rule will not result in $2n$ independent columns if and only if $p_6=0$ and $q_3=0$. This means, that if we represent all pairs (p_6,q_3) by $\mathbb{R}^2$, then step *3* will not be skipped for only one point in the plane.
The second statement about genericity follows from a similar though little more involved reasoning. $\qquad\square$

As well known in the literature the set of all controllable systems with a specified sequence of controllability indices is isomorphic to the vector space $\mathbb{R}^k$, where k equals the number of (non structural) parameters in the associated first Luenberger controllable canonical form (Denham 1974). In the context of the vso canonical form such an isomorphism exists only for a set, that contains any realizations in $\mathcal{S}_{2n}{}^s$ with some sequence of pairwise selection indices, which satisfies $\nu_i\geq\nu_{i+1}$, $i=1,..,m-1$ and $\Sigma\nu_i=2n$. Note that this condition on the sequence $(\nu_1,\nu_2,..,\nu_m)$ is a generic property for $\mathcal{S}_{2n}{}^s$. As a result of the isomorphism the parameters in the vso canonical form and the pairwise selection

indices together make up a complete set of independent invariants for i/o equivalence on each subset of $\mathcal{S}_{2n}{}^s$, that contains all realizations with prescribed $(\nu_1,\nu_2,..,\nu_m)$, which satisfies the condition. For any other sequence of pairwise selection indices, the corresponding set of realizations possesses a complete set of invariants, which consists of the ν_i's, the non structural parameters in the canonical form and possibly a sequence of odd controllability indices (step *4*). However these invariants (parameters) are not independent as will be clear from the examples 2 and 3.

<u>Example 3.</u>
Let the system (A,B,C,D), n=3, have pairwise selection indices $(2,4)$. Note that the controllability indices may be $(3,3)$ or $(2,4)$. The associated matrices R_c and A_c look like

$$R_c=\begin{bmatrix}0&0&1&0&p_1&0&q_1&0&\cdot&r_1\\0&0&0&1&p_2&0&q_2&0&\cdot&r_2\\0&0&0&0&p_3&0&q_3&1&\cdot&r_3\\1&0&0&0&p_4&0&q_4&0&\cdot&r_4\\0&1&0&0&p_5&0&q_5&0&\cdot&r_5\\0&0&0&0&p_6&1&q_6&0&\cdot&r_6\end{bmatrix}\cdots,\ \ A_c=\begin{bmatrix}p_1&0&r_1&1&0&0\\p_2&0&r_2&0&1&0\\p_3&0&r_3&0&0&1\\p_4&0&r_4&0&0&0\\p_5&0&r_5&0&0&0\\p_6&1&r_6&0&0&0\end{bmatrix}$$

where $q=A_cp$, with e.g. q defined as $q:=[q_1,..,q_6]^T$; r can take any value in $\mathbb{R}^6$, but the indices ν_i resulting from the selection rule will be $(2,4)$ only if the equation $p_3q_6-p_6q_3=0$ is satisfied. $\qquad\square$

VSO COMPATIBILITY: A GENERIC PROPERTY

In this section we enunciate some generic properties of the set $\mathcal{S}_{2n}$. First we will define the concept of genericity and we adopt some metric on $\mathcal{S}_{2n}$. Then we show subsequently the following properties. The set of all controllable realizations is generic within $\mathcal{S}_{2n}$. The condition $\eta_i\neq1$ is almost always met within a set $\mathcal{S}_{2n}$, provided that the number m of independent inputs is smaller than or equal to n. And under these latter conditions the vso compatibility is a generic property of the set of realizations $(A,B,C,D)\in\mathcal{S}_{2n}$.

<u>Definition 3.</u> Given any two sets θ and Ψ such that $\theta\subset\Phi$, then the set θ is called *generic* within Φ if $\theta\neq\Phi$ and the closure θ^+ of θ is equal to the closure of Φ^+ of Φ.

The quadruple (A,B,C,D) can be represented by the vector $\theta\in\mathbb{R}^s$, $s=2n(2n+m+p)+mp$, where θ consists of all elements of the system matrices. By adopting the usual metric on $\mathbb{R}^s$, we solidify the concept of genericity with the set of realizations in $\mathcal{S}_{2n}$.

Let $\mathcal{S}_{2n}(j)$, $j=1,..,2n$ denote the subset of $\mathcal{S}_{2n}$ consisting of realizations with at least j uncontrollable modes. Then we have the inclusion $\mathcal{S}_{2n}(2n)\subset\mathcal{S}_{2n}(2n-1)\subset\cdots\subset\mathcal{S}_{2n}(1)\subset\mathcal{S}_{2n}$. Knowing that $\mathcal{S}_{2n}(j)-\mathcal{S}_{2n}(j+1)$ is open within $\mathcal{S}_{2n}(j)$, it is easy to see, that $\mathcal{S}_{2n}(j+1)$ is non-generic within $\mathcal{S}_{2n}(j)$; i.e. the set of realizations with exactly j uncontrollable modes is generic within the set $\mathcal{S}_{2n}(j)$, containing at least j uncontrollable modes. As a first result the set of all controllable realizations is generic within $\mathcal{S}_{2n}$.

Secondly by a corollary of Wonham (1979, p.121) the set of realizations with all controllability indices $\eta_i\geq2$, $i=1,..,m$ is generic within $\mathcal{S}_{2n}-\mathcal{S}_{2n}(1)$ provided all m inputs are independent and $m\leq n$. The extension to dependent inputs is again trivial. Combination with the first property affirms our second claim. Note that by means of the Kalman decomposition, separating the controllable and strictly uncontrollable subspaces (Van Dooren, 1981), this corollary could be extended for all sets $\mathcal{S}_{2n}(j)-\mathcal{S}_{2n}(j+1)$, $j=1,..,2(n-m)$.

Finally consider realizations (A,B,C,D) with a restricted number $(m\leq n)$ of independent inputs. Summarizing we showed that $\mathcal{S}_{2n}{}^c$, consisting of all controllable realizations, is a generic subset of $\mathcal{S}_{2n}$ and that the condition $\eta_i\neq1$, which determines the set $\mathcal{S}_{2n}{}^s$, holds generically within $\mathcal{S}_{2n}{}^c$; conclusively all closures are equal: $\mathcal{S}_{2n}{}^{s+}=\mathcal{S}_{2n}{}^{c+}=\mathcal{S}_{2n}$. And since all elements of $\mathcal{S}_{2n}{}^s$ are vso compatible, we can make the following proposition.

<u>Proposition 2.</u>
The vso compatibility is a generic property of the set of realizations (A,B,C,D) with $A{\in}\mathbb{R}^{2n{\times}2n}$ and $m{\leq}n$ independent inputs.

Note that we could prove also that vso compatibility is a generic property within $\mathcal{S}_{2n}(1)$, $\mathcal{S}_{2n}(2)$, etc., since as can be seen easily, the sufficiency conditions of theorem 2 are almost always satisfied.

CONCLUSIONS

We investigated into what (A,B,C,D) out of a set of $2n$ dimensional realizations admits it's dynamics to be represented by a matrix- or vector second order differential equation. Sets of necessary and of sufficient conditions on (A,B,C,D) to possess this property have been derived. By means of a new canonical form, we reduced these sets for controllable realizations to one necessary and sufficient condition, which says that all controllability indices must be unequal to 1. Finally we showed, that almost every realization (A,B,C,D) with $A{\in}\mathbb{R}^{2n{\times}2n}$ and $m{\leq}n$ independent inputs possess vector second order dynamics.
We end up mentioning, that the implications of these results in view of behavioural modeling (Willems, 1986), especially for mechanical systems, is currently under investigation.

REFERENCES

Blelloch, P.A. (1987). Perturbation analysis of internal balancing for lightly damped mechanical systems with gyroscopic and circulatory forces. *J.Guidance, 10,* 406-410.

Denham, M.J. (1974). Canonical forms for the identification of multivariable linear systems. *IEEE Trans. Automat.Contr., AC-19,* 646-656.

Glover, K. and J.C. Willems (1974). Parametrizations of linear dynamical systems: canonical forms and identifiability. *IEEE Trans.Automat.Contr., AC-19,* 640-646.

Goyder, H.G.D. (1980). Methods and applications of structural modelling from measured structural frequency response data. *Journal of Sound and Vibration, 68,* 209-230.

Hendricks, S.L., S. Rajaram, M.P. Kamat and J.L. Junkins (1984). Identification of large flexible structures mass/stiffness and damping from on-orbit experiments. *J.Guidance, 7,* 244-245.

Hughes, P.C. and R.E. Skelton (1980). Controllability and observability of linear matrix second order systems. *Trans.ASME J.Appl.Mechanics, 47,* 415-420.

Juang, J.-J. (1987). Mathematical correlation of modal-parameter identification methods via system-realization theory. *Journal of Modal Analysis,* 1-18.

Kano, H. (1986). Identification method for modal analysis of mimo linear dynamical systems of mechanical structures. *Proc. ACC, Seattle, WA,* 1204-1209.

Lancaster, P. and M. Tismenetsky (1985). *The theory of matrices, second edition with applications.* Academic Press, Inc., Orlando, Florida.

Laub, A.J. and W.F. Arnold (1984). Controllability and observability criteria for multivariable linear second-order models. *IEEE Trans.Autom.Contr., AC-29,* 163-165.

Luenberger, D.G. (1967). Canonical forms for linear multivariable systems. *IEEE Trans.Autom.Contr., AC-12,* 290-293.

MacLane, S. and G. Birkhoff (1967). *Algebra.* The MacMillan Company, New York.

Natke, H.G. (1983). *Einfuehrung in Theorie und Praxis der Zeitreihen- und Modalanalyse, Identifikation schwingungsfaehiger elastomechanischer Systeme.* Vieweg, Braunschweig / Wiesbaden.

Roether, F. (1987). Improved identification of mechanical multibody systems from discrete-time data. Differential identification of linear mechanical systems. *Trans. ASME, J.Dyn.Syst., Meas., and Control, 109,* 140-145.

Schrama, R.J.P. (1989). *Dynamics described by second order matrix differential equations.* Internal report nr. N-311; also to be published as MEMT-Delft report.

Shieh, L.S., M.M. Mehio and H.M. Dib (1987). Stability of the second-order matrix polynomial. *IEEE Trans. Automat.Contr., AC-32,* 231-233.

Skelton, R.E. (1988). *Dynamic Systems Control, Linear systems analysis and synthesis.* John Wiley & Sons.

Sridar, B. and J.-N. Aubrun (1986). Comparison of identification methods using jitter beam experimental data. *Proc. ACC, Seattle, WA,* 1060-1066.

Van den Bossche, E., L. Dugard and I.D. Landau (1986). Modelling and identification of a flexible arm. *Proc.ACC, Seattle, WA,* 1611-1616.

Van Dooren, P.M. (1981). The generalized eigenstructure problem in linear system theory. *IEEE Trans.Automat. Contr., AC-26,* 111-129.

Willems, J.C. (1986). From time series to linear system; Part I: Finite dimensional linear time invariant systems. *Automatica, 22,* 561-580.

Williams, J.P. and R.C. Montgomery (1982). Simulation and testing of digital control of a flexible beam. *Proc. of the AIAA Guidance and Control Conf.,* 403-409.

Wonham, W.M. (1979). *Linear multivariable control: a geometric approach.* Springer-Verlag, New York.

Yeh, F.-B. and C.-D. Yang (1987). New time-domain identification technique. *J.Guidance, 10,* 313-316.

APPENDIX: PROOFS.

<u>Proof of lemma 1.</u> First note that any matrix B can be factorized as indicated.
If. Given (A,B_m,C,D) is vso compatible, then there exists a transformation T s.t. $(A_c,B_c,C_c,D):=(TAT^{-1},TB_m,CT^{-1},D)$ is a vso realization with TB_m as in Eq.(3a). By $TB = TB_mL$ obviously (TAT^{-1},TB,CT^{-1},D) is also a vso realization and thus (A,B,C,D) is vso compatible.
Only if. Given (A,B,C,D) is vso compatible, then analogously there exists a non-singular matrix $T{\in}\mathbb{R}^{2n{\times}2n}$ such that (TAT^{-1},TB,CT^{-1},D) is a vso realization with the input matrix TB partitioned as in Eq.(3a). Since $\rho(L)=m$ and $m{\leq}k$, it has a right inverse $L^{\bullet}$: $LL^{\bullet}=I_m$. So $TB_m = TB_mLL^{\bullet} = TBL^{\bullet} = \begin{bmatrix} 0 \\ B_2L^{\bullet} \end{bmatrix}$ and thus (A,B_m,C,D) is also vso compatible. □

<u>Proof of lemma 3.</u>
Only if. Given an A_w-reaching T for A, i.e. $A_w=TAT^{-1}$, and let $E_v{\in}\mathbb{R}^{2n{\times}n}$ be defined as $E_v=\begin{bmatrix} 0 \\ I_n \end{bmatrix}$, then we have the trivial equation $[A_w{\cdot}E_v | E_v]=I_{2n}$. With E uniquely defined by $E_v=TE$, this can be written as $[TAT^{-1}TE | TE]=I_{2n}$ and the property $T^{-1}=[AE | E]$ is evident.
If. Given E such that $[AE | E]$ is invertible, then let $T=[AE | E]^{-1}$ and $A_t=TAT^{-1}$. Consider the right half of the matrix A_t only:
$$\begin{bmatrix} A_{t12} \\ A_{t22} \end{bmatrix} = TAT^{-1}{\cdot}\begin{bmatrix} 0 \\ I \end{bmatrix} = TAE$$
which is equivalent to
$$[AE | E]{\cdot}\begin{bmatrix} A_{t12} \\ A_{t22} \end{bmatrix} = AE.$$
By the uniqueness of the solution $A_{t12}=I$, $A_{t22}=0$ the proof has been completed. □

<u>Proof of lemma 5.</u>
We have to prove the following six facts: T_c is B_v-reaching (1), A_w-reaching (2) and invertible (3); an unique T_c exists for all $(A,B,C,D){\in}\mathcal{S}_{2n}{}^s$ (4); and finally two i/o equivalent realizations are transformed into the same vso realization (5).
(1). Let $B{\in}\mathbb{R}^{2n{\times}k}$, $\rho(B)=m$ be factorized as $B_m{\cdot}L$ as in lemma 1, but with the extra property $B_m=R_r(1)$ (which is always possible). So T_c^{-1} can be written as $T_c^{-1}=[U | B_m | V]$, $U{\in}\mathbb{R}^{2n{\times}n}$, $V{\in}\mathbb{R}^{2n{\times}(n-m)}$ (see also the proof of theorem 1). Thus $T_cB_m=B_c$ as in Eq.(5) and $T_cB=B_cL$, which has the same structure as B_v in Eq.(3).
In the sequel of the proof we assume B to be of full rank. By lemma 1 and the result above, the extension to non full rank input matrices is obvious.
(2). By definition $R(k+1) = AR(k)$ and the pairwise selection indices are even. So from Eqs.(6) (7) we have $E_1=AE_2$ for all $(A,B,C,D){\in}\mathcal{S}_{2n}{}^s$. So $T_c^{-1}=[AE | E]$ and by lemma 3 the matrix T_c is A_w-reaching.

(3). The set $\mathcal{V}_{\mathrm{ps}}$ or possibly the compound set $\mathcal{V}_{\mathrm{ps}} \cup \mathcal{V}_{\mathrm{ss}}$ consists by definition of linearly independent vectors only. So side by side arrangement of these vectors leads to a non singular matrix.

(4). The uniqueness of T_{c} for any (A,B,C,D) is an immediate result of the unambiguously ordered selection rule. Furthermore since the realization is controllable the controllability indices select 2n independent vectors from the controllability matrix R. In step 1 these 2n vectors are separated into two sets: N_{ps} vectors in $\mathcal{V}_{\mathrm{ps}}$ and remainder, which consists of N_{nps} non-pairwise selected vectors and which will be indicated as $\mathcal{V}_{\mathrm{nps}}$. Then in steps 2 and 3 $\mathcal{V}_{\mathrm{ps}}$ is extended by linearly independent vectors only. So arriving at step 4 we have N_{ps} independent vectors in $\mathcal{V}_{\mathrm{ps}}$ and 2n independent vectors in $\mathcal{V}_{\mathrm{ps}} \cup \mathcal{V}_{\mathrm{nps}}$. If N_{ps} is not 2n yet, then step 4 effactually selects 2n independent vectors from this compound set, where all elements of $\mathcal{V}_{\mathrm{ps}}$ are chosen. In Eq.(7) $R(\eta_{\mathrm{g}}+1)^{\mathrm{g}}$ is obviously a linear combination of all selected vectors but $R(\eta_{\mathrm{h}})^{\mathrm{h}}$, since otherwise the pair $(R(\eta_{\mathrm{g}})^{\mathrm{g}},R(\eta_{\mathrm{g}}+1)^{\mathrm{g}})$ would have been pairwise selected. So the addition of $R(\eta_{\mathrm{g}}+1)^{\mathrm{g}}$ to $R(\eta_{\mathrm{h}})^{\mathrm{h}}$ does not affect the independency of the vectors. The same is true for the pair $R(\eta_{\mathrm{g}})^{\mathrm{g}}+R(\eta_{\mathrm{h}}-1)^{\mathrm{h}}$ since they are both in the selected set of independent vectors. This proofs the existence of T_{c}.

(5). The controllability matrices R and R_{c} of respectively any (A,B,C,D) and it's associated $(A_{\mathrm{c}},B_{\mathrm{c}},C_{\mathrm{c}},D)$ are related as $R = T_{\mathrm{c}}^{-1} R_{\mathrm{c}}$. For a system with $\Sigma \nu_{\mathrm{i}} = 2n$ the matrix T_{c}^{-1} consists of columns of R, and thus the 2n corresponding columns of R_{c} are the 2n independent elementary vectors. Since additionally the locations of the elementary vectors are determined by the (invariant) interdependence within R, the canonical controllability matrix R_{c} is unique. Finally all columns of A_{c} are in R_{c}, thus $(A_{\mathrm{c}},B_{\mathrm{c}},C_{\mathrm{c}},D)$ is unique for all i/o equivalent realizations. For systems with $\Sigma \nu_{\mathrm{i}} < 2n$, not all elementary vectors occur in R_{c}. However by postmultiplication of the equation $R = T_{\mathrm{c}}^{-1} R_{\mathrm{c}}$ by an unique orthogonal matrix (determined by the invariant indices) the missing elementary vectors are found and thus the uniqueness follows analogously.

ON THE ORDER AND STRUCTURAL INDICES OF LINEAR SYSTEMS REPRESENTED IN POLYNOMIAL FORM

P. Van den Hof

Lab. Measurement and Control, Department of Mechanical Engineering and Marine Technology, Delft University of Technology, Mekelweg 2, 2628 CD Delft, The Netherlands

Abstract. Linear time–invariant, finite dimensional discrete time systems very often are specified in a polynomial form representation in either a forward or backward shift operator (sometimes called MFD and ARMA form). Considering dynamical systems in terms of their behaviour, i.e. the set of admissible signal trajectories, it is shown that there exists a unifying theory for the determination of McMillan degree and Kronecker indices of systems represented by polynomial matrices in the two shift operators, considering the MFD and ARMA forms as special cases. Moreover the notions of McMillan degree and Kronecker indices can be generalized to noncontrollable and noncausal systems.

Keywords. Multivariable systems; matrix polynomial equations; Kronecker indices; system behaviour; McMillan degree; system theory.

INTRODUCTION

Linear, time–invariant, finite dimensional, discrete time systems can be represented in many different ways.
Descriptions of these type of systems in terms of difference equations are quite common, formulating direct relations between the input and output variables of the system and their time shifts. For a multivariable system with input $u(t)$ and output $y(t)$, two different types of polynomial representations are frequently used:

(1) MFD–form: $P(\sigma)y(t) = Q(\sigma)u(t) \quad \forall\ t \in \mathbb{Z};$ (1)

with $y(t) \in \mathbb{R}^p$, $u(t) \in \mathbb{R}^m$, and P, Q polynomial matrices of size $(p \times p)$, $(p \times m)$ in one indeterminate, and σ the (forward) shift operator: $\sigma w(t) = w(t+1)$.

(2) ARMA–form: $P^*(\sigma^{-1})y(t) = Q^*(\sigma^{-1})u(t) \quad \forall\ t \in \mathbb{Z};$ (2)

with P^*, Q^* similar matrices as P, Q, and σ^{-1} the (backward) shift operator: $\sigma^{-1}w(t) = w(t-1)$.

Both forms are generally applied in systems and control topics, while the choice between the two is more or less dependent on the type of problem that is discussed. In identification, and especially in prediction error identification, ARMA forms are popular because of their natural way of treating output predictions: the output signal $y(t)$ is written as a linear combination of previous output and input signals $y(t-i)$, $u(t-j)$, $i>0$, $j\geq0$; see e.g. Ljung (1987). In this respect the so called monic ARMA forms are very suitable, having $P^*(\sigma^{-1}) = I + P_1\sigma^{-1} + . + P_r\sigma^{-r}$. In the area of control and in parametrization problems the MFD forms are more frequently applied. In parametrization issues, and especially in the construction of canonical forms, this is due to the fact that in MFD forms the McMillan degree and the Kronecker indices of systems can simply be related to determinantial degrees and row degrees of specific MFD forms (Guidorzi, 1975, 1981), whereas for ARMA forms this is much more complex.
Obviously there exist straightforward relations between system representations in either MFD or ARMA form, which can be specified by considering the transfer functions $H(z) = P(z)^{-1}Q(z)$ and $H^*(z) = P^*(z^{-1})^{-1}Q^*(z^{-1})$ as the basic systems' characteristics. Using the equality of transfer functions as a notion of system's equivalence, transformations between MFD and ARMA forms have been shown, see e.g. Wolovich and Elliott (1983). In Gevers (1986) the McMillan degree and Kronecker indices

of ARMA models are studied, showing that also for ARMA models expressions for these phenomena can be given, although more complex than for MFD models. These results have been extended and generalized by Janssen (1988a, 1988b), where parallels between the situations of MFD and ARMA forms have been clearly indicated.
In this contribution it will be shown that there exists a unifying theory for the determination of McMillan degree and structural (Kronecker) indices of models in polynomial form, using polynomial representations that contain both σ and σ^{-1} as indeterminates; MFD and ARMA forms will reduce to special cases in these generalized forms. This generalization actually is due to Willems (1988). Using the basic system characteristic of system behaviour (Willems, 1986) as a more general notion for system equivalence than the transfer function, the concepts of McMillan degree and Kronecker indices will be generalized also to noncontrollable as well as noncausal systems, leading to generalized concepts that will be called system order and left structure indices. First we will introduce the notion of system behaviour as advocated by Willems (1986). In section three some additional definitions and notations related to polynomial matrices will be summarized. Subsequently the concepts of system order and structural indices will be discussed and related expressions will be derived in terms of a polynomial representation.

Some notational conventions: $\mathbb{R}^{p \times m}(z)$ is the field of $(p \times m)$ rational matrices; $\mathbb{R}^{p \times m}[z,z^{-1}]$ is the ring of $(p \times m)$ polynomial matrices in the indeterminates z and z^{-1}; $\det_{\mathbb{R}(z)}(\cdot)$ is the determinant over the field of rational functions; $\mathrm{rank}_{\mathbb{C}}(\cdot)$ is the (ordinary) rank over the field of complex numbers.

SYSTEM REPRESENTATIONS AND SYSTEM BEHAVIOUR

There are many ways of representing linear, time–invariant, finite–dimensional discrete time systems. Apart from the MFD and ARMA forms as mentioned in the introduction, and apart from the transfer function representation, also the state space form is an alternative system description. Obviously these different representations all are representations of the same phenomena: dynamical systems. The transfer function of a dynamical system is very often referred to as its basic characteristic; i.e. two dynamical systems (irrespective of their representation) are equal if and only if they have the same transfer function. However a basic definition of a dynamical system in terms of its transfer function shows a number of shortcomings: (1) a transfer function does not reflect the effect of non-zero initial conditions; and (2) a transfer function does not reflect the effect of noncontrollable modes in the system. A very natural definition of a dynamical system is recently introduced by Willems (1986), employing the notion of system behaviour as the basic characteristic of a dynamical system. Consider a dynamical system with input $u(t)$ and output $y(t)$, $t \in \mathbb{Z}$; then the behaviour $\mathcal{B}$ is defined as the set of all signal trajectories $\binom{y}{u}$: $\mathbb{Z} \to \mathbb{R}^{p+m}$ that are admissible, i.e. that satisfy the restrictions (equations) of the dynamical system. Note that a signal trajectory $\{\binom{y(t)}{u(t)}, t = -\infty, \cdot, \infty\}$ actually is denoted as a mapping from $\mathbb{Z}$ to $\mathbb{R}^{p+m}$, also denoted by $\binom{y}{u} \in (\mathbb{R}^{p+m})^{\mathbb{Z}}$. The behaviour is defined on the level of signals and not by coefficient matrices in specific equations. Therefore the system behaviour is representation independent. It will be illustrated that this behaviour can serve as a proper tool for describing dynamical systems in more detail than e.g. a transfer function representation.

Let us first consider the formal definition of a dynamical system according to Willems (1986).

<u>Definition 1.</u> A *dynamical system* S is defined as a triple: $S = (T, W, \mathcal{B})$, with $T \in \mathbb{R}$ the *time set*, W the *signal set*, i.e. the space in which the (input and output) variables that are related to the system take on their values, and $\mathcal{B} \subset W^T$ the *behaviour* of the system, i.e. the space of all signal trajectories w: $T \to W$ that are compatible with the system. □

The behaviour $\mathcal{B}$ of a dynamical system S will generally be denoted by $\mathcal{B}(S)$. In this paper we will be dealing with discrete–time systems having a time set $T = \mathbb{Z}$. The restriction to linearity, time–invariance and finite dimensionality of the dynamical systems is established by the additional requirements that W is a vector space and that $\mathcal{B}$ is a linear subspace of W^T that is closed in the topology of pointwise convergence, satisfying the shift–invariance property: $w \in \mathcal{B}(S) \Leftrightarrow \sigma w \in \mathcal{B}(S) \Leftrightarrow \sigma^{-1}w \in \mathcal{B}(S)$. In this paper we will consider systems with a fixed number of output and input signals, in such a way that the signals related to the system can be denoted by $w = (y, u)$, with $w \in (\mathbb{R}^{p+m})^{\mathbb{Z}}$, and consequently $\mathcal{B} \subset (\mathbb{R}^{p+m})^{\mathbb{Z}}$.

The basic definition 1 does not distinguish between input and output components of a dynamical system but only refers to system variables (signals). When considering these system's variables as inputs and outputs (causes and effects), a special structure is laid upon the dynamical system, referring to the intuitive appeal that is connected to these notions. The character of an input signal is that it can be chosen freely, it is not bound by the system; the character of an output signal is that it is caused by the input, the system dynamics and possibly a finite number of (initial) conditions. These notions are formalized in Willems (1986, 1988). In this paper we will restrict attention to the class of input–output systems $\Sigma_{p,m}$ with properties as indicated above.

Since the time set $(T = \mathbb{Z})$ and the signal set $(W = \mathbb{R}^{p+m})$ are fixed within this class, two dynamical systems will be equal if and only if they have the same behaviour. Now let us consider the consequences of the introduction of this concept of system behaviour, for polynomial representations of systems. The results as presented in this section are due to Willems (1986, 1988).

<u>Proposition 1.</u> For any system $S \in \Sigma_{p,m}$ there exists a polynomial matrix $T = [P|-Q]$, with $P \in \mathbb{R}^{p \times p}[z, z^{-1}]$, $\det_{\mathbb{R}(z)} P \neq 0$, $Q \in \mathbb{R}^{p \times m}[z, z^{-1}]$ such that
$$\mathcal{B}(S) = \{(y, u) \mid P(\sigma, \sigma^{-1})y - Q(\sigma, \sigma^{-1})u = 0\}. \quad \Box$$

The system S will be said to be induced by the polynomial

matrix T, denoted by $S = M_P(T)$. Using polynomial matrices in two indeterminates, a generalization has been obtained for the MFD and ARMA forms. Note that the condition of P being invertible as a rational matrix is implied by the fact that y is an output variable. The system behaviour will be used for deciding whether two dynamical systems are equal or not in terms of their polynomial representation.

<u>Proposition 2.</u> Let S_1, $S_2 \in \Sigma_{p,m}$ be induced by polynomial matrices T_1, T_2 respectively, as meant in proposition 1. Then $\mathcal{B}(S_1) = \mathcal{B}(S_2)$ if and only if $T_1 = U\,T_2$ with U a polynomial matrix that is unimodular over the ring $\mathbb{R}[z, z^{-1}]$, i.e. $\det_{\mathbb{R}(z)} U = cz^d$ with $c \in \mathbb{R} \setminus \{0\}$ and $d \in \mathbb{Z}$. □

Note that the unimodularity of U has to be considered with respect to the ring $\mathbb{R}[z, z^{-1}]$, in contrast with the situation for MFD or ARMA forms where the rings $\mathbb{R}[z]$ or $\mathbb{R}[z^{-1}]$ are considered. Consequently $U(z, z^{-1})$ has to be invertible within $\mathbb{R}[z, z^{-1}]$, which implies the condition on its determinant as formulated in the proposition. As a result it follows that premultiplication of the polynomial equation with e.g. z or z^{-1} does not change the behaviour of the dynamical system, but only comes down to shifting the equations. For this polynomial description of dynamical systems similar properties can be formulated as are valid in the situation of MFD/ARMA forms:

a. The system's transfer function can be denoted by
$$H(z) = P(z, z^{-1})^{-1}Q(z, z^{-1}).$$

b. This transfer function describes the dynamical system completely if and only if the system is controllable, i.e. a corresponding polynomial representation $T = [P|-Q]$ has the property that P and Q are left coprime with respect to $\mathbb{R}[z, z^{-1}]$ (all left common factors are unimodular with respect to $\mathbb{R}[z, z^{-1}]$), or equivalently $\mathrm{rank}_{\mathbb{C}} T(\lambda, \lambda^{-1}) = p$ for all $\lambda \in \mathbb{C} \setminus \{0\}$.

c. The causality of the system is reflected by the properness of the corresponding transfer function.

It has to be stressed that the notions of controllability and causality are defined as properties of the system behaviour, i.e. on the signal level (Willems, 1986). In this paper we will only refer to the consequences of these properties in specific representations of the behaviour as in the polynomial forms discussed.

Structural properties of these polynomial representations will be elaborated upon in the sequel of this paper. We will briefly pay attention to a second representation of system behaviour in terms of a state space description.

<u>Proposition 3.</u> For any system $S \in \Sigma_{p,m}$ that is causal there exists an integer n and matrices A, B, C, D with $A \in \mathbb{R}^{n \times n}$, $B \in \mathbb{R}^{n \times m}$, $C \in \mathbb{R}^{p \times n}$, $D \in \mathbb{R}^{p \times m}$, such that
$$\mathcal{B}(S) = \{(y, u) \mid \exists x \in (\mathbb{R}^n)^{\mathbb{Z}}, \; \sigma x = Ax + Bu \\ y = Cx + Du\} \quad \Box$$

In this representation x is called the state, and n the state space dimension. Similar to the previous situation the dynamical system S will be said to be induced by the realization (A, B, C, D), denoted by $S = M_{ss}(A, B, C, D)$. A realization of S which has a minimal state space dimension n is called a minimal realization. This notion of minimality has an interpretation that is different from the classical notion of minimal realization, which requires the state space representation to be both controllable and observable. Since controllability is a property of the (external) behaviour of a dynamical system it should not be neglected when considering minimal state space realizations. Consequently minimality with respect to a transfer function to be realized is different from minimality with respect to a system behaviour to be realized.

<u>Proposition 4.</u> Let (A, B, C, D) be a realization of dimension n with $S = M_{ss}(A, B, C, D) \in \Sigma_{p,m}$. Then (A, B, C, D) is a minimal realization of S if and only if

1. (C, A) is an observable pair, i.e. $\mathrm{rank}_{\mathbb{C}} \begin{bmatrix} \lambda I - A \\ C \end{bmatrix} = n$ for all $\lambda \in \mathbb{C}$; and

2. rank $[A \mid B] = n$. □

Minimality of the realization is reflected by observability of the pair (C,A) and by absence of uncontrollable modes in the origin. The latter restriction is required because effects of these modes will not be present in the system behaviour. Note that a nonsymmetrical treatment is created of the classical concepts of controllability and observability. Noncontrollable parts in a state space realization are expressed in the system behaviour, whereas nonobservable parts are not. Additional properties in terms of state space representations are:

a. The system's transfer function equals

$$H(z) = C(zI-A)^{-1}B + D;$$

b. A system S is controllable if and only if a realization (A,B,C,D) of S with dimension n satisfies $\text{rank}_{\mathbb{C}}\,[\lambda I - A \mid B] = n$ for all $\lambda \in \mathbb{C} \setminus \{0\}$.

The minimal state space dimension of a dynamical system will be used in the sequel of this paper when discussing the McMillan degree and the order of dynamical systems.

DEFINITIONS AND NOTATIONS
ON POLYNOMIAL MATRICES

Before we can come to the main goal of this paper: a discussion on the system order and structural indices of dynamical systems in terms of their polynomial representation, we have to introduce some additional notations and definitions on polynomial matrices.

<u>Definition 2.</u> Consider a polynomial matrix $T \in \mathbb{R}^{p \times q}[z,z^{-1}]$. Denote with:

$\delta^{(u)}(T)$:= the maximum power of z in $T(z,z^{-1})$;

$\delta^{(\ell)}(T)$:= the minimum power of z in $T(z,z^{-1})$;

$\delta(T)$:= $\delta^{(u)}(T) - \delta^{(\ell)}(T)$;

T_{i*} := the ith row of $T(z,z^{-1})$, $i \in \underline{p}$; with $\underline{p} := \mathbb{Z} \cap [1,p]$;

$\nu_i^{(u)}(T)$:= the maximum power of z in $T_{i*}(z,z^{-1})$;

$\nu_i^{(\ell)}(T)$:= the minimum power of z in $T_{i*}(z,z^{-1})$;

$\nu_i(T)$:= $\nu_i^{(u)}(T) - \nu_i^{(\ell)}(T)$;

$\Gamma_{hr}(T)$:= the leading row coefficient matrix of T, i.e. the coefficient matrix related to the highest row degree terms in T;

$\Gamma_{lr}(T)$:= the trailing row coefficient matrix of T, i.e. the coefficient matrix related to the lowest row degree terms in T;

$\pi_i^{(u)}(T)$:= the maximum power of z in any $i \times i$–minor of $T(z,z^{-1})$;

$\pi_i^{(\ell)}(T)$:= the minimum power of z in any $i \times i$–minor of $T(z,z^{-1})$. □

The well known property of row properness of polynomial matrices (Wolovich, 1974) has to be extended to the situation of polynomial matrices in two indeterminates.

<u>Definition 3.</u>

A polynomial matrix $T \in \mathbb{R}^{p \times q}[z,z^{-1}]$ is called *row proper with respect to* $\mathbb{R}[z]$ if rank $\Gamma_{hr}(T) = p$; it is called *row proper with respect to* $\mathbb{R}[z^{-1}]$ if rank $\Gamma_{lr}(T) = p$.

T is called *bilaterally row proper* if it is row proper with respect to $\mathbb{R}[z,z^{-1}]$, i.e. rank $\Gamma_{hr}(T) = $ rank $\Gamma_{lr}(T) = p$. □

The notion of bilaterally row properness was first presented by Willems (1986).

MCMILLAN DEGREE, SYSTEM ORDER
AND MINIMAL STATE SPACE DIMENSION

We will start this section by considering the McMillan degree of a rational matrix which is interpreted as the transfer function of a dynamical input–output system $S \in \Sigma_{p,m}$.

<u>Definition 4.</u>

The McMillan degree of a rational matrix $H(z) \in \mathbb{R}^{p \times m}(z)$ is defined as:
$\delta_M(H) := $ the sum of polar degrees at all poles of $H(z)$. □

For finite poles this sum of polar degrees can be determined by the sum of the degrees of the denominator polynomials in the Smith–McMillan form of $H(z)$. The polar degree at infinity has to be added to arrive at the McMillan degree (see e.g. Kailath, 1980). For systems that are controllable and causal there exists a close connection between the McMillan degree of the transfer function and the minimal state space dimension that is required for representing the corresponding input–output system. This is formulated in the following proposition.

<u>Proposition 5.</u>

Let S be a dynamical system $S \in \Sigma_{p,m}$ that is controllable and causal, and let $H(z) \in \mathbb{R}^{p \times m}(z)$ be the corresponding proper transfer function of S. Then
$\delta_M(H) = $ the minimal state space dimension of any realization (A,B,C,D) of S. □

The proposition is mainly a classical result from the theory of dynamical systems, see e.g. Rosenbrock (1970) and Kailath (1980). However note that the classical minimal realization of a transfer function requires also observability in order to match the McMillan degree. The McMillan degree is considered to be a measure of complexity that is related to the transfer function of an input–output dynamical system. Therefore it takes account only of the controllable part of a dynamical system and it neglects any noncontrollability. The dimension of a minimal state space realization is only defined for causal systems, related to proper transfer functions $H(z)$. For this reason it is appropriate to consider a closely related measure of complexity that takes account of both noncontrollable as well as noncausal parts in a system. To this end we will use the following notion of system's order, formulated in terms of the model behaviour (Willems, 1986).

<u>Definition 5.</u>

Let $S = (\mathbb{Z}, W, \mathcal{B})$ be a dynamical system $S \in \Sigma_{p,m}$ with behaviour $\mathcal{B}$. Then define the order of S as:

$$n(S) := \sum_{N=0}^{\infty} (\dim(\mathcal{B}^N) - \dim(\mathcal{B}^{N-1}) - m) \qquad (3)$$

with $\mathcal{B}^N := \mathcal{B}|_{\mathbb{Z} \cap [0,N-1]}$, and $\dim(\mathcal{B}^{-1}) := 0$. □

For the interpretation of (3) we refer to Willems (1986). The connection of $n(S)$ with the McMillan degree and the minimal state space dimension is presented in the following theorem, which also gives an expression for the determination of $n(S)$ based on a polynomial representation.

<u>Theorem 1</u>

Let S be a dynamical system $S \in \Sigma_{p,m}$ being induced by a polynomial matrix $T \in \mathbb{R}^{p \times (p+m)}[z,z^{-1}]$, with $T = [P \mid -Q]$. Then

a. $n(S) = \pi_p^{(u)}(T) - \pi_p^{(\ell)}(T)$; $\qquad (4)$

b. – there exists a permutation matrix $X \in \mathbb{R}^{(p+m) \times (p+m)}$ such that $S^* := M_P(TX) \in \Sigma_{p,m}$ is a causal dynamical system with m inputs and p outputs; and

 – Any such system S^* has a minimal realization with dimension $n(S)$.

c. $n(S) \geq \delta_M(P^{-1}Q)$ with equality if and only if $\text{rank}_{\mathbb{C}} T(\lambda, \lambda^{-1}) = p$ for all $0 \neq \lambda \in \mathbb{C}$, or equivalently S is controllable.

295

d. $n(S) \leq \sum_{i=1}^{p} \nu_i(T)$ with equality if and only if T is bilaterally row proper. □

<u>Proof</u>
Parts (b) and (d) are proved in Willems (1986); the proofs of parts (a) and (c) can be found in Van den Hof (1989).

Note that the order $n(S)$ is invariant for any shift in polynomial matrix T, i.e. for premultiplication of T with z or z^{-1}.
It follows from this theorem that $n(S)$ is a generalization of both the McMillan degree of a transfer function and the minimal state space dimension of a state space realization. If S is causal then the permutation matrix X can be chosen the identity matrix and $n(S)$ equals the corresponding minimal state space dimension. If S is noncausal, the signal variables w=(y,u) can be reordered in such a way that the reordered system S^* is causal and again allows a state space representation with minimal dimension $n(S)$. The order $n(S)$ is invariant for any nonsingular constant column operation on T. Note that $\mathcal{B}(S^*)$ is constructed according to $w\in\mathcal{B}(S^*) \Leftrightarrow Xw\in\mathcal{B}(S)$.
Part c of the theorem shows that $n(S)$ takes account of noncontrollable modes in the system. If S is controllable then $n(S)$ is equal to the McMillan degree of the corresponding transfer function. Part a shows a general expression for $n(S)$ in terms of a polynomial representation. It requires the evaluation of the degrees of all $p\times p$ minors of T.
Bilaterally row properness of a polynomial representation appears to be an important property since it allows the system order to be easily determined from the sum of row degrees in the polynomial matrix, as formulated in part d.

In the following example these results are illustrated.

<u>Example 1.</u>
Consider a dynamical system $S\in\Sigma_{2,1}$, induced by the polynomial matrix $T(z,z^{-1}) = [P|-Q]$, with

$$P(z,z^{-1}) = \begin{bmatrix} z + z^{-2} & z^{-1} \\ z^{-1} & 1 \end{bmatrix}, \quad Q(z,z^{-1}) = \begin{bmatrix} 1 \\ 0 \end{bmatrix}.$$

It can easily be verified that rank $T(\lambda,\lambda^{-1}) = 2$ for all $\lambda\in\mathbb{C}\setminus\{0\}$, which implies that S is controllable. Its transfer function $H(z) = \begin{bmatrix} z^{-1} \\ -z^{-2} \end{bmatrix}$.

The row degrees of T satisfy: $\nu_1^{(u)}=1$, $\nu_1^{(\ell)}=-2$, $\nu_2^{(u)}=0$, $\nu_2^{(\ell)}=-1$ leading to $\nu_1+\nu_2=4$.
In order to check the bilaterally row properness of T we verify that $\Gamma_{hr}(T) = \begin{bmatrix} 1 & 0 & 0 \\ 0 & 1 & 0 \end{bmatrix}$ and $\Gamma_{lr}(T) = \begin{bmatrix} 1 & 0 & 0 \\ 1 & 0 & 0 \end{bmatrix}$.
Since $\Gamma_{lr}(T)$ does not have full row rank, T is not bilaterally row proper and consequently (by theorem 1d) $n(S)<4$.

Evaluating expression (4) shows $\pi_2^{(u)}(T)=1$, $\pi_2^{(\ell)}(T)=-1$ leading to $n(S)=2$ which is equal to the McMillan degree of $H(z)$.
Matrix T can be transformed to a bilaterally row proper form by unimodular premultiplication. Construct $T^* = UT$ with $U(z,z^{-1})=\begin{bmatrix} 1 & z^{-2} \\ 0 & 1 \end{bmatrix}$, it follows that $T^* = \begin{bmatrix} z & 0 & -1 \\ z^{-1} & 1 & 0 \end{bmatrix}$.

T^* is bilaterally row proper with row degrees $\nu_1+\nu_2 = 1+1 = 2 = n(S)$. □

For some specific situations the expression (4) for the system order can be further simplified, as formulated in the following proposition.

<u>Proposition 6</u>
Let S and T be as in theorem 1. Then

$\pi_p^{(u)}(T)= \delta^{(u)}(\det(P))$ if and only if $P^{-1}Q$ has no poles in $z=\infty$;
$\pi_p^{(\ell)}(T)= \delta^{(\ell)}(\det(P))$ if and only if $P^{-1}Q$ has no poles in $z=0$. □

<u>Proof.</u> See Van den Hof (1989).

Note that the absence of poles in $z=\infty$ refers to the situation of a proper transfer function, and consequently to a causal system. The results of theorem 1 and proposition 6 include a generalization of the results on the McMillan degree of MFD and ARMA models as reported in Janssen (1988a, 1988b), where separate results were derived for models represented in one of the shift operators σ or σ^{-1}. A generalized model representation in both shift operators leads to an overall theory that covers both situations of MFD and ARMA representations. The connections with the results of Janssen (1988a, 1988b) are shown in the following corollary.

<u>Corollary 1</u>
Let S be a dynamical system $S\in\Sigma_{p,m}$, represented in MFD or ARMA form by:
$$S=M_P(T_f), \ T_f=T_f(z)=[P_f|-Q_f]\in\mathbb{R}^{p\times(p+m)}[z], \quad \text{and}$$
$$S=M_P(T_b), \ T_b=T_b(z^{-1})=[P_b|-Q_b]\in\mathbb{R}^{p\times(p+m)}[z^{-1}].$$
Then the following statements hold true:

a. If $\text{rank}_{\mathbb{C}}T_f(\lambda)=p$ for $\lambda=0$ then $\pi_p^{(\ell)}(T_f) = 0$;

b. If $\text{rank}_{\mathbb{C}}T_b(\lambda)=p$ for $\lambda=0$ then $\pi_p^{(u)}(T_b) = 0$;

c. If P_f, Q_f are left coprime with respect to $\mathbb{R}[z]$, and $P_f^{-1}Q_f$ has no poles in $z=\infty$ then
$$\delta_M(P_f^{-1}Q_f) = n(S) = \delta^{(u)}(\det(P_f));$$

d. If P_b, Q_b are left coprime with respect to $\mathbb{R}[z^{-1}]$ and $P_b^{-1}Q_b$ has no poles in $z=0$ then
$$\delta_M(P_b^{-1}Q_b) = n(S) = -\delta^{(\ell)}(\det(P_b)).$$ □

<u>Proof</u>
a) If $\pi_p^{(\ell)}(T_f)\neq 0$ then $\pi_p^{(\ell)}(T_f)>0$ and consequently $T_f(\lambda)$ has a zero in $\lambda=0$, which is conflicting with the condition $\text{rank}_{\mathbb{C}}T_f(\lambda)=p$ for $\lambda=0$.

b) If $\pi_p^{(u)}(T_b)\neq 0$ then $\pi_p^{(u)}(T_b)<0$ and consequently $T_b(\lambda)$ has a zero in $\lambda=0$, which is conflicting with the condition $\text{rank}_{\mathbb{C}}T_b(\lambda)=p$ for $\lambda=0$.

c) The statement follows directly from theorem 1c, proposition 6, and part a of this corollary, taking into account that left coprimeness with respect to $\mathbb{R}[z]$ is equivalent with the condition $\text{rank}_{\mathbb{C}}T(\lambda)=p$ for all $\lambda\in\mathbb{C}$.

d) The proof is similar as the proof of c, applying theorem 1c, proposition 6 and part b of this corollary. □

There is a certain symmetry in the expressions for MFD and ARMA representations of systems. However there are still some important differences. Note that in situations c and d as formulated in the corollary, the order of the system can be extracted from the matrices P_b, respectively P_f only, irrespective of the matrices Q_b, Q_f. However this situation requires different restrictions on the dynamical systems dependent on the question whether MFD or ARMA representations are involved. In the case of an MFD representation (situation c), this restriction is the quite natural condition of causality of the system (the system should not contain anticipations). In the case of ARMA representations (situation d) the condition of H(z) having no poles in $z=0$ (the system should not contain any delays) is much more restrictive. This is exactly the reason why model sets of causal models with specified model order n are much easier to parametrize in MFD form than in ARMA form. Absence of poles in the origin, is exactly the restriction that is made by Bokor and Keviczky (1987) in order to come up with a canonical ARMA parametrization for causal systems having a fixed

McMillan degree.

STRUCTURAL INDICES

In the same line of thought as exhibited in the previous section, results can be formulated for the structural indices of dynamical systems, represented in the generalized polynomial forms as discussed. Structural indices constitute a further specification of the order of dynamical systems and are often used for the construction of sets of dynamical systems that allow an identifiable and differentiable parametrization. Let us first consider the different structure indices that will be considered.

<u>Definition 6</u>
Let (A,B,C,D) be a state space realization with dimension n of a dynamical system $S \in \Sigma_{p,m}$, with

$C = [c_1^T \ c_2^T \ \cdot \ \cdot \ c_p^T]^T$. Consider the observability matrix

$\mathcal{O} = [C^T \ A^T C^T \ \cdot \ \cdot \ (A^T)^{n-1} C^T]^T$ and evaluate the linearly independent rows in $\mathcal{O}$ from top to bottom. Order these independent rows as

$$\{c_1, c_1 A, .., c_1 A^{\gamma_1 - 1}, .., c_p, .., c_p A^{\gamma_p - 1}\}$$

then the integers $(\gamma_i)_{i=1,.,p}$ are defined to be the *observability indices* of the realization (A,B,C,D). □

The observability indices of a realization are often called the (left) Kronecker indices of the realization, and actually are equal to the left Kronecker indices of the matrix pencil

$\begin{bmatrix} zI - A \\ C \end{bmatrix}$ (Kailath, 1980). The notation $(\gamma_i)_{i=1,.,p}$ refers to

a mapping $\mathbb{Z} \cap [1,p] \to \mathbb{Z}_+$, which means that the ordering of the observability indices is of importance, (e.g. $(2,1) \neq (1,2)$). When writing $\{\gamma_i\}_{i=1,.,p}$ we will refer to a set with p elements, where consequently any ordering of the elements is not relevant.

In the case of a minimal realization (A,B,C,D) it follows directly that

$$\sum_{i=1}^{p} \gamma_i = n(S). \tag{5}$$

The observability indices are presented being related to state space realizations and consequently their interpretation is restricted to causal systems.
Related structure indices can also be defined on the basis of transfer functions. The definition of these – so called – left dynamical indices will be adopted from Forney (1975), Kailath (1980) and Janssen (1988a), and will not be further elaborated here. For more details the reader is referred to these references.

<u>Definition 7</u>
Let $H(z)$ be a rational matrix, $H(z) \in \mathbb{R}^{p \times m}(z)$, satisfying $\text{rank}_{\mathbb{R}(z)} H = p$. The set of *left dynamical indices* $\{\kappa_i\}_{i=1,.,p}$ of $H(z)$, is defined as the set of row degrees of any minimal polynomial basis in $\mathbb{R}[z]$, for the rational vector space generated by the rows of the matrix $[I_p | H(z)]$.
Minimality of this basis is considered with respect to the sum of the row degrees. □

These left dynamical indices of $H(z)$ are often referred to as the left Kronecker indices of $H(z)$. For reasons that are discussed in Janssen (1988a), the name "left dynamical indices" will be used in order . In Janssen (1988a) it is also proven that these left dynamical indices of $H(z)$ sum up to its McMillan degree:

$$\sum_{i=1}^{p} \kappa_i = \delta_M(H). \tag{6}$$

Note that these left dynamical indices are defined on rational matrices, i.e. on the level of transfer functions. This means that these structure indices do not contain

information on any noncontrollable part of a dynamical system. Again, as in the case of the system order $n(S)$, both sets of structure indices can be related to each other in terms of a polynomial matrix representation, in order to deal with both noncontrollable and noncausal dynamical systems. To this end the following set of structure indices is introduced.

<u>Definition 8</u>
Let S be a dynamical system $S \in \Sigma_{p,m}$, and let T be a polynomial matrix $T \in \mathbb{R}^{p \times (p+m)}[z,z^{-1}]$.
The set of *left structure indices* of S, denoted by $\{\rho_i\}_{i=1,.,p}$, is defined as the set of row degrees $\{\nu_i(T)\}_{i=1,.,p}$ of any bilaterally row proper matrix T that induces S. □

It can be verified that this set of left structure indices is an invariant of the system S, i.e. it is not dependent on the specific bilaterally row proper matrix T that it is induced by. Its relation with the structure indices previously defined, is given in the following proposition.

<u>Proposition 7.</u> Let S be a dynamical system $S \in \Sigma_{p,m}$ being induced by a polynomial matrix $T \in \mathbb{R}^{p \times (p+m)}[z,z^{-1}]$ then the following statements hold true:
a. The set of left structure indices of S is equal to the set of left dynamical indices $\{\kappa_i\}_{i=1,.,p}$ of its transfer function if and only if S is controllable;
b. The set of left structure indices of S is equal to the set of observability indices $\{\gamma_i\}_{i=1,.,p}$ of any minimal state space realization (A,B,C,D) that satisfies

$M_{ss}(A,B,C,D) = M_P(TX)$ and $X \in \mathbb{R}^{(p+m) \times (p+m)}$ a nonsingular matrix. □

<u>Proof.</u>
Part a of this theorem is an extension of results of Forney (1975) and Kailath (1980); for the situation that $T \in \mathbb{R}^{p \times (p+m)}[z]$, it has been shown that the set of row degrees of T is equal to the set of left dynamical indices of $P^{-1}Q$ if and only if P and Q are left coprime and T is row proper with respect to $\mathbb{R}[z]$. Extension of this result to polynomials in $\mathbb{R}[z,z^{-1}]$ follows from the consideration that {left coprimeness and row properness with respect to $\mathbb{R}[z]$} $\Leftrightarrow$ {left coprimeness and bilaterally row properness with respect to $\mathbb{R}[z,z^{-1}]$}.
Part b follows from results of Willems (1986), and the consideration that the set of row degrees of T is invariant for postmultiplication of T with a constant nonsingular matrix. □

The set of left structure indices of S coincides with the set of left dynamical indices in case of controllability, and it coincides with the set of observability indices in case of causality. Note that equality as meant in the proposition refers to the sets of indices, i.e. equality of the indices up to ordering.
The left structure indices as defined act as a further structural specification of dynamical systems with order

$$n(S) = \sum_{i=1}^{p} \rho_i(S).$$

The left structure indices have been defined on the basis of a bilaterally row proper polynomial representation of a system. It has to be stressed that this set of indices also can be formulated purely on the basis of system behaviour (see Willems, 1986). Moreover this set of structure indices can also be shown to be related to the left Kronecker

indices of a matrix pencil $\begin{bmatrix} zE - A \\ C \end{bmatrix}$ generated by a

generalized state space representation.

Similar results as formulated in proposition 6 and corollary 1 with respect to the system order, can now be given for the evaluation of the set of left structure indices in some specific situations. We will briefly pay attention to the situation when the set of left structure indices can be determined on the basis of polynomial matrix P only, as a submatrix of the matrix T that induces S. Moreover the specific situation of MFD and ARMA forms will be taken into account. These results will be presented in the following two propositions. The proofs are omitted.

Proposition 8
Let S be a dynamical system $S \in \Sigma_{p,m}$ induced by a bilaterally row proper polynomial matrix $T \in \mathbb{R}^{p \times (p+m)}[z,z^{-1}]$, with $T=[P|-Q]$, then

a. $\{\nu_i^{(u)}(T)\}_{i=1,..,p} = \{\nu_i^{(u)}(P)\}_{i=1,..,p}$ if and only if $P^{-1}Q$ has no poles in $z=\infty$;

b. $\{\nu_i^{(\ell)}(T)\}_{i=1,..,p} = \{\nu_i^{(\ell)}(P)\}_{i=1,..,p}$ if and only if $P^{-1}Q$ has no poles in $z=0$. $\qquad \square$

Proposition 9
Let S be a dynamical system $S \in \Sigma_{p,m}$, represented in MFD or ARMA form by:
$$S=M_P(T_f),\ T_f=T_f(z)=[P_f|-Q_f] \in \mathbb{R}^{p \times (p+m)}[z], \quad \text{and}$$
$$S=M_P(T_b),\ T_b=T_b(z^{-1})=[P_b|-Q_b] \in \mathbb{R}^{p \times (p+m)}[z^{-1}].$$

Then the following statements hold true:

a. If $\mathrm{rank}_{\mathbb{C}} T_f(\lambda)=p$ for $\lambda=0$ and P_f, Q_f are row proper with respect to $\mathbb{R}[z]$, and $P_f^{-1}Q_f$ has no poles in $z=\infty$ then $\{\rho_i(S)\}_{i=1,..,p} = \{\nu_i^{(u)}(P_f)\}_{i=1,..,p}$;

b. If $\mathrm{rank}_{\mathbb{C}} T_b(\lambda)=p$ for $\lambda=0$ and P_b, Q_b are row proper with respect to $\mathbb{R}[z^{-1}]$, and $P_b^{-1}Q_b$ has no poles in $z=0$ then $\{\rho_i(S)\}_{i=1,..,p} = \{-\nu_i^{(\ell)}(P_b)\}_{i=1,..,p}$. $\qquad \square$

As mentioned before the expressions for the determination of the set of left structure indices shows a clear resemblance with similar expressions for the system order n(S). In order to be able to extract the structure indices of S only from the submatrix P within T, additional conditions of causality (for MFD forms) and absence of delays (for ARMA forms) have to be satisfied. These two symmetrical conditions clearly show the symmetrical treatment of the two different forms.
Finally we will illustrate the results as presented in this section.

Example 2
Consider again the dynamical system $S \in \Sigma_{2,1}$ as introduced in example 1, induced by polynomial matrix $T^*=[P^*|-Q^*]$,
$$T^*(z,z^{-1}) = \begin{bmatrix} z & 0 & -1 \\ z^{-1} & 1 & 0 \end{bmatrix}$$

Since T^* is bilaterally row proper the set of left structure indices $\{\rho_1,\rho_2\}$ is determined by the set of row degrees of T^*, is equal to $\{1,1\}$.
An equivalent MFD form for T^* that induces the same system, can be obtained by premultiplying the second row of T^* by z. This is a unimodular operation on T^*. The resulting matrix is
$$T_1 = [P_1|-Q_1] = \begin{bmatrix} z & 0 & -1 \\ 1 & z & 0 \end{bmatrix}.$$

With this matrix T_1 the conditions of proposition 9a are satisfied and consequently the set of upper row degrees of P_1 is equal to the set of left structure indices of S.
An equivalent ARMA form for T^* that induces the same system, can be obtained by premultiplying the first row of T^* by z^{-1}. This is a unimodular operation on T^*. The resulting matrix is
$$T_2 = [P_2|-Q_2] = \begin{bmatrix} 1 & 0 & -z^{-1} \\ z^{-1} & 1 & 0 \end{bmatrix}.$$

This matrix T_2 satisfies the first two conditions of proposition 9b; however since the transfer function of S,

$$H(z) = \begin{bmatrix} z^{-1} \\ -z^{-2} \end{bmatrix}, \text{ contains poles in } z=0, \text{ the third condition}$$

is not satisfied. It follows that the set of left structure indices of S can not be determined from the row degrees of P_2. $\qquad \square$

CONCLUSIONS

In this paper the system order and the left structure indices have been presented as generalizations of respectively the McMillan degree and the left Kronecker indices of linear systems, being applicable to systems that are not necessarily causal and/or controllable. To this end the concept of dynamical system in terms of its behaviour is adopted. General expressions for the evaluation of system order and structure indices are derived in terms of system representations in polynomial matrix form. The polynomial forms discussed exhibit two shift operators, being a generalization of both MFD and ARMA forms. Consequently a unifying theory results for the evaluation of system order and structure indices of linear systems in polynomial forms, in which MFD and ARMA forms are special cases that are completely symmetrical.

REFERENCES

Bokor, J. and L. Keviczky (1987). ARMA canonical forms obtained from constructability invariants. *Int. J. Control, 45*, 861–873.

Forney, G.D. (1975). Minimal bases of rational vector spaces, with applications to multivariable linear systems. *SIAM J. Control Optimiz., 13*, 493–520.

Guidorzi, R.P. (1975). Canonical structures in the identification of multivariable systems. *Automatica, 11*, 361–374.

Guidorzi, R.P. (1981). Invariants and canonical forms for systems structural and parametric identification. *Automatica, 17*, 117–133.

Janssen, P.H.M. (1988a). *On Model Parametrization and Model Structure Selection for Identification of MIMO-Systems.* Dr. Dissertation, Dept. Electrical Engineering, Eindhoven Univ. Technology, The Netherlands.

Janssen, P.H.M. (1988b). General results on the McMillan degree and the Kronecker indices of ARMA and MFD models. *Int. J. Control, 48*, 591–608.

Kailath, T. (1980). *Linear systems.* Prentice–Hall, Englew. Cliffs, N.J.

Ljung, L. (1987). *System Identification: Theory for the User.* Prentice–Hall Inc., Englewood Cliffs, N.J.

Rosenbrock, H.H. (1970). *State space and multivariable theory.* Nelson, London.

Van den Hof, P.M.J. (1989). *On Residual–Based Parametrization and Identification of Multivariable Systems.* Dr. Dissertation, Dept. Electrical Engineering, Eindhoven Univ. Technology, The Netherlands.

Willems, J.C. (1986). From time series to linear system – Part I: Finite dimensional linear time–invariant systems, *Automatica, 22*, 561–580.

Willems, J.C. (1988). Models for dynamics. In: Kirchgraber, U. and H.O. Walther (Eds.), *Dynamics Reported, Vol. 2*, pp. 171–269. Wiley and Teubner.

Wolovich, W.A. (1974). *Linear Multivariable Systems.* Applied Mathematical Sciences, Vol. 11. Springer Verlag, New York.

Wolovich, W.A. and H. Elliott (1983). Discrete models for linear multivariable systems. *Int. J. Control, 38*, 337–357.

MODAL FRAMEWORK FOR SYSTEMS WITH IDENTICAL UNITS

J. Calazans de Castro

*Center of Technology, Universidade Federal da Paraíba, 58.000 João Pessoa,
Paraíba, Brazil*

Abstract. A theoretical technique for modal study of systems with identical units is
proposed. Under the assumption of identical interconnections, a motivated basic structure
for understanding and analyzing linear time-invariant models of dynamic systems with
identical units is developed. It is shown that for these systems there are only two kinds
of modes whose associated eigenvectors satisfy two basic properties. An algorithm is
presented for determination of eigenvectors and chains of generalized eigenvectors of
the system by using reduced order matrices. It is shown how the technique may be applied
on a system where a group of identical units is connected to other group of different
units.

Keywords. Linear systems, modal control, multivariable systems, eigenvalues, model
reduction.

INTRODUCTION

Modern systems, such as large scale systems, have
evolved into complex dynamic models, giving rise
to new problems which necessitate new techniques
to deal with. In those complex systems, it is not
uncommon to find situations where some components
or units are identical and accomplish the same tasks.
In an electric power plant, for instance, some, if
not all, generators are identical. For dynamic
analysis, these units are usually represented by an
equivalent unit in the system model. However, by
using that simplification, a lot of internal
information is lost and, as a consequence, the
model can not represent some situations of dynamic
instability of the system that may occur.

Due to the practice of using equivalent model, the
publication about identical units is scarce and
usually applied on only two identical units. Schleif
and colleagues (1979) and Crenshaw and colleagues
(1983), study power systems with two identical
generators. They obtained interesting results by
using experimental and heuristic analysis. However,
mathematical justifications for the results are not
presented. Alden, Nolan and Bayne (1977) present a
modal analysis of shaft dynamics in identical
generators. Their study although applied on a
particular problem, gives a good analysis of a
two-unit system, however, the extension to a N
generator system is not clear and some typical
characteristics are not observed.

In this paper, a new analytical framework for
analysis of a system with identical units is
proposed. The study is based on the theory of
linear time-invariant systems and develops a well
founded technique to deal with the problem.

The paper is organized as follows. First, it is
presented the linearized system with its special
matrix structure and some modal definitions with a
heuristic introduction to the problem. It is shown
that the system has two kinds of modes, called
intrasystem and local modes. The eigenvalues
associated with intrasystem modes are invariant with
the introduction of new units, increasing however
their multiplicity. The number of eigenvalues
associated with local modes does not depend on the
number of units. Then, the main results and an
algorithm for determination of the eigenvectors
associated with the repeated eigenvalues of the
intrasystem modes are developed. Finally it is shown
how the technique may be applied on a system where
a group of identical units is connected to other
group of different units.

BASIC CONCEPTS

Consider a system composed of p identical units with
identical interconnections represented by

$$\dot{X} = \bar{A}X + \bar{B}u$$
$$y = \bar{C}X \qquad (1)$$

where

$$\bar{A} = \begin{vmatrix} A & H & \dots & H \\ H & A & \dots & H \\ \dots & \dots & \dots & \dots \\ \dots & \dots & \dots & \dots \\ H & H & \dots & A \end{vmatrix} \qquad \begin{array}{l} \bar{B} = \text{bloc diag (B B } \dots \text{ B)} \\ \\ \bar{C} = \text{Bloc diag (C C } \dots \text{ C)} \end{array}$$

Each unit is a subsystem of n-th order.

It is known (Perez-Arriaga, Verghese and Schweppe,
1982), that the j-th component, g_i^j, of the right
eigenvector g_i represents the activity of the
variable x_j on mode i and the component v_i^j of the
left eigenvector v_i represents the weight of this
activity. Due to the system structure, for some
modes, the components of g_i and v_i associated with
similar variables may have the same value, which
corresponds the same activity and the same weight.
The modes which satisfy this property are defined
as *local modes* , and their eigenvectors can be
represented by

$$g_i^T = \begin{vmatrix} G_1^T \dots & G_p^T \end{vmatrix} \text{ and } v_i^T = \begin{vmatrix} V_1^T \dots & V_p^T \end{vmatrix}$$

with $G_1 = \dots = G_p$ and $V_1 = \dots V_p$. G_j and V_j are
vectors of order n associated with the unit j.

However, in an interconnected system there are some
modes for which the variables of two units may be
180^0 out of fase to each other (Schleif and colleagues,
1979). These modes are defined as *intrasystem modes*.

For these modes, the eigenvalue problem is

$$\begin{vmatrix} A-\lambda_i I & H & \cdots & H \\ H & A-\lambda_i I & \cdots & H \\ \cdot & \cdot & \cdots & \cdot \\ \cdot & \cdot & \cdots & \cdot \\ H & H & & A-\lambda_i I \end{vmatrix} \begin{vmatrix} G_1 \\ G_2 \\ \cdot \\ \cdot \\ G_p \end{vmatrix} = 0 \qquad (2)$$

Summing up these equations results

$$\left| A + (p-1)H - \lambda_i I \right| (G_1 + G_2 + \ldots + G_p) = 0 \qquad (3)$$

Assuming that

$$\text{rank}(A + (p-1)H - \lambda_i I) = n \qquad (4)$$

the unique solution for Eq. (3) is

$$G_1 + G_2 + \ldots + G_p = 0 \qquad (5)$$

Similarly, it can be shown that

$$V_1 + V_2 + \ldots + V_p = 0 \qquad (6)$$

Eq. (4) to Eq. (6) define intrasystem modes.

THE EIGENVALUE PROBLEM FOR LOCAL MODES

The formulation for determination of eigenvalues and eigenvectors for local modes is

$$\begin{vmatrix} A-\lambda_\ell I & H & \cdots & H \\ H & A-\lambda_\ell I & \cdots & H \\ \cdot & \cdot & \cdots & \cdot \\ \cdot & \cdot & \cdots & \cdot \\ H & H & \cdots & A-\lambda_\ell I \end{vmatrix} \begin{vmatrix} G_1 \\ G_1 \\ \cdot \\ \cdot \\ G_1 \end{vmatrix} = 0 \qquad (7)$$

which can be reduced to

$$[A + (p-1)H - \lambda_\ell I]G_1 = 0 \qquad (8)$$

Then, the eigenvalues and eigenvectors for local modes are obtained by using a reduced order matrix. Also, from Eq. (8) it is seen that the local modes are dependent on the number of units.

From Eq. (4) and Eq. (8) it is seen that the system has only local and intrasystem modes and that there are n local and $(p-1)n$ intrasystem eigenvalues for any number of units.

A similar development is used to determine the left eigenvectors associated with local modes, by using the matrix $A^T + (p-1)H^T$. The equivalent model for study of local modes of the composite system, under equal inputs in all units, is

$$\dot{X} = [A + (p-1)H]X + B\eta$$
$$\sigma = cX$$

THE EIGENVALUE PROBLEM FOR INTRASYSTEM
MODES

From Eq. (2) and Eq. (5), results

$$(A - H - \lambda_i I)G_1 = 0 \qquad (9)$$

$G_2, \ldots, G_p$ are also solution of Eq. (9), then, to satisfy Eq. (5), $G_2 = \alpha_2 G_1, \ldots, G_p = \alpha_p G_1$, with $\alpha_1 + \alpha_2 + \ldots + \alpha_p = 0$. Therefore, the eigenvalue problem for intrasystem modes is solved by using the n x m matrix A–H, which has n intrasystem eigenvalues. However, as the system has $(p-1)n$ intrasystem eigenvalues, they must be repeated with multiplicity p-1.

If the inputs are $u_1 = \alpha_1 u$, $u_2 = \alpha_2 u$, ..., $u_p = \alpha_p u$, with $\alpha_1 + \alpha_2 + \ldots + \alpha_p = 0$, only the intrasystem modes affect the response, then the equivalent model to study intrasystem modes of the system, under the above inputs, is

$$\dot{X} = (A-H)X + b\eta$$
$$\sigma = CX$$

<u>Preservation of Structure for Generalized Eigenvectors Associated with Intrasystem Modes</u>

It can be shown that $(\overline{A} - \lambda I)^k$, for $k \geq 0$, is a symmetrical matrix, as is $\overline{A} - \lambda I$.

<u>Theorem 1.</u> Let the matrix $(\overline{A} - \lambda I)^k$ be represented by

$$\overline{A}_k \triangleq (\overline{A} - \lambda I)^k \triangleq \begin{vmatrix} A_k & H_k & \cdots & H_k \\ H_k & A_k & \cdots & H_k \\ \cdot & \cdot & \cdots & \cdot \\ \cdot & \cdot & \cdots & \cdot \\ H_k & H_k & \cdots & A_k \end{vmatrix} \qquad (10)$$

Then, $A_k + (p-1)H_k = [A + (p-1)H - \lambda I]^k$

<u>Proof.</u> Firstly, considering the multiplication of $\overline{A} - \lambda I$ by itself, one obtains $A_2 + (p-1)H_2 = [A + (p-1)H - \lambda I]^2$. Now, considering the following multiplication

$$\overline{A}_j(\overline{A} - \lambda I) \triangleq \begin{vmatrix} A_{j+1} & H_{j+1} & \cdots & H_{j+1} \\ \cdot & \cdot & \cdot & \cdot & \cdot \\ H_{j+1} & H_{j+1} & \cdots & A_{j+1} \end{vmatrix} \qquad (11)$$

it is seen that $A_{j+1} + (p-1)H_{j+1} = [A_j + (p-1)H_j] \times [A + (p-1)H - \lambda I]$. Then, the assertion is proved by induction.

<u>Theorem 2.</u> Consider $(\overline{A} - \lambda I)^k$ represent by Eq. (10). Then,

$$A_k - H_k = (A - H - \lambda I)^k$$

<u>Proof.</u> From $(\overline{A} - \lambda I)^2$, it is easily obtained $A_2 - H_2 = (A - H - \lambda I)^2$. Also, by considering the multiplication of $\overline{A}_j$ by $\overline{A} - \lambda I$ as described by Eq. (11) one obtains $A_{j+1} - H_{j+1} = (A_j - H_j)(A - H - \lambda I)$, proving the assertion.

<u>Theorem 3.</u> If rank $[A + (p-1)H - \lambda I] = n$ Then, rank $[A_k + (p-1)H_k] = n$

<u>Proof.</u> By using Sylvester's inequality, it is seen that rank $[A + (p-1)H - \lambda I]^2 = n$. Also, assuming that rank $[A_j + (p-1)H_j] = n$ for $j > 1$ and using the same inequality the following result is obtained:

$$\text{rank}\{[A_j + (p-1)H_j][A + (p-1)H - \lambda I]\} = n$$
which is valid for any j.

<u>Theorem 4.</u> The generalized eigenvectors associated with an intrasystem eigenvalue satisfy the property given by Eq. (5).

<u>Proof.</u> Firstly consider the problem for determination of the eigenvector of highest rank of a chain of length k, g^k, that is

$$\begin{vmatrix} A_k & H_k & \cdots & H_k \\ H_k & A_k & \cdots & H_k \\ \cdot & \cdot & \cdots & \cdot \\ H_k & H_k & \cdots & A_k \end{vmatrix} \begin{vmatrix} G_1^k \\ G_2^k \\ \cdot \\ G_p^k \end{vmatrix} = 0 \qquad (12)$$

Summing up the matrix equations of Eq. (12), results

$$\left| A_k + (p-1)H_k \right| (G_1^k + G_2^k + \ldots + G_p^k) = 0 \qquad (13)$$

By using Eq. (4) and Theorem (3), it is seen that the unique solution for Eq. (13) is

$$G_1^k + G_2^k + \ldots + G_p^k = 0 \qquad (14)$$

Now, consider that an arbitrary eigenvector of the

300

chain, g^j, satisfy Eq. (5). It follows by definition that

$$g^{j-1} = \begin{vmatrix} A-\lambda_i I & H & \cdots & H \\ H & A-\lambda_i I & \cdots & H \\ \vdots & \vdots & \vdots & \vdots \\ H & H & \cdots & A-\lambda_i I \end{vmatrix} \begin{vmatrix} G_1^j \\ G_2^j \\ \vdots \\ G_p^j \end{vmatrix}$$

Summing up these equations results

$$(G_1^{j-1} + \ldots + G_p^{j-1}) = \left| A-\lambda_i I + (p-1)H \right| G_1^j + \ldots + G_p^j)$$
$$= 0$$

Since g^k and g^j satisfy Eq. (5) for any j, the proof is complete. It can also be shown that the chain of generalized left eigenvectors satisfy Eq. (6).

Theorem 5. If $(A-H-\lambda_i I)G_j = 0$ and $g_i = \begin{vmatrix} a_1 G_j \\ \vdots \\ a_p G_j \end{vmatrix}$, with

$a_1 + \ldots + a_p = 0$. Then, $(\bar{A}-\lambda_i I)g_i = 0$.

<u>Proof</u>. Consider the vector

$$\begin{vmatrix} p_1 \\ \vdots \\ p_p \end{vmatrix} = \begin{vmatrix} A-\lambda_i I & H & \cdots & H \\ H & A-\lambda_i I & \cdots & H \\ \vdots & \vdots & \vdots & \vdots \\ H & H & \cdots & A-\lambda_i I \end{vmatrix} \begin{vmatrix} a_1 G_j \\ a_2 G_j \\ \vdots \\ a_p G_j \end{vmatrix}$$

Then, $p_1 = a_1(A-H-\lambda_i)G_j = 0$. Similarly, it is verified that $p_2 = \ldots = p_p = 0$.

<u>Collorary</u>. If $(A-H-\lambda_i I)G_j \neq 0$, then $(\bar{A}-\lambda_i I)g_i \neq 0$
The proof is similar to that used in the theorem.

Theorem 6. Let λ_i be an eigenvalue of A-H with any multiplicity and let G_i^ℓ be a generalized eigenvector of rank ℓ with $G_i^{\ell-1} = (A-H-\lambda_i I)G_i^\ell$, then,

$$\begin{vmatrix} a_1 G_i^{\ell-1} \\ \vdots \\ a_p G_i^{\ell-1} \end{vmatrix} = (\bar{A}-\lambda_i I) \begin{vmatrix} a_1 G_i^\ell \\ \vdots \\ a_p G_i^\ell \end{vmatrix} \quad \text{for } a_1 + \ldots + a_p = 0 \tag{15}$$

<u>Proof</u>. Consider, by definition, the vector

$$\begin{vmatrix} p_1 \\ \vdots \\ p_p \end{vmatrix} \triangleq \begin{vmatrix} A-\lambda_i I & H & \cdots & H \\ \vdots & \vdots & \vdots & \vdots \\ H & H & \cdots & A-\lambda_i I \end{vmatrix} \begin{vmatrix} a_1 G_i^\ell \\ \vdots \\ a_p G_i^\ell \end{vmatrix}$$

Then $p_1 = a_1(A-H-\lambda_i I)G_i^\ell = a_1 G_i^{\ell-1}$

Similarly it is shown that $p_2 = a_2 G_i^{\ell-1}, \ldots, p_p = a_p G_i^{\ell-1}$, as required. Eq. (15) shown how the generalized eigenvectors of $\bar{A}$ associated with λ_i can be constructed from the chain of generalized eigenvectors of the reduced order matrix A-H by selecting $a_1, a_2, \ldots, a_p$ satisfying (15). If A-H has other chains of generalized eigenvectors associated with λ_i, these chains can be constructed in a similar way (Chen, 1970).

<u>Determination of Generalized Eigenvectors Associated with Intrasystem modes</u>

The eigenvectors associated with intrasystem modes are determined by using the reduced order matrix $A-H-\lambda_i I$ and its powers. The following algorithm presents in detail the steps for determination of these eigenvectors, taking advantage of their special structure.

<u>Algorithm</u>

i) Compute the eigenvalues of the matrix A-H. These are the n intrasystem eigenvalues. Let the set of these eigenvalues be $L = \{\lambda_1, \lambda_2, \ldots, \lambda_s\}$, where some of them may be repeated. Any distinct eigenvalue of this set is a repeated eigenvalue of $\bar{A}$, with multiplicity p-1 and for any eigenvalue of L with multiplicity m, the intrasystem eigenvalue of $\bar{A}$ has multiplicity m(p-1). Since, usually, the matrix A - H is not large, the multiplicity m, when existing, is small.

ii) Consider the eigenvalue λ_1. If this is a distinct eigenvalue on A - H, compute its associated eigenvector, G_1, and proceed to step v. If λ_1 has multiplicity m in A - H, compute $(A-H-\lambda_1 I)^k$ for $k = 1, 2, \ldots$, until rank$(A-H-\lambda_1 I)^k = $ rank$(A-H-\lambda_1 I)^{k+1}$. Then find a generalized eigenvector of rank k, G_1^k, of A - H, such that

$(A-\lambda_1 I)^k G_1^k = 0$
and $(A-\lambda_1 I)^{k-1} G_1^k \neq 0$

iii) Define $G_1^{k-j} = (A-H-\lambda_1 I)G_1^{k-j+1}$, for $j = 1, 2, \ldots, k-1$.

iv) If k = m proceed to step v. If k < m, try to find another linearly independent generalized eigenvector with the largest possible rank; that is, with rank k. If this is not possible, try another generalized eigenvector with rank k-1 and so forth and then go to step (iii). This is done until m linearly independent generalized eigenvectors associated with λ_1 are found. Note that if rank $(A-H-\lambda_1 I) = r$ then there are n-r chains of generalized eigenvectors associated with λ_1 in A - H (Chen, 1970).

v) If λ_1 is a distinct eigenvalue on A - H, construct p-1 eigenvectors associated with λ_1 as follows

$$g_1 = \begin{vmatrix} a_1 G_1 \\ \vdots \\ a_p G_1 \end{vmatrix}, \ldots, g_{p-1} = \begin{vmatrix} s_1 G_1 \\ \vdots \\ s_p G_1 \end{vmatrix}$$

The parameters $a_1, \ldots, a_p, \ldots, s_1, \ldots, s_p$ are arbitrarily selected, satisfying (15) and to have $g_1, \ldots, g_{p-1}$ linearly independent. If λ_1 is not a distinct eigenvalue on A - H, for each chain of generalized eigenvectors, construct p-1 chains for $\bar{A}$ as follows. Let be a chain of lenght k in A - H. The corresponding chains of $\bar{A}$ are

$$\text{chain 1: } g_1^1 = \begin{vmatrix} a_1 G_1^1 \\ \vdots \\ a_p G_1^1 \end{vmatrix} \cdots g_1^k = \begin{vmatrix} a_1 G_1^k \\ \vdots \\ a_p G_1^k \end{vmatrix}$$
$$\vdots \qquad \vdots \qquad \vdots$$
$$\text{chain p-1: } g_{p-1}^1 = \begin{vmatrix} s_1 G_1^1 \\ \vdots \\ s_p G_1^1 \end{vmatrix} \cdots g_{p-1}^k = \begin{vmatrix} s_1 G_1^k \\ \vdots \\ s_p G_1^k \end{vmatrix}$$

The parameters $a_1, \ldots, a_p, \ldots, s_1, \ldots, s_p$ are selected satisfying (15) such that all p-1 eigenvectors $g_1^k, \ldots g_{p-1}^k$ be linearly independent. If k = 1, then p-1 eigenvectors of $\bar{A}$ are constructed from G_1^1 by selecting the parameters $a_1, \ldots, s_p$.

vi) Repeat from step (ii) to step (v) for $\lambda_2, \ldots, \lambda_s$.

<u>Jordan Form of $\bar{A}$</u>

First consider that the eigenvalues of A - H are distinct. Then, from the eigenvector G_i associated with an eigenvalue λ_i in A - H, the complete chain of generalized eigenvectors $g_i^1 \ldots g_i^k$ of $\bar{A}$ is

301

obtained as described above.

It is clear that for any eigenvector of that chain, results

$$\overline{A}g_i^j = \lambda_i g_i^j \tag{16}$$

Consider now the Jordan form of $\overline{A}$, that is

$$\Lambda = G^{-1}\overline{A}G \tag{17}$$

where G is the np x np modal matrix of $\overline{A}$, whose columns are the eigenvectors of $\overline{A}$, and

$$\Lambda = \begin{vmatrix} \Lambda_1 & 0 & 0 & 0 \\ 0 & \Lambda_2 & \cdots & \vdots \\ \vdots & \vdots & \Lambda_i & 0 \\ 0 & 0 \cdots 0 & \cdots & \Lambda_\ell \end{vmatrix} \tag{18}$$

where $\Lambda_1, \Lambda_2, \ldots, \Lambda_i \ldots,$ are the blocs associated with the intrasystem modes and Λ_ℓ is the bloc associated with the local modes.

From Eq. (17) one obtains

$$\overline{A}G = G\Lambda \tag{19}$$

and

$$\overline{A}G = \left| \overline{A}g_1^1 \,\right|\, \overline{A}g_1^2 \,\left|\, \ldots \,\right|\, \overline{A}g_i^1 \,\left|\, \ldots \right| \tag{20}$$

From Eq. (18) to Eq. (20) and Eq. (16) it is seen that the blocs associated with intraplant modes must be diagonal, in the form

$$\Lambda_i = \mathrm{diag}(\lambda_i, \lambda_i, \ldots, \lambda_i)$$

Then, in this case, the repeated intrasystem eigenvalues give rise to repeated modes.

For repeated eigenvalues in A - H, each chain of generalized eigenvectors generates one Jordan bloc whose order is equal to the length of the chain. For instance, for a chain of length k associated with the eigenvalue λ_i, the Jordan bloc is

$$\Lambda_i \atop (k \times k) = \begin{vmatrix} \lambda_i & 1 & 0 & \cdots & 0 \\ 0 & \lambda_i & 1 & \cdots & 0 \\ \cdots & \cdots & \cdots & \cdots & \cdots \\ \cdots & \cdots & \cdots & \lambda_i & 1 \\ 0 & 0 & 0 & \cdots & 0 & \lambda_i \end{vmatrix}$$

Therefore, $\overline{A}$ is always non cyclic if $p \geqslant 2$.

<u>Example</u>

Let be a system with three identical units, where

$$A = \begin{vmatrix} 0 & 1 & 0 \\ 0,5 & 0 & 0,5 \\ 0 & 0 & -1 \end{vmatrix} \quad \text{and} \quad H = \begin{vmatrix} 1 & 1 & 0 \\ 0 & 1 & 0 \\ 0 & 0 & 0 \end{vmatrix}$$

The eigenvalues of the matrix A - H (intrasystem eigenvalues) are $\lambda_i = -1$ with multiplicity 3.

Since rank$(A - H + I) = 1$, there are two chains of generalized eigenvectors associated with λ_i in A - H.

$(A - H + I)^2 = 0$, then $G_1^2 = \left| 2\ 0\ 0 \right|^T$ and $G_1^1 = (A-H+I) G_1^2 = \left| 0\ 1\ 0 \right|^T$ forms the first chain in A - H.

The 2nd chain, which is a unique eigenvector, is obtained from $(A - H + I)G_3^1 = 0$, or $G_3^1 = \left| 1\ 0\ -1 \right|^T$.

The chains of generalized eigenvectors associated with intrasystem modes for $\overline{A}$ are then constructed. They are

1st chain $\begin{cases} g_1^1 = \left| 0\ 1\ 0\,\vdots\,0\ -2\ 0\,\vdots\,0\ 1\ 0 \right|^T \\ g_1^2 = \left| 2\ 0\ 0\,\vdots\,-4\ 0\ 0\,\vdots\,2\ 0\ 0 \right|^T \end{cases}$

2nd chain $\begin{cases} g_2^1 = \left| 0\ -2\ 0\,\vdots\,0\ 1\ 0\,\vdots\,0\ 1\ 0 \right|^T \\ g_2^2 = \left| -4\ 0\ 0\,\vdots\,2\ 0\ 0\,\vdots\,2\ 0\ 0 \right|^T \end{cases}$

3rd chain $\{\ g_3^1 = \left| 1\ 0\ -1\,\vdots\,-2\ 0\ 2\,\vdots\,1\ 0\ -1 \right|^T$

4th chain $\{\ g_3^2 = \left| -2\ 0\ 2\,\vdots\,1\ 0\ -1\,\vdots\,1\ 0\ -1 \right|^T$

The eigenvalues of the matrix A + 2H (local eigenvalues) are $\lambda_{\ell 1} = -1$, $\lambda_{\ell 2} = 3,225$ and $\lambda_{\ell 3} = 0,775$.

The Jordan form for $\overline{A}$ results

$$\overline{A}_J = \mathrm{bloc\ diag} \left\{ \underbrace{\begin{vmatrix} -1 & 1 \\ 0 & -1 \end{vmatrix}}_{\substack{\text{1st} \\ \text{chain}}}, \underbrace{\begin{vmatrix} -1 & 1 \\ 0 & -1 \end{vmatrix}}_{\substack{\text{2nd} \\ \text{chain}}}, \underbrace{\left| -1, -1, \right|}_{\substack{\text{3rd, 4th} \\ \text{chains}}} \right.$$

$$\left. \underbrace{-1,\ 3,225,\ 0,775}_{\text{Local modes}} \right\}$$

EXTENSION FOR IDENTICAL UNITS CONNECTED TO A GROUP OF DIFFERENT UNITS

It is assumed that there are p identical units with identical interconnections whith the rest of the system and among themselves. The composite system, represented by Eq. (1), has now the following $\overline{A}$ matrix:

$$\overline{A} = \begin{vmatrix} A_0 & H_1 & H_1 & \cdots & H_1 \\ H_2 & A & H & \cdots & H \\ \vdots & \vdots & \vdots & \vdots & \vdots \\ H_2 & H & H & \cdots & A \end{vmatrix}$$

and $X = \left| X_0^T\ X_1^T\ \ldots\ X_p^T \right|^T$ is the state vector.

$\dot{X}_0 = A_0 X_0 + \sum\limits_{i=1}^{p} H_1 X_i$ represents the external system.

By using the same reasoning used for identical units only, it is easy to accept that the eigenvectors associated with the system modes may be represented in the following forms:

$$g_\ell = \left| G_0^T\ G_1^T\ \ldots\ G_1^T \right|^T, \quad v_\ell = \left| V_0^T\ V_1^T\ \ldots\ V_1^T \right|^T \quad \text{(for a local mode)}$$

and

$$g_i = \left| G_0^T\ G_1^T\ \ldots\ G_p^T \right|^T, \quad v_i = \left| V_0^T\ V_1^T\ \ldots\ V_p^T \right|^T \quad \text{(for an intrasystem mode)}$$

where, for an intrasystem mode, $G_1 + \ldots + G_p = 0$ and $V_1 + \ldots + V_p = 0$. The eigenvector g_i, associated with an intrasystem mode, is determined by

$$\begin{vmatrix} A - \lambda_i I & H_1 & \cdots & H_1 \\ H_2 & A - \lambda_i I & \cdots & H \\ \vdots & \vdots & \vdots & \vdots \\ H_2 & H & \cdots & A - \lambda_i I \end{vmatrix} \begin{vmatrix} G_0 \\ G_1 \\ \vdots \\ G_p \end{vmatrix} = 0 \tag{21}$$

The first matrix equation of Eq. (21) results

$$(A_0 - \lambda_i I)G_0 = 0 \tag{22}$$

Summing up the last p matrix equations of Eq. (21) results

$$H_2 G_0 = 0 \tag{23}$$

The solution for Eq. (22) and Eq. (23) is

$$G_0 = 0 \tag{24}$$

By using one of the last p matrix equations of Eq.

(21), it is obtained

$$(A - H - \lambda_i I)G_1 = 0$$

Comparing with Eq. (9), it is seen that the intrasystem eigenvalues and eigenvectors are not dependent on the external system and can be obtained exactly as in the former case.

Now, consider a local mode, whose eigenvalue is λ_ℓ. The equation for determination of the eigenvector g_ℓ is

$$\begin{vmatrix} A_0 - \lambda_\ell I & H_1 & \cdots & H_1 \\ H_2 & A - \lambda_\ell I & \cdots & H \\ \vdots & \vdots & \ddots & \vdots \\ H_2 & H & \cdots & A - \lambda_\ell I \end{vmatrix} \begin{vmatrix} G_0 \\ G_1 \\ \vdots \\ G_1 \end{vmatrix} = 0$$

which can be reduced to

$$\begin{vmatrix} A_0 - \lambda_\ell I & p H_1 \\ H_2 & A + (p-1)H - \lambda_\ell I \end{vmatrix} \begin{vmatrix} G_0 \\ G_1 \end{vmatrix} = 0 \qquad (25)$$

Then, there one $n_0 + n$ local eigenvalues, where n_0 is the order of the external system. The local eigenvectors are obtained from the reduced order system of equation described by Eq. (25). By using $\overline{A}^T$ and similar development the reduced order matrix equations for determination of the left eigenvectors of the left eigenvectors are easily obtained.

CONCLUSION

The problem of determination of eigenvalues and eigenvectors of a system composed of a number of identical units was analyzed. Two basic heuristic properties were introduced, making possible the reduction of the problem to smaller matrix equations.

It was shown that the system has only two kinds of modes, named local and intrasystem modes and that the eigenvalues associated with intrasystem modes are repeated for a system with more than two identical units. Techniques for determination of eigenvectors of the system, by using reduced order matrices, were proposed.

It was also shown that the treatment applied on a system where all units are identical can also be applied on a system composed of a group of identical units connected to other group of different units.

The extension for a system composed of more than one group of units, where the units of each group are identical, is straightforward. The developed theory can be easily used for designing and analyzing the application of controllers for selective modal control.

REFERENCES

Alden, R.T.H., P.J. Nolan and J.P. Bayne (1977). Shaft dynamics in closely coupled identical generators. IEEE Trans. PAS, PAS-96, 721-728.

Chen, C.T (1970). Introduction to Linear System Theory. Holt Rinehart and Winston, New York, pp. 35-49.

Crenshaw, M.L., J.M. Cutler, G.F. Wright and W.J. Reid (1983). Power system stabilizer application in a two-unit plant analytical studies and field tests. IEEE Trans. PAS, PAS-102, 267-274.

Perez-Arriaga, I.J., G.C. Verghese and F.C. Schweppe (1982). Selective modal analysis with applications to electric power systems. IEEE Trans. PAS, PAS-101, 3117-3134.

Schleif, F.R., R.K. Feeley, W.H. Phillips and R.W. Torluemke (1979). A power system stabilizer application with local mode cancellation. IEE Trans. PAS, PAS-98, 1054-1059.

A THEORY OF DYNAMIC AGGREGATION BY STOCHASTIC MODELS

G. Picci

*Dipartimento di Elettronica e Informatica, Università di Padova, via Gradenigo 6/A,
35131 Padova, Italy*

ABSTRACT: This paper presents the rudiments of a Theory of Aggregation of Linear Dynamical Systems by Stochastic or "noisy" Models. Aggregation ,as defined here, is essentially a Modelling problem of the same type encountered in the foundations of Statistical Mechanics. Potential applications include Model reduction in Large-scale Systems, dynamic aggregagation in Economic systems and in general Thermodynamic-type description of large Systems .

1. INTRODUCTION

We say that a "large" autonomous linear system

$$\dot{z}(t) = Fz(t) \qquad z(t) \in \mathbb{R}^N$$
$$y(t) = Hz(t) \qquad y(t) \in \mathbb{R}^m \qquad (1.1)$$

assumed irreducible, is *Aggregable* if we are able to generate all of its output trajectories $y(.)$ by means of a smaller dimensional "noisy" dynamical system like

$$\dot{x}(t) = Ax(t) + Kv(t) \qquad x(t) \in \mathbb{R}^n$$
$$y(t) = Cx(t) \qquad (1.2)$$

where $n < N$. The "noise" input v in (1.2) can either be understood as a genuine stochastic white process, in which case some "a priori" probabilistic structure on the state space of (1.1) needs to be postulated, or in the deterministic sense of a *driving input signal* as defined in J.C.WILLEMS [1983, 1986] monumental work. Choosing for the moment the second alternative, the problem can be precisely stated in the following way.

Find linear systems of the type (1.2) with state space of dimension n *strictly smaller than* N, with the property that, for every initial state $z(0)$ of the "large" system (1.1), one can find a corresponding initial state $x(0)$ and a noise sample path $v(.)$ of the "small" system (1.2), such that (1.1) and (1.2) produce *exactly the same output function* $y(.)$. (The two systems will then be indistinguishable for an external

observer).

The *Aggregatin Problem* as posed above is a caricature of the problem of going from microscopic to macroscopic dynamical descriptions in Statistical Mechanics. It was first formulated in PICCI [1986]. It has a lot of potential engineering applications. It can also be formulated in a much more general nonlinear setting (see PICCI [1988]).

In the present linear context it is easy to show that the problem is solvable, in the "exact" sense stated above, only if $N = \infty$, i. e. the "microscopic dynamics" (1.1) is *infinite dimensional* (see PICCI [1989] for a simple proof). A particularly simple and important case that we shall consider here is when $z(t)$ evolves on a infinite dimensional Hilbert space $\mathbb{H}$ and F generates a unitary group $U(t)$. Irreducibility in this context means that,

$$\overline{\text{span}} \ \{ \ U(t)h_k; \ t \in \mathbb{R} \ , \ k=1,\dots,m \ \} = \mathbb{H}$$

$\langle h_1,.\rangle,\dots,\langle h_m,.\rangle$ being the components of the read-out map $H : \mathbb{H} \to \mathbb{R}^m$. This dynamical structure is pretty general and encompasses a large class of conservative systems with a qudratic type of energy function. Also, more generally, unitary evolutions can be obtained by "lossles embedding" of dissipative systems by using dilation procedures. (Unitary dilations are treated extensively e.g. in FUHRMANN [1981]).

In order to discuss aggregability we need to introduce the notion of *Compression Subspace*. In the following P^X denotes the orthogonal projection from H onto a subspace X.

DEFINITION 1.1

*A Subspace $X \subset H$ is a Compression Subspace for
$U(t)$ if*

$$T(t) = P^X U(t)\big|_X \qquad t \geq 0,$$

*forms a one-parameter semigroup of contractions
on X which, together with its adjoint $T(t)^*$, is
strongly asymptotically stable.*

From PICCI [1986] we have the following result.

THEOREM 1.1

The irriducible dynamical system

$$\dot{z}(t) = Fz(t) \qquad z(t) \in H$$
$$y(t) = Hz(t)$$

*is aggregable if and only if the unitary group
$U(t)$ generated by F admits finite dimensional
compression subspaces X containing the vectors
$h_1, \ldots, h_m$.*

*Any such X qualifies as state space of an
aggregate model (1.2) in which A is then the
generator of $T(t)$, $C = H|_X$, K a suitable linear
map from $\mathbb{R}^\nu$ into X and $v(.)$ ranges over the
Lebesgue space $L_\nu^2(\mathbb{R})$, of ν-dimensional square
integrable functions. The integer $\nu \leq m$ is the
multiplicity of the group $U(t)$.*

Necessary and sufficient conditions for the
existence of finite dimensional compression
subspaces can be given in terms of a *Spectral
Density matrix* associated to the generators
$h_1, \ldots, h_m$ for the unitary group $U(t)$. Also the
structure of all solutions to the aggregation
problem can be studied (the main points being the
classification of *minimal* models , the existence
of *foward-backward pairs* of aggregate models,
etc.). The mathematical apparatus is to some
extent borrowed from Stochastic Realization
Theory. See LINDQUIST-PICCI [1985]. Since a
discussion of all of this would take us too far,
we shall have to refer the reader to the cited
literature for any details.

The point of this paper is rather to to relate
the above ideas to a differaent setup. The
question is what to do in presence of control
variables in the model. Is absence of control
really necessary for the aggregation procedure to
carry over? Indeed, the goalof this paper will be
to show that an aggregation theory in presence of
control variables, and, more generally, for
systems with both control and observation
variables, is possible pretty much along the same
ideas and techniques of the autonomous case.

2. AGGREGATION OF LINEAR CONTROL SYSTEMS

For the sake of clarity of exposition we shall
for the moment consider non-observed systems
(without output variables).

We are given a linear control system

$$\dot{z}(t) = F z(t) + G u(t) \qquad (2.1)$$

where $z(t) \in H$, an infinite dimensional Hilbert
space, F a skew-adjoint operator with dense
domain $\mathcal{D}(F) \subset H$, $G = [g_1, \ldots, g_p]$ with $g_k \in H$
$k=1, \ldots p$ and $u(t)$ a piecewise continuous $\mathbb{R}^p$-valued
control function. We still denote by $U(t)$ the
unitary group generated by F.

We shall assume, without loss of generality,
that

$$\overline{\text{span}} \{U(t)g_k; \ t \in \mathbb{R}, \ k=1 \ldots p\} = H \qquad (2.2)$$

(irreducibility). For, if the left hand side was
a proper subspace of H, there would be an
orthogonal complement on which the control
variable had no influence whatsoever and
therefore we could as well have taken as a new
state pace just the left hand side subspace in
(2.2).

We are interested in the existence of
aggregate models for (2.1). By this we mean an
orthogonal decomposition $H = X \oplus X^\perp$ where X is a
finite dimensional subspace such that,
i) the X-component, $x(t):=P^X z(t)$, of the state
   $z(t)$ evolves according to a dynamic equation,

$$\dot{x}(t) = Ax(t) + Bu(t) + Kv(t) \qquad (2.3)$$

on the state space X. We require that $v(\cdot)$ be
a *driving input* (noise) in the sense of
[WILLEMS, 1986, 1988] independent of the
control $u(\cdot)$.
ii) For every initial condition $z_o \in H$ there is a
   corresponding pair $x_o \in X$, $v_o(\cdot)$, such that
   the projection $x(t)=P^X z(t)$, $z(o)=z_o$, is the
   solution of (2.3) with $x(o)=x_o$, $v(\cdot)=v_o(\cdot)$,
   irrespective of the choice of the control
   variable $u(\cdot)$.

REMARK

From an engineering point of view the term
$Kv(t)$ represents a *modelling noise* which in the
aggregate model (2.3) describes the effect of the
"high frequency" dynamics $x^\perp(t)$ coupled to $x(t)$.
The coupling term is unaffected by the control
variable u and can be lumped into an additive
input independent noise in the finite dimensional
model (2.3). Being a *free* variable $v(\cdot)$ *does not*

have internal dynamics and, following [WILLEMS,1986,1988] is called "noise" in this specific sense. As we see, aggregation can be interpreted as a precise mathematical description of a very commonly "hoped for" situation in engineering practice.   □

We want now to show that a characterization of aggregability very much analogous to that given in Theorem 1.1 holds for linear control systems as well.

THEOREM 2.1

The irreducible control system (2.1) is aggregable if and only if the unitary group $U(t)$, generated by F admits finite dimensional compression subspaces X containing the vectors $g_1, \cdots, g_p$.

Any such X qualifies as state space of an aggregate model (2.3) in which A is then the generator of the semigroup $T(t)$, $B = G$, K is a suitable linear map from $\mathbb{R}^\nu$ into X and $v(\cdot)$ ranges over $L^2_\nu(\mathbb{R})$, $\nu \leq p$ being the multiplicity of $U(t)$.

    □

We shall give below a brief outline of the argument leading to this characterization.

Let X be a compression subspace containing $g_1, \cdots, g_p$. Define the "low" and "high" frequency components of $z(t)$ as

$$z(t) = x(t) + x^\perp(t) = P^X z(t) + P^{X^\perp} z(t) \qquad (2.4)$$

By decomposing the initial condition, $z(o)$ as in (2.4) the solution of (2.1) takes the form

$$z(t) = U(t)P^X z(o) + \int_o^t U(t-s)Gu(s)ds +$$
$$+ U(t)P^{X^\perp} z(o) \qquad (2.5)$$

Assume $t \geq 0$ and project now $z(t)$ onto X. Recalling that G maps into X one gets

$$x(t) = T(t)P^X z(o) + \int_o^t T(t-s)Gu(s)ds +$$
$$+ P^X U(t)P^{X^\perp} z(o) \qquad (2.6)$$

which is "almost" the integrated version of (2.3). What needs to be shown is that the last term in this expression can be written as $\int_o^t T(t-s)K\, v(s)ds$ for a suitable K and with $v(\cdot)$ depending only on the "high ferquency" component $P^{X^\perp} z(o)$ of the initial state $z(o)$.

Now let $(S, \bar{S})$ be the *scattering pair* corresponding to X. We have $S = \bar{S}^\perp \oplus X$, $\bar{S} = X \oplus S^\perp$

and
$$H = \bar{S}^\perp \oplus X \oplus S^\perp \qquad (2.7)$$

with $U(t)S^\perp \subset S^\perp$, $t \geq 0$ and $U^*(t)\bar{S}^\perp \subset \bar{S}^\perp$, $t \geq 0$. Using the above invariance properties we have

$$P^X U(t)P^{X^\perp} z(o) = P^X U(t)P^{\bar{S}^\perp} z(o) =$$
$$= P^S U(t)P^{\bar{S}^\perp} z(o) \qquad (2.8)$$

for $t \geq 0$. By WOLD' representation Theorem(see e.g. [PICCI 1986] Thm. 4.1),

$$\bar{S}^\perp = \left\{ \int_{-\infty}^0 f(-\sigma)d\bar{w}(\sigma) \; ; \; f \in L^2_\nu(\mathbb{R}_+) \right\} \qquad (2.9)$$

where the functions f are written as ν-dimensional row vectors and $w(\Delta\,) := \mathrm{col}(\bar{w}_1(\Delta) \ldots, \bar{w}_\nu(\Delta))$ is the quasi isometric measure generating $\bar{S}$ applied to a fixed orthonormal set of generating vectors. By (2.9) each element of $\bar{S}$ (or $\bar{S}^\perp$) can be written as a Wiener integral with respect to $d\bar{w}$ of an anticausal (resp. causal) function in $L^2_\nu(\mathbb{R})$. The correspondence is unitary (it is in fact the translation representation). Let $v(\cdot)$ be the (causal) L^2 function corresponding to $P^{\bar{S}^\perp} z(o)$. We can rewrite the last term in (2.8) as

$$P^S U(t)\int_{-\infty}^0 v(-\sigma)d\bar{w}(\sigma) = P^S\int_{-\infty}^t v(t-\sigma)d\bar{w}(\sigma)$$
$$= P^S\int_0^t v(t-\sigma)d\bar{w}(\sigma) \qquad (2.10)$$

because $\int_{-\infty}^0 v(t-\sigma)d\bar{w}(\sigma) \in \bar{S}^\perp$ and is therefore orthogonal to S.

LEMMA 2.1

For any finite dimensional compression subspace, (in fact, for a much wider class of compression subspaces) the limits,

$$\lim_{h\to 0_+} \frac{1}{h} P^S (\bar{w}_i(h) - \bar{w}_i(o)) := k_i \qquad i=1\ldots\nu$$

exist, $k_i \in X$ for all i, and we have the representation

$$P^S(\bar{w}_i(h) - \bar{w}_i(o)) = \int_0^h P^S U(\sigma)k_i \, d\sigma =$$

$$= \int_0^h T(\sigma)k_i \, d\sigma \qquad (2.11)$$

307

Sketch of the Proof:

We need to refer heavily to [LINDQUIST-PICCI 1985]. The notion of "splitting" in this reference is the same as "compression" in this paper.

Since the structural inner function of any finite dimensional compression subspace X is rational, it is decomposable in the sense of the reference above, p. 33. Then, by Corollary 9.2 $\bar{w}$ is *conditionally Lipschitz with respect to S* and the existence of the limits follows. It is then obvious that the limits are in X $(P^S\bar{S} = X)$. The final formula (2.11) follows by noticing that the H-valued function $h \to \eta_i(h)$ equal to the expression on the left hand side of (2.11) is continuosly differentiable on $h \geq 0$. Finally, by a fundamental invariance condition, $P^S U(t)|_X = P^X U(t)|_X$ $\forall$ $t \geq 0$.

□

By the above lemma we can pass the projection operator P^S in the last term of (2.10), under the integral signal, getting $(k := col(k_1 .. k_\nu))$,

$$P^X U(t) P^{X\perp} z(o) = \int_0^t v(t-\sigma) T(\sigma) K \; d\sigma$$

$$= \int_0^t T(t-\sigma) \sum_1^\nu k_i v_j(\sigma) d\sigma$$

which is exactly the integral representation we wanted to get.

3. SUMMING UP

It is easy to put the two characterizations, Thm 1.1 and Thm 2.1, together and find conditions for aggregability of (infinite dimensional) *systems with both observations and control variables*,

$$\dot{z}(t) = Fz(t) + Gu(t)$$
$$y(t) = Hz(t) \tag{3.1}$$

One sees that (3.1) *will be aggregable, i.e. indistinguishable by input-output experiments from a finite dimensional "noisy" system like*

$$\dot{x}(t) = Ax(t) + Bu(t) + Kv(t)$$
$$y(t) = Cx(t) \tag{3.2}$$

if and only if the unitary group generated by F, admits finite dimensional compression subspaces containing both the "columns" of $G = [g_1, .. g_p]$ *and the "rows"* $(h_1, .. h_m)$ *of* **H**.

Hence finding compression subspaces, even in this more general set up, boils down to the same procedure explained in [PICCI 1986].

To conclude, one should say that the above aggregate model can be made "truly stochastic" by postulating a "thermal equilibrium" condition of the microscopic system. There is an essentially unique "probability" distribution on H which is left invariant by the unitary flow U(t), the *normalized Gaussian distribution* μ_o. By introducing this distribution we can e.g. make y into a bona fide stationary stochastic process (see [PICCI 1986]). Unfortunately μ_o is not quite a "probability" measure because it is not generally countably additive on H. This fact actually agrees with (or is responsable for) white noise processes with L^2 paths. To get "truly white" (or Brownian) noises one has to extend the measure to a larger space.

REFERENCES

FUHRMANN P.A. [1981]. *Linear Systems and Operators in Hilbert Spaces*. Mc Graw Hill N.Y.

LAX p.D., PHILLIPS R.S. [1967]. *Scattering Theory*. Ac. Press N.Y.

LINDQUIST A., PICCI G. [1985a]. Realization Theory for Multivariable Stationary Gaussian Processes. *SIAM J. Control Optim.* 23, 809-857. [1985b]. Forward and Backward Semimartingale Models for Gaussian Processes with Stationary Increments. *Stochastics*, 15, pp. 1-50.

PICCI G. [1986]. Applications of Stochastic Realization Theory to a Fundamental Problem of Statistical Physics. *Modelling Identification and Robust Control*, C.J. Byrnes and A. Lindquist eds. North Holland. [1988]. Stochastic Aggregation. *Linear Circuits Systems and Signal Processing Theory and Applications*, C.J. Byrnes, C. Martin, R. Saeks eds. North Holland. [1989]. Aggregation of Linear Systems in a completely deterministic framework. In: *Three Decades of Mathematical System Theory*. Springer Verlag 1989.

WILLEMS J.C. [1983]. Input-output and state space representations of finite-dimensional linear time-invariant systems. *Lin. Alg. Appl.*, 50, 581-608. [1986]. From time series to linear systems. Part I. Finite Dimensional Linear Time Invariant Systems. *Automatica 22*, 561-580. [1988]. Models for Dynamics. *Dynamics Reported*, 2, Wiley and Teubner.

ON THE MODEL ORDER REDUCTION OF LINEAR MULTIVARIABLE SYSTEMS USING A BLOCK-SCHWARZ REALIZATION

P. Resende and V. V. R. Silva

*Department of Electronics Engineering, Minas Gerais Federal University,
Cx.P. 1294, Belo Horizonte, MG 30160, Brazil*

Abstract. A block-Schwarz state-space realization is utilized to derive a model order
reduction method for linear multivariable systems. It is shown that the proposed method
may be regarded as a frequency-domain continued-fraction technique based upon a block
"alpha-beta" expansion of matrix fraction descriptions. A new algorithm is presented in
order to obtain the block-Schwarz realization from matrix fraction descriptions. Once
the block-Schwarz matrix satisfies certain stability conditions then stable reduced
models are provided. The proposed method constitutes a block version of the popular
Routh approximation.

Keywords. Model reduction; systems order reduction; linear systems; matrix algebra;
multivariable systems.

INTRODUCTION

Model reduction techniques are often required in
control systems analysis for deriving low-order
approximations of high-order systems. Using time-
domain concepts, many reduction methods have been
proposed and recently much work has utilized the
idea of balanced realizations introduced by Moore
(1981).

Considering frequency-domain techniques a wide
variety of reduction methods have been suggested
(see Jamshidi, 1983) with several extensions for
multivariable systems. However, most of these
extensions are only applicable to square transfer
function matrices (Chen, 1974; Goldman and others,
1981; Owens and Fields, 1980; Shieh and Wei, 1975).
In addition, some of these methods require the
determination of the dominant poles of the
complete model (Shieh and Wei, 1975; Shamash, 1975)
and in general the preservation of the stability
is unpredictable (Chen, 1974; Goldman and others,
1981). The desired stability property of the
reduced model derived from a stable complete
model, for the multivariable case, suggests the
approach to be proposed here.

It is shown in this paper that once a block-Schwarz
matrix (of the complete model) satisfies the stability conditions given in Resende and Kaszkurewicz
(1989) then some appropriate partitions of
this matrix provide stable block-Schwarz submatrices (that are associated to the reduced models).
To obtain the block-Schwarz state-space realization
for a matrix fraction description a new realization algorithm is presented, which is derived
from a particular case of a block-tridiagonal
realization algorithm proposed by Resende and
others (1986). A "reverse" application of the
presented algorithm for the retained partitions of
the block-Schwarz realization provides a matrix
fraction description of the reduced model. It is
further shown that this reduction method have a
block continued-fraction representation and constitutes a block version of the Routh approximation
proposed by Hutton and Friedland (1975).

A BLOCK-SCHWARZ REALIZATION

Consider the multivariable system represented in
frequency-domain by the relationship

$$Y(s) = G(s) \, U(s) \qquad (1)$$

where $U(s)$, $m \times 1$, and $Y(s)$, $p \times 1$, are the Laplace
transforms of $u(t)$ and $y(t)$, the input and output
vectors, respectively. The $p \times m$ transfer function
matrix $G(s)$ is assumed to be given by a right
matrix fraction description:

$$G(s) = N(s) \, D^{-1}(s) \qquad (2)$$

$$D(s) = Is^n + D_{n-1}s^{n-1} + D_{n-2}s^{n-2} + \ldots + D_1 s + D_o \qquad (3)$$

$$N(s) = N_{n-1}s^{n-1} + N_{n-2}s^{n-2} + \ldots + N_1 s + N_o \qquad (4)$$

where D_i, $m \times m$, and N_i, $p \times m$, $(i=0,1,\ldots,n-1)$, are
real constant matrices; I is the $m \times m$ identity
matrix. It is also assumed that $\det(D_{n-1}) \neq 0$.

From the system description (1)-(4) it is intended
to obtain the following block-Schwarz realization

$$\dot{x}(t) = \begin{bmatrix} -M_1 & -M_2 & 0 & \cdots & 0 & 0 \\ I & 0 & -M_3 & \cdots & 0 & 0 \\ 0 & I & 0 & \cdots & 0 & 0 \\ \vdots & \vdots & \vdots & & \vdots & \vdots \\ 0 & 0 & 0 & \cdots & 0 & -M_n \\ 0 & 0 & 0 & \cdots & I & 0 \end{bmatrix} x(t) + \begin{bmatrix} I \\ 0 \\ 0 \\ \vdots \\ 0 \\ 0 \end{bmatrix} u(t) \qquad (5a)$$

$$y(t) = \begin{bmatrix} K_1 & K_2 & K_3 & \cdots & K_{n-1} & K_n \end{bmatrix} x(t) \qquad (5b)$$

with the state-vector $x' = (x_1', x_2', \ldots, x_n')'$, where
x_i, $(i=1,2,\ldots,n)$, are m-vectors; the block-
elements M_i, $m \times m$, and K_i, $p \times m$, $(i=1,2,\ldots,n)$, are
real constant matrices.

TABLE 1 Sequence for Solving Equations (6)

k	i	obtained matrix
k = n	i = n-2	M_2
	i = n-4	$Q_{n-2,n-4}$
	i = n-6	$Q_{n-2,n-6}$
	$\vdots$	$\vdots$
k = n-1	i = n-3	M_3
	i = n-5	$Q_{n-3,n-5}$
	i = n-7	$Q_{n-3,n-7}$
	$\vdots$	$\vdots$
$\vdots$	$\vdots$	$\vdots$
k = 2	i = 0	M_n

TABLE 2 Sequence for Solving Equations (18)

k	i	obtained matrix
k = 2	i = 0	$\hat{Q}_{20}$
$\vdots$	$\vdots$	$\vdots$
k = r-1	i = r-7	$\hat{Q}_{r-1,r-7}$
	i = r-5	$\hat{Q}_{r-1,r-5}$
	i = r-3	$\hat{Q}_{r-1,r-3}$
k = r	i = r-6	$\hat{Q}_{r,r-6}$
	i = r-4	$\hat{Q}_{r,r-4}$
	i = r-2	$\hat{Q}_{r,r-2}$

Using the result of Lemma 1 in Resende and Kaszkurewicz (1989) with a particular case of Lemma 2 in Resende and others (1986), it is easy to show that the block-Schwarz form (5) and the matrix fraction description (2)-(4) are related by the equations (6)-(11), below

$$Q_{k,i} = Q_{k-1,i-1} + M_{n-k+2}Q_{k-2,i}$$

$$((k+i)\text{ even}; k=2,3,\ldots,n; i=0,1,\ldots,k-1) \quad (6)$$

where

$$Q_{k,i} = I \quad \text{if } k=i>0 \quad (7a)$$

$$Q_{k,i} = 0 \quad \text{if } k<i \quad \text{or } k<0 \quad \text{or } i<0 \quad (7b)$$

$$Q_{n,i} = D_i , \quad ((i+n)\text{ even}; i=0,1,\ldots,n-1) \quad (8)$$

$$Q_{n-1,i} = D_{n-1}^{-1}D_i , \quad ((i+n)\text{ odd}; i=0,1,\ldots,n-2) \quad (9)$$

$$M_1 = D_{n-1} \quad (10)$$

and

$$K_i = \begin{cases} N_{n-i} - \sum_{\substack{j=1 \\ j \text{ odd}}}^{i-2} K_j Q_{n-j,n-i} , & (i \text{ odd}) \\[2em] N_{n-i} - \sum_{\substack{j=2 \\ j \text{ even}}}^{i-2} K_j Q_{n-j,n-i} , & (i \text{ even}) \end{cases}$$

$$(i=1,2,\ldots,n) \quad (11)$$

The matrices $Q_{i,j}$, $((i+j)$ even; $i=2,3,\ldots,n$; $j=0,1,\ldots,i-1)$, are m×m real matrices. The algorithm given below provides a simple recursive manner to solve the system of matrix equations (6), to obtain matrices M_i, $(i=2,3,\ldots,n)$.

Algorithm. The equations (6) can be solved recursively, since for each appropriate pair (k,i) one equation with only one unknown matrix is provided. Table 1 shows the sequence of indices k and i to be taken and the unknown matrix to be obtained in the respective step. To obtain the matrices K_i $(i=1,2,\ldots,n)$, equations (11) are solved recursively by setting i = 1 to i = n.

MODEL ORDER REDUCTION

Considering the block-Schwarz state matrix in equation (5) it follows that if RM_1, RM_1M_2, $\ldots$, $RM_1M_2\ldots M_n$ are symmetric positive definite matrices (for some m×m matrix R, such that $R + R'>0$) then (5) is asymptotically stable. This is proven by Resende and Kaszkurewicz (1989) in Theorem 1. Therefore, it is easy to note that if (5) satisfies this stability condition, then the reduced block-Schwarz form below is also stable.

$$\dot{\hat{x}}(t) = \begin{bmatrix} -M_1 & -M_2 & 0 & \ldots & 0 & 0 \\ I & 0 & -M_3 & \ldots & 0 & 0 \\ 0 & I & 0 & \ldots & 0 & 0 \\ \vdots & \vdots & \vdots & & \vdots & \vdots \\ 0 & 0 & 0 & \ldots & 0 & -M_r \\ 0 & 0 & 0 & \ldots & I & 0 \end{bmatrix} \hat{x}(t) + \begin{bmatrix} I \\ 0 \\ 0 \\ \vdots \\ 0 \\ 0 \end{bmatrix} u(t) \quad (12a)$$

$$\hat{y}(t) = \begin{bmatrix} K_1 & K_2 & K_3 & \ldots & K_{r-1} & K_r \end{bmatrix} \hat{x}(t) \quad (12b)$$

where $1 \leq r < n$. The r-th reduced model for (2)-(4) is proposed to be the right matrix fraction description corresponding to (12), $\hat{Y}(s)=G_r(s)U(s)$,

$$G_r(s) = \hat{N}(s) \, \hat{D}^{-1}(s) \quad (13)$$

$$\hat{D}(s) = Is^r + \hat{D}_{r-1}s^{r-1} + \hat{D}_{r-2}s^{r-2} + \ldots + \hat{D}_1 s + \hat{D}_o \quad (14)$$

$$\hat{N}(s) = \hat{N}_{r-1}s^{r-1} + \hat{N}_{r-2}s^{r-2} + \ldots + \hat{N}_1 s + \hat{N}_o \quad (15)$$

The matrix coefficients $\hat{D}_i$, $\hat{N}_i$, $(i=1,2,\ldots,r-1)$, can be obtained from the equation of the form (6)-(11) rewritten below:

$$\hat{D}_{r-1} = M_1 \quad (16)$$

$$\hat{D}_i = \hat{Q}_{r,i} , \quad ((i+r)\text{ even}; i=0,1,\ldots,r-1) \quad (17a)$$

$$\hat{D}_i = M_1\hat{Q}_{r-1,i} , \quad ((i+r) \text{ odd}; i=0,1,\ldots,r-2) \quad (17b)$$

with

$$\hat{Q}_{k,i} = \hat{Q}_{k-1,i-1} + M_{r-k+2} \hat{Q}_{k-2,i}$$

$$((k+i)\text{ even}; k=2,3,\ldots,r; i=0,1,\ldots,k-1) \quad (18)$$

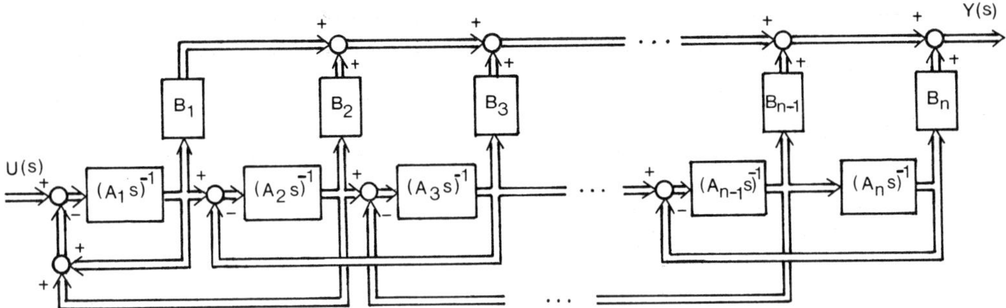

Fig. 1. Diagram of block alpha-beta expansion.

where

$$\hat{Q}_{k,i} = I \quad \text{if} \quad k=i>0 \tag{19a}$$

$$\hat{Q}_{k,i} = 0 \quad \text{if} \quad k<i \quad \text{or} \quad k<0 \quad \text{or} \quad i<0 \tag{19b}$$

and

$$\hat{N}_{r-i} = \begin{cases} K_i + \displaystyle\sum_{\substack{j=1 \\ j \text{ odd}}}^{i-2} K_j \hat{Q}_{r-j,r-i} \ , & (i \text{ odd}) \\[2em] K_i + \displaystyle\sum_{\substack{j=2 \\ j \text{ even}}}^{i-2} K_j \hat{Q}_{r-j,r-i} \ , & (i \text{ even}) \end{cases}$$

$$(i=1,2,\ldots,r) \tag{20}$$

The system of matrix equations (18) can be solved recursively using a "reverse" form of the Algorithm by the sequence shown in Table 2.

BLOCK ALPHA-BETA EXPANSION

Consider the block-Schwarz state-space realization (5) and the following block transformation matrix

$$T = \text{block-diag}(L_0, L_1, \ldots, L_{n-1}) \tag{21}$$

where L_i, $(i=0,1,\ldots,n-1)$ are $m \times m$ matrices defined as follows

$$L_0 = I \tag{22a}$$

$$L_1 = M_1 \tag{22b}$$

$$L_i = L_{i-2} M_i \ , \quad (i=2,3,\ldots,n) \tag{22c}$$

Applying the transformation $\overline{x} = T x$, with matrix T given by (21), to the state equation (5) yields

$$\dot{\overline{x}}(t) = \begin{bmatrix} -A_1^{-1} & -A_2^{-1} & 0 & \ldots & 0 & 0 \\ A_1^{-1} & 0 & -A_3^{-1} & \ldots & 0 & 0 \\ 0 & A_2^{-1} & 0 & \ldots & 0 & 0 \\ \vdots & \vdots & \vdots & & \vdots & \vdots \\ 0 & 0 & 0 & \ldots & 0 & -A_n^{-1} \\ 0 & 0 & 0 & \ldots & A_{n-1}^{-1} & 0 \end{bmatrix} \overline{x}(t) + \begin{bmatrix} I \\ 0 \\ 0 \\ \vdots \\ 0 \\ 0 \end{bmatrix} u(t) \tag{23a}$$

$$y(t) = \begin{bmatrix} B_1 A_1^{-1} & B_2 A_2^{-1} & B_3 A_3^{-1} & \ldots & B_{n-1} A_{n-1}^{-1} & B_n A_n^{-1} \end{bmatrix} \overline{x}(t) \tag{23b}$$

where A_i, $m \times m$, B_i, $p \times m$, $(i=1,2,\ldots,n)$, are given by

$$A_i = L_{i-1} L_i^{-1} \ , \quad (i=1,2,\ldots,n) \tag{24}$$

$$B_i = K_i L_i^{-1} \ , \quad (i=1,2,\ldots,n) \tag{25}$$

Using the state-space representation (23) with null initial conditions, it is easy to obtain the diagram of Fig. 1. Thus, when we consider this diagram the block alpha-beta expansion of $G(s)$ is obtained, i.e.

$$G(s) = B_1 F_1(s) + B_2 F_2(s) F_1(s) + \ldots +$$
$$+ B_n F_n(s) F_{n-1}(s) \ldots F_2(s) F_1(s)$$
$$= \Big[\Big[\Big[\ldots\Big[B_n F_n(s) + B_{n-1}\Big] F_{n-1}(s) + \ldots +$$
$$+ B_3\Big] F_3(s) + B_2\Big] F_2(s) + B_1\Big] F_1(s) \tag{26}$$

with

$$F_1(s) = \Big[I + A_1 s + \Big[A_2 s + \Big[A_3 s + \ldots +$$
$$+ \Big[A_{n-1} s + \Big[A_n s\Big]^{-1}\Big]^{-1} \ldots\Big]^{-1}\Big]^{-1}\Big]^{-1} \tag{27a}$$

$$F_i(s) = \Big[A_i s + \Big[A_{i+1} s + \ldots +$$
$$+ \Big[A_{n-1} s + \Big[A_n s\Big]^{-1}\Big]^{-1} \ldots\Big]^{-1}\Big]^{-1}$$
$$(i=2,3,\ldots,n) \tag{27b}$$

The matrices A_i, B_i, $(i=1,2,\ldots,n)$, given in (24)-(25) correspond to the alpha and beta matrix parameters. It is interesting to note that (26) is a block version of the alpha-beta expansion given by Hutton and Friedland (1975) which considers scalar alpha parameters.

Considering the reduced block-Schwarz form in (12), the block alpha-beta expansion of $\hat{G}_r(s)$ is

obtained in a similar manner:

311

$$G_r(s) = B_1\hat{F}_1(s) + B_2\hat{F}_2(s)\hat{F}_1(s) + \ldots +$$
$$+ B_r\hat{F}_r(s)\hat{F}_{r-1}(s)\ldots\hat{F}_2(s)\hat{F}_1(s)$$
$$= \Big[\Big[\Big[\ldots\Big[B_r\hat{F}_r(s) + B_{r-1}\Big]\hat{F}_{r-1}(s) + \ldots +$$
$$+ B_3\Big]\hat{F}_3(s) + B_2\Big]\hat{F}_2(s) + B_1\Big]\hat{F}_1(s) \qquad (28)$$

with

$$\hat{F}_1(s) = \Big[I + A_1 s + \Big[A_2 s + \Big[A_3 s + \ldots +$$
$$+ \Big[A_{r-1}s + \Big[A_r s\Big]^{-1}\Big]^{-1}\ldots\Big]^{-1}\Big]^{-1}\Big]^{-1} \qquad (29a)$$

$$\hat{F}_i(s) = \Big[A_i s + \Big[A_{i+1}s + \ldots +$$
$$+ \Big[A_{r-1}s + \Big[A_r s\Big]^{-1}\Big]^{-1}\ldots\Big]^{-1}\Big]^{-1}$$
$$(i=2,3,\ldots,r) \qquad (29b)$$

where matrices A_i, B_i, $(i=1,2,\ldots,r)$ are defined in (24)-(25).

Therefore, the r-th reduced model in (13)-(15) corresponds to the truncation of the block alpha-beta expansion of the complete model in (2)-(4). As described above, the r-th reduced model retains only the A_i and B_i, $(i=1,2,\ldots,r)$ terms of the complete block alpha-beta expansion.

It is mentioned in Hutton and Friedland (1975) that reduced models based on alpha-beta expansions preserve high-frequency behavior. However, for control applications it is preferable a low-frequency approximation. This can be achieved applying the proposed reduction method for the reciprocal transfer function matrix $H(s)=G(1/s)/s$ and then considering the reduced model as $G_r(s)=H_r(1/s)/s$.

EXAMPLE

Consider the two-input two-output system given by the representation (1) with

$$G(s)=\Big[N_o s^3+N_1 s^2+N_2 s+N_3\Big]\Big[D_o s^4+D_1 s^3+D_2 s^2+D_3 s+I\Big]^{-1}$$

$$D_o = \begin{bmatrix} 176 & 144 \\ 176 & 464 \end{bmatrix}, \quad D_1 = \begin{bmatrix} 88 & 132 \\ 132 & 288 \end{bmatrix}, \quad D_2 = \begin{bmatrix} 70 & -2 \\ 8 & 90 \end{bmatrix},$$

$$D_3 = \begin{bmatrix} 4 & 6 \\ 6 & 14 \end{bmatrix}, \quad N_o = \begin{bmatrix} 329 & 280 \\ 231 & 210 \end{bmatrix}, \quad N_1 = \begin{bmatrix} 123 & 102 \\ 81 & 78 \end{bmatrix},$$

$$N_2 = \begin{bmatrix} 26 & 20 \\ 16 & 16 \end{bmatrix}, \quad N_3 = \begin{bmatrix} 2 & 1 \\ 1 & 1 \end{bmatrix} \qquad (30)$$

It is desired to find the second (r=2) reduced model, preserving low-frequency behavior, for this system. The reciprocal transfer function matrix is given by

$$H(s)=\Big[N_3 s^3+N_2 s^2+N_1 s+N_o\Big]\Big[I s^4+D_3 s^3+D_2 s^2+D_1 s+D_o\Big]^{-1}$$
$$(31)$$

From (8) and (9), we have $Q_{40}= D_o$, $Q_{42}= D_2$, $Q_{31}= D_3^{-1}D_1$. The block-Schwarz realization algorithm, based on (6), utilized as indicated in Table 1, gives:

(k=4, i=2): $Q_{42}= Q_{31}+ M_2$; $\quad M_2= \begin{bmatrix} 48 & -8 \\ 8 & 72 \end{bmatrix}$

(k=4, i=0): $Q_{40}= M_2 Q_{20}$; $\quad Q_{20}= \begin{bmatrix} 4 & 4 \\ 2 & 6 \end{bmatrix}$

(k=3, i=1): $Q_{31}= Q_{20}+ M_3$; $\quad M_3= \begin{bmatrix} 18 & 2 \\ -2 & 12 \end{bmatrix}$

(k=2, i=0): $Q_{20}= M_4$; $\quad M_4= \begin{bmatrix} 4 & 4 \\ 2 & 6 \end{bmatrix}$

and by (10), $M_1= D_3$. From (11) we obtain: $K_1= N_3$, $K_2= N_2$,

$$K_3 = N_1 - K_1 Q_{31} = \begin{bmatrix} 79 & 72 \\ 59 & 54 \end{bmatrix} ,$$

$$K_4 = N_o - K_2 Q_{20} = \begin{bmatrix} 185 & 56 \\ 135 & 50 \end{bmatrix}$$

It is interesting to note that D(s) verifies the stability condition given by Resende and Kaszkurewicz (1989), since RM_1, $RM_1 M_2$, $RM_1 M_2 M_3$ and $RM_1 M_2 M_3 M_4$ are symmetric positive definite matrices (for example, take R = I). Therefore all reduced models, that are obtained using the proposed method, are stable for the matrix fraction description of this Example.

The right matrix fraction description corresponding to the reciprocal reduced model $H_2(s)$ is given by

$$H_2(s) = \Big[\hat{N}_1 s+\hat{N}_o\Big]\Big[I s^2+\hat{D}_1 s+\hat{D}_o\Big]^{-1} \qquad (32)$$

From (16)-(17), $\hat{D}_1 = M_1$ and $\hat{D}_o = \hat{Q}_{20}$; using (18) $\hat{Q}_{20} = M_2$. Thus $\hat{D}_o = M_2$. From (20), we have $\hat{N}_o = K_2$ and $\hat{N}_1 = K_1$.

Applying again the reciprocal transformation, yields the desired reduced model

$$G_2(s)= \Big[\hat{N}_o s+\hat{N}_1\Big]\Big[\hat{D}_o s^2+\hat{D}_1 s+I\Big]^{-1}$$

$$\hat{D}_o = \begin{bmatrix} 48 & -8 \\ 8 & 72 \end{bmatrix}, \quad \hat{D}_1 = \begin{bmatrix} 4 & 6 \\ 6 & 14 \end{bmatrix},$$

$$\hat{N}_o = \begin{bmatrix} 2 & 1 \\ 1 & 1 \end{bmatrix}, \quad \hat{N}_1 = \begin{bmatrix} 26 & 20 \\ 16 & 16 \end{bmatrix} \qquad (33)$$

which provides four poles at: $-0.0116\pm0.1378i$, -0.0983 and -0.1512 (the complete model (30) has poles at $-0.0064\pm0.1199i$, $-0.0765\pm0.2109i$, $-0.0999\pm0.8070i$, -0.0878 and -0.4216).

If it is desired to obtain the block alpha-beta expansion for H(s), i.e.

$$H(s)=\Big[\Big[\Big[B_4 F_4(s)+B_3\Big]F_3(s)+B_2\Big]F_2(s)+B_1\Big]F_1(s)$$

equations (27) are used to define $F_i(s)$, $(i=1,2,3,4)$, where A_i, B_i, $(i=1,2,3,4)$ are given by (24)-(25) and (22).

As shown in this example, the proposed reduction method is computationally attractive and few matrix operations were utilized.

CONCLUSION

A model reduction method for systems represented
by matrix fraction descriptions has been derived
by using a block-Schwarz realization algorithm.
The reduced models correspond to the truncation
of the block alpha-beta expansion of the
complete model. The proposed method constitutes
a generalization to matrix fraction descriptions
of the Routh approximation method given by
Hutton and Friedland (1975). It is noted that the
present method is not restricted to systems with
an equal number of inputs and outputs and the
stability of the reduced models is predictable. As
shown in the Example the present method is computationally attractive to use.

Considering the Routh approximation with scalar
alpha parameters (Hutton and Friedland, 1975) some
modifications were proposed by Lucas (1988) and
Lucas and Davidson (1983). It is the belief of the
authors that similar modifications can be applied
to the present method, then widening its use. An
modification would be to obtain the reduced model
from the retained block alpha beta parameters
multiplied by a scaling parameter which is chosen
to find an optimal scaled reduced model. Another
modification would be to obtain the reduced
denominator matrix from the retained block alpha
parameters and then time-moments are matched (see
Shamash, 1975) to obtain the reduced numerator
matrix.

ACKNOWLEDGMENT

The authors acknowledge the valuable comments of
Prof. E. Kaszkurewicz concerning this paper.

REFERENCES

Chen, C.F.(1974). Model reduction of multivariable
control systems by means of continued fraction.
Int. J. Control, 20, 225-238.
Goldman, M.J., W.J. Porras, and C.T. Leondes
(1981). Multivariable systems reduction via
Cauer forms. Int. J. Control, 34, 623-650.
Hutton, M.F., and B. Friedland (1975). Routh
approximation for reducing order of linear,
time-invariant systems. IEEE trans. Autom.
Control, 20, 329-337.
Jamshidi, M. (1983). Large-Scale Systems Modeling
and Control.North Holland, New York.
Lucas, T.N. (1988). Scaled impulse energy
approximation for model reduction. IEEE trans.
Autom. Control, 33, 791-793.
Lucas, T.N., and A.M. Davidson (1983). Frequency
domain reduction of linear systems using
Schwarz approximation. Int. J. Control, 37,
1167-1178.
Moore, B.C. (1981). Principal Component analysis
in linear systems: controllability,
observability and model reduction. IEEE Trans.
Autom. Control, 26, 17-32.
Owens, D.W., and A.D. Fields (1980). Nested-Feed-
back-loop decomposition for reduction of
linear multivariable systems. IEE Proc., Part D,
127, 103-114.
Resende, P., and E. Kaszkurewicz (1989). A
sufficient condition for the stability of
matrix polynomials. IEEE Trans. Autom.Control,
34, 539-541.
Resende, P., E. Kaszkurewicz, and L. Hsu (1986).
On the absolute stability of multivariable
systems using a block-tridiagonal state-space
representation. Int. J. Control, 43, 643-655.
Shamash, Y. (1975). Multivariable systems
reduction via modal methods and Padé
approximations. IEEE Trans. Autom.Control, 20,
815-817.
Shieh, L.S., and Y.J. Wei (1975). A mixed method
for multivariable system reduction. IEEE
Trans. Autom. Control, 20, 429-432.

IDEAL MODEL-MATCHING

V. Strejc

*Institute of Information Theory and Automation, Czechoslovak Academy of Sciences,
182 08 Prague 8, Czechoslovakia*

Abstract. The linear continuous basic time-invariant control system consisting of a generator of a command variable, of an ideal model response and of the control process is fitted up by three controllers: one in the feedback and two in the feedforward. The objective is to provide a good match between the responses of the process and of the ideal model by minimalization of a quadratic cost function.The concept of the ideal model- matching can be applied for SISO as well as for MIMO systems. By means of appropriate weighting factors relating to three possible control errors it is possible to attain the required control strategies.Because the dimension of this kind of systems can be very high,the synthesis is based on the application of eigenvectors of the Euler or Hamiltonian matrix which are calculated by the well known QR algorithm.

Keywords. Linear systems;Extended state-space systems;State feedback and feedforward;Control theory.

1   Introduction

In industrial applications of process control we can find a very large variety of control structures using auxiliary variables, i.e. variables measured on the control process,on the model of the process or applied as auxiliary controlling variables.In all these cases the final aim is to gain the best possible control quality. The auxiliary measured variables provide additional information about the dynamics of the process while the auxiliary controlling variables speed up the action of controllers influencing only a part of the controlled process in front of the output, i.e. in front of the main controlled variable. Of course, such a type of control system structure is possible only if auxiliary variables are accessible to measurements or if they admit the acting on the process, respectively.

Besides the main variables of the closed control loop with the controller in the feedback we have, when auxiliary variables are applied, additional or auxiliary control loops. Each auxiliary control loop can be equiped by its own controller irrespective whether the controller is acting in the feedback or in the feedforward configuration.

In this way we can create a great number of intricate control loops. For the analysis as well as for the synthesis of this kind of control loops the state space is a very convenient approach. The main reason sticks in the fact that the mathematical tools are for the single-input-single -output (SISO) as well as for the multi-input-multi-output (MIMO) systems the same, i.e. not needing to modify the design of controllers for the multiloop structure. Of course, the multiloop structure does not concern only MIMO systems corresponding to generally accepted definition of this kind of systems, but it may concern in particular SISO systems extended by auxiliary variables, as it was mentioned above. A single restriction may concern nonlinearities needing special adequate treatment.

The aim of this contribution is to draw attention to general algebraic procedures based on state space theory of linear systems if the objective function applied for the controller design is quadratic one. The design concept will be displayed on a continuously working SISO system belonging to the category of model following and tracking systems. The structure of the system selected for this purpose need actually three controllers,one in the feedback and two in the feedforward,as it is evident from Fig.1. Nevertheless, all these three controllers can be obtained as a single controller gain-matrix and moreover the same procedure can be applied for MIMO systems as well.

2   Problem formulation

Consider a stabillizable and detectable continuous linear timeinvariant SISO process described by the following state equations

$$\dot{x}_p(t) = F_p x_p(t) + g_p u(t)$$
$$y_p = c_p^T x_p(t), \tag{1}$$

a simple model-etalon for the ideal transient response of the controlled process

$$\dot{x}_m(t) = F_m x_m(t) + g_m u_m(t)$$
$$y_m(t) = c_m^T x_m(t), \tag{2}$$

the generator of the command variable

$$\dot{x}_w(t) = F_w x_w(t)$$
$$y_w(t) = c_w^T x_w(t) \tag{3}$$

and the structure of the system outlined in Fig.1 representing a possible type of model following system.

The dynamics of the system is characterized by the correcting variables $u_p \in R$ and $u_m \in R$, the command variable $y_w \in R$, the controlled process and model process outputs $y_p \in R$ and $y_m \in R$ and by the state vectors $x_p \in R, x_m \in R$ and $x_w \in R$ of the dimensions n_p, n_m and n_w, respectively. Let $n = n_p + n_m + n_w$. Matrices (1),(2) and (3) have compatible dimensions and are of the general

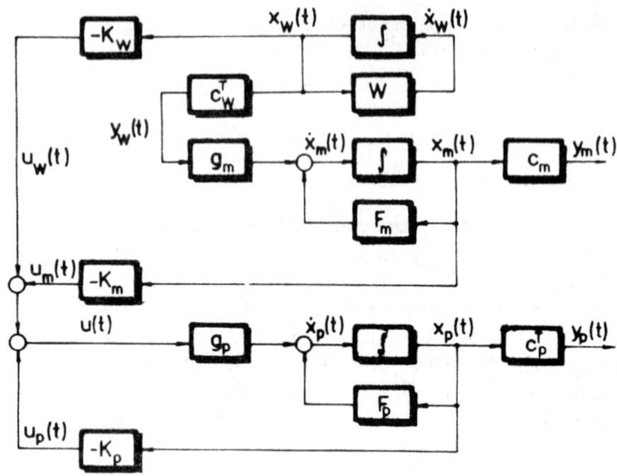

Fig.1 Block diagram of the model following control system

canonical form

$$
F = \begin{bmatrix} -f_{n-1} & 1 & 0 \cdots & 0 \\ -f_{n-2} & 0 & 1 \cdots & 0 \\ \vdots & & & \\ -f_0 & 0 & 0 \cdots & 0 \end{bmatrix} \qquad g = \begin{bmatrix} g_{n-1} \\ g_{n-2} \\ \vdots \\ g_0 \end{bmatrix}
$$

$$
c^T = \begin{bmatrix} 1 & 0 \dots & 0 \end{bmatrix} \tag{4}
$$

corresponding to the differential equation

$$
\sum_{i=0}^{n} f_i y^{(i)}(t) = \sum_{i=0}^{n-1} g_i u^{(i)}(t), f_n = 1. \tag{5}
$$

It is possible to note that the model of the controlled process does not represent a simplified model of the real system to be controlled but for a given command variable it yields the ideal output of the controlled process . The aim is to have the real output y_p as close as possible to the model output y_m.

By inspection of Fig.1 it is clear that

$$
\begin{aligned}
u(t) &= u_p(t) + u_m(t) + u_w(t) \\
&= -[K_p x_p(t) + K_m x_m(t) + K_w x_w(t)] \tag{6}
\end{aligned}
$$

where K_p, K_m and K_w are feedback and feedforward gains of controllers.

Defining the combined state vector $z(t)$ of the dimension n and output vector $y(t)$

$$
z(t) = \begin{bmatrix} x_p(t) \\ x_m(t) \\ x_w(t) \end{bmatrix} \qquad y(t) = \begin{bmatrix} y_p(t) \\ y_m(t) \\ y_w(t) \end{bmatrix} \tag{7}
$$

we can express the overall system by the single set of state equations

$$
\begin{aligned}
\dot{z}(t) &= Fz(t) + gu(t) \\
y(t) &= C^T z(T) \tag{8}
\end{aligned}
$$

where

$$
F = \begin{bmatrix} F_p & 0 & 0 \\ 0 & F_m & g_m c_w^T \\ 0 & 0 & F_w \end{bmatrix} \quad g = \begin{bmatrix} g_p \\ 0 \\ 0 \end{bmatrix} \quad C^T = \begin{bmatrix} c_p^T & 0 & 0 \\ 0 & c_m^T & 0 \\ 0 & 0 & c_w^T \end{bmatrix} \tag{9}
$$

Let the quality of control satisfies the minimum of the quadratic objective function

$$
J = \Theta[z(t)] \mid_{t=0}^{t=\infty} + \int_0^\infty [z^T(t)Qz(t) + u^T(t)Ru(t)]dt \tag{10}
$$

where Q is a symmetric positive semidefinite weighting matrix and R is a symmetric positive definite matrix. In the case of SISO systems R is a weighting constant.

Let the first term in the integral of the objective function (10) has the general form

$$
\begin{aligned}
z^T(t)Qz(t) &= q_1[y_p(t) - y_w(t)]^2 + q_2[y_p(t) - y_m(t)]^2 \\
&+ q_3[y_m(t) - y_w(t)]^2 \tag{11}
\end{aligned}
$$

It is easy to verify that the matrix Q must be expressed as

$$
Q = C \begin{bmatrix} q_1 + q_2 & -q_2 & -q_1 \\ -q_2 & q_2 + q_3 & -q_3 \\ -q_1 & -q_3 & q_1 + q_3 \end{bmatrix} C^T \tag{12}
$$

From practical point of view it is recommendable to apply for the representation of the state matrices (9) the canonical form given by Eqns. (4) and derived for example by Strejc (1981). Hence,the matrix Q attains the final form

$$
Q = \begin{bmatrix}
q_1 + q_2\ 0 \dots 0 & -q_2\ 0 \dots \dots 0 & -q_1\ 0 \dots \dots 0 \\
0 \dots \dots \dots 0 & 0 \dots \dots \dots 0 & \dots \dots \dots 0 \\
-q_2\ 0 \dots \dots 0 & q_2 + q_3\ 0 \dots 0 & -q_3\ 0 \dots \dots 0 \\
0 \dots \dots \dots 0 & 0 \dots \dots \dots 0 & 0 \dots \dots \dots 0 \\
-q_1\ 0 \dots \dots 0 & -q_3\ 0 \dots \dots 0 & q_1 + q_3\ 0 \dots 0 \\
0 \dots \dots \dots 0 & 0 \dots \dots \dots 0 & 0 \dots \dots \dots 0
\end{bmatrix} \tag{13}
$$

By appropriate selection of the three weighting factors q_1, q_2, q_3 and by minimization of the objective function (10) it is possible to satisfy different requirements of system control.

The aim of this contribution is to outline the main steps of the synthesis not needing the formulation and the solution of the Riccati equation relating to the proposed evidently not simple structure of the system.

3 General solution

The problem formulated in section 2 requires a solution of an intricate control loop. One possible approach is the application of the classical Kalman's filtering modified for a deterministic system. It necessitates the solution of the Riccati equation possibly of the overall system. In this case and namely in MIMO systems the dimension of the matrices may be very high and the solution of the Riccati equation can meet serious numerical difficulties. In order to avoid these inconveniences, it is possible to prefer the procedure via the Euler or Hamiltonian matrix which is the starting point for the very qualified numerical algorithms.

Let us recall briefly the derivation of the Euler matrix for a continuous system. Introducing the method of Lagrange multipliers to adjoin the system of differential equality constraint to the objective function and adding the general function of boudary conditions, we have

$$J = \Theta[z(t),t]|_{t=0}^{t=\infty} + \int_0^\infty \{z^T(t)Qz(t) + u^T(t)Ru(t) + $$
$$+ \lambda^T(t)[Fz(t) + Gu(t) - \dot{z}(t)]\}dt \qquad (14)$$

With the Hamiltonian

$$H(t) = z^T(t)Qz(t) + u^T(t)Ru(t) + \lambda^T(t)[Fz(t) + Gu(t)] \qquad (15)$$

the objective function becomes

$$J = \Theta[z(t),t]|_{t=0}^{t=\infty} + \int_0^\infty [H(t) - \lambda^T(t)\dot{z}(t)]dt \qquad (16)$$

Integration by parts of the last term in the integrand of (16) yields

$$J = \{\Theta[z(t),t] - \lambda^T(t)z(t)\}|_{t=0}^{t=\infty} + \int_0^\infty [H(t) - \dot{\lambda}^T(t)z(t)]dt \qquad (17)$$

Now, to find the minimum of (17) with respect to $z(t)$ and $u(t)$, we introduce the first variation of J. For the sake of brevity and clarity let us omit the argument t in the next relations. We obtain

$$\delta J = \{\delta z^T[\frac{\partial \Theta}{\partial z} - \lambda]\}|_{t=0}^{t=\infty} + \int_0^\infty \{\delta z^T[\frac{\partial H}{\partial z} + \dot{\lambda}] + \delta u^T[\frac{\partial H}{\partial u}]\}dt \qquad (18)$$

The necessary condition for a minimum is that the first variations in J vanish for arbitrary δz and δu. Hence, the necessary coditions have the form

$$\delta z^T[\frac{\partial \Theta}{\partial z} - \lambda] = 0; t = 0, \infty \qquad (19)$$

$$\dot{z} = \frac{\partial H}{\partial \lambda} \qquad \dot{\lambda} = -\frac{\partial H}{\partial z} \qquad (20)$$

$$\frac{\partial H}{\partial u} = 0 \qquad (21)$$

If the value of any variable is specified, the corresponding variation vanishes and the respecting conditions (19) — (21) do not apply. Particularly for given $z(0)$ and $z(\infty)$, the corresponding boundary conditions are satisfied by $\delta(0) = \delta(\infty) = 0$.

For the problem formulated in section 2 with specified $z(0)$ and $z(\infty)$, eqns.(20) and (21) yield

$$\frac{\partial H}{\partial u} = Ru(t) + g^T\lambda(t) = 0$$

$$u(t) = -R^{-1}g^T\lambda(t) \qquad (22)$$

$$\frac{\partial H}{\partial \lambda} = \dot{z}(t) = Fz(t) + gu(t) \qquad (23)$$

$$-\frac{\partial H}{\partial z} = \dot{\lambda}(t) = -Qz(t) - F^T\lambda(t) \qquad (24)$$

Equations (22) — (24) define a system of the dimension $2n$

$$\begin{bmatrix} \dot{z}(t) \\ \dot{\lambda}(t) \end{bmatrix} = \begin{bmatrix} F & -GR^{-1}G^T \\ -Q & -F^T \end{bmatrix} \begin{bmatrix} z(t) \\ \lambda(t) \end{bmatrix} \qquad (25)$$

where $\lambda(t)$ may be considered as a costate or adjoint vari-able with respect to $z(t)$ and the matrix on the right hand side is the so called Euler or Hamiltonian matrix E. It is an element of the symplectic algebra satisfying the relation

$$E^T J^* + J^* E = 0 \qquad (26)$$

where

$$J^* = \begin{bmatrix} 0 & -I \\ I & 0 \end{bmatrix} \qquad (27)$$

and I is the identity matrix.

Linear symplectic algebra implies transformations retaining relations of incidence. It is possible to take advantage of this fact in order to transform the matrix E to a matrix of eigenvalues, S, using the so-called commutating square

$$(28)$$

where V is the transforming matrix consisting of eigenvectors of E. Let us split up the matrix V as well as the matrix S to blocks of the dimension of F

$$V = \begin{bmatrix} V_{11} & V_{12} \\ V_{21} & V_{22} \end{bmatrix} \qquad S = \begin{bmatrix} S_1 & 0 \\ 0 & -S_1 \end{bmatrix} \qquad (29)$$

where S_1 includes n unstable eigenvalues of E. Eigenvalues of E must be symmetric with respect to the imaginary axis of the complex plane with no pure imaginary eigenvalues. Hence, $-S_1$ contains n stable eigenvalues of E. It holds according to (28) that

$$EV = VS \qquad (30)$$

For example, the first row of E yields

$$FV_{11} - GR^{-1}G^TV_{21} = V_{11}S_1$$
$$F - GR^{-1}G^TV_{21}V_{11}^{-1} = V_{11}S_1V_{11}^{-1} \qquad (31)$$

Matrix $V_{21}V_{11}^{-1} = P$ is the solution of the corresponding Riccati equation . Evidently, $R^{-1}G^TP = K$ is the weighting gain-matrix of the controller and $F - GK = F^*$ is the matrix of dynamics of the closed control loop. Calculation of the Riccati matrix P or calculation of the gain- matrix K of the controller is transformed to the calculation of eigenvectors of the Euler matrix E.

Let us recall that by means of (25) we can also find the optimum solution for finite time interval T of the cost function. Using (29), solution of (25) can be written in the form

$$\begin{bmatrix} z(t) \\ \lambda(t) \end{bmatrix} = V \begin{bmatrix} e^{S_1 t} & 0 \\ 0 & e^{-S_1 t} \end{bmatrix} V^{-1} \begin{bmatrix} z(0) \\ \lambda(0) \end{bmatrix} \qquad (32)$$

where

$$V^{-1} = \begin{bmatrix} V_{22}^T & -V_{12}^T \\ -V_{21}^T & V_{11}^T \end{bmatrix}$$

With the transversality conditions

$$\lambda(0) = P_0 z(0) \qquad (33)$$

and

$$\lambda(t) = P_t z(t) \qquad (34)$$

Eqn.(32) yields the desired relation for P_t provided that P_0 is known. If $P_0 = 0$ we get

$$P_t = (V_{21}e^{S_1 t}V_{22}^T - V_{22}e^{-S_1 t}V_{21}^T) \cdot$$
$$\cdot (V_{11}e^{S_1 t}V_{22}^T - V_{12}e^{-S_1 t}V_{21}^T)^{-1} \qquad (35)$$

The result holds for any t and consequently also for $t = T$.

As $t \to \infty$, the terms $e^{-S_1 t} \to 0$ and the relation (35) simplifies to $P = V_{21}V_{11}^{-1}$ as mentioned earlier.

The calculation of eigenvectors of a matrix is a problem already solved in a satisfactory way enabling to deal with matrices of very high dimensions. Let us recall some possibilities. The standard or direct method published for example by Vaugham (1969) is based on the well known definition of eigenvectors needing a simultaneous calculation of eigenvalues and generalized eigenvalues, respectively.

More ingenieus is the application of the so-called Schur vectors as proposed by Laub (1979). This approach provides a unifying methodology applicable for standard as well as singular control weighting matrices, cross-weighting matrices, singular transition matrices and generalized state space models. Moreover, it yields reliable numerical solutions.

The original idea is the so-called QR algorithm of Francis (1961,1962) and Kublanovskaya (1961) using shifts according the following equations

$$Q_i(E_i - k_i I) = R_i \qquad (36)$$
$$E_{i+1} = R_i Q_i^T + k_i I \qquad (37)$$

giving

$$E_{i+1} = Q_i E_i Q_i^T \qquad (38)$$

where Q_i is orthogonal, R_i is upper triangular, k_i is the shift of origin and $E_i, i = 1$, is the initial matrix of upper Hessenberg form. For $i = 1, 2, ...$ we obtain a sequence of matrices which converge to a quasi-upper-triangular real Schur form (RSF). It has all eigenvalues isolated on the diagonal or they are the eigenvalues of the 2x2 diagonal submatrices. Rapid convergence is achieved if the shifts are close to an eigenvalue of E_1 and hence of all E_i.

Because E_1 can have complex eigenvalues, it would be necessary to apply complex shifts, too, and consequently E_{i+1}, obtained in two steps, would be in general a complex matrix. The transformations may be accompanied by serious errors. Francis proposed an economical method using two real values of k_i and k_{i+1} or two complex conjugate values,but does not apply the complex arithmetic. The method is based on the relation

$$EQ = QH \qquad (39)$$

with unitary Q and upperHessnberg H. If H has positive subdiagonal elements, then the whole Q and H are determined by the first column of Q. After two steps of QR we have

$$E_{i+2} = QE_iQ^T \qquad (40)$$

where $Q = Q_{i+1}Q_i$. The QR algorithm modified to generalized eigenvectors is in the technical literature denoted as QZ algorithm.

4 Multi-input-multi-output systems

MIMO systems are usually represented as multivariable systems or as interconnected systems. Multivariable systems cannot be splitted up into subsystems while in the case of interconnected systems the blocks of proper processes are situated along the main diagonal of the matrix of dynamics of the whole process and the interactions between the processes are expressed explicitely by elements in the offdiagonal blocks, respectively.

Let us consider the multivariable system first and let us assume that it has s inputs and s outputs. Let the dimension of the overall system be $(N; N)$, where $N = N_p + N_m + N_w$. It is possible to apply the structure indicated by Fig.1 in the straightforward way. The single difference with respect to SISO systems refers to the dimensions of the individual blocks. $\mathrm{Dim}F_p$ is $(N_p; N_p)$, $\mathrm{dim}F_m$ is $(N_m; N_m)$ and $\mathrm{dim}F_w$ is $(N_w; N_w)$. The input matrices are now G of dimension $(N; s)$, G_p of dimension $(N_p; s)$ and G_m of dimension $(N_m; s)$. The individual output matrices are C^T of $\dim(3s; N)$, C_p^T of $\dim(s; N_p)$, C_m^T of $\dim(s; N_m)$ and C_w^T of $\dim(s; N_w)$.

It is evident from (12) that now q_1, q_2 and q_3 in the weighting matrix Q are blocks of the dimension $(s; s)$ and $\dim R$ is (s, s), too.All variables are vector variables of compatible dimensions. The main aim of the application of the model following concept is to reach the most convenient dynamic behaviour of the overall system.

The structure introduced in section 2 generalized for multivariable systems can be applied also for the suboptimal decoupling of a multidimensional system provided that the weighting matrices Q and R strongly accentuate the diagonal elements only.

It is possible to design a process model F_m of dimension $(N_m; N_m)$ as well as a generator F_w of dimension $(N_w; N_w)$ of the command variables, both yielding s outputs. Zero blocks are set according to the pattern given by equations (9). Hence, the outputs $y_i, i = 1, 2, ..., s$ follow expressively the outputs x_{wi} and x_{mi} of the block-diagonal matrices F_m and F_w, respectively.

It is clear that the suboptimal decoupling cannot remove completely the influence of the mutual interconnections. The result of this kind of suboptimal decoupling depends on the properties of the MIMO system and on the selected parameters of Q and R,respectively.

Different situation occurs in the case of interconnected processes. Let us consider the matrix of dynamics of the whole process in the form (41), where $f_{pij}, i, j = 1, 2, ..., s,$ $i \neq j$ are column matrices of elements expressing the inteconnections between the proper processes $F_{pi}, i = 1, 2, ..., s$. The model-state variables as well as the the command variables of each diagonal block do not influence other diagonal blocks.

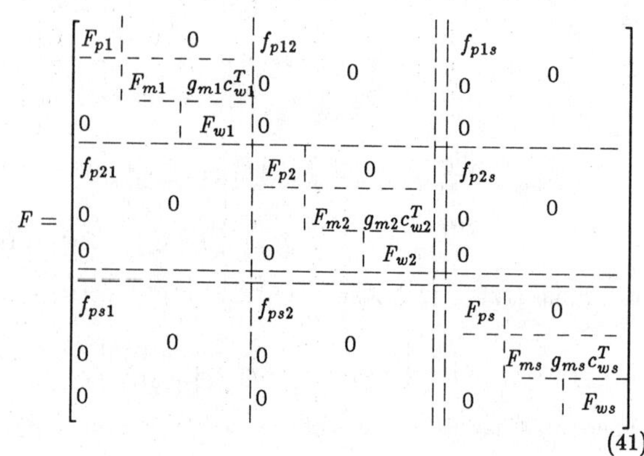

$$F = \begin{bmatrix} F_{p1} & 0 & f_{p12} & & f_{p1s} & \\ & F_{m1}\ g_{m1}c_{w1}^T & 0 & 0 & 0 & 0 \\ 0 & F_{w1} & 0 & & 0 & \\ \hline f_{p21} & & F_{p2} & 0 & f_{p2s} & \\ 0 & 0 & & F_{m2}\ g_{m2}c_{w2}^T & 0 & 0 \\ 0 & & 0 & F_{w2} & 0 & \\ \hline f_{ps1} & & f_{ps2} & & F_{ps} & 0 \\ 0 & 0 & 0 & 0 & & F_{ms}\ g_{ms}c_{ws}^T \\ 0 & & 0 & & 0 & F_{ws} \end{bmatrix}$$
$$(41)$$

The input matrix G may have all its elements nonzero except those belonging to models of the control responses and to generators of command variables so that the inputs $u_i, i = 1, 2, ..., s$ do not influence the ideal response models and the command variables.

The output matrix has the form

$$C^T = \begin{bmatrix} C_1^T & 0\ldots & 0 \\ 0 & C_2^T\ldots & 0 \\ \vdots & \ddots & \vdots \\ 0 & 0\ldots & C_s^T \end{bmatrix} \qquad (42)$$

Now, the overall system is described by the state equations

$$\begin{aligned} x(k+1) &= Fx(k) + Gu(k) \\ y(k) &= C^T x(k) \end{aligned} \qquad (43)$$

where dimensions of the individual matrices are $F(N; N)$, $G(N; s)$, $C^T(3s; N)$ and

$$N = \sum_{i=1}^{s} n_i \qquad (44)$$

n_i being the dimensions of the subsystems $S_i, i = 1, 2, ..., s$.

The vectors $f_{pij}, i, j = 1, 2, ..., s, i \neq j$ describe physical interconnections of the system in question and must be determined by identification or mathematical modeling in the same way as the elements of the diagonal blocks F_{pi}.

In the case of strong interconnections it may happen that the overall system is unstable.Hence, it is necessary to verify the stability of the overall system. If instability occurs, the increase of q_{1i} and q_{2i} values, $i = 1, 2, ..., s$, may remove this inperfection. In the opposite case the desired solution is not possible.

Suboptimal decoupling of interconnected systems in the sense of quadratic cost function can be achieved by similar means as in the case of multivariable systems.

The reader cann find the discrete version of the described concept in the publication by Strejc (1989).

5 Example

Let us consider a two-dimensional interconnected process described by the following matrices

$$F = \begin{bmatrix} -7 & 1 & 0 & 0 & 0 & 10 & 0 & 0 & 0 & 0 \\ -12.96 & 0 & 1 & 0 & 0 & 0 & 0 & 0 & 0 & 0 \\ -14.8 & 0 & 0 & 0 & 0 & 0 & 0 & 0 & 0 & 0 \\ 0 & 0 & 0 & -2 & 2 & 0 & 0 & 0 & 0 & 0 \\ 0 & 0 & 0 & 0 & -5 & 0 & 0 & 0 & 0 & 0 \\ 12 & 0 & 0 & 0 & 0 & -7.5 & 1 & 0 & 0 & 0 \\ 0 & 0 & 0 & 0 & 0 & -13.5 & 0 & 1 & 0 & 0 \\ 0 & 0 & 0 & 0 & 0 & -5 & 0 & 0 & 0 & 0 \\ 0 & 0 & 0 & 0 & 0 & 0 & 0 & 0 & -2 & 2 \\ 0 & 0 & 0 & 0 & 0 & 0 & 0 & 0 & 0 & -5 \end{bmatrix}$$

$$G = \begin{bmatrix} 2 & 0.1 \\ 0.5 & 0 \\ 0 & 0 \\ 0 & 0 \\ 0 & 0 \\ 0 & 1 \\ 0.3 & 1 \\ 0 & 0 \\ 0 & 0 \\ 0 & 0 \end{bmatrix}$$

$$C^T = \begin{bmatrix} 1 & 0 & 0 & 0 & 0 & 0 & 0 & 0 & 0 & 0 \\ 0 & 0 & 0 & 1 & 0 & 0 & 0 & 0 & 0 & 0 \\ 0 & 0 & 0 & 0 & 1 & 0 & 0 & 0 & 0 & 0 \\ 0 & 0 & 0 & 0 & 0 & 1 & 0 & 0 & 0 & 0 \\ 0 & 0 & 0 & 0 & 0 & 0 & 0 & 0 & 1 & 0 \\ 0 & 0 & 0 & 0 & 0 & 0 & 0 & 0 & 0 & 1 \end{bmatrix}$$

Let the initial state vector x be

$$x^T = \begin{bmatrix} 0 & 0 & 0 & 0 & 5 & 0 & 0 & 0 & 0 & 5 \end{bmatrix}$$

Both processes are exposed to the action of the following command variables

$$1 - exp(-5)$$

and the ideal response models have the form

$$1 - exp(-2).$$

The first process is strongly oscillating while the second one is overdamped. The required steady state values are $y_1 = 1$ and $y_2 = 0.8$ for $t \to \infty$.

Because $x(0)$ end $x(\infty)$ are determined, only the weighting matrices Q and R in the quadratic matrix of the cost function (10) are needed. Let us have $q_{11} = 1$, $q_{12} = 5$, $q_{13} = 1$, $q_{21} = q_{11}$, $q_{22} = q_{12}$ and $q_{23} = q_{13}$.

$$R = \begin{bmatrix} 3 & 0 \\ 0 & 3 \end{bmatrix}$$

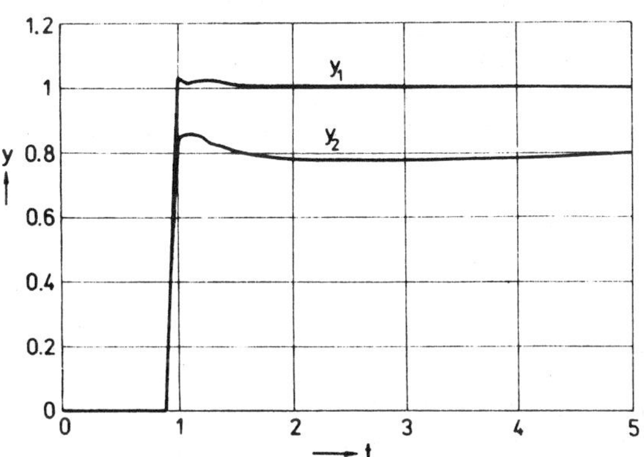

Fig.2 Control step responses of two interconnected systems of the third order

The first number in the indices indicates the correspondence to the process 1 and 2. The second number relates to the pattern (13).

The interaction by elements $f_{16} = 10$ and $f_{61} = 12$ of the matrix F is very strong in comparison with the parameters of both processes. Other interaction is caused by elements G_{12} and $G_{7,1}$ of the input matrix G.

The controller gain-matrix calculated by means of Schur vectors has the form

$$K = \begin{bmatrix} 3.7439 & 0.0830 & -0.2086 & -0.1332 & -0.0151 \\ 1.7737 & 0.0928 & -0.0816 & -0.0154 & 0.0002 \end{bmatrix}$$

$$\begin{bmatrix} 3.1903 & 0.0522 & -0.1815 & -0.0730 & -0.0050 \\ 1.6163 & 0.0773 & -0.0792 & -0.1305 & -0.0279 \end{bmatrix}$$

The closed control loop has following eigenwalues

-17.6041	-0.4369
$-2.4802+i3.1370$	-2
$-2.4802-i3.1370$	-2
$-0.4573+i0.6265$	-5
$-0.4573-i0.6265$	-5

The result of simulation is shown in Fig.2. The transient responses have been calculated by application of $exp((F - GK)t)$ with time interval $\Delta t = 0.1$. Total run time $t = 5$.

It is evident, that the transient responses of both processes are very fast and follow the command variables with negligible errors. The influence of interactions is not visible.

6   Conclusion

The described procedure of multidimensional control system design proves the usefulness of the application of the state space approach. It enables a uniform solution of control systems of different complexity. The selected structure

of the subsystems represents the ideal model-matching control concept yielding a high control quality illustrated by a numerical example. It is recommended to solve the high dimensional controller gain-matrices by means of Euler matrix and by appropriate numerical procedures.

REFERENCES

Francis J.G.F.: The QR transformation—a unitary analogue to the LR transformation. I.Comput.J.4, (1961), 265–271, II.Comput. J. 4, (1962), 332–345.

Kublanovskaya V.N.: On some algorithms for the solution of the complete eigenvalue problem. Ž. Vyčisl. Mat. i Mat. Fiz. 1, (1961), 555–570.

Laub A.J.: A Schur method for solving algebraic Riccati equations. IEEE Trans. Automat. Contr., AC–24, 6, (1979), 913–921.

Strejc V.: State Space Theory of Discrete Linear Control. John Wiley and Sons, Chichester, 1981.

Strejc V.: One possibility of multidimensional control system design. Kybernetika 25, (1989), 6.

Vaugham D.R.: A negative exponential solution for the matrix Riccati equation. IEEE Trans. Automat. Contr., AC–14, (1969), 72–75.

THE LINEAR CONSTRAINED REGULATION
PROBLEM FOR DISCRETE-TIME SYSTEMS

G. Bitsoris* and M. Vassilaki**

**Control Systems Laboratory, Electrical Engineering Department, University of Patras,
26500 Patras, Greece
**AMBER S.A., Computer Systems, P.O. Box 3500 Athens, Greece*

Abstract:In this paper the Linear Constrained Regulation Problem for discrete-time Systems is studied. The first part of the paper deals with the
problem of existence of linear state-feedback control laws that transfer
asymptotically to the origin all initial states belonging to a polyhedral
subset of state space while linear state and control constraints are respected. Then an eigenstructure assignment technique is developed for derivation of a solution to this problem. The results presented in the paper
are based on the properties of systems possessing polyedral positively invariant sets.

Keywords:Constraint theory;discrete time systems;eigenvalue assignment,
invariance;linear systems;stability.

INTRODUCTION

Constrained control problems are usually
solved by first determining the solution of
an unconstrained control problem and then by
modifying this solution so that the control
constraints are respected (Gutman and Hagander 1985). Another approach is the application of optimal control techniques. Both
approaches, however, lead to complicated
nonlinear algorithms which can not be implemented easily. For this reason it is very
important to develop techniques for the derivation of simple solutions to constrained
control problems.

In this paper the constrained regulation
problem for discrete-time systems is studied. The problem consists in the determination of control laws such that all initial states belonging to a polyhedral set of the
state space are transferred to origin while
linear state and control constraints are respected. In Feuer and Heyman (1976) existence conditions for an open-loop solution were
given, while in Gutman and Cwinkel (1987)
an algorithm for the derivation of such a
solution was established. In this paper linear state-feedback controls are considered.

The paper is organised an follows:In the
first section, the LCRP is formulated. In
the next section, conditions guaranteeing
the existence of a solution to this problem
are established. Finally, in the last section, it is shown that the LCRP can be viewed
as an eigenstructure assignment problem and
an algorithm for the derivation of a solution is presented.

PROBLEM FORMULATION

Throughout the paper, upper case letters denote real matrices, lower case letters denote column vectors or scalars, $R^n(C^n)$ denote
the real (complex) n-space and $R^{n \times m}$ $C^{(n \times m)}$
denote the set of all nxm real (complex) ma-
trices. For two real vectors $x = \begin{bmatrix} x_1 & x_2 \dots x_n \end{bmatrix}^T$
and $y = \begin{bmatrix} y_1 & y_2 & \dots & y_n \end{bmatrix}^T$, $x < y$ $(x \leq y)$ is equivalent to $x_i < y_i$ $(x_i \leq y_i)$ i=1,2,...,n. For a
matrix $H \in R^{m \times m}$, $H^{\pm} = (h_{ij}^{+})$ with $h_{ij}^{+} = max(h_{ij}, 0)$
and $H^{-} = (h_{ij}^{-})$ with $h_{ij}^{-} = max(-h_{ij}, 0)$. Finally,
given a square matrix H, $\lambda(H)$ denotes its
eigenvalues.

We consider discrete-time linear systems
described by the difference equation

$$(k+1) = Ax(k) + Bu(k) \qquad (S)$$

321

with $A \in R^{n \times n}$, $B \in R^{n \times m}$, $x \in R^n$, $u \in R^m$ and $t \in T$
where T is the time set $T = \{0,1,2,\ldots,\}$.

The control vector $u(k)$ must satisfy linear
constraints of the form

$$-d_2 \leqslant u \leqslant d_1 \qquad (1)$$

where d_1 and d_2 are real vectors with positive components.

The state vector is also subject to linear
constraints of the form

$$Dx \leqslant c \qquad (2)$$

where $D \in R^{p \times n}$ and $c > 0$.

In most of the practical regulation problems
the upper bounds of the disturbances acting
on the system are known. So, the initial
states of the system are assumed to belong
to a bounded polyhedron defined by the linear
inequality

$$Gx \leqslant w \qquad (3)$$

where $G \in R^{q \times n}$, $\text{rank} G = n$ and $w > 0$.

The regulation problem to be studied is formulated as follows:Determine a linear state-
feedback control law

$$u(k) = Fx(k)$$

such that all initial states belonging to
the polyhedron defined by ineq. (3) are
transferred to the origin aymptotically, while the constraints (1) and (2) on the control and the state vectors respectively,
are respected. Due to the linearily of the
system, of the control law and of the constraints, this problem is termed the <u>Linear
Constrained Regulation Problem</u> (LCRP).

It is clear that the LCRP is well posed if
all initial states belonging to the polyhedron (3) satisfy the state constraints (2).
However, the well posedness of the problem
does not guarantee the existence of a solution even in the case where the open-loop
system is asymptotically stable.

EXISTENCE CONDITIONS

Let us associate with each state feedback
control law $u = Fx$ the polyhedral set

$$Q(F,d_1,d_2) \overset{\Delta}{=} \{x \in R^n : -d_1 \leqslant Fx \leqslant d_2\}$$

which is the region of the state space of
the closed-loop system where the control vector satisfies constraints (1). Let also

$$P(D,c) \overset{\Delta}{=} \{x \in R^n : Dx \leqslant c\}$$

be the region where the state constraints
(2) are respected. According to this notation,
$P(G,w)$ denotes the set of initial states
defined by inequality (3).

It is obvious that a control law $u = Fx$ is a
solution to the LCRP if and only if the resulting closed-loop system

$$x(k+1) = (A+BF)x(k) \qquad (4)$$

is asymptotically stable and no trajectory
$x(k;x_0)$ emanating from the region $P(G,w)$
leaves the regions $Q(F,d_1,d_2)$ and $P(D,c)$ for
any instant $k \in T$. This condition is expressed
as follows:

<u>Proposition 1</u>:

The control law $u = Fx$ with $F \in R^{m \times n}$ is a solution to the LCRP if and only if
i) the eigenvalues λ_i $i = 1,2,\ldots,n$ of the
matrix $A+BF$ are in the open disk $|\lambda_i| < 1$ and
ii) there exists a positively invariant set
$\Omega \subset R^n$ of the resulting closed-loop system (4)
such that

$$P(G,w) \underline{\subset} \Omega \underline{\subset} P(D,c)$$
and
$$\Omega \underline{\subset} Q(F,d_1,d_2)$$

Proof:

a) <u>Necessity</u>:

It is evident that hypothesis (i) is a necessary condition for $u = Fx$ to be a solution
to the LCRP. On the other hand, by definition, the set

$$\Omega = \{x \in R^n : (\exists x_0 \in P(G,w), k \in T : x = x(k;x_0))\} \qquad (5)$$

is a positively invariant set of system (3)
and $P(G,w) \underline{\subset} \Omega$. Furthermore $\Omega \underline{\subset} P(D,c)$ and
$\Omega \underline{\subset} Q(F,d_1,d_2)$ because otherwise there would
exist a $x_0 \in P(G,w)$ and a $k \in T$ such that
$x(k;x_0) \notin P(D,c)$ or/and $x(k;x_0) \notin Q(F,d_1,d_2)$.
This would imply that the state constraints

would not be satisfied, which would contradict the hypothesis that u=Fx is a solution to the LCRP. Therefore hypothesis (ii) is also a necessary condition.

b) <u>Sufficiency</u>:

If x_0 satisfies inequality (3), that is if $x_0 \in P(G,w)$ then $Dx(k;x_0) \leq c$ and $-d_2 \leq Fx(k;x_0) \leq d_1$ for all $k \in T$ because x_0 belongs to the positively invariant set Ω defined by (5), $\Omega \subseteq P(D,c)$ and $\Omega \subseteq Q(F,d_1,d_2)$. Furthermore, condition (i) guarantees the convergence of any state x_0 to the origin. □

By combining this result with well known methods of analysis and synthesis of linear control systems we can establish the following iterative procedure for the determination of a solution to the LCRP:
First, determine a stabilizing control law u=Fx by applying well known eigenvalue assignment methods. Then examine whether there exists a quadratic Lyapunov function $v(x)=x^T Px$ for the closed-loop system such that

$$P(G,w) \subseteq \Omega(v,\rho) \subseteq P(D,c) \qquad (6)$$
and
$$\Omega(v,\rho) \subseteq Q(F,d_1,d_2) \qquad (7)$$

where $\Omega(v,\rho)$ is the positively invariant set defined by the relation

$$\Omega(v,\rho) \overset{\Delta}{=} \{x \in R^n : v(x) \leq \rho\}$$

If it is not the case, then repeat the design procedure to determine a stabilizing control with a "lower gain" F. □

Since it is not easy to examine whether there exists a quadratic Lyapunov function satisfying relations (6) and (7), the application of this procedure to the determination of a solution to the LCRP is a very difficult work. One, however, could observe that it would not be necessary to investigate the existence of such a quadratic Lyapunov function, if one of the polyhedral sets P(G,w) or P(D,c) was positively invariant. Then it would be sufficient to examine whether

$$P(G,w) \subseteq Q(F,d_1,d_2)$$
or
$$P(D,c) \subseteq Q(F,d_1,d_2)$$

These cases, where the stabilizing control u=Fx is computed so that the polyhedron P(G,w) or

P(D,c) is a positively invariant set of the resulting closed-loop system have been reported in Vassilaki et al (1988). Here we consider the special case where no constraints on the state vector are imposed and F is computed so that $Q(F_1,d_1,d_2)$ is positively invariant:

<u>Proposition 2</u>

If F is a real mxn matrix such that
i) $Q(F,d_1,d_2)$ is a positively invariant set of the closed-loop system (4),
ii) $P(G,w) \subseteq Q(F,d_1,d_2)$ and
iii) the eigenvalues of the matrix A+BF are in the open disk $|\lambda|<1$, then the control law u=Fx is a solution to the LCRP. □

The development of an algorithm for the derivation of a solution to the LCRP is possible, if conditions i and ii of Proposition 2 are replaced by equivalent algebraic relations.

Condition i can be replaced by an algebraic one using the following result relative to the existence of positively invariant polyhedral sets for linear systems (Bitsoris 1988b):

<u>Proposition 3</u>

The polyhedral set $Q(F,d_1,d_2)$ is a positively invariant set of the unforced system

$$x(k+1)=Ax(k)$$

if there exists a matrix $H \in R^{mxm}$ such that

$$FA-HF=0$$
and
$$\begin{bmatrix} H^+ & H^- \\ H^- & H^+ \end{bmatrix} \begin{bmatrix} d_1 \\ d_2 \end{bmatrix} < \begin{bmatrix} d_1 \\ d_2 \end{bmatrix}$$

Condition ii can also be replaced by a set of linear inequalities. To this end, let $x^{(i)}$ i=i=1,2,...,r be the vertices of the closed polyhedron P(G,w). The vertices $x^{(i)}$ are solutions of the linear equations

$$G^{(i)} x^{(i)} = w^{(i)} \qquad i=1,2,...,r$$

where $G^{(i)} \in R^{nxn}$ are nonsingular submatrices of G and $w^{(i)}$ are the corresponding subvectors of w. The following fundamental result of the linear programming is an equivalent algebraic version of condition ii of Proposition 2:

Proposition 4

The polyhedral sets $P(G,w)$ and $Q(F,d_1,d_2)$ satisfy the relation

$$P(G,w) \subseteq Q(F,d_1,d_2)$$

if and only if

$$-d_2 \leq Fx^{(i)} \leq d_1 \qquad i=1,2,\ldots,r$$

Now, by combining Propositions 2,3 and 4 we establish the following result:

Proposition 5

If F is a mxn real matrix and there exists a matrix $H \in R^{m \times m}$ such that

$$\text{i)} \quad FA+FBF=HF \tag{8}$$

$$\text{ii)} \quad \begin{bmatrix} H^+ & H^- \\ H^- & H^+ \end{bmatrix} \begin{bmatrix} d_1 \\ d_2 \end{bmatrix} \leq \begin{bmatrix} d_1 \\ d_2 \end{bmatrix} \tag{9}$$

$$\text{iii)} \quad -d_1 \leq Fx^{(i)} \leq d_2 \qquad i=1,2,\ldots,r \tag{10}$$

iv) the eigenvalues of matrix A+BF are in the open disk $|\lambda|<1$,

then the control law u=Fx is a solution to the LCRP. □

There are many ways to use this result to derive a solution to the LCRP. For example, this can be done, by first choosing a matrix H that satisfies inequalities (9) and then by solving the matrix equation (8). It should be noted that given a matrix H, equation (8) possesses at least the trivial solution F=0. Nevertheless, it is not sure that there exists a solution F such that u= Fx is a stabilizing control law. So the following questions arise:

1) Given A and B does there exist a matrix H for which eq. (8) possesses a solution F such that the eigenvalues of the matrix A+BF are in the open disk $|\lambda|<1$?"

2) If such a matrix H exists,how can be chosen?

3) How the solution of eq. (8) can be obained?

The object of the following section is to give answers to these questions.

AN EIGENSTRUCTURE ASSIGNMENT APPROACH

It is well known that the stabilization by linear state feedback can be considered as an eigenstucture assignment problem. From this point of view, relations (8) and (9) expresse implicitly the conditions that the assigned eigenvalues and eigenvectors must satisfy for the control u=Fx not to violate the constraints $-d_2 \leq u \leq d_1$. This will be clarified below:

Proposition 6

The equation (8) is equivalent to the of relations

$$\begin{bmatrix} U \\ V \end{bmatrix} J = \begin{bmatrix} A & B \\ 0 & H \end{bmatrix} \begin{bmatrix} U \\ V \end{bmatrix} \tag{11}$$

$$F=VU^{-1} \tag{12}$$

where $J \in C^{n \times n}$ is a matrix in Jordan form, $V \in C^{m \times n}$ and $U \in C^{n \times n}$ with det $U \neq 0$.

Proof

From (11) and (12) it follows that

$$FA+FBF=VU^{-1}(A+BVU^{-1})$$

$$=VU^{-1}(AU+BV)U^{-1}$$

$$=VU^{-1}UJU^{-1}=VJU^{-1}=HVU^{-1}$$

$$=HF$$

Conversely, assume that $F \in R^{m \times n}$ is a solution of eq. (8) and let $U \in C^{n \times n}$ be the transformation matrix by which the matrix A+BF is transformed to its Jordan canonical form:

$$J=U^{-1}(A+BF)U$$

Now, setting V=FU, we obtain

$$UJ=(A+BF)U=AU+BV \tag{13}$$

and

$$VJ=FUJ=FUU^{-1}(A+BF)U=HFU=HV \tag{14}$$

Remarks

1) Since J is a matrix in Jordan canonical form, from (13) it follows that its diagonal elements λ_i are the eigenvalues of the

324

closed-loop matrix $A+BF$ and the columns u_i of the matrix U are the corresponding eigenvectors or generalised eigenvectors. On the other hand, from (14) it follows that the nonzero columns v_i of the matrix V are eigenvectors or generalised eigenvectors of the matrix H associated with the same eigenvalue λ_i.

2) Since

$$\lambda_i u_i + u_{i-1} = A u_i + B v_i$$
$$\vdots$$
$$\lambda_i u_{i-r-2} + u_{i-r-1} = A u_{i-r-2} + B v_{i-r-2}$$

$$\lambda_i u_{i-r-1} = A u_{i-r-1} + B v_{i-r-1}$$

$$\lambda_i v_i + v_{i-1} = H v_i$$
$$\vdots$$
$$\lambda_i v_{i-r-2} + v_{i-r-1} = H v_{i-r-2}$$

$$\lambda_i v_{i-r-1} = H v_{i-r-1}$$

where r is the multiplicity of the eigenvalue λ_i, we conclude that if $v_i = 0$ then $v_{i-1} = v_{i-2} = \ldots = v_{i-r-1} = 0$, which in turn implies that (λ_i, u_{i-j}) $j = 0, 1, \ldots, r-1$ are pairs of open-loop eigenvalues with the corresponding eigenvectors or generalised eigenvectors.

Now, we are in a position to give an answer to the first question:

Proposition 7

There exists a matrix H for which eq. (8) has a solution F such that the closed-loop system (4) is asymptotically stable, if and only if the pair (A,B) is stabilisable and the matrix A has no more than m unstable eigenvalues.

a) Necessity

It is evident that the stabilisability of the pair (A,B) is a necessary condition for the existence of such a pair (H,F). On the other hand, from the second remark it follows that if (λ_i, u_i) is a pair of an eigenvalue with a corresponding eigenvector or generalised eigenvector assigned by the control low $u=Fx$, then the corresponding column $v_i \neq 0$ of the matrix V is an eigenvector or generalised eigenvector of the matrix H associated with the same eigenvalue. If $u=Fx$ is a stabilising control, then the number of the assigned distinct pairs (λ_i, u_i), and accordingly the number of the distinct pairs (λ_i, v_i) with $v_i \neq 0$, is at least equal to the number of the unstable eigenvalues of the matrix A. This number can not exceed m because $H \in R^{m \times m}$.

b) Sufficiency

Due to space limitations, only the case where the matrix A has $r < m$ distinct unstable eigenvalues will be considered.

Let $V = [v_1, v_2, \ldots, v_r, 0, 0, \ldots, 0]$, $V \in R^{m \times n}$ where $v_i \in R^m$ $i = 1, 2, \ldots, r$ are linearly independent vectors such that $u_{A,i}^T B v_i \neq 0$, $u_{A,i}^T$, $i = 1, 2, \ldots, r$ being the reciprocal eigenvectors of the matrix A associated with unstable eigenvalues λ_i^* $i = 1, \ldots, r$. Such vectors exist because from the stabilisability of the pair (A,B) it follows that $u_{A,i}^T B \neq 0$. Let also $U = [u_1, u_2, \ldots, u_r, u_{r+1}^*, \ldots, u_n^*]$, $U \in R^{n \times n}$ where $u_{r+1}^*, u_{r+2}^*, \ldots, u_r^*$ are the eigenvectors or generalised eigenvectors of the matrix A associated with the stable eigenvalues and

$$u_i = (\lambda_i \mathbf{1} - A)^{-1} B v_i \qquad i = 1, 2, \ldots, r \qquad (15)$$

Here λ_i $i = 1, 2, \ldots, r$ are real numbers such that $|\lambda_i| < 1$ and $\det S \neq 0$ where $S = (s_{ij})$ with

$$s_{ij} = \prod_{\substack{j=1 \\ j \neq i}}^{r} (\lambda_j - \lambda_i^*)\, u_{A,i}^T B v_i \qquad (16)$$

Such numbers exist because $u_{A,i}^T B v_i \neq 0$.

Then taking into account that $\det S \neq 0$, one can easily prove that the matrix U is nonsingular.

By construction, the matrices U and V satisfy relation (13) with

$$J = \begin{bmatrix} J_1 & 0 \\ 0 & J_2 \end{bmatrix}$$

where $J_1 = \operatorname{diag}(\lambda_1, \lambda_2, \ldots, \lambda_r)$ and J_2 is the Jordan block corresponding to the stable eigenvalues of the matrix A. On the other hand, relation (14) is also satisfied by any matrix $H \in R^{m \times m}$ for which $\lambda_i v_i = H v_i$ $i = 1, 2, \ldots, r$. Such a matrix exist because $r \leq m$. Now, taking into account the results stated in Proposition 6 and Remark 1, we conclude that $F=VU^{-1}$ and H satisfy eq. (8) and the eigenvalues of the matrix $A+BF$ are in the open disk $|\lambda_i|<1$. □

The answers to the other questions are also given in the proof of this proposition. By combining these results with those concerning the existence of matrices H satisfying inequalities (9) (Bitsoris 1988a), we can establish the following algorithm for the derivation of a solution to the LCRP.

<u>Step 1</u>:Determine r pairs $(\lambda_i,v_i)\in R_+ xR^m$ such that

a) the matrix $H\in R^{mxm}$, $H=[H_1 \,|\, 0]$ with

$$[v_1 \; v_2 \ldots v_r] \operatorname{diag}(\lambda_1,\lambda_2,\ldots,\lambda_r)=$$

$$=H_1[v_1 \quad v_2 \ldots v_r]$$

satisfies the algebraic conditions for which inequality (9) is verified

b) det $S\neq 0$, where the elements s_{ij} of S are given by (16).

<u>Step 2</u>:Determine the gain matrix $F=VU^{-1}$ where

$$V=[v_1 \; v_2 \ldots v_r \; 0 \ldots 0] \text{ and}$$

$$U=[u_1 \quad u_2 \ldots u_r \quad u_{r+1}^* \ldots u_n^*]$$

with u_i $i=1,2,\ldots,r$ given by the relations (15) and $u_{r+1}^*,\ldots,u_n^*$ being the eigenvectors or generalised eigenvectors of the matrix A associated with the stable eigenvalues.

<u>Step 3:</u> Examine whether F satisfies inequalities (10). If not, then repeat the algorithm by assigning greater eigenvalues λ_i $i=1,2,\ldots,r$.

<u>Example</u>

Let us consider the second order system S with

$$A = \begin{bmatrix} 1 & 0.5 \\ 2 & 0 \end{bmatrix}, \; b = \begin{bmatrix} 1 \\ 0 \end{bmatrix}$$

and a convex polyhedral set of initial states defind by the vertices

$$x_{(1)}=[1 \; 1]^T, \; x_{(2)}=[-1 \; 1]^T, \; x_{(3)}=[0 \; -1]^T$$

The control vector is constrained to satisfy the inequalities

$$-1.5 \leqslant u \leqslant 1$$

The hypotheses of Proposition 7 are satisfied because the pair (A,b) is controllable and only one of the open-loop eigenvalues $\lambda_1^*=1.618$ and $\lambda_2^*=-0.618$ is unstable.

Let us try to assign the stable eigenvalue $\lambda_1=0.8$. Setting $V=[v_1 \; 0]$ and $U=[u_1 \; u_2^*]$ where $v_1=1$, $u_1=(\lambda_1 1 -A)^{-1} bv_1$ and $u_2^*=[-0.618 \quad 2]^T$ is the eigenvector of the matrix A associated with the stable eigenvalue $\lambda_2^*=-0.618$, we obtain $h=0.8$ and

$$F=VU^{-1}=[-0.818 \quad -0.253]$$

It is a simple task to verify that inequality (9) is satisfied with $h^+=0.8$, $h^-=0$, $d_1=1$, $d_2=1.5$, and F satisfies inequalities (10). Therefore the control law $u=-0.818x_1-0.253x_2$ is a solution to the LCRP for the system under consideration

CONCLUSION

In this paper the Linear Constrained Regulation Problem has been formulated and studied. First, conditions for the existence of a solution to this problem have been established. Then an eigenstructure assignment technique for the derivation of a solution has been developed.

REFERENCES

Bitsoris, G.(1988a).Spectral properties of matrices with stable upper bounds. <u>Technical Report UPCSL 88005</u>, University of Patras-Greece.

Bitsoris,G.(1988b). On the positive invariance of polyhedral sets for discrete-time systems. <u>Syst.Control Letters</u>,<u>11</u>,243-248.

Feuer,A.and M. Heyman (1976). Ω-invariance in control systems with bounded controls. <u>J.Math.Anal.Appl.</u>,<u>53</u>, 266-276.

Gutman P.O.and M.Cwinkel (1987). An algorithm to find maximal state and constraint sets for discrete-time systems with bounded control and states. <u>IEEE Trans.Aut. Control, 32</u>, 251-254.

Gutman P.O and P.Hagander (1985). A new design of constrained controllers for linear systems. <u>IEEE Trans.Aut.Control, 30</u>, 22-33

Vassilaki M., J.C.Hennet and G. Bitsoris (1988). Feedback control of linear discrete-time systems under state and control constraints, <u>Int. J.Control, 47</u>, 1727-1735.

INVARIANT REGULATORS FOR A CLASS OF CONSTRAINED LINEAR SYSTEMS

J.-C. Hennet and J.-P. Beziat

*Laboratoire d'Automatique et d'Analyse des Systèmes, 7 avenue du Colonel Roche,
31077 Toulouse, France*

Abstract. Stable dynamic systems admit positively invariant domains associated to their Lyapunov
functions. Conversely, some domains can be made positively invariant for systems with state feedback controllers designed in such a way that some associated definite non-negative functions are
bound to decrease. In particular, this approach can be used to establish conditions on the gain
matrix for Linear Constrained Regulation Problems (LCRP). We propose a fixed and a variable
regulator easy to compute through linear programming, for a class of contrained linear systems.

Keywords. Discrete-time linear systems, Linear Constrained Regulation Problem, Stability,
Positive invariance, Variable regulator, Linear programming.

INTRODUCTION

Technical control limitations have long been considered
as a major problem in control engineering. And actually most control schemes do not integrate constraints in
their design. In practice, control laws often have to be
complemented by adequate control limiting devices. But
saturated linear controllers may fail to stabilize unstable
linear systems.

For some sets of initial conditions,stabilizing saturated linear controllers can be designed by the method of P.O. Gutman and P.Hagander 1985 . Their approach reposes on
the construction of an elliptic positively invariant domain
included in the polyhedral domain of constraints and including the set of initial states. But since the shape of
the invariant domains is directly induced by the selected
Lyapunov functions, the choice of classical quadratic Lyapunov functions is not the most efficient for the Linear
Constrained Regulation Problem (LCRP); it does not maximize the size of the domain of initial states for which a
stabilizing constrained regulator can be computed.

This limitation can be overcome by the use of non-quadratic
Lyapunov functions of the type introduced by H.N. Rosenbrock 1963 , G.Bitsoris and C.Burgat 1977 . In this line,
some authors (M.Vassilaki, J.C.Hennet and G.Bitsoris
1988, A. Benzaouia and C.Burgat 1988) have proposed
methods for constructing polyhedral positively invariant
sets better fitted or even perfectly matching the domain
of linear constraints.

In particular, for a linear state feedback,the convex polyhedron generated in the state space by the set of control constraints can be made invariant by specific matrix
conditions. A simplified version of the invariance conditions is proposed in this paper. For an easy on-line implementation of the control scheme with possible extensions
to adaptive cases (J.P.Béziat and J.C.Hennet 1988), the
non-linear relation to be verified by the gain matrix can
be replaced by a set of linear sufficient conditions. The

control law can then be obtained by adding to the set of
constraints a linear objective function. The selected criterion to be maximized corresponds to a convergence speed
index. This method allows for a fast determination of a
rapidly converging control law; computation of the gain
matrix can also be frequently updated to accelerate the
convergence speed by taking into account the current state
of the system. Global stability of the variable regulating
scheme is proven under a local stability condition.

POSITIVE INVARIANCE
OF POLYHEDRAL SETS

Consider the discrete-time linear dynamical system described by the equation:

$$X_{k+1} = A_0.X_k \qquad (1)$$

with $X_k \in \Re^n$ for any $k \in N$, $A_0 \in \Re^{n*n}$.
Let $R(G, \omega)$ be a not-empty convex polyhedral set defined
by :

$$R(G, \omega) = \{X \in \Re^n \ : \ G.X \leq \omega\} \qquad (2)$$

$G \in \Re^{g*n}$ and ω is a vector in $\Re^g$.
According to the selected couple (G, ω), $R(G, \omega)$ can be
any type of polyhedral set (bounded or unbounded, including or not the origin point) . The case of propre cones
is also included in this representation , for $\omega = 0$.
Existence of positively invariant polyhedral sets for system (1) is a generic property which covers different types
of dynamical behaviours.
In particular, if rank $G = n$ and $(\omega)_i > 0$ for
$i = 1, ..., g$, positive invariance of $R(G, \omega)$ implies the stability of system (1). This last property can easily be shown
as in G.Bitsoris 1988 by selecting as a Lyapunov function
of the system :

$$V(X) = \max_i \{\frac{|(G.X)_i|}{(\omega)_i}\} \qquad (3)$$

But the following proposition is valid for any type of polyhedral set $R(G,\omega)$. It provides a necessary and sufficient condition for $R(G,\omega)$ to be a positively invariant set of system (1).

Proposition 1

The convex polyhedral set $R(G,\omega)$ is positively invariant for system (1) if and only if there exists a matrix $H \in \Re^{g*g}$ such that:

$$H_{ij} \geq 0, \text{ for } i = 1,...,g \; ; \; j = 1,...,g \quad (4)$$
$$HG = GA_0 \quad (5)$$
$$H.\omega \leq \omega \quad (6)$$

Proof

By definition, $R(G,\omega)$ is said to be positively invariant for system (1) if and only if:

$$X_r \in R(G,\omega) \Longrightarrow X_{r+k} = A_0^k.X_r \in R(G,\omega)$$
$$\forall \, r \in \mathcal{N} \, , \, \forall \, k \in \mathcal{N}$$

A necessary and sufficient condition for $R(G,\omega)$ to be positively invariant for system (1) is:

$$GA_0.X \leq \omega, \forall X \in \Re^n \; : \; G.X \leq \omega \quad (7)$$

Condition (7) is a special case of inclusion of a polyhedral convex set in an other polyhedral convex set. The extended Farkas' lemma (J.C.Hennet 1989) provides an algebraic characterization of such an inclusion.

Any point of $R(G,\omega)$ also satisfies the set of linear inequalities $P.X \leq \psi$ with $P \in \Re^{p*n}$ and $\psi \in \Re^p$ if and only if there exists a (dual) matrix H of $\Re^{p*g}$ with non-negative coefficients satisfying conditions $H.G = P$ and $H.\omega \leq \psi$.

This result can easily be proven by concatenation of necessary and sufficient conditions related to each line P_i of matrix P :

By the standard Farkas' lemma (see e.g. A.Schrijver 1986), a necessary and sufficient condition for :

$$P_i.X \leq \psi_i, \quad \forall X \; : \; G.X \leq \omega$$

is :

$$\exists H_i \in \Re^{1*g} \; : \; H_i G = P_i \, ,$$
$$H_i.\omega \leq \psi_i \, , \, (H_i)_j \geq 0 \, , \, \forall j = 1,..,g.$$

The simultaneous respect of all these elementary conditions for any point of $R(G,\omega)$ is equivalent to the existence of p line-vectors H_i satisfying the same types of condition as above. The extended Farkas' lemma is simply obtained by constructing matrix H from these p line-vectors.

In particular, by setting $P = GA$ and $\psi = \omega$, the necessary and sufficient condition for $R(G,\omega)$ to be positively invariant can be written :

$$\exists H \in \Re^{g*g} \; : \; HG = GA_0 \, , \, H.\omega \leq \omega$$
$$H_{ij} \geq 0 \text{ for } i = 1,...,g \; ; \; j = 1,...,g.$$

$\square$

INVARIANCE AND STABILITY UNDER INPUT CONSTRAINTS

Now, the considered multivariable linear systems can be represented by state-space equations of the following type:

$$X_{k+1} = A.X_k + B.U_k \quad (8)$$

$X_k \in \Re^n$ being the state vector, $U_k \in \Re^m$ the control vector, $A \in \Re^{n*n}, B \in \Re^{n*m}$.
The control vector is subject to constraints :

$$-Um \leq U_k \leq UM \quad (9)$$

with $(Um)_i \geq 0, (UM)_i \geq 0$,for $i = 1,...,m$ and $k = 0,1,2,...$
The selected control law is a linear state feedback, with $F \in \Re^{m*n}$:

$$U_k = F.X_k \quad (10)$$

The closed-loop evolution of the system is described by :

$$X_{k+1} = A_0.X_k \text{ with } A_0 = A + B.F \quad (11)$$

The general problem called LCRP (Linear Constrained Regulation Problem) consists of determining a matrix F such that the state vector of system (11) converges to 0 while respecting constraints (9) and relations (8),(10).
From the work of R.E.Kalman and J.E.Bertram 1960, it is well-known that if system (11) has all its eigenvalues in the unit circle of the complex plane, it admits elliptic positively invariant domains associated to its quadratic Lyapunov functions . And more recently, a spectral characterization of linear systems having certain types of positively invariant polyhedral sets has been established by G. Bitsoris 1988.
The convex polyhedron generated in the phase plane by constraints (9) is :

$$R[F,Um,UM] = \{X \in \Re^n : -Um \leq FX \leq UM\}$$

A specialized version of invariance conditions (4),(5),(6) is provided by the following corollary of proposition 1 :

Corollary 1

A necessary and sufficient condition for $R[F,Um,UM]$ to be positively invariant is the existence of a couple of non-negative matrices $(H^+ \in \Re^{m*m}, H^- \in \Re^{m*m})$ such that:

$$F.(A + BF) = HF \quad (12)$$
$$\tilde{H}.\rho \leq \rho \quad (13)$$

with $H = H^+ - H^-$
$$\tilde{H} = \begin{pmatrix} H^+ & H^- \\ H^- & H^+ \end{pmatrix} \, , \, \tilde{H} \in \Re^{2m*2m}$$
and $\rho = \begin{pmatrix} UM \\ Um \end{pmatrix} \, , \, \rho \in \Re^{2m}$

The proof of this corollary is elementary. It is a direct application of proposition 1 to system (11) with:

$$G = \begin{bmatrix} +F \\ -F \end{bmatrix} \text{ and } \omega = \rho$$

$\square$

The LCRP can be solved whenever it is possible to find (H^+, H^-, F) such that :

• Matrix $(A + BF)$ has all its eigenvalues located in the unit circle of the complex plane.
• Positive invariance conditions (12),(13) are verified.
• The initial state vector, $X_0 \in R[F,Um,UM]$.

The algebric formulation of this last condition is :

$$- Um \leq F.X_0 \leq UM \tag{14}$$

A general but heavy technique for solving conditions (12) and (13) is proposed by M.Vassilaki 1987. It consists of choosing H such that (13) is verified and determining F by a classical technique for solving quadratic matrix equations. Stability of the closed-loop system then has to be verified , and some iterations of this process may be necessary to get a stabilizing gain matrix F for which relation (14) is satisfied.

In this paper, we propose a method for finding an easily computable couple (H, F) belonging to a subclass of solutions of equation (12) .

AN ALGORITHM FOR
SOLVING THE LCRP

A possible way to simplify the stability analysis is to impose as a positively invariant set a polytope $R(G, \omega)$ with $R(G, \omega) \subset R[F, Um, UM], \omega > 0$ and rank$G = n$ (M. Vassilaki, J.C.Hennet and G.Bitsoris 1988) . Under this assumption, the function $V(X)$ defined by relation (3) is positive definite and can be choosen as a candidate Lyapunov function for system (11).
In this paper, the polytope to be maintained invariant will be directly constructed by completing $R[F, Um, UM]$ under the assumption : $(Um)_i > 0$, $(UM)_i > 0$, for $i = 1, \ldots, m$.
If the number of control variables , m , is greater or equal to n, some control variables can be suppressed without modifying the controllability properties of the system. Then, for any system in its minimal representation, the assumption rank $B = m$ with $m \leq n$ is quite general.
It is then possible, in the case $m < n$ to add dummy control variables $(U_k)_{m+1}, ..., (U_k)_n$, to add $n - m$ independent columns in B so that rank$B = n$, and to impose the following constraints :

$$- (Um)_{m+i} \leq (U_k)_{m+i} \leq (UM)_{m+i} \tag{15}$$

with, for instance, $(Um)_{m+i} = (UM)_{m+i} = \theta$, for $i = 1, ..., n - m$, and $\theta > 0$ as small as desired. Under these extra conditions, it can be assumed that X_k and U_k have the same dimension, n, and that rank $B = n$.
In this case , the resolution of the LCRP can be directly obtained from the following proposition :

Proposition 2

If $m = n$ and rank$B = n$, conditions (16),(17) and (18) guarantee the positive invariance of $R[F, Um, UM]$ and the stability of the controlled system (11).

$$B^{-1}ABH = HB^{-1}AB \tag{16}$$

$$F = HB^{-1} - B^{-1}A \tag{17}$$

$$\tilde{H}.\rho < \rho \tag{18}$$

Proof :

* From relations (16) and (17),we can derive condition (12) :

$$F(A + BF) = (HB^{-1} - B^{-1}A)BHB^{-1}$$
$$= H(HB^{-1} - B^{-1}A)$$
$$= HF$$

Condition (12) is satisfied.
* Under condition (17), the dynamic matrix of the closed-loop system is :

$$A_0 = BHB^{-1} \tag{19}$$

Matrices A_0 and H being similar, asymptotic stability of the controlled system (11) is equivalent to the asymptotic convergence to 0 of the control sequence, which satisfies, for $k = 0, 1, ..$, the recurrent relation:

$$U_{k+1} = H.U_k \tag{20}$$

Constraint (18) consists of the two inequalities :

$$H^+.UM \mid H^-.Um < UM$$
$$H^-.UM + H^+.Um < Um$$

The summing up of these two inequalities yields :

$$(H^+ + H^-).(UM + Um) < UM + Um$$

and therefore,matrix $\mid H \mid = ((\mid H_{ij} \mid))$ is such that :

$$\mid H \mid .W \leq (H^+ + H^-).W$$
$$< W$$

where :

$$W = UM + Um \tag{21}$$

W is a positive vector of $\Re^n$. Then $(\mathbb{1} - \mid H \mid)$ is an M-Matrix, and from a classical result presented in J.P.Lasalle 1976 and J.Chegancas 1985, it is a necessary and sufficient condition for system (20) to be asymptotically stable.
$\square$

The coefficients of matrices H^+ and H^- can be taken as the unknown variables of a linear programming problem, denoted problem (P) and formulated as follows :

$$\min C = \epsilon \tag{22}$$

subject to :

$$\tilde{H}.\rho \leq \epsilon.\rho \tag{23}$$

$$B^{-1}AB(H^+ - H^-) = (H^+ - H^-)B^{-1}AB \tag{24}$$

$$B^{-1}A.X_0 - Um \leq (H^+ - H^-)B^{-1}.X_0 \leq B^{-1}A.X_0 + UM \tag{25}$$

An efficient way of solving the LCRP can be derived from the following corollary of proposition 2.

Corollary 2

If problem (P) has a solution (H^+, H^-, ϵ) with $\epsilon < 1$, then the LCRP is solved by using the control:

$$U_k = F.X_k$$

with :

$$F = (H^+ - H^-)B^{-1} - B^{-1}A \tag{26}$$

From proposition 2, positive invariance of $R[F, Um, UM]$ and asymptotic stability of the closed-loop system derive from the respect of conditions (24) and (23) when $\epsilon < 1$; and condition (25) in problem (P) guarantees:

$$X_0 \subset R[F, Um, UM]$$

$\square$

If the optimal value of ϵ , ϵ^* , is strictly smaller than 1, compute F by (26) and set $U_k = FX_k$.
If $\epsilon^* \geq 1$, feasibility of the control and stability of the closed-loop system cannot be conjointly guaranteed by this algorithm.

The efficiency of this algorithm can be improved when the current state of the system can be observed. A variable regulating scheme can then be implemented. The gain matrix is periodically updated by solving a linear problem denoted (P_k), similar to problem (P) except for relations (25), in which the initial state vector, X_0, is replaced by the current state vector, X_k :

$$\min C_k = \epsilon_k \qquad (27)$$

subject to :

$$\tilde{H}_k.\rho \leq \epsilon_k.\rho \qquad (28)$$

$$B^{-1}AB(H_k^+ - H_k^-) = (H_k^+ - H_k^-)B^{-1}AB \qquad (29)$$

$$B^{-1}A.X_k - Um \leq (H_k^+ - H_k^-)B^{-1}.X_k \leq B^{-1}A.X_k + UM \qquad (30)$$

The variable regulating law is :

$$U_k = F_k.X_k \qquad (31)$$

and F_k is computed from the optimal solution $(H_k^+, H_k^-, \epsilon_k)$ of problem (P_k) by relation :

$$F_k = (H_k^+ - H_k^-)B^{-1} - B^{-1}A \qquad (32)$$

The closed-loop evolution of the system is described by :

$$X_{k+1} = (A + BF_k).X_k$$

Replace F_k by its expression (32). It yields

$$X_{k+1} = BH_kB^{-1}.X_k \qquad (33)$$

The eigenvalues of H_k are also the eigenvalues of BH_kB^{-1}.

Proposition 3

If there exists an integer r such that problem (P_r) has the optimal solution (H_r, ϵ_r) with $\epsilon_r < 1$, then the optimal solution of problem (P_k) exists and verifies $\epsilon_k < 1$ for any integer $k \geq r$, and the controlled system is asymptotically stable.

Proof

* From corollary 2, relations (16),(17) and (18) are satisfied whenever the optimal solution of problem (P_k) is such that $\epsilon_k < 1$. Then, from proposition 2, $X_{k+1} \in R[F_k, Um, UM]$, and (H_k, ϵ_k) is also a feasible solution of problem (P_{k+1}). Consequently, the set of feasible solutions of problem (P_{k+1}) is not empty and by the choice of the objective function, we must have :

$$\epsilon_{k+1} \leq \epsilon_k$$

And , by induction, $\epsilon_r < 1$ implies $\epsilon_k < 1$ for any integer $k \geq r$.
* Note that in the case of time-varying linear systems, local stability conditions $\epsilon_k < 1$ for $k \geq r$ do not automatically imply the global stability of the system.
Asymptotic stability of the system under the variable feedback law can be prooven by showing the existence of a Lyapunov function for the closed-loop system.

Since vector W defined by relation (21) has positive components, the function L(X), defined as follows, is positive definite :

$$L(X) = \max_i \{ \frac{| (B^{-1}.X)_i |}{W_i} \} \qquad (34)$$

By the same argument as in the proof of proposition 2, it is easy to show that relation (28) implies $| H_k | .W \leq \epsilon_k.W$. And this inequality can be used to majorate $L(X_{k+1})$. The difference between two successive values of this function is :

$$\delta L_k = L(X_{k+1}) - L(X_k)$$

From relation (33), derive:

$$B^{-1}.X_{k+1} = H_kB^{-1}.X_k$$

Then, majorate $L(X_{k+1})$ using (28) :

$$L(X_{k+1}) = \max_i \{ \frac{| (B^{-1}.X_{k+1})_i |}{W_i} \}$$

$$\frac{| (B^{-1}.X_{k+1})_i |}{W_i} = \frac{| (H_kB^{-1}.X_k)_i |}{W_i}$$

$$= \frac{1}{W_i} | \sum_{j=1}^{n} \frac{(H_k)_{ij}W_j(B^{-1}.X_k)_j}{W_j} |$$

$$\leq \frac{(| H_k | .W)_i}{W_i} \max_l \frac{| (B^{-1}.X_k)_l |}{W_l}$$

$$\leq \epsilon_k.\max_l \frac{| (B^{-1}.X_k)_l |}{W_l}$$

Then, $L(X_{k+1}) \leq \epsilon_k.L(X_k)$ and since $\epsilon_k < 1$ for $k \geq r$, $\delta L_k < 0$.
L(X) is a Lyapunov function of the sequence of state vectors, which asymptotically converges to 0:

$$\lim_{k \to \infty} X_k = 0$$

$\square$

EXAMPLE

Consider the second order system described by the state-space equation :

$$X_{k+1} = \begin{pmatrix} 1.7 & -3.3 \\ 1.3 & 0.3 \end{pmatrix}.X_k + \begin{pmatrix} 3.0 & 2.0 \\ -2.0 & 2.0 \end{pmatrix}.U_k$$

The control vector is subject to the following constraints:

$$-\begin{pmatrix} 1.5 \\ 1.5 \end{pmatrix} \leq U_k \leq \begin{pmatrix} 1.0 \\ 1.4 \end{pmatrix}$$

Initial state vector is $X_0 = \begin{pmatrix} -5.0 \\ -3.5 \end{pmatrix}$

The unforced system is unstable. The eingenvalues of matrix A are : $\lambda_{1,2} = 1 \pm j1.95$
• The constant regulator gives the optimal values :

$$\epsilon^* = 0.93$$

$$H = \begin{pmatrix} 0.38 & -0.35 \\ 0.93 & 0.0 \end{pmatrix}$$

and the control law $U_k = FX_k$ is :

$$U_k = \begin{pmatrix} -0.076 & 0.536 \\ -0.544 & 0.384 \end{pmatrix}.X_k$$

Simulation results are presented on fig. 1.

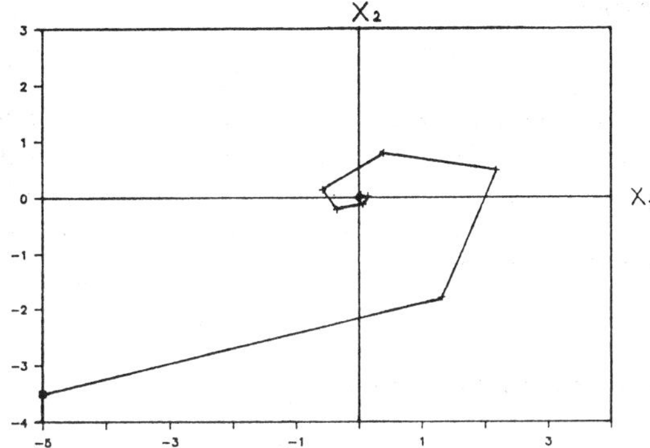

Figure 1 : Trajectory of the state-vector X_k with the constant regulator.

- The same problem can be solved using the variable regulator. The simulation results presented on fig. 2 show that the variable regulator considerably increases the speed of convergence.

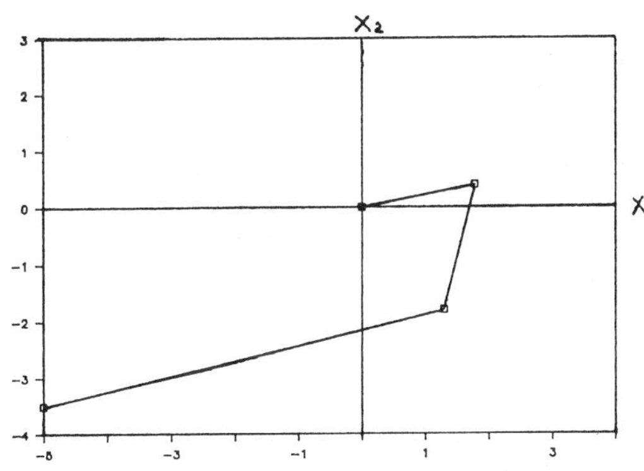

Figure 2 : Trajectory of the state-vector X_k with the variable regulator.

- The same system now has to be controlled from a different initial state vector : $X_0 = \begin{pmatrix} 3.5 \\ 3.0 \end{pmatrix}$
The successive values of the rate of convergence ϵ_k obtained by the linear programming algorithm are :

$\epsilon_0 = 1.53, \epsilon_1 = 1.68, \epsilon_2 = 0.96, \epsilon_3 = 0.92, \epsilon_4 = 0.82, \epsilon_5 = 0.6$

We can see on fig. 3 that although $\epsilon_0 > 1$, the system converges since we get $\epsilon_k < 1$ after two steps.

Since $\epsilon_0 > 1$, the constant regulator is unable to stabilize the system. On the contrary, the variable regulating scheme can be satisfactorily applied.

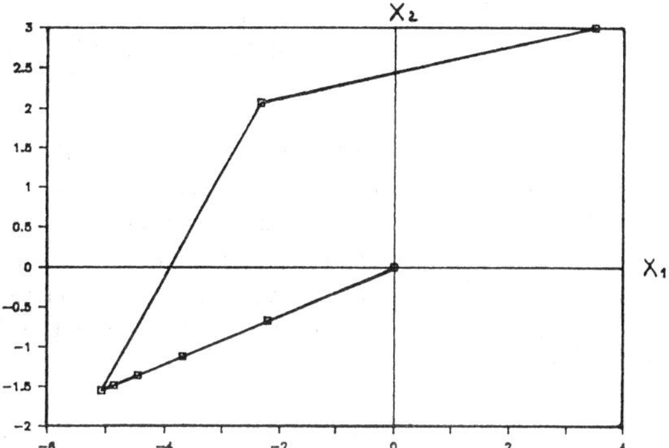

Figure 3 : Trajectory of the state-vector X_k with the variable regulator in a case $\epsilon_0 > 1$.

CONCLUSION

Many linear constrained regulation problems can be solved by constructing positively invariant domains associated with Lyapunov functions. The invariance conditions obtained by this approach can be used as constraints on the gain matrix of the control law. A linear simplified version of these conditions is presented in this paper. The LCRP can then be solved by a standard linear program. The selected objective function to be maximized is denoted ϵ_0. It measures the rate of convergence of the system to the origin.

Two types of regulators are proposed. In the fixed scheme, the same gain matrix is applied at each period, while in the variable scheme, a new gain matrix is computed at each period, from the knowledge of the current state of the system.

If $\epsilon_0 < 1$, stability of the closed-loop system is guaranteed with the fixed and with the variable controller, but the variable regulator may considerably increase the speed of convergence.

If $\epsilon_0 \geq 1$, the fixed regulator is unable to stabilize the system. The variable regulator generates a feasible control as long as ϵ_k remains bounded. At each instant of computation, feasibility of the current control vector is imposed by respect of of the problem constraints . Stability of the closed-loop system can often be obtained by getting $\epsilon_r < 1$ after some periods of time.

This variable control law can also be very efficient in an adaptive context, with a-priori unknown parameters. Then, at each updating time k, the best current estimates of matrices A and B are updated by a recursive identification algorithm and the computation of the gain matrix also directly uses the information on the current state of the system through resolution of problem (P_k).

REFERENCES

A.Benzaouia - Ch.Burgat (1988) : *The regulator problem for a class of linear systems with constrained control.* Systems and Control Letters, vol. 10, N° 5 , pp.357-363.

J-P.Béziat - J-C.Hennet (1988) : *Stability and invariance conditions in Generalized Predictive Control.* IMACS Int. Symposium on 'SMS',Cetraro Italy, pp.163-167.

G.Bitsoris - Ch.Burgat (1977) : *Stability analysis of complex discrete systems with locally and globally stable subsystems.* Int. Journal of Control, vol. 25, N° 3, pp. 413-424.

G.Bitsoris (1988) : *Positively invariant polyhedral sets of discrete-time linear systems.* Int. Journal of Control, vol 47 , pp. 1713-1726.

J.Cheganças (1985) : *Sur le concept d'invariance positive appliqué à l'étude de la commande contrainte des systèmes dynamiques.* Doctorate of University Paul Sabatier,Toulouse, France .

P.O.Gutman - P.Hagander (1985) : *A new design of constrained controllers for linear systems.* IEEE Trans. on Automatic Control, vol AC-30 , pp 22-33.

J.C.Hennet (1989) : *Une extension du lemme de Farkas et son application au problème de régulation linéaire sous contraintes* C.R.Acad.Sciences, t. 308, série 1, pp.415-419.

R.E.Kalman - J.E.Bertram (1960) : *Control systems analysis and design via the second method of Lyapunov.*Trans. A.S.M.E, D 82 , pp.394-400.

J.P.Lasalle (1976) : *The stability of dynamical systems.* SIAM Regional Conference Series in Applied Mathematics.

H.N.Rosenbrock : *A method of investigating stability.* IFAC proceedings, 1963, pp 590-594.

A.Schrijver (1986) : *Theory of linear and integer programming.* Wiley

M.Vassilaki (1987) : *Etude de systèmes sous contraintes; Application à des processus de Production.* Doctorate of Université Paul Sabatier, Toulouse, France .

M.Vassilaki - J.C.Hennet - G.Bitsoris (1988) : *Feedback control of linear discrete-time systems under state and control constraints.* Int. Journal of Control, vol 47 , pp. 1727-1735.

GEOMETRIC CONDITIONS OF STABILITY FOR LINEAR SYSTEMS WITH INVARIANT CONES

S. Tarbouriech and C. Burgat

Laboratoire d'Automatique et d'Analyse des Systèmes du C.N.R.S., 7 avenue du Colonel Roche. 31077 Toulouse Cedex, France

Abstract : The objective of this paper is to give a geometrical characterization of the stability properties (i.e., asymptotic stability, stability, instability) for a linear autonomous system defined by $\dot{x} = A\,x$, without the knowledge of the spectrum of matrix A. This approach only concerns matrices $\Phi_t = e^{tA}$ admitting a proper cone K as a positively invariant set, i.e., $\Phi_t\,K \subset K$. For this class, the proposed necessary and sufficient conditions are directly related to cone K. Moreover, these results allow determination of some bounded domains of invariance and stability for the trajectories of the system, without using Lyapunov functions.

Keywords : Linear systems, stability, cones, positive invariance, geometric conditions.

1. INTRODUCTION

Consider the linear time invariant autonomous system described by :

$$\dot{x} = A\,x \ , \ t \in \mathbb{R}_+ \tag{1}$$

where $x \in \mathbb{R}^n$ and A is a real matrix, $A \in \mathbb{R}^{n \times n}$.

It is well-known that :

(i) this system is asymptotically stable if every eigenvalue of matrix A satisfies $\mathfrak{Re}(\lambda_i(A)) < 0$, for all i ;

(ii) it is stable if every eigenvalue of matrix A satisfies $\mathfrak{Re}(\lambda_i(A)) \not> 0$ for all i and the eigenvalue with $\mathfrak{Re}(\lambda_i(A)) = 0$ are distinct ;

(iii) finally, it is unstable if there exists at least one eigenvalue such that $\mathfrak{Re}(\lambda_i(A)) > 0$ or some eigenvalues not distinct with $\mathfrak{Re}(\lambda_i(A)) = 0$.

The case where there exists at least one eigenvalue with $\mathfrak{Re}(\lambda_i(A)) = 0$, the other eigenvalue being such that $\mathfrak{Re}(\lambda_i(A)) \not> 0$, is often referred to as the critical stability, since from (ii) and (iii) the system can be either stable or unstable according to the multiplicity of these eigenvalues (Lasalle, 1976).

The solution of system (1) can be written as :

$$x(t;x_0) = \Phi(t,t_0)\,x_0 = e^{(t-t_0)A}\,x_0$$

but in the sequel we denote $e^{tA} = \Phi_t$, for $t_0 = 0$. Throughout this paper, it is assumed that matrix e^{tA} has property of leaving positively invariant a proper cone, i.e.,

$$\Phi_t\,K \subset K \tag{2}$$

or equivalently that for any initial value $x_0 \in K$ we have $\Phi_t\,x_0 \in K$ for any $t \in [0\ \infty[$. It clearly appears that the considered class includes, as a particular case, the class of nonnegative matrices $\Phi_t \geq 0$, for which $\Phi_t\,\mathbb{R}^n_+ \subset \mathbb{R}^n_+$ (Berman and Plemmons, 1979, Willems, 1976). This property has a physical meaning and applies to a large class of systems. Indeed, this property becomes particularly interesting when some positively invariant and asymptotically stable polyhedral sets has to be determined, as in the case of the constrained control problem of dynamical systems.

The purpose of this paper is :

(i) to characterize, in geometrical form, the various stability properties of system (1), i.e., asymptotic stability, stability, instability. Necessary and sufficient conditions are given.

(ii) to shown, by means of the preceeding results, under what conditions a bounded set obtained from the intersection of translated cones, is positively invariant and asymptotically stable.

The paper is organized as follows : Section 2 recalls some definitions and notations concerning both proper cones K and matrices having the property of leaving such a cone K positively invariant. Section 3 gives preliminary results, and, in Section 4, the results concerning stability property of system (1) are stated. Section 5 gives conditions guaranteeing the positive invariance and stability of some bounded set containing the origin.

In this section, some definitions that are needed in the sequel are given.

Definition 2.1 : (Berman and Plemmons, 1979) A set K, in real Euclidean n-space, is said to be a cone if :

(i) K is nonempty,
(ii) K is a closed subset of $\mathbb{R}^n$,
(iii) $K + K \subseteq K$,
(iv) $\alpha K \subseteq K$ for all $\alpha > 0$,
(v) int $K \neq \varnothing$,
(vi) $K \cap (-K) = \{0\}$.

Property (iii) means that K is convex. In property (v), int K denote the interior of K ; in other words, K is solid. Property (vi) specifies that K is pointed. A closed pointed solid convex cone K is called a proper cone (Berman and Plemmons, 1979, Vandergraft, 1968, Barker, 1972).

Throughout this paper, ∂K stands for the boundary of K and int K represents the interior of K ; we also denote $K_+ = K$ and $K_- = -K$.

Definition 2.2 : A nonempty subset of $\mathbb{R}^n$, K, is positively invariant relative to a matrix Φ if :
$$\forall x \in K \text{ then } \Phi x \in K, \text{ i.e., } \Phi K \subset K.$$

Definition 2.3 : (Burgat and others, 1988)
(i) A nonempty subset $K \subset \mathbb{R}^n$ is positively invariant with respect to motions of (1) if $\forall x_0 \in K$, $x(t,x_0) \in K$, $\forall t \in \mathbb{R}_+$, i.e., $e^{tA} x_0 \in K$.

(ii) A nonempty subset $K \subset \mathbb{R}^n$ is positively invariant and asymptotically stable if $\forall x_0 \in K$ $x(t,x_0) \in K$, for any $t \in \mathbb{R}_+$, and further $x(t,x_0)$ tends to origin as $t \longrightarrow \infty$.

(iii) A nonempty bounded set $K \subset \mathbb{R}^n$ is strictly contractive if for any $x_0 \in \partial K$, $\lambda > 0$, $\lambda < 1$, we get $x(t_1;x_0) \in \lambda_1 \partial K$, $0 < \lambda_1 < \lambda$ for any $t_1 > 0$, and further for any $\epsilon > 0$ and $t_2 = t_1 + \epsilon$ we get $x(t_2;x_0) \in \lambda_2 \partial K$, $0 < \lambda_2 < \lambda_1$.

Definition 2.4 gives some of the properties allowing characterization of a matrix having the property of leaving K positively invariant.

Definition 2.4 : (Vandergraft, 1968) For a proper cone K, if a matrix Φ is such that $\Phi K \subset K$ then :
(i) $\rho(\Phi)$, the spectral radius of Φ, is an eigenvalue of Φ,

(ii) the degree of $\rho(\Phi)$ is not less than that of any other eigenvalue having the same modulus,

(iii) K contains an eigenvector corresponding to $\rho(\Phi)$.

Lemma 2.5 : (Schneider and Vidyasagar, 1970).
$\Phi_t K_+ \subset K_+$ if and only if $A \in \Sigma(K_+)$, i.e., A is a cross-positive matrix.

In this case, matrix A has a real eigenvalue $\lambda_i(A) = \mu \in \sigma(A)$ (the spectrum of A) such that $\mu > \Re e(\lambda_j(A))$, $j \neq i$, associated with an eigenvector $v_0 \in K_+$ (see also Schröder, 1978).

The purpose of this section is first to obtain a characterization of the positive invariance property for a class of sets generated by translated proper cones K.
For a vector b of $\mathbb{R}^n$ and a proper cone K, we define the translated cone by :

$$K + b = \{x \in \mathbb{R}^n \mid x = y + b, \forall y \in K\}$$

In the sequel, we denote by C_+ the set defined by

$$C_+ = \left\{x \in \mathbb{R}^n \mid -Ax \in K_+\right\} \qquad (3)$$

Regarding the invariance of a translated cone relative to trajectories of system (1), we get :

Lemma 3.1 : Let systems (1) with (2), then the set $K_- + a$ (resp. $K_+ - b$) is positively invariant if and only if $-Aa \in K_+$ (resp. $-Ab \in K_+$).

Proof : (if). Let $x_0 \in K_- + a$. Consequently, x_0 can be written $x_0 = x_- + a$, $x_- \in K_-$. To prove sufficiency, it suffices to show that if $-Aa \in K_+$ then, for any $x_0 \in K_- + a$ and any $t \in \mathbb{R}_+$, we get $e^{tA} x_0 \in K_- + a$. We have $x(t,x_0) = e^{tA} x_- + e^{tA} a$, and from (2) $e^{tA}x_- \in K_-$. Now from $t = \epsilon$ chosen sufficiently small, $e^{tA} a = a + \epsilon A a$ and if $-Aa \in K_+$ then $a + \epsilon A a \in K_- + a$. Clearly, $x(\epsilon;x_0) = a + \epsilon A a + e^{tA} x_- \in K_- + a$. This process can be repeated for $x(\epsilon)$ taken as initial value and we can conclude to the positive invariance of $K_- + a$ for any $t \in \mathbb{R}_+$.
(only if) Assume that $K_- + a$ is positively invariant whereas $-Aa \notin K_+$. Choose as vector x_0 the particular vector $a \in K_- + a$. Clearly, $x(t;a) = e^{tA} a$; for $t = \epsilon$ chosen small enough, we get $x(\epsilon;a) \simeq a + \epsilon A a$. Since by assumption $K_- + a$ is positively invariant, then $a + \epsilon A a \in K_- + a$, i.e., $A a \in K_-$, which contradicts the assumption $-Aa \notin K_+$.

Lemma 3.2 : The set $K_- + a$ is positively invariant and contractive for system (1) if and only if :
$$-Aa \in \text{int } K_+.$$

Proof : As in the proof of lemma 3.1, it is sufficient to prove that this result holds for any $t = \epsilon$ chosen small enough, then the process is repeated.
(if) Let $x_0 \in \lambda\partial(K_- + a) = \partial K_- + \lambda a$ or $x_0 = x_- + \lambda a$, $x_- \in \partial K_-$; we get $x(t;x_0) = e^{tA} x_- + \lambda e^{tA} a$. For $t = \epsilon$, it follows that $x(\epsilon;x_0) = e^{\epsilon A} x_- + \lambda(a + \epsilon A a)$ and if $-Aa \in \text{int } K_+$, then $e^{tA} x_- + \lambda \epsilon A a \in \text{int } K_-$ since $e^{tA} x_- \in K_-$. Finally $x(\epsilon;x_0) \in \text{int } K_- + \lambda a$, i.e., $x(\epsilon;x_0) \in \text{int}(\partial K_- + \lambda a) = \partial K_- + \lambda_1 a$, $\lambda_1 < \lambda$. A strict contraction of the motion on time $t = \epsilon$ is obtained. And through process repetition the proposed result can be obtained.
(only if). The proof is established by contradiction and follows exactly that of lemma 3.1, with the following initial assumptions : $K_- + a$ is contractive and $-Aa \notin \text{int } K_+$.

Lemma 3.3 : If $\Phi_t K_+ \subset K_+$, then Φ_t leaves the cone C_+, defined in (3) positively invariant, that is, $\Phi_t C_+ \subset C_+$.

Proof : Let $x(t;x_0) = e^{tA} x_0$ then $\dot{x}(t;x_0) = A e^{tA}$ $x_0 = e^{tA} A x_0$. If $x_0 \in C_+$ then from (3), $- A x_0 \in$ K_+ and $\dot{x}(t;x_0) \in K_-$. Equivalently, $A e^{tA} x_0 \in K_-$ or $- A e^{tA} x_0 \in K_+$, it follows that $e^{tA} x_0 \in C_+$ for any $x_0 \in C_+$, then $\Phi_t \, C_+ \subset C_+$.

<u>Lemma 3.4</u> :

    (i) if $x_0 \in \partial K_+$ then $- A x_0 \notin$ int K_+.
    (ii) if $x_0 \in K_+$ then $- A x_0 \in K_- + x_0$.

Proof : (Part (i)) Let us assume $x_0 \in \partial K_+$ and $- A x_0 \in$ int K_+. From the previous lemma 3.2, $K_- + x_0$ is a positively invariant and contractive set ; and therefore $e^{tA} x_0 \in \text{int}(K_- + x_0)$. This contradicts property $\Phi_t \, K_+ \subset K_+$ and then $- A x_0 \notin$ int K_+.

(Part (ii)) If $x_0 \in K_+$, then $e^{tA} x_0 \in K_+$. We can write $e^{tA} x_0$ as $x_0 + \epsilon \, A \, x_0$ for t being sufficiently small, $t = \epsilon > 0$; therefore $x_0 + \epsilon \, A$ $x_0 \in K_+ \implies \epsilon A x_0 \in K_+ - x_0$, $\epsilon > 0$, hence $A x_0 \in K_+ - x_0$ and $- A x_0 \in K_- + x_0$.

<u>Lemma 3.5</u> : If matrix A of system (1) is asymptotically stable, i.e., $\mathfrak{Re}(\lambda_i (A)) < 0$ for all i, then :

$$C_+ \subset K_+.$$

Proof : Assume that matrix A is asymptotically stable. Let a vector $x_0 \in C_+$, $x_0 \neq 0$, but such that $x_0 \notin K_+$.
Since $x_0 \in C_+$ then $- A x_0 \in K_+$. From lemma 3.1 it follows that the set $K_- + x_0$ is positively invariant ; further from lemma 3.3, $\Phi_t \, x_0 \in C_+$. Hence, we get $x(t;x_0) \in C_+ \cap (K_- + x_0)$, but the assumption x_0 implies $\{0\} \not\subset K_- + x_0$, and consequently $x(t;x_0)$ cannot reach the origin. This contradicts the fact that $\mathfrak{Re}(\lambda_i (A)) < 0$ for all i, i.e., $e^{tA} x_0 \to 0$, $t \to \infty$, hence any $x_0 \in C_+$ must belong to K_+.

<u>Lemma 3.6</u> : If $\mathfrak{Re}(\lambda_i (A)) < 0$ for all i, then there exists a vector $x_0 \in$ int K_+ such that $- A x_0 \in$ int K_+.

Proof : If $\mathfrak{Re}(\lambda_i (A)) < 0$ for all i, A is a nonsingular matrix. Clearly, from definition (3) of set C_+ and because int $K_+ \neq \emptyset$, we know that int $C_+ \neq \emptyset$. From lemma 3.5, int $C_+ \subset$ int K_+, consequently for any $x_0 \in$ int $C_+ \subset$ int K_+, we have $- A x_0 \in$ int K_+.

4. MAIN RESULTS ON STABILITY PROPERTIES

In the previous Section, we have given lemmas concerning the property of positive invariance for translated proper cones. We are now able to give results regarding the stability properties of system (1), stated in a geometrical form.

With respect to the asymptotic stability of the equilibrium $x = 0$ the following theorem can be formulated :

<u>Theorem 4.1</u> : The equilibrium $x = 0$ is asymptotically stable if and only if
    (i) C_+ is a solid cone
    (ii) $C_+ \subseteq K_+$.

Proof : (if). C_+ is solid (i) means that int C_+ is nonempty ; there exists $x_0 \in$ int C_+ and from (ii) $x_0 \in$ int K_+. From lemma 3.3, if $x_0 \in C_+$ then $e^{t_1 A} x_0 = \Phi_{t_1} x_0$ belongs to C_+ and also to K_+ (from (ii)). Now, $x_0 \in$ int C_+ means that $- A x_0 \in$ int K_+ then $\Phi_{t_1} x_0 \in \text{int}(K_- + x_0)$, hence $\Phi_{t_1} x_0 = x_1 \in C_+ \cap \text{int}(K_- + x_0)$ or equivalently $x_1 \in C_+ \cap \lambda(K_- + x_0)$, with $0 < \lambda < 1 \iff x_1 \in C_+ \cap (K_- + \lambda x_0)$. Repeating this process then $x(t;x_0) \to 0$ as $t \to \infty$.

(only if) Let us assume that the equilibrium $x = 0$ is asymptotically stable but $C_+ \not\subseteq K_+$. Clearly these assumptions are contradicted by lemma 3.5.

<u>Example 4.2</u> : Let us give an illustrative example of the preceding theorem, where one has :
$A = \begin{bmatrix} -0.6 & -0.3 \\ -0.5 & -0.8 \end{bmatrix}$, $\sigma(A) = \{- 0.3 ; - 0.1\}$ (Fig. 1).

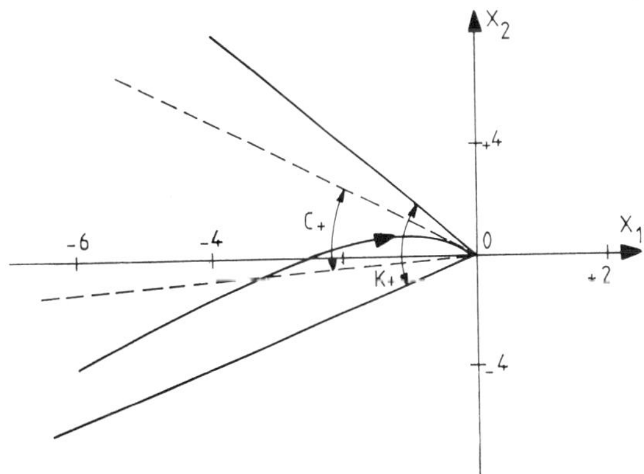

Fig. 1

<u>Theorem 4.3</u> :

    (i) For system (1), $x = 0$ is critically stable $(\mathfrak{Re}(\lambda_i (A)) \not> 0$ and there exists at least one index j such that $\mathfrak{Re}(\lambda_j (A)) = 0)$ if and only :
      $((C_+ \cap C_-)\backslash\{0\}) \cap K_+ \neq \emptyset$.

    (ii) $x = 0$ is unstable with $\mathfrak{Re}(\lambda_i (A)) > 0$ if and only if :
      $C_+ \cap K_+ = \emptyset$.

Proof : (Part (i)) (if). $((C_+ \cap C_-)\backslash\{0\}) \cap K_+ \neq \emptyset$ means that there exists x_0 such that $x_0 \neq 0 \in (C_+ \cap C_-) \cap K_+$. Since $x_0 \in C_+$, then $- A x_0 \in K_+$ but also $x_0 \in C_-$, then $A x_0 \in K_+$ (and $x_0 \in K_+$).
From definition 2.1 K_+ is pointed, therefore $- y$ and y, except 0, cannot belong simultaneously to K_+ and hence $- A x_0 = A x_0 = 0$, i.e., 0 is an eigenvalue of matrix A and the system is critically stable. Then all $x_0 \in (C_+ \cap C_-) \cap K_+$ are such that $x_0 \in$ Ker A, where Ker A is the kernel of matrix A, or equivalently $C_+ \cap C_- =$ Ker A.
(only if). Let us assume that A admits $\mathfrak{Re}(\lambda_i) = 0$ as eigenvalue and that condition (i) of theorem does not hold. From lemma 2.5 this eigenvalue is strictly real, i.e., $\lambda_i = 0$ and associated with the eigenvector $v_0 \in K_+$. Then, it follows $A v_0 = 0 \in K_+$ then $- A v_0 = 0 \in K_+$ implies $v_0 \in C_+$ and $v_0 \in C_-$, $v_0 \neq 0$ and since $v_0 \in K_+$, this contradicts the assumptions made.

(Part (ii)) (if). Let x_0 and $- x_0$ which belong to K_+ and K_- respectively. Let the set $\mathfrak{D} \supset \{0\}$ defined by $\mathfrak{D} = \{x \in (K_+ - x_0) \cap (K_- + x_0)\}$. From condition (ii), $x_0 \notin C_+$, then $- A x_0 \notin K_+$ and with

335

lemma 3.1, $\mathfrak{D}$ is not positively invariant. Clearly, $e^{tA} y$, $y \in \partial\mathfrak{D}$ is such that $e^{tA} y \in \mathbb{R}^n \setminus \mathfrak{D}$. Consequently, for any ϵ, $e^{\epsilon A} y \in \lambda\mathfrak{D}$, $\lambda > 1$. By repeating this process $e^{tA} y \longrightarrow \infty$ as $t \longrightarrow \infty$.

(only if). Let us assume $\exists \, \mathfrak{Re}(\lambda_i(A)) > 0$ for some i and that condition $(C_+ \cap K_+)\setminus\{0\} \neq \varnothing$ is not satisfied or equivalently that for $v_0 \in K_+$, $A v_0 \in K_+$. From lemma 2.5, there exists $\lambda_j - \mu > 0$ such that $\mu \geq \mathfrak{Re}(\lambda_i(A))$, $j \neq i$. This eigenvalue is associated with an eigenvector $v_0 \in K_+$ and $-A v_0 = -\mu v_0 \in K_-$; leading to a contradiction.

<u>Example 4.4</u> : Let an illustrative example of this previous theorem, where $A = \begin{bmatrix} 0.07 & -0.4 \\ -0.6 & -0.12 \end{bmatrix}$ with $\sigma(A) = \{0.4740 ; -0.5240\}$ (Fig. 2).

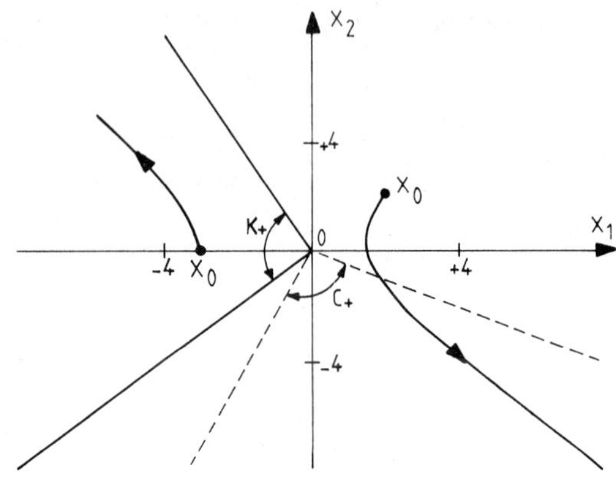

Fig. 2

Let us specify, inside the critical case, the conditions under which the equilibrium is stable or unstable.

<u>Lemma 4.5</u> : If system (1) is unstable for the critical case there cannot exist a vector $x_0 \in \text{int}$ K_+ such that $-A x_0 \in K_+$, or equivalently $C_+ \cap \text{int} K_+ = \varnothing$.

Proof : Consider the set $\mathfrak{D} = (K_- + x_0) \cap (K_+ - x_0)$ containing the origin in its interior, that is, $x_0 \in \text{int} K_+$. Clearly, if the system is unstable it cannot exist $-A x_0 \in K_+$, if not, by virtue of lemma 3.1, the set $\mathfrak{D}$ is positively invariant. This conclusion contradicts the assumption of instability.

<u>Theorem 4.6</u> : The critical case $\mathfrak{Re}(\lambda_i(A)) = 0$ is :
    (i) stable if and only if :
      $((C_+ \cap C_-)\setminus\{0\}) \cap \text{int} K_+ \neq \varnothing.$
    (ii) unstable if and only if :
      $((C_+ \cap C_-)\setminus\{0\}) \cap \text{int} K_+ = \varnothing,$
or equivalently $((C_+ \cap C_-)\setminus\{0\}) \cap \partial K_+ \neq \varnothing.$

Proof : (Part (i)). (if). If condition (i) is satisfied, then there exists $x_0 \in \text{int} K_+$ such that $-A x_0 = 0$; from lemma 3.1, $-A x_0 \in K_+$ and the set $\mathfrak{D}$ previously defined in the proof of lemma 4.5 is positively invariant and $\{0\} \subset \mathfrak{D}$. Stability property is then obtained.
(only if). Assume condition (i) holds and the system is unstable. From (i), there exists $x_0 \in$ int K_+ such that $-A x_0 = 0$. But by virtue of lemma 4.5, this is impossible if the system is

unstable, leading to a contradiction.
(Part (ii)) In the critical case, the equilibrium is unstable if it is not stable, then condition (ii) readily follows ; in other words, from theorem 4.3 we have $((C_+ \cap C_-)\setminus\{0\}) \cap K_+ \neq \varnothing$ and here $((C_+ \cap C_-)\setminus\{0\}) \cap \text{int} K_+ = \varnothing$, therefore the condition reduces to $((C_+ \cap C_-)\setminus\{0\}) \cap \partial K_+ \neq \varnothing.$

<u>Example 4.7</u> :
    case 1 : Let us give an illustrative example of condition (i) of theorem 4.6 ; we choose $A = \begin{bmatrix} 0 & 1 \\ 0 & -1 \end{bmatrix}$, $\sigma(A) = \{0 ; -1\}$ (Fig. 3).

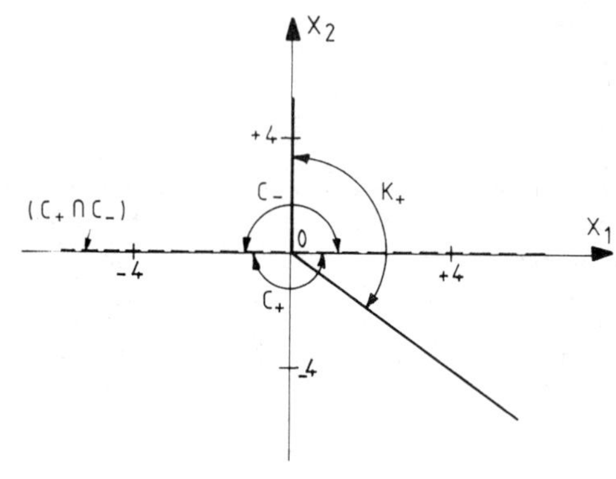

Fig. 3

Case 2 : For an illustrative example of condition (ii) of the previous theorem the classical example of double integrator is chosen, with $A = \begin{bmatrix} 0 & 1 \\ 0 & 0 \end{bmatrix}$, $\sigma(A) = \{0 ; 0\}$ (Fig. 4).

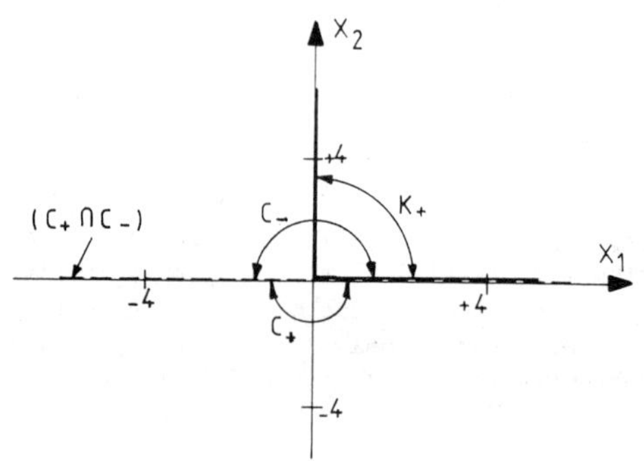

Fig. 4

5. INVARIANCE AND ASYMPTOTIC STABILITY OF SOME BOUNDED SETS

Sets under consideration are obtained from the intersection of translated cones which can be defined as follows, with $a \in \mathbb{R}^n$:

$$K_+ + a = \left\{ y \in \mathbb{R}^n \mid y = x + a, \; x \in K_+ \right\}$$

The set with vertices a, $b \in \mathbb{R}^n$ is described by :

$$\mathcal{D}(K;a,b) = (K_- + a) \cap (K_+ - b) \quad (4)$$

In the sequel, the following question is adressed : under what condition the set $\mathcal{D}(K;a,b)$ is a positively invariant and asymptotically stable subset for motions of system (1).
Obviously, from the previous results of Section 4, we are able to specify the stability property of set $\mathcal{D}(K;a,b)$ without knowing the spectrum of matrix A.

Since the stability properties of the origin are considered, set $\mathcal{D}(K;a,b)$ must contain it ; then we have :

Lemma 5.1 : For $\mathcal{D}(K;a,b)$, the following holds :

    (i) $\{0\} \subset \mathcal{D}(K;a,b)$ if and only if $a \in K_+$, $b \in K_+$.

    (ii) $\{0\} \subset \partial \mathcal{D}(K;a,b)$ if and only if at least a or $b \in \partial K_+$.

    (iii) $\{0\} \subset$ int $\mathcal{D}(K;a,b)$ if and only if $a \in$ int K_+, $b \in$ int K_+.

Proof : (i) Follows from the fact that if $0 \in \mathcal{D}(K;a,b)$, $0 \in K_- + a$, $0 \in K_+ - b$, i.e., $b \in K_+$ and $a \in K_+$. The same process is used for (ii) and (iii).

The following lemma 5.2 gives those conditions on a, b, under which $\mathcal{D}(K;a,b)$ is a nonempty set.

Lemma 5.2 : For the set $\mathcal{D}(K;a,b)$ the following propositions hold :

    (i) $\mathcal{D}(K;a,b) \neq \varnothing$ if and only if $a+b \in K_+$.

    (ii) int $\mathcal{D}(K;a,b) \neq \varnothing$ if and only if $a+b \in$ int K_+.

    (iii) $\mathcal{D}(K;a,b) \neq \varnothing$ and has more than one element if and only if $a+b \in K_+ \setminus \{0\}$.

Proof : (i) It is based on the following fact : if $x \in \mathcal{D}(K;a,b)$ then $x = a + y = - b + z$, $y \in K_-$ and $z \in K_+$; then $a+b \in K_+$. The "only if" part is contradicted. The same process is used to prove (ii) and (iii).

Theorem 5.3 : The set $\mathcal{D}(K;a,b)$, a, $b \in K_+$, is positively invariant for system (1) if and only if $- A a$, $- Ab \in K_+$.

Proof : This follows from lemma 3.1, taking into account that the intersection of positively invariant sets is also a positively invariant set.

Obviously, invariance implies stability but the asymptotic stability is not distinguished. This is specified below.

Theorem 5.4 : For domain $\mathcal{D}(K;a,b)$ such that a, b $\in K_+$ and $- A a$, $- A b \in K_+$ the following properties hold :

    (i) if $- A a$, $- A b \in$ int K_+ then $\mathcal{D}(K;a,b)$ is positively invariant, contractive and asymptotically stable.

    (ii) If, at least, $- A a \in$ int K_+ or $- Ab \in$ int K_+ then $\mathcal{D}(K;a,b)$ is positively invariant and asymptotically stable.

Proof : (Part (i)). From lemma 3.2, $\mathcal{D}(K;a,b)$ is positively invariant and contractive therefore it is also asymptotically stable.
(Part (ii)). The proof is the same as in point (i) ; the only difference is that if $- A b \in K_+$, $- A b \notin$ int K_+ then the contractivity of $\mathcal{D}(K;a,b)$ cannot be obtained.

Remark 5.5 : It must also be noted that even if a or $b \in \partial K_+$, the asymptotic stability of $\mathcal{D}(K;a,b)$ can be obtained. This occurs when $C_+ \subset K_+$ and $C_+ \cap \partial K_+ \neq \varnothing$.

Example 5.6 : This illustrates remark 5.5. Assume $A = \begin{bmatrix} -0.5 & 0 \\ -0.5 & -0.75 \end{bmatrix}$ and $\sigma(A) = \{- 0.5 ; - 0.75\}$, with $a = \begin{bmatrix} -2 \\ 1 \end{bmatrix}$ and $b = \begin{bmatrix} -1 \\ 2 \end{bmatrix}$, obviously $a \in$ int K_+, $b \in \partial K_+$ (see Fig. 5).

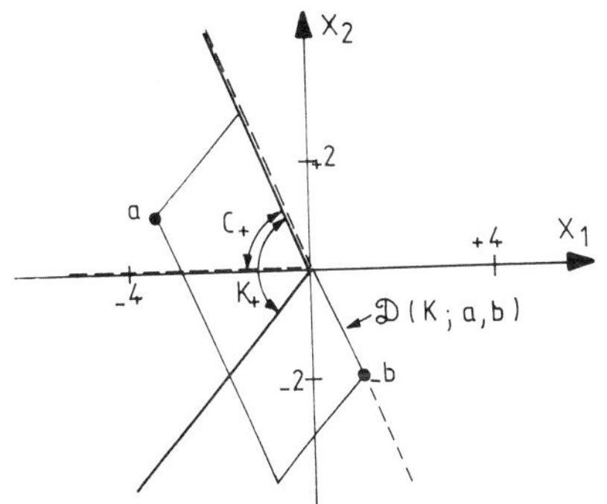

Fig. 5

Condition (ii) of theorem 5.4 is satisfied and $\mathcal{D}(K;a,b)$ is a positively invariant and asymptotically stable subset of $\mathbb{R}^n$ as can be seen in Fig. 5.

Remark 5.7 : Clearly, the asymptotically stable set $\mathcal{D}(K;a,b)$ cannot be obtained from a Lyapunov function $\nu(x)$ since this function must satisfy $\nu(x) = 0 \Rightarrow x = 0$, but this is impossible in this case since $0 \in \partial \mathcal{D}$.

Lemma 5.8 : The set $\mathcal{D}(K;a,b)$, with a, $b \in K_+$, and $- A a$, $- A b \in \partial K_+$, is positively invariant. It is further :

    (i) asymptotically stable if and only if there exists a vector $x_0 \in C_+ \cap$ int K_+ such that $- A x_0 \in$ int K_+, else

    (ii) it is stable and this case corresponds to the stable critical case (theorem 4.6).

Proof : (Part (i)). (if) If there exists $x_0 \in C_+ \cap$ int K_+ such that $- A x_0 \in$ int K_+ then, from lemma 3.2, there exists a contractive domain $\mathcal{D}(K;x_0,x_0)$ which is asymptotically stable ; hence system (1) is asymptotically stable and consequently $\mathcal{D}(K;a,b)$.
(only if) If $\mathcal{D}(K;a,b)$ is a domain of asymptotic stability then from lemma 3.6, there exists x_0 such that $x_0 \in$ int K_+ and $- A x_0 \in$ int K_+, hence there exists a vector $x_0 \in C_+ \cap$ int K_+ such that $- A x_0 \in$ int K_+.
(Part (ii)) If there does not exist $x_0 \in C_+ \cap$ int K_+ such that $- A x_0 \in$ int K_+, then from theorem

4.6, we know that system (1) is stable in the critical case. Then the set $\mathcal{D}(K;a,b)$ is simply positively invariant.

Lemma 5.9 : Let $\mathcal{D}(K;a,b)$ with a, b $\in K_+$ and - A a, - A b $\in K_+$. If , at least -A a = 0, a $\neq$ 0 (or - A b = 0, b $\neq$ 0) then $\mathcal{D}(K;a,b)$ is positively invariant (stable).

Proof : If - A a = 0 then a $\in$ (($C_+ \cap C_-$)\{0\}) $\cap K_+$ and from theorem 4.3, we get the critical case. Since - A a, - A b $\in K_+$, $\mathcal{D}(K;a,b)$ is positively invariant (and stable).

Remark 5.10 : In the unstable critical case, the only possible positively invariant set $\mathcal{D}(K;a,b)$ is such that - A a = - A b = 0. This follows from condition (ii) of theorem 4.6.

Example 5.11 : Let an illustrative case of the foregoing remark regarding the so-called double integrator (e.g., Gutman and Hagander, 1985), with matrix A = $\begin{bmatrix} 0 & 1 \\ 0 & 0 \end{bmatrix}$. When a = $\begin{bmatrix} 1 \\ 0 \end{bmatrix}$ and b = $\begin{bmatrix} 3 \\ 0 \end{bmatrix}$, we get the set $\mathcal{D}(K;a,b)$ positively invariant by virtue of theorem 5.3 (see Fig. 6)

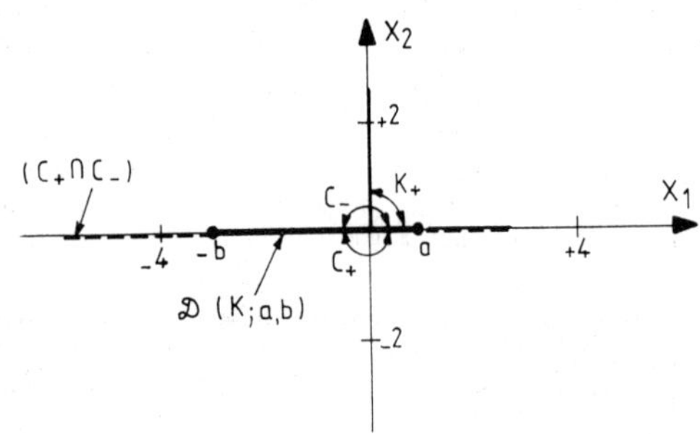

Fig. 6

6. CONCLUSION

Geometrical conditions guaranteeing any of stability properties of system (1) have been given. These properties concern system (1) with a positively invariant cone K. Clearly, if this cone is of a polyhedral nature, simplicial or not, these results can be translated into an algebraic form. Finally, these results can be apply to characterize the stability properties of bounded sets obtained from the intersection of translated sets. When these sets are of a polyhedral nature, algebraic conditions can also be obtained to characterize their properties without the knowledge of the spectrum of matrix A. This is particularly interesting in the case of the saturated state feedback regulator problem.

REFERENCES

Barker, G.P. (1972). On matrices having an invariant cone. Czech. Math. J., 22, 97, pp.49-68.

Berman, A. and R.J. Plemmons (1979). Nonnegative matrices in the mathematical sciences. Academic Press.

Burgat, C., A. Benzaouia and S. Tarbouriech (1988). Positively invariant sets for discrete-time systems with constrained inputs. Report research LAAS n°87187 .

Burgat, C. and A. Benzaouia (to appear). Stability properties of positively invariant linear discrete-time systems. Journal of Math. Anal. Appl.

Gutman, P.O. and P. Hagander (1985). A new design of constrained controllers for linear systems. I.E.E.E trans. A.C.,30, 1, pp.22-33.

Lasalle, J.P. (1976). The stability of dynamical systems. SIAM RegionalConferences Series in Applied Mathematics.

Schneider, H. and M. Vidyasagar (1970). Cross-positive matrices. SIAM J.Numer. Anal. , 7, 4, pp.508-519.

Schröder, J. (1978). M-matrices and generalization using an operator theory approach. SIAM Review., 20, 2, pp.213-244.

Vandergraft, J.S. (1968).Spectral properties of matrices which have invariant a cone. SIAM J.Appl., 16, pp.1208-1222.

Willems, J.C. (1976). Lyapunov functions for diagonally dominant systems. Automatica, 12, pp.519-523.

338

RICCATI EQUATIONS IN THE STABILIZATION OF UNCERTAIN SYSTEMS: A COMPARISON BETWEEN THE METHODS OF CHANG-PENG AND PETERSEN-HOLLOT

M. R. Pickering and I. R. Petersen

Department of Electrical Engineering, University College, University of New South Wales, Australian Defence Force Academy, Canberra ACT 2600, Australia

ABSTRACT This paper considers the problem of determining the stabilizability of a linear system with uncertain parameters contained in the system matrix only. The existing stabilizability approaches of Petersen-Hollot and Chang-Peng are compared and equivalence relations between the two approaches are determined for both single and multiple rank-one uncertainty. It is shown that for the single rank-one uncertainty case the two approaches are identical for one specific value of the constant ε. For the case of multiple rank-one uncertainties the Chang and Peng method is shown to correspond to a particular case of the more general modified Petersen and Hollot method.

KEYWORDS   Feedback control; Linear optimal control; Lyapunov methods; Robust control; Robustness; Robust control; Stability; Stability criteria; State feedback.

I   INTRODUCTION

This paper deals with the problem of determining the stabilizability of an uncertain linear system. In particular, consideration is limited to those systems which contain time-varying unknown-but-bounded uncertainties in the system matrix.

A comparison is made between two existing techniques for determining the stabilizability (via state feedback control) of an uncertain system. These are the methods of Petersen and Hollot [1] and Chang and Peng [2]. The comparison is made for the particular case in which the uncertainty occurs as a rank-one matrix; see [2] and [3].

In both references [1] and [2] the solution to a certain algebraic Riccati equation is required in order to determine the stabilizability of the uncertain system. However the methods differ in the particular form of the Riccati equation. It is shown in section 3 that for certain cases the Riccati equations constructed using each method are equivalent. Furthermore, it is shown that for the case of rank-one uncertainty, the method proposed in [2] can be regarded as a special case of the method proposed in [1].

II   SYSTEM AND DEFINITIONS

The class of linear uncertain systems considered in this report is described by the state equations:

$$\dot{x}(t) = \left[A_0 + \sum_{i=1}^{k} A_i r_i(t) \right] x(t) + B_0 u(t), \qquad (\Sigma)$$

$$\left| r_i(t) \right| \leq \bar{r}, \quad i = 1, 2, \ldots, k$$

where $x(t) \in R^n$ is the state, $u(t) \in R^m$ is the control and $r_i(t) \in R$ is the ith component of the vector of uncertain parameters $r(t) \in R^k$. The matrix A_i is assumed to be a rank-one matrix of the form $A_i = d_i e_i'$.

The stabilizability of the linear uncertain system (Σ) is to be determined by the construction of a quadratic Lyapunov function. That is, it is required to determine a quadratic Lyapunov function which has a negative derivative for all possible states of the uncertain system. This motivates the following definition, see also [4].

DEFINITION 2.1   The system (Σ) is said to be *quadratically stabilizable* if there exists a continuous linear feedback control $u(x) = Kx$, an $n \times n$ positive-definite symmetric matrix P and a constant $\alpha > 0$ such that the following condition is satisfied: given any admissible uncertainties $r(\cdot)$, the Lyapunov derivative corresponding to the quadratic Lyapunov function $V(x) = x'Px$, and the closed loop system satisfies the inequality

$$(x,t) = x' \left[\left(A_0 + \sum_{i=1}^{k} A_i r_i(t) \right)' P + P \left(A_0 + \sum_{i=1}^{k} A_i r_i(t) \right) \right] x$$

$$+ 2x'PB_0 u(x)$$

$$\leq -\alpha \|x\|^2 \qquad (2.1)$$

for all non-zero $x \in R^n$ and all $t \in [0, \infty)$. In order to prove this inequality, [1] and [2] use an approach referred to as the 'quadratic bound method'. This approach involves constructing a quadratic form which is an upper bound for the Lyapunov derivative. This upper bound leads to an augmented algebraic Riccati equation similar to that found in the linear quadratic regulator problem; see [1] and [2]. For the case of systems with zero uncertainty, both augmented Riccati equations found in [1] and [2] reduce to the ordinary Riccati equation which arises in the LQR problem. As in the LQR problem, these Riccati equations depend on positive-definite symmetric weighting matrices $Q \in R^{n \times n}$ and $R \in R^{m \times m}$. These matrices are chosen by the designer.

The two methods described in [1] and [2] use different upper bound expressions for (2.1) and hence obtain different Riccati equations. Both approaches rely on obtaining a positive-definite solution to their respective Riccati equations. The positive-definite solution to the Riccati equation is then used to construct a quadratic Lyapunov function. This quadratic Lyapunov function has a derivative that satisfies inequality 2.1.

The main results of [1] and [2] are now presented.

THE PETERSEN AND HOLLOT METHOD (See [1])

DEFINITION 2.2   The algebraic Riccati equation found in the ordinary LQR problem is augmented in [1] by the addition of the matrix function

$$v(P) = \sum_{i=1}^{k} \left(\varepsilon\, P\, d_i\, d_i'\, P + \frac{1}{\varepsilon}\, e_i\, e_i' \right) \tag{2.2}$$

where $d_i\, e_i' = A_i$; $i = 1,2,3,...,k.$

THEOREM 2.1 (see [1] for proof) Let $Q \in R^{n \times n}$ and $R \in R^{m \times m}$ be given positive-definite symmetric weighting matrices and suppose there exists a constant $\varepsilon > 0$ such that the Riccati equation

$$P A_0 + A_0' P - P B_0 R^{-1} B_0' P + Q + v(P) = 0 \tag{2.3}$$

has a positive-definite symmetric solution P. Then the uncertain system (Σ) is quadratically stabilizable using the continuous linear feedback control $u(x) = -R^{-1} B_0' P x$.

THE CHANG AND PENG METHOD (see [2])

DEFINITION 2.3 The algebraic Riccati equation found in the original LQR problem is augmented in [2] by the addition of the matrix function

$$w(P) = \sum_{i=1}^{k} S_i \left| \Lambda_i \right| S_i' \tag{2.4}$$

where S_i denotes the orthonormal transformation that diagonalizes the symmetric matrix $G_i = P A_i + A_i' P$. That is $\Lambda_i = S_i' G_i S_i$ is the diagonal matrix of eigenvalues of G_i. Furthermore, $|\Lambda_i|$ denotes the diagonal matrix formed by taking the absolute value of each element in the diagonal matrix Λ_i.

THEOREM 2.2 (see [2] for proof) Let $Q \in R^{n \times n}$ and $R \in R^{m \times m}$ be given positive-definite symmetric weighting matrices and suppose that there exists a positive-definite symmetric solution P to the augmented Riccati equation

$$P A_0 + A_0' P - P B_0 R^{-1} B_0' P + Q + w(P) = 0. \tag{2.5}$$

Then the uncertain system (Σ) is quadratically stabilizable using the continuous linear feedback control $u(x) = -R^{-1} B_0' P x$.

III A COMPARISON BETWEEN THE METHODS OF CHANG AND PENG AND PETERSEN AND HOLLOT

In this section a comparison will be made between the methods of [1] and [2] for the cases of single rank-one uncertainty and multiple rank-one uncertainty.

The following theorem establishes the relationship between each method for the case of a single rank-one uncertainty.

THEOREM 3.1 Given the uncertain system (Σ) with a single rank-one uncertainty, the Riccati equation obtained by the Chang and Peng method is equivalent to the Riccati equation obtained by the Petersen and Hollot method for one specific value of the constant ε.

PROOF To prove the equivalence of the respective Riccati equations it is necessary to prove that for a particular value of ε, $w(P) = v(P)$.

From Theorems 2.1 and 2.2 it is obvious that if $k = 1$

$$G_1 = P d_1 e_1' + e_1 d_1' P \tag{3.1}$$

Thus, the matrix G_1 will have rank less than or equal to 2 and therefore has only two non-zero eigenvalues.

To determine $w(P)$ in terms of d_1 and e_1 it is useful to express $w(P)$ in the following form

$$w(P) = \left| \lambda_1 \right| \frac{ff'}{f'f} + \left| \lambda_2 \right| \frac{gg'}{g'g} \tag{3.2}$$

where λ_1, λ_2 are the non-zero eigenvalues of G_1 and $\dfrac{f}{\sqrt{f'f}}, \dfrac{g}{\sqrt{g'g}}$ are the normalized eigenvectors of G_1.

To determine these eigenvalues and eigenvectors let $x = P d_1$ and $y = e_1$. We now postulate that there will be two eigenvectors for G_1 of the form $\alpha x - \beta y$.

It can be shown (see Appendix 1) that two eigenvectors of this form do exist and are given by

$$f = \sqrt{y'y}\ x - \sqrt{x'x}\ y \tag{3.3a}$$
$$g = -\sqrt{y'y}\ x - \sqrt{x'x}\ y \tag{3.3b}$$

and

$$\lambda_1 = x'y - \sqrt{y'y\, x'x} \tag{3.4a}$$
$$\lambda_2 = x'y + \sqrt{y'y\, x'x} \tag{3.4b}$$

are the eigenvalues corresponding to these eigenvectors.

To verify equations 3.3 and 3.4 it can be easily shown that

$$G_1 f = \lambda_1 f \tag{3.5a}$$
$$G_1 g = \lambda_2 g \tag{3.5b}$$

if G_1 is expressed in terms of x and y.

To determine $w(P)$ we take the absolute values of λ_1 and λ_2 and obtain

$$\frac{\left| \lambda_1 \right|}{f'f} = \frac{1}{2\sqrt{y'y\, x'x}} \tag{3.6a}$$

$$\frac{\left| \lambda_2 \right|}{g'g} = \frac{1}{2\sqrt{y'y\, x'x}} \tag{3.6b}$$

Using equation 3.2 and substituting in equations 3.3 and 3.6 we obtain

$$w(P) = \frac{\sqrt{y'y}}{\sqrt{x'x}}\ x\, x' + \frac{\sqrt{x'x}}{\sqrt{y'y}}\ y\, y' \tag{3.7}$$

Now since $x = P d_1$ and $y = e_1$ and if we define ε by the equation

$$\varepsilon = \sqrt{\frac{e_1' e_1}{d_1'\, PP d_1}} \tag{3.8}$$

then $w(P)$ can be expressed in the form

$$w(P) = \varepsilon\, P\, d_1\, d_1' P + \frac{1}{\varepsilon}\, e_1\, e_1' \tag{3.9}$$

Therefore $w(P) = v(P)$ provided ε is defined as in (3.8). $\nabla\nabla\nabla$

Using Theorems 2.1, 2.2 and 3.1, the following corollary is obtained.

COROLLARY 3.1 If there exists a positive-definite solution to the Chang and Peng Riccati equation (2.5) then there must exist a positive-definite solution for the Petersen and Hollot Riccati equation (2.3).

REMARKS   This corollary also follows from the fact that the Petersen and Hollot method provides a necessary and sufficient condition for quadratic stabilizability in the single rank-one uncertainty case.

The converse to Corollary 3.1 does not hold for all cases. The solution to the Chang and Peng Riccati equation must be found by means of an iterative algorithm. Indeed, there exist uncertain systems which can be stabilized using the Petersen and Hollot method, and yet the Chang and Peng algorithm is divergent. The following is an example of one such system.

EXAMPLE 3.1   Consider the uncertain system of the form (Σ) defined by the following matrices

$$A_0 = \begin{bmatrix} 0 & 1 & 0 \\ 0 & 0 & 1 \\ -24 & -26 & -9 \end{bmatrix} \quad A_1 = \begin{bmatrix} 0 & 0 & 0 \\ 0 & 0 & 0 \\ 0 & 0 & 5 \end{bmatrix}$$

$$B_0 = \begin{bmatrix} 0 \\ 0 \\ 1 \end{bmatrix} \quad d_1 = \begin{bmatrix} 0 \\ 0 \\ 1 \end{bmatrix} \quad e_1 = \begin{bmatrix} 0 \\ 0 \\ 5 \end{bmatrix}$$

$$Q = \begin{bmatrix} 1 & 0 & 0 \\ 0 & 1 & 0 \\ 0 & 0 & 1 \end{bmatrix} \quad R = 1$$

For this example, the Chang and Peng iterative algorithm is divergent.

However, the Petersen and Hollot method provides an infinite number of solutions to the stabilization problem depending on the value of ε chosen. For a solution which minimizes the norm of the linear feedback control matrix a value of $\varepsilon = 27$ is chosen. This value of ε corresponds to the matrix

$$P = \begin{bmatrix} 5.3002 & 4.2348 & 0.0210 \\ 4.2348 & 5.3785 & 0.2026 \\ 0.0210 & 0.2026 & 0.1725 \end{bmatrix}$$

The result stated in Theorem 3.1 suggests that if an extra variable is introduced into the Chang and Peng bounding expression w(P) then the two methods will be equivalent in all cases.

DEFINITION 3.1   The modified Chang and Peng bounding expression w'(P) is defined by

$$w'(P) = \delta_1 \frac{\sqrt{e_1' e_1}}{\sqrt{d_1' PP d_1}} P d_1 d_1' P$$

$$+ \frac{1}{\delta_1} \frac{\sqrt{d_1' PP d_1}}{\sqrt{e_1' e_1}} e_1 e_1' \qquad (3.10)$$

LEMMA 3.1 (see Appendix 2 for proof)   Let the matrix S_1 and the constants λ_1 and λ_2 be defined as in the definition of w(P).

If we define $w_1 = -\sqrt{\lambda_1}$, $w_2 = \sqrt{\lambda_2}$ and $\tilde{x} = S_1 x$ then

$$x'[w'(P)]x = \left(\frac{\delta_1}{2} + \frac{1}{2\delta_1} \right) w_1^2 \tilde{x}_1^2$$

$$+ \left(\frac{1}{\delta_1} - \delta_1 \right) w_1 w_2 \tilde{x}_1 \tilde{x}_2$$

$$+ \left(\frac{\delta_1}{2} + \frac{1}{2\delta_1} \right) w_2^2 \tilde{x}_2^2 \qquad (3.11)$$

LEMMA 3.2   The modified Chang and Peng bound w'(P) satisfies the equation

$$x'[w'(P)]x \geq x'G_1 x \qquad (3.12)$$

and therefore w'(P) is established as a valid upper bound expression for the Lyapunov derivative of inequality 2.1.

PROOF   If w_1, w_2 and $\tilde{x}$ are defined as in Lemma 3.1 then

$$x'[w'(P)]\,x = \left(\frac{\delta_1}{2} + \frac{1}{2\delta_1} \right) w_1^2 \tilde{x}_1^2$$

$$+ \left(\frac{1}{\delta_1} - \delta_1 \right) w_1 w_2 \tilde{x}_1 \tilde{x}_2$$

$$+ \left(\frac{\delta_1}{2} + \frac{1}{2\delta_1} \right) w_2^2 \tilde{x}_2^2$$

$$= \frac{\delta_1}{2} \left(w_2 \tilde{x}_2 - w_1 \tilde{x}_1 \right)^2$$

$$+ \frac{1}{2\delta_1} \left(w_2 \tilde{x}_2 + w_1 \tilde{x}_1 \right)^2$$

$$\geq \left| \left(w_2 \tilde{x}_2 - w_1 \tilde{x}_1 \right) \left(w_2 \tilde{x}_2 + w_1 \tilde{x}_1 \right) \right|$$

$$= w_2^2 \tilde{x}_2^2 - w_1^2 \tilde{x}_1^2$$

$$\geq w_2^2 \tilde{x}_2^2 - w_1^2 \tilde{x}_1^2$$

$$= \lambda_1 \tilde{x}_1^2 + \lambda_2 \tilde{x}_2^2$$

$$= \tilde{x}' \Lambda_1 \tilde{x}$$

$$= x'G_1 x . \qquad \nabla\nabla\nabla$$

THEOREM 3.2   The modified Chang and Peng Riccati equation is equivalent to the Petersen and Hollot Riccati equation.

PROOF   Since the addition of δ allows w'(P) to vary over all values of v(P) it follows that given any $\varepsilon > 0$, there exists a $\delta_1 > 0$ such that

$$w'(P) = v(P). \qquad (3.13)$$

Therefore the Riccati equations are equivalent. $\nabla\nabla\nabla$

We now recall that the Petersen and Hollot method is both a necessary and sufficient condition for quadratic stabilizability given a single rank-one uncertainty. Furthermore we have just shown the modified Chang and Peng method is equivalent to the Petersen and Hollot method. Thus we obtain the following corollary.

COROLLARY 3.2 The modified Chang and Peng method provides a necessary and sufficient condition for the quadratic stabilizability of an uncertain system with a single rank-one uncertainty.

We now compare the methods of [1] and [2] for the case of multiple rank-one uncertainties.

OBSERVATION 3.1   The Chang and Peng bounding

expression w(P) for the case of multiple rank-one uncertainty is defined by the following equation

$$w(P) = \sum_{i=1}^{k} \left(\frac{\sqrt{e_i' e_i}}{\sqrt{d_i' P P d_i}} \, P d_i d_i' P + \frac{\sqrt{d_i' P P d_i}}{\sqrt{e_i' e_i}} \, e_i e_i' \right) \tag{3.14}$$

The Petersen and Hollot bounding expression v(P) for this case is given by equation 2.2. We observe that each term in the summation of equation 3.14 has a different scaling factor according to the values of d_i and e_i. Therefore unlike the single uncertainty case the Chang and Peng bounding expression is not a special case of the Petersen and Hollot bounding expression. To overcome this problem a modification to v(P) is proposed. This modified Petersen and Hollot bounding expression is defined as follows.

DEFINITION 3.2   The modified Petersen and Hollot bounding expression v'(P) for multiple rank-one uncertainty is defined as

$$v'(P) = \sum_{i=1}^{k} \left(\xi_i \, \varepsilon \, P d_i d_i' P + \frac{1}{\xi_i \varepsilon} \, e_i e_i' \right)$$

$$\xi_i > 0 \quad i = 1, 2, \ldots, k. \tag{3.15}$$

THEOREM 3.3   Let $Q \in P^{n \times n}$ and $R \in R^{m \times m}$ be given positive-definite symmetric weighting matrices and suppose there exists a constant $\varepsilon > 0$ and constants $\xi_i > 0$; $i = 1,2,3\ldots k$, such that the Riccati equation

$$A_0' P + P A_0 - P B_0 R^{-1} B_0' P + Q + v'(P) = 0 \tag{3.16}$$

has a positive-definite symmetric solution P. Then the uncertain system (Σ) is quadratically stabilizable with the continuous linear feedback control $u(x) = R^{-1} B_0' P x$.

PROOF   It can be shown that given any $\{\xi_i : \xi_i > 0 \ \ i = 1,2,\ldots,k\}$

$$2 \sum_{i=1}^{k} \left| x' P d_i e_i' x \right| \leq x' [v'(P)] \, x \tag{3.17}$$

Hence a simple modification to the proof of Theorem 3.1 in [1] establishes this   theorem. $\nabla\nabla\nabla$

COROLLARY 3.3   Given any uncertain system with multiple rank-one uncertainty, there exists a set of positive constants $\{\xi_i : \xi_i > 0 \ \ i = 1,2,\ldots,k\}$ such that the modified Petersen and Hollot bounding expression v'(P) is equal to the Chang and Peng bounding expression w(P).

OBSERVATION 3.2 Using the above corollary, we can see that for the case of multiple rank-one uncertainty, the Chang and Peng method is a special case of the modified Petersen and Hollot method. However, it should be pointed out that the modified Petersen and Hollot method is not computationally feasible at present. This is due to the fact that there is no systematic way to choose the constants $\xi_1, \xi_2, \ldots, \xi_k$. In contrast, the constant $\varepsilon > 0$ in the standard Petersen and Hollot method was simply chosen to be sufficiently small.

IV   CONCLUSIONS

The results presented in this paper show that for a single rank-one uncertainty the Chang and Peng method is a special case of the Petersen and Hollot method. More precisely, the Chang and Peng method corresponds to the Petersen and Hollot method for a particular value of the constant ε. Furthermore the modified Chang and Peng method is shown to provide a necessary and sufficient condition for quadratic stabilizability as it is equivalent to the Petersen and Hollot method for all values of ε.

For the case of multiple rank-one uncertainties the Chang and Peng method is shown to correspond to a particular case of the more general modified Petersen and Hollot method. However this modified Petersen and Hollot method provides a sufficient condition only for the quadratic stabilizability of the defined uncertain system.

V   REFERENCES

[1]   PETERSEN I.R. and HOLLOT C.V., "A Riccati Equation Approach to the Stabilization of Uncertain Linear Systems", Automatica, Vol. 22, No. 4, pp. 397-411, 1986.

[2]   CHANG S.S.L. and PENG T.K.C., "Adaptive Guaranteed Cost control of Systems with Uncertain Parameters", IEEE T-AC, Vol. AC-17, No. 4, pp. 474-483, August 1972.

[3]   PETERSEN I.R., "A Stabilization Algorithm for a Class of Uncertain Linear Systems", Systems and Control Letters, Vol. 8, pp. 351-357, 1987.

[4]   BARMISH B.R., "Necessary and Sufficient Conditions for Quadratic Stabilizability of an Uncertain Linear System", J. Opt Theory Applic., Vol. 46, p. 399, 1985.

APPENDIX 1

LEMMA A1.1   Given the matrix G_1 defined by

$$G_1 = P d_1 e_1' + e_1 d_1' P \tag{A1.1}$$

if we let $x = P d_1$ and $y = e_1$, then there exists two eigenvectors of G_1 given by

$$f = \sqrt{y'y} \, x - \sqrt{x'x} \, y \tag{A1.2(a)}$$
$$g = -\sqrt{y'y} \, x - \sqrt{x'x} \, y \tag{A1.2(b)}$$

and

$$\lambda_1 = x'y - \sqrt{y'y \, x'x} \tag{A1.3(a)}$$
$$\lambda_2 = x'y + \sqrt{y'y \, x'x} \tag{A1.3(b)}$$

are the eigenvalues corresponding to these eigenvectors.

PROOF   If we assume there exist eigenvectors of G_1 of the form $\alpha x - \beta y$ then we can write the equation

$$(x y' + y x') (\alpha x - \beta y) = \gamma (\alpha x - \beta y) \tag{A1.4}$$

where γ represents the corresponding eigenvalue.

From equation A1.4 we obtain two simultaneous equations in the variables α and β as follows. Expand equation A1.4

$$\alpha \, y'x \, x - \beta \, y'y \, x + \alpha \, x'x \, y - \beta \, x'y \, y = \gamma \alpha x - \gamma \beta y. \tag{A1.5}$$

Equating the coefficients of x and y

$$\alpha \, y'x - \beta \, y'y = \gamma \alpha \tag{A1.6(a)}$$
$$\alpha \, x'x - \beta \, x'y = -\gamma \beta. \tag{A1.6(b)}$$

We now represent these equations in matrix form :

$$\begin{bmatrix} y'x & -y'x \\ -x'x & x'y \end{bmatrix} \begin{bmatrix} \alpha \\ \beta \end{bmatrix} = \gamma \begin{bmatrix} \alpha \\ \beta \end{bmatrix} \tag{A1.7}$$

This allows the values of γ to be calculated from the characteristic equation

$$(y'x - \gamma)(x'y - \gamma) - x'x\, y'y = 0 \qquad \text{A1.8}$$

Expanding equation A1.8 and evaluating the quadratic equation provides the following expressions for the non-zero eigenvalues of G_1

$$\lambda_1 = x'y - \sqrt{y'y\, x'x} \qquad \text{A1.9(a)}$$
$$\lambda_2 = x'y + \sqrt{y'y\, x'x} \qquad \text{A1.9(b)}$$

Substituting the value of λ_1 into equation A1.6(a) we obtain the following equation relating α to β

$$\alpha = \beta\, \frac{\sqrt{y'y}}{\sqrt{x'x}} \qquad \text{A1.10}$$

Similarly substituting λ_2 into equation A1.6(a) provides the following equation

$$\alpha = -\beta\, \frac{\sqrt{y'y}}{\sqrt{x'x}} \qquad \text{A1.11}$$

Now consider the eigenvector corresponding to λ_1 which is given by

$$f = \alpha\, x - \beta\, y \qquad \text{A1.12}$$

Substituting equation A1.10 into A1.12 and letting $\beta = \sqrt{x'x}$ we obtain

$$f = \sqrt{y'y}\, x - \sqrt{x'x}\, y \qquad \text{A1.13}$$

Similarly the eigenvector corresponding to λ_2 is given by

$$g = \alpha\, x - \beta\, y \qquad \text{A1.14}$$

and by substituting A1.11 into A1.14 and letting $\beta = \sqrt{x'x}$ we obtain

$$g = -\sqrt{y'y}\, x - \sqrt{x'x}\, y. \qquad \nabla\nabla\nabla$$

APPENDIX 2

PROOF OF LEMMA 3.1 Convert the expression given in the right hand side of equation 3.11 into a quadratic expression given by

$$\tilde{x}'\Phi\,\tilde{x} = \tilde{x}'\begin{bmatrix} w_1^2\left(\dfrac{\delta_1}{2}+\dfrac{1}{2\delta_1}\right) & w_1 w_2\left(\dfrac{1}{2\delta_1}-\dfrac{\delta_1}{2}\right) & \\[2ex] w_1 w_2\left(\dfrac{1}{2\delta_1}-\dfrac{\delta_1}{2}\right) & w_2^2\left(\dfrac{\delta_1}{2}+\dfrac{1}{2\delta_1}\right) & 0 \\[2ex] \hline & 0 & 0 \end{bmatrix}\tilde{x} \qquad \text{A2.1}$$

To prove equation 3.11 holds it is sufficient to show the following is true

$$w'(P) = S_1'\Phi S_1 \qquad \text{A2.2}$$

The orthonormal transformation matrix S_1 which diagonalizes G_1 is of the form

$$S_1 = \begin{bmatrix} \dfrac{f'}{\sqrt{f'f}} \\[2ex] \dfrac{g'}{\sqrt{g'g}} \\[1ex] x'_3 \\ x'_4 \\ \cdot \\ \cdot \\ \cdot \\ x'_n \end{bmatrix} \qquad \text{A2.3}$$

Now expand the right hand side of equation A2.2

$$S_1'\Phi S_1 = w_1^2\left(\frac{\delta_1}{2}+\frac{1}{2\delta_1}\right)\frac{ff'}{f'f}$$

$$+\, w_1 w_2\left(\frac{1}{2\delta_1}-\frac{\delta_1}{2}\right)\frac{gf'}{\sqrt{g'g f'f}}$$

$$+\, w_1 w_2\left(\frac{1}{2\delta_1}-\frac{\delta_1}{2}\right)\frac{fg'}{\sqrt{f'f g'g}}$$

$$+\, w_2^2\left(\frac{\delta_1}{2}+\frac{1}{2\delta_1}\right)\frac{gg'}{\sqrt{g'g}} \qquad \text{A2.4}$$

Substituting into equation A2.4 the expressions for f, g, w_1 and w_2 and a considerable amount of algebraic manipulation yields the desired result. $\nabla\nabla\nabla$

PERFORMANCE LIMITATIONS OF NON-MINIMUM PHASE SYSTEMS*

L. Qiu and E. J. Davison

*Department of Electrical Engineering, University of Toronto, Toronto, Ontario
M5S 1A4, Canada*

Abstract. This paper studies the cheap regulator problem and the cheap servomechanism problem for non-minimum phase systems. Some well-known properties of the "perfect regulation" (Francis, 1979; Kwakernaak and Sivan, 1972; Scherzinger and Davison, 1985) and the "perfect tracking" (Davison and Scherzinger, 1987) problem of minimum phase systems are generalized to systems which may be non-minimum phase. It is shown that a fundamental limitation exists re the speed of tracking and disturbance rejection for a non-minimum phase system, and that this limitation is completely characterized by the location of the unstable transmission zeros of the system. Furthermore, this limitation provides a quantitative measure of the "degree of difficulty" which is inherent in the control of such non-minimum phase systems.

Key Words. Linear optimal regulator; Servomechanisms; Non-minimum phase systems; Transmission zeros; Cheap control; Inner-outer factorization.

1 INTRODUCTION

It has long been realized that minimum phase systems have certain advantages over non-minimum phase systems; in particular, minimum phase systems have the desirable property that their response can be made arbitrarily fast with no "peaking" occurring in the output. This, however, is not the case for non-minimum phase systems. See, e.g. (Davison and Scherzinger, 1987; Francis, 1979; Kwakernaak and Sivan, 1972a; Scherzinger and Davison, 1985). In particular, in studying the "perfect control problem" for a "high gain servo controller" and the "robust servo controller" (Davison and Scherzinger, 1987), it is shown that a necessary condition which must be satisfied in order to obtain "perfect control" is that the system must be minimum phase. On the other hand, it is to be recognized that not all non-minimum phase systems behave the same; for example, some non-minimum phase systems produce results which are "almost as good" as minimum phase systems, whereas other non-minimum phase systems are indeed "almost impossible" to control.

In this paper, we first study the cheap quadratic regulator problem for non-minimum phase systems, which is the limiting case of the optimal quadratic regulator problem when the weight on the input energy of the performance index goes to zero. When a special state space realization is adopted, it is shown that a simple expression for the limiting performance can be obtained. This result is then applied to the study of the cheap optimal servomechanism problem, in which the original servomechanism problem is transformed to a linear quadratic regulator problem and the cheap control of this regulator problem is then studied. Two control schemes are considered. One is the high gain servomechanism controller (Davison and Scherzinger, 1987); the other is the robust servomechanism controller (Davison, 1976b). It is shown that for any non-minimum phase system there exists a fundamental limitation on the resultant optimal cost which characterizes the transient behavior of the closed loop system. This limitation can be completely characterized by the number and the locations of the transmission zeros of the system contained in the open right half of the complex plane, and it provides a quantitative measure of the "degree of difficulty" which is inherent in the control of a non-minimum phase system.

*This work has been supported by the Natural Sciences and Engineering Research Council of Canada under grant no. A4396.

2 PRELIMINARIES

Throughout this paper, a transfer matrix means a proper real rational matrix. Let $F(s)$ be a $r \times m$ transfer matrix. $F(s)$ is said to be tall (or wide, square) if $r \geq m$ (or $r \leq m$, $r = m$). By a realization of $F(s)$, we mean a 4-tuple of matrices over the real numbers (A, B, C, D) such that $D + C(sI - A)^{-1}B = F(s)$. The following notation is used to divide the complex plane into three parts: $\mathbf{C}^{+} = \{s \in \mathbf{C} : \Re(s) > 0\}$, $\mathbf{C}^{0} = \{s \in \mathbf{C} : \Re(s) = 0\}$, $\mathbf{C}^{-} = \{s \in \mathbf{C} : \Re(s) < 0\}$. A transfer matrix is said to be stable if all of its poles are contained in $\mathbf{C}^{-}$, and a square real matrix is said to be stable if all of its eigenvalues are contained in $\mathbf{C}^{-}$.

The transmission zeros of a realization (A, B, C, D) are defined to be the complex numbers λ which make the matrix

$$\begin{bmatrix} A - \lambda I & B \\ C & D \end{bmatrix}$$

reduce rank. The transmission zeros of a transfer matrix $F(s)$ are defined to be the transmission zeros of its minimal realization. It can be shown that the transmission zeros of $F(s)$ are also the roots of the numerator polynomials of the diagonal elements of the Smith-McMillan form of $F(s)$. $F(s)$ is said to be minimum phase if it has no transmission zeros in $\mathbf{C}^{+}$; otherwise it is said to be non-minimum phase. For more details about transmission zeros, see (Davison and Wong, 1974).

Associated with any realization (A, B, C, D) in which A is stable, there are two Lyapunov equations:

$$AP + PA' = -BB' \tag{1}$$

$$A'Q + QA = -C'C. \tag{2}$$

The solutions P, Q to equation (1)-(2) are called the controllability gramian and the observability gramian of (A, B, C, D) respectively. A minimal realization (A, B, C, D) of a stable transfer matrix $F(s)$ is called a balanced realization if the solution P, Q to equations (1)-(2) are diagonal and equal. The diagonal elements of such P or Q are called the Hankel singular values of $F(s)$. It is shown in (Moore, 1981) that every transfer matrix has a balanced realization. A procedure to find a balanced realization from any minimal realization of a transfer matrix is also given in (Moore, 1981).

A stable transfer matrix $F(s)$ is called inner if $F'(-s)F(s) = I$. An inner transfer matrix must be tall. All the transmission zeros of an inner transfer matrix must be located in $\mathbf{C}^+$.

Lemma 1 (Glover, 1984) *Let (A, B, C, D) be a minimal realization of an inner transfer matrix $F(s)$ and let P, Q be the solutions to equations (1)-(2). Then*

(a) $PQ = I$,

(b) $D'D = I$,

(c) $D'C + B'Q = 0$, $DB' + CP = 0$.

Corollary 1 *Let (A, B, C, D) be a balanced realization of an inner transfer matrix $F(s)$ and let P, Q be the solutions to equations (1)-(2). Then*

(a) $P = Q = I$,

(b) $D'D = I$,

(c) $D'C + B' = 0$, $DB' + C = 0$.

A stable transfer matrix $F(s)$ is outer if $F(s)$ has full row rank for every $s \in \mathbf{C}^+$. The concept of outer matrices is not as important as that of inner matrices in our development. Instead, we are more interested in the class of transfer matrices which are minimum phase and wide. Clearly, outer matrices belong to this class, but a matrix in this class does not have to be stable.

The following factorization result, which serves as a foundation for our development, is obtained. It is remarked here that when the poles and/or zeros of two transfer matrices are compared in the following, we consider not only their values but also their multiplicities in the Smith-McMillan sense.

Lemma 2 *A transfer matrix $F(s)$ can always be factorized as $F(s) = F_1(s)F_2(s)$ such that $F_1(s)$ is inner, $F_2(s)$ is minimum phase and wide, and the unstable poles of $F_2(s)$ are equal to the unstable poles of $F(s)$.*

Proof: In this proof, we will use the notation $\mathbf{RH}_\infty$ to denote the set of all stable transfer matrices and will use some knowledge of the factorization theory in $\mathbf{RH}_\infty$ (Francis, 1987). It is known that $F(s)$ can be decomposed as $F(s) = F_s(s) + F_a(s)$ where $F_s(s) \in \mathbf{RH}_\infty$ and $F_a(s)$ is strictly unstable, i.e. all its poles are contained in $\mathbf{C}^0 \cup \mathbf{C}^+$. The first factorization result we have to use is the coprime factorization result which says that any transfer matrix $G(s)$ can be factorized as $G(s) = N(s)M^{-1}(s)$ where $N(s), M(s) \in \mathbf{RH}_\infty$ and there exist $X(s), Y(s) \in \mathbf{RH}_\infty$ such that

$$X(s)N(s) + Y(s)M(s) = I$$

Moreover, $M(s)$ can be chosen so that $M^{-1}(s)$ is proper (Francis, 1987). Let $N(s)M^{-1}(s)$ be such a coprime factorization of $F_a(s)$. Then the transmission zeros of $M(s)$ are the poles of $F_a(s)$. Let $L(s)$ be any matrix (compatible in size) in $\mathbf{RH}_\infty$ with the property that $L^{-1}(s) \in \mathbf{RH}_\infty$ and $M(s)L^{-1}(s), N(s)L^{-1}$ are proper. The existence of such $L(s)$ can be verified simply by choosing $L(s) = I$. Then

$$\begin{aligned} F(s) &= F_s(s) + N(s)L^{-1}(s)L(s)M^{-1}(s) \\ &= [F_s(s)M(s)L^{-1}(s) + N(s)L^{-1}(s)]L(s)M^{-1}(s) \end{aligned}$$

in which $F_s(s)M(s)L^{-1}(s) + N(s)L^{-1}(s)$ belongs to $\mathbf{RH}_\infty$, $L(s)M^{-1}(s)$ is minimum phase, square and its unstable poles are the unstable poles of $F(s)$.

The second factorization result we have to use is the inner-outer factorization result which says that any $G(s) \in \mathbf{RH}_\infty$ can be factorized as $G(s) = G_i(s)G_o(s)$ where $G_1(s)$ is inner and $G_o(s)$ is outer (Chen, 1987). Let the inner outer factorization of $F_s(s)M(s)L^{-1}(s) + N(s)L^{-1}(s)$ be $F_i(s)F_o(s)$ and let $F_1(s) = F_i(s)$ and $F_2(s) = F_o(s)L(s)M^{-1}(s)$. Then we immediately have that $F(s) = F_1(s)F_2(s)$ and $F_1(s)$ is inner. For all $s \in \mathbf{C}^+$, $F_o(s)$ has full row rank and $L(s)M^{-1}(S)$ is a nonsingular square matrix. Therefore $F_2(s)$ must have full row rank for all $s \in \mathbf{C}^+$, which means that $F_2(s)$ must be minimum phase and wide. $\square$

Given a transfer matrix $F(s)$, let $F(s) = F_1(s)F_2(s)$ be the factorization described in Lemma 2. Let (A_1, B_1, C_1, D_1) be a

balanced realization of $F_1(s)$ and (A_2, B_2, C_2, D_2) be any stabilizable and detectable realization of $F_2(s)$. Then a stabilizable and detectable realization of $F(s)$ is given by

$$A = \begin{bmatrix} A_1 & B_1C_2 \\ 0 & A_2 \end{bmatrix} \quad B = \begin{bmatrix} B_1D_2 \\ B_2 \end{bmatrix} \tag{3}$$
$$C = [C_1 \ D_1C_2] \quad D = D_1D_2.$$

This realization is called a factorized realization of $F(s)$ and plays an important role in the development.

A class of transfer matrices called right-invertible transfer matrices deserves special treatment. A transfer matrix $F(s)$ is said to be right-invertible if it has full row rank for at least one $s \in \mathbf{C}$. If (A, B, C, D) is a realization of $F(s)$, then the right-invertibility of $F(s)$ is equivalent to the fact that

$$\begin{bmatrix} A - \lambda I & B \\ C & D \end{bmatrix}$$

has full row rank for at least one $\lambda \in \mathbf{C}$. The following lemma gives some useful properties of right-invertible transfer matrices with respect to the factorization described by Lemma 2.

Lemma 3 *Let $F(s)$ be a right-invertible transfer matrix and let $F(s) = F_1(s)F_2(s)$ be the factorization described in Lemma 2. Then $F_1(s)$ is square, the transmission zeros of $F_1(s)$ are equal to those transmission zeros of $F(s)$ contained in $\mathbf{C}^+$, and the poles of $F_1(s)$ are equal to the negative of the transmission zeros of $F_1(s)$.*

Proof: Let $F(s)$ be $r \times m$. Then $F_1(s)$ has r rows and at most r columns. If $F_1(s)$ has less than r columns, then the rank of $F_1(s)$ is less than r for all $s \in \mathbf{C}$, which implies that the rank of $F(s)$ is less than r for all $s \in \mathbf{C}$. This contradicts with the fact that $F(s)$ is right-invertible. Therefore $F_1(s)$ must have r columns, i.e. it must be square.

The transmission zeros of $F_1(s)$ are the poles of $F_1^{-1}(s)$. A minimal realization of $F_1^{-1}(s)$ is given by

$$(A_1 - B_1D_1^{-1}C_1, B_1D_1^{-1}, -D_1^{-1}C_1, D_1^{-1}),$$

where D_1^{-1} exists, so that the transmission zeros of $F_1(s)$ are equal to the eigenvalues of $A_1 - B_1D_1^{-1}C_1$. For the proof of the remaining part of the lemma, it is enough to show that

(a) the eigenvalues of $A_1 - B_1D_1^{-1}C_1$ are equal to the negative of the eigenvalues of A_1;

(b) the eigenvalues of $A_1 - B_1D_1^{-1}C_1$ are equal to those transmission zeros of $F(s)$ contained in $\mathbf{C}^+$.

By Corollary 1,

$$\begin{aligned} A_1 - B_1D_1^{-1}C_1 &= A_1 + B_1D_1^{-1}D_1B_1' \\ &= A_1 + B_1B_1' \\ &= A_1 - (A_1 + A_1') \\ &= -A_1'. \end{aligned}$$

This proves (a).

Since the factorized realization is stabilizable and detectable, those transmission zeros of $F(s)$ which are contained in $\mathbf{C}^+$ are the complex numbers $\lambda \in \mathbf{C}^+$ which make the matrix

$$\begin{bmatrix} A_1 - \lambda I & B_1C_2 & B_1D_2 \\ 0 & A_2 - \lambda I & B_2 \\ C_1 & D_1C_2 & D_1D_2 \end{bmatrix}$$

reduce rank. This matrix can be shown to be similar to the matrix

$$\begin{bmatrix} A_1 - B_1D_1^{-1}C_1 - \lambda I & 0 & 0 \\ 0 & A_2 - \lambda I & B_2 \\ D_1^{-1}C_1 & C_2 & D_2 \end{bmatrix}.$$

This proves (b). $\square$

3   CHEAP LQR PROBLEM

Let $F(s)$ be a transfer matrix and let (A, B, C, D) be a stabilizable and detectable realization of $F(s)$. Consider the linear time-invariant system defined by (A, B, C, D):

$$\begin{aligned}
\dot{x} &= Ax + Bu, \qquad x(0) = x_0 \qquad (4)\\
y &= Cx + Du
\end{aligned}$$

and consider the optimal linear quadratic regulator problem of this system with respect to the cost

$$J_\epsilon = \int_0^\infty (y'y + \epsilon x'x)dt, \qquad \epsilon > 0. \qquad (5)$$

The problem concerning the limit of J_ϵ as $\epsilon \searrow 0$ is called the cheap LQR problem. It is known that the optimal control which minimizes J_ϵ is given by $u = -(\epsilon I + D'D)^{-1}(B'P_\epsilon + D'C)x$, the optimal value of J_ϵ is given by $x_0'P_\epsilon x_0$, where P_ϵ is the unique positive semi-definite solution to the algebraic Riccati equation (ARE)

$$\begin{aligned}
[A - B(\epsilon I + D'D)^{-1}D'C]'P_\epsilon + P_\epsilon[A - B(\epsilon I + D'D)^{-1}D'C]\\
+ C'[I - D(\epsilon I + D'D)^{-1}D']C - P_\epsilon B(\epsilon I + D'D)^{-1}B'P_\epsilon = 0 \quad (6)
\end{aligned}$$

It is easy to show that P_ϵ is a monotonically increasing function of $\epsilon > 0$, and so the limit of P_ϵ as $\epsilon \searrow 0$ exists.

Lemma 4 (Scherzinger and Davison, 1985) *Let (A, B, C, D) be a stabilizable and detectable realization of a transfer matrix $F(s)$, and let P_ϵ be the unique positive semi-definite solution to ARE (6). Then $P_\epsilon \to 0$ as $\epsilon \searrow 0$ if $F(s)$ is minimum phase and wide.*

The condition in Lemma 4 is also necessary if matrices B and C are assumed to have full column rank and full row rank respectively. The same result for the case when $D = 0$ is obtained in (Francis, 1979; Kwakernaak and Sivan, 1972a). The main purpose of this section is to give a result on the limit of P_ϵ as $\epsilon \searrow 0$ for systems which do not satisfy the condition given in Lemma 4.

Lemma 5 *Let (A, B, C, D) be a factorized realization of any transfer matrix $F(s)$. Then the unique positive semi-definite solution P_ϵ to ARE (6) is given by $\begin{bmatrix} I & 0 \\ 0 & P_{\epsilon 2} \end{bmatrix}$, where $P_{\epsilon 2}$ is the unique positive semi-definite solution to the ARE*

$$\begin{aligned}
[A_2 - B_2(\epsilon I + D_2'D_2)^{-1}D_2'C_2]'P_{\epsilon 2}\\
+ P_{\epsilon 2}[A_2 - B_2(\epsilon I + D_2'D_2)^{-1}D_2'C_2]\\
+ C_2'[I - D_2(\epsilon I + D_2'D_2)^{-1}D_2]C_2\\
- P_{\epsilon 2}B_2(\epsilon I + D_2'D_2)^{-1}B_2'P_{\epsilon 2} = 0.
\end{aligned}$$

Let $x = \begin{bmatrix} x_1 \\ x_2 \end{bmatrix}$ be partitioned accordingly with the partition of A given by (3). Then the optimal control of system (4) under cost (5) is given by $u = -(\epsilon I + D_2'D_2)^{-1}(B_2'P_{\epsilon 2} + D_2'C_2)x_2$.

The proof of the first statement of this lemma is obtained simply by checking that the given solution indeed satisfies ARE (6). The second statement follows by using Corollary 1. The details of the proof are omitted.

Direct application of Lemma 5 and Corollary 1 leads to the following corollary.

Corollary 2 *If (A, B, C, D) is a balanced realization of an inner transfer matrix, then the unique positive semi-definite solution P_ϵ to the ARE (6) is $P_\epsilon = I$ and the optimal control of system (4) under cost (5) is $u = 0$.*

Since any minimal realization of a transfer matrix is similar to a balanced realization, the corollary implies that if (A, B, C, D) is a minimal realization of an inner transfer matrix, then the optimal control of the system (4) under cost (5) is always zero and thus the optimal performance is independent of ϵ. This is consistent with the well-known fact that cheap control asymptotically puts all the poles of the closed loop system to the mirror points of the system's unstable zeros with respect to the imaginary axes.

Since an inner matrix already has this property, no control is needed to make it optimal.

Since (A_2, B_2, C_2, D_2) is a stabilizable and detectable realization of a minimum phase and wide system, the following theorem is obtained immediately from Lemma 4-5. (In the statement of the theorem, we assume that P_ϵ is partitioned accordingly with the partition of A given in (3).)

Theorem 1 *Let (A, B, C, D) be a factorized realization of a transfer matrix $F(s)$, and let P_ϵ be the unique positive semi-definite solution to ARE (6). Then $P_\epsilon \to \begin{bmatrix} I & 0 \\ 0 & 0 \end{bmatrix}$ as $\epsilon \to 0$.*

4   HIGH GAIN SERVOMECHANISM CONTROLLER

Consider a plant described by the state space equation:

$$\begin{aligned}
\dot{x} &= Ax + Bu + B\omega, \qquad x(0) = 0 \qquad (7)\\
y &= Cx + Du + D\omega + \eta\\
e &= y_{\text{ref}} - y.
\end{aligned}$$

where ω is the input disturbance, η is the output disturbance and y_{ref} is the output reference. Assume that ω, η and y_{ref} are constant signals. A control problem which often arises is to design a controller for system (7) such that the closed loop system is stable, and such that the tracking error $e \to 0$ as $t \to \infty$ for arbitrary ω, η and y_{ref}. In order for this to be possible, the following assumption is necessary (Davison, 1976a).

Assumption 1 *Assume that system (7) satisfies the following condition:*

(a) *(A, B, C, D) is stabilizable and detectable;*

(b) *$\begin{bmatrix} A & B \\ C & D \end{bmatrix}$ has full row rank.*

Assumption 1(b) implies that $F(s) = D + C(sI - A)^{-1}B$ is right-invertible and has no transmission zeros at the origin.

Under Assumption 1, it can be shown that the following controller can accomplish the required task:

$$\begin{aligned}
u &= [D - (C + DK)(A + BK)^{-1}B]^\dagger e\\
&+ \{K - [D - (C + DK)(A + BK)^{-1}B]^\dagger(C + DK)\}x \quad (8)
\end{aligned}$$

where K is any matrix which makes $A + BK$ stable and $M^\dagger$ denotes the Moore-Penrose pseudo-inverse of matrix M. Assumption 1 implies that $M = D - (C + DK)(A + BK)^{-1}B$ has full row rank so that $M^\dagger = M'(MM')^{-1}$. It is easy to check that the closed loop state matrix of (7)-(8) is given by $A + BK$ which is stable. The fact that the error e goes to zero is proved in (Davison and Scherzinger, 1987).

Now assume controller (8) is applied to system (7), and assume that the steady-state values of the input and the state are given by $\bar{u}$ and $\bar{x}$ respectively. Let $v := u - \bar{u}$ and $z := x - \bar{x}$. Then $\bar{u}$, $\bar{x}$ must satisfy equations

$$\begin{aligned}
0 &= A\bar{x} + B\bar{u} + B\omega\\
y_{\text{ref}} &= C\bar{x} + D\bar{u} + D\omega + \eta,
\end{aligned}$$

and as a result, system (7) can be written as

$$\begin{aligned}
\dot{z} &= Az + Bv \qquad (9)\\
e &= Cz + Du.
\end{aligned}$$

Equation (9) suggests that

$$J_\epsilon = \int_0^\infty (e'e + \epsilon v'v)dt \qquad (10)$$

can be used as a performance index for the closed loop system, where the stabilizing gain matrix K is chosen to minimize the performance index J_ϵ. The purpose of this section is to investigate the behavior of J_ϵ as $\epsilon \searrow 0$. It is obvious that J_ϵ in general depends on y_{ref}, ω and η. It is shown in (Davison and Scherzinger, 1987) that $\lim_{\epsilon \searrow 0} J_\epsilon = 0$ for all y_{ref}, ω and η, if (A, B, C, D) is minimum phase. We would like to determine what the limit of $\lim_{\epsilon \searrow 0} J_\epsilon$ is when (A, B, C, D) is non-minimum phase. Before stating our main result, we review some results on quadratic forms.

Let $\mathcal{W}$ be an Euclidean space with inner product $\langle \cdot, \cdot \rangle$. A quadratic form Q on $\mathcal{W}$ is a function $\mathcal{W} \to \mathbf{R}$ mapping w to $Q(w) = \langle w, Hw \rangle$ where H is a given (real) Hermitian operator on $\mathcal{W}$. The quadratic form Q uniquely determine the matrix H and is completely characterized by H. The collection of all quadratic forms on $\mathcal{W}$ is a linear space and clearly is isomorphic to the linear space of all Hermitian operators on $\mathcal{W}$. A quadratic form Q is said to be positive semi-definite if $Q(w) \geq 0$ for all $w \in \mathcal{W}$. A quadratic form is positive semi-definite if and only if its corresponding Hermitian operator is positive semi-definite. A partial ordering can then be defined in the space of all quadratic forms on $\mathcal{W}$: $Q_1 \geq Q_2$ if $Q_1 - Q_2$ is positive semi-definite. Define the norm of a quadratic form to be the norm of its corresponding Hermitian operator. In the following, we always use the operator norm which is given by the sum of all of the singular values of the operator. For positive semi-definite quadratic forms, which are the type we are interested in, this norm is equal to the trace of the corresponding Hermitian operator and it has following two properties:

Lemma 6 (Levine and Athans, 1970) *Let Q be a positive semi-definite quadratic form on Euclidean space $\mathbf{R}^p$ with the usual inner product $\langle w_1, w_2 \rangle = w_1' w_2$, and let w be a random vector in $\mathbf{R}^p$ with $E(w) = 0$ and $E\{ww'\} = I$. Then $E\{Q(w)\} = \|Q\|$.*

The notation $E(\cdot)$ in Lemma 6 denotes the expectation operator. Lemma 6 can be interpreted that the norm of Q is the "average" value of $Q(w)$ over all w on a sphere with radius $\sqrt{p}$.

Let $\mathcal{W}_1$ be a subspace of $\mathcal{W}$, and let Q be a quadratic form on $\mathcal{W}$. Then the restriction of Q on $\mathcal{W}_1$, denoted by $Q|\mathcal{W}_1$, is a quadratic form on $\mathcal{W}_1$.

Lemma 7 *Let $\mathcal{W}_1$, $\mathcal{W}_2$ be mutually orthogonal subspaces of an Euclidean space $\mathcal{W}$, and let Q be a positive semi-definite quadratic form on $\mathcal{W}$. Then $\|Q|\mathcal{W}_1\| + \|Q|\mathcal{W}_2\| = \|Q|(\mathcal{W}_1 + \mathcal{W}_2)\|$.*

Proof: Let H be the corresponding operator of Q and let $\mathcal{W}_3$ be the orthogonal complement of $\mathcal{W}_1 + \mathcal{W}_2$. Then H can be represented by the following 3×3 matrix with operator entries:

$$\begin{bmatrix} H_{11} & H_{12} & H_{13} \\ H_{21} & H_{22} & H_{23} \\ H_{31} & H_{32} & H_{33} \end{bmatrix},$$

where $H_{ij} = P_i H | \mathcal{W}_j$ and P_i, $i = 1, 2, 3$ is the projection of $\mathcal{W}$ onto $\mathcal{W}_i$. In this case, H_{11} and H_{22} are just the corresponding operators of $Q|\mathcal{W}_1$ and $Q|\mathcal{W}_2$, and $\begin{bmatrix} H_{11} & H_{12} \\ H_{21} & H_{22} \end{bmatrix}$ is the corresponding operator of $Q|(\mathcal{W}_1 + \mathcal{W}_2)$. The positive semi-definiteness of H implies the positive semi-definiteness of H_{11}, H_{22} and $\begin{bmatrix} H_{11} & H_{12} \\ H_{21} & H_{22} \end{bmatrix}$. The lemma then directly follows from the fact that $\mathrm{tr} H_{11} + \mathrm{tr} H_{22} = \mathrm{tr} \begin{bmatrix} H_{11} & H_{12} \\ H_{21} & H_{22} \end{bmatrix}$. $\square$

If $\mathcal{W}$ has an orthogonal decomposition $\{\mathcal{W}_1, \mathcal{W}_2, \ldots, \mathcal{W}_q\}$, i.e. $\mathcal{W}_1, \mathcal{W}_2, \ldots, \mathcal{W}_q$ are mutually orthogonal and their sum is $\mathcal{W}$, then $\|Q\|$, as well as any $\|Q| \sum_{i \in \mathcal{I}} \mathcal{W}_i\|$ where $\mathcal{I} \subset \{1, 2, \ldots, q\}$, is completely determined by all $\|Q|\mathcal{W}_i\|$, $i = 1, 2, \ldots, q$.

Now let $\mathcal{W}$ be the space of all vectors of the form $\begin{bmatrix} \omega \\ \eta \\ y_{\mathrm{ref}} \end{bmatrix}$.

The inner product on $\mathcal{W}$ is defined as usual, e.g. $\langle w_1, w_2 \rangle = w_1' w_2$. Define subspaces

$$\mathcal{W}_\omega = \left\{ \begin{bmatrix} \omega \\ \eta \\ y_{\mathrm{ref}} \end{bmatrix} \in \mathcal{W} : y_{\mathrm{ref}} = 0, \ \eta = 0 \right\}$$

$$\mathcal{W}_\eta = \left\{ \begin{bmatrix} \omega \\ \eta \\ y_{\mathrm{ref}} \end{bmatrix} \in \mathcal{W} : y_{\mathrm{ref}} = 0, \ \omega = 0 \right\}$$

$$\mathcal{W}_y = \left\{ \begin{bmatrix} \omega \\ \eta \\ y_{\mathrm{ref}} \end{bmatrix} \in \mathcal{W} : \omega = 0, \ \eta = 0 \right\}.$$

Then $\mathcal{W}_y$, $\mathcal{W}_\omega$ and $\mathcal{W}_\eta$ form an orthogonal decomposition of $\mathcal{W}$.

Theorem 2 *Given the system (7), assume that Assumption 1 holds; then the performance index J_ϵ given by (10) is a positive semi-definite quadratic form on $\mathcal{W}$, and J_ϵ is a monotonically increasing function of ϵ. Let $\lim_{\epsilon \searrow 0} J_\epsilon = J_0$, and let the transmission zeros of (A, B, C, D) which are contained in $\mathbf{C}^+$ be given $\lambda_1, \lambda_2, \ldots, \lambda_l$; then*

$$\|J_0|\mathcal{W}_\omega\| = 0$$
$$\|J_0|\mathcal{W}_\eta\| = 2(\frac{1}{\lambda_1} + \frac{1}{\lambda_2} + \cdots + \frac{1}{\lambda_l})$$
$$\|J_0|\mathcal{W}_y\| = 2(\frac{1}{\lambda_1} + \frac{1}{\lambda_2} + \cdots + \frac{1}{\lambda_l}).$$

This result states that the high gain servomechanism controller (8) produces perfect control, i.e. the optimal performance goes to zero as $\epsilon \searrow 0$, for the case when input disturbances are only present, even if the system (7) is non-minimum phase. However for the case when output disturbances or/and nonzero reference are present, perfect control cannot be obtained for non-minimum phase systems, and the optimal performance J_ϵ is now bounded from below by $2 \sum_{i=1}^{l} \frac{1}{\lambda_i}$. This result shows that $\sum_{i=1}^{l} \frac{1}{\lambda_i}$ can be considered as a quantitative measure of the degree of difficulty in the control of non-minimum phase systems. This result also emphasizes the fact that not all non-minimum phase systems behave the same. For example, a plant with one unstable transmission zero at θ has a degree of difficulty equal to $\frac{1}{\theta}$ which is large if θ is small, whereas a plant with two unstable transmission zeros at $\theta \pm j\sigma$ has a degree of difficulty equal to $\frac{2\theta}{\theta^2 + \sigma^2}$ which is small if $\sigma \gg \theta$.

The proof of Theorem 2: Since the initial condition of system (7) is assumed to be zero, the input-output relation of the system (7) is determined solely by its transfer matrix $F(s) = D + C(sI - A)^{-1}B$. Thus (A, B, C, D) can be assumed to be the factorized realization of $F(s)$ which is of the form of (3). Let $F_1(s) = D_1 + C_1(sI - A_1)^{-1}B_1$ and $F_2(s) = D_2 + C_2(sI - A_2)^{-1}B_2$. Since $F(s)$ is assumed to be right-invertible, it follows from Lemma 4 that $F_1(s)$ must be square and the poles of $F_1(s)$ are $-\lambda_1, -\lambda_2, \ldots, -\lambda_l$.

It is known that $J_\epsilon = z'(0)P_\epsilon z(0)$ where P_ϵ is the unique positive semi-definite solution of ARE (6) and

$$z(0) = x(0) - \bar{x} = -\bar{x}.$$

Since $x(0)$ is zero, the closed loop system can be considered as a linear system with input $\begin{bmatrix} \omega \\ \eta \\ y_{\mathrm{ref}} \end{bmatrix}$ and output x. Hence there exist a linear transformation T such that $\bar{x} = T \begin{bmatrix} \omega \\ \eta \\ y_{\mathrm{ref}} \end{bmatrix}$. It therefore follows that $J_\epsilon = \bar{x}' P_\epsilon \bar{x}$ is a positive semi-definite quadratic form on $\mathcal{W}$ for all $\epsilon > 0$. The fact that J_ϵ is monotonically increasing with respect ϵ follows directly from the monotonicity of P_ϵ.

By Theorem 1, $P_\epsilon \to \begin{bmatrix} I & 0 \\ 0 & 0 \end{bmatrix}$ as $\epsilon \searrow 0$. Let $\bar{x}$ be partitioned as $\bar{x} = \begin{bmatrix} \bar{x}_1 \\ \bar{x}_2 \end{bmatrix}$ according to the partition of A given by (3). Then

$$J_0 = \lim_{\epsilon \searrow 0} J_\epsilon = \bar{x}' \begin{bmatrix} I & 0 \\ 0 & 0 \end{bmatrix} \bar{x} = \bar{x}_1' \bar{x}_1.$$

Now assume that the system (7), under control (8), is at steady-state. The following relation exists

$$y_{\text{ref}} - \eta = F(0)(\bar{u} + \omega) = F_1(0)F_2(0)(\bar{u} + \omega).$$

Since

$$\bar{x}_1 = -A_1^{-1}B_1 F_2(0)(\bar{u} + \omega),$$

where A_1^{-1} is well defined, it follows that

$$\bar{x}_1 = -A_1^{-1}B_1 F_1^{-1}(0)(y_{\text{ref}} - \eta).$$

Therefore, the following expression is obtained for J_0:

$$
\begin{aligned}
J_0 &= (y_{\text{ref}} - \eta)' F_1(0)' B_1' A_1'^{-1} A_1^{-1} B_1 F_1(0)(y_{\text{ref}} - \eta) \\
&= \begin{bmatrix} \omega \\ \eta \\ y_{\text{ref}} \end{bmatrix}' \begin{bmatrix} 0 & 0 & 0 \\ 0 & H & -H \\ 0 & -H & H \end{bmatrix} \begin{bmatrix} \omega \\ \eta \\ y_{\text{ref}} \end{bmatrix},
\end{aligned}
$$

where H is used to denote $F_1(0)' B_1' A_1'^{-1} A_1^{-1} B_1 F_1(0)$. It becomes clear that $\|J_0|\mathcal{W}_\omega\| = 0$ and

$$\|J_0|\mathcal{W}_\eta\| = \|J_0|\mathcal{W}_y\| = \|F_1(0)' B_1' A_1'^{-1} A_1^{-1} B_1 F_1(0)\|.$$

By using the fact that (A_1, B_1, C_1, D_1) is a balanced realization of square inner function $F_1(s)$, we obtain:

$$
\begin{aligned}
\|J_0|\mathcal{W}_\eta\| &= \|J_0|\mathcal{W}_y\| \\
&= \text{tr}[F_1(0)' B_1' A_1'^{-1} A_1^{-1} B_1 F_1(0)] \\
&= \text{tr}[F_1(0)F_1(0)' B_1' A_1'^{-1} A_1^{-1} B_1] \\
&= \text{tr}[A_1^{-1} B_1 B_1' A_1'^{-1}] \quad \text{(since } F_1(s) \text{ is square)} \\
&= -\text{tr}[A_1^{-1}(A_1 + A_1')A_1'^{-1}] \quad \text{(by Corollary 1)} \\
&= -\text{tr}(A_1'^{-1} + A_1^{-1}) \\
&= -2\text{tr}(A_1^{-1}) \\
&= 2\left(\frac{1}{\lambda_1} + \frac{1}{\lambda_2} + \cdots + \frac{1}{\lambda_l}\right) \quad \text{(by Lemma 3)}.
\end{aligned}
$$

$\square$

5 ROBUST SERVOMECHANISM CONTROLLER

Consider system (7), where ω, η and y_{ref} are constant signals. It is now desired to apply a controller to solve the robust servomechanism problem (Davison, 1976b) for (7). Assume that Assumption 1 holds. In this case, a controller which solves the problem must include a servo-compensator, and so consider now the following controller for (7):

$$
\begin{aligned}
\dot{z} &= e, \qquad\qquad z(0) = 0 \qquad (11) \\
u &= K_0 x + K z
\end{aligned}
$$

where $[K_0 \ K]$ is chosen to stabilize the following matrix

$$\begin{bmatrix} A & 0 \\ C & 0 \end{bmatrix} + \begin{bmatrix} B \\ D \end{bmatrix}[K_0 \ K]. \qquad (12)$$

Controller (11) has the significant advantage over controller (8) in that tracking and disturbance rejection occur for all perturbations of the system parameters (A, B, C, D) and controller parameters $[K_0 \ K]$, provided only that the perturbed closed loop system remains stable.

The augmented system (with input u and output z), on combining the original system and the integrator, is then described by the state-space equation:

$$
\begin{bmatrix} \dot{x} \\ \dot{z} \end{bmatrix} = \begin{bmatrix} A & 0 \\ C & 0 \end{bmatrix} \begin{bmatrix} x \\ z \end{bmatrix} + \begin{bmatrix} B \\ D \end{bmatrix}(u + \omega) + \begin{bmatrix} 0 \\ I \end{bmatrix}(\eta - y_{\text{ref}})
$$

$$
\begin{bmatrix} x(0) \\ z(0) \end{bmatrix} = 0
$$

$$
z = [0 \ I]\begin{bmatrix} x \\ z \end{bmatrix}.
$$

Differentiate all variables once, and define new variables:

$$\tilde{x} := \dot{x}, \quad \tilde{z} := \dot{z}, \quad \tilde{u} := \dot{u}.$$

On noticing that $\dot{z} = e$, the augmented system then becomes

$$\begin{bmatrix} \dot{\tilde{x}} \\ \dot{\tilde{z}} \end{bmatrix} = \begin{bmatrix} A & 0 \\ C & 0 \end{bmatrix} \begin{bmatrix} \tilde{x} \\ \tilde{z} \end{bmatrix} + \begin{bmatrix} B \\ D \end{bmatrix} \tilde{u} \qquad (13)$$

$$\begin{bmatrix} \tilde{x}(0) \\ \tilde{z}(0) \end{bmatrix} = \begin{bmatrix} B & 0 & 0 \\ D & I & I \end{bmatrix} \begin{bmatrix} \omega \\ \eta \\ y_{\text{ref}} \end{bmatrix}$$

$$e = [0 \ I]\begin{bmatrix} \tilde{x} \\ \tilde{z} \end{bmatrix}.$$

where $\left(\begin{bmatrix} A & 0 \\ C & 0 \end{bmatrix}, \begin{bmatrix} B \\ D \end{bmatrix}, [0 \ I], 0 \right)$ is stabilizable and detectable if Assumption 1 holds.

This suggests that

$$J_\epsilon = \int_0^\infty (e'e + \epsilon \tilde{u}'\tilde{u})dt \qquad (14)$$

can be used as a performance index of the closed loop system, where the matrix $[K_0 \ K]$ is chosen to minimize the performance index J_ϵ. The same question to the one studied in last section arises: what is the limit of J_ϵ as ϵ goes to zero? It is known that if (A, B, C, D) is minimum phase, then $\lim_{\epsilon \searrow 0} J_\epsilon = 0$ (Davison and Scherzinger, 1987). In this section, we will obtain a result similar to Theorem 2 for the case of non-minimum phase systems.

Assume that $\mathcal{W}$, $\mathcal{W}_\omega$, $\mathcal{W}_\eta$ and $\mathcal{W}_y$ have the same meaning as used in the last section, and that the same norm of quadratic forms is used.

Theorem 3 *Given the system (7), assume that Assumption 1 holds; then the performance index J_ϵ given by (14) is a positive semi-definite quadratic form on $\mathcal{W}$, and J_ϵ is a monotonically increasing function of ϵ. Let $\lim_{\epsilon \searrow 0} J_\epsilon = J_0$, and let the transmission zeros of (A, B, C, D) which are contained in $\mathbb{C}^+$ be given by $\lambda_1, \lambda_2, \ldots, \lambda_l$; then*

$$
\begin{aligned}
\|J_0|\mathcal{W}_\omega\| &= 0 \\
\|J_0|\mathcal{W}_\eta\| &= 2\left(\frac{1}{\lambda_1} + \frac{1}{\lambda_2} + \cdots + \frac{1}{\lambda_l}\right) \\
\|J_0|\mathcal{W}_y\| &= 2\left(\frac{1}{\lambda_1} + \frac{1}{\lambda_2} + \cdots + \frac{1}{\lambda_l}\right).
\end{aligned}
$$

Since the same type of result, as obtained for the high gain controller case, is also obtained in this case, the discussion following Theorem 2 equally applies to this case.

The proof of Theorem 3: Since the initial condition of system (7) is assumed to be zero, the input-output relation of the system (7) is determined solely by its transfer matrix $F(s) = D + C(sI - A)^{-1}B$. Then (A, B, C, D) can be assumed to be the factorized realization of $F(s)$ which is of the form of (3). Consequently (13) can be rewritten as

$$
\begin{bmatrix} \dot{\tilde{x}}_1 \\ \dot{\tilde{x}}_2 \\ \dot{\tilde{z}} \end{bmatrix} = \begin{bmatrix} A_1 & B_1 C_2 & 0 \\ 0 & A_2 & 0 \\ C_1 & D_1 C_2 & 0 \end{bmatrix} \begin{bmatrix} \tilde{x}_1 \\ \tilde{x}_2 \\ \tilde{z} \end{bmatrix} + \begin{bmatrix} B_1 D_2 \\ B_2 \\ D_1 D_2 \end{bmatrix} \tilde{u} \quad (15)
$$

$$e = [0 \ 0 \ I]\begin{bmatrix} \tilde{x}_1 \\ \tilde{x}_2 \\ \tilde{z} \end{bmatrix}.$$

Let

$$T := \begin{bmatrix} A_1 & B_1 & 0 \\ 0 & 0 & I \\ C_1 & D_1 & 0 \end{bmatrix} \quad \text{and} \quad \begin{bmatrix} \tilde{x}_1 \\ \tilde{x}_2 \\ \tilde{z} \end{bmatrix} := T \begin{bmatrix} \hat{x}_1 \\ \hat{z} \\ \hat{x}_2 \end{bmatrix}.$$

Then (15) becomes

$$
\begin{bmatrix} \dot{\hat{x}}_1 \\ \dot{\hat{z}} \\ \dot{\hat{x}}_2 \end{bmatrix} = \begin{bmatrix} A_1 & B_1 & 0 \\ 0 & 0 & C_2 \\ 0 & 0 & A_2 \end{bmatrix} \begin{bmatrix} \hat{x}_1 \\ \hat{z} \\ \hat{x}_2 \end{bmatrix} + \begin{bmatrix} 0 \\ D_2 \\ B_2 \end{bmatrix} \tilde{u} \quad (16)
$$

$$e = [0 \ 0 \ I]\begin{bmatrix} \hat{x}_1 \\ \hat{z} \\ \hat{x}_2 \end{bmatrix}.$$

and the initial condition becomes

$$\begin{bmatrix} \hat{x}_1(0) \\ \hat{z}(0) \\ \hat{x}_2(0) \end{bmatrix} = T^{-1} \begin{bmatrix} \tilde{x}_1(0) \\ \tilde{x}_2(0) \\ \tilde{z}(0) \end{bmatrix} = T^{-1} \begin{bmatrix} B_1 D_2 & 0 & 0 \\ B_1 & 0 & 0 \\ D - 1D_2 & I & I \end{bmatrix} \begin{bmatrix} \omega \\ \eta \\ y_{\text{ref}} \end{bmatrix}. \tag{17}$$

Let P_ϵ be the unique semi-definite solution to the following ARE

$$\begin{bmatrix} A_1' & 0 & 0 \\ B_1' & 0 & 0 \\ 0 & C_2' & A_2' \end{bmatrix} P_\epsilon + P_\epsilon \begin{bmatrix} A_1 & B_1 & 0 \\ 0 & 0 & C_2 \\ 0 & 0 & A_2 \end{bmatrix} + \begin{bmatrix} C_1' \\ D_1' \\ 0 \end{bmatrix} [C_1 \; D_1 \; 0]$$

$$-\frac{1}{\epsilon} P_\epsilon \begin{bmatrix} 0 \\ D_2 \\ B_2 \end{bmatrix} [0 \; D_2' \; B_2'] P_\epsilon = 0.$$

Then $J_\epsilon = [\hat{x}_1'(0) \; \hat{z}'(0)\hat{x}_2'(0)]P_\epsilon \begin{bmatrix} \hat{x}_1(0) \\ \hat{z}(0) \\ \hat{x}_2(0) \end{bmatrix}$. It follows from (17) and the monotonicity of P_ϵ that J_ϵ is a quadratic form on $\mathcal{W}$ for all $\epsilon > 0$ and is monotonically increasing with respect to ϵ.

Note that (16) is a factorized realization with (A_1, B_1, C_1, D_1) being inner and ($\begin{bmatrix} 0 & C_2 \\ 0 & A_2 \end{bmatrix}, \begin{bmatrix} D_2 \\ B_2 \end{bmatrix}, [0 \; I], 0$) being wide and minimum phase. By Theorem 1, $\lim_{\epsilon \searrow 0} P_\epsilon = \begin{bmatrix} I & 0 & 0 \\ 0 & 0 & 0 \\ 0 & 0 & 0 \end{bmatrix}$. Hence

$$J_0 = \begin{bmatrix} \omega \\ \eta \\ y_{\text{ref}} \end{bmatrix}' \begin{bmatrix} B_1 D_2 & 0 & 0 \\ B_1 & 0 & 0 \\ D_1 D_2 & I & I \end{bmatrix}' T'^{-1}$$

$$\cdot \begin{bmatrix} I & 0 & 0 \\ 0 & 0 & 0 \\ 0 & 0 & 0 \end{bmatrix} T^{-1} \begin{bmatrix} B_1 D_2 & 0 & 0 \\ B_1 & 0 & 0 \\ D_1 D_2 & I & I \end{bmatrix} \begin{bmatrix} \omega \\ \eta \\ y_{\text{ref}} \end{bmatrix}.$$

Direct calculation shows that

$$J_0 = \begin{bmatrix} \omega \\ \eta \\ y_{\text{ref}} \end{bmatrix}' \begin{bmatrix} 0 & 0 & 0 \\ 0 & H & -H \\ 0 & -H & H \end{bmatrix} \begin{bmatrix} \omega \\ \eta \\ y_{\text{ref}} \end{bmatrix}$$

where $H = F_1(0)'B_1'A_1'^{-1}A_1^{-1}B_1F_1(0)$ and $F_1(0) = D_1 - C_1A_1^{-1}B_1$.

The rest of the proof now proceeds in exactly the same way as in the last part of the proof of Theorem 2. $\square$

6 CONCLUSION

This paper considers the cheap regulator problem and the cheap optimal servomechanism problem for systems which may be non-minimum phase. The basic tool used is a factorization which factorizes an arbitrary system transfer matrix into the product of an inner transfer matrix and a wide minimum phase transfer matrix. Based on this factorization, the study of an arbitrary system can be decomposed into the study of a system with an inner transfer matrix and the study of a minimum phase and wide system. The cheap control problem of a system with an inner transfer matrix becomes easy to analyze by exploiting various properties of inner matrices, while the cheap control problem of a wide minimum phase system has been intensively studied.

A novel contribution of this paper is the establishment of the fact that the number and the locations of the system's transmission zeros in the open right half of the complex plane, are crucial factors of a system which determines the best attainable closed loop system performance. In another words, we have shown that the design limitations on the closed loop system performance, for the servomechanism problem, can be completely characterized by the number and the locations of the system's open loop transmission zeros in the open right half of the complex plane. This design limitation can be used to evaluate an open loop system, i.e. to determine whether the system is inherently hard to control, and to assess a given closed loop design, i.e. to determine

how near the closed loop system's performance is from the best attainable.

The servomechanism problem considered in this paper is only for the case of constant reference and disturbances. The extension of the results obtained to more general reference and disturbance signals offers a direction for future research.

References

Chen, T. (1987). *Spectral and Inner-Outer Factorizations of Rational Matrices.* Master's thesis, Dept. of Electrical Eng., Univ. of Toronto, Toronto, Canada.

Davison, E. J. (1976a). The steady-state invertibility and feedforward control of linear time-invariant systems. *IEEE Trans. Automat. Contr.*, AC-21:529–534.

Davison, E. J. (1976b). The robust control of a servomechanism problem for linear time invariant multivariable systems. *IEEE Trans. Automat. Contr.*, AC-21:25–34.

Davison, E. J. and Scherzinger, B. M. (1987). Perfect control of the robust servomechanism problem. *IEEE Trans. Automat. Contr.*, AC-32:689–702.

Davison, E. J. and Wong, S. H. (1974). Properties and calculation of transmission zeros of linear multivariable systems. *Automatica*, 10:634–658.

Francis, B. A. (1979). The optimal linear-quadratic time-invariant regulator with cheap control. *IEEE Trans. Automat. Contr.*, AC-24:616–621.

Francis, B. A. (1987). *A Course in H_∞ Control Theory.* Springer-Verlag, Berlin.

Glover, K. (1984). All optimal hankel-norm approximations of linear multivariable systems and their l^∞-error bounds. *Int. J. Control*, 39:1115–1193.

Kwakernaak, H. and Sivan, R. (1972a). The maximal achievable accuracy of linear optimal regulators and linear optimal filters. *IEEE Trans. Automat. Contr.*, AC-17:79–86.

Kwakernaak, H. and Sivan, R. (1972b). *Linear Optimal Control Systems.* Wiley-Interscience, New York.

Levine, W. S. and Athans, M. (1970). On the determination of the optimal constant output feedback gains for the linear multivariable systems. *IEEE Trans. Automat. Contr.*, AC-15:44–48.

Moore, B. C. (1981). Principle component analysis of linear systems: controllability, observability, and model reduction. *IEEE Trans. Automat. Contr.*, AC-26:17–32.

Scherzinger, B. M. and Davison, E. J. (1985). The optimal LQ regulator with cheap control for not strictly proper systems. *Optimal Control Application & Method*, 6:291–303.

SYNTHESIS OF ROBUST LQ CONTROLLERS

A. M. Valderhaug, D. Di Ruscio and J. G. Balchen

*Division of Engineering Cybernetics, The Norwegian Institute of Technology,
N7034 Trondheim, Norway*

Abstract. The robustness problem for a control system and different ways of describing the model
uncertainty are discussed, and an iterative synthesis algorithm for robust controllers is proposed.
The LQ optimization is combined with an optimization of a robustness measure in order to find a
robust controller. The robustness measure, chosen to be the maximum structured singular value
(μ), is minimized with respect to the weight matrices in the LQ criterion. In the algorithm it is
also possible to minimize the robustness measure directly onto the elements in a controller with a
prescribed structure. The algorithm is presented for an LQ controller combined with integral action,
and for direct optimization.

Keywords.   Linear optimal control; robust control; robustness; multivariable control systems;
optimization.

INTRODUCTION

The robustness of model uncertainty is a very critical factor
for every practical control system. It has been the topic of a
large number of studies, and a comprehensive survey of some
of the proposed robustness methodologies has been given by
e.g., Lunze (1989).

In this paper the LQ optimization is chosen as part of a
robust controller synthesis.
The ordinary LQ state controller is given by:

$$\min_{\underline{u}} J = \int_0^\infty \left(\underline{x}^T Q \underline{x} + \underline{u}^T P \underline{u} \right) dt \tag{1}$$

where the weight matrices Q and P are symmetric and respectively n-dimensional and nonnegative, and r-dimensional
and positive definite. The system is described by:

$$\dot{\underline{x}} = A\underline{x} + B\underline{u} \tag{2}$$
$$\underline{y} = D\underline{x} \tag{3}$$

where $\underline{x}$ is the n-dimensional state vector, $\underline{u}$ the r-dimensional
control input vector, and $\underline{y}$ the m-dimensional measurement
vector, which here is assumed to be identical to the state
vector (i.e., the matrix $D = I$). The optimal control action
according to the criterion is:

$$\underline{u} = G\underline{x} \tag{4}$$
$$G = -P^{-1}B^T R \tag{5}$$

where R is the positive definite solution of the algebraic Riccati equation:

$$RA + A^T R - RBP^{-1}B^T R + Q = 0 \tag{6}$$

The objective of this paper is to propose a synthesis algorithm for a robust controller which iteratively combines the
optimization of an LQ criterion with a minimization of a ro-

bustness measurement. The paper is organized as follows: in
the first section the robust synthesis algorithm is described,
then robustness analysis is treated, the algorithm is finally
examined in an example.

A ROBUST LQ CONTROLLER ALGORITHM

The task of determining a controller that has some prescribed
robustness properties and that is optimal in the sense of
a linear quadratic performance index, can be obtained by
solving the following optimization problem:

$$\min_{P,Q} \mu \tag{7}$$

given that

$$RA + A^T R - RBP^{-1}B^T R + Q = 0 \tag{8}$$
$$G = -P^{-1}B^T R \tag{9}$$

are satisfied. Equations(8) and (9) are the solution to
the ordinary LQ problem, and Eq.(7) is a function defining
the robustness measure. The structured singular value (μ),
which is described in the next section, has been chosen as the
robustness measure in the algorithm, but other robustness
measures could be implemented instead. The prescribed robustness is achieved by a minimization of Eq.(7) with respect to the weight matrices P and Q. The weight matrices
can be chosen to be diagonal or full, and one of them could
be fixed throughout the optimization. The optimization can
be done for example with respect to the state weight matrix
Q with a given structure, while the control weight matrix P
is fixed.

An alternative way to determine a robust controller would be
to optimize the robustness measure directly with respect to
the feedback matrix G with a prescribed structure. This optimization problem can be illustrated by the following task:

$$\min_G \mu \tag{10}$$

<u>The optimization algorithm.</u>   The robust LQ algorithm can be described as follows, where k is the iteration index. It will only be illustrated for the case when the weight matrix P is fixed, and the criterion in Eq.(7) is optimized with respect to the elements in the full state weight matrix Q. Similar algorithms have been developed for other combinations of the weight matrices, and for the direct optimization of Eq.(10) with respect to the controller G.

1. Given an initial value of the state weight matrix Q_k.

2. Determine G_k for the given Q_k as the solution of the LQ problem given in Eqs.(8) and (9):

$$R_k A + A^T R_k - R_k B P^{-1} B^T R_k + Q_k = 0$$
$$G_k = -P^{-1} B^T R_k$$

3. Determine the value of the robustness measure given in Eq.(7):

$$\mu\left(Q_k\right)$$

A search technique for determining the optimal state weight matrix Q can now be employed. In the algorithm the Sequential Quadratic Programming-method for solving nonlinear constrained problems (SQP) is used (Schittkowski, 1984). The matrix Q is transformed to a parameter vector $\underline{p}$, which is used in the optimization.

The proposed algorithm can be used in various optimization tasks. Some possible problems that will be further described by numerical examples in the examples section, are:

<u>1.</u>   Fixing the control input weight matrix P. Finding the state weight matrix Q that gives a feedback matrix G that minimizes the robustness measure. The structure of the state weight matrix Q could be chosen as a) diagonal or b) full. In this way it will be possible to prescribe some "economical" weights on the control inputs, and robustness is found by varying the state weight matrix.

<u>2.</u>   Fixing the state weight matrix Q. Finding the diagonal control input weight matrix P that gives a feedback matrix G that minimizes the robustness measure. In this matter some weights on the states can be prescribed, and the robustness is found by varying the control weight matrix. This will be a useful approach if a controller is adopted with weight on the control input velocity (Balchen, and Mumme, 1988), because it then is difficult to choose a priori a reasonable weight matrix P.

<u>3.</u>   Minimizing the robustness measure directly with respect to the feedback matrix G with a given structure. This approach will provide the optimal controller with respect to robustness, given the structure of the controller.

ROBUSTNESS ANALYSIS

Modelling is a very important issue with respect to robustness, because decreased model uncertainty will improve the robustness margins. Considerable effort should therefore be made in the modelling, and all available a priori knowledge about the plant should be used. The model uncertainty usually consists of time-varying plant parameters, non-linearities and neglected model dynamics. An intuitive way of modelling these uncertainties is to model them in the time domain based on the first-principles knowledge of the plant, but a drawback with time-domain description is that some classes of uncertainties, e.g., unknown time delays and neglected dynamics, may be difficult to describe. A robust control synthesis is to find a controller that ensures system stability and satisfies some performance specifications when the uncertainty is limited by some prescribed bounds. Although the modelling of the uncertainty in the time domain is physically reasonable, it may become a formidable task, and at present there are few robustness measures that can be used to analyze a system with a time-domain description. Therefor, a frequency domain description robustness measure is chosen in the algorithm.

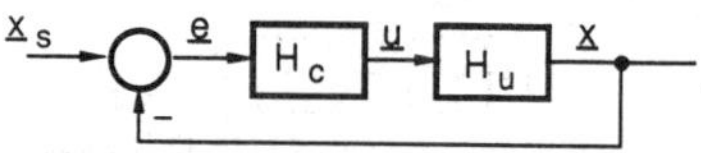

Fig. 1. Nominal model in frequency domain.

<u>Robustness Analysis in Frequency Domain.</u>

An equivalent frequency domain description of the nominal system given in Eqs.(2) - (4) is:

$$H_u = D\left(sI - A\right)^{-1} B \tag{11}$$
$$H_c = -G \tag{12}$$

The model is shown in Fig.(1). The nominal model description (system with no uncertainty) is separated from the perturbed model description (system with uncertainty), which actually is a <u>set</u> of possible models, by the following nomenclature:
- H : the nominal transfer matrix.
- $\tilde{H}$: the perturbed transfer matrix.

The uncertainty in a frequency domain description can be stated by additional perturbation blocks to the nominal system. Each perturbation block contains two different parts; a description of the uncertainty magnitude, and a normalized perturbation block. In order to avoid too conservative stability conditions, the perturbation blocks should be modelled where the uncertainty occurs (Skogestad, Morari, and Doyle, 1988). In addition to uncertainty description it will also be necessary to state some performance specification. A common performance specification is to state some bounds on the closed-loop transfer function between the external input and error signals. The robustness problem can be stated in the following way: The perturbed system must be stable <u>and</u> satisfy the performance specification, i.e., the worst case response of the system should satisfy the desired performance specification.

The uncertainty and performance description in the frequency domain is illustrated by considering a system with uncertainty in the input channels, as shown in Fig.(2). In the example a similar system is considered. The perturbed model can be described as:

$$\tilde{H}_u = H_u\left(I + W_I \Delta_I\right) \tag{13}$$

W_I describes the magnitude of the uncertainty in the input

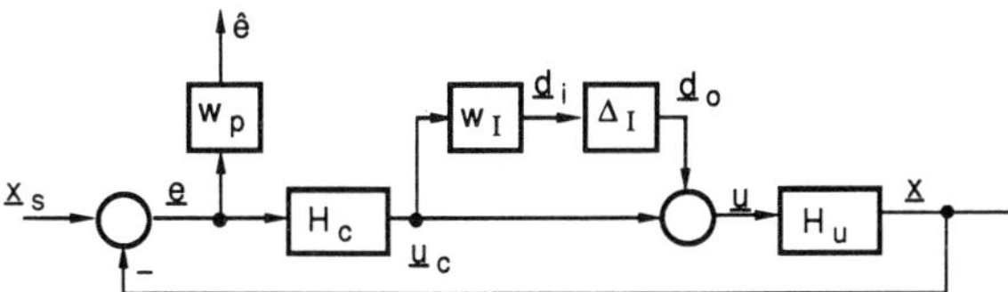

Fig. 2. Perturbed model with performance weight in frequency domain.

channels, and Δ_I is a normalized perturbation that satisfies $\overline{\sigma}(\Delta_I) \leq 1 \; \forall \omega$. Performance is specified by a bound on the error signal $\hat{\underline{e}}$, and W_P describes the frequency range where the error is specified as "small".

A comprehensive survey for uncertainty modelling and robustness analysis in the frequency domain is given by Morari and Zafiriou (1989).

Robustness analysis with the structured singular value.

The structured singular value (μ) (Doyle, Wall, and Stein, 1982) has been shown to be a useful tool for robustness analysis when the uncertainty is modelled as described above (Skogestad, Morari, and Doyle, 1988), and it is chosen as the robustness measure in the robust LQ algorithm. In μ-analysis the following subobjectives are examined:
Nominal stability (NS). The nominal closed-loop system of Fig.(1) must be stable.
Nominal performance (NP). The nominal system should satisfy some performance specifications. The performance specification for the system in Fig.(2) is:

$$NP \Leftrightarrow \overline{\sigma}\left(W_P N\right) < 1, \quad \forall \omega \qquad (14)$$

where $N = W_P \left(I + H_u H_c\right)^{-1}$ is the weighted sensitivity matrix.
Robust stability (RS). The perturbed closed-loop system shown in Fig.(2) must be stable.
Robust performance (RP). The perturbed closed-loop system should satisfy the performance specifications. The corresponding RP specification to the NP specification in Eq.(14) is:

$$RP \Leftrightarrow \overline{\sigma}\left(W_P \tilde{N}\right) < 1, \quad \forall \omega, \; \forall \tilde{H}_u \qquad (15)$$

where $\tilde{N} = \left(I + \tilde{H}_u H_c\right)^{-1}$ is the perturbed weighted sensitivity matrix.

System representation. When the μ-concept is used, the system should be rearranged into the structure shown in Fig.(3). The new system consists of two blocks; an interconnection matrix S, and a perturbation matrix Δ, containing the different perturbation matrices. The rearrangement is illustrated by the system in Fig.(2). The interconnection matrix S becomes:

$$S = \begin{bmatrix} -W_I H_c N H_u & W_I H_c N \\ -W_P N H_u & W_P N \end{bmatrix} \qquad (16)$$

and the Δ-matrix will contain the normalized perturbation block Δ_I. The signal $\hat{\underline{v}} = \underline{x}_s$ is the normalized input signal, $\hat{\underline{e}} = W_P \underline{e}$ is the normalized output signal, $\underline{d}_i = W_I \underline{u}_c$ is the normalized output perturbation, and $\underline{d}_o$ is the normalized input perturbation.

The RP specification for the system in Fig.(3) is on the general form:

$$RP \Leftrightarrow \overline{\sigma}\left(E\right) < 1, \quad \forall \omega, \; \forall \Delta_I \qquad (17)$$

where

$$\hat{\underline{e}} = E\hat{\underline{v}}, \quad E = S_{22} + S_{21}\Delta_I \left(I - S_{11}\Delta_I\right)^{-1} S_{12}$$

The NP condition in Eq.(14) is equal to this condition when E is chosen at the weighted sensitivity matrix and $\Delta_I = 0$.

Condition for robust stability. If NS is satisfied (S is stable) and the matrix Δ_I is stable ($\overline{\sigma}(\Delta_I) \leq 1 \; \forall \omega$), instability can only occur because of additional feedback paths through Δ_I. A condition for robust stability can be found by the use of the generalized small gain theorem applied to the system in Fig.(3). An equivalent condition is achieved by the use of the structured singular value, μ:

$$\mu_{\Delta_I}(S_{11}) \, \overline{\sigma}(\Delta_I) \quad < \quad 1, \quad \forall \omega, \; \forall \Delta_I, \; \overline{\sigma}(\Delta_I) \leq 1 \qquad (18)$$
$$\Updownarrow$$
$$\mu_{\Delta_I}(S_{11}) \quad < \quad 1, \quad \forall \omega \qquad (19)$$

Condition for robust performance. A condition for robust performance is derived by transforming the system in Fig.(3) to the structure of Fig.(4). The perturbation block $\bar{\Delta}$ in Fig.(4) contains two different perturbation blocks; the uncertainty block Δ_I, and a fictitious perturbation block Δ_P of the same dimension as E, which ensures that the RP condition in Eq.(17) is satisfied.

$$\bar{\Delta} = \begin{pmatrix} \Delta_I & 0 \\ 0 & \Delta_P \end{pmatrix} \qquad (20)$$

RP of the system in Fig.(3) is equivalent to RS of the system in Fig.(4). The RP condition is:

$$RP \Leftrightarrow \mu_{\bar{\Delta}}(S) < 1, \quad \forall \omega \qquad (21)$$

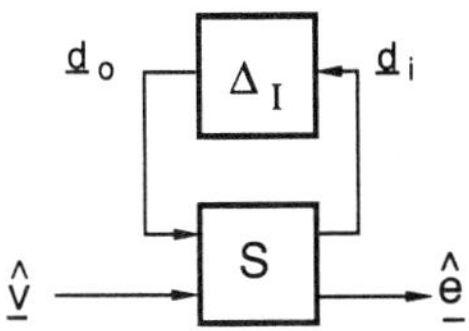

Fig. 3. General representation of a system with uncertainty Δ_I.

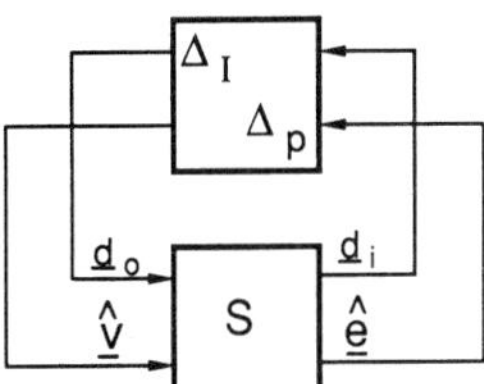

Fig. 4. System representation for RP analysis.

EXAMPLE

The robust LQ algorithm, together with direct optimization on the controller, is tested on a simplified model of a binary distillation column with six trays and a relative volatility of 3.0. The model is given by:

$$\dot{\underline{x}} = A\underline{x} + B\underline{u} + C\underline{v}, \quad \underline{x}_s = 0 \qquad (22)$$

with the following matrices:

$$A = \begin{bmatrix} -3.721 & 2.946 \\ 3.054 & -3.612 \end{bmatrix} 10^{-2}, \quad B = \begin{bmatrix} 0.602 & -0.672 \\ 0.624 & -0.555 \end{bmatrix} 10^{-2}$$

$$C = \begin{bmatrix} 0.475 & 0.721 \\ 0.138 & 0.612 \end{bmatrix} 10^{-2}$$

The states, control inputs and disturbances in the model are:

x_1: the bottom product concentration
x_2: the distillate concentration
u_1: the reflux flow, (mol/min)
u_2: the vapour flow (the "boil up"), (mol/min)
v_1: feed flow, (mol/min)
v_2: feed concentration

and the model is linearized around the steady state values:

$$x_1 = 0.04 \quad u_1 = 2.0 \text{ mol/min} \quad v_1 = 1.0 \text{ mol/min}$$
$$x_2 = 0.96 \quad u_2 = 2.5 \text{ mol/min} \quad v_2 = 0.5$$

The model is rather ill-conditioned with a steady-state RGA of 44.5. The LQ state controller combined with integral action will be used in the robust LQ algorithm.

LQ Controller with Integral Action.

The LQ state controller combined with integral action (e.g., Balchen and others, 1973) is given by:

$$\min_{\underline{u}} J = \int_0^\infty \left(\underline{\bar{x}}^T Q \underline{\bar{x}} + \underline{u}^T P \underline{u} \right) dt \qquad (23)$$

where $\underline{\bar{x}}$ is the augmented state vector, $\underline{\bar{x}} = [\underline{x}, \underline{z}]^T$. The augmented system for the example is:

$$\begin{bmatrix} \dot{\underline{x}} \\ \dot{\underline{z}} \end{bmatrix} = \begin{bmatrix} A & 0 \\ -D & 0 \end{bmatrix} \begin{bmatrix} \underline{x} \\ \underline{z} \end{bmatrix} + \begin{bmatrix} B \\ 0 \end{bmatrix} \underline{u} + \begin{bmatrix} C \\ 0 \end{bmatrix} \underline{v} \qquad (24)$$

All states are assumed to be known ($D = I$). The optimal control action is:

$$\underline{u} = G_1 \underline{x} + G_2 \underline{z} \qquad (25)$$

which is a multivariable PI controller. This control structure is shown in Fig.(5).

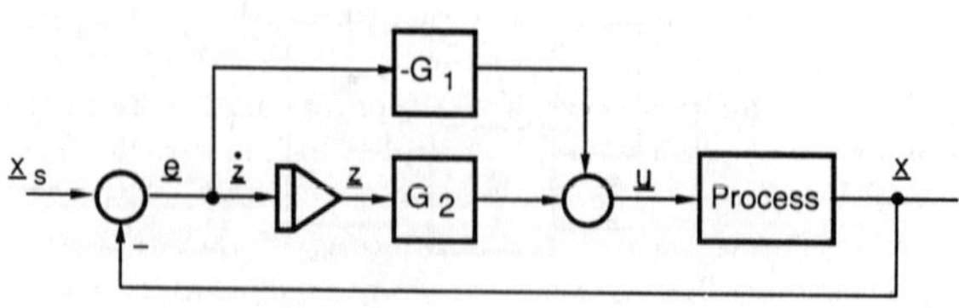

Fig. 5. Multivariable PI controller.

Uncertainty and Performance Specifications.

Because μ is used as a robustness measure in the algorithm, some model uncertainty and performance bounds have to be stated. The bounds that are used in the example, are the same that were used by Skogestad, Morari, and Doyle (1988).

Uncertainty description. It is assumed that there are unknown variations in the plant input flows because of uncertainty in the control valves. The perturbed model can be described by Eq.(13). The magnitude of relative uncertainty in each of the control input channels is chosen to be equal and is given by W_u. The uncertainty block Δ_I is chosen to be diagonal, because the valve uncertainties are assumed to be independent.

$$W_I = 0.2 \frac{1 + 5s}{1 + 0.5s} \qquad (26)$$

This uncertainty weight will correspond to a low frequency dynamic error of 20%. At higher frequencies the uncertainty will increase and reach 100% at a frequency at about 1 min^{-1}, which can describe neglected high-frequency dynamics.

Performance specification. The setpoint $\underline{x}_s$ is chosen as the external input $\hat{\underline{v}}$, and the weighted errors is given as $\hat{\underline{e}} = W_P \underline{e}$.

$$W_P = 0.5 \frac{1 + 10s}{10s} \qquad (27)$$

The performance weight W_P requires use of integral action in the controller and a maximum disturbance amplification of 2 at high frequency ($\overline{\sigma}(\tilde{N})_{\max} = 2$). The desired closed-loop bandwidth is stated as 0.1 min^{-1}.

Conditions for stability and robustness. The interconnection matrix S in the RP representation is equal to Eq.(16), and the conditions for stability and performance according to Eqs.(14), (19) and (21) are:

$$\text{NP} \quad \Leftrightarrow \quad \mu_{NP} = \sup_\omega \overline{\sigma}(W_P N) < 1 \qquad (28)$$

$$\text{RS} \quad \Leftrightarrow \quad \mu_{RS} = \sup_\omega \mu_{\Delta_I}(-W_I H_c N H_u) < 1 \qquad (29)$$

$$\text{RP} \quad \Leftrightarrow \quad \mu_{RP} = \sup_\omega \mu_{\Delta}(S) < 1 \qquad (30)$$

The structured singular value with respect to robust performance μ_{RP} is used as the robustness measure in the algorithm. The optimal controller is then found by minimizing the maximal μ_{RP}.

Examination of the Robust LQ Algorithm.

The algorithm will be examined by the following cases:

Case 1. Optimization with respect to the state weight matrix Q with two different structures; a) diagonal, and b) full Q. The control weight matrix P is fixed:

$$P = diag \begin{bmatrix} 1.0 & 0.62 \end{bmatrix}$$

Case 2. Optimization with respect to a diagonal control weight matrix P. The state weight matrix Q is fixed:

$$Q = diag \begin{bmatrix} 1600 & 1600 & 30 & 30 \end{bmatrix}$$

Case 3. Optimization with respect to the feedback matrix $G = [G_1, G_2]$ for two different structures; a) full, and b) diagonal G.

Cases 1a - 3a yield multivariable PI controllers, and case 3b yields a diagonal PI controller. The optimizations produce μ-optimal controllers, given the structures on the controllers and the weight matrices. The results are:

Case 1a.
$$\max \mu_{RP} = 1.02$$
$$Q = diag \begin{bmatrix} 1500.12 & 1499.95 & 35.63 & 24.41 \end{bmatrix}$$
$$G = \begin{bmatrix} 11.83 & -56.61 & -0.21 & 4.94 \\ 70.59 & -7.70 & -7.56 & -0.22 \end{bmatrix}$$

Some insight about the choice of the state weight matrix can be given by this result. A common guideline in choosing the elements in diagonal weight matrices is (Balchen, and Mumme, 1988):
$$q_{ii} = \frac{1}{(\Delta \hat{x}_i)^2}$$
where $\Delta \hat{x}_i$ is maximum "allowable" deviations of the states. The result of case 1a implies that it is optimal to have approximately equal weights on the states, ($q_{11} \approx q_{22} \approx 1500$), corresponding to a maximum "allowable" deviation: $\Delta \hat{x}_i = 0.025$, which seems to be reasonable.
The weights on the integrated errors are usually difficult to choose. The optimal weights from case 1a yield almost equal weights on the integrated errors, ($q_{33} \approx q_{44} \approx 30$), corresponding to a maximum "allowable" deviation: $\Delta \hat{z}_j = 0.18$. This result is interpreted by the following ad hoc argumentation:
$\Delta \hat{z}_j$ is the integral of the deviations of x_i:
$$\Delta \hat{z}_j = \int_0^\infty \Delta x_i \, dt \approx \Delta \hat{x}_i \int_0^\infty e^{-\frac{t}{T}} \, dt = \Delta \hat{x}_i T$$
where T is the most dominant time constant of the "proportional-part" of the closed-loop system. The eigenvalues of the closed-loop system are: -0.52, -0.15, $-0.06 \pm j0.03$, where the complex conjugated pair corresponds to the "integral-part" of the system. The most dominant time constant of the "proportional-part" is then: $T = 6.7$. This corresponds to a maximum deviation: $\Delta \hat{z}_j = 0.17$, which is approximately equal to the given result.

Case 1b.
$$\max \mu_{RP} = 0.95$$
$$Q = \begin{bmatrix} 3784.56 & 3859.84 & 0 & -773.80 \\ 3859.84 & -12126.67 & -89.99 & 0 \\ 0 & -89.99 & 10.45 & -1.06 \\ -773.80 & 0 & -1.06 & 103.79 \end{bmatrix}$$
$$G = \begin{bmatrix} 8.41 & -39.78 & 0.70 & 9.87 \\ 57.81 & -13.50 & -4.00 & 3.20 \end{bmatrix}$$

Case 2.
$$\max \mu_{RP} = 1.015$$
$$P = diag \begin{bmatrix} 0.84 & 0.85 \end{bmatrix}$$
$$G = \begin{bmatrix} 8.17 & -64.07 & 0.05 & 5.97 \\ 61.11 & -10.97 & -5.95 & 0.05 \end{bmatrix}$$
It is optimal to have equal weights on the control inputs.

Case 3a.
$$\max \mu_{RP} = 0.95$$
$$G = \begin{bmatrix} 8.41 & -39.78 & 0.73 & 9.87 \\ 57.82 & -13.50 & -4.01 & 3.23 \end{bmatrix}$$
It should be noticed that this controller is approximately equal to the controller in case 1b.

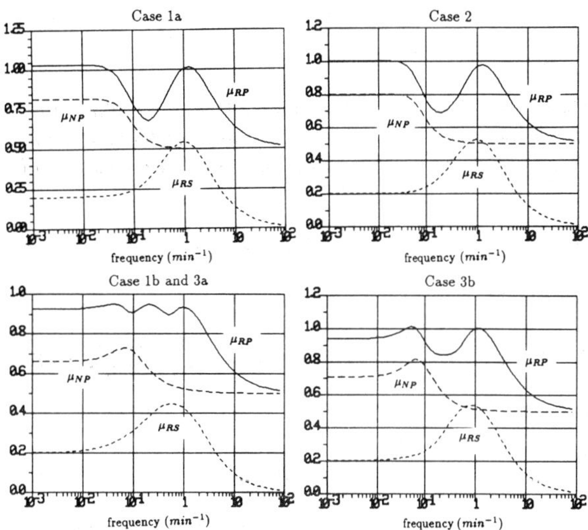

Fig. 6. μ-plots for the examples.

Case 3b.
$$\max \mu_{RP} = 1.005$$
$$G = \begin{bmatrix} 0 & -40.62 & 0 & 11.93 \\ 59.60 & 0 & -5.24 & 0 \end{bmatrix}$$
The μ-optimal multivariable PI controller in cases 1b and 3a has somewhat better robustness properties than the μ-optimal diagonal PI controller in case 3b.
The results show that the robust LQ algorithm can be used in order to synthesize robust controllers. Although only the controllers from cases 1b and 3a satisfy the RP condition, all the different controllers have almost similar robustness properties. The achieved robustness depends on the degree of freedom in the optimization, e.g., optimization with respect to a full state weight matrix yields a more robust controller than optimization with respect to a diagonal matrix. The μ-plots for the different cases are shown in Fig.(6).

Discussion.

The optimization of the robustness measure with respect to the weight matrices in the LQ criterion has some advantages compared to the direct optimization of the controller:
(i)   Closed-loop stability is guaranteed during every iteration with the robust LQ algorithm, due to the closed-loop stability properties of LQ controllers. With direct optimization stability must be checked separately.
(ii)   The robust LQ algorithm yields insight in the choice of the weight matrices in the LQ criterion, especially when the matrices are chosen to be diagonal.
(iii)   If the weight matrices are chosen to be diagonal, or one of them is fixed, the optimization task will become smaller with the robust LQ algorithm. The parameter vector will then have a smaller dimension (i.e., $\dim(\underline{p}_{LQ}) < \dim(\underline{p}_G)$).
(iv)   In the example it was noticed that the robust LQ algorithm converged rapidly at the beginning of the optimization, but somewhat slower near the optimum. The optimization was not sensitive to the choice of the initial value of the weight matrices. The direct optimization had an opposite nature; it was sensitive to the choice of the initial value of the controller, but it converged more rapidly when the optimum was close.
(v)   If the robust LQ algorithm is used, it will be possible

to state physically interpreted weights on states and control inputs, something that hardly can be done with direct optimization.

A possible synthesis procedure is to combine the two optimization methods, the robust LQ algorithm and direct optimization on the controller.
(i)   First an optimization with respect to the weight matrices in the LQ criterion.
(ii)   Then direct optimization of the controller.
Using this approach the optimization will not be sensitive to the initial value, the closed-loop stability will be guaranteed, and the optimization will converge rapidly when the optimum is close.

The SQP optimization algorithm guarantees only local minima, and a possible arrival at a global minimum depends on the initial value of the optimization.

CONCLUSION

By combining the LQ controller synthesis with the μ-concept, an iterative algorithm has been developed for a synthesis of a robust controller. The robustness properties in terms of μ, are optimized with respect to the weight matrices in the LQ criterion. It is also possible to optimize the robustness directly with respect to the controller. The LQ optimization has several advantages compared to direct optimization of the controller. Closed-loop stability is guaranteed in every iteration, the optimization task may be decreased due to a smaller dimension in the optimization parameter vector, and it will be possible to weight states and control inputs. The robust LQ algorithm also yields insight in choosing the weight matrices in the LQ criterion.

The extension of this synthesis methodology to a practical control system including a state estimator, is a subject for further research.

REFERENCES

Balchen, J. G., T. Endresen, M. Fjeld, and T. O. Olsen (1973). Multivariable PID estimation and control with biased disturbances. Automatica, 9 295-307.

Balchen, J. G., and K. I. Mumme (1988). Process control: Structures and applications. Van Nostrand Reinhold Company Inc., New York. Chap 2.11, pp. 63-65.

Doyle, J. C., J. E. Wall, G. Stein (1982). Performance and robustness analysis for structured uncertainty. Proc. 21th IEEE Conf Decision and Control, Orlando, Florida, 629-636

Lunze, J. (1989). Robust multivariable feedback control. Prentice Hall International (UK) Ltd, London.

Morari, M., E. Zafiriou (1989). Robust process control. Prentice Hall, Inc, Englewood Cliff, New Jersey. Chap 10 and 11, pp 205-292.

Schittkowski, K. (1984). NLPQL : A Fortran subroutine solving constrained nonlinear programming problems. Working paper, University of Stuttgart.

Skogestad, S., M. Morari, and J. Doyle (1988). Robust Control of ill-conditioned plants: High-purity distillation. IEEE Trans. Aut. Control, AC-33, 1092-1105.

STATE SPACE DECOUPLING OF NON-MINIMUM PHASE SYSTEMS

B. Lohmann

*Institut für Regelungs- und Steuerungssysteme, Universität Karlsruhe,
Kaiserstraße 12, D-7500 Karlsruhe, FRG*

Abstract: The decoupling of the input-output behaviour of linear multivariable systems by *constant* state-feedback generally requires the compensation of all invariant zeros, which causes instability in the case of non-minimum phase systems. The paper presents a method for partial but stable decoupling, ensuring without compensations coupling of only *one* output with *several* inputs. The application of a *dynamic* state-feedback however allows complete and stable decoupling with the disadvantage of an increased number of disagreeable transmission zeros.

Keywords: Decoupling, constant and dynamic state feedback, non-minimum phase systems, invariant zeros.

1. PROBLEM STATEMENT

Consider an nth order linear time-invariant multivariable system

$$\dot{\underline{x}}(t) = \underline{A}\,\underline{x}(t) + \underline{B}\,\underline{u}(t)\,,\quad \underline{y}(t) = \underline{C}\,\underline{x}(t)\,,\qquad (1.1)$$

with the $(n,1)$-state vector $\underline{x}(t)$, the $(m,1)$-input vector $\underline{u}(t)$, the $(m,1)$-ouput vector $\underline{y}(t)$ and the matrices $\underline{A}$, $\underline{B}$, $\underline{C}$ of conformal dimensions. *Input-output decoupling* is achieved if one can find a constant (m,n)-controller matrix $\underline{R}$ and a constant (m,m)-prefilter $\underline{F}$ such that via the state feedback law

$$\underline{u}(t) = -\underline{R}\,\underline{x}(t) + \underline{F}\,\underline{w}(t)\qquad (1.2)$$

every output y_i, $i = 1,...,m$, is only affected by the corresponding w_i. Hence the transfer-function matrix

$$\underline{G}_w(s) = \underline{C}(s\underline{I} - \underline{A} + \underline{B}\,\underline{R})^{-1}\underline{B}\,\underline{F}\qquad (1.3)$$

of the closed-loop system must be *diagonal*, i.e.

$$\underline{G}_w(s) = \mathrm{diag}[g_{11}(s),...,g_{mm}(s)]\,.\qquad (1.4)$$

Falb and Wolovich (1967) first gave a solution to this problem, Roppenecker and Lohmann (1988) achieved decoupling by the design method of "Complete Modal Synthesis". If non-minimum phase systems are decoupled by these methods, the compensation of the so called *invariant zeros* (MacFarlane, Karcanias, 1976) in the right half of the complex plane lead to unstable poles. In the third section it will be shown how the design steps of the complete decoupling can be modified in the case of an "unstable" invariant zero, in order to obtain a *partial* but *stable* decoupling of the form

$$\underline{G}_w(s) = \begin{bmatrix} g_{11}(s) & & 0 \\ & \ddots & \\ g_{j1}(s) \cdots g_{jj}(s) \cdots g_{jm}(s) \\ & \ddots & \\ 0 & & g_{mm}(s) \end{bmatrix}\,.\qquad (1.5)$$

With the transfer-function matrix (1.5) the partial decoupling is an advantage compared to the triangular decoupling, where a greater or equal number of elements of $\underline{G}_w(s)$ are non-zero (see e.q. Koussiouris (1970)). The choice of the coupled row in $G_w(s)$ and the influence of remaining design parameters are treated in the sections 4 and 5. The design of a dynamic state-feedback controller which allows complete and stable decoupling is presented in section 6. The "unstable" invariant zero appears as transmission zero in several (generally all) elements of the diagonal matrix $\underline{G}_w(s)$ in this case, causing undershooting of the referring outputs.

2. COMPLETE DECOUPLING BY CONSTANT STATE-FEEDBACK

The design method of *Complete Modal Synthesis* by Roppenecker (1986) is based on the fact that every state-feedback controller $\underline{R}$ is related to a set of closed-loop eigenvalues λ_μ and *invariant parameter vectors* $\underline{p}_\mu$ by the equation

$$\underline{R} = [\underline{p}_1,...,\underline{p}_n]\cdot[\underline{v}_1,...,\underline{v}_n]^{-1}\,,\ \text{where}\qquad (2.1)$$

$$\underline{v}_\mu = (\underline{A} - \lambda_\mu\underline{I})^{-1}\underline{B}\,\underline{p}_\mu\,,\ \mu = 1,...,n\,.\qquad (2.2)$$

In order to determine the free parameters λ_μ and $\underline{p}_\mu$ such that a diagonal closed-loop transfer-function matrix is achieved, we first apply the modal transformation

$$(\underline{A} - \underline{B}\,\underline{R}) = \underline{V}\,\underline{\Lambda}\,\underline{V}^{-1}\,,\qquad (2.3)$$

where $\underline{V} = [\underline{v}_1,...,\underline{v}_n]$ is the matrix of closed-loop eigenvectors and $\underline{\Lambda}$ is the diagonal matrix of closed loop eigenvalues, to eq.(1.3) resulting in

$$\underline{G}_w(s) = \underline{C}\,\underline{V}(s\underline{I}-\underline{\Lambda})^{-1}\underline{V}^{-1}\underline{B}\,\underline{F} = \sum_{\mu=1}^{n} \frac{\underline{C}\,\underline{v}_\mu\,\underline{w}_\mu^T\underline{B}\,\underline{F}}{s-\lambda_\mu}\,.\qquad (2.4)$$

The transposed vectors $\underline{w}_\mu^T$ are the rows of $\underline{V}^{-1}$.
Now the elements $g_{ii}(s)$ in the desired diagonal transfer-function matrix (1.4) are set up as

$$g_{ii}(s) = \frac{\prod_{k=1}^{\delta_i}(-\lambda_{ik})}{(s-\lambda_{i1})\cdot\ldots\cdot(s-\lambda_{i\delta_i})}\,,\, i = 1,...,m.\qquad (2.5)$$

The degree δ_i of the denominator is called the *difference order* of the output y_i and is defined as

$$\delta_i = \min_{\mu=1,\ldots,m}\delta_{i\mu}\qquad (2.6)$$

where $\delta_{i\mu}$ is the difference between the degrees of denominator and numerator of the element $g_{i\mu}$ of $\underline{G}_w(s)$ from eq. (1.3), calculated with arbitrary $\underline{R}$ and arbitrary, regular $\underline{F}$. To ensure the difference orders in the decoupled system, the numerators in eq. (2.5) must be set up as constants and were chosen to avoid steady state error.
Comparing eq. (2.4) to eqns. (1.4) and (2.5) we get conditions that must be satisfied by the eigenvectors: If the

eigenvalue λ_{ik} appears only in the ith element $g_{ii}(s)$ of $\underline{G}_w(s)$, then the corresponding closed loop eigenvector must satisfy

$$\underline{C}\,\underline{v}_{ik} = \underline{e}_i\,, \quad i = 1,\ldots,m,\quad k = 1,\ldots,\delta_i \qquad (2.7)$$

(the indices of the $\underline{v}$ are adapted to those of the eigenvalues). $\underline{e}_i$ denotes the ith unit-vector. Eq. (2.7) guarantees the strict connection of every eigenvalue λ_{ik} to *one row* of $\underline{G}_w(s)$ since every dyadic product $\underline{C}\,\underline{v}\,\underline{w}^T\underline{B}\,\underline{F}$ in eq. (2.4) is an (m,m)-matrix in which only the ith row is unequal $\underline{0}^T$. Combining eqns. (2.7) and (2.2) to

$$\begin{bmatrix} \underline{A} - \lambda_{ik}\underline{I} & \underline{B} \\ \underline{C} & \underline{0} \end{bmatrix} \cdot \begin{bmatrix} \underline{v}_{ik} \\ -\underline{p}_{ik} \end{bmatrix} = \begin{bmatrix} \underline{0} \\ \underline{e}_i \end{bmatrix}, \quad \begin{array}{l} i = 1,\ldots,m \\ k = 1,\ldots,\delta_i \end{array} \qquad (2.8)$$

we can calculate the vectors $\underline{v}_{ik}$ and $\underline{p}_{ik}$, if the eigenvalues λ_{ik} are prescribed.
By eq. (2.8) only the

$$\delta = \delta_1 + \delta_2 + \ldots + \delta_m \qquad (2.9)$$

poles of the elements of $\underline{G}_w(s)$ are transformed to conditions on the closed loop eigenvectors. For the remaining $n-\delta$ eigenvalues (it is $n-\delta \geq 0$, see Falb, Wolovich (1967) or Roppenecker, Lohmann (1988)), which do *not* appear in $\underline{G}_w(s)$, it also follows from eq. (2.4) that

$$\underline{C}\,\underline{v}_\nu\underline{w}_\nu^T\underline{B}\,\underline{F} = 0\,, \quad \nu = \delta+1,\ldots,n\,. \qquad (2.10)$$

Assuming controllability of the system[1], i.e. $\underline{w}_\nu^T\underline{B}\,\underline{F} \neq \underline{0}^T$, $\nu = 1,\ldots,n$, eq. (2.10) can only be satisfied if

$$\underline{C}\,\underline{v}_\nu = \underline{0}\,, \quad \nu = \delta+1,\ldots,n\,. \qquad (2.11)$$

Together with eq. (2.2) we then get

$$\begin{bmatrix} \underline{A} - \lambda_\nu\underline{I} & \underline{B} \\ \underline{C} & \underline{0} \end{bmatrix} \cdot \begin{bmatrix} \underline{v}_\nu \\ -\underline{p}_\nu \end{bmatrix} = \underline{0}\,. \qquad (2.12)$$

Non-null solutions $\underline{v}_\nu$, $\underline{p}_\nu$ of this equation exist if the eigenvalues λ_ν are chosen such that

$$\det \begin{bmatrix} \underline{A} - \lambda\underline{I} & \underline{B} \\ \underline{C} & \underline{0} \end{bmatrix} = 0\,. \qquad (2.13)$$

Since the solutions λ of eq. (2.13) just define the *invariant zeros* of the system (MacFarlane, Karcanias, 1976) we can state: Eq. (2.12) is solvable if the λ_ν are chosen equal to the invariant zeros of the system, whereas eq. (2.8) is solvable for any other value of λ_{ik}.
We can now summarize the *steps of calculation of the controller matrix* $\underline{R}$: To every pole of the elements of the desired $\underline{G}_w(s)$ (eqns. (1.4), (2.5)) corresponding vectors $\underline{v}_{ik}$, $\underline{p}_{ik}$ are determined via eq. (2.8) which is solvable if all δ poles are chosen unequal to the invariant zeros of the system. The remaining $n-\delta$ eigenvalues are chosen equal to the invariant zeros which ensures solvability of eq. (2.12). Necessarily the system must have *at least* $n-\delta$ zeros. The so found n pairs of vectors $\underline{v}$ and $\underline{p}$ determine the controller matrix $\underline{R}$ via eq. (2.1). It can be shown that the required inverse in eq. (2.1) exists if the system has *not more* than $n-\delta$ invariant zeros.

How is the precompensator $\underline{F}$ to be chosen? In the desired transfer functions $g_{ii}(s)$ of eq. (2.5) the numerators

avoid steady state error; hence the precompensator must satisfy the wellknown relation

$$\underline{F} = \lim_{s \to 0} \left[\underline{C}(s\underline{I} - \underline{A} + \underline{B}\,\underline{R})^{-1}\underline{B}\right]^{-1}. \qquad (2.14)$$

Actually, this choice of $\underline{F}$ ensures decoupling (for *all* s) as the following consideration shows: With $\underline{F}$ of eq. (2.14) all non-diagonal elements $g_{ik}(s)$, $i \neq k$, of $\underline{G}_w(s)$ disappear at $s = 0$. The controller $\underline{R}$ guarantees by eqns. (2.7), (2.11) δ_i poles in the elements $g_{ik}(s)$, $i = 1,\ldots,m$. Because the δ_i don't change under feedback, the numerators of the $g_{ik}(s)$, $i \neq k$, are of degree zero, i.e. they don't depend on s. Hence, to avoid steady state error, these numerators are equal zero and thus $g_{ik}(s) \equiv 0$, $i \neq k$, which ensures the desired decoupling.
Necessary and sufficient condition for the existence of a decoupling pair $\underline{R}$, $\underline{F}$ is: *The number of finite invariant zeros of the system must be $n-\delta$.* The condition is equivalent to that one given by Falb, Wolovich (1967) (detailed prove by Roppenecker, Lohmann (1988)).

3. PARTIAL DECOUPLING

The decoupling of non-minimum phase systems results in unstable closed-loop eigenvalues since eq. (2.12) requires the choice of the λ_ν equal to invariant zeros.
Generally stability can only be achieved if these compensations and also the complete decoupling are renounced. The design steps for a *partially decoupling, but stabilizing controller* are derived in the following. The diagonal elements $i \neq j$ of the desired $\underline{G}_w(s)$ in eq. (1.5) are chosen as in eq. (2.5), the functions $g_{j1}(s),\ldots,g_{jm}(s)$ are left undefined for the present. Comparing eq. (2.4) to eq. (1.5) we find for the poles of the $g_{ii}(s)$, $i \neq j$ (in correspondence to eq. (2.7)) the relations

$$\underline{C}\,\underline{v}_{ik} = \underline{e}_i + a_{ik}\underline{e}_j\,, \quad \begin{array}{l} i = 1,\ldots,m,\ i \neq j, \\ k = 1,\ldots,\delta_i\,. \end{array} \qquad (3.1)$$

The free parameters a_{ik} are introduced since the appearance of the poles λ_{ik}, $i \neq j$, in elements of the jth row of $\underline{G}_w(s)$ must be allowed explicitly. Together with eq. (2.2) we thus get for the vectors $\underline{v}_{ik}$ and $\underline{p}_{ik}$ the condition

$$\begin{bmatrix} \underline{A} - \lambda_{ik}\underline{I} & \underline{B} \\ \underline{C} & \underline{0} \end{bmatrix}\begin{bmatrix} \underline{v}_{ik} \\ -\underline{p}_{ik} \end{bmatrix} = \begin{bmatrix} \underline{0} \\ \underline{e}_i \end{bmatrix} + a_{ik}\begin{bmatrix} \underline{0} \\ \underline{e}_j \end{bmatrix} \begin{array}{l} i=1,\ldots,m, \\ i \neq j, \\ k=1,\ldots,\delta_i. \end{array} \qquad (3.2)$$

Real eigenvalues λ_{ik} demand real a_{ik}, self-conjugate λ_{ik} demand the choice of self conjugate a_{ik}. Supposing the system to have one "unstable" zero we may only compensate the remaining $n-\delta-1$ "stable" zeros by satisfying

$$\begin{bmatrix} \underline{A} - \lambda_\nu\underline{I} & \underline{B} \\ \underline{C} & \underline{0} \end{bmatrix} \cdot \begin{bmatrix} \underline{v}_\nu \\ -\underline{p}_\nu \end{bmatrix} = \underline{0}\,, \quad \nu = \delta+2,\ldots,n\,, \qquad (3.3)$$

where the λ_ν have to be chosen equal to these $n-\delta-1$ zeros. By eqns. (3.2) and (3.3)

$$\delta_1 + \ldots + \delta_{j-1} + \delta_{j+1} + \ldots + \delta_m + n - \delta - 1 = n - \delta_j - 1 \qquad (3.4)$$

vectors $\underline{v}$ and $\underline{p}$ are determined. The remaining $\delta_j + 1$ pairs of vectors $\underline{v}$, $\underline{p}$ must satisfy the relation

$$\underline{C}\,\underline{v}_{jk} = \underline{e}_j \qquad (3.5)$$

since all poles in rows $i \neq j$ of $\underline{G}_w(s)$ are already considered by eq. (3.2). With eq. (2.2) this yields

---

[1] This assumption can be dropped without eqns. (2.8) and (2.12) loosing their suffiency for decoupling.

$$\begin{bmatrix} \underline{A} - \lambda_{jk}\underline{I} & \underline{B} \\ \underline{C} & \underline{0} \end{bmatrix} \begin{bmatrix} \underline{v}_{jk} \\ -\underline{p}_{jk} \end{bmatrix} = \begin{bmatrix} \underline{0} \\ \underline{e}_j \end{bmatrix}, \quad k = 1,...,\delta_j + 1, \qquad (3.6)$$

where the λ_{jk} can be chosen arbitrarily but unequal to all invariant zeros. With the solutions $\underline{v}$ and $\underline{p}$ of the n linear eqns. (3.2), (3.3), (3.6) the controller matrix $\underline{R}$ can be calculated from eq. (2.1). The precompensator $\underline{F}$ (eq. 2.14) guarantees the desired $\underline{G}_w(s)$ since the proof of decoupling of the rows $i \neq j$ of $\underline{G}_w(s)$ is possible in the same way as in the last section. For the existence of $\underline{R}$ the system must be stabilizable, i.e. must not have uncontrollable eigenvalues in the right half of the complex plane. $\underline{F}$ exists if there is no invariant zero equal to zero. Both conditions are satisfied by all systems appropriate to be controlled. Furtheron note the following condition when choosing $\underline{G}_w(s)$.

4. CHOICE OF THE COUPLED CHANNEL

The choice of the coupled row j of $\underline{G}_w(s)$ in eq. (1.5) to allow partial decoupling of a system with the "unstable" zero η is restricted:

Coupling can be prescribed optionally in one of the rows $j \in [1,...,m]$ that satisfy

$$\underline{q}^T \underline{e}_j \neq 0, \qquad (4.1)$$

where the vector $\underline{q}^T$ is defined via the solution of

$$[\underline{r}^T,\underline{q}^T] \cdot \begin{bmatrix} \underline{A} - \eta\underline{I} & \underline{B} \\ \underline{C} & \underline{0} \end{bmatrix} = \underline{0}^T. \qquad (4.2)$$

It is $\underline{q}^T \neq \underline{0}^T$ since the matrix in eq. (4.2) is singular (see eq. (2.13)) and the block $[\underline{A} - \eta\underline{I},\underline{B}]$ is, stabilizable systems assumed, of full rank[2]. Hence there always exists at least *one* $j \in [1,...,m]$ satisfying eq. (4.1) and thus allowing partial decoupling.

In order to proof the *necessity* of condition (4.1) we first multiply eq. (4.2) with a regular matrix:

$$[\underline{r}^T,\underline{q}^T] \begin{bmatrix} \underline{A} - \eta\underline{I} & \underline{B} \\ \underline{C} & \underline{0} \end{bmatrix} \cdot \begin{bmatrix} \underline{I} & \underline{0} \\ -\underline{R} & \underline{F} \end{bmatrix} =$$

$$= [\underline{r}^T,\underline{q}^T] \cdot \begin{bmatrix} \underline{A} - \underline{B}\,\underline{R} - \eta\underline{I} & \underline{B}\,\underline{F} \\ \underline{C} & \underline{0} \end{bmatrix} = \underline{0}^T. \qquad (4.3)$$

Again multiplying with a suitable matrix we find an equation containing the transfer–function matrix $\underline{G}_w(\eta)$:

$$[\underline{r}^T,\underline{q}^T] \begin{bmatrix} \underline{A} - \underline{B}\,\underline{R} - \eta\underline{I} & \underline{B}\,\underline{F} \\ \underline{C} & \underline{0} \end{bmatrix} \begin{bmatrix} \underline{I} & -(\underline{A} - \underline{B}\,\underline{R} - \eta\underline{I})^{-1}\underline{B}\,\underline{F} \\ \underline{0} & \underline{I} \end{bmatrix} =$$

$$= [\underline{r}^T,\underline{q}^T] \begin{bmatrix} \underline{A} - \underline{B}\,\underline{R} - \eta\underline{I} & \underline{0} \\ \underline{C} & \underline{G}_w(\eta) \end{bmatrix} = \underline{0}^T \qquad (4.4)$$

(the inverse $(\underline{A} - \underline{B}\,\underline{R} - \eta\underline{I})^{-1}$ exists, since η is not a closed-loop eigenvalue). From eq. (4.4) we have

$$\underline{q}^T \cdot \underline{G}_w(\eta) = \underline{0}^T. \qquad (4.5)$$

which must be satisfied for *all obtainable* $\underline{G}_w(s)$. Putting the *desired* $\underline{G}_w(s)$ (eq. (1.5)) in eq. (4.5) and denoting

---

[2] From the controllability criterion of Hautus follows: rank $[\underline{A} - \eta\underline{I},\underline{B}]$ = n, see Föllinger (1985).

the elements of $\underline{q}^T$ by $q_1,...,q_m$ we can write

$$\underline{q}^T\underline{G}_w(\eta) = \left[q_1 g_{11}(\eta) + q_j g_{j1}(\eta),...,q_j g_{jj}(\eta),..., \right.$$

$$\left. q_m g_{mm}(\eta) + q_j g_{jm}(\eta) \right] = \underline{0}^T. \qquad (4.6)$$

Suppose now $\underline{G}_w(s)$ to injure condition (4.1), i.e. $q_j = 0$. Then, with a (always existing) $q_i \neq 0$, the ith element of $\underline{q}^T\underline{G}_w(\eta)$ reads $q_i g_{ii}(\eta)$, an expression which can never be equal to zero since $g_{ii}(\eta) \neq 0$, $i = 1,...,m$, $i \neq j$ from eq. (2.5). Hence, with $q_j = 0$ and the transfer-function matrix (1.5), eq. (4.6) cannot be satisfied. Therefore, partially decoupling matrices $\underline{R}$ and $\underline{F}$ can only exist if the coupled channel in the desired $\underline{G}_w(s)$ is chosen such that $\underline{q}^T\underline{e}_j \neq 0$. The proof of sufficiency of condition (4.1) is here omitted due to space.

From eq. (4.6) we immmediately find the relation $g_{jj}(\eta) = 0$ i.e. the invariant zero η appears in the partially decoupled matrix $\underline{G}_w(s)$ as zero of the jth diagonal element. This element reads

$$g_{jj}(s) = \frac{(s-\eta)}{(s-\lambda_{j1})\cdot\,...\,\cdot(s-\lambda_{j\delta_j+1})} \cdot \frac{\prod\limits_{\nu=1}^{\delta_j+1}(-\lambda_{j\nu})}{(-\eta)}. \qquad (4.7)$$

The time response $y_j(t)$ to a unit step function $w_j(t) = \sigma(t)$ shows the *undershooting phenomenon* (caused by the numerator $s-\eta$ with $\eta > 0$) whereas the elements $g_{ii}(s)$, $i \neq j$ guarantee an agreeable behaviour of the referring $y_i(t)$ (by eq. 2.5).

5. CHOICE OF THE REMAINING DESIGN PARAMETERS

Not only the arbitrarily prespecified eigenvalues but also the parameters a_{ik} determine the controller (see eq. (3.2)). A suitable choice of these a_{ik} demands a relation describing the influence of the a_{ik} on the elements $g_{ji}(s)$, $i \neq j$ of $\underline{G}_w(s)$. If we assume $a_{ik} = 0$ the nondiagonal elements of $\underline{G}_w(s)$ are

$$g_{ji}(s) = \frac{s \cdot f_{ji}}{(s-\lambda_{j1})\cdot\,...\,\cdot(s-\lambda_{j\delta_j+1})}, \quad \begin{array}{l} i = 1,...,m, \\ i \neq j, \end{array} \qquad (5.1)$$

where the "s" in the numerator avoids steady state error. Only the poles $\lambda_{j1},...,\lambda_{j\delta_j+1}$ appear in $g_{ji}(s)$ since all other eigenvalues are strictly connected to their rows $i \neq j$ (see section 3). Because of the difference order δ_j the degree of the numerator is equal to *one*, hence f_{ji} is not depending on s. Evaluating eq. (4.6) element by element with regard to eq. (5.1) we find

$$f_{ji} = -\frac{1}{\eta}\frac{q_i}{q_j} g_{ii}(\eta) \cdot (\eta - \lambda_{j1}) \cdot\,...\,\cdot (\eta - \lambda_{j\delta_j+1}) \qquad (5.2)$$

and obviously $g_{ji}(s) \equiv 0$ if $q_i = 0$. $\qquad (5.3)$

In words: if an element q_i of $\underline{q}^T$ is equal to zero, $g_{ji}(s) \equiv 0$ can be achieved by choosing $a_{ik} = 0$, $k = 1,...,\delta_i$. If $\underline{q}^T$ contains only *one* element unequal to zero, *complete decoupling* can be achieved by choosing $a_{ik} = 0$, $i = 1,...,m$, $i \neq j$, $k = 1,...,\delta_i$. In this special case the *partially* decoupling $\underline{G}_w(s)$ from eq. (1.5) is reduced to the *completely* decoupling $\underline{G}_w(s)$ (eq. 1.4) with the diagonal elements $g_{ji}(s)$, $i \neq j$ following eq. (2.5) and the element $g_{jj}(s)$ from eq. (4.7). Figuratively speaking the influence of η is restricted to the output y_j ("non–interconnecting zero", Koussiouris (1970)) and allows complete decoupling.

But back to the general case, where the formula for the non–diagonal elements of $\underline{G}_w(s)$ reads (proved by Windpassinger 1989):

$$g_{ji}(s) = \frac{-s}{\displaystyle\prod_{\nu=1}^{\delta_{j+1}} (s-\lambda_{j\nu})} \left[\sum_{k=1}^{\delta_i} \left\{ \frac{r_{ik}}{(s-\lambda_{ik})} \cdot \frac{a_{ik}(s-\eta)}{\lambda_{ik}(\eta-\lambda_{ik})} \cdot \prod_{\mu=1}^{\delta_j+1} (\lambda_{ik}-\lambda_{j\mu}) \right\} + \frac{1}{\eta} \frac{q_i}{q_j} g_{ii}(\eta) \prod_{\mu=1}^{\delta_j+1} (\eta-\lambda_{j\mu}) \right] \begin{array}{l} i = 1,\dots,m, \\ , i \neq j. \end{array}$$

$$(5.4)$$

Eq. (5.1) results from eq. (5.4) for $a_{ik} = 0$. The r_{ik} in eq. (5.4) are the residues of the known functions $g_{ii}(s)$, defined via

$$g_{ii}(s) = \sum_{k=1}^{\delta_i} \frac{r_{ik}}{s-\lambda_{ik}}, \quad \begin{array}{l} i = 1,\dots,m, \\ i \neq j. \end{array}$$

$$(5.5)$$

By suitable choice of the a_{ik} the numerator degrees of the functions $g_{ji}(s)$ can be minimized, causing low transmission at high frequencies. Alternatively one can minimize the energy function

$$J = \int_0^\infty h_{ji}^2(t)dt .$$

$$(5.6)$$

$h_{ji}(t)$ is the response of $g_{ji}(s)$ to a unit step function.

Systems with *several*, say γ, "unstable" invariant zeros can be treated by the presented method as well. If there exists a channel j such that $q_1^T \underline{e}_j \neq 0,\dots,q_\gamma^T \underline{e}_j \neq 0$, a design with coupling only in the jth row of $\underline{G}_w(s)$ is possible, otherwise the number of coupled channels increases. The equations for the controller calculation (3.2), (3.3), (3.6) are easy to modify, details of the non–diagonal elements $g_{ji}(s)$ are omitted here.

Complete decoupling will occur if *all* q_i^T, $i = 1,\dots,\gamma$ satisfy the mentioned condition, i.e. if they contain only one non–zero element respectively.

6. COMPLETE DECOUPLING BY DYNAMIC STATE–FEEDBACK

From the last section we know that complete decoupling without compensation of the "unstable" zero η is possible if q^T from eq. (4.2) contains just *one* element q_j

$\neq 0$. Actually let's suppose the first α elements of q^T to be unequal to zero. The idea of complete and stable decoupling is the following: add a dynamic system $\underline{A}'$, $\underline{B}'$, $\underline{C}'$ of order $\alpha-1$ to the existing $\underline{A}$, $\underline{B}$, $\underline{C}$ such that the resulting system $\tilde{A}$, $\tilde{B}$, $\tilde{C}$ has the zero η with multiplicity α and corresponding vectors $q^T,\dots,q_\alpha^T$ satisfying the above–mentioned condition for *complete* decoupling by *constant* state–feedback. An appropriate structure of the extended system is shown in fig. 1 (see also Cremer 1973).

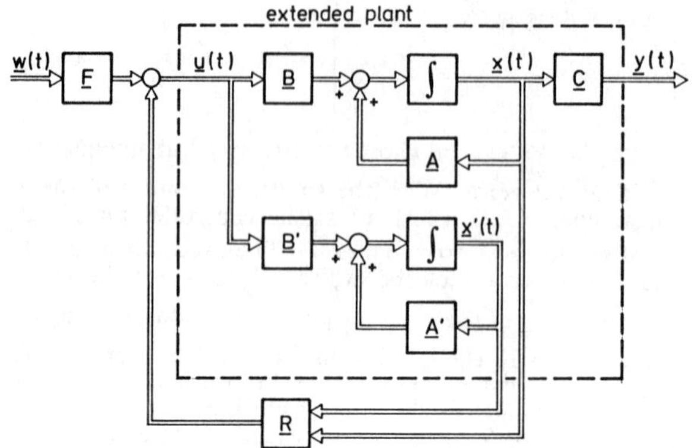

Fig. 1: Extended plant with controller $\underline{R}$ and prefilter $\underline{F}$

The state space description of the extended plant $\tilde{\underline{A}}$, $\tilde{\underline{B}}$, $\tilde{\underline{C}}$, is

$$\tilde{A} = \begin{bmatrix} \underline{A}' & 0 \\ \underline{0} & \underline{A} \end{bmatrix} \begin{array}{l} \alpha-1 \\ n \end{array} , \quad \tilde{B} = \begin{bmatrix} \underline{B}' \\ \underline{B} \end{bmatrix} \begin{array}{l} \alpha-1 \\ n \end{array} , \quad (6.1)$$

$$\tilde{C} = [\, \underline{C}', \underline{C} \,] \quad \text{with} \quad \underline{C}' = \underline{0} .$$

With the choice $\underline{A}' = \eta\underline{I}$ the invariant zeros of the extended system are the solutions s of

$$\det \begin{bmatrix} \tilde{\underline{A}}-s\underline{I} & \tilde{\underline{B}} \\ \tilde{\underline{C}} & \underline{0} \end{bmatrix} = \det \begin{bmatrix} \eta\underline{I}-s\underline{I} & \underline{0} & \underline{B}' \\ \underline{0} & \underline{A}-s\underline{I} & \underline{B} \\ \underline{0} & \underline{C} & \underline{0} \end{bmatrix} =$$

$$= \det(\eta\underline{I}-s\underline{I}) \det \begin{bmatrix} \underline{A}-s\underline{I} & \underline{B} \\ \underline{C} & \underline{0} \end{bmatrix} = 0 . \quad (6.2)$$

Obviously the multiplicity of the zero $s = \eta$ is equal to α, as desired, since η is assumed to be a single zero of the original system $\underline{A}$, $\underline{B}$, $\underline{C}$.

If we can find a $\underline{B}'$ such that α nontrivial solutions $\left[\, \underline{r}_i^T, \underline{q}_i^T \,\right]$, $i = 1,\dots,\alpha$ of

$$\underbrace{\left[\, \underline{r}_{iL}^T , \underline{r}_{iR}^T , \underline{q}_i^T \,\right]}_{\underline{r}_i^T} \begin{bmatrix} \eta\underline{I}-\eta\underline{I} & \underline{0} & \underline{B}' \\ \underline{0} & \underline{A}-\eta\underline{I} & \underline{B} \\ \underline{0} & \underline{C} & \underline{0} \end{bmatrix} = \underline{0} \quad (6.3) \quad , i=1,\dots,\alpha$$

satisfying $\underline{q}_i^T = \underline{e}_i^T$ exist, then we can achieve complete decoupling of the system $\tilde{\underline{A}}$, $\tilde{\underline{B}}$, $\tilde{\underline{C}}$ as described in section 5. In order to determine $\underline{B}'$ we combine the α equations (6.3) to

$$\alpha[\, \underline{R}_L , \underline{R}_R , I , \underline{0} \,] \cdot \begin{bmatrix} \underline{0} & \underline{0} & \underline{B}' \\ \underline{0} & \underline{A}-\eta\underline{I} & \underline{B} \\ \underline{0} & \underline{C} & \underline{0} \end{bmatrix} \begin{array}{l} \alpha-1 \\ n \\ m \end{array} = \underline{0}$$

$$(6.4)$$

$$\text{with } R_L = \begin{bmatrix} \underline{r}_{1L}^T \\ \vdots \\ \underline{r}_{\alpha L}^T \end{bmatrix} , \quad \underline{R}_R = \begin{bmatrix} \underline{r}_{1R}^T \\ \vdots \\ \underline{r}_{\alpha R}^T \end{bmatrix} . \quad (6.5)$$

Evaluation of the second column block

$$\underline{R}_R \cdot (\underline{A} - \eta\underline{I}) + [\,\underline{I}\,,\,\underline{0}\,]\,\underline{C} = \underline{0} \qquad (6.6)$$

leads to a solution

$$\underline{R}_R = -\,[\,\underline{I}\,,\,\underline{0}\,]\cdot\underline{C}\cdot(\underline{A}-\eta\underline{I})^{-1} \qquad (6.7)$$

if η is not eigenvalue of the original system[3].
Evaluation of the third column block yields

$$\underline{R}_L \cdot \underline{B}' = -\,\underline{R}_R\underline{B} \,. \qquad (6.8)$$

A solution $\underline{R}_L$ of this equation exists if

$$\mathrm{Rank}\begin{bmatrix} \underline{B}' \\ \underline{R}_R\,\underline{B} \end{bmatrix} = \mathrm{Rank}\,[\,\underline{B}'\,]\,, \qquad (6.9)$$

i.e. for solvability of eq. (6.8) the α rows of $\underline{R}_R\underline{B}$ must
be linear combinations of the $\alpha-1$ rows of $\underline{B}'$. If we can
show that Rank $[\,\underline{R}_R\cdot\underline{B}\,] < \alpha$ then we can choose $\underline{B}'$
simply equal to the linear independent rows of $\underline{R}_R\cdot\underline{B}$
and hence guarantee the existence of $\underline{R}_L$, $\underline{R}_R$ satisfying
eq. (6.4).

Proof of Rank $[\,\underline{R}_R\cdot\underline{B}\,] < \alpha$:
We first multiply eq. (4.2) with a suitable regular matrix

$$\begin{bmatrix}\underline{r}^T,\underline{q}^T\end{bmatrix}\begin{bmatrix}\underline{A}-\eta\underline{I} & \underline{B} \\ \underline{C} & \underline{0}\end{bmatrix}\cdot\begin{bmatrix}\underline{I} & -(\underline{A}-\eta\underline{I})^{-1}\underline{B} \\ \underline{0} & \underline{I}\end{bmatrix} = \begin{bmatrix}\underline{r}^T,\underline{q}^T\end{bmatrix}\cdot$$

$$\begin{bmatrix}\underline{A}-\eta\underline{I} & \underline{0} \\ \underline{C} & \underline{C}(\eta\underline{I}-\underline{A})^{-1}\underline{B}\end{bmatrix} = \underline{0}^T \qquad (6.10)$$

and find that $\underline{q}^T\underline{C}(\eta\underline{I}-\underline{A})^{-1}\underline{B} = \underline{0}^T$. Since the last $m-\alpha$
elements of $\underline{q}^T$ are equal to zero, we can write

$$\underline{q}^T\cdot[\underline{I},\,\underline{0}]\underline{C}(\eta\underline{I}-\underline{A})^{-1}\underline{B} = \underline{0}^T\,, \qquad (6.11)$$

which shows that Rank $\left\{[\underline{I},\,\underline{0}]\underline{C}(\eta\underline{I}-\underline{A})^{-1}\underline{B}\right\} < \alpha$. From
eq. (6.6) we have just $\underline{R}_R\underline{B} = [\underline{I},\,\underline{0}]\underline{C}(\eta\underline{I}-\underline{A})^{-1}\underline{B}$ and
thus Rank $[\underline{R}_R\underline{B}] < \alpha$.

7. EXAMPLE

Consider the simplified model of a rear–axle–gear test
stand

$$\underline{A} = \begin{bmatrix} -0.00633 & -2.49 & 0 & 0 \\ 95.61 & 0 & -48.2 & -48.2 \\ 0 & 0.456 & -0.0137 & 0 \\ 0 & 0.456 & 0 & -0.0137 \end{bmatrix},$$

$$\underline{B} = \begin{bmatrix} 2.48 & 0 & 0 \\ -23.2 & -2.0 & -2.0 \\ 0 & 0 & -0.462 \\ 0 & -0.462 & 0 \end{bmatrix},\quad \underline{C}^T = \begin{bmatrix} 0 & 0 & 0 \\ 1 & 0 & 0 \\ 0 & 1 & 0 \\ 0 & 0 & 1 \end{bmatrix}$$

with the difference orders $\delta_1 = \delta_2 = \delta_3 = 1$ and the invariant zero $\eta = +10.2$. From eq. (4.2) we calculate $\underline{q}^T$
$= [\,1,\,0.1043,\,0.1043\,]$ which allows prescription of a
$\underline{G}_w(s)$ with coupling in row 1,2 or 3. We choose the
transfer–function matrix with coupling in channel 1

$$\underline{G}_w(s) = \begin{bmatrix} \dfrac{0.49\,(s-10.2)}{(s^2+4s+5)} & g_{12}(s) & g_{13}(s) \\ 0 & \dfrac{10}{s+10} & 0 \\ 0 & 0 & \dfrac{10}{s+10} \end{bmatrix}\,.$$

From eq. (5.4) we find the non–diagonal elements

$$g_{1i}(s) = \frac{s^2(3.2a_{i1} - 0.76) + s(-3.28a_{i1} - 7.6)}{(s^2+4s+5)(s+10)}\,,\quad i = 2,3.$$

$a_{i1} = 0$ yields (design A)

$$g_{12}(s) = g_{13}(s) = \frac{-0.76s}{(s^2+4s+5)}$$

whereas $a_{i1} = -0.215$ minimizes the cost function (5.6)
and yields (design B)

$$g_{12}(s) = g_{13}(s) = \frac{-1.45s^2 - 0.51s}{(s^2+4s+5)(s+10)}\,.$$

The controller $\underline{R}$ is calculated from eq. (2.1) via eq.
(3.2) with $\lambda_{21} = -10$, $a_{21} = -0,215$, $\lambda_{31} = -10$, $a_{31} =$
$-0,215$ and via eq. (3.6) with $\lambda_{11} = -2+j$, $\lambda_{12} = -2-j$.

$$\underline{R} = \begin{bmatrix} -2.5 & -0.27 & 3.2 & 3.2 \\ 0 & -0.99 & 0 & -21.6 \\ 0 & -0.99 & -21.6 & 0 \end{bmatrix},$$

$$\underline{F} = \begin{bmatrix} 0.021 & 1.93 & 1.93 \\ 0 & 0 & -21.6 \\ 0 & -21.6 & 0 \end{bmatrix},$$

$\underline{F}$ is the precompensator from eq. (2.14).

The design of a completely decoupling controller requires the extension of the system (section 6) since $\underline{q}^T$
contains *more* than one non–zero element (see eq. 5.3).
In order to determine $\underline{B}'$ of the added system we calculate from eq. (6.1) (it is $\alpha = 3$):

$$\underline{R}_R\cdot\underline{B} = \underline{C}\cdot(\eta\underline{I}-\underline{A})^{-1}\cdot\underline{B} = \begin{bmatrix} 0 & 0.005 & 0.005 \\ 0 & 2.1\cdot10^{-4} & -0.045 \\ 0 & -0.045 & 2.1\cdot10^{-4} \end{bmatrix}$$

and choose $\underline{B}' = \begin{bmatrix} 0 & 1 & 0 \\ 0 & 0 & 1 \end{bmatrix}$ which quarantees solvability of eq. (6.8). The extended system $\tilde{A}$, $\tilde{B}$, $\tilde{C}$ (eq.
6.1) with $A' = \mathrm{diag}\,[10.2,\,10.2]$ can be completely decoupled (special case of sections 3, 4, 5) by constant
state–feedback via the solutions of eq. (3.6)

$$\begin{bmatrix}\underline{A}-\lambda_{jk}\underline{I} & \underline{B} \\ \underline{C} & \underline{0}\end{bmatrix}\cdot\begin{bmatrix}\underline{v}_{jk} \\ -\underline{p}_{jk}\end{bmatrix} = \begin{bmatrix}\underline{0} \\ \underline{e}_j\end{bmatrix}\quad\begin{matrix}j = 1,\dots,3 \\ k = 1,2\end{matrix}$$

and eq. (2.1). The choice $\lambda_{11} = -2+j$, $\lambda_{12} = -2-j$, λ_{21}
$= \lambda_{31} = -10+5j$, $\lambda_{22} = \lambda_{32} = -10-5j$ yields (design C):

$$\underline{G}_w(s) =$$

$$\begin{bmatrix} \dfrac{0.49\,(s-10.2)}{(s^2+4s+5)} & 0 & 0 \\ 0 & \dfrac{-12.25(s-10.2)}{s^2+20s+125} & 0 \\ 0 & 0 & \dfrac{-12.25(s-10.2)}{s^2+20s+125} \end{bmatrix}\,.$$

---

[3] This assumption is not restricitive, since a pole $\lambda = \eta$
can be shifted before the design ($\underline{A}$, $\underline{B}$, $\underline{C}$ is stabilisable!).

Figures 2-4 show the time responses of the closed-loop system to unit-step functions $\sigma(t)$. The disadvantage of the *completely* decoupled system is the undershooting of all outputs. In the *partially* decoupled system the influence of the inputs w_2 and w_3 on output y_1 is relatively small (designs A, B), i.e. design B is to prefer in the considered example.

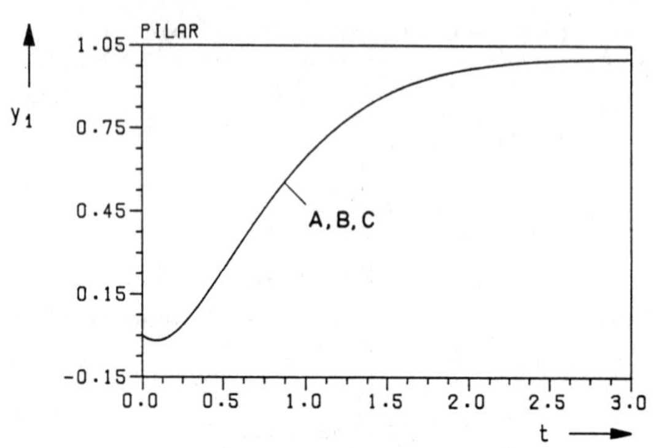

Fig. 2: $y_1(t) = g_{11}(t) \star \sigma(t)$ for design A, B and C.

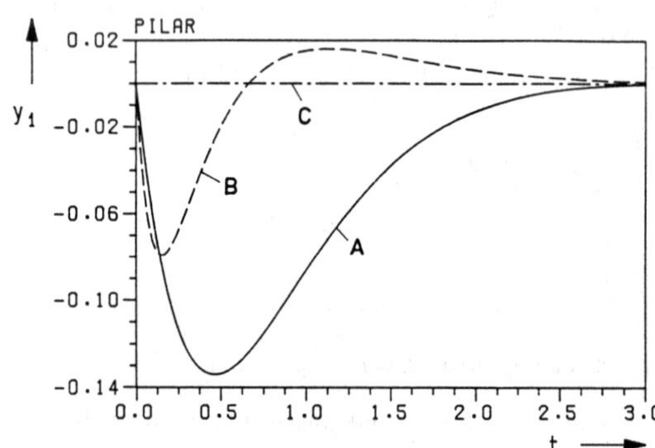

Fig. 3: $y_1(t) = g_{12}(t) \star \sigma(t)$, design A, B, C.

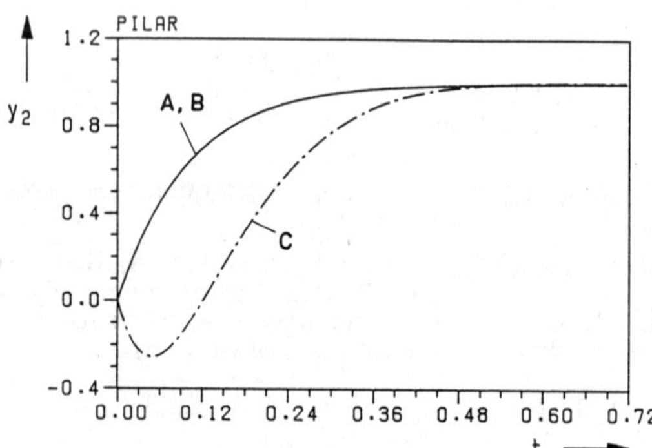

Fig. 4: $y_2(t) = g_{22}(t) \star \sigma(t)$, design A, B, C.

8. CONCLUSIONS

The introduced state-space methods allow *partial* decoupling by *constant* state-feedback or *complete* decoupling by *dynamic* state-feedback in the case of a non-minimum phase system having $n-\delta$ finite invariant zeros. The practical application of such controller is above all the realization of model based feedforward controllers generating the input signal $\underline{u}(t)$, furtheron multivariable compensators can be realized.

The design steps of section 3 even allow partial decoupling of systems having *less* than $n-\delta$ invariant zeros. However the choice of the coupled channel(s) is not possible by the condition (4.1) in this case and will be treated in future works.

REFERENCES

Cremer, M. (1973). Festlegen der Pole und Nullstellen bei der Synthese linearer entkoppelter Mehrgrößenregelkreise. Regelungstechnik und Prozeß-Datenverarbeitung, 21, p. 146-150, p. 195-199.

Falb, P.L. and Wolovich, W.A. (1967). Decoupling in the Design and Synthesis of Multivariable Control Systems. IEEE Trans. Automat. Control, 12, p. 651-659.

Föllinger, O. (1985). Regelungstechnik. 5. Auflage, Hüthig-Verlag Heidelberg.

Koussiouris, T. (1970). A frequency domain approach to the block decoupling problem. Int. J. Control, 32, p. 443-464.

Litz, L. (1979). Reduktion der Ordnung linearer Zustandsraummodelle mittels modaler Verfahren. Hochschulverlag, Freiburg.

MacFarlane, A.G.J. and Karcanias, N. (1976). Poles and Zeros of linear multivariable systems: a survey of the algebraic, geometric and complex-variable theory. Int. J. Control, 24, p. 33-74.

Roppenecker, G. (1986). On Parametric State Feedback Design. Int. J. Control, 43, p. 793-804.

Roppenecker, G. and Lohmann, B. (1988). Vollständige Modale Synthese von Entkopplungsregelungen. Automatisierungstechnik, 36, p. 434-441.

Roppenecker, G. and Preuss, H.-P. (1982). Nullstellen und Pole linearer Mehrgrößensysteme. Regelungstechnik, 30, p. 219-225, 255-263.

Sogaard-Andersen, P. (1986). Eigenstructure and residuals in multivariable state-feedback design. Int. J. Control, 44, p. 427-439.

Windpassinger, R. (1989). Implementierung und Erprobung eines modalen Entwurfsverfahrens zur Führungsentkopplung. Diplomarbeit, Institut für Regelungs- und Steuerungssysteme, Universität Karlsruhe.

AUTHOR INDEX

KEYWORD INDEX

CONTENTS OF VOLUMES I - VI

VOLUME I

VOLUME II

VOLUME III

VOLUME IV

VOLUME V

VOLUME VI

<h1 style="text-align:center">CUMULATIVE AUTHOR INDEX, VOLUMES I - VI</h1>

CUMULATIVE KEYWORD INDEX, VOLUMES I-VI

Factory automation, V:327
Failure detection, II:61, 67, IV:13, 19, 89,
    303, 361, 367, 373
Farming, VI:287
Fast PID-controller optimization, II:443
Fault detection, IV:1, 75, 459
Fault-tolerance, IV:129, 303
Fedbatch process, VI:253
Feedback, III:287
Feedback control, I:173, 235, 277, 339, III:165,
    213, 351, 459, V:477, VI:143
Feedforward, II:421, V:65
Feedforward control, V:127, VI:143
Fermentation processes, VI:241, 265
Field oriented control, IV:539
Field theory, VI:535
Finite automata, III:507, 523
Finite spectrum assignment, I:187
Finite wordlength effects, II:17
First passage times, II:43
Fisher information matrix, VI:241
Fixed accuracy identification, II:179
Fixed modes, III:287
Fixed-interval smoother, II:117
Flexible jointed robots, V:175
Flexible joints, III:361
Flexible links, III:361
Flexible manufacturing, I:71, IV:123,
    V:27, 31, 47, 53
Flexible manufacturing systems, V:37, 181,
    VI:477
Flexible structure, III:111
Flexible system, III:249
Flight control, III:75, 87
Flotation modelling, IV:31
Flow control, III:477
Flowmeters, IV:483, 557
Force control, IV:335, V:169, 199, 255,
    275
Forecasting, VI:325
Four block problem, I:229
4-quadrant torque-speed curves, IV:557
Fractal boundary, III:379
FRENDA project, IV:289
Frequency and voltage control, VI:121
Frequency control, II:331, VI:115
Frequency domain, I:223, II:191, 209,
    III:23, 263, IV:89, 373, VI:417
Frequency domain design method, III:107
Frequency estimation, IV:483
Frequency response, II:209
Frequency response design, II:427
Friction, V:223
Full state feedback, IV:317
Function generators, VI:291
Functional analysis, II:13

Functional programming, V:349, VI:435
Fuzzy control, IV:193, 205
Fuzzy integral, IV:57
Fuzzy modeling, IV:199
Fuzzy moving average models, II:111
Fuzzy program, IV:69
Fuzzy reasoning, V:105
Fuzzy set theory, II:111, IV:63, VI:61
Fuzzy sets, IV:199
Fuzzy stochastic systems, II:111
Fuzzy systems, IV:57, 175

Gain control, II:331
Gain scheduling, I:173
Game theory, VI:365, 393
General embedded problem, VI:359
General least-squares estimation, II:179
General orthogonal polynomials, II:129
Generalized predictive control, II:49
Generalized state-space systems, I:247
Generalized stochastic Petri Nets, V:321
Geometric approach, IV:505
Geometric conditions, I:333
GKS, IV:253
Global asymptotic stability, I:205
Global convergence, II:161, 279
Global models, VI:505
Goodness of fit, II:185
Graph theory, III:511
Graphical interface, V:77
Gross-fine manipulation, V:191
Ground support systems, III:29, 47
Group, III:269
Group-projects, I:39
Guidance systems, III:11, V:111
Gydrodines, III:11

Hard real time, V:379
Harmonic relation, VI:371
Heading measurement, IV:471
Heat systems, II:461
Heliostats, VI:321
Heuristic programming, IV:111
Hierarchially intelligent control, IV:389
Hierarchical control, V:181
Hierarchical control architectures, V:321
Hierarchical control structure, V:297
Hierarchical decision making, I:87, III:47,
    427, V:473, VI:309, 379,
Hierarchical systems, I:131, III:529, IV:51,
    165, V:71, 281, 373, 429
Hierarchically intelligent control, V:31
High level languages, IV:123
Higher level adaptation mechanism, II:325
Hot strip, VI:143
Human centred, VI:487

SYMPOSIA VOLUMES

ADALI & TUNALI: Microcomputer Application in Process Control

AKASHI: Control Science and Technology for the Progress of Society, World Congress 1981, 7 volumes

ALBERTOS & DE LA PUENTE: Components, Instruments and Techniques for Low Cost Automation and Applications

ALONSO CONCHEIRO: Real Time Digital Control Applications *

AMOUROUX & EL JAI: Control of Distributed Parameter Systems (1989)

ATHERTON: Multivariable Technological Systems

BABARY & LE LETTY: Control of Distributed Parameter Systems (1982)

BALCHEN: Automation and Data Processing in Aquaculture

BANKS & PRITCHARD: Control of Distributed Parameter Systems (1977)

BAOSHENG HU: Analysis, Design and Evaluation of Man-Machine Systems (1989)

BARKER & YOUNG: Identification and System Parameter Estimation (1985)

BASANEZ, FERRATE & SARIDIS: Robot Control "SYROCO '85"

BASAR & PAU: Dynamic Modelling and Control of National Economies (1983)*

BAYLIS: Safety of Computer Control Systems (1983)*

BEKEY & SARIDIS Identification and System Parameter Estimation (1982)

BINDER & PERRET: Components and Instruments for Distributed Computer Control Systems*

CALVAER: Power Systems, Modelling and Control Applications

Van CAUWENBERGHE: Instrumentation and Automation in the Paper, Rubber, Plastics and Polymerisation Industries (1980)*(1983)

CHEN HAN-FU: Identification and System Parameter Estimation (1988)

CHEN ZHEN-YU: Computer Aided Design in Control Systems (1988)

CHRETIEN: Automatic Control in Space (1985)

CHRISTODULAKIS: Dynamic Modelling and Control of National Economies (1989)

COBELLI & MARIANI: Modelling and Control in Biomedical Systems

CUENOD: Computer Aided Design of Control Systems*

DA CUNHA: Planning and Operation of Electric Energy Systems

DE CARLI: Low Cost Automation

De GIORGIO & ROVEDA: Criteria for Selecting Appropriate Technologies under Different Cultural, Technical and Social Conditions*

DEVANATHAN: Intelligent Tuning and Adaptive Conrol

DUBUISSON: Information and Systems

EHRENBERGER: Safety of Computer Control Systems (SAFECOMP'88)

ELLIS: Control Problems and Devices in Manufacturing Technology (1980)*

FERRATE & PUENTE: Software for Computer Control (1982)

FLEISSNER: Systems Approach to Appropriate Technology Transfer

FLORIAN & HAASE: Software for Computer Control (1986)

GEERING & MANSOUR: Large Scale Systems: Theory and Applications (1986)

GENSER et al: Control in Transportation Systems (1986)

GERTLER & KEVICZKY: A Bridge Between Control Science and Technology, World Congress 1984, 6 volumes*

GHONAIMY: Systems Approach for Development (1977)

HAIMES & KINDLER: Water and Related Land Resource Systems

HARDT: Information Control Problems in Manufacturing Technology (1982)

HERBST: Automatic Control in Power Generation Distribution and Protection

HRUZ & CICEL: Automatic Measurement and Control in Woodworking Industry-Lignoautomatica'86

HUSSON: Advanced Information Processing in Automatic Control

ISERMANN: Automatic Control, World Congress 1987, 10 volumes

ISERMANN: Identification and System Parameter Estimation (1979)

ISERMANN & KALTENECKER: Digital Computer Applications to Process Control*

ISIDORI: Nonlinear Control Systems Design

JAAKSOO & UTKIN: Automatic Control, World Congress 1990, 6 volumes

JANSSEN, PAU & STRASZAK: Dynamic Modelling and Control of National Economies (1980)

JELLALI: Systems Analysis Applied to Management of Water Resources

JOHANNSEN & RIJNSDORP: Analysis, Design, and Evaluation of Man-Machine Systems*

JOHNSON: Adaptive Systems in Control and Signal Processing

JOHNSON: Modelling and Control of Biotechnological Processes

KAYA & WILLIAMS: Instrumentation and Automation in the Paper, Rubber, Plastics and Polymerization Industries (1986)

KLAMT & LAUBER: Control in Transportation Systems (1984)*

KOPACEK & GENSER: Skill Based Automated Production

KOPACEK, TROCH & DESOYER: Theory of Robots

KOPPEL: Automation in Mining, Mineral and Metal Processing (1989)

KUMMEL: Adaptive Control of Chemical Processes (ADCHEM'88)

LARSEN & HANSEN: Computer Aided Design in Control and Engineering Systems

LEININGER: Computer Aided Design of Multivariable Technological Systems*

LEONHARD: Control in Power Electronics and Electrical Drives (1977)*

LESKIEWICZ & ZAREMBA: Pneumatic and Hydraulic Components and Instruments in Automatic Control*

LINKENS & ATHERTON: Trends in Control and Measurement Education

MACLEOD & HEHER: Software for Computer Control (SOCOCO'88)

MAHALANABIS: Theory and Application of Digital Control

MANCINI, JOHANNSEN & MARTENSSON: Analysis, Design and Evaluation of Man-Machine Systems (1985)

MARTOS, PAU & ZIERMANN: Dynamic Modelling and Control of National Economies (1986)

McGREAVY: Dynamics and Control of Chemical Reactors and Distillation Columns

MLADENOV: Distributed Intelligence Systems: Methods and Applications

MUNDAY: Automatic Control in Space (1979)

NAJIM & ABDEL FATTAH: System Approach for Development (1980)*

NIEMI: A Link Between Science and Applications of Automatic Control, World Congress 1978, 4 volumes*

NISHIKAWA & KAYA: Energy Systems, Management and Economics

NISHIMURA: Automatic Control in Aerospace

NORRIE & TURNER: Automation for Mineral Resource Development

NOVAK: Software for Computer Control (1979)

O'SHEA & POLIS: Automation in Mining, Mineral and Metal Processing (1980)

OSHIMA: Information Control Problems in Manufacturing Technology (1977)

PAUL: Digital Computer Applications to Process Control (1985)

PERRIN: Control, Computers, Communications in Transportation

PONOMARYOV: Artificial Intelligence

PUENTE & NEMES: Information Control Problems in Manufacturing Technology (1989)

RAMAMOORTY: Automation and Instrumentation for Power Plants

RANTA: Analysis, Design and Evaluation of Man-Machine Systems (1988)

RAUCH: Applications of Nonlinear Programming to Optimization and Control*

RAUCH: Control of Distributed Parameter Systems (1986)*

REINISCH & THOMA: Large Scale Systems: Theory and Applications (1989)

REMBOLD: Robot Control (SYROCO'88)

RIJNSDORP: Case Studies in Automation Related to Humanization of Work

RIJNSDORP et al: Dynamics and Control of Chemical Reactors (DYCORD'89)

RIJNSDORP, PLOMP & MÖLLER: Training for Tomorrow-Educational Aspects of Computerized Automation

ROOS: Economics and Artificial Intelligence
SANCHEZ: Fuzzy Information, Knowledge Representation and
  Decision Analysis*
SAWARAGI & AKASHI: Environmental Systems Planning, Design
  and Control
SINHA & TELKSNYS: Stochastic Control
SMEDEMA: Real Time Programming (1977)*
STRASZAK: Large Scale Systems: Theory and Applications (1983)
SUBRAMANYAM: Computer Applications in Large Scale Power
  Systems
TAL': Information Control Problems in Manufacturing Technology
  (1986)
TITLI & SINGH: Large Scale Systems: Theory and Applications
  (1980)*

TROCH, KOPACEK & BREITENECKER: Simulation of Control
  Systems
UHI AHN: Power Systems and Power Plant Control (1989)
VALADARES TAVARES & EVARISTO DA SILVA: Systems
  Analysis Applied to Water and Related Land Resources
Van WOERKOM: Automatic Control in Space (1982)
WANG PINGYANG: Power Systems and Power Plant Control
WESTERLUND: Automation in Mining, Mineral and Metal
  Processing (1983)
YANG JIACHI: Control Science and Technology for Development
YOSHITANI: Automation in Mining, Mineral and Metal Processing
  (1986)
ZWICKY: Control in Power Electronics and Electrical Drives (1983)

WORKSHOP VOLUMES

ASTROM & WITTENMARK: Adaptive Systems in Control and Signal
  Processing
BOULLART et al: Industrial Process Control Systems
BRODNER: Skill Based Automated Manufacturing
BULL: Real Time Programming (1983)*
BULL & WILLIAMS: Real Time Programming (1985)
CAMPBELL: Control Aspects of Prosthetics and Orthotics*
CHESTNUT: Contributions of Technology to International Conflict
  Resolution (SWIIS)
CHESTNUT et al: International Conflict Resolution using Systems
  Engineering (SWIIS)
CHESTNUT et al: Supplemental Ways for Improving International
  Stability*
CICHOCKI & STRASZAK: Systems Analysis Applications to
  Complex Programs
CRESPO & DE LA PUENTE: Real Time Programming (1988)
CRONHJORT: Real Time Programming (1978) *
DI PILLO: Control Applications of Nonlinear Programming and
  Optimization
ELZER: Experience with the Management of Software Projects
GELLIE & TAVAST: Distributed Computer Control Systems (1982)*
GENSER et al: Safety of Computer Control Systems (SAFECOMP'89)
GOODWIN: Robust Adaptive Control
HAASE: Real Time Programming (1980)
HALME: Modelling and Control of Biotechnical Processes*
HARRISON: Distributed Computer Control Systems (1979)
HASEGAWA: Real Time Programming (1981)*
HASEGAWA & INOUE: Urban, Regional and National Planning -
  Environmental Aspects
JANSEN & BOULLART: Reliability of Instrumentation Systems for
  Safeguarding and Control
KOTOB: Automatic Control in Petroleum, Petrochemical and
  Desalination Industries
LANDAU, TOMIZUKA & AUSLANDER: Adaptive Systems in
  Control and Signal Processing
LAUBER: Safety of Computer Control Systems (1979)
LOTOTSKY: Evaluation of Adaptive Control Strategies in Industrial
  Applications

MAFFEZZONI: Modelling and Conrol of Electric Power Plants
  (1984)*
MARTIN: Design of Work in Automated Manufacturing Systems*
McAVOY: Model Based Process Control
MEYER: Real Time Programming (1989)*
MILLER: Distributed Computer Control Systems (1981)
MILOVANOVIC & ELZER: Experience with the Management of
  Software Projects (1988)
MOWLE & ELZER: Experience with the Management of Software
  Projects (1989)
NARITA & MOTUS: Distributed Computer Control Systems (1989)
OLLUS: Digital Image Processing in Industrial Applications - Vision
  Control
QUIRK: Safety of Computer Control Systems (1985)(1986)
RAUCH: Control Applications of Nonlinear Programming*
REMBOLD: Information Control Problems in Manufacturing
  Technology (1979)
RODD: Artificial Intelligence in Real Time Control (1989)
RODD: Distributed Computer Control Systems (1983)
RODD: Distributed Databases in Real Time Control
RODD & LALIVE D'EPINAY: Distributed Computer Control
  Systems (1988)
RODD & MULLER: Distributed Computer Control Systems (1986)
RODD & SUSKI: Artificial Intelligence in Real Time Control
SIGUERDIDJANE & BERNHARD: Control Applications of
  Nonlinear Programming and Optimization
SINGH & TITLI: Control and Management of Integrated Industrial
  Complexes*
SKELTON & OWENS: Model Error Concepts and Compensation
SOMMER: Applied Measurements in Mineral and Metallurgical
  Processing
SUSKI: Distributed Computer Control Systems (1985)
SZLANKO: Real Time Programming (1986)
TAKAMATSU & O'SHIMA: Production Control in Process Industry
UNBEHAUEN: Adaptive Control of Chemical Processes
VILLA & MURARI: Decisional Structures in Automated
  Manufacturing

*Out of stock · High quality repro print available. Details of prices sent on request from the IFAC Publisher.

IFAC related titles

BROADBENT & MASUBUCHI: Multilingual Glossary of Automatic Control Technology

EYKHOFF: Trends and Progress in System Identification

NALECZ: Control Aspects of Biomedical Engineering